PHYSICO-CHEMICAL BEHAVIOUR OF ATMOSPHERIC POLLUTANTS

Commission of the European Communities

PHYSICO-CHEMICAL BEHAVIOUR OF ATMOSPHERIC POLLUTANTS

Proceedings of the Third European Symposium
held in Varese, Italy,
10-12 April 1984

Organized within the framework
of the Concerted Action COST 61 A bis

Edited by

B. VERSINO and G. ANGELETTI

Commission of the European Communities
Directorate-General Science, Research and Development

D. REIDEL PUBLISHING COMPANY

A MEMBER OF THE KLUWER  ACADEMIC PUBLISHERS GROUP

DORDRECHT / BOSTON / LANCASTER

Library of Congress Cataloging in Publication Data
Main entry under title:

Physico-chemical behaviour of atmospheric pollutants.

       Organized by the Commission of the European Communities, Joint
Research Centre Ispra, Directorate-General, Science, Research and Development, Brussels.

       Includes index.

       1. Air–Pollution–Congresses. 2. Atmospheric chemistry–Congresses.
3. Pollutants–Congresses. 4. Aerosols–Environmental aspects–Congresses.
5. Acid deposition–Environmental aspects–Congresses. I. Versino, B.,
1935-   . II. Angeletti, G.,   1943-   . III. Commission of the
European Communities. Joint Research Centre. IV. Commission of the
European Communities. Directorate-General for Science, Research and
Development.
TD881.P484   1984   628.5'3   84-18192
ISBN-13: 978-94-009-6507-2   e-ISBN-13: 978-94-009-6505-8
DOI: 10.1007/978-94-009-6505-8

The symposium was organized by the
Commission of the European Communities
Joint Research Centre Ispra
Directorate-General Science, Research and Development, Brussels

Publication arrangements by
Commission of the European Communities
Directorate-General Information Market and Innovation, Luxembourg

EUR 9436
© 1984, ECSE, EEC, EAEC, Brussels and Luxembourg
Softcover reprint of the hardcover 1st edition 1984

Published by D. Reidel Publishing Company
P.O. Box 17, 3300 AA Dordrecht, Holland

Sold and distributed in the U.S.A. and Canada
by Kluwer Boston Inc.,
190 Old Derby Street, Hingham, MA 02043, U.S.A.

In all other countries, sold and distributed
by Kluwer Academic Publishers Group,
P.O. Box 322, 3300 AH Dordrecht, Holland

P R E F A C E

The Commission of the European Communities presents in this book
the proceedings of the third European Symposium on "Physico-Chemical
Behaviour of Atmospheric Pollutants", held in Varese (Italy) from
10 to 12 April 1984.*

The symposium was organized within the framework of the Concerted
Action COST 61a bis**, which has the same name and is included in the
third R & D Programme on Environment of the Commission of the European
Communities - Indirect and Concerted Actions - 1981 to 1985.

The scope of the Concerted Action is to co-ordinate all research in
the area executed in the participating countries and to collect and
disseminate the results.

The aim of the symposium was to review the progress and results achieved
during the past two and a half years, since the second symposium held
in Varese in October 1981.

The programme of the symposium consisted of papers and posters covering
different areas related to the physico-chemistry of air pollutants,
including identification and analysis of pollutants, chemical and
photochemical reactions, aerosols characterisation, pollutant cycles,
transport and modelling. Particular emphasis was put on the physico-
chemical aspects of acid deposition.

We think the book contributes substantially to a better understanding
of some of the causes which affect our ecosystems today and demonstrates
the opportunity for a follow-up programme in this field.

Brussels/Ispra, May 1984

      B. VERSINO   G. ANGELETTI

*The proceedings of the first European Symposium on physico-chemical
behaviour of atmospheric pollutants have been published by the
Commission of the European Communities in 1980 as report EUR 6621
and may be obtained on request from the Commission of the European
Communities, Directorate General for Science, Research and Development,
200, rue de la Loi, B-1049 Brussels. The proceedings of the second
European Symposium held in Varese, Italy, September 29 - October 1,
1981, have been published by D. Reidel Publishing Company in 1982 as
report EUR 7624.

**COST 61a bis : Scientific and Technical Cooperation among European
Community Member Countries and Non-Member Countries (Austria, Switzerland,
Sweden and Yugoslavia) in the field of "Physico-Chemical Behaviour of
Atmospheric Pollutants".

CONTENTS

Preface   v

SESSION 1 - IDENTIFICATION AND ANALYSIS OF POLLUTANTS

Summary by the Chairman
   A. LIBERTI   2

Field measurements of tropospheric OH by long-path UV laser
absorption
   J. HÄGELE, R. PASCHKE and R. ZELLNER, Institut für Physikalische Chemie, Universität Göttingen, Federal Republic of
   Germany   5

Evaluation of atmospheric acidity - Sampling and analytical
techniques
   I. ALLEGRINI, F. DE SANTIS, A. FEBO, A. LIBERTI and
   M. POSSANZINI, Istituto Inquinamento Atmosferico del CNR,
   Area della Ricerca di Roma, Monterotondo Stazione, Italy   12

Attempted measurement of gaseous H_2O_2 in the ambient atmosphere
   H.M. TEN BRINK, Chemistry Department, Netherlands Energy
   Research Center ECN, Petten, The Netherlands; T.J. KELLY,
   Y.N. LEE and S.E. SCHWARTZ, Environmental Chemistry Div-
   ision, Department of Energy and Environment, Brookhaven
   National Laboratory, Upton, N.Y., USA   20

Properties, formation and detection of peroxyacetyl nitrate
   U. SCHURATH, U. KORTMANN and S. GLAVAS, Institut für Physi-
   kalische Chemie der Universität Bonn, Federal Republic of
   Germany   27

Cryogenic sampling and analysis of peroxyacetyl nitrate in the
atmosphere
   H. MEYRAHN, J. HAHN, G. HELAS and P. WARNECK, Max-Plank-
   Institut für Chemie (Otto-Hahn-Institut), Mainz, Federal
   Republic of Germany; S.A. PENKETT, AERE Harwell, Environmen-
   tal and Medical Sciences Division, Oxfordshire, United
   Kingdom   38

Sampling and analysis of acetaldehyde in tropospheric air
   B. SCHUBERT, U. SCHMIDT and D.H. EHHALT, Institut für
   Chemie 3: Atmosphärische Chemie der Kernforschungsanlage
   Jülich GmbH, Jülich, Federal Republic of Germany   44

Optimisation des méthodes de prélèvement et d'analyse des hydrocarbures aromatiques polycycliques et de leurs dérivés azotés - Détermination de leur stabilité dans l'atmosphère
    M.A. BRESSON, S. BEYNE, P. MASCLET and G. MOUVIER, Université Paris VII, Laboratoire de Physico-Chimie Instrumentale, Paris, France   53

Sampling, identification and quantitative determination of biogenic and anthropogenic hydrocarbons in forestal areas
    P. CICCIOLI, E. BRANCALEONI, M. POSSANZINI, A. BRACHETTI and C. DI PALO, Istituto Inquinamento Atmosferico del CNR, Area della Ricerca di Roma, Monterotondo Stazione, Italy   62

Determination of hydrogen peroxide in cloud and rain water
    F.G. RÖMER, A.A. VELDKAMP, P. VAN GALEN, Environmental Department, NV KEMA, Arnhem, The Netherlands   74

Bestimmung von Schwefelsäure und Sulfaten in der Luft
    M. BUCK, Landesanstalt für Immissionsschutz, Essen, Federal Republic of Germany   83

La mesure en continu des sulfates par photométrie de flamme
    M. PAYRISSAT, B. NICOLLIN and H. STANGL, Commission of the European Communities, Joint Research Centre, Ispra Establishment, Italy   90

Plants as monitoring samplers of airborne PAH
    E. BRORSTRÖM-LUNDEN and L. SKÄRBY, Swedish Environmental Research Institute, Göteborg, Sweden   101

Preparation of diffusion denuder tubes for collection of ammonia or acidic gases in air - Equipment for coating and extraction
    E.E. LEWIN, K. FUGLSANG and K.A. HANSEN, National Agency of Environmental Protection, Air Pollution Laboratory, Roskilde, Denmark   111

The PIXE analytical technique and its application to environmental problems
    I.V. MITCHELL, Commission of the European Communities, Directorate-General Science, Research and Development, Brussels, Belgium   120

SESSION 2 - CHEMICAL AND PHOTOCHEMICAL REACTIONS

Summary by the Chairman
    R.A. COX   141

Kinetic study of reactions of OH radicals with organic sulfur compounds
    J.L. JOURDAIN, H. MAC LEOD, G. POULET and G. LE BRAS, Centre de Recherches sur la Chimie de la Combustion et des Hautes Températures, CNRS, Orléans, France   143

Reactions of OH radicals with reduced sulfur compounds under atmospheric conditions
I. BARNES, V. BASTIAN, K.H. BECKER and E.H. FINK, Physikalische Chemie/FB 9, Bergische Universität-GH Wuppertal, Federal Republic of Germany   149

LIF studies of formation and kinetics of primary radical products in OH-oxygenated hydrocarbon reactions
K. LORENZ, D. RHÄSA and R. ZELLNER, Institut für Physikalische Chemie, Universität Göttingen, Federal Republic of Germany   158

The temperature dependence of the forward-backward reactions of the addition of OH to benzene, aniline and nitrobenzene
F. WITTE and C. ZETZSCH, Lehrstuhl für physikalische Chemie I, Ruhr-Universität Bochum, Federal Republic of Germany   168

Absolute rate constant measurements of OH reactions under atmospheric conditions by laser photolysis/dye laser fluorescence
V. SCHMIDT, GUI-YUN ZHU, K.H. BECKER and E.H. FINK, Physikalische Chemie - FB 9, Bergische Universität - Gesamthochschule Wuppertal, Federal Republic of Germany   177

Photooxidation of acetaldehyde
G.N. BAGNALL and H.W. SIDEBOTTOM, Chemistry Department, University College, Dublin, Ireland   188

A FTIR spectroscopic study of the photooxidation of acetaldehyde in air
G.K. MOORTGAT and R.D. McQUIGG, Max-Planck-Institut für Chemie, Air Chemistry Division, Mainz, Federal Republic of Germany   194

Absorption spectrum and kinetics of the NO_3 radical
R.A. COX, R.A. BARTON, E. LJUNGSTRÖM and D.W. STOCKER, Environmental and Medical Sciences Division, AERE, Harwell, United Kingdom   205

Hydroxyl radical concentration in ambient air at a semirural site estimated from $C^{13}O$ oxidation
J. HJORTH, G. OTTOBRINI, F. CAPPELLANI, G. RESTELLI and H. STANGL, Commission of the European Communities, Joint Research Centre, Ispra Establishment, Italy; C. LOHSE, University of Odense, Chemistry Department, Denmark.   216

Temperature dependence of the reactions $SO + O_3$ (1) and $SO + O_2$ (2)
U. SCHURATH and H.J. GOEDE, Institut für Physikalische Chemie der Universität Bonn, Federal Republic of Germany   227

A study of N_2O_5 and NO_3 chemistry in the photolysis of N_2O_5 mixtures
J.P. BURROWS, G.S. TYNDALL and G.K. MOORTGAT, Max-Planck-Institut für Chemie, Mainz, Federal Republic of Germany   240

A study of the reaction between ClO and NO_2 using matrix iso-
lation FTIR spectroscopy and UV-visible spectroscopy
   J.P. BURROWS, G.S. TYNDALL, G.K. MOORTGAT and
   D.W.T. GRIFFITH, Max-Plank-Institut für Chemie, Mainz,
   Federal Republic of Germany   249

Oxidation of methylchloroform
   L. NELSON, J.J. TREACY and H.W. SIDEBOTTOM, Chemistry
   Department, University College, Dublin, Ireland   258

Transformation of reactive PAH on particles by exposure to
oxidized nitrogen compounds
   A. LINDSKOG, E. BRORSTRÖM-LUNDEN and A. SJÖDIN, Swedish
   Environmental Research Institute   264

The kinetic coefficient of the C_2H_4 + O reaction over extended
pressure and temperature ranges
   V. FONDERIE, D. MAES and J. PEETERS, Department of Chemistry, Katholieke Universiteit Leuven, Belgium   274

Kinetics of the reaction of OH with ethane and a series of Cl-
and F-substituted methanes at 300–400 K, studied by pulse
radiolysis combined with kinetic spectroscopy
   O.J. NIELSEN, P. PAGSBERG and A. SILLESEN, Chemistry Department, Risø National Laboratory, Roskilde, Denmark   283

The role of freons in the chemistry of the upper atmosphere
   L. BATT, Department of Chemistry, University of Aberdeen,
   United Kingdom   293

SESSION 3 - AEROSOLS

Summary by the Chairman
   J.G. MADELAINE   298

Mesure de particules fines dans une zone polluée
   M.L. PERRIN, G. MADELAINE, C. FRAMBOURT, Laboratoire de
   Physique et Métrologie des Aérosols, Commissariat à l'Ener-
   gie Atomique, Fontenay-aux-Roses, France   300

Design and performance of an aerosol reactor for photochemical
studies
   W. HULLÄNDER, W. BEHNKE, W. KOCH and G. POHLMANN,
   Fraunhofer-Institute for Toxicology and Aerosol Research,
   Münster-Roxel, Federal Republic of Germany   309

Physical and chemical characteristics of suspended particulates
during smog conditions in Berlin(W)
   G.W. ISRAËL, H.-W. BAUER and K. WENGENROTH, Fachgebiet Luftreinhaltung im Institut für Technischen Umweltschutz, Technische Universität Berlin, Federal Republic of Germany   320

A study of the concentration of sulfates in the particulate
matter
    J. DE LA SERNA, R. FERNANDEZ PATIER and F. PEREZ CARLES,
    Departamento de Sanidad Ambiental, Escuela Nacional de Sani-
    dad, Madrid, Spain   322

Aerosol neutralization by atmospheric ammonia
    A.G. CLARKE, M.J. WILLISON and E.M. ZEKI, Department of Fuel
    and Energy, Leeds University, United Kingdom   331

Elemental composition and size distribution of atmospheric
aerosols during long range transport
    H.W. GEORGII, Institut für Meteorologie und Geophysik;
    P. METTERNICH, id. + Institut für Kernphysik;
    K.O. GROENEVELD, Institut für Kernphysik, Johann Wolfgang-
    Goethe-Universität, Frankfurt/Main, Federal Republic of
    Germany   339

Characterization of suspended particulate matter in a lead
smeltery area
    M. FUGAS, J. HRSAK and K. SEGA, Institute for Medical
    Research and Occupational Health, Zagreb. P. SOUVENT, Lead
    Mine and Smeltery Mezica, Yugoslavia   348

The concentration of sulfate in broken cloud layers
    H.M. TEN BRINK, Chemistry Department, Netherlands Research
    Center, ECN, Petten, The Netherlands; P.H. DAUM and
    S.E. SCHWARTZ, Environmental Chemistry Division, Department
    of Applied Science, Brookhaven National Laboratory, Upton,
    USA   356

Combined photolytic and radiolytic aerosol formation in a SO_2
-NO_2-air mixture
    F. RAES and A. JANSSENS, Rijksuniversiteit Gent, Nuclear
    Physics Laboratory, Gent, Belgium   364

Measurement of gaseous halogenated hydrocarbons in ambient air
    J. MÜLLER and F. RIEDEL, Umweltbundesamt, Pilotstation
    Frankfurt, Federal Republic of Germany   373

SESSION 4 - POLLUTANT CYCLES

Summary by the Chairman   380
    S. BEILKE

NO_x background mixing ratios in surface air over Europe and the
Atlantic Ocean
    A. BROLL, G. HELAS and P. WARNECK, Max-Planck-Institut für
    Chemie, Otto-Hahn-Institut, Mainz; K.J. RUMPEL, Umwelt-
    bundesamt, Messtelle Deuselbach, Federal Republic of Germany   390

Regional background concentrations of NO$_2$ in Sweden
A. SJÖDIN and P. GRENNFELT,, Swedish Environmental Research
Institute, Göteborg, Sweden   401

Nitrous acid in polluted air masses - Sources and formation
pathways
C. KESSLER and U. PLATT, Kernforschungsanlage Jülich GmbH,
Institut für Chemie 3: Atmosphärische Chemie, Jülich,
Federal Republic of Germany   412

Study of the chemical characteristics of wet and dry deposition
in Switzerland
J. FUHRER, Institute of Plant Physiology, University of
Bern, Switzerland   423

Vertical and horizontal profiles of hydrogen chloride in the
Mediterranean region
B. VIERKORN-RUDOLPH, J. RUDOLPH and F.X. MEIXNER, Institut
für Chemie 3: Atmosphärische Chemie der Kernforschungsanlage
Jülich GmbH; K. BÄCHMANN and B. SCHWARZ, FB 8 (Anorganische
Chemie und Kernchemie) Technische Hochschule Darmstadt,
Federal Republic of Germany   433

Bilan ionique et acidité de la précipitation antarctique
M. LEGRAND, R.J. DELMAS and F. ZANOLINI, Laboratoire de
Glaciologie et Géophysique de l'Environnement, St. Martin
d'Hères, France   441

Etude de l'influence d'une source locale naturelle intense de
composés organosoufrés sur la chimie de la troposphère en
milieu non pollué
P. CARLIER, C. LUCE, R. GIRARD, G. MOUVIER, J. MORELLI,
L. GIRARD-REYDET, T. MARCHAL and S. CADENE, Université Paris
VII, Laboratoire de Physico-Chimie instrumentale et Laboratoire de Chimie minérale des milieux naturels, Paris, France   451

Fluctuations temporelles fines de la composition chimique de
l'aérosol côtier
J. MORELLI, L. GIRARD-REYDET and T. MARCHAL, Laboratoire de
Chimie Minérale des Milieux Naturels, ERA CNRS 889, Univer-
sité Paris VII; P. MASNIERE, Direction des Etudes et Recher-
ches, Electricité de France, Chatou; M. FEDOROFF,
J.-C. ROUCHAUD, P. CARLIER and L. DEBOVE, Laboratoire de
Physico-Chimie Instrumentale - ERA CNRS 889, Université
Paris VII, France   461

Toxic metals and metalloids in high altitude alpine glaciers
snow and ice
F.M. BATIFOL and C.F. BOUTRON, Laboratoire de Glaciologie et
Géophysique de l'Environnement du CNRS, Grenoble, France   471

Results of many years' analyses of precipitation chemistry on
samples obtained simultaneously at 3.0 km, 1.8 km and 0.7 km ASL
R. REITER, K. PÖTZL and K. MUNZERT, Fraunhofer Institute for
Atmospheric Environmental Research, Garmisch Partenkirchen,
Federal Republic of Germany   480

Measurements of the latitudinal distribution of light hydrocarbons and halocarbons over the Atlantic
   J. RUDOLPH, C. JEBSEN, A. KHEDIM and F.J. JOHNEN, Institut
   für Chemie 3: Atmosphärische Chemie der Kernforschungsanlage
   Jülich GmbH, Federal Republic of Germany   492

SESSION 5 - TRANSPORT AND MODELLING FIELD EXPERIMENTS

Summary by the Chairman   504
   A.J. ELSHOUT

Composition and origin of cloudwater solutes
   G.P. GERVAT, A.S. KALLEND and A.R.W. MARSH, Central Electricity Research Laboratories, Leatherhead, United Kingdom   506

Air sampling flights round the British Isles at low altitudes :
SO_2 oxidation and removal rates
   D.J. BAMBER, P.G. HEALEY, A.F. TUCK and G. VAUGHAN, Meteorological Office, Bracknell; P.A. CLARK, G.M. GLOVER,
   A.S. KALLEND and A.R.W. MARSH, Central Electricity Generating Board, Central Electricity Research Laboratories,
   Leatherhead, United Kingdom   517

Oxidation of NO to NO_2 in flue gas plumes of power stations
   A.J. ELSHOUT, N.V. KEMA, Arnhem, The Netherlands; S. BEILKE,
   UBA, Frankfurt, Federal Republic of Germany   535

Campagne experimentale en vue de la modélisation de panache
réactif
   G. MAFFIOLO, E. JOOS and A.E. SAAB, Electricité de France,
   Direction des Etudes et Recherches, Chatou, France   544

The atmospheric significance of liquid phase oxidation of SO_2
to sulphate by O_3 and H_2O_2
   T. BØHLER, Norwegian Institute for Air Research, Lillestrøm;
   I.S.A. ISAKSEN, Institute of Geophysics, University of Oslo,
   Norway   554

Some possibilities of modelling data from the "FOS" European
field experiment (June 6/15, 1983)
   H. AUGUSTIN, Délégation aux Risques Majeurs, Paris;
   P. BESSEMOULIN, Etablissement d'Etudes et de Recherches
   Météorologiques, Toulouse, France   565

Atmospheric tracer dispersion experiments (at the Mol Nuclear
Energy Research Center)
   B. VANDENBORGHT, I. MERTENS and J. KRETZSCHMAR, Nuclear
   Energy Research Center, SCK/CEN, Mol, Belgium   573

A sampling network for the assessment of the heavy metal pol-
lution originating from a municipal incinerator plant in
Belgium
   F. CANDREVA and R. DAMS, Institute for Nuclear Sciences,
   University of Gent, Belgium   583

Evaluation of the information from a continuously working precipitation monitor
    P. WINKLER, Deutscher Wetterdienst, Meteorologisches Obser-
    vatorium Hamburg, Federal Republic of Germany   590

Production et transfert d'ozone sur le bassin de Fos-Berre
    P. PERROS and G. TOUPANCE, Laboratoire de Physico-Chimie de
    l'Environnement, Université de Paris Val de Marne, Créteil,
    France   596

Comparison between the scavenging ratios for nitrate and sulphate at a rural site
    M. FERM, Swedish Environmental Research Institute, Göteborg,
    Sweden   607

Identification des sources d'hydrocarbures aromatiques polycycliques particulaires dans l'atmosphère urbaine
    P. MASCLET, K. NICOLAOU et G. MOUVIER, Laboratoire de
    Physico-Chimie Instrumentale, Université de Paris VII,
    France   616

Photochemical air pollution in Denmark. Weekday effects and
evidence of large-scale formation
    J. FENGER, Air Pollution Laboratory, National Agency of
    Environmental Protection, Risø National Laboratory,
    Roskilde, Denmark   626

Classement automatique des trajectoires du panache de l'Etna -
Etude climatologique
    D. MARTIN and D. CHEYMOL, Etablissement d'Etudes et Recher-
    ches Météorologiques, Magny les Hameaux; M. IMBARD and
    B. STRAUSS, Service Météorologique Métropolitain, Paris,
    France   635

What is the source of acid in clouds?
    F.G. RÖMER, N.V. KEMA, Arnhem; H.F.R. REIJNDERS, National
    Institute of Public Health and Environmental Hygiene, Bilt-
    hoven, The Netherlands.   649

LIST OF PARTICIPANTS   657

INDEX OF AUTHORS   664

SESSION 1

IDENTIFICATION AND ANALYSIS OF POLLUTANTS

Summary by the Chairman
A. LIBERTI

Field measurements of tropospheric OH by long-path UV laser absorption

Evaluation of atmospheric acidity - Sampling and analytical techniques

Attempted measurement of gaseous H_2O_2 in the ambient atmosphere

Properties, formation and detection of peroxyacetyl nitrate

Cryogenic sampling and analysis of peroxyacetyl nitrate in the atmosphere

Sampling and analysis of acetaldehyde in tropospheric air

Optimisation des méthodes de prélèvement et d'analyse des hydrocarbures aromatiques polycycliques et de leurs dérivés azotés - Détermination de leur stabilité dans l'atmosphère

Sampling, identification and quantitative determination of biogenic and anthropogenic hydrocarbons in forestal areas

Determination of hydrogen peroxide in cloud and rain water

Bestimmung von Schwefelsäure und Sulfaten in der Luft

La mesure en continu des sulfates par photométrie de flamme

Plants as monitoring samplers of airborne PAH

Preparation of diffusion denuder tubes for collection of ammonia or acidic gases in air - Equipment for coating and extraction

The PIXE analytical technique and its application to environmental problems

IDENTIFICATION AND ANALYSIS OF POLLUTANTS.

Summary by the Chairman

A. Liberti
Istituto Inquinamento Atmospherico CRN - Roma

A. Liberti
Istituto Inquinamento Atmospherico CRN - Roma

The activity related to identification and analysis of pollutants covers an extensive field with a variety of contributions which can be classified according to the following lines :
a) development of sampling techniques to prevent artifacts and realize reliable sampling of traces components;
b) development of analytical procedures to evaluate species of major interest in the conversion of pollutants and of their cycle in the atmosphere
c) evaluation of new means of investigation for the determination of atmospheric pollutants.

The problem related to chemical artifacts, which affect the accuracy of the procedure for sampling atmospheric pollutants, is strictly connected with gas-particles interaction as in most cases a filter is set in front of the collection device to eliminate particulated matter. The use of diffusion tubes (denuders) permits the collection of the aerosol particles leaving the denuder without sampling artifacts, providing suitable and efficient systems are selected to act as a sink for the sampled gas.

The major limitations of these devices, namely efficiency, capacity and ruggedness, have been overcome by introducing a new annular design, which is capable of sampling rates corresponding to several liters per minute, with a very high efficiency.

The use of the denuder technique allows precise and accurate measurement of minor components in the atmosphere such as HCl, HNO_3, NH_3, at ppb levels and might supply useful information on the dissociation equilibrium of ammonium salts, which are the main components of atmospheric particulates.

The use of denuders may supply a large amount of information on atmospheric acidity as most acid and basic species, responsible for this state, are in a dynamic equilibrium and their application may ensure the collection of atmospheric samples with the same physical and chemical properties as the atmosphere being sampled.

Their application to monitor ammonia and ammonium ion without mutual interference may clean up the action of this system in determining atmospheric acidity as ammonia is the main neutralizing species able to counterbalance the effect of strong acids.

There is a general feeling that the use of denuders might replace conventional procedures for the sampling of gases and aerosols.

The determination of atmospheric sulphuric acid and sulphate which is of importance in connection with atmospheric acidity has been extensively studied and it has been shown that a commercial flame photometric detector can be adapted to continuous sulphate measurement. This determination might help to define haze formation due to ammonium sulphate production.

The role exerted in the atmosphere by free radicals and oxidizing agents has been fully recognized and research has been carried out to determine OH, H_2O_2 and peroxyacetyl nitrate. The previously applied laser absorption methods for the determination of tropospheric OH have been improved by the use of sub-Doppler resolution analysis laser lines as well as a reference laser beam.

The measurement of gaseous hydrogen peroxide in the ambient atmosphere has been attempted with a variety of analytical procedures in an effort to determine this constituent, which is the main species responsible for oxidation in the droplet phase.

The presence of ozone interferes and it appears that surface reactions related to the sampling procedures determine artifacts which prevent various methods being employed with confidence..

The determination of peroxyacetyl nitrate (PAN) has been further investigated and a cryosampling method wich can be applied on board aircraft has been developed; it has been thus possible to evaluate the PAN mixing ratio as a function of altitude. A new method for PAN determination has been worked out consisting in the conversion to NO and its detection by a chemiluminescence detector. The detection limit of this procedure is 20 ppt.

The sampling and analysis of acetaldehyde, which is produced predominantly during the combustion of ethane and which plays a significant role in the formation of photochemical smog, has been carried out by stripping this compound into aqueous 2-4 dinitrophenylhydrazine solution and further analysis by HPLC.

The determination of various classes of hydrocarbons in the atmosphere has been the object of several studies: highly sensitive and fast methods for analyzing biogenic and anthropogenic hydrocarbons in forestal areas and polycyclic aromatic hydrocarbons (PAH) in urban areas have been developed. The presence of nitro-PAH in the atmosphere has also been reported and it has been shown that the levels of these compounds are about two orders of magnitude lower than the concentration of the most common carcinogenic PAH.

Kale and elm leaves can be used as monitoring samplers for airborne PAH. The possibilities and the usefulness in the analysis of aerosols by means of the non-destructive energetic ion beam technique of PIXE (particle induced X-ray emission) has been presented for its application to atmospheric studies.

The wide variety of problems examined and the large number of analytical techniques which have been used for their solution indicate the importance of the research activity related to the identification and analysis of pollutants. The contribution to the characterization of atmospheric acidity and to the sampling techniques for its determination are worth noting.

It is suggested that some achievements which have been realized should be converted into analytical methods to be used in the evaluation of air quality in addition to standard procedures used at the present for the determination of conventional pollutants. Indexes such as rate of conversion of SO_2 in a certain area, aerosol sulphate and PAH concentration, level of atmospheric acidity, the extent of sulphate dry deposition, might supply additional information on the status of the atmosphere of a certain area.

A basic problem which needs to be properly solved and to which is called attention deals with the sampling of atmospheric aerosols for chemical analysis. This aspect, which has been previously overlooked, can be a source of serious errors and affect analytical results. Sampling artifacts, can be responsible either for a loss or an enrichment of the analytes on the filter medium and may yield different collection efficiencies of the sampling inlets. The use of properly designed inlets with characteristics largely independent of the wind speed and direction, coupled to high precision samplers with constant flow capabilities, reduces such errors to acceptable values. There is, however, an urgent need for the standardization of the collection devices in order to have reproducible mesurements as the basis for any intercomparison work. This is particularly true when size classifying devices such as cyclones, impactors, dichotomous etc., have to be used.

FIELD MEASUREMENTS OF TROPOSPHERIC OH
BY LONG-PATH UV LASER ABSORPTION

J. HÄGELE, R. PASCHKE and R. ZELLNER
Institut für Physikalische Chemie, Universität Göttingen
3400 Göttingen, FRG

Summary

Due to its predicted concentration and high reactivity
the OH radical is generally accepted as the most important driving agent of chemical conversion and transformation in the troposphere. Moreover, since its concentration level is strongly coupled to solar radiation
and the mixing ratios of a large number of atmospheric
trace constituents (which are all variable in space and
time), a direct measurement of OH provides one of the
most sensitive tests of tropospheric chemical models. In
the present paper we report on a double-beam long-path
UV-laser absorption technique recently developed in our
laboratory. This technique improves on previously applied
laser absorption methods by the use of sub-Doppler resolution analysis laser lines as well as a reference
laser beam. First tests show that our method provides the
required detection sensitivity.

1. INTRODUCTION

Tropospheric OH is formed mainly as a result of ozone
photolysis near 300 nm:

$$(1) \quad O_3 + h\nu \quad (\lambda \leq 310 \text{ nm}) \rightarrow O(^1D) + O_2$$

$$(2) \quad O(^1D) + H_2O \rightarrow 2 \text{ OH}$$

At steady state its formation rate is balanced by chemical
consumption

$$(3) \quad OH + A \rightarrow \text{products}$$

where $A \equiv CO$, CH_4, O_3, and other trace constituents [1, 2].
For the free troposphere and also for most pollution situations the CO reaction is dominant, leading to a 'chemical'
lifetime of OH of ~2.5 s (for 100 ppb CO). It is a uniqueness
of the CO reaction ($OH + CO \rightarrow CO_2 + H$) that its products re-
cycle OH via the coupling with NO_x (Fig. 1). Hence, the actual
OH level depends on the rates of reactions (1) - (3) as well
as (in a more complex way [3]) on the concentration level of
NO_x. Model calculations predict annual average OH concentrations during a 24-hour period of ~ 5×10^5 cm^{-3} [4]. Regional
day-time concentrations may be expected in the order of
several $10^6 - 10^7$ cm^{-3} [5, 6]. The accuracy of such predictions cannot be tested except by in-situ measurements. Hence,
attempts to determine tropospheric OH have begun as long as
10 years ago; their results, however, have been largely con-

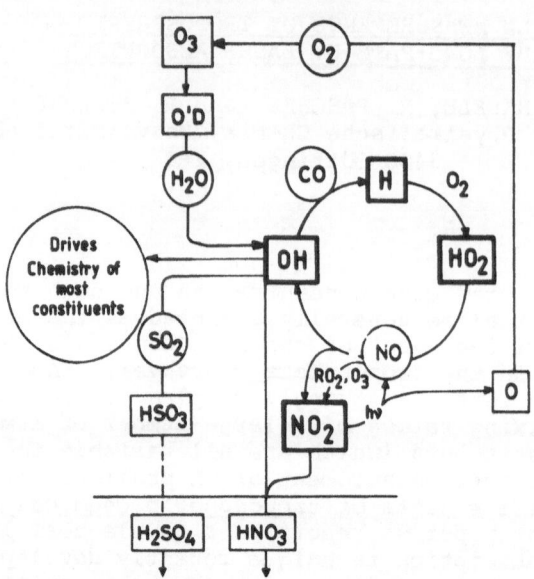

Fig. 1: Main tropospheric cycle of HO_x and its coupling with CO, NO_x, and SO_2

troversial.

Due to its strong $A^2\Sigma^+-X^2\Pi$ UV transition the OH radical is a suitable candidate for spectroscopic detection. The technique of laser induced fluorescence (LIF) was the first to be applied to this problem [7, 8]. However, until now results are less than satisfactory mainly because of a high OH generation rate by the laser beam (intensities typically 50 $\mu J/cm^2$) itself.

An alternative method is the use of long path laser absorption. This technique, which has been pioneered by Perner et al. [9, 10] is essentially devoid of OH generation problems since it applies laser intensities ($\sim$1 $\mu J/cm^2$) considerably below the corresponding integrated solar intensity for the near UV region. However, due to the low optical densities of OH pathlengths of several km have to be used. For $[OH] = 10^6$ cm^{-3} the expected absorption of the $Q_1(2)$-line (307.995 nm) at 10 km pathlength and using a laser system with sub-Doppler bandwidth ($\Delta\lambda = 0.0002$ nm) is $\sim 2.5 \times 10^{-4}$. This cannot be determined in a single pulse experiment, in particular since air turbulence leads to beam displacement and strong intensity fluctuations of the transmitted laser light. Hence, averaging techniques have to be used.

By improvement of their original experimental approach [9], Hübler et al. [10, 11] have recently demonstrated the feasibility of the laser absorption technique. In their work a broad band laser light source is used and the integrated line absorption is determined by tuning rapidly across the line by means of a high resolution ($\Delta\lambda = 0.0027$ nm) monochromator. Averaging of the signals is achieved by repetitive scanning with a 6.6 kHz frequency. For two locations in Germany average noon-time concentrations of OH on sunny days during the summer of $\sim 1.6 \times 10^6$ cm^{-3} were reported.

Despite this successful application the Hübler et al. method suffers of spectrally resolved light intensity, which prevents the use of increased absorption pathlengths and hence still higher detection sensitivities. A modified absorption technique which potentially meets these requirements has recently been developed in our laboratory.

We report in this paper an absorption technique with sub-Doppler laser line width and hence high narrow band intensities. By use of a reference beam which probes the same atmosphere the standard deviation of a single pulse measurement is essentially reduced to the accuracy of the detection system.

2. EXPERIMENTAL ARRANGEMENT

The essential components of the experimental set-up are shown schematically in Fig. 2. A photograph is provided in Fig. 3. An excimer laser (LAMBDA, EMG 102, E $\approx$ 150 mJ) serves as pump source for a frequency-doubled dye laser (LAMBDA, FL 2002 E). A splitted portion of the direct excimer laser beam is delayed by 100 ns ($\hat{=}$ 30 m) and joined to the output of the frequency-doubled dye laser as a reference beam.

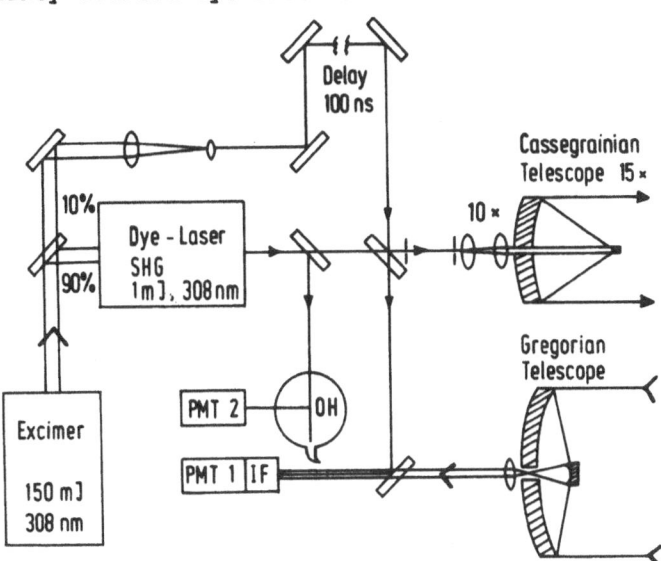

Fig. 2: Schematic representation of double-beam laser absorption experiment

Both beams, consisting of pulses of the same duration ($\sim$10 ns) and nearly the same energy ($\sim$1 mJ) probe the same volume of the atmosphere. Whereas one of them (dye laser) has sub-Doppler OH line bandwidth ($\Delta\lambda$ = 0.0002 nm) and is tuned to the center of the $Q_1(2)$-line at 307.995 nm, the second one (delayed excimer laser) consists of two broad lines with bandwidths $\Delta\lambda \sim 0.08$ nm and centers at 307.9 and 308.2 nm [12]. Hence, only the former will be attenuated by OH. Due to the proximity of the absolute wavelengths, however, both beams are

Fig. 3: Photograph of experimental arrangement

subject to the same intensity loss by physical interaction with the atmosphere, i.e. Mie scattering. Hence, the reference beam serves to eliminate this part of the attenuation from the analysis laser beam.

The "double-beam" is directed to a 10 x beam expander and further to a 15 x Cassegrainian telescope. This expands the beam to nearly 30 cm_2 diameter and simultaneously reduces its intensity (< 1 $\mu J/cm^2$) and divergence (7 μrad). A 50 cm plane mirror with a surface flatness of $\lambda/10$ serves as a reflector at a distance of 6.4 and 15.7 km. The reflected beam (d $\sim$ 40 - 50 cm) is collected by a 65 cm Gregorian telescope and directed via an interference filter (SCHOTT, UV DIL, $\Delta\lambda$ = 6.9 nm) to a fast photomultiplier (EMI QB 9815). No monochromator is used. The intensity of the outgoing beam is also measured by the same photomultiplier. In order to control the wavelength stability of the dye laser, a fraction of the frequency-doubled light is used for OH-fluorescence excitation in a reference cell. The fluorescence arising from this is detected by a second photomultiplier.

The interval between outgoing and incoming double-pulse is $\geq$ 43 μs. As the excimer laser normally runs at 20 Hz, the next pulse sequence follows 50 ms later. A fast analog electronic circuit provides subsequent treatment of the measured pulses to an accuracy of better than 0.5 %. The digitized data are processed by a 64-kByte microcomputer (CBM 8032) and transferred to a floppy-disc (CBM 8050). The computer can be used for simultaneous and later data reduction. Provided that the signal intensities show random fluctuations, averaging over 2500 signals leads to an accuracy of 10^{-4} corresponding to an OH concentration of about 2 x 10^5 cm^{-3}. Since beam displacements cause data losses up to 80 % and absorption measurements will be made on and off the OH-line in order to re-

duce interferences by other species (SO_2, CH_2O), a single
measurement period takes about 20 minutes. Further details of
the technique have been described elsewhere [13].

3. TESTS OF METHOD

An important criterion of the applicability of a technique with a single pulse accuracy far below the required
final accuracy is the normal (Gaussian) statistical behaviour
of pulse fluctuations. As shown by Killinger and Menyuk [14]
this is not generally the case in field experiments but has to
be tested.

Fig. 4 shows the measured standard deviation of outgoing
and incoming intensity I_{out}/I_{in} for the dye laser beam for two
different weather conditions.

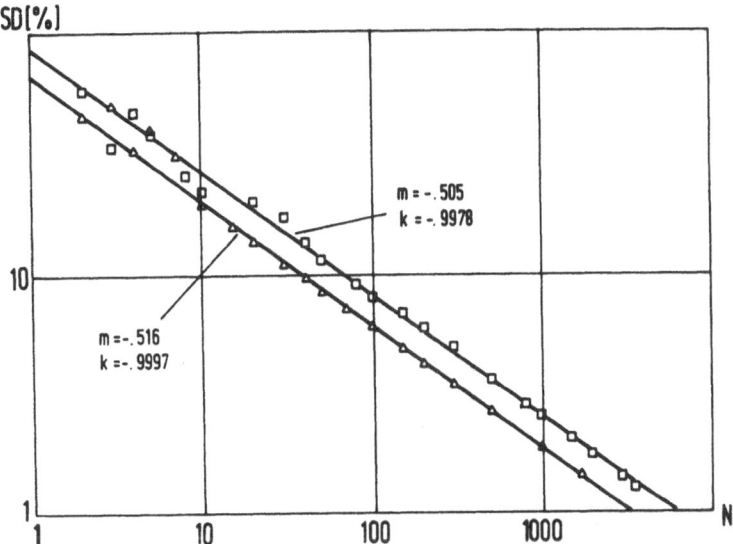

Fig. 4: Standard deviation of I_{out}/I_{in} for a narrow band dye
laser beam: m = mean slope, k = correlation coefficient, N = number of pulses

The extrapolated standard deviation (SD) of a single
pulse is on the order of 60 - 80 %, where the larger number
corresponds to stronger winds. This is the effect of atmospheric turbulence. As can be seen, however, SD is proportional to $N^{-0.5}$ and hence behaves normally statistical.

This is the behaviour of the field. We have performed the
same kind of analysis for the ratios of outgoing analysis
laser and reference laser intensities I_{out}/I_{out}^R as well as the
incoming ratios (I_{in}/I_{in}^R). Both show the same standard deviations (SD = 8 - 12 % for N = 1) which results from fluctuations of the adjustment of the optical components in the laboratory. Note, that ratioing the signals by using the reference
beam immediately reduces the SD from the field (Fig. 4) to the
one imposed by the laboratory arrangement. Therefore, the required accuracy of 10^{-4} for the field measurement is available

with a pulse number in the order of 2500.

The work of Huebler et al. [10, 11] has presented the problem of interference of the OH measurement with other atmospheric trace gases that have a structured absorption in the near UV (SO_2, CH_2O, CS_2). The effect of these trace gases is to add to the total "line" absorption and to modify the OH base line. Obviously, the amount of interference then depends on the concentration level of these trace gases and the resolution of the experiment.

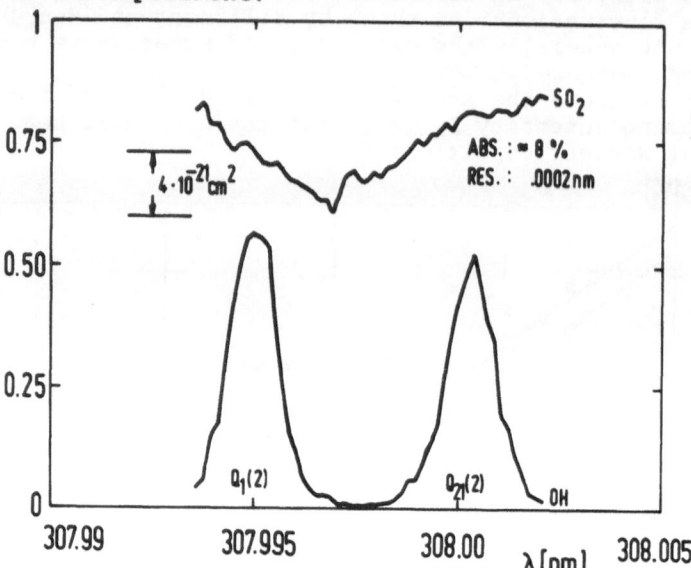

Fig. 5: High resolution SO_2 spectrum and OH line reference around 308 nm

Fig. 5 shows a section of a high resolution ($\Delta\lambda$ = 0.0002 nm) absorption spectrum of SO_2 at 1 atm of air, measured in our laboratory using the narrow band laser system. The figure also contains fluorescence excitation signals of the $Q_1(2)$ line and its satellite ($Q_{21}(2)$) of the OH radical excited with the same laser. As can be seen the SO_2 absorption cross sections vary by roughly 4×10^{-21} cm^2 between 307.992 and 308.002 nm. Hence, an intended measurement of OH on and off its line could be obscured by an absorption of as much as 1×10^{-3} for a background SO_2 of 10 ppb. This effect, however, can be compensated by taking OH measurements in the center of the line as well as to the left and right from it. In between the SO_2 absorption cross section can be approximated as to change linearly. Moreover, a line can be chosen (i.e. $Q_{21}(2)$) across which σ_{SO_2} varies only slightly anyway.

In summary, we feel that due to a higher resolution of our experiment interference problems will also be reduced. We expect to have actual OH data available shortly.

ACKNOWLEDGEMENT

Financial support of this work by "Bundesministerium für

Forschung und Technologie" (BMFT) is gratefully acknowledged.

REFERENCES

[1] B. Weinstock, Science 166, 224 (1969)

[2] H. Levy II, Science 173, 141 (1971)

[3] S. Hameed, J.P. Pinto, R.W. Stewart, J. Geophys. Res.84, 763 (1979)

[4] P.J. Crutzen in "Atmospheric Chemistry", E.D. Goldberg (Ed.), Springer, 1982

[5] B. Weinstock, H. Niki, T.Y. Chang, Adm. Environm. Sci. Technol. 10, 221 (1980)

[6] J.A. Logan, M.J. Prather, S.C.Wolfsy, M.R. McElroy, J. Geophys. Res. 86, 7210 (1981)

[7] C.C. Wang, L.I. Davis, Phys. Rev. Lett. 32, 349 (1974), C.C. Wang, L.I. Davis, P.M. Selzer, R. Munoz, J. Geophys. Res. 86, 1181 (1981)

[8] D.D. Davis, W. Heaps, T. McGee, Geophys. Res. Lett. 3, 331 (1976); D.D. Davis, W. Heaps, D. Philen, T. McGee, Atm. Environm. 13, 1197 (1979); D.D. Davis, W. Heaps, D. Philen, M. Rodgers, T. McGee, A. Nelson, A.J. Moriarty, Rev. Sci. Instr. 50, 1505 (1979)

[9] D. Perner, D.H. Ehhalt, H.W. Pätz, U. Platt, E.P. Röth, A. Volz, Geophys. Res. Lett. 3, 466 (1976)

[10] G. Huebler, D.H. Ehhalt, H.W. Pätz, D. Perner, U. Platt, J. Schröder, A. Toennissen, in "Physico-chemical Behaviour of Atmospheric Pollutants", B. Versino, H. Ott (Eds.), D. Reidel (1982) p. 2

[11] G. Huebler, D. Perner, U. Platt, A. Toennissen, D.H. Ehhalt, J. Geophys. Res., to be published

[12] Data sheet of Lambda Physik GmbH, Göttingen

[13] R. Zellner, J. Hägele in "Optical methods for the remote sensing of air pollution", S. Sandroni (Ed.), Elsevier, 1984, p. 351

[14] D.K. Killinger, N. Menyuk, IEEE J. Quant. Electr. 17, 1917 (1981), N. Menyuk, D.K. Killinger, C.R. Menyuk in "Optical and laser remote sensing"; D.K. Killinger, A. Mooradian (Eds.), Springer Ser. Opt. Sci. 39, 185 (1983)

EVALUATION OF ATMOSPHERIC ACIDITY - SAMPLING AND ANALYTICAL TECHNIQUES

I. Allegrini, F. De Santis, A. Febo, A. Liberti and M. Possanzini

Istituto Inquinamento Atmosferico del C.N.R. - Area della Ricerca di Roma
Via Salaria Km. 29,300 - C.P. 10 - 00016 Monterotondo Stazione (Roma) ITALY

Summary

A full methodology for the evaluation of atmospheric acidity based upon
properly designed instrumentations for sampling selected gaseous species,
suspended particulate matter and rain water has been developed.
It includes newly developed sampling heads and high performance annular denuders, high precision constant flow electronic samplers and solar powered
wet-dry deposimeters equipped with real time sensors of pH and conductivity
reading. A set-up of three annular denuders has been used for the absorption of HNO_3, HCl, SO_2 and NH_3 before collecting aerosol particles on a
back up filter. Measurement data of the chemical composition of rain water
and aerosol particles collected in parallel at three different locations
are presented and the results discussed.

INTRODUCTION

The characterization of atmospheric acidity is currently under extensive investigation in an effort to obtain a correct understanding of wet and
dry deposition. Most species responsible of atmospheric acidity are in the
atmosphere in a dynamic equilibrium with other air components and, in most
cases, at levels not directly measurable. Therefore, to obtain the real con
centration of the various species, a highly reliable measurement methodology is required. The speciation of atmospheric strong acids, namely H_2SO_4,
HNO_3 and HCl is very difficult as they have to be detected in the ambient
air simultaneously with their precursors (SO_2 and NO_x), neutralizing agents
(NH_3) and other compounds coming from sea-spray and soil erosion. Therefore,
in order to obtain a detailed insight into the complex acid-base-salt system responsible of the atmospheric acidity, a combination of a selected de
tection system with an appropriate sampling and sample treatment procedure
is needed. This investigation reports a detailed analytical procedure for
the determination of the major contributors to wet and dry deposition and
outlines the requirements which have to be met in order to draw a correct
sample.
Emphasis is given to the sampling technique, as this operation has to
ensure that a representative sample should be collected, i.e. the physical
and chemical properties of the sample should not differ from those of the
atmosphere being sampled. It is stressed that, though most analytical
methods utilized   for the evaluation of selected components are able to
yield satisfactory results with a proper precision and accuracy, the sampling
step has been largely overlooked. It can be a source of serious errors
which reflect unrealistic environmental situations. These errors arise from
the interaction between gas and particle phases during sampling resulting
in positive or negative chemical artifacts (1-3).

INSTRUMENTATION AND METHODS

Sampling train for particulate matter

The collection of particulate matter in the atmosphere should be perfor
med isokinetically i.e. with an intake velocity equal to the actual veloci
ty of the particles. This condition is however difficult to be fulfilled in
open atmosphere because of the high variability in wind speed and direction.
In order to overcome this limitation,a sampling head has been assembled as
suggested by Liu (4). It is basically built around a flat surface with a
100 mm diameter entrance which meets the Zebel conditions for a very effi
cient collection of particles smaller than 20 μm for wind speeds up to 10
Km/h. Downstream, an impactor provides a cut size of 10 μm aerodynamic diameter at a flow rate of 20 1/min. A cover protects the sampling head from
fallout and precipitation.

Particles which are not collected by the impactor flow through a series of newly developed High Efficiency Annular Denuders (HEAD), which remo
ve selected gaseous components, and are collected on a Teflon filter (Gelman
TF-1000,1.0 μm pore size, 47 mm diameter). A back-up denuder is used to col
lect ammonia released from the filter in order to correctly estimate the
concentration of particulate ammonia (NH_4^+) (5).

The sampling train includes also a novel device for the measurement and
control of the sampling inlet velocity rate which affects the cut size of
the impactor. This device, known as Real Time Flow Control (RTFC) uses a
high precision standard flow rate, an electronic transducer and a digital
valve which puts in the sampling line a programmed aerodynamic impedance.
In this way a positive feedback on the inlet velocity field is directly pro
vided in order to compensate for any change in the sampling train due to
filter clogging, changes in temperature and in the characteristics of the
pump. The RTFC device,which is controlled by a microprocessor, measures the
standard volume with precision and accuracy better than ± 1%, while the
flow rate regulation is within ± 2% of the nominal value (6).

High efficiency annular denuders (HEAD)

Diffusion tubes (denuders) have been used for sampling and separating
gaseous species from particles. They basically consist of glass tubes the
inner walls of which are coated with a suitable substance able to react
with a selected gas. In a laminar airstream through a coated tube the gas
molecules diffuse to the walls, which act as a perfect sink, while the particles, owing their low diffusion coefficients, will go through the tube
unaffected.

The use of diffusion tubes has been so far strongly limited by some
factors such as low capacity and low flow rate for achieving a satisfactory
efficiency. An additional drawback is the practical difficulty in assem-
bling several tubes in series to remove more than one reacting gas.

In order to overcome these limitations a new geometry design for diffusion tubes has been developed (7). These tubes consist of two coaxial cylin
ders so that,when air passes through the annular space,selective absorption
takes place on the coated annulus walls.

According to the theory of diffusion it has been shown that in laminar
flow through an annular denuder the term Δ in the general equation

$$\frac{C}{C_0} = A \exp\left(-\alpha \Delta\right) + B \exp\left(-\beta \Delta\right) + \ldots\ldots \tag{1}$$

which describes the fractional penetration of a species in terms of the outlet (C) to inlet (C_0) concentration ratio, can be expressed as

$$\Delta_a = \frac{\pi DL}{4F} \frac{d_1 + d_2}{d_2 - d_1} \tag{2}$$

where D is the diffusion coefficient of the gas, L the length of the denuder, F the operating flow rate and d_1, d_2 the inner and outer diameters of
the annulus, respectively. By comparing eq. (2) with the following

$$\Delta_c = \frac{\pi DL}{4F} \tag{3}$$

which applies to a cylindrical denuder, it appears that values of Δ_a much
larger than those corresponding to Δ_c can be obtained by a proper choice
of d_2 and d_1. In this way, a very high sorption efficiency can be achieved
even at high flow rates with short annular denuders.

Fig. 1 shows a comparison of the performances of both cylindrical and
annular SO_2 denuders.

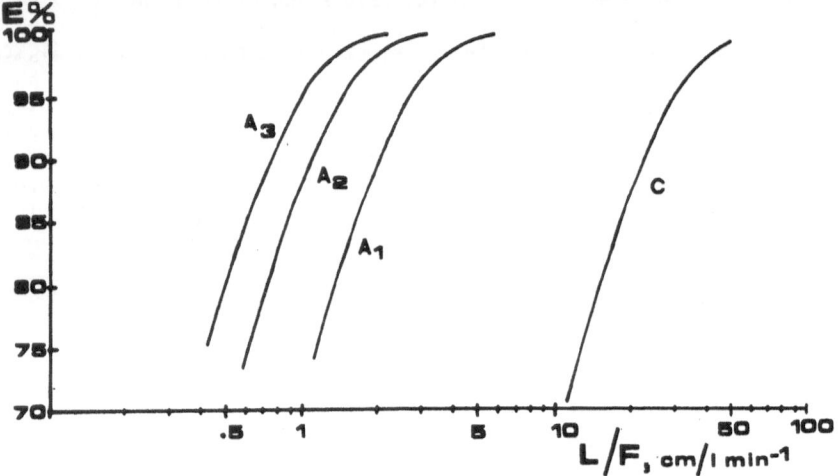

Figure 1 - SO_2 sorption efficiency curves as a function of the length/flow-
rate ratio for a cylindrical denuder (C) and varying size annular denuders ($d_1(A_1)$ = 1 cm; $d_1(A_2)$ = 2 cm; $d_1(A_3)$ = 3 cm) with
the same equivalent diameter ($d_2 - d_1$ = 0.32 cm).

By assuming for SO_2 a diffusion coefficient of 0.136 $cm^2 s^{-1}$ efficiencies larger than 95% are ensured for L/F of about 1. This means that an annular denuder 20 cm long with d_1 = 3 cm and $d_2 - d_1$ = 0.3 cm is effective
at flow rates larger than 15 l min^{-1} typical of the sampling of suspended
particulate matter. Annular denuders are easily coupled by means of flanges,
so that with a simple arrangement the collection of different gases is
achieved. Another peculiar characteristic of the device is its rather high
capacity. Measurements carried out for the sampling of SO_2 show that, if a
proper coating is adopted, pollutant amounts larger than 5 mg can be trap-

ped (8).

Wet-dry collector

The collection of precipitations has been performed by using a wet-dry collector which consists of two jars alternatively exposed to atmospheric fallout. The collectors have been assembled according to an ISO recommendation (9). The rain-drop sensor has been built by using a gold plated grid which acts as a capacitor. The capacity controls the frequency of an oscillator which changes when water droplets are deposited over the sensor. The output frequency is compared with a reference frequency from a local oscillator by means of a PLL (Phase Locked Loop) whose digital output enables a reversible DC motor. This will expose the wet collector during precipitation events and will close it again during dry periods preventing contamination by dry fallout. The PLL output enables in addition a relay to actuate exter nal devices, such as pH and electrical conductivity meters.

The collector is small in size and its low quiescent current (5 mA at 6 V) is such that the unit can be powered by a little solar panel to permit its operation even in remote areas.

EXPERIMENTAL

A speciation of the gaseous atmospheric acidity has been achieved through the separate determination of hydrogen chloride, nitric acid and am monia. A set of three denuders in series has been assembled as shown in Fig. 2.

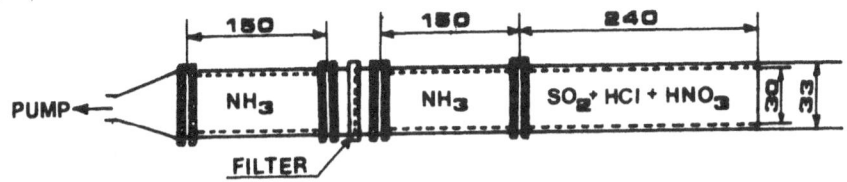

Figure 2 - Filter and denuder assembly for sampling gaseous strong acids, ammonia and particulate matter in ambient air. Units are mm.

The molecules of HCl and HNO_3 diffuse to the walls of a sodium carbonate coated denuder, where SO_2 also is trapped, whilst NH_3 is absorbed in a shorter denuder, coated with oxalic acid. Particles are collected on the fil ter, while the back-up denuder collects the ammonia which might be released from the filter. A typical sampling lasts for about 24 hrs at 15 L min^{-1} for a total air volume of about 20 m^3. After sampling, the tubes are washed out with 10 ml water and analyzed by ion-chromatography for Cl$^-$, NO$_3^-$ and by a selective electrode for NH$_4^+$. As the detection limits for the ion chromatographic and potentiometric procedures which have been used for the analysis are almost the same (0.2 μg/ml of Cl$^-$, NO$_3^-$ and NH$_4^+$), the final sensitivity of the method is 0.1 μg/m^3 for HCl, HNO$_3$ and NH$_3$. Such a high sensitivity level is a prerogative of the sampling system which has been adopted.

The filters have been extracted with 10 ml water in an ultrasonic bath

at 60°C and the solution analyzed for the above ions and $SO_4^=$. In addition
pH and electrical conductivity of filter extracts have been also measured.
Rainwater samples have been analyzed with the same procedure.

RESULTS AND DISCUSSION

In order to assess the extent of wet and dry deposition, field measurements of selected gaseous species and ionic compounds in aerosols have been
performed during fall 1983. Atmospheric particles, nitric acid, hydrogen
chloride and ammonia were collected at the Institute which is located in a
semirural area. For the sake of comparison the same measurements have also
been carried out in an urban area (Rome) and in a coastal site. Rain samples
have been collected in the semirural area only.

In table 1 are shown the arithmetic mean and the range of the concentration data for each component.

TABLE 1 - MEAN, MINIMUM AND MAXIMUM LEVELS ($\mu g/m^3$) OF GASEOUS POLLUTANTS AND
IONIC SPECIES IN AEROSOL.

Site		$SO_4^=$	Cl^-	NO_3^-	NH_4^+	pH	HNO_3	HCl	NH_3
rural	min	2.7	0.2	0.3	0.6	4.3	0.1	0.1	0.8
n=30	mean	5.9	0.5	2.2	2.5	5.2	0.3	0.5	3.8
	max	12.0	1.0	5.6	5.1	6.8	0.9	0.9	6.1
urban	min	3.7	0.4	0.7	0.8	4.9	0.4	0.3	0.5
n=30	mean	7.7	0.8	3.5	2.3	5.5	0.8	1.0	2.0
	max	11.4	1.1	5.9	6.7	6.6	1.1	1.8	3.4
coastal	min	1.8	0.5	0.4	0.8	5.2	0.3	0.6	0.6
n=15	mean	5.4	5.7	2.0	1.3	5.7	0.9	1.7	1.0
	max	11.5	32.3	4.0	1.9	6.6	1.3	2.5	1.8

With the exception of ammonia, in all sites the concentration levels are
quite similar. This is, at least in part, an expected result because ammonia
is mainly produced by biogenic activity in the soil as result of the decompo
sition of organic material. The contribution of Cl^- to the ionic composition
of atmospheric particles near the sea is, of course, considerably higher
than at the other sites. It is worthnoting that in this case the concentration of particulate Cl^- is highly variable because of the strong influence
of wind direction during sampling.

From Table 1 it can be inferred that the ground-level concentrations of
gaseous acids are rather low and their effect in the atmosphere is usually
counteracted by ammonia and, likely, by alkaline particles. These are respon
sible of the relatively high pH values of filter extracts. An additional indication on the alkaline nature of airborne matter in the areas examinated
derives from the large difference in rain acidity (up to 2 units of pH)
between the bulk and wet-only collector. This has to be ascribed both to
the nature of the soil abounding in sedimentary rocks and to dust trasport
by winds blowing from the South.

The small amounts of HCl found (< 2 μg/m3) might be attributed to the reaction of marine aerosols with acid components of the atmosphere (10).

Due to the stripping of volatile ammonium salts from the filter, appreciable amounts of ammonia have been found in the back-up denuder (5).

The results relative to pH and conductivity of the water extracts of the filters and rain samples are reported in Figure 3.

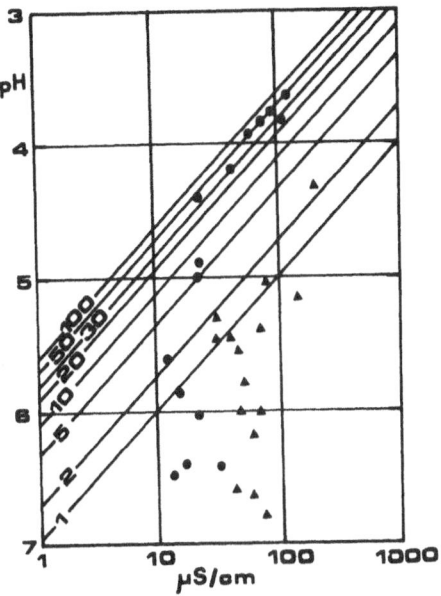

Figure 3 - Simultaneous measurements of pH and electrical conductivity in aerosol leaching solutions (▲) and precipitation samples (•) at the semirural station. The inclined lines give the fraction of free acid contributing to the dissolved material (11).

As it has been shown by Winkler (11), the pH/conductivity diagram is a direct measurement of the free acidity percentage of total water soluble matter. The plot of Fig. 3 shows that free acid content of aerosols is absent or very small (1-2%) while most data for the rain samples indicate a noticeable acid fraction (30-50%). According to these findings, it can be concluded that the acid deposition is mainly due to wet deposition.

In Fig. 4 are plotted the pH of single events during the period Oct.-Dec. 1983 together with the relative precipitation rate.

The results on the temporal evolution of wet deposition are depicted in Fig. 5a and b. The former shows that the relative contribution of sulphate is about 50-60%, that of nitrate 20-30% and that of chloride is 15-20% on an equivalent basis. These values agree well with the measurements carried out in FRG (12) and in the USA (13). Fig. 5b shows the average ionic balance of rain water samples with a pH lower than 4.5 for the same period. Sea salt, soil dust and ammonium are assumed to contribute to rain composition; the net acidity depends therefore upon the extent of neutralization of strong acids (HNO_3 and H_2SO_4) by ammonia and alkaline matter (Ca^{++}) once accounting for the contribution of sea salt (Mg^{++} and Na^+ associated with Cl^-).

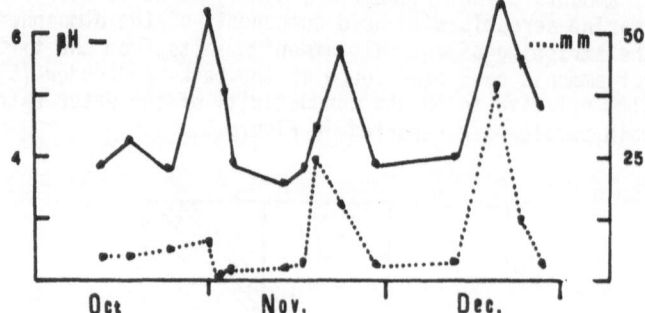

Figure 4 - Temporal variation of pH and precipitation rate for rain samples collected during the period October to December 1983.

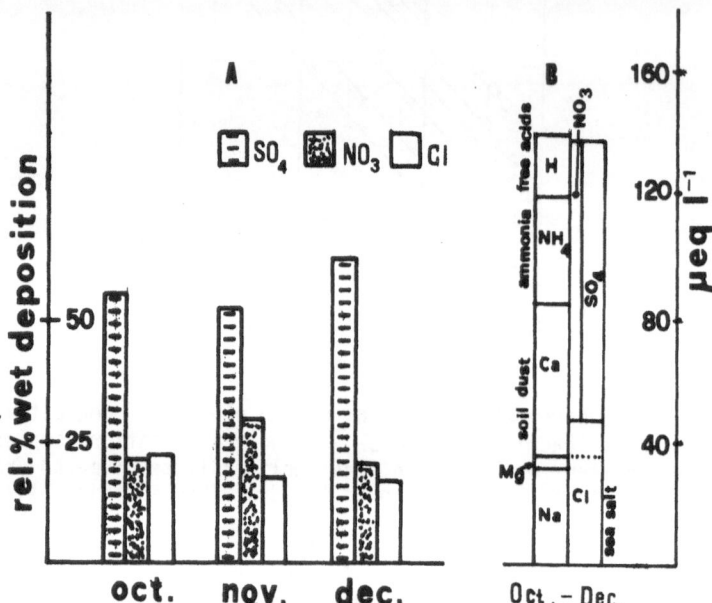

Figure 5 - a) Relative contribution of the wet $SO_4^=$, NO_3^-, Cl^- deposition to total wet deposition.
   b) Average composition of rain expressed in terms of different source contributions.

CONCLUSIONS

It has been shown how the measurements of various chemical components known to contribute to atmospheric acidity can be carried out only by the use of efficient sampling devices in connection with reliable analytical procedures. The use of proper sampling heads for particulate matter, high efficiency annular denuders for gaseous species and wet-only rain collectors permits to follow a methodology which may clarify and quantify wet and dry deposition.

Acknowledgment

The authors wish to thank V. Di Palo and R. Tappa for their assistance in field measurements and G. Calogero and M. Giusto for their contribution to the development of sampling instrumentations.

REFERENCES

1) Klockow D., Tablonski B. Niessner R., Atmos. Environ. 13, 1665-1676, (1979).
2) Pierson W.R., Brachaczek W.W., Korniski T.J., Truex T.J., Butler J.W., J. Air Pollut. Control Assoc., 30, 30-34, (1980).
3) Ferm M., Paper presented at the EMEP Expert Meeting on Chemical Matters Geneva 10-12 March 1982 EMEP/CCC Report 1/82.
4) Liu B.Y.H., Pui D.Y.H., Atmos. Environ., 15, 589-600, (1981).
5) Allegrini I., De Santis F., Di Palo V., A. Liberti. J. Aerosol Sci., in press.
6) Liberti A., Febo A., Allegrini I.. Proceedings of VIth World Congress on Air Quality, Paris, 16-20 May 1983, Vol. 1 pp. 49-56, SEPIC (Paris, 1983).
7) Possanzini M. Febo A. Liberti A., Atmos. Environ. 17, 2605-2610, (1983).
8) De Santis F., Febo A., Possanzini M., unpublished results.
9) Qualité de l'air - Mesurage des retombées particulaires. Méthode à collecteur horizontal. Project de norme internationale ISO/DIS 4222.2 (1980).
10) Hitchcock D.R., Spiller L.L., Wilson W.E., Atmos. Environ. 14, 165-182, (1980).
11) Winkler P., J. Geophys. Res. 85, C8, 4481-4486, (1980).
12) Perseke C., In: Deposition of Atmospheric Pollutants, Georgii H.W., Pankrath (Eds) pp. 77-87, Reidel Publ. Comp. Dordrecht, 1982.
13) Wilson J., Mohnen J., ASRC Publication 796, State University of NY, Albany, (1980).

ATTEMPTED MEASUREMENT OF GASEOUS H_2O_2 IN THE AMBIENT ATMOSPHERE

H.M. TEN BRINK*
Chemistry Department
Netherlands Energy Research Center ECN; Petten 1755 ZG, The Netherlands

T.J. KELLY, Y.N. LEE and S.E. SCHWARTZ
Environmental Chemistry Division
Department of Energy and Environment
Brookhaven National Laboratory; Upton, New York 11973, USA

Summary

The concentration of H_2O_2 observed in cloud water is indicative of gas phase concentrations of 0.1 to 1 ppb of H_2O_2 in the atmosphere. Recent direct measurements of gaseous H_2O_2 in the atmosphere proved to be unsuccessful because of artifact formation of H_2O_2 during sampling and analysis. In the existing methods for the determination of the gas, ambient air is passed through impingers to collect H_2O_2 in the aqueous phase. Attempts were made in the present study to overcome the artifact formation of H_2O_2 by trying different types of sampling devices and by using a new analysis technique for the measurement of dissolved H_2O_2. It was found that the artifact H_2O_2 formation occurred irrespective of the type of collector and analysis method. This artifact H_2O_2 formation is evidenced by the presence of comparable concentrations of H_2O_2 in collectors placed in series during air scrubbing. Artifact H_2O_2 was highest in highly photochemically active air masses and only on "clean" days did the amount of artifact H_2O_2 appear to be low. In laboratory studies generation of artifact H_2O_2 was found to require the presence of O_3 and one or more other unidentified precursors. In an attempt to remove O_3 from ambient air by an aluminum trap artifact H_2O_2 was found. The present observations all indicate the participation of surface reactions in the artifact H_2O_2 formation. The results establish that aqueous sampling of H_2O_2 in ambient air cannot be confidently employed until a means for elimination of artifact H_2O_2 formation is found. Further investigations on the formation of H_2O_2 from O_3 in aqueous solution are in progress.

1. INTRODUCTION

H_2O_2 is an important constituent of cloud water since it rapidly oxidizes dissolved SO_2 to H_2SO_4 and it is evidently the only oxidizing species that keeps its reactivity in an acidifying cloud droplet [1, 2]. The concentrations of H_2O_2 in cloud water can be translated to a concentration of about 0-3 ppb of gaseous H_2O_2 in the atmosphere, which is in agreement with the amount predicted from model calculations of atmospheric reactions [3]. In recent years H_2O_2 gas was measured by the method developed by Kok and co-workers [4, 5]. However Heikes et al. [6] demonstrated that this method is totally unreliable; in collecting ambient air in successive aque-

---

* The present study was conducted while on leave at Brookhaven National Laboratory

ous traps they found comparable quantities of H_2O_2 even though the collection effiency of a trap for H_2O_2 gas was greater than 99%. However, in their study the luminol chemiluminescence method was used to measure the dissolved H_2O_2 and it is well-known that this analysis method is not specific for H_2O_2. Very recently Kok and co-workers [7] modified an existing specific enzyme method [8] for the detection of H_2O_2 and improved its sensitivity. So as a first aim in the present study this enzyme technique was used for the measurement of gaseous H_2O_2 after trapping the gas in water.

From the observation that artifact H_2O_2 is formed when ambient air is contacted with water Heikes et al. [6] hypotethized that an analogous process could occur in clouds, viz that H_2O_2 is produced from the contact of air constituents and cloud water. It is in this respect that the artifact formation of H_2O_2 was further investigated since this would lead to a better understanding of the origin of H_2O_2 in cloud water. It is known from literature that H_2O_2 can be formed as a product of the decomposition of O_3 in water [6] and Hoigné and co-workers [9] found that the formation of H_2O_2 is p_H dependent. Therefore the formation of artifact H_2O_2 was investigated as a function of p_H and special laboratory experiments were performed to investigate the production of H_2O_2 from O_3 in the aqueous phase.

To examine the possibility that H_2O_2 could be formed by photochemical processes, a possibility suggested by the calculations of Chameides and Davis [10] for the formation of H_2O_2 in cloud water, measurements of ambient H_2O_2 gas were performed both at night and in the day time and in some day time tests the collection device was darkened.

2. EXPERIMENTAL

Atmospheric measurements

Ambient air was sampled by passing air through various water traps. A train of midget glass impingers (50 ml volume) filled with 10 ml of deionized water was used, analogous to the set-up chosen by Heikes et al. [6]. The air flow rate was typically 2 l. min^{-1} and the collection time thirty minutes. To check the efficiency of the samplers H_2O_2 gas in air was generated by passing air through a H_2O_2 solution. This test showed that H_2O_2 gas was trapped with an efficiency greater than 99%.

Also, H_2O_2 was collected in a cooled glass trap by dissolution in the condensing atmospheric water vapor, a method which resembles that of Farmer and Dawson [11], who used a static cooled plate to collect the water vapor and H_2O_2 gas. To study possible artifact formation of H_2O_2 in this method two condensers were used in series, in which the second one was cooled to a lower temperature. The amount of collected water is of the order of 1 ml. More than 85% of the H_2O_2 gas in the sampled air is collected in a condenser of this type.

During collection of ambient air the O_3 concentration was monitored by a standard chemilumiscent ozone monitor. Measurements were made in the period of February to September 1983 on over forty days. On several days parallel trains of samplers were used in which in one of the trains modifications were made to further investigate the artifact H_2O_2 formation. On occasion the second collection system was darkened to search for a possible direct photochemical source of the artifact H_2O_2 in the water phase. Furthermore the second collection device was proceeded by an aluminum trap which removed O_3 from the incoming ambient air, as discussed in the next section in this chapter. To further investigate the formation of H_2O_2 from the decomposition of O_3 in solution, which is p_H dependent [9], solutions of different p_H values were used in a train of impingers to search for minimum artifact H_2O_2 formation.

Laboratory experiments

Laboratory studies were performed, not with the purpose of a systematic analysis of the artifact H_2O_2 formation, but in order to better characterize the artifact associated with collection of H_2O_2 gas in ambient air measurements and to identify the important parameters in the artifact formation. In these studies O_3 was led through the same impingers as used in the ambient measurements. O_3 was produced via photolysis of O_3-free house air and air obtained from bottles of breathing air, which were chosen to simulate most closely ambient air. After the O_3 was passed through the impinger any residual O_3 in the water was purged by leading N_2 gas through the impingers. O_3 concentrations in the range of 0.1 to 3 ppm were used and the sample rate was equal to that in the ambient air measurements. The collection time was mostly 30 minutes. Aluminum was tested for the specific removal of O_3 from air. The detailed set-up of this system will be described elsewhere [12].

Analysis of dissolved H_2O_2

After sampling of air containing H_2O_2 in the aqueous traps the concentration of aqueous H_2O_2 was determined by the peroxidase catalyzed dimerization of p-hydroxyphenylacetic acid (PHOPAA) followed by fluorescent detection of the dimer [7, 12]. To distinguish H_2O_2 from organic peroxides the enzyme catalyse is added which preferentially destroys the hydrogen peroxide, leaving the organic peroxides. Thus H_2O_2 is determined in a difference method. Via flow injection 1 ml of the aqueous sample is injected into the reagent stream containing PHOPAA in a concentration of 150 mg . 1^{-1}. In addition 7.2 mg of horseradish peroxidase is present in the reagent solution which is buffered to p_H 8.5 by a TRIS buffer. The reaction between the peroxides and the reagent takes place in a delay line of 0.75 ml in which the flow rate is 1 ml-min^{-1} and the concentration of the product dimer is measured in a standard spectro fluorimetor (Perkin-Elmer 204 S) equipped with a small flow-through cell. For the elimination of the hydrogen peroxide 6 ppmv of bovine liver catalase is added to the aqueous sample some twenty seconds prior to injection of the sample into the flow line. Further details of the PHOPAA-technique will be described in the forthcoming paper [12]. The limit of detection of this method is 1.2 * 10^{-8}M (0.3 ppbm) which is equivalent to a detection limit for H_2O_2 gas of 0.03 ppbv in air using impingers at the standard collection conditions (10 ml of water; 2 1.min^{-1} air flow rate; 30 minutes sampling time); linearity is excellent up to a concentration of 1 * 10^{-4} M. In figure 1 the experimental trace obtained in a typical measurement is shown, which indicates base-line stability and repeatibility of the present method. Occasional determination of aqueous H_2O_2 were made with the luminol chemiluminescent method [4, 6] with an equivalent LOD. However large background signals appeared in an analysis sequence using the luminol technique, which were absent in the PHOPAA method. It is in this respect that the PHOPAA technique is preferred to the luminol system for the analysis of low concentrations of H_2O_2.

It was observed that only H_2O_2 and no organic peroxides were measured in the present study, with the exception of the occasions that polyvinyl chloride (PVC) impingers were tried. The organic peroxides found during the sampling of ambient air are presumably formed in a reaction of the wall material and O_3 and the results using plastic impingers were disregarded.

3. RESULTS AND DISCUSSION

Atmospheric measurements

The results obtained in the present extended study using the PHOPAA technique as analysis method for H_2O_2 are analogous to the rather limited set of results reported by Heikes et al. [6]. On most days the concentration of H_2O_2 measured in the second and third impingers was comparable to and occasionally even greater than the concentration in the first impinger, see e.g. table I. The H_2O_2 observed in subsequent impingers is evidently due to artifact formation. It appears that the presence of H_2O_2 in the second and third impinger in a series of impingers is real and not an analysis artifact, since Zika et al. [13] have also indicated the artifact using a different analysis technique for H_2O_2. The artifact formation is generally higher in summer than in the winter/spring time, as evidenced in table I. Several causes for the artifact formation could be excluded by the experiments described in the previous section. Artifact H_2O_2 is not attributable to an individual contaminant and it is not photochemically induced. Also it was found that the artifact formation of H_2O_2 was not reduced by a change of the p_H of the collection solution. Whereas O_3 was indeed removed from the ambient air by an aluminum trap it appeared that additional H_2O_2 gas was formed, which suggests that this compound is formed from a reaction of an air constituent and the aluminum surface.

When ambient air was processed via a cooled trap (condenser) the concentration of dissolved H_2O_2 in the condensate was over one order of magnitude greater than the concentration of H_2O_2 in the impingers. However the absolute amount of H_2O_2 gas trapped in a condenser was on the average two to three times less than that in the first impinger of an impinger system which was used in parallel. As will be shown in full detail in a next paper [12] the experiments using condensers give the strongest indication of the participation of wall reactions in the artifact H_2O_2 formation.

Laboratory studies

When O_3-free air was passed through impingers no measurable amount of H_2O_2 was found. However when the air was ozonated typically an aqueous concentration of 20 ppbm of H_2O_2 was found at the standard collection conditions, irrespective of the O_3 concentrations. These results demonstrate that artifact H_2O_2 requires the presence of O_3 but a kinetic calculation using literature values for the various parameters shows that this artifact is not the result of the homogeneous decomposition of aqueous O_3 [12].

The fact that the concentration of dissolved H_2O_2 is independent of the O_3 concentration is strongly indicative that the formation of H_2O_2 is limited by precursors other than O_3.

4. GENERAL DISCUSSION AND CONCLUSIONS

Attempts to collect and measure atmospheric H_2O_2 gas by the trapping of this compound in water were confounded by the formation of artifact H_2O_2 in the samplers. On some sampling days the artifact formation of H_2O_2 was relatively small as indicated by the small amount of H_2O_2 in the second impinger relative to that in the first impinger of a series, see table I. It might go too far to interpret these data as showing that the first impinger indeed collected atmospheric H_2O_2 gas. It is of importance however for further studies of the artifact H_2O_2 formation to mention the observation that on these days with low H_2O_2 artifacts the air masses were of a "clean" nature, as indicated by the wind direction, the low O_3 concentration and the good visibility. It is advisable to further quantify the composition of the ambient atmosphere also with respect to organic compounds in future studies.

Whereas laboratory studies show that O_3 is necessary for the artifact production of H_2O_2 it was found in the atmospheric measurements that a good correlation between artifact H_2O_2 and O_3 concentration is absent, see table I. Therefore it is concluded that precursors other than O_3 are required for the artifact H_2O_2 formation. This assumption is supported by the results of the laboratory experiments where the H_2O_2 formation in ozonated air is presumably limited by the concentration of other precursors. Since it was found that the artifact formation of H_2O_2 is caused by contact of ambient air and surfaces it appears improbable that a method to seperate O_3 and H_2O_2 can be developed based on a surface reaction. Methods to separate O_3 via a selective gas-phase reaction seem to be left as possible means of avoiding sampling artifacts.

ACKNOWLEDGMENTS

We thank personnel of the National Center for Atmospheric Research, Boulder CO, especially A.L. Lazrus and J. Lind for assistance in setting up instrumentation for H_2O_2 measurement and B.G. Heikes for his helpful discussions in the current experiments.
The help of P.H. Daum in the interpretation of the results using the condensers is gratefully acknowledged.

This research was performed under the auspices of the United States Department of Energy under Contract No. DE-AC02-76CH00016.

REFERENCES

[1]   a. L.W. Richards, J.A. Anderson, D.L. Blumenthal, J.A. McDonald and
       G.L. Kok and A.L. Lazrus;
       Atmospheric Environment 17 (1983), 671.

       b. P.H. Daum, S.E. Schwartz and L. Newman, in:
       "Precipitation Scavenging Dry Deposition and Resuspension, vol. 1";
       H.R. Pruppacher et al. eds, Elsevier;
       New York/Amsterdam (1983), 31-52.

[2]   L.R. Martin and D.E. Damschen;
       Atmospheric Environment 15 (1981), 1615.

[3]   L. Kleinman;
       Brookhaven National Laboratory;
       BNL report 33367 (1983).

[4]   G.L. Kok, T.P. Holler, M.B. Lopez, H.A. Nachtrieb and M. Yuan;
       Environ. Sci. Techn. 12 (1978), 1072.

[5]   T.N. Das, P.N. Moorthy and K.N. Rao;
       Atmospheric Environment 17 (1983), 79.

[6]   B.G. Heikes, A.L. Lazrus, G.L. Kok, S.M. Kunen, B.W. Landrud,
       S.N. Gitlin and P.D. Sperry;
       J. Geophys. Res. 87 (1982), 3045.

[7]   A.L. Lazrus, G.L. Kok, J.A. Lind and P.D. Sperry;
      Abstract in EOS, Trans. Amer. Geophys. Union, 64 (1983), 670.
      R.E. Schetter, A.L. Lazrus, J.A. Lind and G.L. Kok;
      Abstract in EOS, Trans. Amer. Geophys. Union, 64 (1983), 670.

[8]   G.G. Guilbault, P.J. Brignac, Jr and M. Juneau;
      Anal. Chem. 40 (1968), 1256.

[9]   J. Staehelin and J. Hoigné;
      Environ. Sci. Technol. 16 (1982), 676 and ref's therein

[10]   W.L. Chameides and D.D. Davis;
      J. Geophys. Res. 87 (1982), 4863.

[11]   J.C. Farmer and G.A. Dawson;
      J. Geophys. Res. 87 (1982), 8391.

[12]   H.M. ten Brink, T.J. Kelly, Y.N. Lee and S.E. Schwartz;
      Manuscript in preparation for Atmospheric Environment.

[13]   R.G. Zika and E.S. Saltzman;
      Geophys. Res. Lett. 9 (1982), 231.

Date	Time	Apparent H_2O_2 gas concentration, ppb 1	2	3	Analysis method	O_3 conc., ppb
2-17	10:00	0.3	0.5	0.2	PHOPAA	20
	15:00	0.4	0.7	0.3		35
3-10	11:00	0.7	0.5	0.3	luminol	20
	12:00	0.4	0.6	0.4		30
	20:00	0.5	0.6	0.2		30
3-18	15:00	0.7	0.2	0.1	luminol	25
4-26	14:30	1.7	0.7		PHOPAA	85
6-15	17:00	1.1	0.9		PHOPAA	105
		1.1	* 0.8			
	22:00	0.9	0.4		PHOPAA	85
7-20	16:00	4.1	2.4		PHOPAA	100
	21:00	1.1	1.2			110

Table I   Apparent gasphase H_2O_2 concentrations during sampling of ambient air for 30 minutes via a series of impingers, data collected in 1983. Numbers indicate successive impingers.
* Parallel collection trains

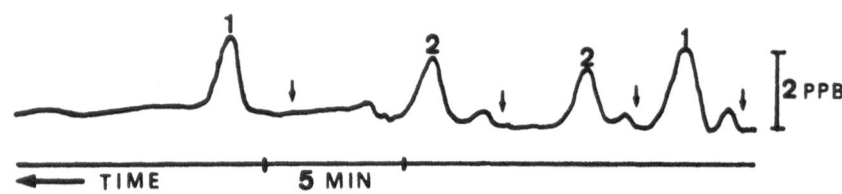

Figure 1   Experimental signal trace of a measurement of aqueous H_2O_2 via the PHOPAA analysis technique. Arrows indicate switches of the injection valve of the flow injection system, see text. The numbers indicate the sample that is being analyzed.

PROPERTIES, FORMATION AND DETECTION OF PEROXYACETYL NITRATE

U. SCHURATH, U. KORTMANN and S. GLAVAS
Institut für Physikalische Chemie der Universität
D-5300 Bonn 1   Wegelerstr. 12

Summary

The properties of PAN in the gas phase are reviewed. The convertibility to NO of organic nitrogen compounds on a hot carbon-molybdenum catalyst forms the basis for a sensitive and specific GC detector which detects PAN at the 50 pg level. When a correction is applied for heterogeneous decay on the GC column, the detector measures absolute PAN concentrations when calibrated with NO. The detector was used in a smog chamber study to determine the fractional conversion of NO2 to PAN in irradiations with individual hydrocarbons. It is concluded that a few reactive compounds are mainly responsible for high PAN concentrations in photochemical smog. By combination with a cryotrapping technique the detection limit of the GC-NO detector could be extended to 20 ppt (0.5 pg). The effect of liquid water on PAN in the cryotrap was rationalized by measurements of the deposition velocity on liquid water surfaces, which is in the order of 0.006 cm/s below pH 7.

1. INTRODUCTION

Peroxyacetyl nitrate (PAN) has been extensively studied for over two decades, mostly because of its phytotoxic and eye-irritating properties (1). However, the exceedingly high concentrations frequently observed in Southern Californian smog (1-5) have not been approached elsewhere. Average summer concentrations in the low ppb range are typical of the European countries, where the 10 ppb level is rarely exceeded (1,6-10). The main loss mechanism for PAN is monomolecular decomposition (11-13), in reversal of its formation in the atmosphere:

$$(1) \quad CH_3CO_3NO_2 \longrightarrow CH_3CO_3 + NO_2 \; ; \quad k_1, \; k_{-1}.$$

$$\log k_1(1/s) = 15.8 - 5715/T \quad (13a)$$

$$\log k_{-1}(\text{molecular units}) = -11.2 \quad (14).$$

The reaction is reversible unless the CH3CO3 radical is removed by reaction with NO or other radicals. Scavenging of the radical by reactive hydrocarbons has been studied at 335 K (13), but the reactions are strongly temperature dependent (13a), becoming negligibly slow at ambient temperature (rate constants in molecular units):

$$CH_3CO_3 + \text{ethene} \qquad \log k = -11.7 - 2200/T \; ;$$
$$+ \text{cis-2-butene} \qquad \log k = -11.6 - 1450/T \; ;$$
$$+ \text{2.3-dimethyl-2}$$
$$\text{-butene} \qquad \log k = -11.9 - 670/T \; .$$

According to k1 the stability of PAN increases dramatically with decreasing temperature, particularly at high NO2 : NO ratios.

PAN is therefore a candidate for long range transport under suitable conditions (15). This suggestion was made even before its reversible monomolecular decay had been elucidated, on account of wind tunnel experiments which yielded low deposition velocities por PAN on natural surfaces (16). This is surprising since heterogeneous loss poses a serious problem in sampling and analysis of the compound. Although the deposition velocity on water surfaces was too slow to measure in the wind tunnel, other authors stated more recently that PAN is soluble and decomposes in rain water (17), suggesting that wet removal is a loss mechanism in the atmosphere. - New efforts towards improved techniques of PAN analysis in the sub-ppb-range were stimulated by Singh and Hanst who postulated (18) that a seizable fraction of odd nitrogen in the troposphere might exist in the form of PAN, and thus be photochemically inert. Their simple steady-state estimates were borne out by one-dimensional photochemical model calculations (19,20), one of which predicts that PAN contributes 17 % to the integrated column of odd nitrogen in the natural troposphere, and up to 50 % above 2 km (20). The presence of PAN in the higher troposphere was qualitatively confirmed by air-borne measurements above the Pacific (21,22) with a mobile GC-ECD analyzer. Novel calibration techniques have also been worked out (9,23,24) which extend the calibration range down to ambient PAN concentrations. - Due to the instability of PAN, suitable techniques for trapping and storage of PAN for later analysis in the laboratory have only recently been developed (25). Such techniques are, however, needed for field investigations at feasible costs.

2. ANALYTICAL INSTRUMENTATION

In all experiments 1 - 5 ml air samples were injected with syringes into a HP 5710A GC, modified for ambient and sub-ambient temperature operation. Two identical quartz columns (60 cm x 1.5 mm i.d., 4.8 % QF-1 + 0.18 % diglycerol on chromosorb G-AW-DMCS, 80-100 mesh) were used at 30.0 ^{o}C with 30 ml/min carrier gas. Under these conditions PAN appeared well separated 2.1 min after injection. The sensitivity and low noise of the ECD would have allowed to detect less than 0.1 ppb in clean air with Ar/5 % methane as carrier gas, but a limit was set by unidentified background peaks in the polluted air near the laboratory. The other column (nitrogen carrier gas) was fitted to a specially designed chemiluminescence NO detector with 0.15 s reaction time constant (0.3 s chamber passage time at 0.65 torr pressure) via a thermal converter and gas blender for pressure equalization at the input orifice of the NO detector. Thermal conversion occurred at 330 ^{o}C on ground ultra pure graphitized carbon mixed with molybdenum powder in a quartz tube of 204 mm length which caused negligible peak broadening. Figure 1 shows conversion efficiencies as function of converter temperature for various organic nitrogen compounds. The efficiencies remained unchanged during one year of continuous operation at 335 ^{o}C. Although any convertible organic nitrogen compound in air can be used for calibrating the NO detector, NO in argon was used for convenience. The detection limit amounted to 50 pg PAN (2 ppb injected in 5 ml air). Figure 2 shows a log-log plot of peak areas versus calculated PAN concentration (1 ml injections

for the ECD, 5 ml injections for the NO detector). Initially ca.
1 ppm PAN in a 425 liter glass reactor (26) was   diluted
stepwise by partially evacuating and replenishing the vessel
with synthetic air. Both the NO detector and ECD at 35 ^{o}C are
linear within experimental error of the dilution procedure. At
100 ^{o}C the sensitivity and linearity of the ECD are considerably reduced.

Since the calibrated NO detector measures the amount of
PAN exiting from the column, a correction must be applied for
the fractional loss of PAN on the column material. This was
evaluated by varying the contact time of the sample with the
column material via the carrier gas flow rate. The peak areas
for repeated injections of PAN are plotted in figure 3 as function of contact time for two column temperatures. The decay is
non-exponential at short contact times, asymptotically approaching the calculated rate for homogeneous decomposition at 20
and 30 ^{o}C at longer contact times. A correction factor of 1.25
± 0.10 was estimated for standard working conditions (30.0 ^{o}C,
2.1 min contact time) by extrapolating the peak areas to zero
time.

The ECD could be calibrated conveniently with any test gas,
provided it contained some PAN in the ppb range, by measuring
the PAN concentration simultaneously with the GC-NO detector.
No purification of the test gas, which was prepared by photolysis of NO2 in air in the presence of trans-2-butene, was
required. The ECD was surprisingly stable over long periods of
time when operated at 35 ^{o}C. The precision of ± 15 % in these
experiments seemed to be limited by irregular PAN losses in the
syringes, and could have been improved with an automatic sampling system.

3. PAN YIELDS IN IRRADIATIONS OF NO2 WITH HYDROCARBONS

With the dual detector system, absolute PAN yields could be
determined in smog chamber irradiations of NO2 (photolysis frequency 0.355 ± 0.010 1/s) with individual hydrocarbons in synthetic air (for a description of the smog chamber facility see
reference 26). The peroxyacetyl radical is created predominantly
in reactions of OH with acetyldehyde and a few other keto and
diketo compounds (MEK, methylglyoxal), which are photooxidation
products of specific hydrocarbons. The PAN yield must thus depend on the hydrocarbon reactivity and structure (27), the NO2
: HC ratio, the concentrations, and the light intensity, in a
complex way.

Trans-2-butene was selected for a pilote study, because of
its high reactivity which brings about short irradiation times,
reduces smog chamber effects, and allows several irradiations to
be carried out per day. Figure 4 summarizes the PAN yields obtained as function of both reactant concentrations. All yields
are expressed as % NO2 converted to PAN. The yields are approximately constant for fixed HC : NO2 ratios in the concentration
range studied. A maximum conversion of about 50 % occurs at
HC : NO2 = 2.5 : 1. The concentration dependence of the PAN
yield for this fixed ratio was measured over an extended concentration range. The results are plotted on a logarithmic concentration scale in figure 5. The PAN yield does not change by more
than a factor 2 for a 400fold change in concentration, although

several hours of irradiation were required below 100 ppb NO2 to
reach the PAN maximum, and heterogeneous losses may have reduced the yield. The weak concentration dependence of the fractional conversion of NO2 to PAN could not be reproduced by a model
calculation based on a fairly detailed kinetic mechanism of the
trans-2-butene/NO2/light system.

A limited study of the light intensity effect was also
carried out. The rate of PAN formation depended linearly on
light intensity, whereas the PAN maximum decreased only by 30 %
for a 4fold reduction in light intensity.

With other less reactive hydrocarbons, higher HC : NO2
ratios were required for maximum NO2-to-PAN conversions. Figure
6 shows PAN growth curves for 5 individual hydrocarbons at a
HC : NO2 ratio of 6 : 1, except for n-butane which was used at
a ratio of 20 : 1. The PAN maxima were not reached for the less
reactive hydrocarbons. However, at higher HC concentrations,
NO2-to-PAN conversions of 35 % for propene, and 6 % for toluene,
have been measured, in fair agreement with literature data (28,
29). The low but reproducible rate of PAN formation from ethene
must be due to minor acetaldehyde yields from the addition of
oxygen atoms, and/or OH radicals, to the double bond at atmospheric pressure.

We conclude that the rapid formation of PAN in photochemical smog depends on the availability of relatively few reactive
species, particularly the butenes, propene, and acetaldehyde.
The highest PAN concentration which has been measured at Bonn
was in fact coincident with an exceptionally high propene concentration in the atmosphere (10).

4. CRYOTRAPPING OF PAN AND ITS INTERACTION WITH LIQUID WATER

The existing need of a PAN sampling technique for field
use has been outlined in the introduction. Obviously, cryotrapping is the only feasible method. After an ineffectual search
for a drying agent which does not affect PAN, a surprisingly
simple technique led to success (30):

In a cylindrical pyrex flask fitted with a teflon stopcock
and a silicone rubber septum, 5 to 10 liter air are liquefied
within about 20 minutes when the flask is immersed in liquid
nitrogen. The liquid is then removed by carefully controlled
pumping at liquid nitrogen temperature. The sample volume is
measured with a gas meter at the pump exhaust. About 40 minutes
are required for evacuation. The evacuated sample can be stored
in liquid nitrogen without detectable loss of PAN.

For analysis the flask must be warmed to room temperature
as quickly as possible with cold running tab water, and filled
with pure nitrogen to ambient pressure. At least six samples of
5 ml each can be withdrawn through the septum for GC analysis
in rapid succession. PAN was found to decay at a rate of about
2.5E-4 1/s in the wet flask. Figure 7 shows a chromatogram obtained with the NO detector. The peak corresponds to an ambient
PAN concentration of 170 ppt, which was amplified 55 fold by
the trapping technique.

The reliability of the technique was tested by irradiating
ambient air in the smog chamber until enough PAN was formed for
direct analysis with the ECD. 1 - 2 % of the chamber air were
then cryotrapped. A total of 12 samples, collected at PAN concentrations of 1 - 45 ppb in the chamber, were analyzed as de-

scribed. A plot of PAN recovered versus PAN in the chamber was linear, indicating however, that about 12 % PAN were lost in the procedure. - In another test, ambient air was sampled simultaneously with 5 cryosamplers on a well ventilated roof. Analyses with the NO detector and with the ECD yielded ambient PAN mixing ratios of 43 ± 13 and 44 ± 9 ppt (one standard deviation is given). The results of a few unsystematic field measurements in Bonn, Athens and Patras have been reported elsewhere (30).

Spicer et al. have reported (17) that PAN is soluble in acidic water and actual rain water, and is slowly decomposed in solution. Since the persistence of gaseous PAN in the presence of liquid water is crucial for the cryotrapping technique (each warmed sample contains ca. 0.1 ml liquid water), the loss rate of PAN on a water surface was investigated separately. Experiments were carried out in a pyrex cylinder of ca. 4.5 liter volume, fitted with a teflon stopcock, a septum, a teflon tube leading to the bottom of the reactor, and a miniature stainless steel fan for rapid convection of the gas in the reactor. The vessel was thoroughly rinsed with distilled water and pumped dry before each experiment, filled with synthetic air, and fully immersed in a thermostated water bath. PAN in air from a storage tank was injected with a 100 ml syringe to yield about 150 ppb in the reactor. The PAN concentration was measured with the ECD every 5 minutes. About an hour later, 200 ml distilled water which had been buffered to the desired pH was added through the teflon tube to cover the bottom of the flask. The PAN measurements were continued for at least one hour. Semilog plots of the PAN concentration versus time were linear both before and after the addition of liquid water. The change in slope yielded effective first order rate constants for the loss of PAN to the 130 cm^2 water surface. These rate constants k could be converted to deposition velocities v_D using the formula (16) $v_D = F/c = k \cdot V/A$, where F is the flux to the water surface of area A, and V the volume of the well-mixed reactor. The results are plotted as function of pH in figure 8. The deposition velocity is constant and extremely low, in the order of 0.006 cm/s, in the pH range of cloud and rain water. It increases under alkaline conditions with increasing pH, presumably due to hydrolysis in the liquid phase. Under the conditions of the measurements the overall resistance $r = 1/v_D$ was so large that the resistance r_a for absorption and/or reaction of PAN at the water surface was clearly rate controlling. In the range 276.5 to 304.5 K at pH 2.7, 283.5 to 302 K at pH 4.1, and 283.5 to 297 K at pH 6.8, the deposition velocity was independent of temperature within experimental scatter. It should be noted that the rate constant k_1 for the homogeneous decomposition of PAN varies by two orders of magnitude in the range studied here.

In summary, the deposition velocity of PAN on liquid water surfaces is extremely low under all practical conditions. Deposition of PAN on water surfaces as well as rainout must be quite inefficient loss mechanisms for PAN in the atmosphere. The results also rationalize the slow decomposition of PAN in the warm cryosamplers during their analysis, confirming the feasibility of the technique.

ACKNOWLEDGEMENTS

This work was supported by the Minister für Wissenschaft und Forschung des Landes Nordrhein-Westfalen. S.G. wishes to thank the International Bureau in Jülich for a grant.

The valuable contributions of K. Nußbaumer, W. Wendler and Brigitte Wosnitza are acknowledged.

5. REFERENCES

1   P.J. Temple and O.C. Taylor, Atmos. Environ. 17,1583-1587 (1983)

2   E.C. Tuazon, A.M. Winer and J.N. Pitts, Jr., Environ. Sci. Technol. 15,1232-1237(1981)

3   P.L. Hanst, N.W. Wong, J. Bragin, Atmos. Environ. 16,969-981(1982)

4   E. Peake and H.S. Sandhu, Can. J. Chem. 61,927-935(1983)

5   D. Grosjean, Environ. Sci. Technol. 17,13-14(1983)

6   F.J. Sandalls, S.A. Penkett and B.M.R. Jones, Atmos. Environ. 9,139-141(1975)

7   H. Nieboer and J. van Ham, Atmos. Environ. 10,115-120(1976)

8   S.A. Penkett, F.J. Sandalls and B.M.R. Jones, VDI-Berichte 270,47-54(1979)

9   P. Bruckmann und W. Mülder, Schriftenreihe der Landesanstalt für Immissionsschutz des Lndes Nordrhein-Westralen, 47,30-41(1979)

10 J. Löbel, V. Wipprecht und U. Schurath, Staub-Reinhalt. Luft 40,243-244(1980)

11 D.G. Hendry and R.A. Kenley J. Am. Chem. Soc. 99,3198-3199(1977)

12 R.A. Cox and M.J. Roffey, Environ. Sci. Technol. 11,900-906(1977)

13 U. Schurath and V. Wipprecht, 1st Europ. Sympos. Physico-chemical Behaviour of Atmospheric Pollutants, Ispra, Oct. 1979. Proceedings edited by B. Versino & H. Ott, 1980

13a U. Schurath, re-evaluation of preliminary data in ref. 13, supplemented by unpublished results

14 C.M. Addison, J.P. Burrows. R.A. Cox and R. Patrick, Chem. Phys. Lett. 73,2-5(1980)

15 T. Nielsen, U. Samuelsson, P. Grennfelt and E.L. Thomsen, Nature 293,553-555(1981)

16 J.A. Garland and S.A. Penkett, Atmos. Environ. 10,1127-1131(1976)

17 C.W. Spicer, M.W. Holdren and G.W. Keigley, Atmos. Environ. 17,1055-1058(1983)

18 H.B. Singh and P.L. Hanst, Geophys. Res. Lett. 8,941-944 (1981)

19 A.C. Aikin, J.R. Herman, E.J. Maier and C.J. McQuillan, J. Geophys. Res. 87,3105-3118(1982)

20 D.A. Brewer, T.R. Augustsson and J.S. Levine, J. Geophys. Res. 88,6683-6695(1983)

21 H.B. Singh and L.J. Salas, Nature 302,326-328(1983)

22 H.B. Singh and L.J. Salas, Atmos. Environ. 17,1507-1516 (1983)

23 W.A. Lonnemann, J.J. Bufalini and G.R. Namie, Environ. Sci. Technol. 16,655-660(1982)

24 T. Nielsen, A.M. Hansen, E.L. Thomsen, Atmos. Environ. 16, 2447-2450(1982); see also "Discussions" and "Author's Replies" in Atmos. Environ. 16,2755-2757(1982) and Atmos. Environ. 17,1855-1858(1983)

25 H. Meyrahn, J.Hahn. G. Helas. P. Warneck and S.A. Penkett, published in this volume

26 K. Henrich, H. Lippmann, U. Schurath and W. Wendler, pp. 218-227 in: Physico-Chemical Behaviour of Atmospheric Pollutants, 2nd Europ. Sympos., Varese 1981. Proceedings edited by B Versino & H. Ott, D. Reidel Publishing Co., Dordrecht 1982

27 R.A. Cox, R.G. Derwent and M.R. Williams, Environ. Sci. Technol. 14,57-61(1980)

28 H. Akimoto, H. Bandow, F. Sakamaki, G. Inoue, M. Hoshino and M. Okuda, Environ. Sci. Technol. 14,172-179(1980)

29 R. Atkinson. A.C. Lloyd and L. Winger, Atmos. Environ. 16, 1341-1355(1982)

30 S. Glavas and U. Schurath, Chimika Chronika, New Series, 12,89-97(1983)

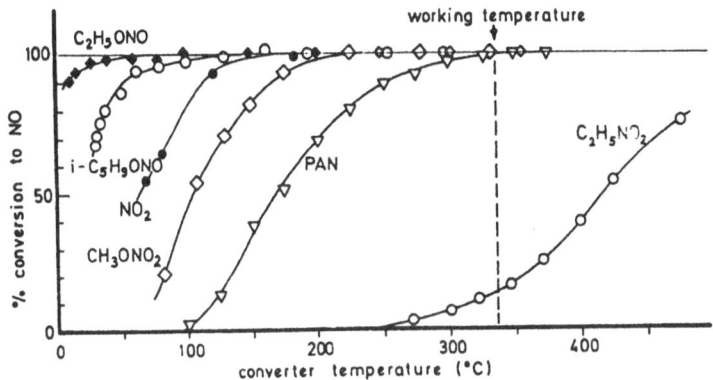

Figure 1: Conversion of NO$_2$, PAN, and some other nitrogen compounds to NO (in %) by the carbon/molybdenum catalyst, as function of temperature.

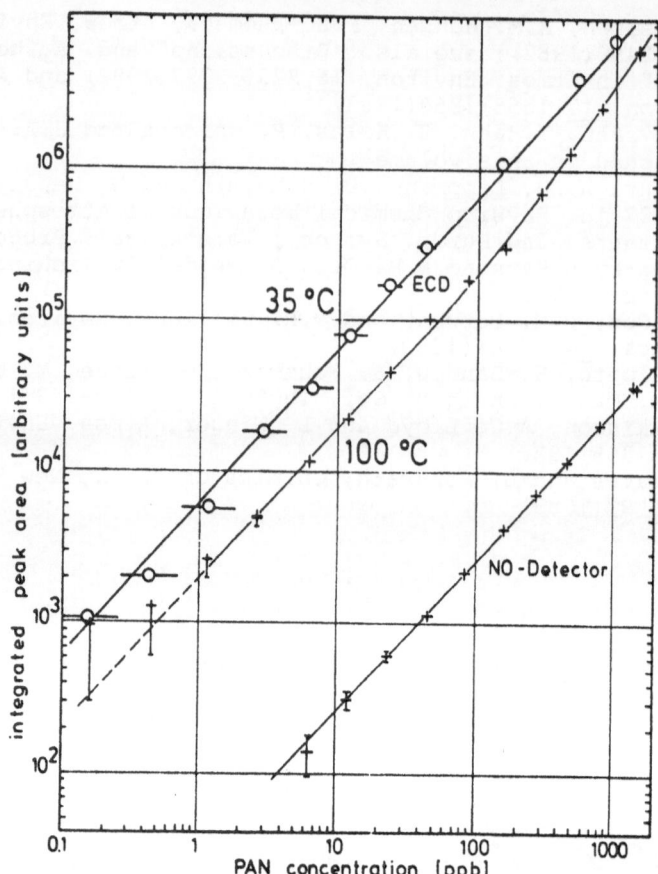

Figure 2: Linearity test of the NO detector and ECD (2 detector temperatures) for PAN. The concentrations were calculated from the pressure ratios in the pump-and-fill cycles, as described in the text.

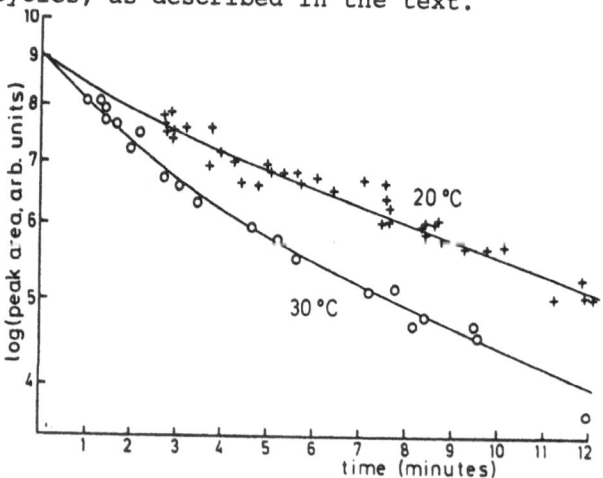

Figure 3: Semilog plot of PAN detected with the NO analyzer, as function of contact time with the column material at two temperatures.

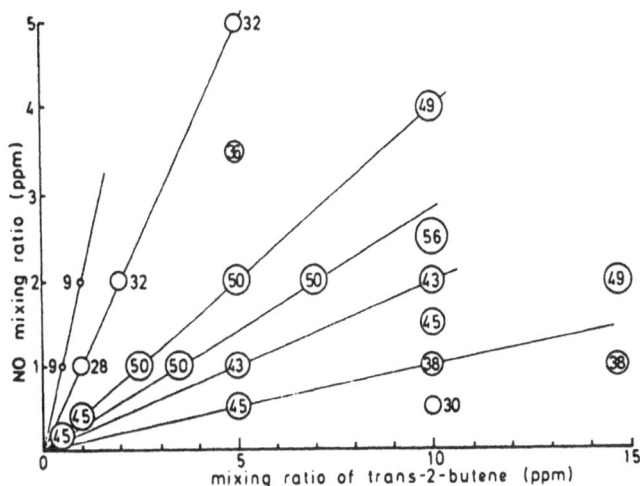

Figure 4: NO₂-to-PAN conversions in irradiations of NO₂ with trans-2-butene in synthetic air. The numbers in the diagram refer to % of initial NO₂ converted to PAN (circle diameters proportional to % conversion).

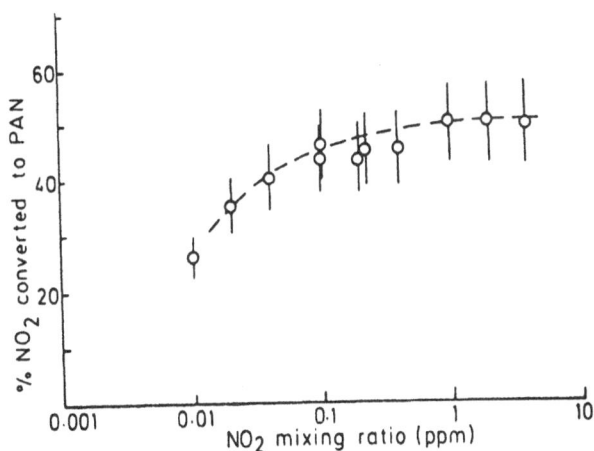

Figure 5: NO₂-to-PAN conversions, as in figure 4, but for fixed molar ratios of trans-2-butene : NO₂ = 2.5 : 1.

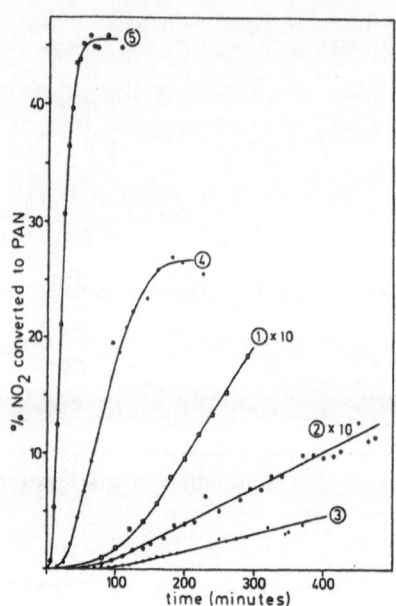

Figure 6: PAN growth curves, as % of initial NO$_2$ converted to PAN, for (1) 2 ppm n-butane; (2) 600 ppb ethene; (3) 600 ppb toluene; (4) 600 ppb propene; (5) 600 ppb trans-2-butene. Initial NO$_2$ was 100 ppb in all experiments.

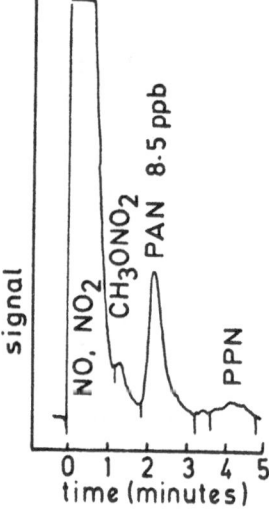

Figure 7: NO-detector-chromatogram of PAN in ambient air after cryotrapping. The first broad peak is due to NO + NO$_2$ which is also cryotrapped. Two other nitrogen compounds appear just above the detection limit and are tentatively identified by retention times. The amplification factor of the trapping procedure was 55.

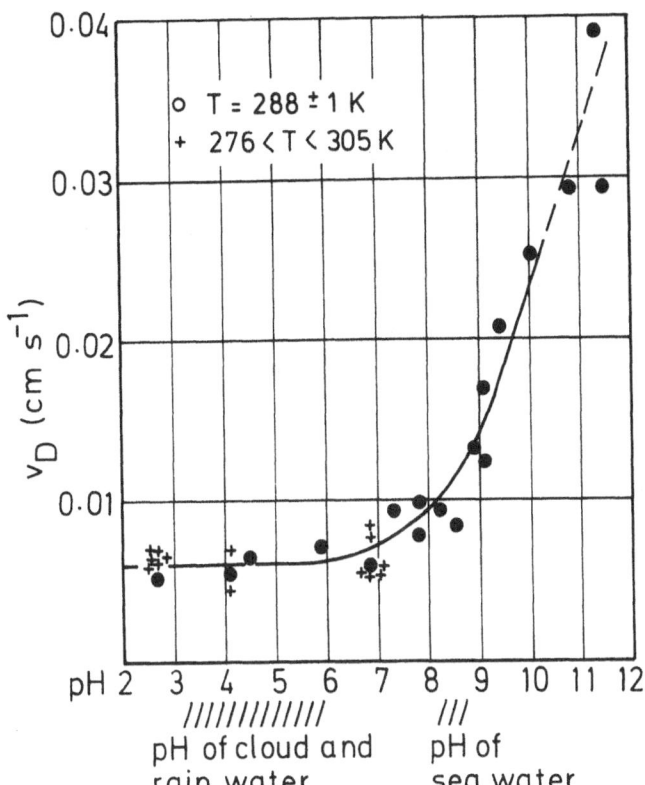

Figure 8: Deposition velocity of PAN on liquid water as function of pH, at 288 ± 1 K. Temperature was varied at pH 2.7, pH 4.1, and pH 6.8, as described in the text.

CRYOGENIC SAMPLING AND ANALYSIS OF PEROXYACETYL NITRATE

IN THE ATMOSPHERE

H. Meyrahn, J. Hahn, G. Helas, and P. Warneck
Max-Planck-Institut für Chemie (Otto-Hahn-Institut)
6500 Mainz, F.R.G.

S. A. Penkett
AERE Harwell, Environmental and Medical Sciences Division
Oxfordshire OX11 ORA   England

Summary

A cryosampling method has been developed which was
designed to collect peroxyacetyl nitrate (PAN) in air
samples onboard small air craft. Some exploratory
measurements were performed over southern Germany and an
altitude profile up to 5 km was obtained.

1. Introduction

Peroxyacetyl nitrate (PAN) is an important constituent of
photochemical air pollution. PAN is a peroxydic nitrogen
compound originating in the atmosphere from the irradiation of
certain hydrocarbons in the presence of nitrogen oxides. It
was recognized by Stephens et al. (1) to be one of the
products of photochemical smog causing plant damage and eye
irritation. As a thermally unstable compound, PAN decomposes
in the atmosphere according to the following reactions

$$CH_3C(O)O_2NO_2 \longrightarrow CH_3C(O)O_2 + NO_2$$

$$CH_3C(O)O_2 + NO \longrightarrow CH_3COO + NO_2$$

$$CH_3COO \longrightarrow CH_3 + CO_2$$

Crutzen (2) computed the residence time for PAN as a
function of altitude and noted that it is relatively stable in
the cold upper troposphere. Singh and Hanst (3) further
suggested that PAN should be present in the entire troposphere
because it is generated as a by-product during the oxidation
of ethane and propane, two ubiquitous atmospheric hydrocarbons.

In the present paper, we describe a cryogenic sampling procedure for the collection of PAN in air onboard an airplane, the subsequent analysis of the samples by gas chromatography, and the results of some measurements made at altitudes up to 5 km over southern Germany.

2. Experimental

a. Sampling procedure

Samples were collected onboard a propeller-driven twin-engine airplane (DO 28). Outside air flowed through a 3 cm o.d. teflon tube which was led through the cabin and served as a sampling manifold. Samples were taken with a calibrated syringe and then injected, one at a time, into U-shaped glass tubes fitted with shut-off teflon valves. The U-tubes were immersed and kept in liquid nitrogen until subsequent analysis in the laboratory. Storage for about 20 hours did not change the results.

b. Analysis

Analysis was performed immediately after the flights in a laboratory next to the airfield. PAN was determined by gas chromatography (gc) with an electron capture detector (ECD). Separation from other species was achieved using a 40 cm x 3 mm o.d. glass column packed with 10 percent Carbowax on Chromosorb WAW, 80-100 mesh. Detector and column were operated at 30 ^{o}C; the carrier gas flow was 35 cc. per minute of nitrogen. Under these conditions, the retention time of PAN was 6.5 minutes.

Identification of ambient PAN was achieved by comparison with chromatograms obtained from PAN synthesized in the laboratory. The synthesis followed the procedure by Nielsen et al. (4). The following tests confirmed the identification by retention time

1. The PAN signal disappeared when the air sample was passed through a scrubber, filled with aqueous 1 percent NaOH.
2. The PAN signal disappeared after heating the air sample for 4 minutes at 100 ^{o}C.
3. The PAN signal was reduced when the air sample was passed through a brass tube.
4. The PAN signal decreased with increasing ECD temperature.

c. Transfer of samples

A sampling tube was connected to the gas sampling valve which served as the gc inlet, keeping the tube immersed in liquid nitrogen. The air inside the tube was gently pumped off until the pressure was less than 1 torr. The cooling agent was then removed, the gas sampling valve turned to carrier gas flow, and the U-tube warmed up with water to a temperature of 15 ^{o}C. In this way, the content of the tube was flushed onto the gc column. A number of experiments were performed to determine the extent of losses during this procedure. The cryotrap efficiency was found to be 100 percent, no PAN was detectable in the air which was pumped off, but the amount of PAN recovered was only 85 percent, on average, of that introduced. The air craft data were corrected for this effect.

d. Calibration

The thermal instability of PAN and the high sensitivity of the ECD coupled with its limited range of linearity require special procedures for calibration. We have used two independent analytical schemes to characterize the PAN content of air samples prepared for calibration.

Synthesized PAN was used to prepare PAN-air mixtures at mixing ratios of 100-1000 ppm(v). These mixtures were passed through an aqueous 1 percent NaOH solution. Hydrolysis converts PAN quantitatively to nitrite which was in turn analysed using an azo-coupling reaction and colorimetry (5). The PAN-air mixtures were also analysed by gas chromatography using a flame ionisation detector (FID), calibrated with a PAN-air standard obtained by the above procedure. The lowest mixing ratio which the FID can detect is 10 ppm(v).

Another technique for the determination of PAN is to use a chemiluminescent NO_x analyser, fitted with a molybdenum converter which is heated to 350 ^oC. The converter reduces PAN nearly quantitatively to NO. With this technique, PAN is separated from other NO_x species by means of a short chromatographic column (30 cm x 6 mm o.d., packed with 10 percent Carbowax 400 on Diatomite CQ, 100-120 mesh). The PAN signal observed with the NO_x analyser is calibrated utilising a 4 ppm(v) mixture of NO in nitrogen. For the same mixture of PAN in air, the two analytical methods gave 102 $\pm$ 7 ppm(v) using the FID and 95 $\pm$ 2 ppm(v) using NO_x chemiluminescence detection.

The calibration of the ECD requires more dilute mixtures which were prepared by static dilution. With a sample injection volume of 5 cc., the performance of the ECD was linear for mixing ratios up to 150 ppb(v). This is below the detection limits of both the FID and the NO_x analyser used (the latter instrument was fit to analyse a 1 ppm(v) mixture in a 5 cc. sample volume). Direct comparison between both instruments was made possible without saturating the ECD by reducing to 30 microliters the sampling volume injected onto the gc column used with the ECD. In this way, a cross-check of the dilution procedure was achieved.

In the field, an NO_x analyser was not available at the time of the measurements. Therefore, a different procedure was followed taking advantage of the continuous decay of a calibrated PAN sample: A PAN-air mixture of approximately 100 ppm(v) was put into a 1 liter flask and its decomposition was recorded for about a week prior to field measurements. After that time, the PAN mixing ratio was in the range of 5 ppm(v), suitable for calibration in the field. After return to the laboratory, the decay of PAN was followed up again. The mixing ratios observed were found to fall on the extrapolated decay curve which was established earlier. Interpolation then provided the mixing ratios valid at the time of the field measurements.

3. Results of test flights

In order to test the airborne operation of the cryosampling device and to obtain some data on the vertical distribution of PAN in the lower troposphere, we participated in a series of flights which were carried out in September and October 1983 in southern Germany, starting from Oberpfaffenhofen about 40 km south of Munich. Samples were collected during day time between 11 a.m. and 2 p.m. Sample volumes were varied and ranged from 50 to 300 cc. PAN data obtained for samples taken at different altitudes are listed in Table I. With few exceptions, three to four samples were obtained at each flight level. Only the averages and the standard deviations are given in the table. The average scatter of the individual data is 20 percent except for flight no. 2 at 4.6 km altitude where two groups of values (1.49 ± 0.20 and 0.24 ± 0.06 ppb(v)) were found, indicating that different air masses were encountered at the same flight level. In the laboratory, overall precision was 8 percent (i.e. for the complete sampling and analysing procedure). Vertical profiles of PAN mixing ratios obtained from each flight are summarized in Figure 1. The general tendency is a decrease with altitude with a scale height of about 1.5 km for the first 2.5 km, then leveling off to a more moderate slope or even a constant value.

4. Discussion

From the multiple samples taken at each flight level, it appears that the reproducibility of the procedure described for sampling and analysing PAN is satisfactory. The data obtained indicate a considerable variability of PAN mixing ratios over Bavaria with values varying over at least an order of magnitude. This feature is not unexpected in view of the instability of PAN.

Backward trajectories for the 850, 700 and 500 mbar pressure levels show that on September 7 and 8 and on October 12, the air masses encountered were advected with north-westerly winds from southern England and northern France. On October 13 and 14, air mass transport occurred from southern France and Spain. On all days of the flights, the air masses sampled must be classified as continental.

Although all flights were performed during fair weather periods, October 13 was an exceptionally clear and cloudless day. The meteorological situation was characterised by subsidence of air into a high pressure region so that it is possible that the air masses sampled originated from higher altitudes than actually reached by our aircraft. The low PAN mixing ratios observed on this day suggest that values of 50 to 200 ppt(v) exist in the upper troposphere.

Our data may be compared to those of Singh and Salas (6) who made airplane measurements of PAN over the Pacific ocean west of the Californian coast. The majority of their data indicate mixing ratios from 10 to 100 ppt(v) which are lower than ours by an order of magnitude. Singh and Salas also used cryogenic sampling and found a similarly high variability of PAN mixing ratios from day to day. The difference in absolute

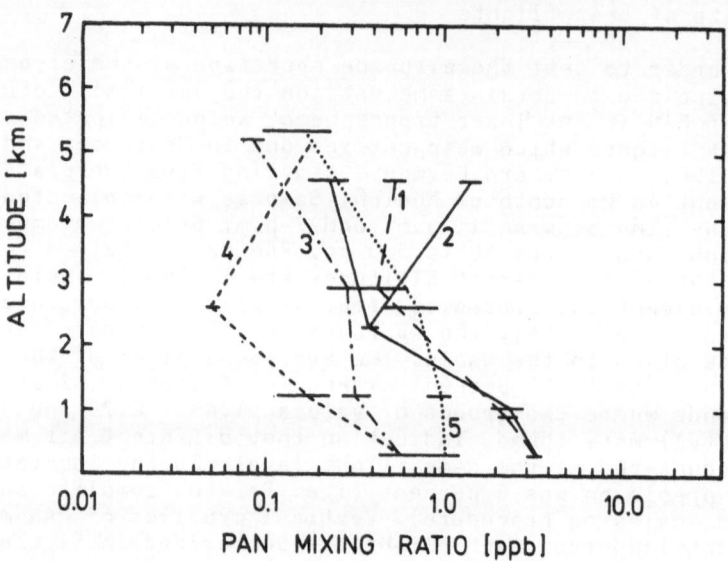

Figure 1 : Results of the test flights
PAN mixing ratios in ppb (v) as a function of altitude

Flight No (date)	Altitude (km)	PAN mixing ratio (ppbv)	Number of samples
1 (07.09.1983)	0.30	3.46 ± 0.25	3
	0.91	2.45 ± 0.25	3
	2.90	0.45 ± 0.21	3
	4.57	0.55 ± 0.08	3
2 (08.09.1983)	0.30	3.69 ± 0.20	2
	1.06	2.51 ± 0.51	4
	2.29	0.40 ± 0.04	4
	4.57	1.49 ± 0.20	2
	4.57	0.24 ± 0.06	2
3 (12.10.1983)	0.30	0.60 ± 0.12	4
	1.22	0.34 ± 0.06	4
	2.90	0.30 ± 0.07	4
	5.18	0.088 ± 0.013	4
4 (13.10.1983)	0.30	0.62 ± 0.16	4
	1.22	0.18 ± 0.06	3
	2.59	0.053 ± 0.004	3
	5.18	0.19 ± 0.004	3
5 (14.10.1983)	0.30	1.07 ± 0.24	5
	1.22	1.02 ± 0.15	4
	2.59	0.74 ± 0.26	4
	5.33	0.17 ± 0.007	4

Table I Results of test flights

values may be ascribed to the fact that the continents are source areas of PAN, whereas the oceans presumably are not. The decline of mixing ratios with height as demonstrated in Figure 1 shows that PAN originates in near-surface air and that it is partially destroyed as it is mixed upwards to higher levels. Even at 5 km altitude, the mixing ratios which we observed were still much higher than those found by Singh and Salas over the Pacific ocean. If PAN were completely stable at the 5 km height level and above, it should spread to produce a uniform background mixing ratio throughout the upper troposphere. Tentatively, we must conclude, therefore, from the difference in mixing ratios found by Singh and Salas and by us that a loss process exists whereby the mixing ratio of PAN is lowered as air masses from continental areas proceed over the oceans.

The lowest PAN mixing ratio of 53 ± 4 ppt(v) observed by us on October 13 at the 2.5 km flight level falls well into the range of data reported by Singh and Salas. It is an isolated point which cannot be taken to bridge the difference between the two sets of data. It may be more representative of the situation in the upper troposphere as explained above. If so, the upper troposphere would indeed contain a fairly uniform mixing ratio of PAN of about 50 ppt(v). Further observations are necessary to yield a better understanding of the behaviour of PAN in the troposphere.

Acknowledgements

We thank D. Stein for providing us with the air mass trajectory analysis. Research at Harwell was performed as part of an air pollution research programme sponsored by the U.K. Department of the Environment. Research at Mainz was supported in part by the Deutsche Forschungsgemeinschaft within the programme Sonderforschungsbereich 73 "Atmosphärische Spurenstoffe".

5. References

1) E. R. Stephens, Adv. Environ. Sci. 1,119 - 146 (1969)
2) P. J. Crutzen, Ann. Rev. Earth Planet. Sci.7,443 - 472 (1979)
3) H. B. Singh and P. L. Hanst, Geophys. Res. Lett. 8,941 - 944 (1981)
4) T. Nielsen, A. N. Hansen, and E. L. Thomsen, Atmos. Environ. 16,2447 - 2450 (1982)
5) F. J. Sandalls, S. A. Penkett, B. M. R. Jones, AERE report AERE-R 7807, Harwell, England.
6) H. B. Singh and L. J. Salas, Atmos. Environ. 17,1507 - 1516 (1983)

SAMPLING AND ANALYSIS OF ACETALDEHYDE IN TROPOSPHERIC AIR

B. SCHUBERT, U. SCHMIDT, and D.H. EHHALT
Institut für Chemie 3: Atmosphärische Chemie der
Kernforschungsanlage Jülich GmbH, P.O.Box 1913, D-5170 Jülich, F.R.G.

Summary

A method already used in field experiments to determine the formaldehyde mixing ratio in the atmosphere has been modified for the additional measurement of acetaldehyde. The technique consists of two basic steps: the stripping of acetaldehyde from about 200 l of air into an aqueous 2.4-dinitrophenylhydrazine solution at 3-4° C followed by separation and determination of the resulting reaction product - CH_3CHO-2.4-dinitrophenylhydrazone - by high performance liquid chromatography. Some details of the derivatization and critical parameters of the sampling procedure are discussed. Including an enrichment step by means of a precolumn, the lower detection limit is 0.03 ppbv. Laboratory experiments were performed for the determination of the collection efficiency that was found to be 82 ± 6 %. For CH_3CHO mixing ratios of 0.1 ppbv the precision of this sampling technique is about 20 %. First results from measurements in moderately polluted air are presented.

1. INTRODUCTION

Besides formaldehyde (HCHO) the most abundant aldehyde in the atmosphere is acetaldehyde (CH_3CHO); both are involved in air pollution chemistry and play a significant role during the formation of photochemical smog, for example acetaldehyde is one of the precursors of the well known smog compound peroxyacetyl nitrate (PAN) (1, 2).

The major anthropogenic source of CH_3CHO is the combustion of fossil fuels. The photochemical oxidation of hydrocarbons in the atmosphere leads to the formation of aldehydes as intermediate compounds and appears to be the main natural source.

Acetaldehyde is produced predominantly during the oxidation of ethane (C_2H_6) according to the following suggested reaction mechanism (2).

C_2H_6 is destroyed primarily by OH radicals,

$$CH_3CH_3 + OH\cdot \longrightarrow CH_3CH_2\cdot + H_2O \qquad (R1)$$

$$CH_3CH_2\cdot + O_2 + M \longrightarrow CH_3CH_2OO\cdot + M \qquad (R2)$$

$$CH_3CH_2OO\cdot + HO_2\cdot \longrightarrow CH_3CH_2OOH + O_2$$

$$CH_3CH_2OOH + h\nu \longrightarrow CH_3CH_2O\cdot + OH\cdot \qquad (R3a)$$

$$CH_3CH_2OO\cdot + NO \longrightarrow CH_3CH_2O\cdot + NO_2 \qquad (R3b)$$

$$CH_3CH_2O\cdot + O_2 \longrightarrow CH_3CHO + HO_2\cdot \qquad (R4)$$

Similar reaction chains lead from hydrocarbons having carbon numbers larger than 2 to the formation of higher aldehydes as well as to HCHO and CH_3CHO. In the vicinity of vegetation, for example the oxidation of isoprene and the terpenes is of some importance (3, 4). However, heterogeneous processes are more important since the larger radicals are rather soluble and therefore are removed from the photochemical system.

Destruction of aldehydes occurs mainly through reaction with OH radicals or photolysis. The removal of HCHO is fairly well understood (5). Aldehydes are also removed to a lesser extent by dry and wet deposition. Photolysis of CH_3CHO yields a methyl and a formyl radical (R5) but it is slower than in the case of HCHO. The preferred destruction process is reaction with OH radicals (R6 - R9). During this reaction chain PAN is generated to an extent of about 25 % (R8a) whereas the major reaction path leads to the formation of a methyl radical and CO_2 (R8b, R9).

$$CH_3CHO + h\nu \longrightarrow CH_3\cdot + CHO\cdot \quad\quad (R5)$$

$$CH_3CHO + OH\cdot \longrightarrow CH_3CO\cdot + H_2O \quad\quad (R6)$$

$$CH_3CO\cdot + O_2 \longrightarrow CH_3C(O)OO\cdot \quad\quad (R7)$$

$$CH_3C(O)OO\cdot + NO_2 \rightleftharpoons CH_3C(O)OONO_2 \quad\quad (R8a)$$

$$CH_3C(O)OO\cdot + NO \longrightarrow CH_3C(O)O\cdot + NO_2 \quad\quad (R8b)$$

$$CH_3C(O)O\cdot \longrightarrow CH_3\cdot + CO_2 \quad\quad (R9)$$

These reaction sequences show that eventually almost every CH_3CHO molecule yields a methyl radical. We know from the methane oxidation chain reactions that every $CH_3\cdot$ yields a HCHO molecule if the NO_x concentration is large enough ($\approx$ 100 pptv). Photochemical destruction of formaldehyde leads to an enhancement of the HO_x radical concentration (5) and therefore mixing ratios of the two most abundant aldehydes may be interpreted as an indirect measurement of atmospheric reactivity.

In addition to the gas phase chemistry of aldehydes, aqueous phase reactions are of interest. HCHO and CH_3CHO have been found in precipitation (6, 7) and they can react to form carboxylic acids. Rather large amounts of formic acid and acetic acid were observed in precipitation (8, 9). Some analyses have even suggested that weak organic acids may account for 63 % of the total acidity in rain (9).

The interest in measurements of aldehyde concentrations in the atmosphere is twofold: a) to investigate their atmospheric distribution, b) to estimate their impact on photochemical processes in the atmosphere. While there is a reasonable data base on HCHO (6, 10, 11, 12) - its mixing ratio has even been measured in clean marine air - the data base for CH_3CHO is rather scarce. Mixing ratios ranging from 1 to 35 ppbv were observed in polluted air (10, 11, 13). A model calculation of hydrocarbon oxidation predicts a CH_3CHO mixing ratio of about 20 pptv (14) at 2 km altitude. This calculation assumed a C_2H_6 mixing ratio of 1.5 ppbv and an OH radical concentration of 10^6 cm^{-3}. Since the oxidation of higher hydrocarbons also yields CH_3CHO it may be that the actual mixing ratios are larger, too.

Many of the analytical methods employed by various investigators are based on the derivatization of the aldehydes with 2.4-dinitrophenylhydrazine (2.4-DNPH). The reaction products, the 2.4-dinitrophenylhydrazones of the respective aldehydes, are determined by means of gas chromatography (12, 15) or high performance liquid chromatography (HPLC) (6, 10, 11, 16, 17). We have investigated whether this technique may also be used to determine the CH_3CHO mixing ratio of unpolluted air and have tried to modify the sampling and analytical technique described by Lowe et al. (6).

They passed 1-2 m³ of air at a flow rate of 40 l/min through a glass sampling tube filled with Raschig rings. The sampling tube contains about 100 cm³ of an aqueous H_2SO_4 acidic 2.4-DNPH solution. The sampler is rotated during sampling to achieve 95-100 % sampling efficiency even at these high flow rates. After separation of the hydrazones on a reversed phase (RP 18) column by means of HPLC, the derivatives are quantified by UV absorption at 254 nm. The chromatograms of air samples frequently also showed small peaks that could be identified as $CH_3CHO-2.4-DNPH$.

This investigation aimed at an attempt to extend the technique described above for simultaneous quantitative measurements of CH_3CHO and CH_2O in the atmosphere.

2. EXPERIMENTAL

2.1 HPLC analysis of $CH_3CHO-2.4.-DNPH$

Aqueous calibration standards of $CH_3CHO-2.4-DNPH$ were either prepared by adding aqueous solutions of CH_3CHO to a 2.4-DNPH solution, adjusted to pH 2.4 by sulfuric acid or by dissolution of pure $CH_3CHO-2.4-DNPH$. Such standards agree within 5 %. Usually standard solutions may be stored for several months, whereas solutions which have been used for air sampling sometimes deteriorate within a few days (6). For CH_3CHO concentrations up to about 110 ng/ml the calibration curve is linear.

We used a sample loop of 1 ml volume for direct sample injection and obtained a lower detection limit of 0.5 ng/ml. Sensitivity can be improved to 0.1 ng/ml by enriching the hydrazones from about 5 ml of the sampling solution by means of a short precolumn. With a typical air volume of about 200 l and 100 cm^3 sampling solution the lowest detectable quantity is 0.03 ppbv.

2.2 Reactions of CH_3CHO with an aqueous 2.4.-DNPH solution

Laboratory tests on the efficiency of CH_3CHO enrichment from air samples using the same technique and sampling parameters as for HCHO showed much smaller efficiencies of about 20 % to 30 % at a flow rate of 40 l/min and of only about 60 % at a flow rate as low as 2 l/min. Trapping efficiency was studied with a second sampling tube in series. To optimize the sampling conditions the individual steps of dissolution and derivatization of CH_3CHO in the aqueous 2.4-DNPH solution had to be investigated in detail.

In an aqueous solution the following equilibrium reactions must be considered (18, 19):

$$CH_3CH(OH)_2$$

$$(H^+) \Updownarrow H_2O, K_2$$

$$CH_3CHO_{gas} \underset{K_1}{\overset{+H_2O}{\rightleftharpoons}} CH_3CHO_{aq}$$

$$(H^+) \Updownarrow + O_2N\text{--}\langle O \rangle\text{--}NH\text{-}NH_2 \overset{+H^+}{\rightleftharpoons} O_2N\text{--}\langle O \rangle\text{--}NH\text{-}\overset{+}{N}H_3 \quad (NO_2)$$

$$O_2N\text{--}\langle O \rangle\text{--}NH\text{-}N{=}C\begin{smallmatrix}CH_3\\H\end{smallmatrix} + H_2O \quad (NO_2)$$

The equilibrium for physical dissolution of CH_3CHO in water is achieved instantaneously.

Hydration of physically dissolved CH_3CHO is catalyzed by the H_3O^+ ions present in the solution. The rate constant of the hydration reaction is $k_2 = 125$ $mol^{-1}s^{-1}$ at 0.3° C. The hydration constants, K_2, and the Henry constants, corresponding to the total solubility, K_3, are reported by Kurz (18). Table I lists the equilibrium constants for physical dissolution (K_1), hydration (K_2), and the total solubility (K_3).

$$K_1 = \frac{[CH_3CHO]_{aq}}{[CH_3CHO]_{gas}} = \frac{K_3}{K_2 + 1} \qquad (R10)$$

$$K_2 = \frac{[CH_3CH(OH)_2]}{[CH_3CHO]_{aq}} \qquad (R11)$$

$$K_3 = \frac{[CH_3CHO]_{aq} + [CH_3CH(OH)_2]}{[CH_3CHO]_{gas}} \qquad (R12)$$

We have determined experimental values K_3^* of the effective Bunsen solubility by calculating the ratio of the CH_3CHO concentrations measured in an aqueous CH_3CHO solution and in the gaseous phase that was equilibrated with this solution in a closed glass flask.

Table I: Equilibrium constants for the dissolution of CH_3CHO in water

$T[°C]$	K_1	K_2	K_3	K_3^*
0	318	2.52	1122	1933 ± 423
4	-	2.17	-	-
20	150	1.25	337	620 ± 55

The solubility increases significantly at lower temperature. The solubility of HCHO is much larger with $K_3^* = 1.40 \pm 0.25 \times 10^5$ at 20° C and $K_3^* = 4.41 \pm 2.98 \times 10^5$ at 10° C (20).

Accordingly, sampling at low temperatures should increase the efficiency and we investigated the reaction time of CH_3CHO with 2.4-DNPH in an aqueous solution at different temperatures. The results are plotted in Figure 1. They show that the reaction proceeds much more slowly at low temperatures, however, derivatization is completed after about 80 min at all temperatures. The same kinetics were previously observed for HCHO in this reaction (6).

Therefore the rate limiting step of the reaction sequence in the sampling solution is the acid catalyzed derivatization reaction. The reaction rate goes through a maximum with pH, because a too high H_3O^+ concentration leads to an increased salt formation of the reagent, and thus reactivity will be reduced. Laboratory experiments showed that the gain in the range pH 2.5 to 2.0 is rather small. We decided to adjust the sampling solution to pH 2.4 as a standard condition. Lower pHs would affect the chromatographic system.

The fact that both HCHO and CH_3CHO react with 2.4-DNPH at almost the same rate leaves different solubilities as the only explanation for the much smaller sampling efficiency for acetaldehyde. It appears that HCHO

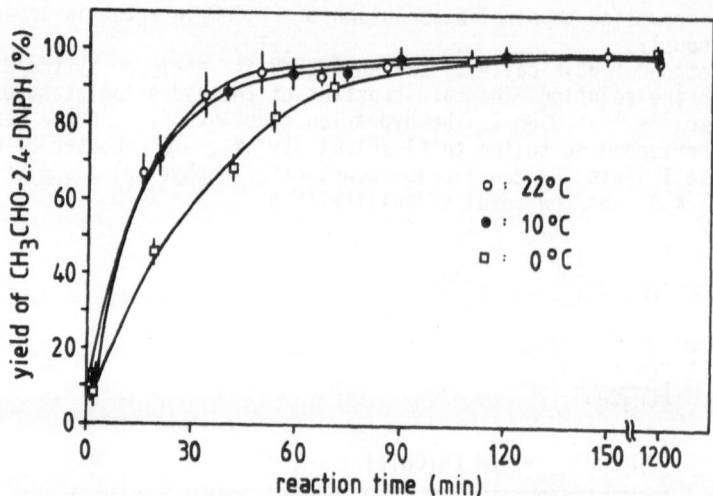

Fig. 1: CH_3CHO-2.4-DNPH reaction yield of the derivatization in aqueous
solutions as a function of time at various temperatures

is stripped at about 100 % efficiency, because it is hydrated very rapidly
to more than 99 %, i.e. the effective total solubility of HCHO is much
greater than that of CH_3CHO.

Though the derivatization equilibrium with 2.4-DNPH favours the reaction product, the degree of dissociation, α, is rather large at low concentrations as expected for atmospheric samples. Various amounts of pure
CH_3CHO-2.4-DNPH were dissolved in methanol and standard solutions were prepared by diluting 1 ml of the stock solution in 1 l of water. The standards
were immediately analysed after preparation and subsequently we monitored
the decrease of the CH_3CHO-2.4-DNPH concentration with time. At pH 6 the
dissociation proceeds very slowly, equilibrium is not reached before about
two days. However at pH 2.4 the equilibrium of dissociation is established
after only 4 hours at room temperature. The results of the experiments are
listed in Table II. From these data we calculated the equilibrium constant,
K_D, for the derivatization reaction

$$K_D = c_0 \cdot \frac{\alpha^2}{1 - \alpha} \, ,$$

where c_0 is the initial CH_3CHO-2.4-DNPH concentration in the solution. For
temperatures of about 20° C we obtained $K_D = 2.0 \pm 0.7 \times 10^{-5}$ and for 0° C,
$K_D = 3.3 \pm 0.1 \times 10^{-6}$

Table II: Degrees of dissociation and dissociation constants for CH_3CHO-
2.4-DNPH dissolved in water

T [°C[	c_0 [mol/l]	α	K_D [mol/l]
20	4.6×10^{-6}	0.83	1.9×10^{-5}
	5.5×10^{-6}	0.89	2.8×10^{-5}
	4.5×10^{-5}	0.43	1.4×10^{-5}
0	4.7×10^{-6}	0.56	3.4×10^{-6}
	5.5×10^{-6}	0.61	3.3×10^{-6}

At lower temperatures the equilibrium is further shifted towards hydrazone formation. The CH_3CHO-2.4-DNPH is regenerated by adding the 2.4-DNPH reagent to the dissociated solution. The signal is diminished by about 10 % relative to the initial signal. The time of reagent addition, elapsed since the beginning of the dissociation reaction, does not influence the peak height of the CH_3CHO-2.4-DNPH. Thus, the degree of dissociation in the sampling solution is about 10 % at 20° C.

These results suggested that more favourable sampling conditions would result if the sampling solution was thermostated at 3 - 4° C. At lower temperatures the 2.4-DNPH reagent sometimes recrystallized.

2.3 Collection efficiency studies

Passing air through a sampler spiked with HCHO-2.4-DNPH and CH_3CHO-2.4-DNPH, the HCHO concentration remains stable, but the concentration of CH_3CHO decreases. At 20° C about 30 - 50 % and at 3° C about 5 - 10 % of the initial amount of CH_3CHO are lost. This loss is probably due to the degree of dissociation occurring in aqueous solutions. Therefore a sampler kept at room temperature and containing CH_3CHO-2.4-DNPH solution can act as a calibration source. In order to determine the quantitative sampling efficiency we used a series of three rotated sampling tubes. The first sampler, loaded with standard solution of CH_3CHO-2.4-DNPH was kept at room temperature. The other two samplers were cooled to 3 - 4° C. Synthetic air was passed through the sampling device at a flow rate of 2 l/min for about 90 minutes. In addition to the trapping efficiency the performance of the cooled sampler could be checked by determining the recovery for the CH_3CHO stripped from the source. These values agree with the efficiency to within 10 - 15 %. In Figure 2 we have plotted the sampling efficiency for various amounts of CH_3CHO lost by the calibration source. Up to 4500 ng, an amount corresponding to a CH_3CHO mixing ratio of about 12 ppbv in the air sample, the efficiency does not depend on the amount of CH_3CHO passing through the sampler.

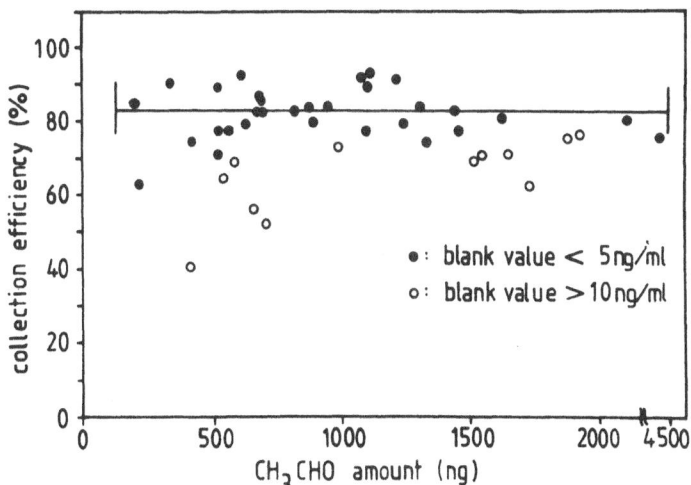

Figure 2: Collection efficiency plotted versus the amount of CH_3CHO lost by a calibration source

However, there are notable exceptions throughout the whole concentration range, with values indicating strongly reduced efficiencies. For these experiments the cooled samplers were loaded with sampling solutions having rather large blank concentrations of > 10 ng/ml. In this case the blank value was already reduced significantly by stripping CH_3CHO off the cooled sampling tube, but this loss cannot be quantitated when calculating the CH_3CHO concentration by the difference between sample and blank value. Excluding these measurements with high blank values, the collection efficiency was found to be 82 ± 6 %.

It has been suggested that addition of organic solvents such as iso-octane, cyclohexane or n-hexane to the aqueous sampling solution could improve sampling efficiency by in situ extraction of the hydrazones (16, 21). However, laboratory tests showed that the overall sampling efficiency could not be increased. Values of about 70 ± 10 % were obtained.

Collection efficiencies obtained by applying this technique have been reported in some detail by Grosjean (22) for the concentration range of about 15 to 40 ppbv CH_3CHO. Addition of a cyclohexane-isooctane mixture (9:1 v/v) to a 2.4-DNPH solution, led to a trapping efficiency of 82 ± 7 % for humid air (50 % r. h.), but only to 63 ± 5 % for dry air (< 1 % r. h.). Air was passed through impingers at a flow rate of 1 l/min.

Admixture of organic solvents does not seem to offer advantages over our technique employing cooled samplers, and implies a higher risk of introducing impurities.

3. FIELD MEASUREMENTS

A few measurements in moderately polluted areas were performed by using a sampling device with two cooled samplers in series. So efficiency could be determined in all cases.

Table III: HCHO and CH_3CHO measurements in moderately polluted regions of West Germany

Date	Time	CH_3CHOexp. [ppbv]	Eff. [%]	CH_3CHOcorr. [ppbv]	CH_2O [ppbv]
Measurements in Jülich					
02-10-83	15:15 - 16:15	3.06	86	3.56	*
02-11-83	10:30 - 11:30	1.74	94	1.85	3.21
02-16-83	11:55 - 12:55	1.94	70	2.77	*
02-17-83	12:00 - 13:15	2.60	90	2.89	5.03
01-03-84	13:30 - 15:00	0.15	100	0.15	0.01
01-05-84	14:30 - 16:00	0.32	100	0.32	1.04
01-05-84	16:20 - 17:40	0.16	72	0.22	0.87
01-06-84	12:20 - 14:20	0.44	85	0.52	1.27
01-06-84	14:50 16:50	0.24	77	0.31	0.96
Measurements in Deuselbach, Hunsrück					
10-13-83	17:45 - 18:15	0.78	76	1.02	3.22
10-13-83	21:00 - 22:30	0.33	38	0.87	3.84
10-14-83	13:50 - 15:10	0.79	55	1.44	3.57
10-14-83	16:15 - 17:45	0.26	100	0.26	2.27
10-14-83	18:30 - 20:00	0.32	68	0.48	*

* = The CH_2O concentrations have not been measured.

Experiments conducted in the Hunsrück region of West-Germany show significant variations in collection efficiency , which could not be explained. Cooling of samplers was performed with a water bath and temperature variations could not be excluded.

The measurements at Jülich were performed either using a transportable refrigerator, or simply at the ambient temperature of 2 °C in February 1983.

Table III shows the CH_3CHO mixing ratio actually found (ppbv$_{exp}$) and the concentration (ppbv$_{corr}$) corrected for collection efficiency. Simultaneously determined HCHO concentrations are also reported.

Very low CH_3CHO concentrations were found in January 1984 under conditions of strong westerly winds accompanied by clouds and rainshowers whereas the weather was quite calm during the measurements in February 1983. As expected, a correlation of the HCHO and CH_3CHO concentrations has been found.

The precision of the sampling technique is about 20 % for a CH_3CHO mixing ratio of 0.1 ppbv.

4. CONCLUSIONS

Extending the sampling technique employed for HCHO to simultaneous measurements of CH_3CHO is only possible with some restrictions. Sampling parameters have to be altered to a flow rate of 2 l/min and thus an additional enrichment step was necessary for determining CH_3CHO concentrations in the sub-ppbv range.

To avoid irreproducible losses of CH_3CHO during the sampling procedure, it is necessary to maintain sampling conditions exactly. Above all, the blank values have to be controlled and maintained constant. It seems that the solubility of aldehydes in the aqueous reaction medium is of great importance and thus temperature has to be kept constant too. Nevertheless, trapping efficiency could not be increased to more than 82 ± 6 %.

We will continue measurements in the moderately polluted area of Jülich. Additional experiments are planned in marine air to obtain data on the diurnal variation and latitudinal dependence of acetaldehyde concentrations in unpolluted regions.

5. REFERENCES

(1) Graedel, T.E., L.A. Farrow, T.A. Weber, Kinetic studies of the photochemistry of the urban troposphere. Atmos. Environ. 10, 1095-1116 (1976)

(2) Aikin, A.C., J.R. Herman, E.J. Maier, C.J. McQuillan, Atmospheric chemistry of ethane and ethylene. J. Geophys. Res. 87, 3105-3118 (1982).

(3) Graedel, T.E. Terpenoids in the atmosphere. Rev. Geophys. Space Phys. 17, 937-947 (1979)

(4) Zimmerman, P.R., R.B. Chatfield, J. Fishman, P.J. Crutzen, P.L. Hanst, Estimates on the production of CO and H_2 from the oxidation of hydrocarbon emissions from vegetations. Geophys. Res. Lett. 5, 679-682 (1978)

(5) Calvert, J.G., The homogeneous chemistry of formaldehyde generation and destruction within the atmosphere. Federal Aviation Agency Report No. FAA-EE-80-20, 153-190 (1980)

(6) Lowe, D.C., U. Schmidt, D.H. Ehhalt, The tropospheric distribution of formaldehyde. Berichte der Kernforschungsanlage Jülich, No. Jül-1756, 5170 Jülich, F.R.G. (1981)

(7) Thompson, A.M., Wet and dry removal of tropospheric formaldehyde at a coastal site. Tellus 32, 376-383 (1980)

(8) Chameides, W.L., D.D. Davies, Aqueous-phase source of formic acid in clouds. Nature 304, 427-429 (1983)

(9) Keene, W.C., J.N. Galloway, J.D. Holden, Jr., Measurements of weak organic acidity in precipitation from remote areas of the world. J. Geophys. Res. 88, 5122-5130 (1983)

(10) Grosjean, D., Formaldehyde and other carbonyls in Los Angeles ambient air. Environ. Sci. Technol. 16, 254-262 (1982)

(11) Kuwata, K., M. Uebori,Y. Yamasaki, Determination of aliphatic and aromatic aldehydes in polluted airs as their 2.4-dinitrophenylhydrazones by high performance liquid chromatography. J. Chromatogr. Sci. 17, 264- 268 (1979)

(12) Neitzert, V., W.Seiler, Measurements of formaldehyde in clean air. Geophys. Res. Lett. 8, 79-82 (1981)

(13) Hoshika, Y., Simple and rapid gas-liquid-solid chromatographic analysis of trace concentrations of acetaldehyde in urban air. J. Chromatogr. 137, 455-460 (1977)

(14) Singh, H.B., P.L. Hanst, Peroxyacetyl nitrate in the unpolluted atmosphere: An important source for nitrogen oxides. Geophys. Res. Lett. 8, 941-944 (1981)

(15) Papa, L.J., L.P. Turner, Chromatographic determination of carbonyl compounds as their 2.4-dinitrophenylhydrazones. I. Gas Chromatography, J. Chromatogr. Sci. 10, 744-747 (1972)

(16) Grosjean, D., K. Fung, R. Atkinson, Measurements of aldehydes in the air environment. Paper No. 80-50.4, 73rd Annual Meeting of the Air Pollution Control Association, Montreal, Canada, June 23-27 (1980)

(17) Johnson, L., B. Josefson, P. Marstorp, G. Eklund, Determination of carbonyl compounds in automobile exhausts and atmospheric samples. Intern. J. Environ. Anal. Chem. 9, 7-26 (1981)

(18) Kurz, J.L., The hydration of acetaldehyde. I. Equilibrium thermodynamic parameters, J. Am. Chem. Soc. 89, 3524-3528 (1967)

(19) Roberts, J.D., M.C. Caserio, Basic principles of organic chemistry, W.A. Benjamin Inc., New York, Amsterdam (1965)

(20) Lowe, D.C., personal communication (1981)

(21) Selim, S., Separation and quantitative determination of traces of carbonyl compounds as their 2.4-dinitrophenylhydrazones by high pressure liquid chromatrography. J. Chromatogr. 136, 271-277 (1977)

(22) Grosjean, D., K. Fung, Collection efficiencies of cartrigdes and microimpingers for sampling of aldehydes in air as 2.4-dinitrophenyl-hydrazones. Anal. Chem. 54, 1221-1224 (1982)

OPTIMISATION DES METHODES DE PRELEVEMENT ET D'ANALYSE

DES HYDROCARBURES AROMATIQUES POLYCYCLIQUES ET DE LEURS DERIVES

AZOTES - DETERMINATION DE LEUR STABILITE DANS L'ATMOSPHERE

M.A. BRESSON, S. BEYNE, P. MASCLET et G. MOUVIER[*]
Université PARIS VII - Laboratoire de Physico-
Chimie Instrumentale - 75251 PARIS CEDEX 05. -

Summary

A highly sensitive and fast method for analyzing polycyclic aromatic
hydrocarbons by HPLC with fluorimetric detection was developed. This
method permits to execute daily several samplings and treat them
without any risk of degradation during filter collection and analysis,
so that it is possible to observe significantly the diurnal variation
of these compounds. From this study, a scale of PAH reactivity in
atmosphere was determined.

De nombreux hydrocarbures aromatiques polycyliques (HAP) formés dans
différentes combustions (bois, charbon, essence, gas oil, fuels ...) sont
mutagènes ou cancérigènes. Leurs teneurs dans l'atmosphère ne sont pas très
élevées ; aussi est-il fondamental de bien maitriser les techniques d'échantillonnage et d'analyse. Nous avons recherché une méthode rapide et très
sensible de manière à pouvoir limiter le temps de prélèvement. Avec cet
outil, il a été possible de suivre les variations journalières de différents
HAP et en tirer des conclusions intéressantes sur leurs comportements physico-chimiques dans l'atmosphère et en particulier sur leur stabilité relative.

METHODES D'ECHANTILLONNAGE ET D'ANALYSE

Les HAP sont des composés très réactifs susceptibles de se transformer
sous l'action de la lumière et d'autres polluants. Pour limiter les pertes
lors du prélèvement, il suffit d'effectuer ces prélèvements pendant des
temps courts : avec la filtration à haut volume, on obtient des quantités
mesurables en 1 heure pour des atmosphères urbaines et 3 heures pour les
atmosphères peu polluées. L'appareil utilisé est un HIVOL Sierra 305 -
2000 X, permettant un débit ajustable et constant de 50 m3/h. Les particules
sont retenues sur un filtre en fibre de verre Whatman GF/A de 20 x 25 cm,
et les gaz sont ensuite piégés sur un copolymère de styrène et de vinyl-
benzène commercialisé sous le nom de résine XAD 2. L'extraction des filtres
et de l'adsorbant XAD 2 peut se faire suivant différentes techniques utilisant des solvants et des temps variables suivant les auteurs.
L'efficacité de l'extraction par ultrason et de celle par Soxhlet a
été comparée, différents mélanges de solvants polaires et non polaires ont
été testés et l'influence du temps d'extraction a été étudiée (6).

Ces différents essais ont conduit à retenir l'extraction au Soxhlet par un mélange cyclohexane - chlorure de méthylène (2/3) pendant 3 heures. Les solutions d'extraction sont évaporées à sec et le résidu est repris dans 2 ml de méthanol. Cette méthode s'est révélée tout aussi efficace pour les filtres que pour les XAD 2.

Parmi les méthodes d'analyse d'HAP : chromatographie gazeuse, spectro-fluorimétrie à basse température, chromatographie liquide à haute perfor-mance (5), cette dernière avec détection fluorimétrique a été choisie essen-tiellement à cause de sa grande sensibilité. D'autre part, cette méthode permet d'analyser les échantillons sans séparation préalable. Certains auteurs (1) ont constaté que pour une même phase mobile, les colonnes de silice greffée en C 18 de différents fabricants ont des sélectivités diffé-rentes. Cette comparaison nous a fait choisir une colonne Merck Lichospher C 18, 5 μ pour l'analyse des HAP parents et une colonne Vydac 210 TP, 5 μ pour l'analyse des azaarènes. Une phase mobile permettant, dans le temps le plus court possible, une bonne séparation des HAP parents, de l'anthra-cène au coronène, a été recherchée en comparant différents mélanges de sol-vants couramment utilisés : eau - méthanol ou eau - acétonitrile. Les sépa-rations les plus délicates sont celles d'une part du triphénylène, benzo a anthracène, chrysène, d'autre part du benzo b fluoranthène, benzo k fluoran-thène, benzo e pyrène. Nos essais ont montré que le mélange eau - méthanol était favorable à la séparation chrysène benzo a anthracène, alors que le mélange eau - acétonitrile était favorable à la séparation BbF, BkF, BeP. D'où la mise au point d'un gradient ternaire eau - méthanol - acétonitrile qui donne une séparation de tous les HAP en une vingtaine de minutes (Voir Figure 1). Cette séparation est affinée par le choix de deux couples de longueurs d'onde : 297-430 et 313-390. Ainsi, la quantification de 14 HAP est obtenue en 1 heure 30.

Les HAP gazeux et piégés sur la résine XAD 2 sont principalement des composés à 2 ou 3 cycles et leurs dérivés méthylés : la rapidité d'élution du gradient ternaire ne permet pas une bonne séparation de ces composés. Avec des isocratiques eau - méthanol ou eau - acétonitrile, à forte propor-tion d'eau, la séparation est complète, mais l'élution lente. Le mélange eau - acétonitrile a été retenu pour sa meilleure résolution. Un gradient de pente assez faible permet d'accélérer l'élution sans trop perdre sur la sélectivité. En 15 minutes, une bonne séparation de ces composés est obte-nue comme le montre la figure 1.

L'analyse des azaarènes est souvent effectuée après séparation entre fractions acide, neutre et basique (3). Nous avons constaté qu'une extrac-tion acide simplifiait effectivement l'analyse, mais que son rendement était variable suivant les composés (Tableau 1). Le gradient mis au point, eau ammoniaquée - méthanol avec un choix de longueurs d'onde approprié per-met d'identifier des azaarènes sans préparation initiale (Voir Figure 2).

Les mesures réalisées dans les émissions de véhicules diesels ont montré que les azaarènes étaient présents en quantité extrèmement faibles (Tableau 1). On constate par ailleurs l'absence de dérivés aminés primaires. Aussi est-il possible de procéder à la réduction de l'échantillon en pré-sence de borohydrure de sodium, suivant la méthode de Gibson : les dérivés nitrés deviennent ainsi des dérivés aminés primaires que l'on peut analyser en chromatographie liquide haute performance. Des essais sont en cours pour étudier le rendement.

Cette méthode permet d'effectuer dans une journée plusieurs prélèvements qui sont analysés rapidement, limitant les risques de dégradation des
échantillons. On peut ainsi suivre les variations journalières de chacun
des HAP particulaires. Ces profils ont été déterminés dans des conditions
météorologiques très différentes. Nous présentons les résultats pour des
prélèvements effectués un jour d'hiver avec une température moyenne de 9 °C
et un jour d'été avec une température de 28 °C. Ces prélèvements ont été
réalisés au Centre de Paris, à 20 m d'altitude où l'on peut considérer que
les sources prépondérantes de HAP sont les automobiles en toute saison et
le chauffage en hiver.

L'allure générale des profils journaliers en hiver et en été est à peu
près similaire (Figure 3). On observe un maximum entre 8 h. et 9 h. ;
7 heures plus tard, et cela pour tous les composés, la courbe passe par un
minimum; le soir un autre maximum moins important apparaît. Il vient tout
de suite à l'idée que ces maxima pourraient correspondre aux pointes journalières de la ciculation. Cependant, si les variations journalières des
HAP étaient dues uniquement aux variations de circulation, les courbes de
HAP devraient toutes être identiques les unes aux autres, et devraient également être similaires à la courbe de variation du trafic dans la journée.
Comme cela n'est pas le cas, il faut bien admettre que ces variations ont
d'autres causes, notamment la transformation physico-chimique de ces
composés.

Pour étudier cette transformation, nous avons introduit le rapport
entre la concentration maximale de la journée et la concentration minimale
qui le suit. Afin d'évaluer la réactivité des HAP à partir de ce rapport,
il faut dans un premier temps s'affranchir de la variation des sources. Les
statistiques indiquent qu'à Paris la quantité d'énergie dépensée pour le
chauffage au cours d'une journée, notamment entre 8 et 20 h., est pratiquement constante. Donc, on peut considérer que la contribution du chauffage
domestique constitue une sorte de bruit de fond constant de la courbe de
variation journalière hivernale des HAP. Pour les véhicules, les comptages
montrent que la diminution des véhicules entre l'heure du maximum de nos
courbes et l'heure du minimum est de 20 % aussi bien en été qu'en hiver.
Il est donc possible de calculer un rapport R(HAP) corrigé de la variation
de la circulation véhiculaire (Tableau 2). On peut alors considérer que
cette diminution est due :

- aux réactions dans l'atmosphère avec des oxydants, favorisées essentiellement par la lumière, mais également par la température.
- à la volatilisation des HAP par déplacement de l'équilibre dans l'
aérosol entre phase gazeuse et phase particulaire. Elle dépend de la température et de la masse moléculaire de chaque HAP (voir Tableau 3).
- à la dilution dans l'atmosphère par augmentation de la hauteur de la
couche de mélange qui dépend des conditions météorologiques, mais agit de la
même manière sur tous les HAP.

Le tableau 2 mettant en évidence une différence très nette de comportement pour les différents HAP analysés, nous avons pensé établir une échelle
de réactivité (7). Pour s'affranchir des facteurs agissant de la même manière sur tous les HAP, nous avons défini un indice de stabilité relative
(ISR). Le fluoranthène a été pris comme référence : en effet, la présence
d'un pseudo plateau dans sa variation journalière indique qu'il est le
composé le plus stable. Donc l'indice de stabilité relative est défini par
la formule :

$$ISR = R(HAP)/R(FLA)$$

Etant donné que les maxima et les minima des courbes de variation jour-nalière sont simultanés pour tous les composés, il est intéressant de cal-culer un temps de vie moyen relatif. En 7 heures, les concentrations passent de leur valeur maximale à leur valeur minimale, d'où :

$$TVR = \frac{7}{ISR} \quad \text{(en heures)}$$

Ces valeurs, reportées au tableau 4 sont évaluées à 20 % près.
On observe que l'indice de stabilité relative est en général 2 fois plus élevé en été qu'en hiver, et que sa variation n'est pas corrélée avec la variation de masse molaire. Ces remarques montrent que l'effet de la conver-sion gaz - particules est très faible dans la transformation des HAP, et que les HAP évoluent essentiellement par réaction avec d'autres polluants oxydants.

Il faut bien souligner qu'il s'agit d'un indice relatif : ainsi le fait qu'un composé ait le même indice en été et en hiver ne signifie pas que la réactivité est la même en été et en hiver, mais seulement qu'il se comporte de la même manière que le fluoranthène.

Notre but a été d'établir une échelle de réactivité in situ. D'autres auteurs ont réalisé des études sur la réactivité des HAP, sous des condi-tions atmosphériques simulées : par exemple, l'étude de Butler et Crossley (2) sur des HAP adsorbés sur des particules de suie dans l'air contenant 10 ppm de NO_x a donné l'échelle suivante :

BaP > BghiP > BaA > Pyr > BeP > CHR > FLA > COR

Katz et al. (4), en faisant réagir dans l'obscurité des HAP avec de l'ozone, ont obtenu l'échelle suivante :

BaP > Ant > DiBahAnt > BaA > BeP > Pyr > BkF > BbF

On constate qu'à 3 composés près, l'échelle "in situ" est en accord avec ces deux échelles. Ce résultat confirme l'hypothèse suivant laquelle la transformation des HAP est essentiellement due aux réactions avec d'au-tres polluants.

L'intérêt de cette étude réside dans le fait que nous avons établi une échelle de réactivité in situ, qui tient compte de toutes les transfor-mations possibles des HAP. A l'avenir, nous nous proposons d'établir une variation de l'indice de stabilité en fonction des divers paramètres inter-venant dans les réactions photochimiques, de suivre la conversion gaz - particules et l'évolution des HAP, notamment en dérivés nitrés.

BIBLIOGRAPHIE

1. K.D. BARTLE, M.L. LEE, S.A. WISE. Chem. Soc. Rev. 10, 113 (1981).

2. J.D. BUTLER, P. CROSSLEY. Atm. Env. 15(1), 91 (1981).

3. M. DONG, D.C. LOCKE. J. Chrom. Sci. 15, 32 (1977).

4. M. KATZ, C. CHAN, H. TOSINE, T. SAKUMA. Third Int. Symp. on Polynuclear Aromatic Hydrocarbons, P.W. JONES, P. LEBER, Ed., Ann Arbor Science, p. 171 (1979).

5. M. LAMOTTE, P. MASCLET. 4th. Int. Conference Chemistry for Protection of the Environment, TOULOUSE (France), Sept. 1983.

6. P. MASCLET, K. NIKOLAOU, G. MOUVIER. Poll. Atm. 95, 175 (1982).

TABLEAU 1

RENDEMENT DE L'EXTRACTION ACIDE ET CONCENTRATIONS D'AZAARENES DANS LES EMISSIONS PARTICULAIRES DES VEHICULES DIESELS

COMPOSES	1AzFla	1AzPyr	Acr	B(a)Acr	B(c)Acr	Me2B(a)Acr	Me7B(c)Acr	DiBahAcr	Fla	Pyr	BaP
Rendement d'Extraction %		21	68	21	12	81	85				
Teneur Emission diesel (ng/m3)	0,58	0,48	7,00	6,10	5,80	1,12	0,49	0,82	620	625	245

TABLEAU 2

RAPPORT ENTRE MAXIMA ET MINIMA D'UNE JOURNEE R(HPA)

	BaP	BaAnt	Cor	DiBahAnt	BghiPer	InPyr	BbF	BhF	Ant	Pyr	Chrys	BeP
Hiver	11	9	8	7	5	5	5	5	5	5	4	4
Eté	21	9	10	14	13	12	9	9	9	5	9	8

TABLEAU 3

REPARTITION DES HAP ENTRE PHASE GAZ ET PHASE PARTICULE

PHASE GAZ/PARTICULES	Ant	Fla	Pyr	BaAnt	BbF	BkF	BaP
Prélèvement atmosphérique t° = 9 °C	0,9	0,32	0,60	0,17	0,01	0,01	particules uniquement
Prélèvement émission Centrale charbon t° = 110 °C		0,54	2,22	0,38	1,00	0,60	
Prélèvement émission automobile – t° = 200 °C		8,00	17,00	5,00	0,60	0,30	0,50

TABLEAU 4

INDICE DE STABILITE RELATIVE ET TEMPS DE VIE MOYEN RELATIFS DES HAP PARTICULAIRES DANS L'ATMOSPHERE URBAINE

		BaP	BaAnt	Cor	DiBahAnt	BghiPer	IndPyr	BbF	BhF	Ant	Pyr	Chrys	BeP
Hiver	ISR	3,7	3,0	2,7	2,3	1,7	1,7	1,7	1,7	1,7	1,7	1,3	1,3
(en h.)	TVR	1,9	2,3	2,6	3,0	4,2	4,2	4,2	4,2	4,2	4,2	5,3	5,3
Eté	ISR	7,0	3,0	3,3	4,7	4,3	4,0	3,0	3,0	3,0	1,7	3,0	2,7
(en h.)	TVR	1,0	2,3	2,1	1,5	1,6	1,8	2,3	2,3	2,3	4,2	2,3	2,6

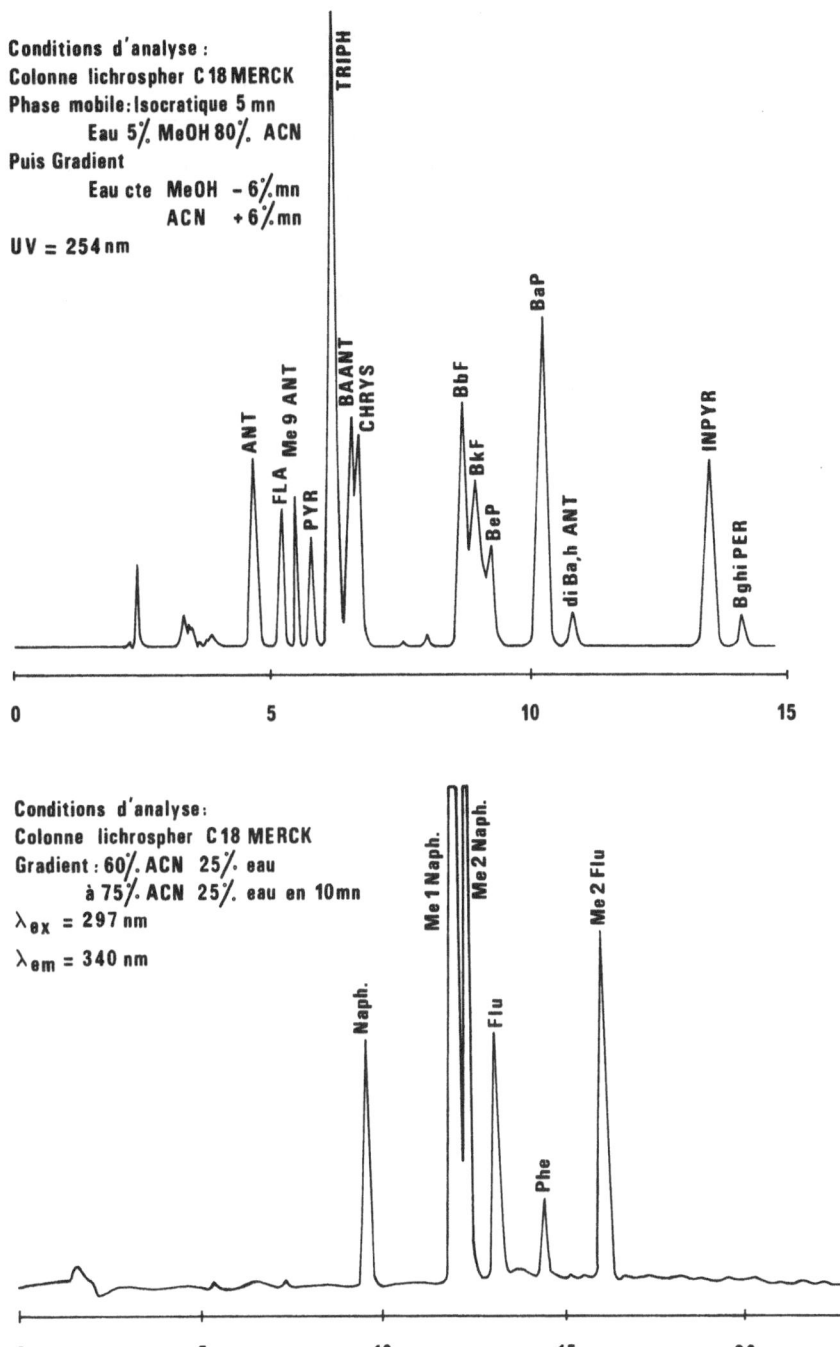

Conditions d'analyse :
Colonne lichrospher C18 MERCK
Phase mobile: Isocratique 5 mn
        Eau 5% MeOH 80% ACN
Puis Gradient
        Eau cte MeOH – 6%/mn
                 ACN   + 6%/mn

UV = 254 nm

TRIPH
ANT
FLA Me9 ANT
PYR
BAANT
CHRYS
BbF
BkF
BeP
BaP
diBa,h ANT
INPYR
Bghi PER

0   5   10   15

Conditions d'analyse:
Colonne lichrospher C18 MERCK
Gradient : 60% ACN 25% eau
         à 75% ACN 25% eau en 10mn
λ_{ex} = 297 nm
λ_{em} = 340 nm

Me1Naph.
Me2Naph.
Me2 Flu
Naph.
Flu
Phe

0   5   10   15   20

Fig. 1   HPLC of PAH in AIRBORNE PARTICULATES

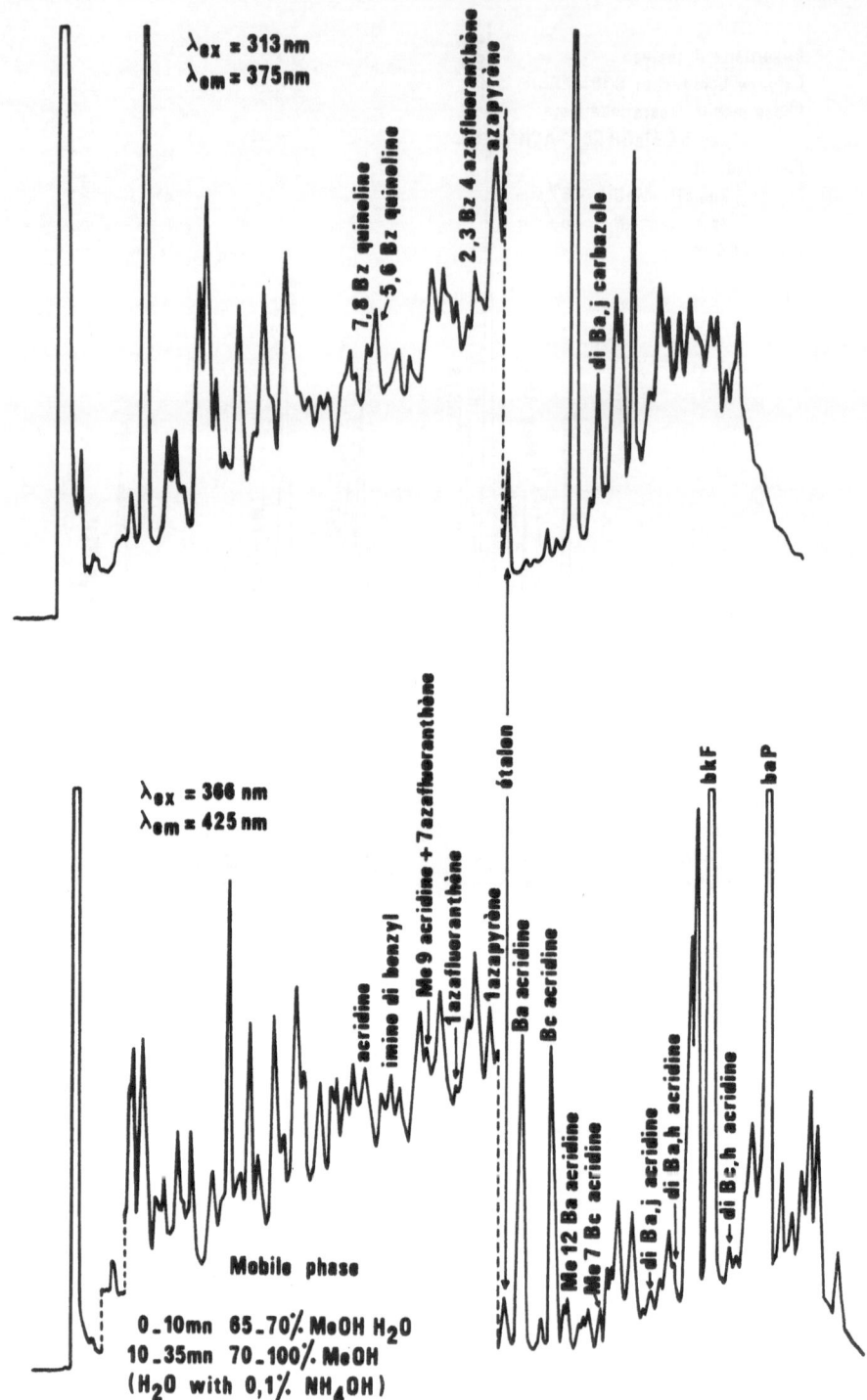

Fig. 2 <u>HPLC</u> OF AZAARENES AND AMINO – PAH IN AIRBORNE PARTICULATES

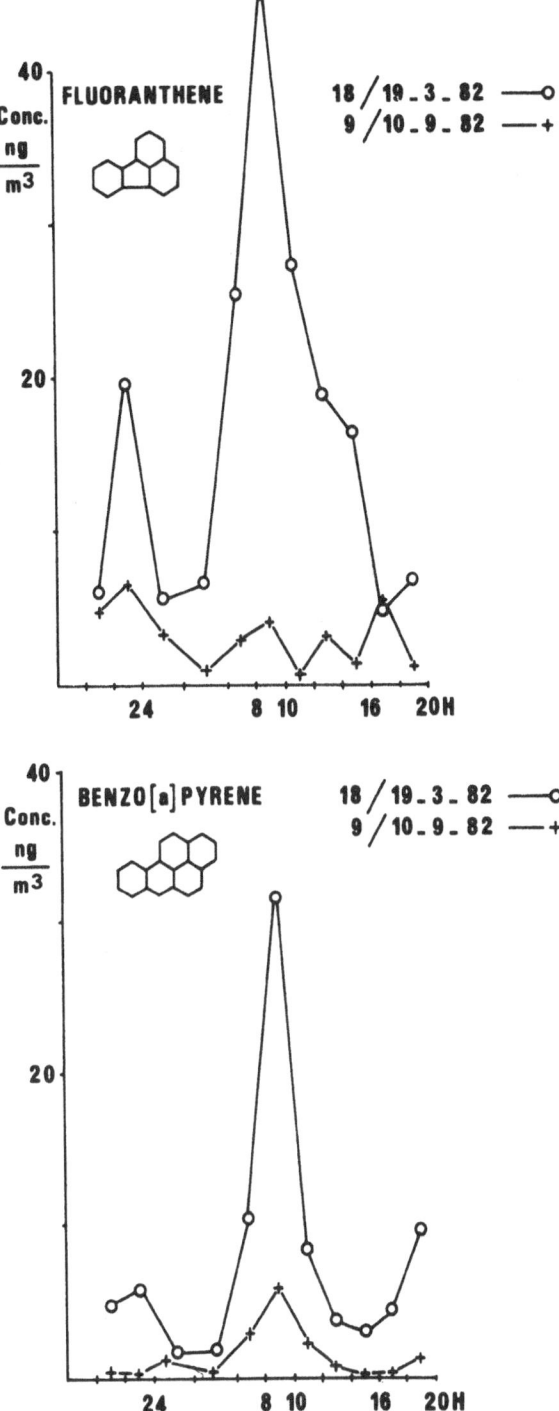

Fig. 3 Variation de la concentration des HAP au cours de 2 journées
typiques de l'hiver (o) et de l'été (+) à Paris

SAMPLING, IDENTIFICATION AND QUANTITATIVE DETERMINATION OF BIOGENIC AND ANTHROPOGENIC HYDROCARBONS IN FORESTAL AREAS

P. Ciccioli, E. Brancaleoni, M. Possanzini, A. Brachetti and C. Di Palo

Istituto Inquinamento Atmosferico del C.N.R. - Area della Ricerca di Roma
Via Salaria km. 29,300 - C.P. 10 - 00016 Monterotondo Stazione (Roma) ITALY

Summary

An analytical procedure for the collection, analysis and quantitative determination of biogenic and anthropogenic hydrocarbons (HC's) from C_5 to C_{10} in forestal areas is presented. The samples are collected on adsorption traps packed with Carbopack B which is a highly inert material with respect to oxidizing agents and allows a complete recovery of the various components by programmed thermal desorption. Monoterpene and alkylbenzene HC's present in a pine forest at ppt level can be quantitatively determined by using a GC-MS apparatus operating in the Selected Ions Detection (SID) mode. Diurnal variations of the HC's concentration within a pine forest are presented together with the variation of the meteorological parameters and the O_3 and NO_2 concentration. The measurements were carried out during sunny days when photochemical smog episodes were observed.

INTRODUCTION

In recent years, increasing attention has been devoted to find out whether possible environmental effects may arise from biogenic HC's emitted from vegetation. The results of this research can be summarized as follows:

a) The most abundant natural components which can participate to atmospheric reactions are isoprene and monoterpenes emitted from tree foliage (1)

b) These compounds rapidly react with OH radicals and O_3 to form peroxy radicals which, in turn, can polymerize into particles or be converted into gaseous pollutants (2).

c) Global emission of biogenic HC's is estimated to be of the order of several Mtons per year and appears to be higher than that arising from anthropogenic and vehicular sources (3).

d) The amount of isoprene and monoterpenes measured in ambient air is small and becomes significant only within the forest canopy (4).

Based on these results, it has been suggested that a substantial build-up of particulate matter and photochemical oxidants should be observed in forestal areas when the solar radiation, temperature and emission rate of HC's reach a high value. The formation of blue-haze sorrounding forests during windless, warm and sunny days could be thus explained by gas into particle conversion of biogenic HC's (5, 6). Recently, the contribution of biogenic HC's to the O_3 and particle formation has been questioned (7). It has been pointed out that not only the amount of biogenic HC's surviving atmospheric reactions with O_3 and OH radicals is too low when compared to the global emission but, more, important the ratio between anthropogenic and natural HC's measured in remote areas is higher than that predicted on the basis of emission estimate and reactivity. It has been suggested that overestimation of biogenic sources is coupled to underestimation of the anthropogenic contribution and the environmental impact of natural HC's might be neglegible ($\leq$ 10%). Computer models developed for simulating a multi-days photochemical smog episode (8) are in agreement with these conclusions. However, most of the present uncertainties still remains because of the lack of measurements and the insufficient specificity and sensitivity of the present analytical procedures. Such methods might be also affected by sampling artifacts. The availability of analytical procedures capable to specifically detect both anthropogenic and natural HC's in the ppb-ppt range would permit a more accurate determination of the emission rates from air measurements and a better evaluation of the anthropogenic impact. In this paper a method for the simultaneous determination of alkylbenzenes and monoterpenes is presented. The method is used for measuring diurnal variations of the HC's within the forest canopy when photochemical smog episodes are observed. Hourly variations in the HC's concentration are reported together with the change of the meteorological parameters and the O_3 and NO_2 concentration.

SAMPLING OF ANTHROPOGENIC AND BIOGENIC HC's IN AIR.

To minimize possible sampling artifacts arising from leaks, diffusion through the walls and valves or decomposition of the polymeric materials, the use of plastic bags for the sampling of HC's was avoided and enriching traps packed with solid adsorbents preferred. Five different adsorbents, selected on the basis of their chemical nature (non specific) and surface area,were tested. The choice of the best adsorbent was made on the basis of the following criteria:
  i) sampling efficiency for HC's ranging from C_6 - C_{10},
 ii) chemical stability to the oxidizing agents (mainly O_3)
iii) possibility of analyzing as much sample as possible in a single run.
Standard NIOSH traps (5 cm x 4 mm i.d.) were used for evaluating the adsorption features of Active Charcoal whereas tubes of 15 cm x 4 mm i.d. were filled with the other adsorbents. The traps were all in glass and protected from sunlight with aluminium foil. The adsorption efficiencies were measured by injecting into traps packed with different adsorbents known amounts of alkylbenzenes and highly reactive olefines (including some naturally emitted compounds) from C_6 to C_{10} and by measuring the amount of each compound recovered when 2,5 and 10 L of air containing 1.9 ppm of CH_4 and 0.5 ppm of HC's from C_2 to C_4 were passed through the adsorbent. The injection of the sample was made under constant flow of air (100 ml/min) by avoiding condensation of the gaseous components into the injector. Standard gas mixtures were prepared by using the head-space technique. The experiments were repeated by adding 0.5 ppm of O_3 to the eluting air. The compounds were recovered either by liquid extraction (0.5 cc CS_2 with Active Charcoal and Carbopack B, 0.5 cc CH_3OH with Porapack T) or programmed thermal desorption (the rate was 50°C/min, the final temperature 250°C). Thermal decomposition of the olefines was prevented by using glass coated metal tubes in the construction of the desorption apparatus (9). Trapping efficiencies measured for the various adsorbents are shown in Figure 1. Similar results were obtained when 0.5 ppm of O_3 was added to the eluting air. Although Active Charcoal and Porapack T give high percent recoveries, only 1% of the total sample can be injected into the column (the extract is 0.5 cc) and the sensitivity of the method is so poor that large volumes of air need to be sampled to obtain detectable peaks. Tenax GC, which can be ther-

mally desorbed, gives lower recoveries for natural compounds than Carbopack B but, more important, undergoes a chemical decomposition in the presence of O₃. The products formed which are shown in Figure 2, can interfere with the species to be determined. Based on these observations, Carbopack B appears to be the most suitable adsorbent for the determination of natural and anthropogenic HC's. For ambient air measurements, traps of 25 cm x 6 mm i.d. were used in order to collect HC's with 5 carbon atoms and to prevent sample losses when high relative humidity is measured in the atmosphere. The aspirating pump used was controlled by an electronic device developed

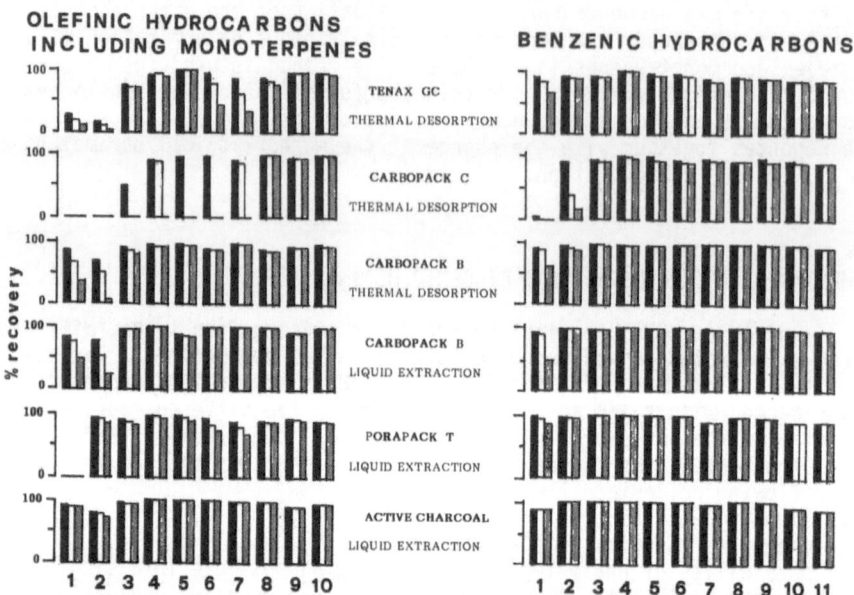

Figure 1 - Percent recovery of olefinic and benzenic HC's measured on various adsorbents when different volumes of air (■ 2 L, ▭ 5 L, ▨ 10 L) were passed through the trap. Olefines: 1) hexene 1,2) bicycloheptadiene,3)heptene 1,4)vynilcyclo-hexene,5) octene 1,6)cis,cis octadiene,7) α- pynene,8)care-ne,9) α -terpinene,10)limonene. Benzenic HC's: 1)benzene,2) toluene,3)ethylbenzene,4)o-xylene,5)m-xylene,6)p-xylene,7) 1,3,5 trimethylbenzene,8) n-butylbenzene,9)1,3 diethylbenze ne,10)1,3,4,5 tetramethylbenzene,11)1,2,3,4 tetramethylben‾ zene.

in our Institute ;it allows the measure of the volume sampled with a preci-sion better than ± 1% in the range of temperatures between 5 and 50°C. To minimize the effect of particulate matter on the sampling of gaseous HC's, a Teflon filter (0.25 μm) was placed at the trap inlet. The filter was pro-tected from rain and large particles with a PTFE shield especially designed.

SELECTION OF THE BIOGENIC AND ANTHROPOGENIC INDICATORS

To better evaluate the performances of our method, a forestal site suf-

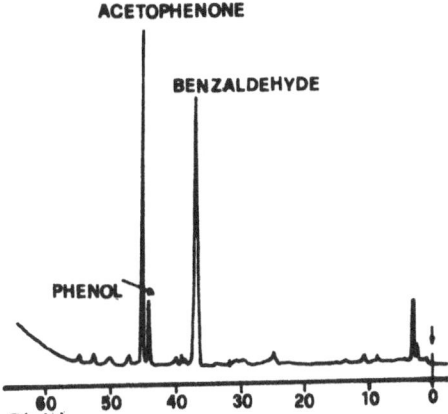

Figure 2 - GC-MS determination of the decomposition products of the Tenax GC polymeric matrix obtained by passing 5 L of air containing 0.5 ppm of O_3 through the traps.

ficiently close to the city but not directly polluted by industrial emissions was selected for the sampling of HC's. The site was located in the countryside 30 Km NE of Rome. The altitude was 30 m above the sea level and the prevalent winds were blowing from W-SW (30%). The average concentrations of the major pollutants measured in the closest sampling station (200 m,SW) are reported in Table I.

TABLE I - AVERAGE DAILY CONCENTRATION OF THE MAIN POLLUTANTS MEASURED NEAR THE SAMPLING SITE.

SO_2 ($\mu g/m^3$)	NO_2 ($\mu g/m^3$)	HNO_3 ($\mu g/m^3$)	NH_3 ($\mu g/m^3$)	O_3 (ppb)	CH_4 (ppm)	THC (ppm)
2-4$_s$ 7-12$_w$	20-30	0.3	5	15$_w$ 60$_s$	1.9	0.07-0.18

Particulate matter: 20 - 40 $\mu g/m^3$

Alkanes: ca. 1 $\mu g/g$; PAH: ca. 40 ng/g; $SO_4^=$: ca. 5 $\mu g/m^3$; NO_3^-: ca. 2 $\mu g/m^3$.

w = winter;   s = summer

The sampling site was within a small forest (15,000 m^2) of evergreen 10 - 15 years old. The species present in the forest were, in order of abundance, Pinus Silvestris, Cupressus Arizonica Conica, Pinus Pinea, Pinus Nigra and Pinus Aleppensis. The forest was sorrounded by a farmland were sparse eucaliptus trees and oaks were present. Figure 3 shows the typical chromatographic profiles observed when a 10 L sample collected below the forest canopy (1 m from the ground) was analyzed with the multidetection unit described elsewhere (10).
The separation of the various components was carried out on a 3mx2mm GC column packed with Carbopack B 80-100 mesh coated with 0.5% PPE 20. After 1 minute at 70°C the temperature of the column was raised up to 235°C using a linear program at a rate of 3°C/min. Figure 4 shows the HC's distribution when the same air volume was analyzed by GC-MS. The various traces are relative to the different classes of HC's selectively detected by plotting the change of the intensity of specific ions   vs.   time (reconstructed chro-

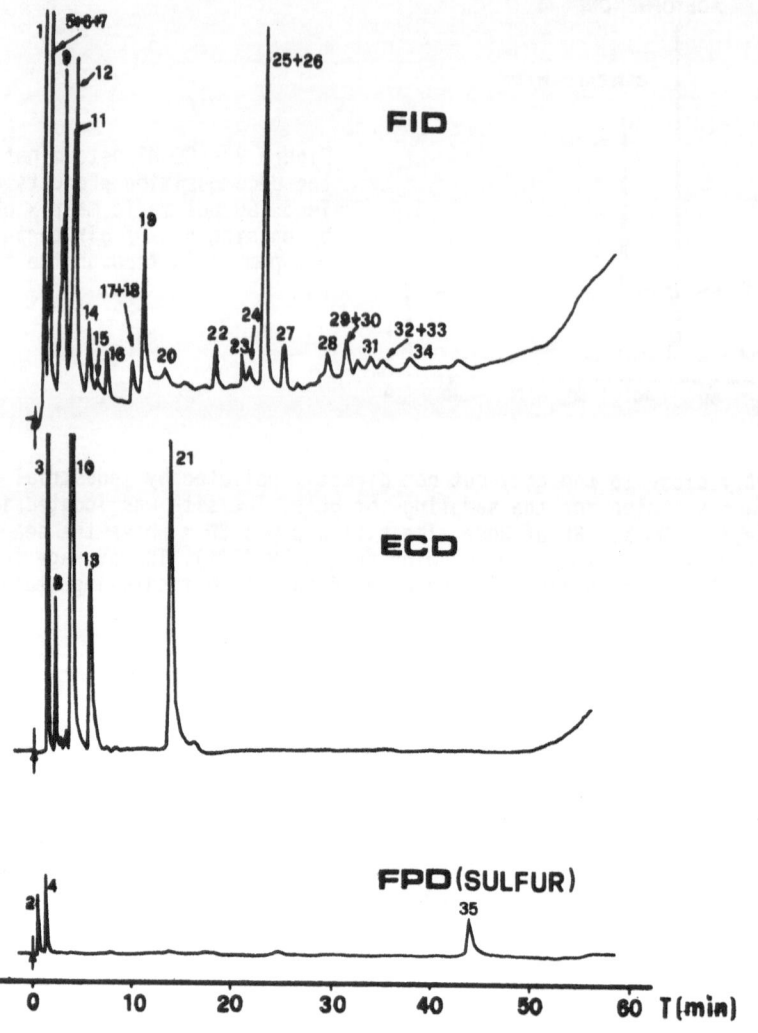

Figure 3 - GC analysis of a 10 L air sample collected in a Pine forest.
Simultaneous traces were obtained by using a multidetection
unit equipped with three detectors (FID, ECD and FPD). For
the peak number see   Table II.

matogram). These preliminary experiments were carried out during winter
season (October 20). By combining the information of Figure 3 and 4 most of
the compounds desorbed from the trap were identified and quantified; the re
sults are shown in Table II.

From these results it appears clearly that, within the range of carbon
atoms fully recovered from the trap, alkylbenzene and monoterpene HC's are
the most suitable indicators for the identification of anthropogenic (mainly vehicular) and biogenic sources. Other possible indicators (such a linear or branched olefins emitted from automobile exhausts) are present in
such a small concentration that their monitoring by GC or GC-MS is dif-

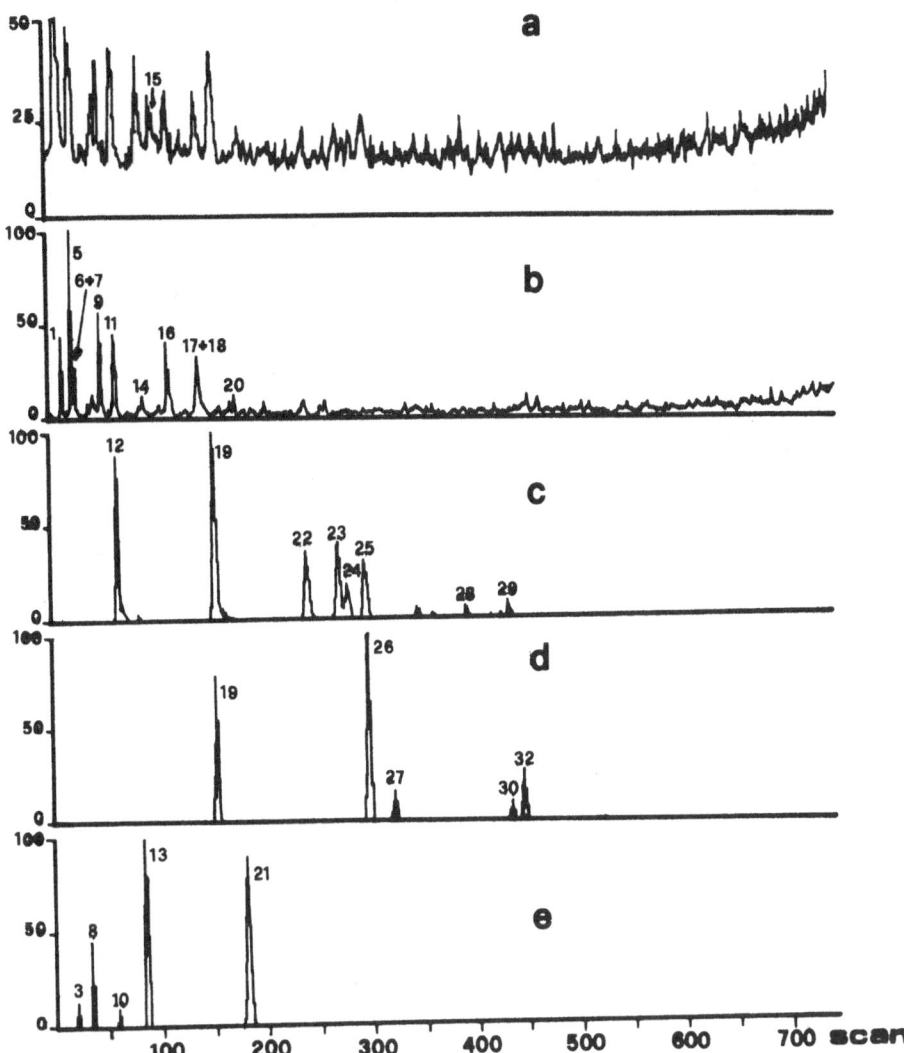

Figure 4 - Reconstructed chromatograms of a 10 L air sample collected in a Pine forest. a) Total Ion Current, b) Alkanes, c) Benzenic HC's, d) Monoterpenes, e) Chlorofluorocarbons. For the identification see Table II.

ficult to be performed with sufficient accuracy. For air monitoring, the specificity and sensitivity of the method was greatly improved by using the MS apparatus (a VG - Micromass 70-70 F) in the Selected Ions Detection (SID) mode. Figure 5 shows a comparison between the SID traces recorded during the analysis of an air sample collected in the forest (A) and those obtained from the vapours emitted from Pinus Silvestris (B)and Cupressus Ari zonica Conica (C). Chromatogram B and C were obtained by injecting directly into the GC column 0.5 cc of vapours sampled from a 50 cc glass vial contai

TABLE II - CONCENTRATION OF THE HC's IDENTIFIED IN THE AIR SAMPLES SHOWN IN FIGURE 3 and 4 (see text).

Peak N.	Compound name	Conc. (ppb)	Peak N.	Compound name	Conc. (ppb)
1	n-butane	—	19	toluene	0.65
2	$H_2S + SO_2$	—	20	n-heptane	0.06
3	CCl_3F	0.13*	21	C_2Cl_4	0.05
4	CS_2	—	22	ethylbenzene	0.17
5	isopentane	0.16	23	m-xylene	0.13
6	trimethylpropane	0.08	24	p-xylene	0.08
7	n-pentane		25	o-xylene	0.1
8	$C_2Cl_3F_3$	0.03	26	α-pinene	1.5
9	cyclohexane	0.15	27	β-pinene	0.18
10	CCl_3-CH_3	0.11	28	trimethylbenzene	—
11	2methylpentane+3methylpentane	0.14	29	2 ethyltoluene	0.07
12	benzene	0.5	30	carene	0.06
13	$CCl_2=CHCl$	0.05	31	p-cymene	0.046
14	n-hexane	0.064	32	limonene	0.04
15	heptene	0.06	33	1,2,4 trimethylbenzene	0.02
16	dimethylpentane	0.18	34	1,4 diethylbenzene	0.05
17	2 methylhexane	0.16	35	not identified (sulfur)	—
18	3 methylhexane	0.065			

* Compound quantified with the cold trap technique.

ning 1 g of pine needles. The good correlation existing between monoterpene HC's measured in the air and the vapours emitted from these species confirm the reliability of our sampling method. The sensitivity allows the quantitation of natural and anthropogenic HC's present in air at ppb-ppt levels.

DIURNAL VARIATIONS OF BIOGENIC AND ANTHROPOGENIC HC's WITHIN A FOREST DURING PHOTOCHEMICAL SMOG EPISODES.

Figure 6a and b reports hourly concentrations of anthropogenic and biogenic HC's measured within the forest canopy respectively in June and July when photochemical smog episodes were observed. For sake of clarity the curves of alkylbenzenes from vehicular sources have been separated from those of monoterpenes and alkylbenzene HC's (p-cymene) emitted from trees. To account for the influence of the meteorological conditions, temperature, solar radiation, relative humidity and, when necessary, wind speed were also reported. In both cases, the prevalent wind was blowing from W-SW. To evidentiate the influence of pollutants transported from the city, O_3 and (in one case) NO_2 were also measured. In general, daily variations of natural HC's follow a regular trend. The concentration increases after sunset and reaches a maximum value during the night. In the absence of dry and wet deposition, this high concentration is maintained until late morning when a significant, sudden drop is observed. Episodically, dramatic increases in concentration can be observed during the day if, due to the peculiarity of the meteorological conditions, photochemical activity dies out. The trend of natural HC's can be explained as follows. In the late afternoon, when the

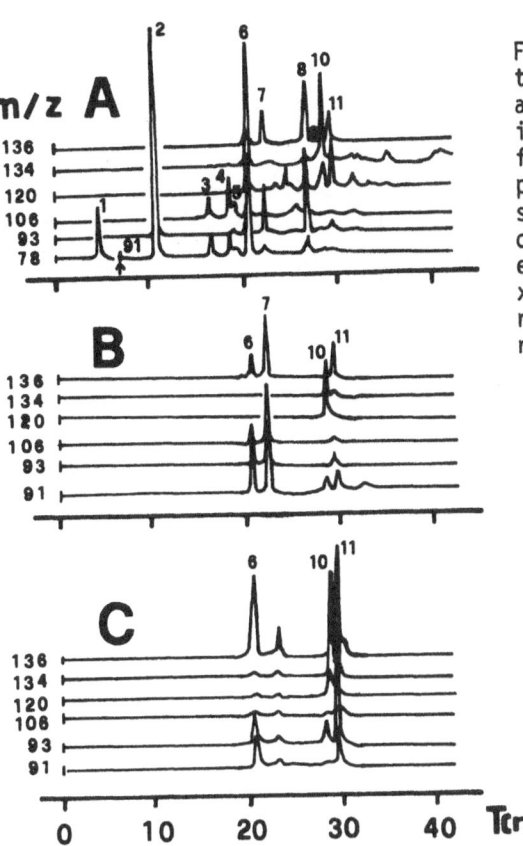

Figure 5 - A) SID traces from the chromatographic analysis of an air sample (IOL) collected in a pine forest. SID traces from the GC analysis of the vapours emitted from Pinus Silvestris (B) and Cupressus Arizonica.(C) 1) benzene, 2) toluene, 3) ethylbenzene, 4) m-xylene, 5) p-xylene, 6) α -pinene, 7) β -pinene, 8) carene, 9) phellandrene, 10) p-cymene, 11) limonene.

soil below the canopy starts to cool, convective stirring dies out, stable layers are formed near the ground and low mixing occurs with the overlaying layers. Due to the nocturnal inversion, natural HC's (which are continuously emitted from trees but not consumed by photochemical reactions) accumulate within the forest. During the night, only a small portion of organic species can be removed from the forest. The dominating sinks are the slow reaction with O_3 and, occasionally, dry and wet deposition. The minimum at 6 a.m. observed in Figure 6a indicates that wash-out caused by condensation of water droplets deposited at ground may be a quite efficient process for the removal of both anthropogenic and natural HC's. During the day, when the soil below the canopy is heated by the sun, deeper and deeper layers become well mixed and convectional stirring takes place (in July data from the nearest airport indicated that the mixing depth was about 2 Km). Natural HC's can thus diffuse over the canopy and be rapidly consumed by the fast reaction with OH radicals. The effectiveness of this reaction is well evidentiated in Figure 6a. When, due to light rain and cloudy weather, photochemical reactions were reduced (note the decrease of O_3 and anthropogenic HC's occurring at 5 p.m.) a drastic increase in natural HC's within the forest was observed. When one hour later, the rain ceased and the weather became sunny, the photochemical production of the OH radicals was partly restored (the increase in the concentration of O_3 and anthropogenic HC's observed at 7 p.m. well evidentiates this process)   and the concentra-

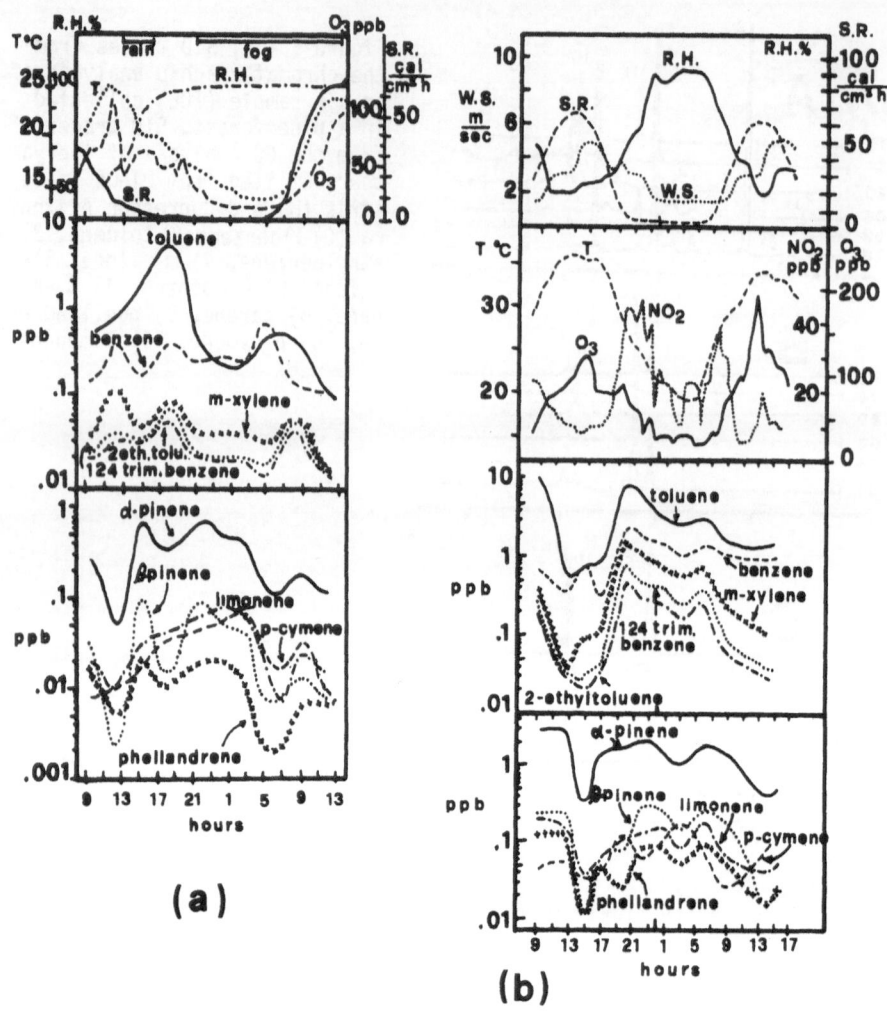

Figure 6 - a) Hourly variations of natural and anthropogenic HC's measured
in June (21-22) during a photochemical smog episode. b) Same
plots of Figure a. The measurements were carried out in July
(21-22) during a photochemical smog episode. In both cases, the
main metereological parameters were monitored together with some
pollutants T: temperature, R.H.: Relative Humidity, S.R.: Solar
Radiation, W.S.: Wind speed.

tion of natural HC's dropped down again. Our statement that the determining
process for the consumption of biogenic HC's emitted in the forest is the
reaction with the OH radicals is supported by the results shown in Figure 7
where the product between the concentration of ⍺ -pinene (which represents
more than 60% of the total biogenic content) and O_3 is plotted together
with the temperature. Data are taken from Figure 6b because the meteorological conditions established in July were more representative of the typical
anticyclonic weather occurring in Central Italy during summer. By assuming

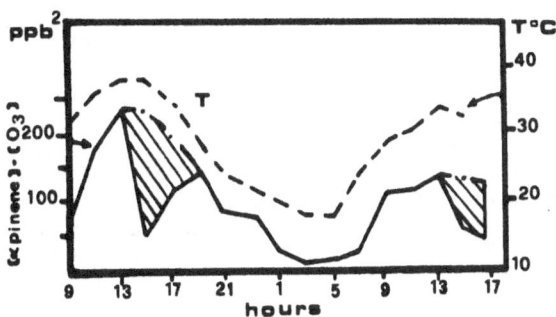

Figure 7 - Plots of the product between the concentration of α-pinene (ppb) and O_3 (ppb) (solid line). Data are relative to the measurements carried out in July (see Figure 6b). The dotted line is relative to the temperature variations measured in the same days. For the meaning of the shadowed area see the text.

that the concentration of α-pinene in the forest is exclusively controlled by the reaction with O_3, the emission rate E can be described by the following equation (11):

$$E = K_{O_3} \left[\alpha - \text{pinene} \right] \cdot \left[O_3 \right] \qquad (1)$$

where K_{O_3} is the second order rate constant for the reaction between O_3 and α-pinene. As the emission rate from trees is a function of the temperature (12), deviations of the product on the right side of eq. 1 from the temperature profile can be used for evidentiating the contribution of the OH radicals to the consumption of α-pinene. The results of Figure 7 indicate that, at night and during most part of the day, the emission rate is fairly described by eq. 1 and the main sink for natural HC's is represented by the reaction with O_3. In the late morning, however, the emission rate as defined by eq. 1 largely deviates from the temperature profile. The extent of such deviation, estimated by assuming that the emission rate at noon is the same as that at 2 p.m. (this is reasonable because in both cases T = 38°C), is shown in Figure 7 by the shadowed area. By assuming that other processes (diffusion, adsorption, wet and dry deposition) do not contribute for more than 20-30% to the decrease in the α-pinene concentration, the deviation of the product $[\alpha$-pinene$] \cdot [O_3]$ from the temperature profile can be used for evaluating the increase in concentration in OH radicals necessary to balance the emission rate equation. Simple calculations, carried out by adding the term K_{OH} $[\alpha$-pinene$] \cdot [OH]$ in the right side of eq. 1, indicate that a two order of magnitude increase in the concentration of the OH radicals should be necessary to account for the disappearence of natural HC's. This value well correlates with some theoretical predictions reported in the literature (13).

The results of Figure 6a and b indicate that the time available to natural HC's for reacting with OH radicals is quite limited (no more than 3-5 hours per day) and much shorter than that available to the anthropogenic HC's emitted near the forest. This point is illustrated well in Figure 6b where the consumption of alkylbenzenes starts earlier in the morning and dies out later in the afternoon. As the reactivity of monoterpenes toward OH radicals is equal or higher than that of alkylbenzenes, this behaviour can be only explained by the small difference in meteorological conditions

established inside and outside the forest. Very likely, the farmland sorrounding the forest (which is directly exposed to the sun) is heated more
rapidly than the soil below the canopy where the trees shadow shields the
solar radiation. When, in the morning, convectional stirring starts to be
effective below the canopy, a large portion of the anthropogenic HC's emitted near the forest are already transported in the upper layers and
consumed by the reaction with OH radicals. A similar effect can be observed
in the afternoon. Natural HC's starts to accumulate within the forest when
anthropogenic HC's are still reacting with the OH radicals and diluted in
the overlaying layers. These observations suggest that the amount of natural
HC's effectively involved in the formation of photochemical oxidants and
particulate matter can be much smaller than that estimated by emission in-
ventories where the density of trees, their height from the ground and the
metereological situation are not taken into account.
    The changes in concentration of anthropogenic HC's shown in Figure 6a
and b are more complex than those relating to natural HC's as two different
trends can be observed. By looking at Figure 6a we can see that during the
night anthropogenic HC's emitted from local sources (roads and houses) do
not diffuse into the forest. It is likely that the nocturnal inversion or
the emission rate are so low that the anthropogenic HC's present into the
canopy are only those left from the overlying layers when convectional stir
ring dies out. In this case, the concentration measured at night is very
small when compared to that of natural HC's and lower than that observed du
ring the day when HC's together with other pollutants emitted in the city
can diffuse in the upper layers and be transported into the forest. Transport phenomena seem to well account for the trend of O_3 and some anthropogenic HC's observed during the two days measurements carried out in June.
By looking at Figure 6b we can see that the concentration of anthropogenic
HC's measured at night is very high and follows a trend similar to that of
natural HC's. It is likely that, due to the relatively high nocturnal inver
sion (100-200 m), HC's emitted from local sources can diffuse and accumulate within the forest. Similarly to what has been noted for natural HC's, con
vective stirring and reaction with the OH radicals eliminate in the late mor
ning the bulk of anthropogenic HC's accumulated at night and a large drop
in concentration is observed when higher is the photochemical reactivity.
This decrease, is only partly counterbalanced by the diffusion of anthropogenic HC's from the overlaying layers. The extent of this contribution can
be evaluated by the small increase in concentration centered at 3 p.m. and
shown in Figure 6b. Also in this case, anthropogenic HC's diffuse into
the forest together with O_3. Due to the occurrence in the same site of such
different situations, it is difficult to establish to what extent anthropogenic HC's contribute to the formation of photochemical oxidants and particulate matter in forestal areas. We can only say that this contribution
can be quite high depending upon the meteorological conditions occurring
over the forest and the presence of emission sources near to the site.

CONCLUSIONS

    The method described here allows an accurate measure of biogenic and
anthropogenic HC's in forestal areas. The contribution of natural HC's to
the formation of photochemical pollutants appears to be lower than that
estimated from laboratory experiments as the time available for the diffusion and reaction with oxidizing agents is ranging between 3-5 hours per
day. The major sinks of natural HC's appears to be the reaction with OH ra-

dicals during the day and the reaction with O_3 during the night. The concentration of anthropogenic HC's in the forest can be significant and, in many instances, comparable to that of natural HC's. Local emission sources and, to a lesser extent, transport from the city account for the presence of anthropogenic species.

REFERENCES

1) P.R. Zimmermann, EPA Report No.904/9 - 77 - 028 (1979).
2) R.R. Arnts and B.W. Gay, EPA Report No.600/3 - 79 - 081 (1979).
3) P.R. Zimmermann, EPA Report No.450/4 - 4 - 79 - 004 (1979).
4) R.A. Rasmussen, R.B. Chatfield, M. Holdren and E. Robinson, Technical Report to coordinating Research Council Inc. 219 Perimeter Center Parkway, Atlanta, GA 30346, U.S.A..
5) B.W. Gay, R.R. Arnts, EPA Report No. 600/3 - 77 - 001 (1977).
6) F.W. Went, Nature (London), 187, (1960), 641.
7) A.P. Altshuller, Atm. Environ., 17, (1983), p. 2131-2165.
8) R.G. Derwent, D. Hov, on "Physico-Chemical Behaviour of Atmospheric Pollutants" Ispra, 16-18 October 1979, B. Versino and H. Ott Editors. p. 367-382.
9) P. Ciccioli, M. Possanzini, E. Brancaleoni, S. Brachetti and R. Tappa. 6° World Congress on Air Quality, I.U.A.P.P.A., 16-20 May 1983. Poster n. 2, p. 7.
10) M. Possanzini, P. Ciccioli, E. Brancaleoni, R. Tappa; A. Brachetti. Physico-Chemical Behaviour od Atmospheric Pollutants, 29 Sept.-1° Oct. 1981, Varese. B. Versino and H. Ott Editors, p. 76.
11) Y. Yokouchi, M. Okaniwa, Y. Ambe and K. FUWA, Atm. Environ. 17, (1983), 743.
12) D.T. Tingery, M. Manning, H.C. Ratsch, W.F. Burns, L.C. Grothaus and R.W. Field, EPA Report No. 904/9 - 78 - 013 (1978).
13) A. Lopez, G. Lecouteux, S. Prieur and J. Fontan, J. Aerosol. Sci., 14, (1983), 99.

DETERMINATION OF HYDROGEN PEROXIDE IN CLOUD AND RAIN WATER

F.G. RÖMER, A.A. VELDKAMP and P. van GALEN
Environmental Department, N.V. KEMA, Arnhem, The Netherlands

Summary

Hydrogen peroxide may be a major oxidant for sulphur dioxide in the water phase. Concentration levels found are variable and depend on sampling location as well as on weather conditions. The highest concentrations were found in summer in clouds. The application of the copper catalyzed chemiluminescent reaction of luminol to determine hydrogen peroxide was studied in order to be optimized and applied to a flow injection system. The interferences of trace metals are eliminated by a cation exchanger before final determination. The cation exchange was carried out in the analytical system itself. Analytical characteristics of the system are discussed. The stability of hydrogen peroxide solutions was established. A summary is given of the concentration levels observed.

1. INTRODUCTION

The concentration and composition of acid in cloud and rain water is determined mainly by the absorption of oxidation products of sulphur and nitrogenoxides and by the amount of oxidant in the liquid phase forming acid from dissolved oxides. Concerning the oxidation of sulphur dioxide in the water phase, the reaction with hydrogen peroxide is believed to be a chemical pathway which is important in a quantitative respect. This hypothesis is supported both by laboratory [1, 2] and field experiments [3, 4, 5] and from a mechanistic point of view [6].

An appropriate method for the analysis of hydrogen peroxide is a first step to study the conditions (season, sampling location, pollutant concentrations) which are favourable for the formation of this substance in the gas and/or liquid phase. Because methods for the direct determination of gaseous hydrogen peroxide are not available and sampling from the gas phase suffers from artefacts [7] our attention was given mainly to the determination in the liquid phase. Amongst other methods the chemiluminescent method with luminol in a flow system seemed to be practicable [8, 9, 10]. This method, however, is sensitive to interferences. Therefore its accuracy in real samples has to be determined. For our experiments the determination of hydrogen peroxide has to be fast enough to allow sequential sampling (e.g. plume washout experiments, ~ 100 samples) and must be applicable on small samples (cloud water, volume << 1 ml).

The principle of the method is widely discussed in literature [8-11]. It is based on the measurement of the photon flux (chemiluminescence) from the reaction of luminol and hydrogen peroxide. This reaction catalyzed by e.g. copper ions is carried out in an alkaline solution.

This paper deals with experiences obtained from a modified method described by Kok et al. [8] concerning the determination of hydrogen peroxide in the ambient atmosphere. Because of the wide concentration range, from less than 1 µg/l up to several mg/l [3, 12] of hydrogen peroxide in cloud and rain water, a method with a large dynamic range is desirable.

2. EXPERIMENTAL

2.1. Apparatus

In essence the flow scheme used in our method is the same as the method used by Kok et al. [8]. For our purpose the method has been adapted to water samples. Therefore modification in the flows and reagent concentrations have been made. The final scheme is depicted in figure 1 in which the several components are indicated and assigned with a number.

Figure 1. Analytical system for the determination of H_2O_2

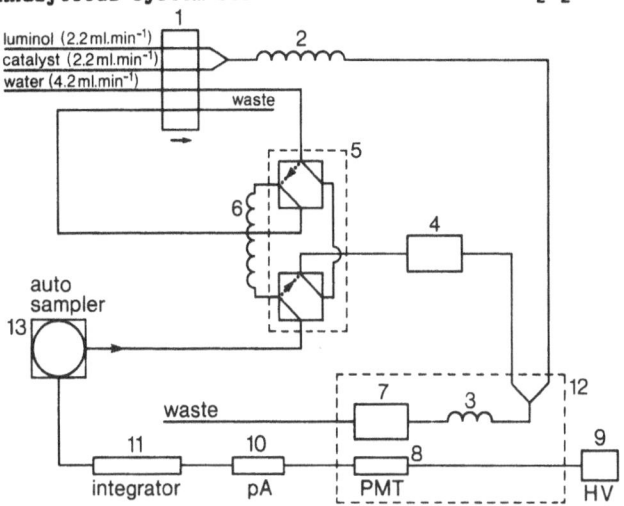

Reagents are pumped with a peristaltic pump (1: Gilson minipuls 2 with variable gear) through tubes (tygon). The luminol and catalyst reagents are mixed in a coil (2: 1 m teflon tube with an inner diameter of 0.8 mm). To this stream distinct volumes of samples are added after passing an in-line cation exchanger column (4: 50 mm glass tube with 3 mm inner tube diameter). Distinct volumes are obtained by assimilating samples from the autosampler (13: Cenco 34517-040) and filling a sample loop (6: 30-200 µl) of an eight-ports slider injection valve (5: Flarefit no. 0300521 pressure activated, 500 kPa). Reagent and sample streams are mixed in a coil (3: 20 mm teflon tube with an inner diameter of 0.8 mm filled with teflon wool) and through a flow through cuvet (7: spiral shaped cell with a volume of 0.4 ml). The effluent of the cell is wasted. The photon flux from the cuvet is measured with a photomultiplier tube (8: RCA Ip 21). To the photomultiplier tube a high voltage is applied which is obtained from a high voltage source (9: Thermoelectron Corp. H 10120, 950 V). The cuvet is positioned in such a way that the distance to the photomultiplier is minimal. The side of the cuvet opposite to the photomultiplier is provided with a reflective shield. Cuvet and photomultiplier are placed in a light tight housing (12). The current from the photomultiplier is measured with a pico-amperometer (10: Keithley 414S or 614). The signal of the pico-amperometer is recorded and integrated (11: Shimadzu C-R1A).

2.2. Reagents

Luminol was obtained from Fluka (09253). The stock solution contains 5.6×10^{-2} M/l luminol in 0.15 M/l NaOH. The stock solution was stabilized by settling for three days and can be used for two months. From the stock solution working solutions are prepared containing 2.8×10^{-3} M/l and 2.8×10^{-4} M/l luminol with a pH of 13. The catalyst solution was made of copper nitrate of Merck (pro analysi 2753) and potassium bromide of Merck (suprapur 4904). A solution was prepared containing 2×10^{-4} M/l Cu^{2+} and 9×10^{-2} M/l Br^-. Hydrogen peroxide (30%) was obtained from Merck (pro analysi 7209). From this solution a 1% solution is obtained which can be used for two months. This solution is standardized weekly against potassium permanganate. For calibration of the method dilutions of the standardized hydrogen peroxide solution were made daily. Cation exchange resin was obtained from Baker (CGC 241X8, 200-400 mesh). The capacity of the column (figure 1: 4) is about 1 mmol Na^+. Demineralized water with a specific resistance of at least 18 M ohm/cm was prepared with a water demineralizing system of Millipore Corp. (Milli RO-Milli Q).

3. RESULTS

3.1. Calibration graph

First we optimized the method as described originally by Kok et al. [8]. For optimization we applied a simplex method [13] taking into account the influence of the pH and of the luminol, copper and bromide concentrations. The influence of bromide ion on the reaction was studied because in literature an increase in photon yield is claimed [11]. Although it turned out that hydrogen peroxide could be determined reproducible in small quantities (lower limit of detection ~ 0.3 µg/l; signal to noise ratio 3) the time per determination was too long and the sample volumes (2 ml) needed was too large. Besides the calibration graph was not linear over a wide range. Using the system depicted in figure 1 with a sample loop of 30 µl and accepting a difference between calculated (using linear regression) and measured concentration of 5 percent of the calculated value, five linear ranges could be defined (tabel I).

Table I Linear and non-linear concentration ranges for the determination of hydrogen peroxide with the copper-luminol reaction (see text)

linear ranges (mg/l)	non-linear ranges (mg/l)
0 −0.15	0 −0.15
0.15−0.50	0.10−1.10
0.50−1.00	0.50−3.50
1.00−2.00	
2.00−3.50	

The relative standard deviation in the concentration range from 20 to 2000 µg/l was 2% (n = 4 per concentration). For the 5% difference as mentioned before and using a quadratic equation ($y = ax^2+bx+c$) - which gave better results than applying an exponential equation ($y = ax^b$) - the number of calibration curves could be limited to three graphs (table I). We studied the characteristics of the system using a low luminol concentration (2.8×10^{-4} M/l) combined with various sample loop volumes (30, 50, 100, 200 µl). Despite this effort we could only describe the calibration curve over the complete range with three different graphs of which always one was non-linear. Without changing the analytical system and using three instead of five graphs a relative standard deviation of less than 5% is found with the modified method (figure 1, loop 30 µl).

Some results for different concentrations and different operational conditions are given in table II.

Table II Results of the determination of hydrogen peroxide in standard solutions

concentration (µg/1)	relative stand.dev.(%)	number of detn.	remarks
10	0.0	8	
20	1.0	5	loop: 30 µl
40	0.2	5	lum.: 2.8×10^{-3} M/l
60	0.7	5	
120	1.2	5	linear curve
140	0.3	5	
400	0.7	5	loop: 50 µl
500	1.6	5	
1 500	1.4	6	lum.: 2.8×10^{-4} M/l
2 500	0.5	4	
3 500	1.3	5	linear curve
4 000	0.6	3	
10 000	0.5	4	

3.2. The influence of trace metal ions

It is known that rather low concentrations of several cations and anions will catalyze the reaction of hydrogen peroxide and luminol (e.g. Cu(II), Co(II), Ni(II), $S_2O_8{}^{2-}$, Fe(CN)$_6$ $^{3-}$, Cr(III), Fe(II), Fe(III), Mn(II) [8-11, 14-16], or may inhibit the reaction (Fe(II), Mn(II)) when copper is used as catalyst [16].

The occurence of several trace metal ions in rain as well as in cloud water is established [17] and concentrations of 10^{-6} M/l are no exception. Recently we measured in cloud water Fe concentrations up to 2.5×10^{-5} M/l and Mn concentratins up to 1.3×10^{-6} M/l. So it may be expected that concentrations of metal ions are sufficient to be effective as catalyst. Therefore the presence of several trace metals in samples will cause a change in the optimum chemical conditions for which the calibration curve holds. In general the extent of signal change due to metals is unknown and will lead to random errors. The extent of the influence of metal ions can be seen from figure 2 in which the effect of Ca, Co(II), Mn(II), Fe(II), Fe(III) on the copper-luminol reaction is shown.

Figure 2 The effect of trace metal ion concentration on the copperluminol reaction

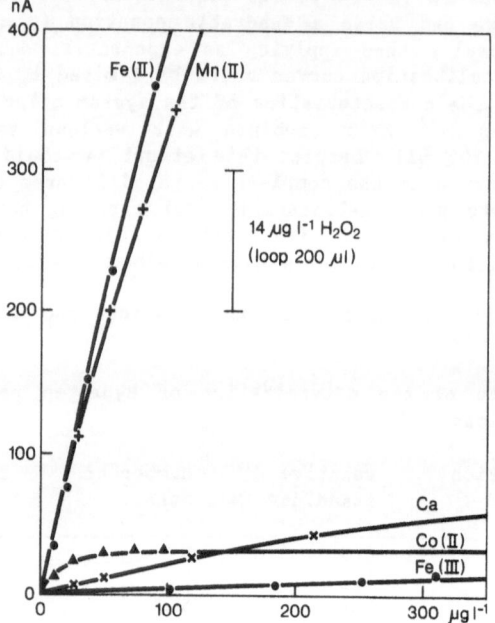

To remove trace metals a small in-line cation exchanger column was inserted into the analytical system (paragraph 2.1. and 2.2.). This column may be considered as an additional element causing peak broadening. Especially when low concentrations have to be determined the signal to noise (S/N) ratio has to be considered. We tried to compensate for this effect by applying a RC-network (R = 22 kΩ, C = 47 μF) to the input signal of the integrator. The improvement of the S/N ratio on introducing the RC-network is shown in table III.

Table III Signal to noise ratio (loop 200 μl, $[H_2O_2]$ = 20 μ/1, n = 10)

column	electronic filter	
	−	+
−	15.8 ± 1.3	20.4 ± 1.7
+	5.1 ± 0.9	6.9 ± 0.8

Inserting the column into the analytical system resulted in a decrease of the sensitivity of the standard calibration curves (e.g. 30% decrease for 0-25 μg/1 H_2O_2, loop 200 μl). Despite this decrease in sensitivity the analytical system is useful in our experiments.

As the trace metal contents in the samples of interest are very low compared to ion exchanger column capacity, only after a very large number of samples the complete capacity is used (at an average total metal ion concentration of 400 μeq/1 and using a loop of 50 μl, 50% of the capacity is used after determining 25 000 samples).

3.3. Standard addition

For rain water samples a comparison is made between results obtained using calibration curves and results obtained in applying standard addition curves (table IV). The ionic composition (main components) of these samples was normal [17]. The differences between the results using both methods were less than 1 µg/l.

In table IV the results from determinations in samples collected in the same period at the same site are presented.

Table IV Hydrogen peroxide concentrations (µg/l) calculated using normal calibration curves and on applying standard addition (range 0–15 µg/l)

sample		calibration curve	standard addition
83–10–08	1	5.6	5.5
	2	5.8	5.4
	3	5.1	4.8
	4	4.7	4.1
83–10–11	5	3.9	4.4
	6	2.7	2.7
	7	3.6	4.2
	8	3.4	3.5

To examine the effect of the matrix of rain water samples several standard additions were done. From these results the sensitivity was calculated and compared with the sensitivity of a normal calibration curve. It was found that for concentrations up to 1.2 mg/l the sensitivity of the two methods deviated less than 5%.

For cloud water, in the concentration range up to 15 µg/l hydrogen peroxide, the agreement between the results using calibration curves and standard addition procedures was less than found for rain water. For 14 samples maximum deviations of about 30% were found. For the samples examined so far with the method of standard addition the calculated sensitivity is always higher than the sensitivity of the normal calibration graph. The sensitivity calculated from results of standard addition to cloud water samples up to concentrations of about 1 mg/l showed an agreement within 10% with the sensitivity of a normal calibration curve. An outlier was found for a sample collected downwind Rijnmond industrial area. To this sample 500 µg/l hydrogen peroxide had to be added before the standard addition curve had a sensitivity comparable with the sensitivity of a normal calibration curve.

3.4. Sample storage and stability

Depending on sample type (chemical composition, sampling location) and wall material of the collection bottle, hydrogen peroxide may be decomposed or generated. Therefore the samples have to be analyzed as soon as possible after collection.

The stability of standard solution both in borosilicate glass and polyethylene(pe) bottles (25 ml and 50 ml respectively) was compared for several concentrations. The storage conditions and the results for an initial concentration of 750 µg/l hydrogen peroxide are summarized in table V.

Table V The effect (% recovery) of storage period and conditions on
        the H_2O_2 concentration in standard solutions

days	~ 6°C (dark)		roomtemp.(dark)		roomtemp. (daylight)	
	glass	pe	glass	pe	glass	pe
0	100	100	100	100	100	100
0.9	94	95	95	94	91	95
4.0	90	95	7	52	36	83
5.0	83	92	0	32	9	69

The best results were obtained under standard conditions i.e. sample
storage in polyethylene bottles in the dark at low temperature. It can be
established that for a recovery of at least 90% and storage under
standard conditions analyses must be carried out within 24 hours after
sampling (table VI).

Table VI Percentage recovery for H_2O_2 after storage under standard
         conditions (see text)

days	initial concentration (μg/l)		
	1030	605	190
0	100	100	100
0.2	96	97	98
0.8	93	91	95
1.2	92	84	72

4. HYDROGEN PEROXIDE CONCENTRATIONS IN CLOUD AND RAIN WATER [4]

In table VII the cumulative frequency distribution of hydrogen peroxide concentrations measured in cloud water and collected upwind and
downwind of some industrial areas in The Netherlands during the summer
1983 are summarized.

Table VII Cumulative frequency distribution of concentrations
           of H_2O_2 in cloud water (number of samples 152, maximum
           value 2.2 mg/l)

percentile	mg/l
25	0.04
50	0.12
75	0.35
90	0.70
95	1.05
99	1.95

The highest concentrations usually were found in the upwind regions
(North Sea). Winter values are low (< ~ 10 μg/l).
Hydrogen peroxide concentrations in rain water measured in Arnhem in
the period April–October 1983 are much lower (3–117 μg/l, mean 16
μg/l, number of samples 26, sampling duration $\leq$ 24 h).
In general hydrogen peroxide concentrations in cloud water are
higher than in rain water sampled at ground level.

5. DISCUSSION

A fast automated flow injection method to determine hydrogen peroxide in water samples was developed. The analysis can be carried out in a sample volume of only 30 µl with a lower limit of detection of about 0.3 µg/l (S/N = 3). Interference from dissolved trace metal cations was decreased by insertion of a cation exchanger column in the flow system.

The calibration curve is not linear over the range of hydrogen peroxide concentrations occuring in cloud and rain water samples. Within the desired accuracy, the calibration can be approximated by two linear and one non-linear calibration graph.

The sensitivity calculated from results of standard additions to rain water was in good agreement with the sensitivity of a normal calibration graph. For cloud water the difference in sensitivity was less than 30% for low concentration levels and less than 10% for high concentration levels. However, one has to bear in mind that the experiments were carried out in only a limited number of samples. From the experiments on sampling stability it is concluded that standard solutions have to be stored preferably at low temperature in the dark in polyethylene material. Even then significant losses occur incidentally. Until now no acceptable explanation for this phenomenon was found. Therefore more extensive studies in real samples are in progress.

Using the method described in this paper we found hydrogen peroxide concentrations in rain water up to more than 100 µg/l during the summer. In winter much lower concentration were measured [3]. This might indicate that the presence of hydrogen peroxide is related to photochemical reactions in the atmosphere. Comparable information can be derived from measurements in clouds; the highest concentrations, up to a few milligrams per liter, were always found in less polluted areas.

ACKNOWLEDGEMENT

The authors thank Mr. J.T.M. Hermans and Mr. J. Maaskant for valuable assistance in carrying out part of the experiments and Dr. H.F.R. Reijnders (RIVM, Bilthoven) for useful discussions.

REFERENCES

1. PENKETT, S.A., JONES, B.M.R., BRICE, K.A. and EGGLETON, A.E.J., (1979). The importance of atmospheric ozone and hydrogen peroxide in oxidizing sulphur dioxide at low pH, Atmos.Environ. 13, 123–137

2. MARTIN, L.R. and DAMSCHEN, D.E., (1981). Aqueous oxidation of sulphur dioxide by hydrogen peroxide at low pH, Atmos.Environ. 15, 1615–1621

3. RÖMER, F.G., VILJEER, J.W., VAN DEN BELD, L., SLANGEWAL, H.J., VELDKAMP, A.A. and REIJNDERS, H.F.R., (1982). Preliminary measurements form an aircraft into the chemical composition of clouds, Proc. CEC Workshop September 1982, Berlin, ed. Beilke, S and Elshout, A.J., D. Reidel Publ.Comp., Dordrecht

4. RÖMER, F.G., SLANGEWAL, H.J., VELDKAMP, A.A. and REIJNDERS, H.F.R., (1983). Wolkenwateronderzoek in Nederland. In: Zure regen; oorzaken, effecten en beleid. Proc. Symp. november 1983, 's Hertogenbosch; ed. Adema, E.H. and Van Ham, J., Pudoc-I II, Wageningen

5. DAUM, P.H., SCHWARTZ, S.E. and NEWMAN, L., (1982). Studies of the gas and aqueous phase composition of stratiform clouds. Proc. 4th International Conference on Precipitation Scavenging, Dry Deposition and Resuspension, Santa Monica, CA

6. CALVERT, J.G. and MOHNEN, V.A., (1982). The chemistry of acid rain development. Personal communication from Mohnen, V.A.

7. ZIKA, R.G. and SALZMAN, E.S., (1982). Interaction of ozone and hydrogen peroxide in air. Geophys. Res.Lett. 9, 231-234

8. KOK, G.L., HOLLER, T.P., LOPEZ, M.B., NACHTRIEB, H.A. and YUAN, M, (1978). Chemiluminescent method for the determination of hydrogen peroxide in the ambient atmosphere. Envir.Sci.Technol. 12, 1072-1076

9. BOSTICK, D.T. and HERCULES, D.H., (1975). Quantative determination of blood glucose using enzyme induced chemiluminescence of luminol. Analyt.Chem. 47, 447-452

10. SHAW, F. (1980). Development and construction of an analyser for the determination of the hydrogen peroxide content of natural water using a chemiluminscent reaction. Analyst 105, 11-17

11. CHANG, C.A. and PATTERSON, H.H., (1980). Halide ion enhancement of chromium (III), iron (III) and cobalt (II) catalysis of luminol chemiluminescence. Analyt.Chem. 52, 653-656

12. KOK, G.L., (1980). Measurement of hydrogen peroxide in rain water. Atmos.Environ. 14, 653-646

13. MELDER, J.A. and MEAD, R. (1965). Comp. J., 308

14. AUSES, J.P., COOK, S.L. and MALOY, J.T., (1975). Chemiluminescent enzyme method for glucose. Analyt.Chem. 47, 244-249

15. HAAPAKKA, K.E. and KANKARE, J.J., (1982). Apparatus for mechanistic and analytical studies of the electrogenerated chemiluminescence of luminol. Anal.Chim.Acta 138, 253-262

16. IBUSUKI, T., (1983). Influence of trace metal ions on the determinations of hydrogen peroxide in rain water by using a chemiluminescent techniques. Atmos.Environ. 12, 393-396

17. RIDDER, T.B., REIJNDERS, H.F.R., VAN ESSEVELD, F., WEGMAN, R.C.C. and MOOK, W.G., (1981). Chemical precipitation over The Netherlands. Survey 1978-1980, RIV-KNMI report, KNMI, De Bilt, The Netherlands

BESTIMMUNG VON SCHWEFELSÄURE UND SULFATEN IN DER LUFT

M. BUCK
Landesanstalt für Immissionsschutz, Essen (Deutschland)

SUMMARY

Most of the numerous methods for the determination of atmospheric concentrations of sulfuric acid and sulfates are
assigned to the determination of total sulfate ions and do
not allow separate measurements of the free acid and sulfa-
te-ions bound as salts.

Our investigations were aimed at the development of a method,
by which sulfuric acid will be fixed upon its first contact
with the sampling system. Thus the acid is prevented from
reacting with other substances during and after sampling.

This isolation of the sulfuric acid is achieved by este-
rification with glucose. Separation of the sulfuric acid
ester from simultaneously sampled sulfates is accomplished
by the aid of anhydrous ethanol. After decomposition of
the ester and conversion of the sulfuric acid to barium sulfate excess barium ions are determined photometrically with
disodium rhodizonate. Simultaneously sampled sulfate is
subsequently determined by this disodium rhodizonate method.

For sampling, the air volume containing free sulfuric acid
and sulfates is drawn through a glucose coated glass frit
held at $5o^{o}C$. Under sampling conditions of 2 m^3 of air at
1 m^3/h, the detection limit for sulfuric acid is at
1.5 $\mu g/m^3$ approx.. Standard deviation amounts to approx.
$\pm$ 1o % in the concentration range from 3 to approx. 15 $\mu g/m^3$.

1. EINLEITUNG

In der Literatur ist eine Reihe verschiedener Verfahren zur
Bestimmung von Sulfaten und von Schwefelsäure als Bestand-
teile des Aerosols in der Luft beschrieben worden (Über-
sicht siehe [1]. Einige davon werden in der Praxis eingesetzt, in den meisten Fällen solche, die auf die Bestimmung
des Gehaltes an Gesamtsulfat ausgerichtet sind, ohne daß
zwischen freier Schwefelsäure und ihren Salzen, den Sulfa-
faten, unterschieden wird.

Da jedoch wahrscheinlich die biologische Wirksamkeit der
Schwefelsäure und der Sulfate unterschiedlich ist, wobei
die Schwefelsäure relevanter sein dürfte als die Sulfate -
war es das Ziel unserer Untersuchung, ein Verfahren zur
selektiven Schwefelsäure-Bestimmung zu entwickeln.

Voraussetzung für eine spezifische Erfassung der Schwefelsäure ist es, zu gewährleisten, daß die Schwefelsäure während der Probenahme bereits beim ersten Kontakt mit dem
Sorptionssystem fixiert und an einer Reaktion mit anderen

Stoffen während und nach der Probenahme gehindert wird.

Diese Fixierung gelingt dadurch, daß als Sorbens Glucose gewählt wurde, die mit Schwefelsäure unter Esterbildung reagiert. Von gleichzeitig auf dem Sorbens abgeschiedenen Sulfaten wird der Schwefelsäure-Ester mit Hilfe von wasserfreiem Ethanol abgetrennt. Nach Zerlegung des Esters und Umsetzung der wieder freigesetzten Schwefelsäure zu Bariumsulfat bestimmt man den Barium-Ionen-Überschuß mit Dinatriumrhodizonat photometrisch.

2. VERFAHRENSVORSCHRIFT

2.1 Probenahme

2.1.1 Vorbereitung des Probenahmegefäßes

Das Probenahmegefäß besteht aus einem 15 cm langen Glasrohr (Innendurchmesser: 3o mm) mit aufgesetzter G3-Fritte (Frittenrohr). 55o mg D(+)-Glucose-Monohydrat werden direkt in das Frittenrohr eingewogen und mit 1o - 15 ml wasserfreiem Ethanol versetzt. Um Festanteile der Glucose zu entfernen, wird das Ethanol unter leichtem Schütteln des Frittenrohres abgezogen. Anschließend werden weitere 5 - 6 ml Ethanol auf die Fritte gegeben, und durch mehrmaliges Schütteln wird die Glucose gleichmäßig auf die Fritte verteilt. Man läßt das Ethanol ablaufen und spült evtl. beim Schütteln hochgeschwemmte Glucose mit weiteren ca. 3 ml Ethanol auf die Fritte hinunter. Das Frittenrohr wird auf der der Belegung zugewandten Seite mit einem mit Silikagel gefüllten Silikagelröhrchen versehen, um bei der nachfolgenden Absaugung von restlichem Ethanol mit einer Wasserstrahlpumpe eine Einwirkung der Laborluft auszuschließen. Nach dem Absaugen wird die Fritte im Trockenschrank bei 5oºC z.B. über Nacht getrocknet. Das Frittenrohr wird zum Verbringen an den Probenahmeort an beiden Enden mit Silikonstopfen verschlossen und senkrecht transportiert.

2.1.2 Aufbau der Probenahmevorrichtung

Der Aufbau der Probenahmevorrichtung geht aus der Abb. 1 hervor. Das Frittenrohr wird während der gesamten Probenahmezeit mit Hilfe eines elektrisch beheizten Strahlerofens auf 5oºC beheizt. Das Ausgleichsgefäß dient zum Auffangen von Pumpenstößen; es kann durch einen etwa 1 m langen Gummi- oder Plastikschlauch ersetzt werden (S.Abb.1).

2.1.3 Durchführung der Probenahme

Zur Probenahme werden 2 m^3 Luft mit einer Absauggeschwindigkeit von etwa 1 m^3/h durch die mit Glucose belegte, beheizte Fritte geleitet. Bei hohen Gehalten an Sulfaten und Schwefelsäure ist die Probenahmemenge unter Beibehaltung der Absauggeschwindigkeit entsprechend zu senken. Nach erfolgter Probenahme wird das Frittenrohr an beiden Enden mit Silikonstopfen verschlossen und in das Laboratorium gebracht.

2.2 Analytische Bestimmung

2.2.1 Bereitung der Pufferlösung

Zur Bereitung der Pufferlösung wird 1 l Wasser mit 11,4 ml in 1oo %iger Essigsäure versetzt und mit 1 n Natronlauge unter Verwendung eines ph-Meters auf ph 2,7 eingestellt.

2.2.2 Bereitung der Lösung "A"

Aus einer Titrisolampulle Bariumchloridlösung (Merck) wird durch entsprechende Verdünnung eine 0,005 molare $BaCl_2$-Lösung hergestellt. 3,3 ml dieser Lösung werden mit 7 ml 2 molarer Essigsäurelösung versetzt und mit Ethanol auf 1oo ml aufgefüllt.

2.2.3 Bereitung der Lösung "B"

1oo mg Ascorbinsäure und 5 mg des Dinatriumsalzes der Rhodizonsäure werden in einer braunen Glasflasche in 2o ml Wasser gelöst und anschließend mit 8o ml Ethanol versetzt. Die Lösung ist vor der Einwirkung von Luft zu schützen und täglich frisch herzustellen.

2.2.4 Aufarbeitung der Probe

Das Frittenrohr wird senkrecht, mit der belegten Seite der Fritte nach oben, in eine Halterung eingespannt. Der obere Silikonstopfen des Frittenrohres wird entfernt. Auf die Fritte werden 5,o ml Ethanol gegeben; während einer Einwirkungszeit von 1o min wird das Frittenrohr mehrmals leicht geschüttelt. Anschließend wird die entstandene ethanolische Lösung in ein unter das Frittenrohr gestelltes 5o-ml-Becherglas, in dem sich 3 ml Pufferlösung mit o,5 g darin gelöster Glucose befinden, abgelassen. Zur Entfernung restlicher Lösung aus der Fritte wird mit Hilfe eines Gummi-Handgebläses Luft von oben in das Frittenrohr eingedrückt. Danach wird der Vorgang mit 2,o ml Ethanol wiederholt, jedoch ohne Zwischenschaltung einer Einwirkungszeit. Die im Becherglas gesammelten Extrakte werden in einem Meßkolben auf 1o,o ml mit Ethanol aufgefüllt und bis zur photometrischen Bestimmung beiseite gestellt.

Da nach dieser Extraktion noch geringe Reste an freier Schwefelsäure oder Ammoniumhydrogensulfat auf der Fritte verblieben sein können, wird die Auswaschung des Fritteninhaltes mit 5 und 2 ml Ethanol und unter Vorlage von 3 ml der Lösung von Glucose in Pufferlösung wiederholt. Diese zweite Extraktionslösung wird nicht mit der zuerst erhaltenen vermischt, so daß die Vollständigkeit der Auswaschung überprüft werden kann und um bei sehr geringen Gehalten an freier Schwefelsäure die erste Auswaschlösung nicht unnötig zu verdünnen.

Durch eine weitere Extraktion werden die in Ethanol unlöslichen Sulfate aus dem Fritteninhalt herausgewaschen: Hierzu werden 3,o ml der Pufferlösung auf die Fritte gegeben und das Frittenrohr etwa 3 - 4 min vorsichtig geschüttelt.

Danach wird der Extrakt in ein Becherglas abgelassen und mit dem Handgebläse nachgedrückt. Die in der Fritte verbliebene restliche Pufferlösung wird durch Nachwaschen zuerst mit 5 und nochmals mit 2 ml Ethanol unter Verwendung des Handgebläses verdrängt. Die in dem Becherglas gesammelten Waschlösungen werden in einem Meßkolben auf 1o,o ml mit Ethanol aufgefüllt.

Die photometrische Bestimmung des H_2SO_4- bzw. -Sulfat-Gehaltes der obigen Lösungen wird gemäß den Ausführungen in Abschnitt 2.5.2 vorgenommen, in dem jeweils 3 ml der Lösungen in gleicher Weise wie die Kalibrierlösungen behandelt werden.

2.2.5 Reinigung der Fritte

Nach drei- bis fünfmaligem Gebrauch ist die Fritte zu reinigen, um z.B. organische Begleitkomponenten zu entfernen. Hierzu wird die Fritte über Nacht mit Chromschwefelsäure behandelt. Anschließend wird die Fritte mit heißem Wasser mehrfach gespült und mit heißer, halbkonzentrierter Salzsäure chromfrei gewaschen. Schließlich wird restliche Säure mit heißem Wasser entfernt. Der Reinigungseffekt ist durch die Aufnahme von Blindwerten zu überprüfen.

2.3 Kalibrierung

Die vorläufige Kalibrierung des Meßverfahrens erfolgte ohne Einbeziehung des Probenahmevorgangs. Die Kalibrierung des gesamten Verfahrens unter Einschluß der Probenahme mit Hilfe von künstlich hergestellten Schwefelsäure- und Sulfat-Aerosolen wird derzeit in Angriff genommen.

2.3.1 Herstellung der Kalibrierlösungen

Blindlösung:

o,5 g Glucose werden in 3,o ml Pufferlösung gelöst. Die Lösund wird in einem Meßkolben mit Ethanol auf 1o,o ml aufgefüllt.

Kalibrierlösung I:

o,55 g Glucose werden in 2,o ml Pufferlösung gelöst und mit 1 ml einer mit Pufferlösung verdünnten Schwefelsäurelösung versetzt. Die Lösung wird darauf im Meßkolben mit Ethanol auf 1o,o ml aufgefüllt. Die so erhaltene Lösung enthält 24,5 μg H_2SO_4 pro 1o ml. Die zur Herstellung der Kalibrierlösung I verwendete, mit Pufferlösung verdünnte Schwefelsäurelösung wird hergestellt, indem 1o,o ml einer o,oo5 n H_2SO_4 mit Pufferlösung im Meßkolben auf 1oo,o ml aufgefüllt werden. Die erhaltene o,oo5 n Schwefelsäure enthält 24,5 μg H_2SO_4 pro ml.

Kalibrierlösung II:

1,5 ml der Kalibrierlösung I, enthaltend 3,67 μg H_2SO_4 , werden mit 1,5 ml Pufferlösung verdünnt. Die so erhaltene Kalibrierlösung II enthält 1,23 μg H_2SO_4 pro ml.

2.3.2 Anfärbung der Lösungen und photometrische Bestimmung

Von jeder der anzufärbenden Kalibrierlösungen werden drei Milliliter entnommen und mit 2,5 ml Ethanol und 1,o ml der Lösung "A" im Meßkolben gemischt; von der Kalibrierlösung II wird die ganze hergestellte Menge (3,o ml) eingesetzt. Nach 1o min werden die so vorbereiteten Lösungen mit je 1,5 ml der Lösung "B" vermischt. Nach einer Standzeit von 15 min wird die entstandene Rotviolettfärbung mit einem Elko-Filterphotometer, Filter S 51 E 67, oder mit einem Spektralphotometer bei 5o7 nm in einer 1-cm-Küvette gegen Wasser photometriert.

Obwohl die Farbentwicklung in der zu photometrierenden Lösung, wie aus Abbildung 2 zu ersehen, bereits nach etwa 5 min beendet ist, hat es sich aus Gründen des Arbeitsablaufs bewährt, die Standzeit auf 15 min festzusetzen. Um die Einhaltung der Standzeit zu gewährleisten, ist es zweckmäßig, die einzelnen anzufärbenden Lösungen in zeitlichen Abständen von z.B. 2 min, beginnend mit den Eichlösungen, mit der Lösung "B" zu versetzen. Nach jeweils 15 min Einwirkungsdauer wird dann in den gleichen Abständen von z.B. 2 min die Extinktion gemessen.

2.3.3 Aufnahme der Kalibrierkurve

Zur Auswertung der erhaltenen Meßwerte wird eine Kalibrierkurve aus den Extinktionswerten der Kalibrierlösung erstellt. Es werden die Extinktionen gegen die Gehalte der Kalibrierlösungen in μg pro 3 ml aufgetragen. Da mit steigendem Gehalt der zu messenen Lösungen an Sulfationen die Extinktion abnimmt, kann, um eine ansteigende Kalibrierkurve zu erhalten, der Extinktionsmeßbereich auf der Ordinate in absteigender Reihe angelegt werden. Im zu messenden Konzentrationsbereich bis 8 μg H_2SO_4 pro 3 ml - entsprechend 27 μg H_2SO_4/m^3 unter den in Abschnitt 2.3.3 genannten Probenahmebedingungen - ist der Verlauf der Kalibrierfunktion linear.

Die aus den Kalibrierwerten errechnete Analysenfunktion ergibt sich zu

$$c_L = 33,93 \; A + 21,09$$

c_L = Menge an H_2SO_4 (in μg) in 3 ml der anzufärbenden Probelösung

A = am Photometer abgelesener Extinktionswert.

2.4 Auswertung der Meßergebnisse

Mit Hilfe der Analysenfunktion werden die Konzentrationen
an Schwefelsäure in den Probelösungen in /ug pro 3 ml aus
den am Photometer abgelesenen Extinktionswerten errechnet.
Durch Multiplikation mit dem Faktor 3,33 erhält man daraus
den Schwefelsäuregehalt in dem am Gasmengenmesser abgelesenen Probenvolumen.

Bei der Messung des Sulfatgehaltes werden bei der Anwendung
dieser Berechnungslgeichungen die Meßwerte als Konzentra-
tionswerte an freier Schwefelsäure erhalten. Zur Umrechnung
der Schwefelsäurekonzentrationen in Sulfatkonzentrationen
müssen die Meßwerte deshalb mit dem sich aus dem Verhältnis
der Molekulargewichte von Sulfat-Anion und Schwefelsäure
ergebenden Faktor F = o,979 multipliziert werden.

2.5 Verfahrenskenngrößen

2.5.1 Standardabweichung

Aus einer Anzahl von jeweils 15 Doppelbestimmungen wurden
die Standardabweichung des beschriebenen Verfahrens zur
H_2SO_4- bzw.- Sulfat-Bestimmung auf die übliche Weise berechnet. Bei H_2SO_4-Konzentrationen größer als 3 /ug/m^3 bis ca.
15 /ug/m^3 liegt die Standardabweichung in der Größenordnung
von ± 1o %, bezogen auf den Konzentrations-Klassen-Mittel-
wert von ca. 1o /ug/m^3. Der 95-%-Streubereich liegt dement-
sprechend bei ca. ± 2o %.

Die Standardabweichung für die Sulfat-Bestimmung liegt im
Konzentrationsbereich zwischen 3 und 15 /ug/m^3 in der Grös-
senordnung von ± 15 %, der diesbezügliche 95-%-Streubereich
bei ca. ± 3o %, bezogen auf den Konzentrations-Klassen-
Mittelwert von ca. 1o /ug/m^3.

2.5.2 Nachweisgrenze

Die Nachweisgrenze des Verfahrens wurde aus der Standardabweichung der Blindwerte als NG = 3 · S_{Bl} zu

$$NG_{H_2SO_4} = 1,3 /ug/m^3 \quad und \quad NG_{SO_4{}^{2-}} = 1,2 /ug/m^3$$

berechnet.

2.6 Querempfindlichkeiten

Bei der H_2SO_4-Messung wird Ammoniumhydrogensulfat mit er-
faßt. Die Trennung der Sulfate von der veresterten Schwefelsäure verläuft aufgrund der äußerst geringen Löslichkeit
der Sulfate in wasserfreiem Ethanol nahezu vollständig.
Nitratgehalte, entsprechend NO_x-Konzentrationen in der
Größenordnung von bis zu 5oo /ug/m^3 in der Luft, stören die
H_2SO_4-Bestimmung nicht. Das gleiche gilt für Fluorkonzen-
trationen bis zu 2oo /ug/m^3.

3. ANWENDUNG DES MEßVERFAHRENS

Bei der ersten Anwendung des Meßverfahrens zur Bestimmung des Gehaltes an Schwefelsäure und Sulfate in der Luft zum Zwecke der Routineanpassung des Meßverfahrens an Meßstellen im Ruhrgebiet wurden H_2SO_4-Konzentrationen zwischen 1,3 und 35 µg H_2SO_4/m^3 und Sulfat-Konzentrationen zwischen 1,2 und 3o µg SO_4^{2-}/m^3 ermittelt.

Über die begonnenen systematischen Schwefelsäure- und Sulfatmessungen in Nordrhein-Westfalen wird demnächst berichtet werden.

LITERATURVERZEICHNIS

[1] BEINE, H., R. SCHMIDT und M. BUCK:
     Ein Meßverfahren zur Bestimmung des Schwefelsäure-
     und Sulfatgehaltes in der Luft.
     LIS-Bericht Nr. 31 (1983) ISSN o72o - 8499
     Herausgeber: Landesanstalt für Immissionsschutz, Essen.

[2] TAKIURA, K., H. YUKI, S. HONDA, Y. KOJIMA, L.Y. CHEN:
     Trisulfation of Hexoses by Means of Concentrated
     Sulfuric Acid.
     Chem.Pharm. Bull., __18__ (197o), S. 429/435.

[3] REES, D.A.:
     Bemerkung zur Charakterisierung von Kohlenhydrat-
     sulfaten durch Säurehydrolyse.
     Biochem. J., 88 (1963), S. 343/345.

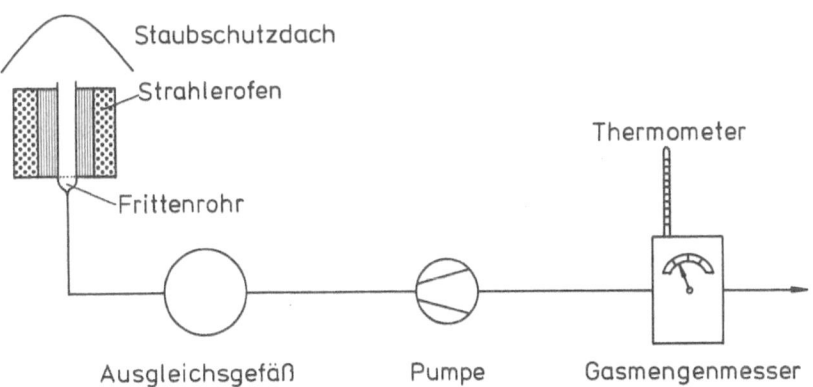

Abb. 1 : Probenahmeanordnung für die Bestimmung von Schwefelsäure und
         Sulfaten in der Luft

LA MESURE EN CONTINU DES SULFATES PAR PHOTOMETRIE DE FLAMME
M. PAYRISSAT, B. NICOLLIN and H. STANGL
Commission of the European Communities
Joint Research Center, Ispra Establishment

Summary

A commercial flame photometric detector (FPD) normally used for SO_2 determinations has been adapted to continuous sulfate measurements in ambient air. As described, the modification consisted in adapting an appropriate inlet system composed of a coated lead dioxide denuder for the removal of gaseous sulfur compounds, and a controlled heater to maintain a steady flame inside the detector. With a detection limit of the order of 4 µg SO_4^{--} /m³, the modified analyser can be favourably used for comparative measurements.

In the Ispra region, the sulfates well correlated with the visibility are one of the constituents of the summer haze. The measurements gave evidence of ammonium sulfate aerosol formation by photochemical conversion of SO_2 emitted by local sources. These measurements are compared with some data from an EEC campaign at the site of Fos-Berre (France), a heavily industrialized and polluted area.

1. INTRODUCTION

Diverses techniques ont été développées au CCR en vue de rechercher les éventuels précurseurs de la formation de brume dans la région d'Ispra et la Vallée du Po. La fraction organique qui constitue 13 à 25 % de l'aérosol total a été analysée mais ses précurseurs gazeux décelables par GCMS (1) ne peuvent être corrélables avec d'autres paramètres mesurables en continue comme la visibilité. Récemment, Altshuller et d'autres auteurs (2) (3) ont indiqué que les sulfates étaient universellement présents en été, même dans les sites ruraux. Par fluorescence X on a donc analysé les résidus soufrés prélevés sur filtre pendant différentes périodes de l'année 1981. Cette fraction qui était pratiquement constante, représentait une moyenne de 6 % S pendant les mois d'été, représentant donc une concentration en sulfates non négligeable. La question qu'on se posait était la suivante: quel est l'impact de cette fraction sur les épisodes de formation de brume? Pour y répondre, on s'est proposé d'effectuer une mesure en continue des sulfates par photométrie de flamme au moyen d'un analyseur normalement utilisé pour le monitorage du SO_2.

Le principe qui exploite la spectrométrie d'émission du soufre dans une flamme réductrice d'H_2 à 394 nm suscite un intérêt récent pour la mesure de l'aérosol. Depuis 1964 où W.L. Crider avait réussi à mesurer 0,1 mg H_2SO_4/m³ d'air en modifiant la configuration de la flamme d'H_2 émise par un tel analyseur, la méthode s'est considérablement améliorée et diversifiée

(4) (5) (6). Grâce aux performances des DPF (Détecteur à Photométrie de Flamme) actuels et en utilisant des "denuder" (*) à diffusion gaseuze (7) (8) (9), on détecte en temps réel quelques μg SO_4^{--}/m^3 d'air.

Le denudeur, placé à l'entrée de l'analyseur, exploite les différences de diffusivités entre gaz et particules. C'est un tube cylindrique dont la paroi interne est enduite d'un parfait absorbant pour les composés gazeux de soufre (SO_2, H_2S). Pour un écoulement laminaire, l'aérosol, à plus faible diffusivité, est seul entraîné par le courant gazeux jusqu'au four à détection. L'étude ci-après présente cette méthode telle quelle a été développée et appliquée au CCR.

2. Développement de la méthode

2.1 Modifications apportées à l'analyseur standard

Le DPF soumis à l'essai est le Monitor Labs 8450 dont le système d'admission comprend les éléments suivants: un filtre particulaire de téflon et un piège H_2S constitué d'une pastille (tamis en Ag) placée dans un four à 110 °C. Progressivement délesté de ces éléments, l'instrument a été soumis à la réponse d'un aérosol de $(NH4)_2SO4$ introduit dans un ballon de téflon de 450 litres rempli d'air pur reconstitué (fig. 1). Cet essai produit un aérosol de granulométrie sensiblement constante mesurée autour de 0,4 μm avec l'analyseur optique Climet.
Introduit dans le ballon pendant un temps d'admission déterminé, la concentration en ppb équivalents SO_2 est mesurée, puis le ballon est vidé. Des mesures successives sont effectuées pour des temps d'admission plus longs. Répété pour des configurations différentes du système d'admission, cet essai permet de tracer une courbe d'étalonnage (fig. 2). La réponse de l'instrument s'avère d'autant meilleure que le système d'admission est dépourvu de ses éléments, que le tube d'échantillonnage (1/8 ") est raccourci et ne présente aucun coude. Etant donné la position verticale du four à réaction, la configuration optimale oblige l'analyseur à échantillonner "par dessous" (fig. 3). Un tube d'admission trop court dé-stabilisant la flamme lorsque la mesure se fait à l'air ambiant, un four de préchauffage a été placé sur une longueur de 25 cm. Afin d'éviter la vaporisation de certains aérosols avant d'être mesurés (comme H_2SO4) et la condensation prématurée des produits ainsi formés (8) (10)(11), une température de l'air de 60 °C ne doit pas être dépassée. Une régulation du four à 90 °C permet dans notre cas d'entretenir la flamme sans que le rapport H_2/O_2 prescrit par le constructeur soit modifié et sans que la température du courant gazeux, mesurée à la sortie du four, dépasse 45 °C. Avec cette configuration (fig. 3), le temps de réponse de l'analyseur est plus court. On l'explique par la diminution du temps mort causée par le raccourcissement du tube d'échantillonnage (8).

(*) francisé en "denudeur" dans le texte

2.2 Choix du "dénudeur"

Les dénudeurs sont des tubes de verre ou d'inox recouverts de PbO_2 avec lequel SO_2 et H_2S entrent en réaction. L'application de ce dépôt a été effectuée selon la technique de Durham (7). Pour améliorer l'adhérence de l'enduit, nous avons utilisé un tube de verre préalablement rodé. Dans le cas d'un écoulement laminaire, obtenu grâce au faible débit de l'analyseur (3 cm^3 s^{-1}), l'efficacité d'absorption est peu influencée par le diamètre du dénudeur car, lorsqu'il varie, les deux composantes perpendiculaires agissant sur la molécule ou la particule (longitudinale dans le sens d'écoulement et transversale de diffusion vers la paroi, fig. 3) varient dans le même sens. Le diamètre choisi est donc assez faible (4 mm) de façon à réduire les pertes par impaction au niveau de la jonction avec le tube d'échantillonnage qui est faite d'un épaulement conique en téflon. La longueur du dénudeur a été préalablement calculée par l'équation de Gormley-Kennedy (1949):

$$\frac{Ci}{C_{oi}} = 0,819 \exp(-14,6272\,\Delta) + 0,0976 \exp(-89,22\,\Delta)$$
$$+ 0,01896 \exp(-212\,\Delta) \text{ , avec } \Delta = \frac{\pi\,Di\,l}{4\,F} \text{ , dans laquelle:}$$

C_{oi} est la concentration à l'entrée du tube de l'espèce diffusante i

Ci est la concentration à la sortie du tube " "

Di est le coefficient de diffusion " "

F est le débit de pompage de l'analyseur

l la longueur du dénudeur.

Le tableau I donne les longueurs de dénudeurs pour divers taux d'absorption de SO_2 (D = 0,13 cm^2 s^{-1}) et d'aérosols de 0,01 μm pour lesquels D $\simeq$ 10^{-3} cm^2 s^{-1} :

Tableau I -   Efficacité d'absorption théorique du dénudeur

longueur du dénudeur pour SO_2 et particules	taux d'absorption $\frac{Co - Ci}{Co} \times 100$			
	99,9	99,5	99	95
$l(SO_2)$	13,5 cm	10,2 cm	8,8 cm	5,6 cm
$l(0,01\ \mu m)$	1752 -	1330 -	1150 -	729 -

On a effectué des essais d'absorption comparatifs, entre dénudeurs et un même tube-étalon sans Pb O_2 pour des longueurs variant de 6 à 15 cm, vis à vis du SO_2 émis par une source à perméation et des sulfates émis par le dispositif de la fig. 1. Vis à vis du SO_2 on a trouvé sensiblement les

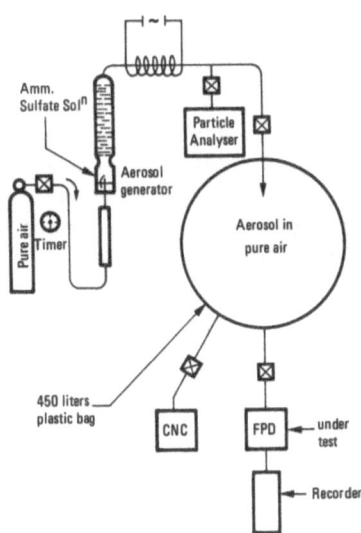

Fig. 1 - Dispositif expérimental
pour l'étalonnage de la méthode

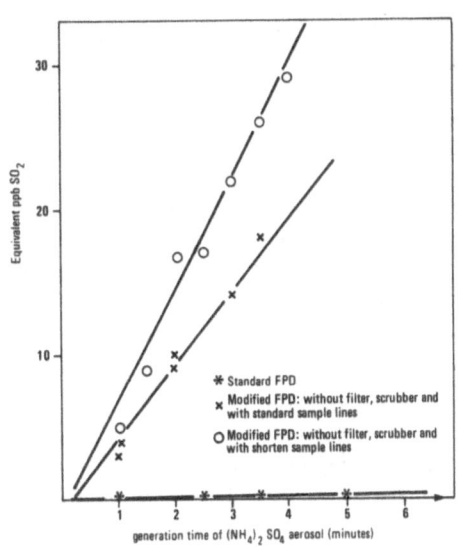

Fig. 2 - Courbe d'étalonnage pour
différentes configurations du DPF

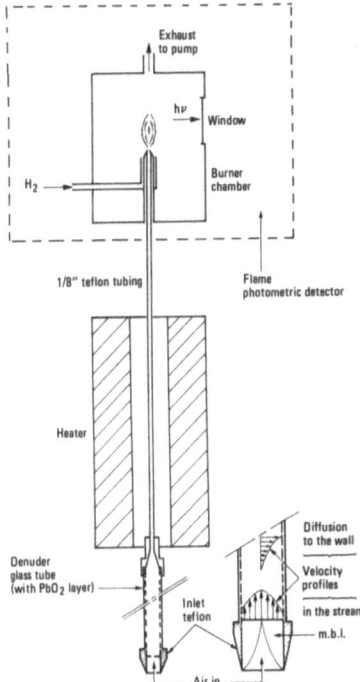

Fig. 3 - Schéma des modifications
apportées à l'analyseur standard

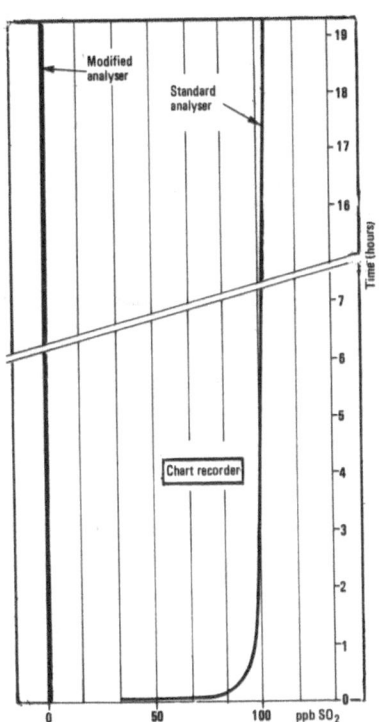

Fig. 4 - Efficacité du dénudeur
en fonction du temps par rapport
au SO_2 émis par perméation

données du tableau I; un dénudeur de 14 cm, retenu pour la suite des essais, démontre (fig. 4) 100 % d'efficacité. Ce même dénudeur, vis à vis de l'aérosol donne environ 15 % de pertes. Il est possible que celles-ci soient concentrées à l'entrée du tube, zone d'écoulement transitoire où la couche limite moléculaire ("molecular boundary layer", m.b.l., fig. 3) n'est pas complètement développée. Cette couche est bien moins résistante à la diffusion des particules vers la paroi. Un dénudeur disposant d'un embout de téflon de longueur $L = 0,05.d.Re = 1,1$ cm (d: diamètre du tube, Re nombre de Reynolds) L étant la longueur de la couche transitoire, est en cours d'expérimentation.

2.3 Etalonnage de la méthode et interférences

Un étalonnage par comparaison vis à vis du SO_4^{--} est difficile. Le DPF a une sensibilité qui dépend de la dimension des particules (8); la spectrométrie X en parallèle dans un ballon (fig. 1) a été tentée mais elle nécessite de grands volumes et un temps trop long pendant lequel la dimension de l'aérosol croît. Des mesures comparatives faites en atmosphère ambiante ont montré une interférence du SO_2 sur le filtre de prélèvement donnant des valeurs en excès de SO_4^{--}.

L'étalonnage du DPF a donc seulement été effectué vis à vis du SO_2. Sa réponse étant proportionnelle à la concentration d'atomes de S, 1 ppb SO_2 (TPN) équivaut à $(32 \times 2,66) / 64 \simeq 1,33$ µg S/m^3 d'air soit à $(96 \times 1,33) /32 \simeq 4$ µg SO_4^{--} /m^3 d'air. Le SO_2 provient d'une source à perméation, le gaz porteur étant de l'air purifié qui évite l'interférence négative du CO_2 qui est de l'ordre de - 12 % sous une concentration moyenne de 330 ppm (12). L'humidité de l'air est un autre interférant dont l'effet se fait sentir sur la dérive du courant zéro; on constate une dérive négative proportionnelle à l'humidité ambiante. De fréquents étalonnages à l'aide d'échantillons d'air purifié et sous degré hygrométrique ambiant sont nécessaires.

3. Applications de la méthode

Le tableau II rassemble quelques exemples d'applications de la méthode: ex. 1 correspond à 4 jours consécutifs d'un épisode de brume d'été à Ispra; ex. 2 à des mesures d'hiver également à Ispra et ex. 3 dans le complexe industriel de Fos-Berre (France), lors de la campagne CEE du 1 au 16 juin 1983.

Les périodes d'observation sont de 24 h et les valeurs ont été moyennées par heure. Le tableau donne les valeurs maximales du rapport SO_4^{--}/SO_2, du % SO_4^{--} d'après $(SO_4^{--}/0,38$ bs$).100$, où bs est le coefficient d'extinction mesuré par néphélométrie, et les coefficients de corrélation linéaires entre $SO_4^{--}-SO_2$, $SO_4^{--}-O_3$ (de 11 h à 19 h) et bs-SO_4^{--}.

Tableau II
Exemples d'application de la méthode à Ispra et Vitrolles

Exemple n°	Date	% $\overline{HR}_o \pm x$ (1)	$\dfrac{SO_4^{--} max}{SO_2 hor.}$	%SO_4^{--} max hor	Coeffic. corrél. linéaire (r)		
					SO4- SO2	SO4-O3(2)	b_S - SO4
1 a	17.9.82	84 ± 17,3	3,8	9,8	<0,15	0,78	0,16
1 b	18.9.82	81 ±12,6	2,9	8,4	-	0,97	0,83
1 c	19.9.82	85 ± 11,7	4,4	8,6	-	0,98	0,68
1 d	20.9.82	87 ± 13,2	5,5	9,0	-	0,71	0,41
$\sum 1 \begin{smallmatrix} d \\ a \end{smallmatrix}$	17-20.9	<70	4,4	9,0	-0,1;-0,62 (3)	0,72	0,93
		>75	5,5	8,3	0,2;0,56	-	0,86
2 a	du 14 - 15.1.83	87,3±14,1	0,32	15,9	0,93	-	0,98
2 b	29.1.83	93.6± 2,4	0,83	9,6	0,11	-	0,83
2 c	30.1.83	88,7± 11	0,51	13,3	0,34	-	0,91
3	9.6.83 (Vitrolles)				0,78	-	-

(1) moyenne de valeurs journalières de l'humidité relative: $x = t_{0,01} \sigma$
où t = 3,11 est le coefficient de Student pour 12 valeurs bihoraires
au cœfficient de sécurité de 99 % et σ l'écart-type.

(2) entre 11 h et 19 h.

(3) valeurs maximales et minimales de valeurs journalières durant l'épisode.

Commentaire:

Exemple 1: Les concentrations maximales de SO_4^{--} sont de l'ordre de 35 µg/m^3.
L'évolution journalière, avec d'autres paramètres simultanément mesurés,
est similaire à la fig. 5 (ex. 1 b). Elle ne présente pas de fluctuation
importante, démontrant une relative stabilité par rapport à SO_2, celui-ci se
trouvant en moyenne très peu concentré. Il s'en suit une distribution régionale homogène, apparemment peu affectée par les sources locales (13).
Toutefois, entre 11 h et 19 h, SO_4 évolue de façon identique à O_3 (qui est
un indicateur photochimique) avec 0,71 < r < 0,98 selon la présence matinale
de NO, puits pour O_3 (fig. 7). On doit être en présence d'une formation
photochimique à partir de faibles niveaux de concentration de SO_2.

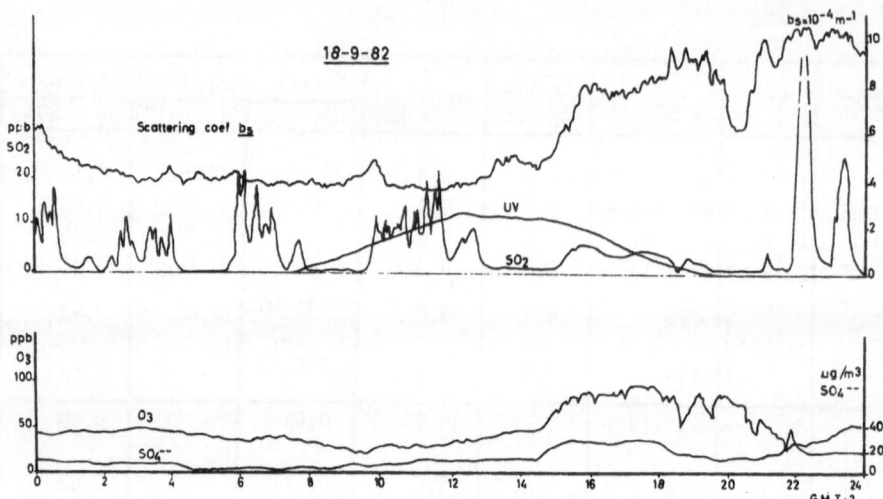

Fig. 5 - Evolution journalière du SO_4^{--} et autres paramètres au cours d'un épisode de brume d'été

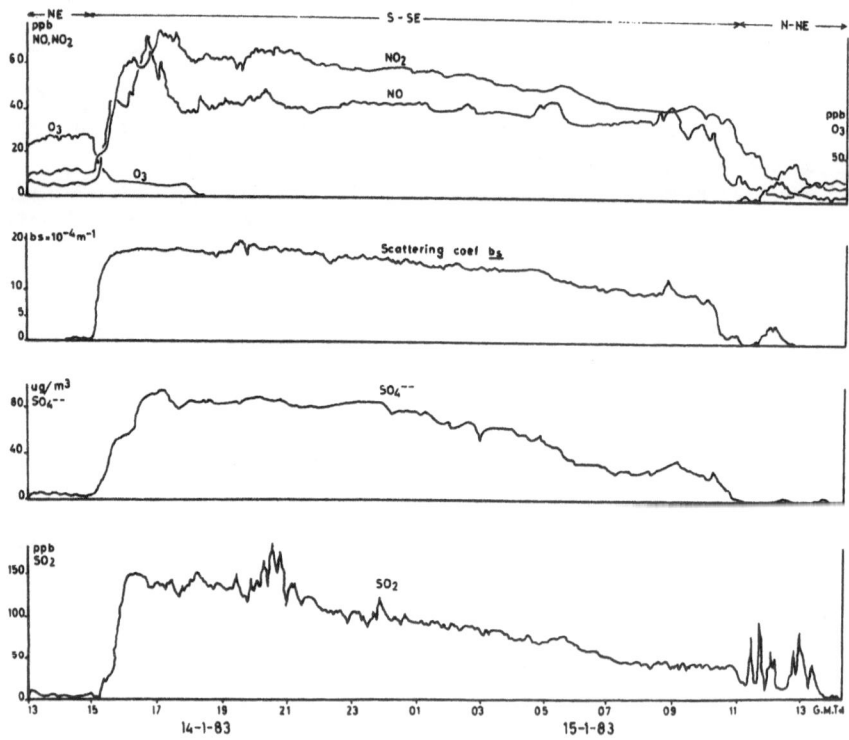

Fig. 6 - Evolution du SO_4^{--} et autres paramètres au cours d'un épisode de transport en hiver

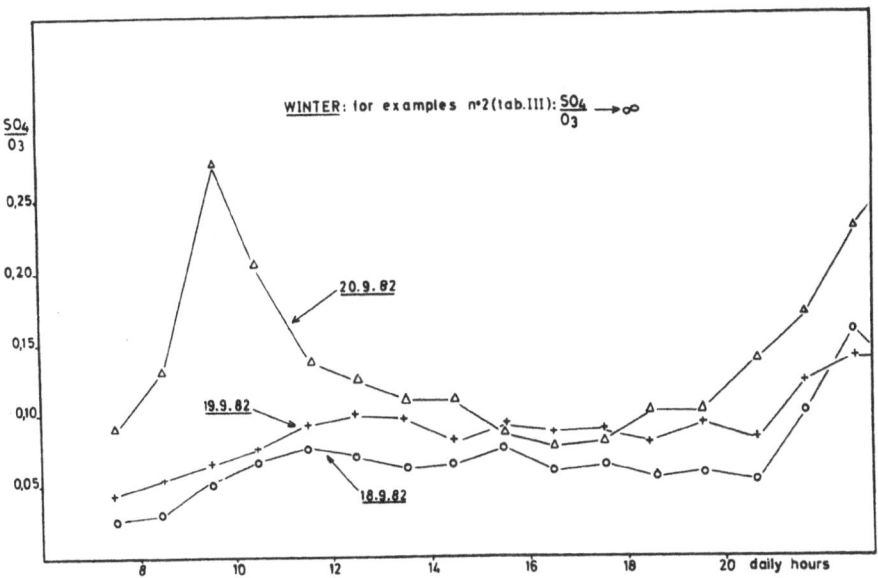

Fig. 7 - Evolution du rapport SO4⁻⁻ / O₃ au cours d'un épisode
de brume d'été

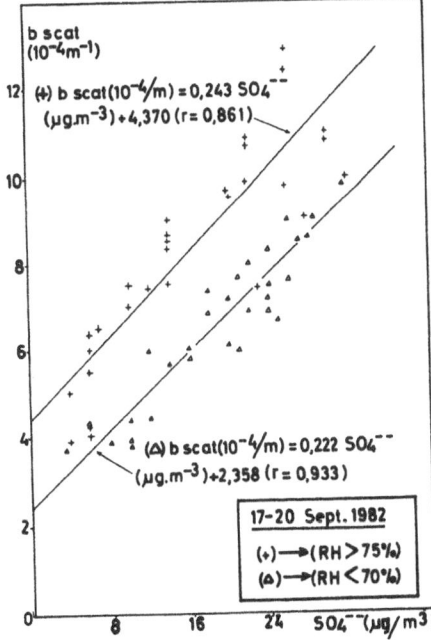

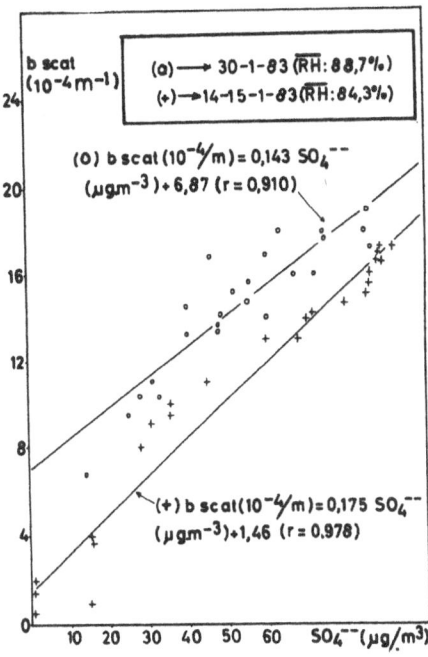

Fig.8 - Influence de la concentration
de SO₄⁻sur la visibilité:
mesures d'été

Fig.9 - Influence de la concentration
de SO₄⁻sur la visibilité:
mesures d'hiver

Cette formation "in situ" s'ajoute au "background" de SO_4^{--} Elle se produit à partir de diverses sources locales de SO_2 disséminées dans la région. Cette conversion gaz-particule est mise en évidence par la formation de noyaux d'Aitken observables le matin (14). Une condensation de molécules gazeuses sur les particules préexistantes s'ajoute à ce processus (14)(15).

Le tableau II ne montre pas une excellente corrélation entre b_S et SO_4^{--} (0,16 < r < 0,83). Deux raisons sont invoquées: D'abord le SO_4^{--} n'est pas le seul paramètre influençant la visibilité, même s'il faut accorder peu de crédit aux valeurs de % SO_4^{--}. Ensuite, on doit considérer les proprié- tés de déliquescence de certains aérosols qui, selon leur composition chi- mique, provoquent sous un degré hygrométrique déterminé, une brusque varia- tion de b_S (16). Soupçonnant le sulfate d'ammonium (2) dont le point de déliquescence est à 75 - 80 % d'humidité relative, on a rassemblé 2 grou- pes de données SO_4^{--}— b_S, sous HR < 70 % et H.R > 75 %. Les corrélations s'avèrent alors effectivement satisfaisantes (fig. 8) identifiant la nature de l'aérosol vraisemblablement $Zn(NH_4)_2SO_4$. Le néphélomètre étant utilisé sans son dispositif de chauffage, on serait tenté d'expliquer les écarts d'ordonnée à l'origine par une condensation d'eau sur des particules autres que les sulfates mais les observations, données à titre d'exemple, sont insuffisantes pour l'affirmer.

$$b_S \ (10^{-4} \ m^{-1}) = 0,222 \ [SO_4^{--}] (\mu g.m^{-3}) + 2,358 \quad (r = 0,933) \text{ pour } r < 70 \ \%$$
$$- \quad = 0,243 \quad - \quad + 4,370 \quad (r = 0,861) \text{ pour } r > 75 \ \%$$

Exemple 2: Les concentrations maximales sont de l'ordre de 90 μg SO_4^{--}/m^3 d'air. Ex. 2 a est un cas typique de transport, observable sur la fig. 6. Simultanément, SO_4^{--}, SO_2 et les NO_x "envahissent" le site de mesure provo- quant la rapide destruction de O_3 par NO. Cet épisode s'observe par coups de vent S-SE, déplaçant les masses d'air stagnantes des zones polluées au- tour de Milan. Dans ce cas, la corrélation de SO_4^{--} avec SO_2 (r = 0,93), suggérant la présence de SO_4^{--} d'origine primaire simultanément émis avec le SO_2, est bonne; de même celle avec la visibilité (fig. 9) où

$$b_S \ (10^{-4}m^{-1}) = 0,175 \ [SO_4^{--}] (\mu g.m^{-3}) + 1,46, \text{ avec } r = 0,978.$$

Sans vent et en période de "fog" (2 b, 2 c), des niveaux comparables de SO_4^{--} ont été mesurés sans être correlés avec SO_2 (r = 0,11 et 0,34) suggérant ici une formation "in situ". Toutefois, la comparaison avec la conversion photochimique d'été montre qu'avec 20 fois plus de SO_2 on ne mesure que 3 fois plus de sulfates. Il est donc probable qu'une fraction de SO_4^{--} pro- vienne d'émissions directes locales. Une autre fraction, catalysée par des éléments simultanément présents avec NO_x, ceux-ci étant alors fort élevés (fig. 7), pourrait provenir d'une conversion "in situ". Cette situation, par rapport aux données d'été, montre une corrélation de SO_4^{--} avec la vi- sibilité bien meilleure pour un ensemble de valeurs journalières. Par ex- emple, pour le 30.1.83 (fig. 9)

$$b_S \ (10^{-4}m^{-1}) = 0,143 \ [SO_4^{--}] \ \mu g.m^{-3} + 6,87, \text{ avec } r = 0,910,$$

On pense qu'on est en présence ici d'un aérosol acide de type H_2SO_4 qui présente un humidogramme $b_S = f (RH)$ monotone, c.à.d. sans point d'in- flexion avec un coefficient b_S moins divergent en fonction de RH.

Pour le prouver, on a essayé de traiter les données de plusieurs jours consécutifs, mais des conditions stables sous HR < 75 % sont rares en hiver. Des données isolées, s'observant sous ensoleillement et bonne visibilité, mettent difficilement en évidence une formation de sulfates qui, d'ailleurs, ne représenterait pas une situation identique à celle de "fog" (2b ou 2c).

Exemple 3: Les mesures effectuées à Fos-Berre dépendant d'une situation topographique et météorologique complexe, les concentrations de SO_4^{--}, mesurées simultanément avec un seul paramètre (SO_2), variant sensiblement d'un site à l'autre.

A Lavera, trop près d'une raffinerie, le panache passait au-dessus du site de mesure, les valeurs étaient négligeables sauf un pic, dans la nuit du 6 ou 7 juin, donnant 230 $\mu g/m^3$ qui est la plus forte concentration mesurée par la méthode.

A Vitrolles (ex. 3), le site était influencé par des émissions situées à 8 - 10 km. Les 7 et 8 juin des niveaux de 13 $\mu g/m^3$ étaient mesurables sans trace de SO_2 entre 12 h et 17 h, pour disparaître pendant la nuit malgré des afflux importants de SO_2. Le 9 juin on mesurait 700 $\mu g/m^3$ de SO_2 pour 60 $\mu g/m^3$ de SO_4^{--} avec un rapport SO_4^{--}/SO_2 faible ($\simeq$ 0,15), mais une corrélation moyenne de 0,78 (fig. 10). Il est possible qu'une partie du SO_4 soit due à une transformation de SO_2 au cours du transport.

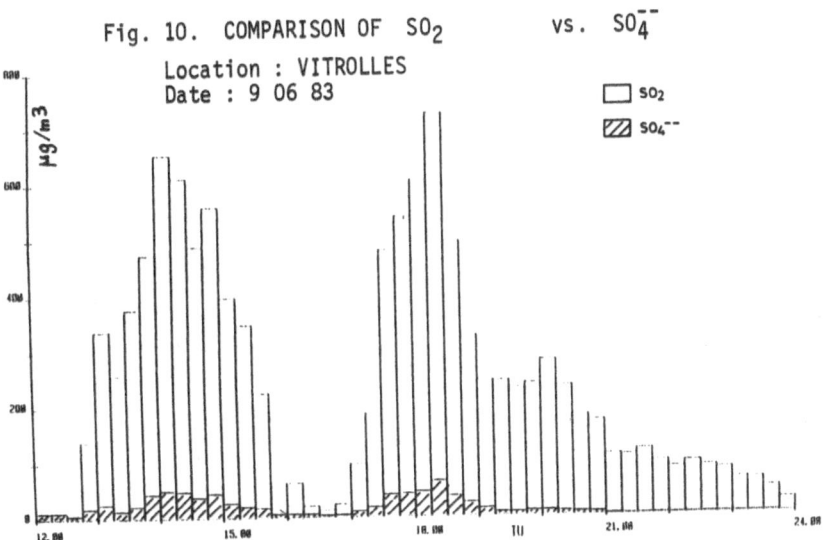

Fig. 10. COMPARISON OF SO_2 vs. SO_4^{--}
Location : VITROLLES
Date : 9 06 83

4. Conclusion

Au moyen de quelques modifications apportées à un détecteur à photométrie de flamme commercial, il a été possible de mesurer les sulfates en continu et en temps réel. La méthode mesure cet aérosol avec un rendement de 85 % lorsqu'il est supérieur à 0,1 μm ce qui la rend intéressante pour

des études affectant la visibilité. Elle a montré que la brume d'été dans
la région d'Ispra est constituée d'un niveau de base régional de sulfates.
A cette concentration s'ajouterait un surplus de nature photochimique, produit
à partir   des sources de SO_2, même faibles,pendant le transport.
Les sulfates mesurés présentent une bonne corrélation avec la visibilité.
D'abord à l'état de brume sèche lorsque l'humidité relative est inférieure
à 70 %, ensuite à l'état déliquescent lorsque $RH > 75$ %. Cette mise en
évidence, caractéristique du sulfate d'ammonium, a permis d'identifier
l'aérosol. On mesure en hiver une concentration 3 fois plus élevée, souvent corrélée avec le SO_2. Une formation "in situ" d'H_2SO_4 corrélant bien
avec la visibilité d'origine catalytique, est aussi très probable.

REFERENCES

1- H.KNOPPEL, B.VERSINO, A.PEIL, H.SCHAUENBURG, H.VISSERS, "Quantitative determination of terpenes emitted by conifers", Proceeding of the 2nd Europ.
Symposium, Doc. EUR 7624, 1981 pp 89-98
2- A.ALTSHULLER "Relationships involv.particle mass + sulfur content at
sites in + around St.Louis, Miss",Atmos.Envir.vol 16,4,pp 837-43, 82
3- W.PIERSON et al. "Ambient sulfate measurem on Alleghany mountains + the
question of atmosph.sulf. in N.E. US" Annuals NYAS,vol "38,pp 145-73, 80
4- W.CRIDER "Hydrogen flame emission spectrophotometry in monit.air for sulf
dioxide + sulf acid aerosol", Analytical Chemistry vol 37 n°13 Dec 1965
5- D.LEAHY et al "Separation + characterization of sulf.aerosol"
Atmosph.Envir; vol 9 pp 219-229, 1975
6- P.ROBERTS et al "Analysis of sulfur in depos.aerosol part. by vaporiz.+
flame photom.detect." Atmos.Envir.vol 10 pp 403-8, 1976
7- J.DURHAM et al "Applic. SO_2 denuder f.contin.measurem of sulfur in sub-
micrometric aerosols" Atmos.Envir. vol 12 pp 883-86, 1978
8- W.COBURN et al "Airborne in situ measurem of particul.sulfur + sulfuric
+ with flame photometry + therm anal." Atmos.Env.12, pp 89 - 98, 1978
9- R.GARBER et al "Determination of amb.aerosol + gaseous sulfur using a
continuous FPD" Atmosph.Envir. vol 17 n° 7, pp 1381-85, 1983
10- R.TANNER et al "Determination of ambient aerosol sulfur using a continuous FPD" Atmosph.Envir. vol 14, pp 121-127, 1980
11- J.HUNTZICKER et al "Cont.measur. + specific. of sulfur-containing aerosols by flame photometry" atmos.Envir. n°12 pp 83-88, 1978
12- E.MUYLE, PEPERSTRAETE "Harmonisation of methods for measurements of SO_2"
EUR report 8052, 1982
13- B.IFADERER et al "Diurnal variations,chem.compos.+ relation to meteorological variables of summer aeros. in NY subr." Atmos.Env. vol 16,9, 1982
14- M.PAYRISSAT, G.OTTOBRINI, H.STANGL "Discussion on partic.formation processes in Po Valley" Proceed. 2nd Eur Symp. Doc EUR 7624, pp.319-333
15- M.LURIA et al "Dynamics of sulfate part.prod. + growth in smog chamber
experiments" Atmos.Envir. vol 16, 4, pp. 697-708, 1982
16- R.J.CHARLSON et al "Sulfuric Acid - Ammonium sulfate aerosol: optical
detection in the St.Louis region,"Science vol 184, 30/11/1973

PLANTS AS MONITORING SAMPLERS OF AIRBORNE PAH

E. BRORSTRÖM-LUNDÉN AND L. SKÄRBY
Swedish Environmental Research Institute

Summary

In this study kale and elm leaves have been used as
samplers of airborne PAH. Kale was cultivated at three
different sites in an urban area, and at two sites with
increasing distance from much frequented roads. Control
plants were grown at a rural site. Elm leaves were collected at two of the urban sites and at the rural site.
Depending on duration of exposure, the PAH content was
increased 10-100 times in plants grown in the urban area,
compared to the control plants. The PAH profiles from the
plants at the different sites with different exposure
times were quite equal between themselves. This may indicate that reactive PAH-components do not further react
after deposition to the plants. Furthermore, the PAH-
profiles were almost identical with an air particle
sample of PAH, taken in the urban area. The PAH-content
in the elm leaves increased also compared to the reference
site, and the PAH profiles were constantly corresponding
to the profiles of kale. The results also showed that
the PAH-content rapidly decreased with increased distance
from the busy roads.

1. INTRODUCTION

Polycyclic aromatic hydrocarbons (PAH) are substances
with a far-reaching distribution in the environment. Beside
a negligible quantity which exists naturally, PAH are delivered
via the atmosphere both in gas phase and bound to particles.
The most common sources of PAH are motor traffic and energy
production. Especially in urban environments high PAH- concentrations in air have been reported (1). PAH are transformed
from the atmosphere to soil and plants. The contribution of
PAH in plants occurs partly through the soil by absorption of
roots, but probably mainly directly from the atmosphere by
adsorption and absorption by the leaves. Then the surface area
of the leaf and its morphology have great importance to the
amount of PAH taken from the air. Plants with large and waxy
leaves like salad and kale have demonstrated a great capacity
to take up and accumulate PAH (2, 3). In studies of different
sources of PAH the content in plants has been analysed concerning quality and quantity of PAH (4). In this study, kale
and elm leaves have been used as indicators in the examination
of PAH deposition in an urban area. Kale was chosen because
of its large leaf area and because its ability to survive in

cold and snow during several months. Elm leaves were used because these trees are common in urban environments and moreover the leaves have a surface covered with fine hairs on which particles are easily deposited.

2. MATERIALS AND METHODS

Sampling sites and exposure times

Two monitoring periods of PAH in plants have been carried out.

The aim of the first period was to study the deposition of PAH to vegetation in an urban area. The monitoring was carried out in a city at the Swedish west coast (Gothenburg) in the autumn and winter of 1980-1981. Kale was cultivated in central parts of the city at three different sites (site 1,2, 3). All of them had high traffic intensity. A reference site was chosen for kale cultivation about 35 km south of the city. Elm leaves were collected close to the sites of kale. The kale leaves were collected at two different occasions. The first sampling took place after three months of exposure (in September) and the second after six months (in January). The collection of the elm leaves was carried out in September which gave an exposure time of about five months.

The aim of the second monitoring period was to study the PAH content in plants cultivated with increasing distance from a busy street in an urban area and a much frequented highway in a rural area, located about 20 km south of the city. The kale was placed 10, 25 and 50 meters from the roads. The exposure time was six months from August 1981 to January 1982.

Cultivation technique

The kale plants were cultivated and grown in pots with controlled soil. When the plants were about 20-30 cm high they were taken outdoors and placed at the sampling sites. The elm leaves were collected 1 or 2 m above ground level. All samples were kept froozen in aluminum foil from the collection to the time of analysis.

Analysis

The plant and leaf samples were prepared according to Grimmer before analysis (5). The analysis was performed with a gas chromatograph equipped with a flame ionisation detector and supplied with a 50 m x 0.38 mm (i.d.) capillary column coated with SE 54. The quantification and identification of PAH was performed with the use of reference substances. All samples were analysed on a fresh weight base.

3. RESULTS AND DISCUSSION

Kale leaves which had been exposed in the city had a much higher PAH content than the reference from the rural site (Table 1).

		Site 1	Site 2	Site 3	Ref.	Site 2	Site 3
Exposure time (months)		3	3	3	3	6	6
				ng/g (fresh weight)			
Biphenyl		1.2	2.1	2.0	<0.5	2.7	2.0
Acenaphtylene		2.2	3.2	7.6	"	11	44
Acenaphtene		<0.5	<0.5	<0.5	"	0.6	4.4
4-Methylbiphenyl		<0.5	<0.5	<0.5	"	0.6	3.9
Dibenzofuran		3.1	5.4	1.9	"	7.6	19
Fluorene		1.8	3.5	11	"	14	52
9-Methylfluorene		<0.5	<0.5	1.8	"	1.9	23
9.10-Dihydroanthracene		<0.5	<0.5	<0.5	"	<0.6	1.6
2-Methylfluorene		1.2	2.5	4.8	"	8.2	44
1-Methylfluorene		<0.5	2.7	2.4	"	0.9	52
Dibenzothiophene		2.0	5.8	0.9	"	13	60
Phenanthrene	PHE	26	67	30	11	220	520
Anthracene	ANT	3.3	6.7	3.4	<0.5	30	84
Acridine		<0.5	<0.5	<0.5	"	<0.6	<1
Carbazole		<0.5	<0.5	<0.5	"	0.6	13
2-Methylanthracene		2.4	4.5	6.4	"	2.0	76
1-Methylphenanthrene	1-MPHE	6.4	13	12	"	47	160
9-Methylanthracene		<0.5	<0.5	<0.5	"	2.5	17
3.6-Dimethylphenanthrene		3.0	6.1	2.4	"	18	80
1.2-Dihydropyrene	1.2DIHP	4.2	12	6.0	"	43	160
Fluoranthene	FTH	29	67	28	9.0	310	760
Pyrene	P	33	76	41	8.5	350	1000
Benzo(a)fluorene	B(a)F	10	14	9.2	<0.5	75	300
Benzo(b)fluorene	B(b)F	5.7	5.8	2.6	"	35	140
1-Methylpyrene	1-MP	2.4	4.1	3.0	"	15	110
Benz(a)anthracene	B(a)ANT	18	38	26	"	76	160
Chrysene ⎫ Triphenylene ⎭	CHR	19	28	16	"	150	230
Benzo(b)fluoranthene	B bjk FTH	2	20	14	"	120	190
Benzo(e)pyrene	B(eP)	1.7	4.8	<0.8	"	41	57
Benzo(a)pyrene	B(a)P	<0.5	2.0	<0.8	"	11	26
Perylene	PER	<0.5	<0.5	<0.8	"	-*	-*
Indeno (1,2,3-cd)pyren	IND	<1.0	<1.0	<1.2	"	19	18
Dibenz(a,h)anthracene		<1.0	<1.0	<1.2	"	<1	<1
Picene		<1.0	<1.0	<1.2	"	<1	<1
1,2,3,4-Dibenzanthracene		<1.0	<1.0	<1.2	"	<1	<1
Benzo(g,h,i)perylene	B(ghi)PER	<1.0	<1.0	<1.2	"	21	38
Anthanthrene		<1.0	<1.0	<1.2	"	<1	<1
Σ identified PAH		187.6	385.2	232.4	28.5	1646	4449

*) high background, cannot be quantified.

Table 1. PAH concentrations in kale.

A comparison of the PAH content in the leaves with the traffic intensity at the different sites shows that the highest traffic intensity resulted in the highest PAH concentration in the kale leaves (Table 2).

Site	Cars/24 hours	Σ identified PAH ng/g fresh weight
1	14.000	187
2	42.000	385
3	22.000	232

Table 2. Traffic intensity at three examinated sites in the city

In spite of the difference in PAH content in the plants between the different sites, the PAH-profiles showed great similarity. The PAH-profiles, e.g. the logaritmes of the relative distribution of selected PAH-compounds, normalized with the concentration of the non-reactive benzo(b,j,k)fluoranthenes as the unit, are shown in Figure 1.

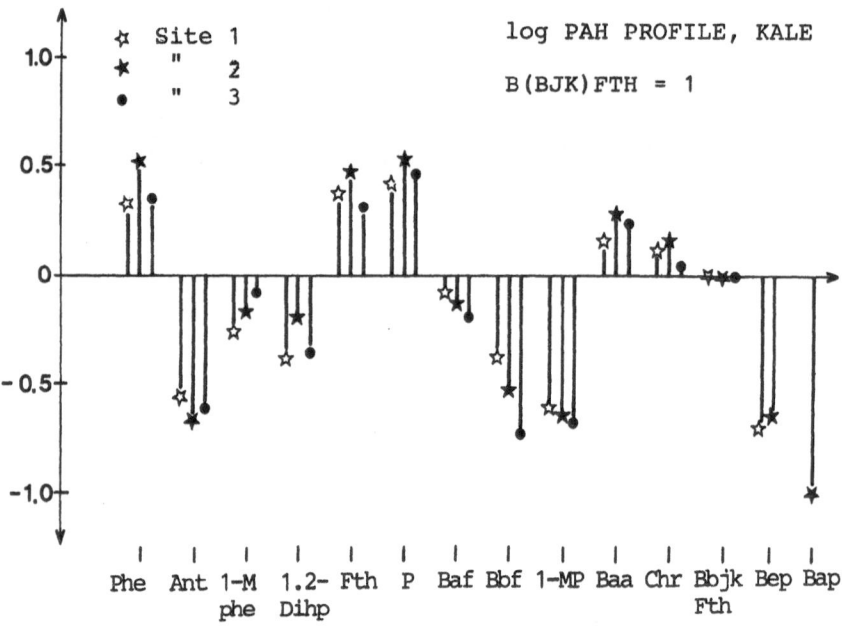

Figure 1. PAH-profiles from kale plants at the different sites in the city

Kale leaves with six months of exposure had a PAH content which was 10 times higher than kale exposed three months. Thus, most of the PAH was deposited to the plants during the last three months. This result agrees with the higher PAH concentrations in air during wintertime (6). It should be noticed that in spite of the long exposure time, and therefore the possibility for transformation or degradation of the PAH, the PAH profiles from September and January showed remarkably good agreement (Figure 2 and 3). This indicates that the PAH compounds are stable, in contrast to PAH adsorbed on air-borne particles(7) after deposition to the leaf surface.

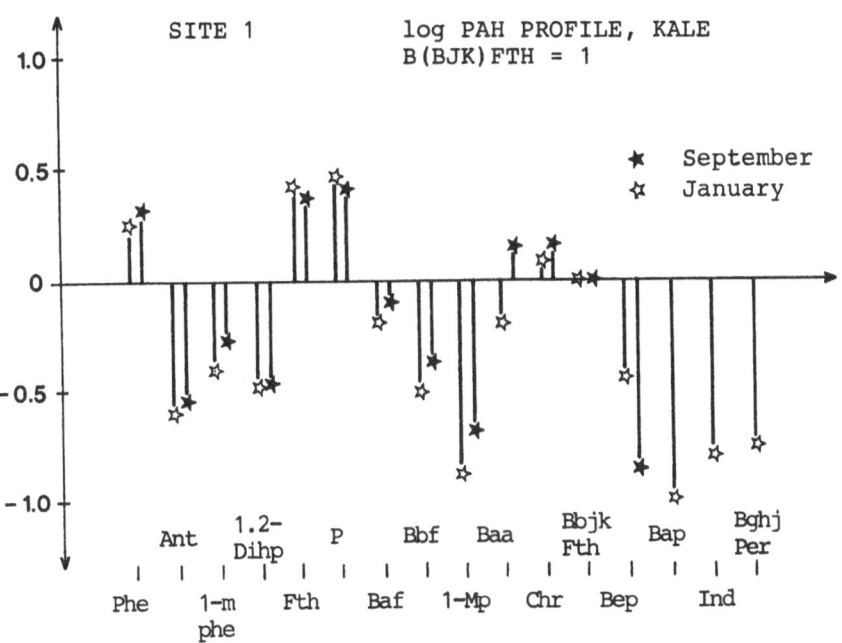

Figure 2. PAH-profiles from kale collected at different exposure times, after three months and six months, respectively.

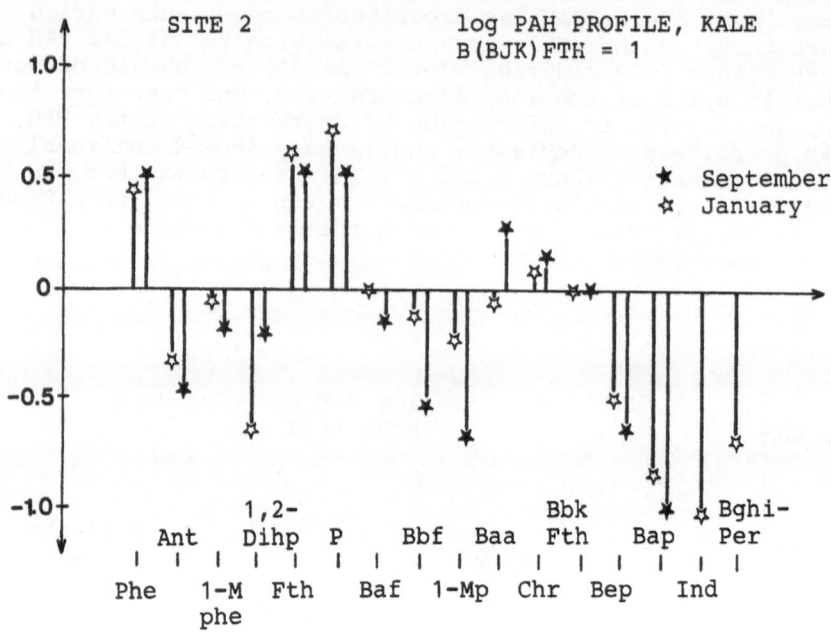

Figure 3. PAH-profiles from kale collected at different exposure times, after three months and six months, respectively.

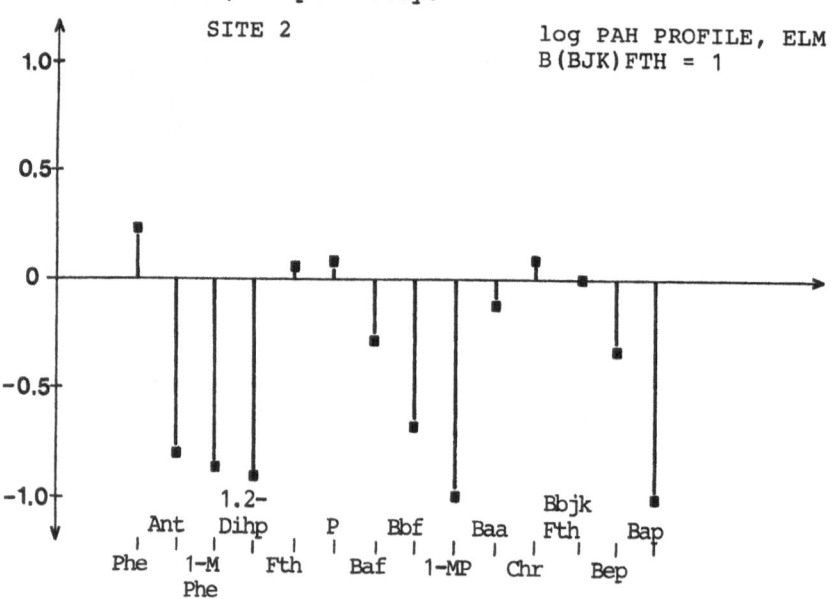

Figure 4. PAH-profiles from elm leaves at site 2.

The PAH-profiles from the elm leaves showed a greater part of nonvolatile PAH compounds (Figure 4). This may be explained by the fact that particles and soot more easily are deposited to the "hairy" surfaces like the elm leaves and also simply by the fact that there were higher PAH concentrations in the elm leaves compared with the kale leaves with the same exposure time (Table 3).

Figure 5 shows a PAH-profile from an HVS air particle sample taken near site 2, 20 meters above ground during a seven days period in February, 1981. The similarity of the PAH profiles between the air sample and the leaf sample is quite evident, but the amount of nonvolatile compounds are much larger in the leaves. This difference may partly depend on the fact that gas phase PAH adsorbs and accumulates in the leaves, partly on the fact that an evaporation of the volatile PAH takes place during high-volume sampling ("blow-off")(1).

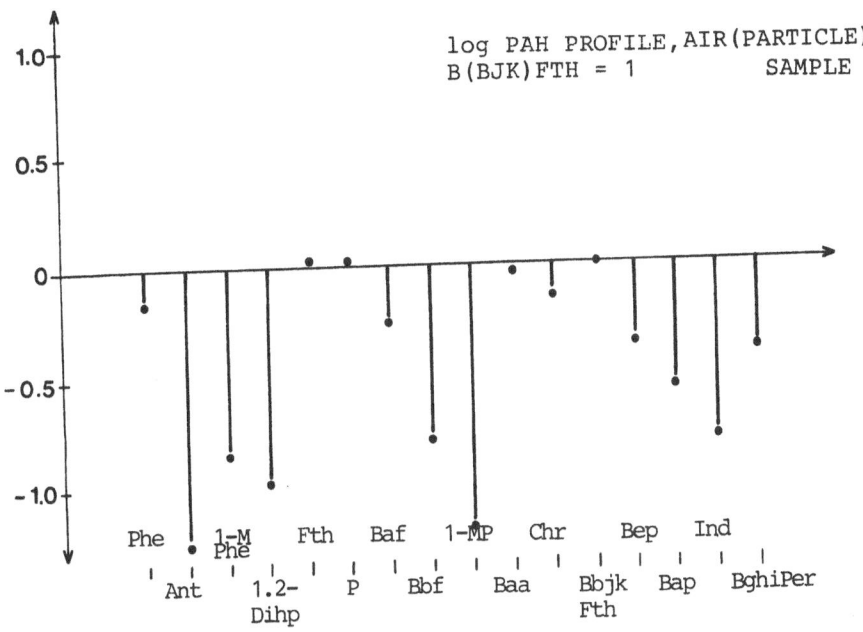

Figure 5. PAH-profile from a seven 24-hours period air particle sample from the city.

The results of the PAH analyses of kale leaves,which were grown with increasing distance from a street in the city, showed that the PAH content decreased with increased distance from the road (Figure 6). The difference in the PAH concentration between 10 and 25 meters was about 1300 ng and between 25 and 50 meters about 560 ng/g. A comparison of the PAH-content in the kale 10 meters from the street in 1981-1982 with the kale from site 1 from 1980-1981, shows that the PAH-content in the kale from the different periods were in the same level, 1600 and 1400 ng/g, respectively. This, in spite of snow covering the plants during 2 months of the last period.

Table 3. PAH concentrations in elm leaves.

Compound	Abbr.	Site 1 ng/g (fresh weight)	Site 2 ng/g (fresh weight)	Ref.
Biphenyl		2.3	4.6	<0.5
Acenaphtylene		4.0	25	<0.5
Acenaphtene		(0.3)	2.4	<0.5
4-Methylbiphenyl		<0.5	0.5	<0.5
Dibenzofuran		2.5	13	<0.5
Fluorene		3.7	13	<0.5
9-Methylfluorene		<0.5	1.1	<0.5
9.10-Dihydroanthracene		(0.2)	<0.5	<0.5
2-Methylfluorene		1.5	8.5	<0.5
1-Methylfluorene		1.7	9.9	<0.5
Dibenzothiophene		2.9	17	<0.5
Phenanthrene	PHE	52	190	10
Anthracene	ANT	5.0	22	<0.5
Acridine		<0.5	<0.5	<0.5
Carbazole		3.9	12	<0.5
2-Methylanthracene		4.4	39	<0.5
1-Methylphenanthrene	1-MPHE	<0.5	<0.5	<0.5
9-Methylanthracene		1.3	16	<0.5
3.6-Dimethylphenanthrene		3.7	35	<0.5
1.2-Dihydropyrene	1.2DIHP	32	160	<0.5
Fluoranthene	FTH	34	180	6.3
Pyrene	P	16	80	4.3
Benzo(a)fluorene	B(a)F	6.7	29	<0.5
Benzo(b)fluorene	B(b)F	3.4	11	<0.5
1-Methylpyrere	1-MP	23	67	<0.5
Benz(a)anthracene	B(A)ANT	34	100	<0.5
Chrysene	CHR	30	110	<0.5
Triphenylene		14	42	<0.5
Benzo(b)fluoranthene	B(b,j,k)FTH	3.1	12	<0.5
Benzo(e)pyrene	B(e)P	-*	-*	<0.5
Benzo(a)pyrene	B(a)P	<1	27	<1
Perylene	PER	<1	<1	<1
Indeno(1,2,3-cd)pyrene	IND	<1	<1	<1
Dibenz(a,h)anthracene		<1	<1	<1
Picene		<1	27	<1
1,2,3,4-Dibenzanthracene		<1	<1	<1
Benzo(g,hi)perylene	B(g,h,i)PER	<1	<1	<1
Anthanthrene		<1	<1	<1
Σ identified PAH		285.6	1.253	20.6

Table 3. PAH concentrations in elm leaves.

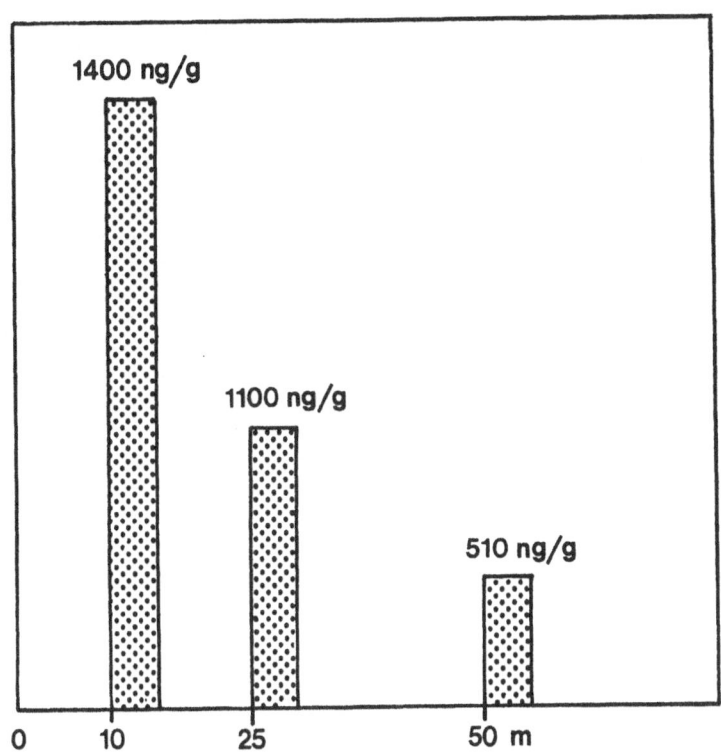

Figure 6. Σ identified PAH in ng/g fresh weight with increasing distance from a street in the city.

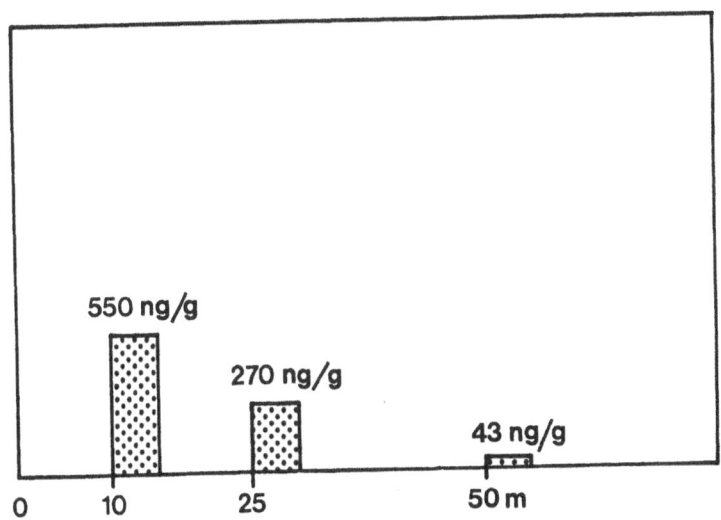

Figure 7. Σ identified PAH in ng/g fresh weight with increasing distance from a highway in a rural area.

The kale plants exposed to traffic at the highway showed increased PAH-concentration in the sampling sites closest to the highway. At a 50 meters distance from the highway the PAH content was down on the background level (Figure 7). A comparison of the PAH-concentrations at the different roads showed four times higher PAH concentrations in the plants that had been exposed in the city.

The results from these investigations indicate that PAH analysis of kale and elm leaves constitute a good complement to conventional PAH air sampling, for instance in studies of local sources or in an inventory of air quality in different environments. Kale and elm as passive samplers of air pollutants of course are not only limited to PAH, but can be extended to other organic compounds with equal properties. Then, PAH could be used as an index for the organic compounds in the air which are emitted, distributed and deposited in a similar manner.

4. REFERENCES

1. Lindskog, A. and Brorström, E. (1981) Determination of polycyclic aromatic hydrocarbons in airborne particulate matter i Göteborg. Nordic PAH-project. Report No. 11

2. Björseth, A. et al. (1978) Analys av polycykliska aromatiska kolväten i sallad från Sverige. Nordic PAH-project. Report No. 2.

3. Kveseth, K., Southland, B., Stovert, M-M. (1981) Polycyclic aromatic hydrocarbons in leafy vegetables. A comparison of the Nordic results. Nordic PAH-project, No. 8.

4. Larsson, B. and Sahlberg (1981) Polycyclic aromatic hydrocarbons in lettuce. Influence of a highway and on aluminum smelter. In Polynuclear aromatic hydrocarbons: Typical and Biological Chemistry. M. Cooke, A. Dennis, G. Fisher (Eds) Battelle Press, Columbus, Ohio (1981) 417-427.

5. Grimmer, G. and Böhnke, H. (1975) Polycyclic Aromatic Hydrocarbon, Profile Analysis of High-Protein Foods, Oils and Fats by Gas Chromatography, Journal of the AOAC.

6. van Vaeck, Brodaoin, G. and van Cauwenberghe, K. (1981) On the relevance of air pollution. Measurements of alphatic and polyaromatic hydrocarbons in ambient particulate matters. Biomodic. Mass Spectrometry Vol. 1, No. S11-12.

7. Brorström,E. Grennfelt, P. Lindskog, A. (1983) The effect of nitrogen dioxide and ozone on the decomposition of particle associated polycyclic aromatic hydrocarbons during sampling from the atmosphere. Atm. Environ. Vol. 17, No. 3.

PREPARATION OF DIFFUSION DENUDER TUBES FOR COLLECTION OF AMMONIA OR ACIDIC GASES IN AIR. EQUIPMENT FOR COATING AND EXTRACTION

Ella E. Lewin, Karsten Fuglsang and Knud A. Hansen
National Agency of Environmental Protection
Air Pollution Laboratory
DK-4000 Roskilde
Denmark

Summary

The construction and preparation of a denuder tube assembly for collection and direct analysis of ammonia or
acidic gases is described. The assembly consists of fifteen glass tubes arranged in a holder; it can be easily
connected with a filter holder for concurrent collection
of particles. Also, an equipment for rapid and convenient coating and extraction of the tubes is described.

1. INTRODUCTION

It has recently become known that the collection of gases
and aerosols on filters can lead to erroneous results. This
is due to the release of volatile compounds from particles and
to the absorption of gas on the collected particles.

In order to avoid this artifact it has been proposed to
remove the gases prior to collection of the aerosol. Such a
separation can be achieved by employing diffusion denuder
tubes. In a diffusion denuder tube a laminar flow of air passes through a hollow tube covered inside with a suitable absorption agent: Gas molecules diffuse to the walls and are retained there, while aerosol particles pass virtually unaltered
because of their much lower diffusion coefficient.

Several proposals for suitable arrangements of denuder
tubes have been made during the last 5-6 years. The methods
presented to date can be classified in three different groups:

- methods involving direct determination of the collected gas
  by extraction and analysis (1, 2, 3, 4),
- methods applying the "denuder difference method", i.e. simultaneous collection of particles on a filter following a
  denuder tube, and collection of gases and particles on a
  filter. The amount of gas is found as the difference between
  the collected amounts on the filters (5,6),
- methods involving thermal desorption of the collected gas
  from the tube wall and detection by means of e.g. chemiluminescence (7).

This paper describes the construction and features of an
equipment for direct determination of ammonia or acidic gases
collected in diffusion denuder tubes.

The demand of laminar flow considerably limits the amount
of material collected in a tube and on the following filter.

In order to overcome this difficulty, the assembly presented consists of 15 tubes arranged in a holder, similar to the device used by Stevens et al. (8).

2. DESCRIPTION OF THE ASSEMBLY

The geometrical dimensions of the diffusion denuder assembly are dictated by the demand of the total separation of gas and particles on one side and enough material for analysis on the other. The amount of gas absorbed is governed by the Gormley-Kennedy equation (9) and is a function of the length of the tube, flow of air and the molecular diffusion coefficient of the gas. The amount of material collected is, of course, proportional to the air flow and the time of collection, but an increase in flow entails an increase in the length of the tube, if the requirement of the total gas removal is to be observed. On the other hand, the increase in the length increases the possibility of particle deposition, and the flow must still be in the laminar region. Therefore, the geometrical dimensions chosen are dictated by a compromise between length, flow, and also convenience in handling. The assembly is shown in fig. 1.   15 quartz glass tubes, each 30 cm long and 0.4 cm i.d. are arranged in a circle and placed in a cylindrical plexiglass protective cover. A choice of 12 $1 \cdot min^{-1}$ flow through the entire system will lead to a Reynolds number of 325, and the requirement for laminar flow is thus fulfilled. The tubes are supported by two silicon rubber plates (A). To connect the assembly with a filter, an open part of a holder (B) for 47 mm Selectron filters is glued to the protective cover (C). At the inlet end, the tubes go into a 0.5 cm Teflon plug (D), which also facilitates connection with the vacuum system during the preparation procedure. To prevent the coating film to deliquesce at high humidities, a warm air can be led into the protective cover to heat the tubes a few degrees above the ambient temperature.

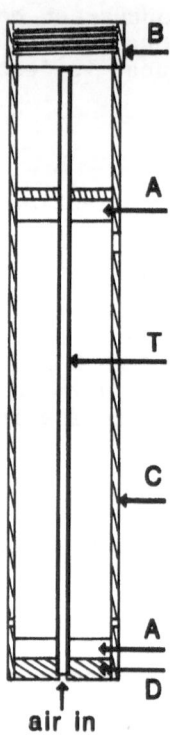

air in

Figure 1. Denuder tube assembly, only one tube shown. A - silicon rubber support plates. B - open part of the Selectron filter holder. C - protective cover. D - Teflon plug. T - denuder tubes.

3. EQUIPMENT FOR COATING AND EXTRACTION

To speed up the coating and extraction procedures and to limit the extraction volume as far as feasible, an equipment - shown schematically in Figure 2 - was constructed. The coating solution (or water for extraction) is sucked into the tubes of the holder by connecting them to a vacuum pump, and the coating film is afterwards dried out with a gentle flow of clean pressurized air. The equipment enables alternative use of sucking (valve E) or blowing (valve G). Every tube is connected to the main line through a separate valve (H). This gives us a possibility of choosing how many tubes we coat or extract simultaneously, depending on the volume of liquid available. The diffusion denuder assembly is connected to the vacuum pump from its inlet end (D, Figure 1).

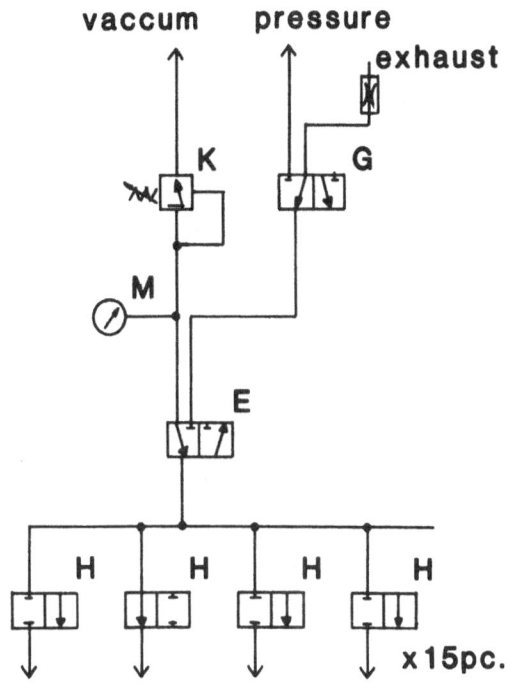

Figure 2. Schematic diagram of connections for vacuum and pressurized air.

  E - connection valve to vacuum
  G - connection valve to pressurized air
  H - connection valves to denuder tubes
  K - valve regulating the height of the liquid in tubes
  M - manometer for controlling the height of the liquid in tubes

A special valve (K) allows regulation and permanent posi-
tioning of the height of the liquid column sucked into the
tubes. The manometer (M) serves as an indicator of this
height. In this way the liquid comes to the same level every
time in successive operations. The equipment, with the assembly mounted for preparation, is shown in Figure 3.

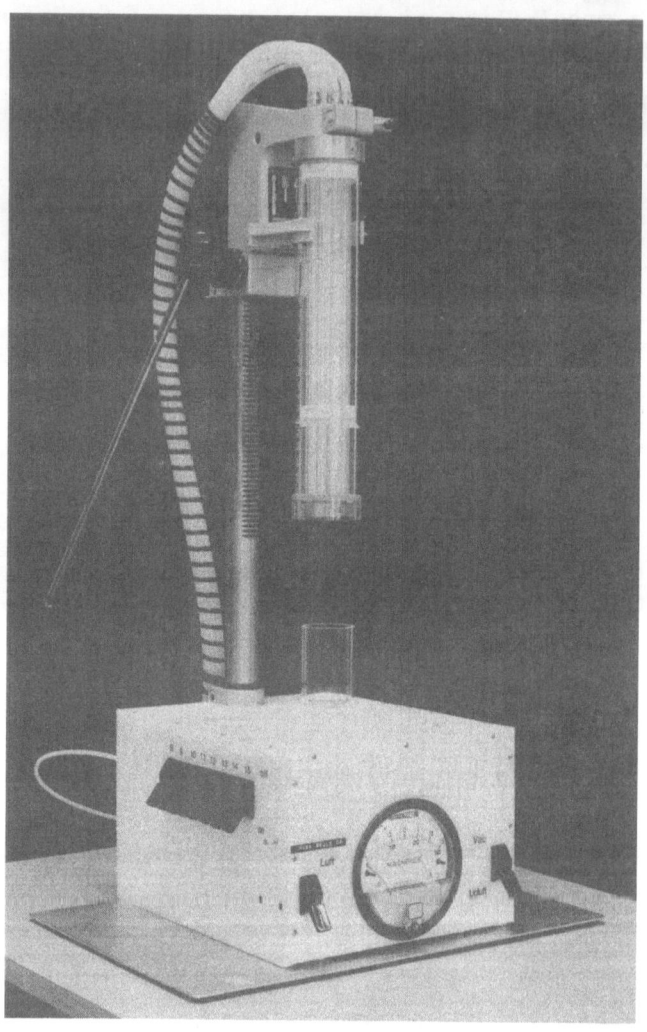

Figure 3. General view of the equipment for coating and extracting the denuder tubes in the assembly.

3.1. COATING

To avoid losses of particles due to impaction caused by turbulence in the inlet part of the tubes, the first few centimeters should remain uncoated. The length required to obtain laminar flow can be calculated according to Ferm (1), and with the given condition the air must pass 6,5 cm to reduce turbulence. Therefore, the first 7 cm from the inlet end remain uncoated.

It is easiest to coat all tubes at once. Following its connection to the equipment, the tubes of the assembly are lowered into the beaker containing the appropriate coating solution. The tubes are pressed gently against the bottom of the beaker, the height of the liquid column is set to the desired level and the passage to the vacuum is opened. After disconnection, the liquid flows slowly out of the tubes and back into the glass. However, when the connection is reestablished, the liquid is sucked again to the same level. To dry out the cover, the assembly is raised and the gentle stream of dry, clean air is blown through the tubes. It seems that the cleaning of tubes is the determining factor for the homogenity of coating. In our experiment, we left the tubes in the detergent solution for 24 hours or more, then rinsed them extensively with deionized water and dried them with hot air. Just before coating, the tubes were flushed with methanol. Blank values for such treated tubes are shown in Tables 1 and 2 for ammonia and acidic gases, respectively. To determine the precision of coating procedure within the system, the tubes were extracted one by one to fresh portions of water (10 ml each) and the amount retained on the walls was determined from pH measurements (NH_3-denuder tubes) or by titration (acid-denuder tubes); the reproducibility of coating between the tubes varied by about 20%.

3.2. EXTRACTION

Extraction is in principle performed in the same way as coating. Only the coated part of the tubes (i.e. containing the retained gas) should be extracted. The same portion of water is used for extracting all tubes and the number of tubes extracted simultaneously depends on the extraction volume. In our case, twenty millimeters of the extract is enough for the necessary analyses colorimetric determination of SO_4^{2-}, NO_3^- and Cl^-, and the complete wash-out of the system is achieved when the tubes are flushed successively with 2 x 10 ml of water. In our case (23 cm of every tube coated) three tubes can be handled in parallel and the complete extraction procedure takes about 10 min. It has not been investigated whether the absorbed material can be completely removed with volumes lower than 20 ml.

4. APPLICATIONS

The assembly can be coated with in principle any absorbant found suitable for the collection and analysis of gases desired. In our case, we chose oxalic acid as an absorbant for the collection of ammonia as described by Ferm (1), and

KOH for the collection of acidic gases. The absorption efficiency can be calculated from the Gormley-Kennedy equation (9). The molecular diffusion coefficient of gaseous HNO_3 is assumed to be 0.16 $cm^2 \cdot s^{-1}$ (6); the diffusion coefficient of gaseous NH_3, according to (10), is 0,236 $cm^2 \cdot s^{-1}$. With an absorption length of 23 cm and a flow of ca. 14 $cm^3 \cdot s^{-1}$ this means a calculated absorption efficiency of 96% for HNO_3 and 99% absorption efficiency for NH_3. Experiments conducted in our laboratory confirmed these calculations.

4.1. AMMONIA

As a coating solution, 1.5% oxalic in ethanol is used (Ferm (1) used methanol, but we found that a smoother film is obtained with ethanol). In order to determine the theoretical capacity of the system, coated but unexposed tubes were flushed with 2 x 10 ml water and the pH of the extract was measured with a calibrated glass electrode. The amount of acid retained on the walls of the tubes was calculated to be 0.89±0.14 meq H^+ (mean of 5 determinations). Ammonia is determined in our laboratory by the automated indophenol method.To determine the blank value and possible interferences in the analysis from the absorption film, the coated tubes were flushed with pure water or solutions containing respectively, 0.2, 0.7 or 1.0 ppm NH_3-N. The results are shown in Table 1; the values are corrected for the blank value of uncoated tubes. Storage of coated tubes sealed with Parafilm in a clean atmosphere did not affect the blank value.

Expected µg N/ml	Found µg N/ml
0	0.020 ± 0.003
0.202	0.209
0.738	0.747
1.022	1.022

Table 1. Denuder tube assembly for collection of NH_3 - blank value and interferences in analysis (indophenol method). Blank value - mean of 5 measurement, others - mean of 2 measurements. The results are corrected for blank values.

4.2. Acidic gases

For collection of acidic gases the tubes are coated with a solution of 1 M KOH in methanol. The drying of the tubes with clean air after coating result in conversion to K_2CO_3. This coating will absorb HNO_3, HCl and also some SO_2, depending on the relative humidity. The collection capacity of the tubes was determined by titration of the extract of the coated, unexposed tubes. It was calculated to be 0.44 ± 0.08 meq OH^-. The blank values of coated tubes for sulphate, nitrate, and chloride are shown in Table 2. The methods of analyses are: sulphonazo (11) for sulphate and sulphanilamid for nitrate. The analyses were performed on an Auto Analyzer.

Also, the ion chromatographic method is used to determine all
three ions in one sample. In the case of the sulphonazo-me-
thod and the ion chromatographic method, KOH is first removed
from the sample with an ion exchanger (Dowex 50-X). To investigate the possible influence of the presence of KOH in the
extract or the ion exchanging step on the analytical results,
the coated tubes were flushed with solutions of sulfate, nitrate and chloride of known concentrations and the extracts
were analyzed. The results are given in Table 2.

	Expected	Found	Method
SO$_4$	0	0	
	0.20	0.20	ion chroma-tography
μg S·ml^{-1}	0.98	1.00	
	1.88	1.89	
	0.20	0.25	
	1.00	0.99	sulphonazo
	4.00	3.80	
NO$_3$	0	0	ion chroma-tography
	0.14	0.12	
μg N·ml^{-1}	0.95	0.85	
	1.90	1.82	
	0	0.020±0.005	
	0.20	0.20	sulphani-lamid
	0.70	0.69	
	1.00	0.97	
Cl	0	0	ion chroma-tography
	0.20	0.20	
μg·ml^{-1}	0.95	0.91	
	1.89	1.85	

Table 2. Denuder tube assembly for collection of acidic gases
– blank values and interferences in analysis. Blank values –
mean of five, others – mean of two. The results are corrected
for blank values.

5. DISCUSSION

Measurements of ambient ammonia concentrations conducted by Ferm (1) in background areas showed concentrations between 0.05 and 0.68 µg $NH_3 \cdot m^{-3}$. For our sampling conditions of 12 $l \cdot min^{-1}$ for 24 h and with an extraction volume of 20 ml, these figures correspond to 0.04-0.5 µg $N \cdot ml^{-1}$ after extraction, which is within the detection range of the method of analysis applied. On the other hand, as the blank value for unexposed (but coated) assembly is 0.02 µg $N \cdot ml^{-1}$, extra precautions must be taken in order to reduce it. As collection of ambient air with NH_3-concentration of 1.0 µg $\cdot m^{-3}$ under the given circumstances will lead to a collected amount of ammonia of about 17 µg NH_3, the shown theoretical capacity of 0.89 meq (i.e. $1.5 \cdot 10^4$ µg NH_3) is fully sufficient.

Concentrations of gaseous HNO_3 cited in (6) lie between 1 and 35 µg $NO_3^- \cdot m^{-3}$. This corresponds in our case to concentrations in the analysis of 0.2-7 µg NO_3-$N \cdot ml^{-1}$, which is within the detection range of the method. In this case the blank value is about 10 times lower than the lowest expected sample concentration and the method can be used without further modifications. The theoretical capacity of the denuder tube assembly for collection of acidic gases was found to be about 0.4 meq. If the concentration of acidic gases is assumed to be 35 µg $NO_3^- \cdot m^{-3}$ and 50 µg $SO_2 \cdot m^{-3}$ (i.e. about 2 µeq $H^+ \cdot m^{-3}$), then the amount of acid collected will be 35 µeq under our circumstances and the cited capacity of 0.44 meq is sufficient.

Losses of particles due to sedimentation can be avoided by placing the tubes vertically during sampling while losses due to diffusion are of importance only for the smallest particles and in the range 3-5% (3).

The dimensions of the tube assembly were partly chosen for reasons of convenience. The primary aim was to find an easy and versatile coating and extraction method. If necessary, the length of the tubes, the flow etc. can be easily changed according to specific requirements.

At present the system is undergoing field trials in respect to its performance under different ambient conditions. Also, it is planned to investigate the applicability of the acid-tube assembly for collection of SO_2 and gaseous HCl.

REFERENCES

(1) M. Ferm. - Method for Determination of Atmospheric Ammonia. Atmos. Environ. 1979, 13, 1385-1393.

(2) J. Slanina, L. v. Lamoen-Doornenbal, W.A. Lingerak, W. Meilof, D. Klockow and R. Niessner. - Application of a Thermo-Denuder Analyzer to the Determination of H_2SO_4, HNO_3 and NH_3 in Air. Intern. J. Environ. Anal. Chem. 1981, 9, 59-70.

(3)   E.E. Lewin and D. Klockow. - Application of the TCM De-
      nuder for SO_2 Collection.   Proceedings of the Second
      European Symposium, Varese, Italy, Sept, 29 - Oct. 1,
      1981.

(4)   M. Ferm. - Method for Determination of Gaseous Nitric
      Acid and Particulate Nitrate in the Atmosphere. EMEP Ex-
      pert Meeting on Chemical Matters, Geneva, 10.-12. March,
      1982.

(5)   R.W. Shaw, Jr., R.K. Stevens, J. Bowermaster, J.W. Tesch
      and E. Tew. - Measurements of Atmospheric Nitrate and Ni-
      tric Acid: The Denuder Difference Experiment. Atmos. En-
      viron. 1982, 16, 845-853.

(6)   J. Forrest, D.J. Spandau, R.L. Tanner, and L. Newman. -
      Determination of Atmospheric Nitrate and Nitric Acid em-
      ploying a Diffusion Denuder with a Filter Pack.   Atmos.
      Environ. 1982, 16, 1473-1485.

(7)   R.S. Braman, T.J. Shelley and W.A. McClenny. - Tungstic
      Acid for Preconstruction and Determination of Gaseous and
      Particulate Ammonia and Nitric Acid in Ambient Air. Anal.
      Chem. 1982, 54, 358-364.

(8)   R.K. Stevens, T.G. Dzubay, G.M. Russwurm and D. Rickel. -
      Sampling and Analysis of Atmospheric Sulphate and Related
      Species. Atmos. Environ. 1978, 12, 55-68.

(9)   P. G. Gormley and M. Kennedy. - Diffusion from a Stream
      Flowing through a Cylindrical Tube. Proc. R. Ir. Acad.
      1949, Sect. A, 52, 163-169.

(10) J.M. Coulson and J.F. Richardson. - Chemical Engineering,
      vol. 1, Pergamon Press, London 1956, p. 239.

(11) K. Keiding and A.-M. Hansen. - SO_2-analyser i det lands-
      dækkende luftkvalitetsmåleprogram, 1983, MST LUFT A-66,
      National Agency of Environmental Protection, Forsøgsanlæg
      Risø, DK-4000 Roskilde, Denmark.

THE PIXE ANALYTICAL TECHNIQUE AND ITS APPLICATION TO ENVIRONMENTAL PROBLEMS

I.V. MITCHELL
Directorate-General for Science, Research and Development
Commission of the European Communities

Summary

The non-destructive energetic (MeV) ion beam technique of PIXE (particle induced X-ray emission) is described and its usefulness in the analysis of aerosols is illustrated by reference to recent examples carried out in environmental science by a range of laboratories throughout the world. A brief review of compatible aerosol samplers and procedures suitable for the optimum exploitation of the PIXE technique for atmospheric studies is also presented.

1. INTRODUCTION

In recent years, many governments and a number of international agencies have been increasingly concerned by the gradual erosion of the quality of the environment by all manner of industrial, man-made and natural pollutants. Environmental quality investigations have indicated that, due to the small samples generally available and the low toxic levels often involved, there is a need for highly sensitive multi-elemental procedures capable of ready application to air, water, soil and biological samples. In particular, concern about the harmful effects of air pollution has stimulated an interest in the origin, conversion and eventual dispersal of particulate matter in the environment. The PIXE (Particle induced X-ray emission) technique has shown itself well suited to precisely this type of broad range analysis, covering almost the entire spectrum of elements whilst requiring only tens of micrograms of total sample. In principle, PIXE is an X-ray analysis technique where X-rays characteristic of the elements of interest are generated by the action of high energy charged particles impinging on a sample.

PIXE is proving to be particularly useful in research on the chemical properties of the atmosphere, perturbed by both natural and anthropogenic causes. The good detection limits of PIXE, where low levels of elemental concentrations can be accurately determined, at the ppm level, together with the speed of analysis and hence the cost effectiveness of analysing large numbers of samples is going far in providing environmental scientists with a very versatile tool. To this must be added the fact that PIXE is, in principle, a non-destructive technique with multi-elemental capability (more than 20-40 elements simultaneously analysed) and where quantitative information at 5-10% levels of accuracy are routinely obtained. It compares very favourably with other more classical techniques and when combined with the complementary ion beam techniques of Rutherford backscattering (RBS) and Nuclear Reaction Analysis (NRA), which can be carried out simultaneously with PIXE, it may often be chosen as the optimum analytical solution.

Added to this, and most important from an environmentalist's point of view, the simplicity and relatively straightforward air sampling equipment needed for PIXE compatibility should not be underestimated.

This having been said it must be stressed, however, that whilst mythological-type pixies have long been recognized as possessing magical powers, the PIXE technique is unhappily not blessed with such supernatural qualities. It should not be considered in isolation of other analytical tools as the unique solution to all trace element analytical problems, but rather as a valuable additional asset to be exploited by atmospheric scientists.

PIXE is not an ultra trace analysis method in the sense of Neutron Activation Analysis (NAA) or some types of mass spectrometry (MS). Its advantages accrue from its simplicity, speed, cost effectiveness and multi-elemental nature. When only a few medium to heavy elements are required to be studied and samples of abundant mass are available over large enough specimen areas, PIXE may not exhibit extra-special advantages over other available modern techniques such as NAA, MS, AAS or XRF.

In the following sections the following questions will try to be answered:
- Where are the needs for trace analysis in aerosol science?
- Who is currently using PIXE for aerosol studies?
- What is PIXE and why is it chosen?

The literature on PIXE-based aerosol studies is now so extensive that only a few selected examples will be presented in the next section to illustrate the diversity and range of institutes and regions where studies are being carried out.

2. THE ROLE OF TRACE ELEMENTS IN AEROSOL POLLUTION

In the field of aerosol pollution, the levels of trace elements are of considerable importance. Two categories of trace elements can be identified: on the one hand, those elements which are known to be or suspected of being hazardous to mankind or to the environment in which we live, and on the other hand, those elements which do not imperil the milieu but which can usefully be used to trace the passage of pollution from its origins to its final resting places in the oceans and on land. These are the main routes by which trace elements can reach mankind – inhalation, eating (through the food chain) and drinking.

Whilst it is not intended to give a detailed description of the sources of airborne trace elements, it is worthwhile, in the discussion which follows, to classify air particulate sources under three broad headings of natural (ambient or rural) sources, anthropogenical (urban and industrial) sources, and localized (work environment) sources.

2.1 Natural sources

The natural sources of trace elements in air are largely from terrestrial dust, radioactive aerosols of terrestrial origin, oceans and sea spray, volcanic emissions and geothermal activity, fires and biomass burning and biosphere emanations.

A number of projects have been carried out recently using the PIXE method to investigate such influences on pollution levels. Winchester (1-3) has carried out PIXE investigations to determine baseline concentrations in rural continental and marine air to try to establish the natural or background exposure levels for humans. By so doing, a reference for comparing concentrations in polluted air could be defined.

A long range transport study on soil-generated and pollution aerosols has been carried out using PIXE analysis by Metternich et al (4, 5), comparing the airborne aerosols over the Mediterranean and mid-Atlantic ocean. Elemental concentrations approximately 10 times higher were measured for the Mediterranean. The significance of the results were correlated with aerosol origin.

A similar comparison has been carried out with cascade impactor samples collected on a sea voyage in the Pacific along the Peru, Ecuador coast and on the voyage between Peru and Hawaii (6).

Maenhaut et al (7, 8) in a series of experiments carried out over the sea surface from a small polyester and wooden sail boat far from local pollution sources, was able to distinguish the transport of aerosol pollutants from terrestrial sources long distances over water without extraneous interferences. Here, the PIXE analysis was essential because its high analytical detectability and panoramic capability allowed the use of economical and compact aerosol samplers and vacuum pumps operated by battery power. Up to 20 elements, in individual particle size fractions, collected by a pair of 6 stage cascade impactors operating over 2-day periods, were quantitatively analysed by PIXE.

Similar measurements have been performed by Maenhaut et al (9), at 4-day intervals, off the coast of Samoa to provide a baseline for comparing other marine atmospheres containing anthropogenically derived substances. Darzi (10) has reported PIXE analyses carried out on fumarolic aerosols from Kilauea volcano in Hawaii. The size fractionated concentrations from approximately 15 elements including Si, P, S, Cl and Fe were compared to measurements of gaseous emissions and other physical parameters made by volcanologists from the French CNRS and USA Geological survey. Differentiation could be made between the primary aerosols converted from the gases and captured ground particles swept into the plume. The effects on ambient pollution of large-scale biomass burning in the form of forest and grassland fires far from urban sites have been investigated by Leslie et al (11, 12) in the Southern Amazon Basin and Central Brazil using ground and airborne samplers and PIXE as the analytical tool. Between 15-20 elements were typically analysed including Mg, Al, Si, P, S, Cl, K, Ca, Ti, V, Cr, Mn, Fe, Ni, Cu, Zn, Br, Pb and Sr. One of the prominent features found was the large flux of particulates less than 2 μm in diameter emanating from these fires particularly of P, S, Cl, Mg, Al, Si and K whilst the crustal elements Ca, Ti, Fe tended to predominate in the large particle mode.
In addition, PIXE analysis has been carried out on ambient aerosols collected from a wide range of different rural and semi-rural land sites in Tasmania and Queensland, Australia (13), South America (14, 14a), China (15, 16), South Africa (17-19), USA (20-22) and Sweden (23, 24) to name but a few.

Time sequence "streaker" filters sampling at 2-4 h time intervals have been extensively deployed at 5 locations in Japan to follow the transport of aerosols generated by an especially strong dust storm originating in China (25). Based on measurements made by PIXE, modelling of aerosol transport on a regional basis was attempted.

2.2 Anthropogenic sources

Most of the major sources of trace elements in air are associated with urban activity; motor vehicle exhaust gases, coal and oil combustion, waste incineration, as well as iron and steel smelting and other industrial processes counting amongst the most important. In contrast with natural source investigations which have tended to be carried out on a small and uncoordinated scale in isolated campaigns, recent PIXE investigations on anthropogenic sources have been characterized by several comprehensive monitoring campaigns, to establish the prevailing concentrations and variability and ranges of trace elements at a network of sites including background, rural, urban and industrial areas . By so doing, the sources of pollution and long term trends have been ascertained with the aim of assessing the possible needs for introducing further controls.

In the USA, an automated PIXE based monitoring network has been established since the late 1970s, at 40 sites covering 2 million square kilometres of landmass (22). The good sensitivity of PIXE for low Z elements (but above Z = 11) has permitted the concentrations of the key element sulphur to be measured as well as a wide range of other elements. This has allowed an estimation of the conversion rates of SO_2 to sulphate aerosols and their long range transport to be established.

In Florida (26, 27) a dedicated PIXE laboratory has been routinely analysing anthropogenic and other aerosols since the mid 1970s collected on modified cascade impactor and time sequence "streaker" filters. Sweden has also been in the forefront of such studies and at the present time, has several multi-station PIXE-based studies underway for both long range aerosol transport and conversion studies (23, 24). South Africa has several coordinated monitoring networks encompassing more than 22 sites (17-19). PIXE analysis has been carried out together with neutron activation analysis (NAA) and atomic absorption spectroscopy (AAS) to determine concentrations of up to 25 elements in each sample.

Recently, PIXE-based networks in Italy (28, 29), Brazil (13, 30) and to a lesser extent West Germany (5, 31-33) have been set up, to supplement existing gas and total suspended particulate (TSP) monitoring stations with size-fractionated multi-element concentration data. These measurement campaigns have confirmed that PIXE is a powerful tool for routine aerosol analyses and can be both cost effective and reliable.

Numerous other examples, although on a smaller scale, have been reported since 1970 on the application of PIXE to anthropogenic pollution studies. Many of these have been reported in the three conference proceedings on PIXE and its analytical applications (34-36) and in recent review articles by Johansson and Johansson (23), Khan and Crumpton (37), Mitchell and Barfoot (38) and Cahill (39).

2.3 Work environmental sources

Work environmental pollution sources are treated here separately although they are clearly industrially-formed and hence anthropogenic by nature. The reason for this separation is that in most cases the aerosol exposure is spatially limited, only having an influence on those in the immediate vicinity of the localized source. A good example is in welding shops and paint sprayshops where sudden and large changes in elemental concentrations occur over relatively short time durations.

Nevertheless, whilst these localized sources tend to leave the general population unaffected, about 25% of a worker's life is spent at his work environment. This will have a significant influence on his total exposure to various chemical substances. Increased knowledge about potential health hazards in industrial plant and at the workbench is required if preventative actions are to be improved. Detailed examinations, including elemental concentrations and compositions, particle sizes and time and site variations are all relevant to the hygenic effects of work environment aerosols. At the present time, the PIXE technique has not been applied to work environment aerosol measurements as much as it has been in the atmospheric studies discussed earlier. This is in some ways surprising, as the special features of work environment aerosol samples, i.e. small size, rapid and wide changes in concentration in short spaces of time, as well as the special interest from a health point of view in medium to heavy elements, make PIXE a very powerful and economical analytical tool for such applications.

Sweden is currently involved in such PIXE-based studies, (23, 38, 40, 41) partly financed by the Swedish Ministry of Industrial Hygiene. A group at the Univerity of Lund have carried out several largescale studies on a variety of occupational and work-oriented situations including a thorough characterization of aerosols formed in industrial high temperature metal processes such as electric arc welding and metal spraying operations where Ni and Cr were of particular interest (42, 43). In one study, (44) about 1000 samples from 500 welding stations, taken in 90 different types of working areas were analyzed. Of all the different analytical methods used, PIXE was ascertained to be the most economical from a commercial point of view when all relevant considerations were taken into account (45).

Time-resolved aerosol composition has been correlated with particulate size by Barfoot et al (46). In a limited trial study, samples were generated in a busy welding shop. Analyses were carried out using PIXE combined with SEM. The "streaker" sampler was modified so that such elemental concentration changes could be continuously monitored for periods of up to 3 days with time resolutions as short as 10 minutes. Figure 1 is shown as an example of the concentration variations observed for some elements in this study.

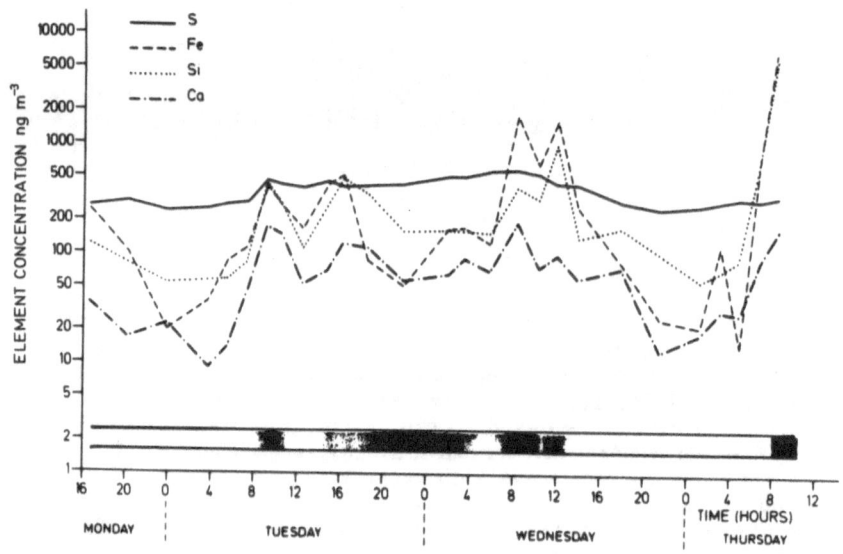

Figure 1: Trace element levels in welding shop particulates for the period 8-11 October 1979. A photograph of the streak sample is also shown for comparison.

In other similarly small scale studies, PIXE has been applied to cascade impactor, size-fractionated samples taken during silver soldering in an art studio (47); to aerosols collected in a lead battery production plant (48), in copper smelter plants (49), in gold mines (50), in telecommunication exchangers (51) and in garbage incinerator plants and plastics milling laboratories (52).

In all these examples, and many others, PIXE has shown itself to be a powerful technique. Several exercises have been carried out to intercompare PIXE, for aerosol investigations, with other analytical techniques such as NAA, XRF, AAS, OES (optical emission spectroscopy), PAA (photon activation analysis) and wet chemistry techniques (18, 31, 33, 53). In some particular circumstances PIXE has been shown to be a unique or at least the optimum tool when such factors as sampling size, economy, sensitivity and multi-elemental capability are all taken into account.

As a conclusion to this section, Table 1 gives a selective list of laboratories and institutes currently applying PIXE to aerosol and other studies.

TABLE 1 - SELECTION OF INSTITUTES INVOLVED IN PIXE-BASED
AEROSOL AND RELATED STUDIES

Institute	Proj. Ion & Energy (MeV)	PIXE applied to	Comments	References ( )
Australia				
Melbourne	p,2.5	O	+RBS,NRA	
Belgium				
Geel	p,α,1-3	St,LV;A,R,W	+RBS,NRA,PAA	(38,46,61)
Gent	p,2.4	CI,HV;A,R,N	+NAA	(6,7,9)
Liege	p,α,3;18	St;A,R	+RBS,NRA	(38)
Namur	p,0.5-3	O	+RBS,NRA	(39)
Brazil				
Sao Paulo	p,α,3-8	CI,LV;R		(30,35)
Canada				
Guelph	p,2	O	+RBS	(35)
Manitoba	p,20-50	O	+NRA	(39)
Chile				
Santiago	p,6			(39)
China				
Beijing	p,3.5	CI;A,R	+RBS	(15,16)
Denmark				
Aarhus	p,α,0.5-5	O	+RBS,BRA	(38)
Copenhagen	p,2-3	A	+RBS	(39)
Finland				
Helsinki	p,2	A,R	+RBS,XRF,NRA	(51)

Institute	Proj. Ion & Energy (MeV)	PIXE applied to	Comments	References ( )
France				
Saclay	p,α,4-22	O	+PAA	(39)
Strasbourg	p,d,α,5-20	O	+RBS,NRA	(39)
Germany				
Bonn	α,30	O	+RBS	(37,38)
Frankfurt	p,1-3	CI,LV;A,R		(4,5)
Heidelberg	p,2	O	+RBS	(38)
Karlsruhe	p,α,2-3	O	+RBS	(35,38)
Marburg	p,2-4	CI,LV,HV;R,A		(31,32,33)
Gr. Britain				
Aston	p,1-3	O		(37)
Harwell	p,1-3	O		(39)
Surrey	p,1-3	St,LV;R,A		(35,38)
Greece				
Athens	p,1-2	O	+RBS	(38,39)
Hungary				
Budapest	p,d,α,1-4	O	+RBS,XRF,NRA	(35,39)
India				
Kanpur	p,2	O	+RBS	(39)
Italy				
Catania	p,α,1-2.5		+RBS,AAS	(39)
Milan	p,2.8	CI,LV	+RBS,XRF	(28,29)
Japan				
Osaka	p,α,1-2	O	+RBS	(39)
Yokohama	p,2.5	St,LV;N,A		(25)
Netherlands				
Eindhoven	p,3	O	+XRF,RBS	(60)
New Zealand				
Lower Hutt	p,2.5	O	+RBS,XRF,NRA	(39)
Poland				
Cracow	p,2.9	O	+NRA	(35,39)
Warsaw	p,2-3	O	+RBS	(35,39)
Portugal				
Sacavem	p,0.2-2	O		(39)
Rumania				
Bucharest	p,4	O	+RBS	(39)
South Africa				
Johannesburg	p,3	St,LV,CI;A,R,W		(17)
Pelindaba	p,2.4	CI,HV;A,R		(18,19)
Faure	p,1-3	O		(39)

Institute	Proj. Ion & Energy (MeV)	PIXE applied to	Comments	References ( )
Sweden				
Lund	p,⍺,2.5	LV,CI,St;A,R,W	+RBS,XRF,NRA	(23,34-36)
Uppsala	p,⍺,1-2	O	+RBS	(36)
USA				
Bell Labs	p,⍺,2-4	O	+RBS,NRA,XRF	(38,39)
Brigham Young	p,2	CI;A,W		(49)
Davis, Calif.	p,⍺,4.5,18	CI,HV,LV,ST; A,R,N		(22,36,39, 56)
Florida St.	p,4;15	CI,St,LV;A,R,N	+RBS	(1,2,3)
Florida Univ.	p,3.8	O		(39)
IBM N.Y.	p,⍺,1-2.5	O	+RBS,NRA,XRF	(37,38,39)
Illinois	p	A	+XRF	(39)
Mississippi	p,3	O		(39)
NRL, Washingt.	p,3	O	+RBS,NRA	(38,39)
Nevada	p	O	+RBS,NRA	(38)
New York	p,3	CI,LV;A		(47,52)
Tallahassee	p,3	CI,LV,St;A,R	+RBS	(10)
Texas	p,⍺,0.5-1.5	O	+RBS,NRA	(37,38)
Yugoslavia				
Ljublijana	p,1		+NRA	(35,48)

O = Other applications than aerosols such as biological, fluids, gases,
    materials science

A = Anthropogenic   CI = Cascade impactor
N = Natural   HV = High volume sampling
R = Rural   LV = Low volume sampling
W = Work environment   St = Streakers

3. THE PIXE TECHNIQUE

3.1 Historical

The phenomonen of X-ray emission brought about when fast ions impinge
on materials was first reported by Chadwick as early as 1912. Since that
time, ion induced X-ray emission (IIX) has been much studied. However, it
was not until little more than a decade ago that it was realised that IIX,
by now rechristened particle induced X-ray emission (PIXE), had not only
theoretical interest but could be applied as an analytical tool for the
observation of trace elements in a wide variety of materials. By 1970,
Johansson and Johansson (23) by making use of the recently developed high
resolution energy dispersive Si(Li) X-ray detector and using MeV proton
beams, showed that PIXE is a highly sensitive method for multi-elemental
analysis. They demonstrated that under suitably favourable conditions
many elements could be detected simultaneously and non-destructively at
the 10^{-12} g level.

Even from the earliest publications, some of the unique advantages of
PIXE to the analyses of airborne particulate matter collected from the
atmosphere was recognised. At the present time, a majority of the groups
carrying out PIXE-based studies include aerosol-related analyses amongst
their most important tasks (39).

3.2 Description of the PIXE technique

X-rays may be produced following the excitation of target atoms induced by an incident ion beam. The main interaction of the ion beam with the target is related to the interaction of the coulomb fields associated with the incoming ions and those of the bound electrons of the target atoms. Thus, when an energetic (⪞ MeV) proton or helium ion impinges on a target atom there is a high probability, typically of the order of hundreds of barns, that an inner shell electron will be excited and removed from the target atom. The excited target atom regains a stable energy state by a cascade process of filling the vacancies created, by outer shell electrons together with the simultaneous emission of characteristic X-rays or Auger electrons. Hence, elemental composition analysis can be accomplished by determining the X-ray emissions emanating from the excited sample. Figure 2 schematically illustrates the physical process involved.

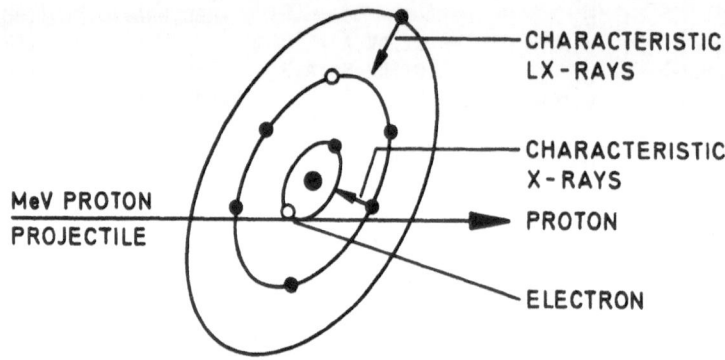

Figure 2: Schematic of physical process involved in particle induced X-ray emission (PIXE).

In most experimental arrangements, charged particle beams of protons or helium of energies between 1-5 MeV are used for PIXE analysis. The exciting projectiles are normally generated by a Van de Graaff (VdG) accelerator or a fixed or variable energy cyclotron. Figure 3 illustrates in a schematic manner the experimental arrangement used with a VdG accelerator.

The most suitable samples (filters) have thicknesses below 1 mg cm^{-2} in which case only minor corrections are required for X-ray attenuation or projectile slowing down effects in the specimen.

Even in the dirtiest work environment, the aerosol concentrations are usually less than 10_2mg . m^{-3} (of air) and under typical sampling conditions (e.g. 10 cm^2 sample, 2 lpm air pump rate, 24h sampling time), the limit of 1 mg cm^{-2} is normally not exceeded (45). On the other hand, a typical urban aerosol concentration is about 80 μg m^{-3}. This sets a lower limit to the use of more traditional analysing techniques, which require perhaps 0.2 g of total deposited material. PIXE generally provides good sensitivity even with total quantities of deposit of less than 0.1 mg!

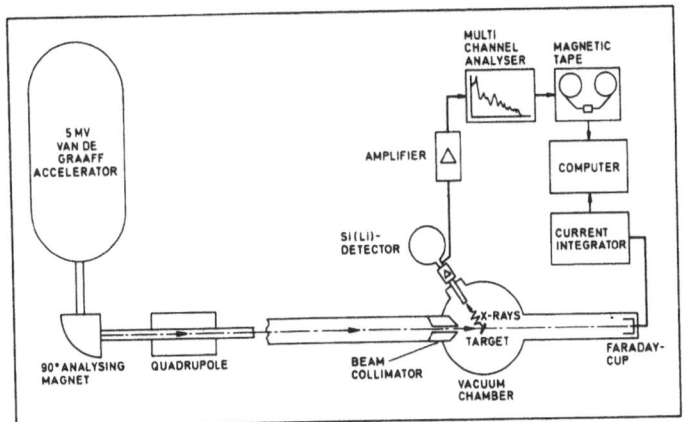

Figure 3: Schematic of the experimental arrangement used for PIXE analysis
with a Van de Graaf accelerator

Typically, PIXE analysis consists of two parts. The first is to
identify the atomic species in the target from the energies of the
characteristic peaks in the X-ray-emission spectrum. The second part is
to determine the amount of a particular element present in the target from
the emission rate of its characteristic X-ray spectrum. If an absolute
calibration approach is chosen, this normally requires knowledge of
ionization cross sections, fluorescence yields and absorption
coefficients. However, in most practical cases, a relative path is chosen
by calibration with thin layer standards of the elements of interest.

Theoretical minimum detectable limits have been calculated for all
elements by Folkmann (1974) et al (54) and others (23) and are illustrated
in Figure 4.

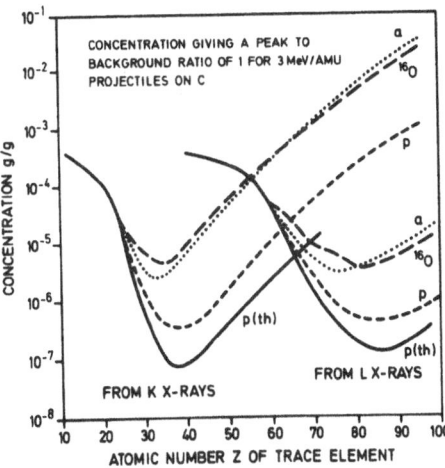

Figure 4: Theoretical minimum detectable concentrations calculated for all
elements (p(th) = theoretical limit for protons).

These theoretical values may be compared with minimum detection limits experimentally determined by Barfoot et al (38, 55) over a more limited element range, shown in Figure 5.

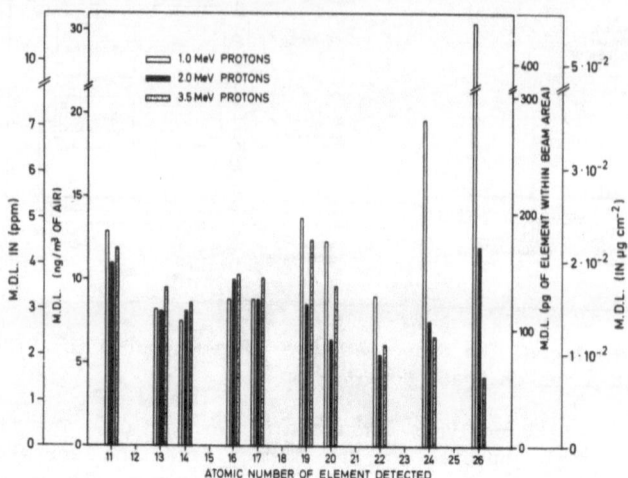

Figure 5: Histogram of experimental minimum detection limits (MDL) for elements with Z = 11-26.

A number of fast and reliable computer codes now exist for locating the different X-ray energy peaks and determining the intensities and hence the mass contents of all trace elements in the X-ray spectrum (26, 56-60). Most modern, small computers can perform peak search, background stripping, area integration and full trace element concentration analysis in a few minutes (32, 45).

In contrast with a number of other chemical and physical spectrometric techniques, the basic advantages of PIXE analysis accrue from the relative simplicity of the characteristic X-ray-emission spectra and from the intrinsic capability for analysis without destructively interfering with the specimen. In addition, it is capable of fast multi-elemental analysis with a good minimum detection capability and typically only requires very small samples. Moreover, the method is essentially insensitive to the chemical state of the atom. In practical aerosol studies detection limits for trace elements are typically as low as 1 ppm which corresponds to amounts of trace element at the 10^{-10} g level.

The lowest detection limit obtainable for a given range of elements depends to some extent on the ion species and energy; so that often an experiment can be 'tailored' to provide the optimum condition for a given range of elements and backings.

Figures 6 a) and b) show PIXE spectra (46) generated from a "streaker" filter sample taken in a welding shop, during night-time (a) and during a welding episode (b). (N.B. the ordinate is plotted on a logarithmic scale.)

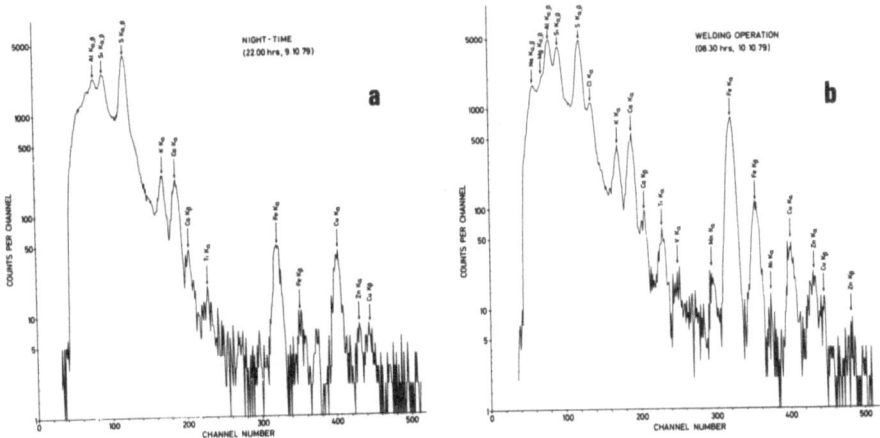

Figures 6 a) and b): 2MeV proton PIXE spectra obtained with a Nuclepore
streak sample of welding shop aerosol during a) night-time and b)
during a welding episode.

PIXE can be carried out simultaneously with the complementary
technique of Rutherford backscattering (RBS) on the same target and using
the same exciting projectile beam. Figure 7 illustrates a vacuum-system
arrangement used for such simultaneous, PIXE, RBS experiments (62).

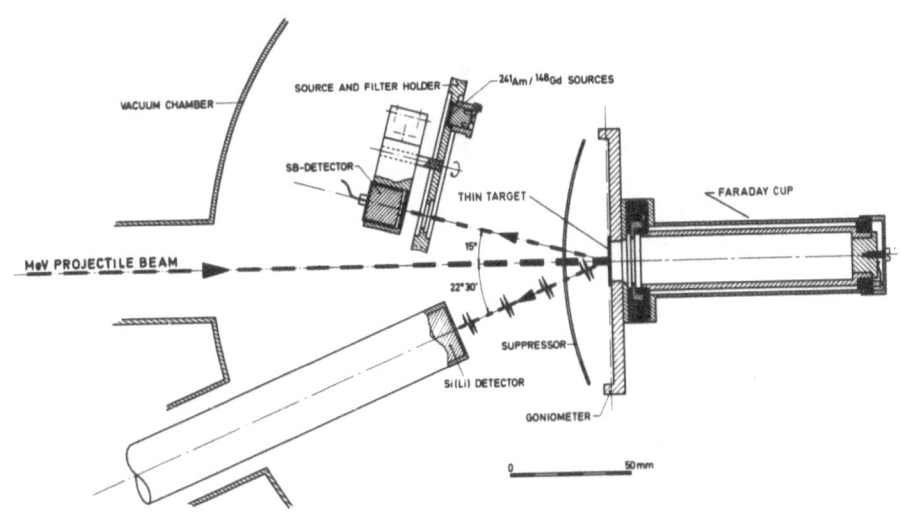

Figure 7: Schematic of a vacuum system arrangement used for simultaneous
PIXE, RBS experiments.

The only experimental addition necessary is the incorporation of a suitable solid-state particle detector and associated electronics.

PIXE supplements the good mass separation but poor sensitivity for low-Z elements of RBS, providing good sensitivity down to elements as low as Z=11 (Na). For high-Z elements, relatively good elemental peak separation is provided by PIXE using the LX-rays rather than the KX-rays. With currently commercially available Si(Li) X-ray detectors, PIXE is not normally useful for elements lighter than sodium. This is because of the strong attenuation of K-shell X-rays from such light elements by the Be vacuum window in front of the detector. Some of these deficiencies can be overcome by using RBS or NRA simultaneously for quantitative analysis of all elements from hydrogen to sodium (38, 61). Recent developments in specialized windowless Si(Li) detectors have permitted PIXE to be used to detect elements (but not at the trace level) as low as boron. However, in aerosol studies, where many difference sized and shaped particulates are often present, the difficulty in quantitatively analysing by PIXE very low Z elements (because of self-absorption effects) remains. In some circumstances nondestructive depth profiling with RBS complements PIXE analysis which is not so well suited to this type of analysis. The use of a Van de Graaff accelerator as the generator of the charged particle beam rather than a radioactive source, or other means, such as photons or fast electrons, is attractive for a number of reasons. The most important of these are: a choice of ion species is available; an accelerator is capable of delivering a wide range of incident ion flux; a simultaneous analysis can be made by PIXE and RBS; favourably high peak-to-background ratios can be expected when compared with electron and photon excitation sources due to a much reduced Bremsstrahlung background.

In principle, PIXE is a nondestructive technique. It is a relatively fast analytical method needing short experimental times, typically of between 1-15 minutes per sample. With the aid of suitable computer programs, PIXE spectra from routine specimens can be analyzed and trace-element contents obtained in a short time(1-3 minutes). Background effects in PIXE are in almost all cases lower than those for comparable X-ray techniques using electrons, and photons (such as X-rays) as the exciting source.

On the other hand, some of the disadvantages of the PIXE method are that, for complicated spectra, interferences between KX rays from light elements and L and M X rays from heavier elements can occur. This is also true for overlaps between the K_α- and K_β- X-ray lines from neighbouring elements. Future improvements in the resolution obtainable from energy dispersive Si(Li) X-ray detectors could go some way towards reducing these problems.

4. SAMPLING TECHNIQUES

Most aerosol samplers employed in airborne pollution monitoring have been developed for use with the traditional environmentalists analytical techniques such as NAA, AAS, XRF, reflectometry, graviometry,wet chemistry, etc. More often than not, they are not completely suitable for PIXE analysis. Consequently for optimal exploitation of the PIXE technique in atmospheric studies, suitable samplers, strategies and procedures have had to be developed.

In a recent comprehensive study, Akselsson (62) reviewed the current state-of-the-art in PIXE-based aerosol samplers. These include modified cascade impactor samplers capable of providing PIXE compatible samples with ten or more discrete ranges of aerosol size fractions from 0.06μm up to 16μm diameter (63, 64). For diameters less than 0.25μm, a nuclepore filter stage is added. Greased impact surfaces have eliminated the bounce-off effects reported in earlier models (65). Concentrically rotating impact plates have reduced the overloading of the smaller size impactor stages (42). Rotating drum compactors designed for air sampling from aircraft have been designed by the Davis Group (62, 65, 66) with up to 9 impactor stages. Solar powered samples have been developed (62) suitable for remote areas and unattended monitoring. Recently, Prodi et al (67, 67a) have developed a new type of size fractionating sampler called an inertial spectrometer, which permits the elemental composition of aerosols to be sampled with a size resolution of 0.08μm for 0 to 10μm diameter particulates. Leslie et al have reported on its use in a PIXE-based study (67b).

Linear time resolution "streaker" samples have been in use for almost 10 years (23, 38, 68). Time resolutions of from 1-2 h can easily be achieved during periods of several weeks unattended monitoring.

A recent development in this area is the PIXE compatible two stage circular "streaker" version (62, 64, 69) which appears to provide a simple, robust and inexpensive means of following temporal variations in elemental concentrations of pollution aerosols, although it has been recently reported (70) that an unpredictable particulate dimension cut-off, due to the low throughput in one of these devices (64), could present problems.

Of particular interest from the industrial hygiene point of view is the recent introduction of a Personal Aerosol Monitoring (PAM) sampler (45, 71) which is a low-weight and low-power microelectronic controlled sampler which may be attached close to the breathing zone area of the wearer. This development should prove to be of great benefit in monitoring working environment aerosols with time resolutions down to a few minutes. Coupled with a very fast and sensitive PIXE analysing system capable of multi-elemental analysis on a 10 second repetition rate at detection limits typically 10% of the currently set hygienic thresholds, this development will guarantee the wider application of this type of system for monitoring in-plant and work environment aerosols. These types of developments at Lund University are proving invaluable for industrial hygienists and indicate that such measurements can be carried out on a commercial basis, and at economic costs per sample that compare very favourably with the more traditional techniques currently in use.

5. CONCLUSION

The nondestructive analytical technique of PIXE has been described and its usefulness in the analysis of aerosols has been illustrated by reference to recent examples carried out in atmospheric research by a range of laboratories throughout the world. One problem still remains however, and this problem is one that is common to all comparatively new and developing analytical techniques. The problem is that of bringing to the attention of those who could exploit the potentials and capabilities of PIXE (and the other energetic MeV ion beam techniques) in helping to solve specific analytical problems. This is particularly so for PIXE which does not at present enjoy the benefits of commercial publicity since PIXE systems are not yet fully available commercially from the manufacturers of analytical systems. On the other hand, under-utilized 1-to-3 MeV Van de Graaff accelerators are now becoming available in Universities and Research Institutes throughout the world as nuclear physicists move to ever higher-energy machines. The vacuum system and ancillary equipment needed to make such analytical measurements are very straightforward. The responsibility therefore falls on those currently engaged in the PIXE analysis fields to give this method the publicity it rightly deserves.

REFERENCES

(1) Winchester J.W., Nucl. Instr. and Meth. 142, (1977), 85-90.
(2) Winchester J.W., Nucl. Instr. and Meth. 181, (1981), 367-381.
(3) Winchester J.W., Proceedings of the 3rd Int. Conf. on particle induced X-ray emission (PIXE) and its analytical applications; Heidelberg, Germany, 18-22 July 1983 to be published in Nucl. Instr. and Meth. (1984).
(4) Metternich P., Georgi H.W. and Groenveld K.O., Proceedings of the 3rd Int. Conf. on particle induced X-ray emission (PIXE) and its analytical applications; Heidelberg, Germany, 18-22 July 1983 to be published in Nucl. Instr. and Meth. (1984).
(5) Metternich P., Georgi H.W. and Groenveld K.O. (this conference).
(6) Raemdonck H., Maenhaut W., Ferek R. and Andreae M.O. Proceedings of the 3rd Int. Conf. on particle induced X-ray emission (PIXE) and its analytical applications; Heidelberg, Germany, 18-22 July 1983 to be published in Nucl. Instr. and Meth. (1984).
(7) Maenhaut W., Selen A., Van Espen P., Van Grieken R. and Winchester J.W., Nucl. Instr. and Meth. 181, (1981), 399-405.
(8) Maenhaut W., Raemdonck H., Selen A., Van Grieken R. and Winchester J.R. J. of Geophysical Res. 88, C9, (1983), 5353-5364.
(9) Maenhaut W., Darzi M. and Winchester J.W., J. of Geophysical Res. 86, C4, (1981), 3187-3193.
(10) Darzi M., Nucl. Instr. and Meth., 181, (1981), 359-365.
(11) Leslie A.C.D., Winchester J.W. and Nelson J.W., 177th American Chem. Soc. Nat. Meeting and 38th Nat. Meeting of the Chem. Soc. of Japan, Honolulu, (1979).
(12) Leslie A.C.D., Nucl. Instr. and Meth., 181, (1981), 345-351.
(13) Andreae M.O. and Barnard W.R., Nucl. Instr. and Meth., 181, (1981), 383-390.
(14) Orsini C.Q., Kaufmann H.C., Akselsson K.R., Winchester J.W. and Nelson J.W., Nucl. Instr. and Meth. 142, (1977), 91-96.
(14a) Adams F., Van Espen P. and Maenhaut W., Atmosph. Environ. 17, (1983), 1521-1536.
(15) Winchester J.W., Wang Ming-Xing, Ren Li-Xin, Lu Wei-Xin, Hansson H.C., Lannefors H., Darzi M and Leslie A.C.D., Nucl. Instr. and Meth., 181, (1981), 391-398.
(16) Winchester J.W., Wang Min-Xing, Lu Wei-Xin, Ren Li-Xin, and Hong Zhong-Shang. Proceedings of the 3rd Int. Conf. on particle induced X-ray emission (PIXE) and its analytical applications; Heidelberg, Germany, 18-22 July 1983 to be published in Nucl. Instr. and Meth. (1984).
(17) Annegarn H.J., Madiba C.C.P., Winchester J.W., Baumann J. and Sellschop J.P.F., Nucl. Instr. and Meth., 181, (1981), 435-440.
(18) Vleggar C.M., Van As D., Watkins J.L., Mingay D.W., Wells R.B., Briggs A.B. and Louw C.M. Report No PEL 274 October (1980), Atomic Energy Board, Pelindaba, Pretoria, S.A.
(19) Mingay D.W., Vleggar C.M. and Watkins J.L., Nucl. Instr. and Meth. 181, (1982), 487- 492.

(20) Cahill T.A., Ashbaugh L.L., Barone J.B., Eldred R.A., Feeney P.J., Flocchini R.G., Goodart C., Shadoan D., and Wolfe G.W., J. Air Pollut. Control Assoc. 27, (1977), 675-678.

(21) Cahill T.A., Nucl. Instr. and Meth. 149, (1978), 431.

(22) Eldred R.A., Cahill T.A., Ashbaugh L.L. and Nasstrom J.S. Proceedings of the 3rd Int. Conf. on particle induced X-ray emission (PIXE) and its analytical applications; Heidelberg, Germany, 18-22 July 1983 to be published in Nucl. Instr. and Meth. (1984).

(23) Johansson S.A.E. and Johansson T.B., Nucl. Instr. and Meth., 137, (1976), 473-516.

(24) Lannefors H., Ph.D Thesis University of Lund, Sweden (1982) No LUTFD/2/CTFKF-1003/1-105.

(25) Hashimoto Y, Tanaka Sh. and Winchester J.W. Proceedings of the 3rd Int. Conf. on particle induced X-ray emission (PIXE) and its analytical applications; Heidelberg, Germany, 18-22 July 1983 to be published in Nucl. Instr. and Meth. (1984).

(26) Johansson T.B., Van Grieken R.E., Nelson J.W. and Winchester J.W., Anal. Chem. 47, (1975), 855-860.

(27) Johansson T.B., Van Grieken R.E. and Winchester J.W., J. Geophys. Res. 81, (1976), 1039.

(28) Caruso E., Braga-Marcazzan G.M. and Redealli P., Nucl. Instr. and Meth. 181, (1981), 425-429.

(29) Caruso E., Braga-Marcazzan G.M. and Redealli P., Proceedings of the 3rd Int. Conf. on particle induced X-ray emission (PIXE) and its analytical applications; Heidelberg, Germany, 18-22 July 1983 to be published in Nucl. Instr. and Meth. (1984).

(30) Orsini C.Q., Netto P.A. and Tabacniks M.H., Proceedings of the 3rd Int. Conf. on particle induced X-ray emission (PIXE) and its analytical applications; Heidelberg, Germany, 18-22 July 1983 to be published in Nucl. Instr. and Meth. (1984).

(31) Bomelka E., Richter F.W., Ries H. and Wätjen U., Proceedings of the 3rd Int. Conf. on particle induced X-ray emission (PIXE) and its analytical applications; Heidelberg, Germany, 18-22 July 1983 to be published in Nucl. Instr. and Meth. (1984).

(32) Wätjen U., Ph. D Thesis, University of Marburg, Germany (1983).

(33) Wätjen U., Bomelka E., Richter F.W., and Ries H., J. Aeros Sci 14, (1983), 305-308.

(34) Proceedings of first international conference on particle induced X-ray emission (PIXE) and its analytical applications, Lund, Sweden August 1976, published in Nucl. Inst. and Meth. 142, (1977).

(35) Proceedings of second international conference on particle induced X-ray emission (PIXE) and its analytical applications, Lund, Sweden 9-12 June 1980, published in Nucl. Inst. and Meth. 181, (1981).

(36) Proceedings of third international conference on particle induced X-ray emission (PIXE) and its analytical applications, Heidelberg, Germany, 18-22 July (1983) to be published in Nucl. Inst. and Meth. (1984).

(37) Khan Rashiduzzaman Md, and Crumpton D. Critical Reviews in Analyt. Chem. May (1981), 103-155 and June (1981), 161-193.

(38) Mitchell I.V., and Barfoot K.M., Nucl. Science Appl., 1, part B, (1981), 99-162.

(39) Cahill T.A., Ann. Rev. Nucl. Part. Sci. 30, (1980), 211-252.

(40) Malmqvist K.G., Ph.D Thesis, Lund University, Sweden (1981), No. LUTFD2/TFKF-1002/1-104.

(41) Johansson G.I., Ph.D Thesis, Lund University, Sweden (1981) No. LUTFDS/TFKF-1001/1-111.

(42) Malmqvist K.G., Johansson G.I., Bohgard M. and Akselsson K.R., Nucl. Instr. and Meth. 181 (1981), 465-471.

(43) Bohgard M., Welinder H. and Akselsson K.R., to be published in Scand. J. Work Environ. and Health, (1983).

(44) Ulfvarson U., Scand. J. Work Environ. and Health, 7, Suppl. 2, (1981), 1.

(45) Malmqvist K.G., Proceedings of the 3rd Int. Conf. on particle induced X-ray emission (PIXE) and its analytical applications; Heidelberg, Germany, 18-22 July 1983 to be published in Nucl. Instr. and Meth. (1984).

(46) Barfoot K.M., Mitchell I.V., Verheyen F. and Babeliowsky T., Nucl. Instr. and Meth. 181, (1981), 449-457.

(47) Williams E.T. and Punyasena L.W., IEEE. Trans. on Nucl. Sci. NS-30 No 2, (1983).

(48) Starc V., Budnar M., Cindro V., Remsak M., Ravnikar M., Smit Z., Modic S. and Cernivec R., Proceedings of the 3rd Int. Conf. on particle induced X-ray emission (PIXE) and its analytical applications; Heidelberg, Germany, 18-22 July 1983 to be published in Nucl. Instr. and Meth. (1984).

(49) Smith T.J., Eatough D.J., Hansen L.D. and Mangelsen N.F., Bull. Env. Cont. and Toxicol., 15, (1976), 651.

(50) Annegarn H.J. (see reference (45), Private communication), (1983).

(51) Raunemaa T., Hantojärvi A., Kauppinen E. and Lannefors V., Proceedings of the 3rd Int. Conf. on particle induced X-ray emission (PIXE) and its analytical applications; Heidelberg, Germany, 18-22 July 1983 to be published in Nucl. Instr. and Meth. (1984).

(52) Williams E.T. and Punyasena L.W., Proceedings of the 3rd Int. Conf. on particle induced X-ray emission (PIXE) and its analytical applications; Heidelberg, Germany, 18-22 July 1983 to be published in Nucl. Instr. and Meth. (1984).

(53) Cahill T.A., Eldred R.A., Shadvan D., Feeney P.J. and Kusko B.H., Proceedings of the 3rd Int. Conf. on particle induced X-ray emission (PIXE) and its analytical applications; Heidelberg, Germany, 18-22 July 1983 to be published in Nucl. Instr. and Meth. (1984).

(54) Folkmann F., Borggren J. and Kjeldgaard A., Nucl. Instr. and Meth., 119, (1974), 117.

(55) Barfoot K.M., Ph. D Thesis, University of Surrey, UK, (1980) and CBNM, Geel, Belgium, Internal Report No CBNM/AS/105/80.

(56) Cahill T.A., New Uses of Ion Accelerators, ed. Ziegler J.F., Plenum Press, N.Y., USA, (1975), 1.

(57) Willis R.D. and Walter R.L., Nucl. Instr. and Meth., 142, (1977), 231.

(58) Van Espen P., Nullens H. and Adams F., Nucl. Instr. and Meth., 142, (1977), 243.

(59) Nass M.J., Lurio A. and Ziegler J.F., Nucl. Instr. and Meth., 154, (1978), 567.

(60) Kivits H.P.M., Ph.D Thesis, University of Eindhoven, Netherlands, (1980).

(61) Mitchell I.V. and Reher D. Int. J. Appl. Radiat. Isot. 34, No 8, (1983), 1277-1290.

(62) Akselsson K.R., Proceedings of the 3rd Int. Conf. on particle induced X-ray emission (PIXE) and its analytical applications; Heidelberg, Germany, 18-22 July 1983 to be published in Nucl. Instr. and Meth. (1984).

(63) Bauman S., Houmere P.D. and Nelson J.W., Nucl. Instr. and Meth., **181**, (1981), 499.

(64) PIXE International Corp. PO Box 2744, Tallahassee, Florida, 32316, USA.

(65) Lawson D.R., Atm. Environ. **14**, (1980), 195.

(66) Shadoan D.J., Barone J.B. and Cahill T.A. Nucl. Instr. and Meth., **181**, (1981), 503.

(67) Prodi V., Melandri C., Tarroni G., De Zaiacomo T., Formignani M., and Hochrainer D., J. Aerosol Science **10**, (1979), 411.

(67a) Prodi V., Melandri C., Tarroni G., De Zaiacomo T., Formignani M., Olivier P., Barilli L. and Oberdörster G., "Aerosols in the Mining and Industrial Work Environments", **3**, (1983), 931, eds. Marple V.A. and Lin B.Y.H., Ann Arbor Publ. Ann Arbor, USA.

(67b) Leslie A.C.D., Kaufmann H.C. and Prodi V, Proceedings of the 3rd Int. Conf. on particle induced X-ray emission (PIXE) and its analytical applications; Heidelberg, Germany, 18-22 July 1983 to be published in Nucl. Instr. and Meth. (1984).

(68) Jensen B. and Nelson J.W. Proc. 2nd Int. Conf. on Nucl. Meth. in Environ. Res. Columbia Missouri, USA (1974), 366.

(69) Kemp K., and Tscherning Moller J., Nucl. Instr. and Meth., **181**, (1981), 481-485.

(70) Wätjen U., private communication (1984).

(71) Bohgard M., Malmqvist K.G., Johansson G.I. and Akselsson K.R., Aerosols in Mining and Industrial Work environments, **3**, (1983), 907, eds. Marple V.A. and Liu B.Y.H., Ann Arbor Sci. Publ. Ann Arbor, USA.

SESSION 2

CHEMICAL AND PHOTOCHEMICAL REACTIONS

Summary by the Chairman
R.A. COX

Kinetic study of reactions of OH radicals with organic
sulfur compounds

Reactions of OH radicals with reduced sulfur compounds
under atmospheric conditions

LIF studies of formation and kinetics of primary radical
products in OH-oxygenated hydrocarbon reactions

The temperature dependence of the forward-backward
reactions of the addition of OH to benzene, aniline and
nitrobenzene

Absolute rate constant measurements of OH reactions
under atmospheric conditions by laser photolysis/dye
laser fluorescence

Photooxidation of acetaldehyde

A FTIR spectroscopic study of the photooxidation of
acetaldehyde in air

Absorption spectrum and kinetics of the NO_3 radical

Hydroxyl radical concentration in ambient air at a
semirural site estimated from $C^{13}O$ oxidation

Temperature dependence of the reactions $SO + O_3$ (1) and
$SO + O_2$ (2)

A study of N_2O_5 and NO_3 chemistry in the photolysis of N_2O_5 mixtures

A study of the reaction between ClO and NO_2 using matrix isolation FTIR spectroscopy and UV-visible spectroscopy

Oxidation of methylchloroform

Transformation of reactive PAH on particles by exposure to oxidized nitrogen compounds

The kinetic coefficient of the $C_2H_4 + O$ reaction over extended pressure and temperature ranges

Kinetics of the reaction of OH with ethane and a series of Cl- and F-substituted methanes at 300–400 K, studied by pulse radiolysis combined with kinetic spectroscopy

The role of freons in the chemistry of the upper atmosphere

CHEMICAL AND PHOTOCHEMICAL REACTIONS

Summary by the chairman

R.A. COX

U.K.A.E.A. - AERE Harwell

Investigations of the kinetics and mechanism of reactions involved in atmospheric transformations has progressed on a broad front on the COST 61a programme during the past $2\frac{1}{2}$ years. Work reported at the present Symposium (both formal and poster sessions) and in the last Discussion Meeting of Working Party 2 held in Dublin in October 1982 can be summarised under the following headings.

OH and HO$_2$ radical reactions

Papers at the Symposium have concentrated on the sulphur compounds for which a number of rate coefficients are presented. Agreement is not always as good as desirable. This is partly due to the complex mechanisms which appear to be involved in these reactions. Complex mechanisms also are a feature of the reactions of radicals with aromatic hydrocarbons and the use of modern optical spectroscopic techniques has been especially beneficial in improving the precision and diagnostic capability of the experimental work in this area.

Photo-oxidation Studies

Work designed to understand the mechanisms of oxidation of hydro-carbons and related organic species has continued to provide useful results. Earlier work on formaldehyde is now being extended to acetaldehyde photolysis. An interesting result to come from these studies is the possible important reaction of HO$_2$ with carbonyl compounds leading eventually to organic acids.

NO$_x$ Chemistry

The most recent work on NO$_x$ chemistry has concentrated on the higher oxides NO$_3$ and NO$_2$O$_5$. These are involved in the non-photochemical conversion of NO$_2$ to nitric acid and there is currently a major effort to improve the kinetic data base for the quantitative understanding of the behaviour of this aspect of NO$_x$ chemistry. The reactive nature of nitrogen oxides in the atmosphere is also borne out by their reaction with polynuclear aromatic hydrocarbons in the condensed phase.

The above research covers the atmospheric chemistry of sulphur and nitrogen species, hydrocarbons, oxygenated hydrocarbons and organic halogen species, areas which have been central to the programme. A new topic is the homogeneous reactions of gaseous mercury compounds, interest in this arising from the requirement of the understanding of the atmos-pheric mercury. The behaviour of volatile compounds of the heavier elements may have significance for other trace element cycles.

In the last two years the acid precipitation issue has come to the forefront and in consequence there has been some shift of emphasis in the programme towards sulphur and nitrogen chemistry. It should be borne in mind, however, that the atmospheric trace gas cycles are highly coupled and hydrocarbon oxidation chemistry may influence rates and mechanisms of transformation of inorganic species (and vice versa) giving rise to complex relationships between oxidation rates, pollutant concentrations and acid deposition. The production of ozone in atmospheric photochemical reactions can also be a factor in pollution effects from acidic species.

Precipitation chemistry is seen to be important for assessing the acid rain problem but at the present time there are very few projects in the working group that are addressing the problems of liquid phase oxidation rates and mechanisms. The main emphasis on the programme remains in the field of photochemically initiated homogeneous reactions, which has seen the more spectacular advances in knowledge related to atmospheric chemistry over the past decade. Recent ideas concerning the origin and role of oxidising species such as ozone and particularly peroxidic species in cloud chemistry, links the photochemical and liquid phase regimes. This may set the framework for a future direction in studies of atmospheric transformations related to precipitation chemistry.

There can be no doubt now about the importance of free radicals in atmospheric transformations. The concentration of OH and related radical species (HO_2, organic radicals) and its spatial and temporal variation, is of paramount importance for quantitative assessment of rates, fluxes and budgets of trace gases as well as for interpretation of atmospheric measurements. Whilst the photochemical radical sources and the sinks in the background atmosphere are probably adequately defined, the more complex boundary layer free radical chemistry requires more attention if the regional transformation rates and mechanisms are to be quantitatively understood. This is also an important direction for future work.

KINETIC STUDY OF REACTIONS OF OH RADICALS WITH ORGANIC
SULFUR COMPOUNDS

J.L. JOURDAIN, H. MAC LEOD, G. POULET and G. LE BRAS
Centre de Recherches sur la Chimie de la Combustion
et des Hautes Températures - CNRS - 45045 Orléans Cedex (France)

Summary

The reactions of OH radicals with CH_3SCH_3, CH_3SH, C_2H_5SH and C_4H_4S were studied using the discharge flow - EPR - mass spectrometric method at total pressures in the range 0.4 - 0.8 Torr. The rate constants measured are respectively :

$$OH + CH_3SCH_3 \quad (1) \qquad k_1 = (9.2 \pm 0.6) \times 10^{-12} \qquad (373\ K)$$
$$k_1 = (7.8 \pm 1) \times 10^{-12} \qquad (573\ K)$$
$$OH + CH_3SH \quad (2) \qquad k_2 = (2.1 \pm 0.2) \times 10^{-11} \qquad (293\ K)$$
$$OH + C_2H_5SH \quad (3) \qquad k_3 = (2.7 \pm 0.2) \times 10^{-11} \qquad (293\ K)$$
$$OH + C_4H_4S \quad (4) \qquad k_4 = (1.3 \pm 0.8) \times 10^{-13}\ exp\ |(1750 \pm 200)/T|$$
$$(293 - 473\ K)$$

(units are $cm^3\ molecule^{-1}\ s^{-1}$).
A product analysis by mass spectrometry for reactions of OH with CH_3SH and C_2H_5SH gave some information about their mechanism.

1. INTRODUCTION

The gas phase oxidation of organic sulfur compounds has become of increasing importance in relation with pollution by SO_2 and tropospheric and stratospheric sulfur containing aerosols. The reactions of OH radicals with these sulfur compounds appear to be the main first oxidation steps and then the determination of the rate constant of these reactions allows the calculation of the atmospheric lifetime of these compounds. We have measured by the discharge flow - EPR technique the rate constants for the reactions of OH with CH_3SCH_3, CH_3SH, C_2H_5SH and C_4H_4S (1) (2). The results obtained are reported and compared with previous data obtained for most of them by photolysis techniques. A mass spectrometric study of the products of the reactions of OH with CH_3SH and C_2H_5SH is also reported. This study has been made in order to obtain some indication on the mechanisms of these reactions. Such data are also needed to know the fate of these sulfur compounds in the atmosphere.

2. EXPERIMENTAL

The discharge flow - EPR - mass spectrometric technique used has been previously described (3). The EPR spectrometer, equiped with an 1 inch cylindrical cavity, was used for gas phase analysis of OH radicals. The reactor, made of quartz or teflon, was 22 mm i.d. and 1 mm thick. OH radicals were produced by reaction between H atoms and NO_2 in excess. H atoms were generated from a microwave discharge in H_2 highly diluted in helium. NO_2 and the sulfur compound were flowed through two sliding injectors kept in constant relative position, NO_2 being introduced into the reactor upstream from the sulfur compound inlet. Flow velocities and pressures in the reactor were respectively in the range 20 - 50 m/s and 0.4 - 2 Torr. The reactor could be heated between 293 and 473 K by means of an electrical furnace. The reaction products were analyzed by a quadrupole mass spectrometer which was interfaced at the end of the reactor, 45 cm downstream from the EPR cavity. The open source of the mass spectrometer was located very close to the effusion sampling hole in order to detect free radicals at low electron ionization energies.

3. RESULTS AND DISCUSSION

a) Rate constant measurements

The rate constants were determined under the pseudo-first order conditions, using a large excess of the sulfur compound compared to OH radicals. k was calculated from the slope of the straightline $-d \ln (|OH|)/dt = f (|RSR'|_o)$ from the EPR decay rate of OH at different initial concentrations $|RSR'|_o$. Typical concentrations were : $|OH| = 0.5 - 2 \times 10^{11}$ cm^{-3}, $|RSR'|_o = 0.2 - 5 \times 10^{13}$ cm^{-3}. The data obtained are reported in table I.

Table I : Rate constants for the reactions of OH

with CH_3SCH_3, CH_3SH, C_2H_5SH and C_4H_4S.

Units are cm^3 $molecule^{-1}$ s^{-1}. Error is 2σ

reaction	T (K)	k
OH + CH_3SCH_3 (1)	373	$(9.2 \pm 0.6) \times 10^{-12}$
	573	$(7.8 \pm 1) \times 10^{-12}$
OH + CH_3SH (2)	293	$(2.1 \pm 0.2) \times 10^{-11}$
OH + C_2H_5SH (3)	293	$(2.7 \pm 0.2) \times 10^{-11}$
OH + C_4H_4S (4)	293	$(5.0 \pm 0.4) \times 10^{-11}$
	333	$(2.2 \pm 0.2) \times 10^{-11}$
	373	$(1.2 \pm 0.2) \times 10^{-11}$
	473	$(0.52 \pm 0.05) \times 10^{-11}$
		$(1.3 \pm 0.8) \times 10^{-13}$ exp $|(1750 \pm 200)/T|$

For OH + CH$_3$SH   reaction, no reliable data could be obtained at room temperature, due to wall effects. These wall effects became less important with increasing temperature and reliable data were obtained at 373 and 573K with teflon and quartz reactor respectively. For reactions of OH with CH$_3$SH and C$_2$H$_5$SH, measurements were made in a teflon reactor and no important wall effects were observed. For reaction OH + thiophene, the quartz reactor was found to be convenient.

These results have been compared with previous data and also very recent determinations which are reported in table II.

Table II : Summary of the rate constant determinations at room temperature for reactions of OH with CH$_3$SCH$_3$, CH$_3$SH, C$_2$H$_5$SH and C$_4$H$_4$S — F.P.: Flash Photolysis — R.F.: Resonance Fluorescence — C.P.: Continuous Photolysis — G.P.C.: Gas Phase Chromatography — D.F.: Discharge Flow.

k $\times$ 10^{12} (cm^3.molecule^{-1}.s^{-1})	Pressure (Torr)	Technique	Reference
CH$_3$SCH$_3$			
8.28 $\pm$ 0.87	50	F.P. – R.F.	(4)
9.8 $\pm$ 1.2	50 – 98	F.P. – R.F.	(5)
9.1 $\pm$ 1.4	760	C.P. – G.P.C.	(6)
4.26 $\pm$ 0.56	50 – 200	F.P. – R.F.	(7)
9.2 $\pm$ 0.6 (373 K)	0.5	D.F. – E.P.R.	This work
7.8 $\pm$ 1 (573 K)	0.5		
CH$_3$SH			
33.9 $\pm$ 3.4	50 – 100	F.P. – R.F.	(8)
90.4 $\pm$ 8.5	760	C.P. – G.P.C.	(6)
33.7 $\pm$ 4.1	40 – 120	F.P. – R.F.	(7)
25.6 $\pm$ 4.4	1.9 – 3	F.P. – R.F.	(9)
21 $\pm$ 2	0.5	D.F. – E.P.R.	This work
C$_2$H$_5$SH			
36.5 $\pm$ 1.8	1.9 – 3	D.F. – R.F.	(9)
27 $\pm$ 2	0.5	D.F. – E.P.R.	This work
C$_4$H$_4$S			
47.7 $\pm$ 6.3	1.63 – 3.04	D.F. – R.F.	(10)
9.58 $\pm$ 0.38	760	C.P. – G.P.C.	(11)
50 $\pm$ 4	0.45 – 2	D.F. – E.P.R.	This work

<u>Reaction OH + CH$_3$SCH$_3$</u> (1) : our value agrees with two previous flash
photolysis measurements (4) (5) and a continuous photolysis one (6). This
agreement would indicate that k_1 is pressure independent between 0.5 and
760 Torr which is the pressure range covered by these four studies. Against
our k_1 value is about two times higher than a more recent flash photolysis
study (7).

<u>Reaction OH + CH$_3$SH</u> (2) : our k_2 value is 50% lower than the two
flash photolysis determinations (7), (8), obtained in the pressure range
40 - 120 Torr. But our value fairly agrees with another discharge-flow measurement (9) at low pressure (1.9 - 3 Torr). The difference between the
flash photolysis and discharge-flow values might be explained by a pressure
dependence of k_2 which would support the assumption of an addition mechanism. This would also be consistent with the higher value of k_2 in ref (6)
which has been tentatively explained by Wine et al (7) to be due to O_2 reaction with the adduct CH$_3$S (OH) H in experiments of ref (6).

<u>Reaction OH + C$_2$H$_5$SH</u> (3) : only one further measurement of k_3, by
the discharge flow method, has been reported (9) and the deviation from
our value remains within the range of uncertainty of the method.

<u>Reaction OH + C$_4$H$_4$S</u> (4) : As for reactions (2) and (3), our value
of k_4 agrees with another discharge-flow determination (10) and moreover
the agreement is excellent. Against, these values are 5 times higher than
a further measurement (11) by a relative method with continuous photolysis
in the presence of air at atmospheric pressure. The higher value obtained
in flow tubes could be due to an additional OH + C$_4$H$_4$S reaction at the wall
of the reactor. However no important wall effect was observed both in study
of ref (10) and in the present work. A tentative explanation can be proposed considering the nature of the mechanism which is an addition step as
shown from the negative temperature dependence observed for k_4. If we consider that the lifetime of the adduct is higher than about 0.5 ms, as for
some OH – aromatic adducts (12), the OH – C$_4$H$_4$S adduct could be partly stabilized at the reactor wall. Then the redissociation of the adduct into
the reactants could be slower in the flow tube compared to the photolysis
cell, leading to an higher value of k_2 by the flow tube method.

b) <u>Product analysis for reactions of OH with CH$_3$SH and C$_2$H$_5$SH</u>

<u>Reaction OH + CH$_3$SH</u>

The mass spectrometric study of reaction 2 was made for similar initial concentrations of OH and CH$_3$SH which were in the range 1.8 - 3 x 10^{12}
cm^{-3}. Under these conditions a peak was observed at mass 64, which is the
parent peak of methylsulfenic acid CH$_3$SOH. The kinetic curves of mass 64
formation showed an induction period. Besides, the decay curves of reactants were monitored by mass spectrometry for CH$_3$SH and by EPR for OH in
two different zones of the reactor during a same experiment. These curves
showed that the OH decay rate was higher than the CH$_3$SH one. The addition
of NO in the reactor at concentrations ranging from 0.6 x 10^{14} to 4 x 10^{14}
cm^{-3} led to the detection of a product at m/e = 77. This mass corresponds
to the parent peak of methylthionitrite CH$_3$SNO. Against, no new peak was
observed by addition of O_2 at concentration near 10^{15} cm^{-3}. Our experimental observations can be explained by the following chain mechanism :

$$(2) \quad OH + CH_3SH \longrightarrow H_2O + CH_3S$$

$$(5) \quad CH_3S + CH_3S \longrightarrow CH_3SSCH_3$$

$$(6) \quad CH_3S + CH_3S \longrightarrow CH_3SH + CH_2S$$

$$(7) \quad OH + CH_3SSCH_3 \longrightarrow CH_3SOH + CH_3S$$

$$(8) \quad CH_3S + NO \ (+M) \longrightarrow CH_3SNO \ (+M) \ \text{(in the presence of NO)}$$

Reactions (5) (6) and (7) are fast enough to be significant in the time scale of our experiment, with the reactant concentrations used. The rate constant of reaction (6) was found to be 4×10^{-11} at room temperature (13), with a similar value for k_5 at the low pressures used in this study (14). The rate constant of reaction (7) was also found to be very high (7) ($k_7 = 2 \times 10^{-10}$ at 298 K). The mechanism of reaction (7), leading to CH_3S and CH_3SOH formation, suggested in ref (7) has been experimentally verified (15). Then, the chain mechanism (2) + (5) + (6) + (7) explains the different experimental observations : i) the CH_3SOH formation with an induction period ; ii) the faster decay of OH, compared to CH_3SH, by the chains (2) + (6) and (5) + (7) ; iii) the CH_3SNO formation in the presence of NO by reaction (8).

This mechanism apparently favors an H atom transfer mechanism for the initial step (2). This assumption disagrees with the conclusions of the recent work of Hatakeyama and Akimoto (15) who suggested an addition step to explain the CH_3SNO formation in their system by the reaction : CH_3SH + $RCH_2ONO \longrightarrow CH_3SNO + RCH_2OH + O$. However, these authors indicated that the adduct $CH_3S(OH)H$ decomposes into CH_3S and H_2O in the absence of effective receptors such as RCH_2ONO. Then, our assumption is not necessarily in contradiction with the conclusion of these authors if we suppose that in our flow tube the reaction between OH and CH_3SH proceeds via an adduct which rapidly decomposes into CH_3S and H_2O. If this adduct exists, the absence of detection at mass 65 in our system, corresponding to the parent peak of CH_3S (OH)H, may signify that this adduct is either short living or too unstable to have a parent peak.

Reaction OH + C_2H_5SH :

The mass spectrometric study of reaction (3) was made under similar conditions as for reaction (2) with initial concentrations of C_2H_5SH ranging from 0.9 to 3×10^{12} cm^{-3}. As for reaction (2), the decay rate of OH was found higher than the decay rate of C_2H_5SH. When NO was added, the peak at mass 91 was observed, corresponding to $C_2H_5SNO^+$. A more intense peak at mass 61 was also produced, which probably represented $C_2H_5S^+$, which would be a cracking peak of C_2H_5SNO. Then, C_2H_5SNO was produced in the presence of NO. Against, O_2 addition did not lead to any new product as in reaction (2). These experimental results can be explained by a mechanism similar to that proposed for reaction (2).

c) Atmospheric implications

The rate constant measurements for reactions of OH with CH_3SCH_3, CH_3SH C_2H_5SH and C_4H_4S allow one to estimate the lifetime of these sulfur compounds in the atmosphere, since these reactions appear to be the main sink for these compounds. If we consider the upper and lower limits of OH concentration in the lower troposphere to be 10^7 and 3×10^5 cm^{-3}, the following lifetime ranges are obtained for CH_3SCH_3, CH_3SH, C_2H_5SH and C_4H_4S :

3 - 100, 1.3 - 44, 1 - 34 and 0.5 - 18.5 hours respectively. For CH_3SH and CH_3SCH_3 for instance, these values would be upper limits because it has been observed that these compunds would react faster with OH in the presence of air at atmospheric pressure (6) (16). For C_4H_4S, the calculated lifetime would represent a lower limit if it is confirmed that the rate constant of OH + C_4H_4S reaction is, under atmospheric conditions, 5 times lower than the discharge - flow determinations. Besides, the product analysis for reactions of OH with CH_3SH and C_2H_5SH gives partial information on the nature of the initial oxidation step of these compounds. But further work, and specially quantitative study of the products, is needed to precise these mechanisms.

Acknowledgement : This work was supported by the "Secrétariat d'Etat chargé de l'Environnement et de la Qualité de la Vie" through grant n° 83117.

References :

1) H. MAC LEOD, G. POULET and G. LE BRAS - J. Chim. Phys. 80 (1983) 287.

2) H. MAC LEOD, J.L. JOURDAIN and G. LE BRAS - Chem. Phys. Lett. 98 (1983) 381.

3) J.L. JOURDAIN, G. LE BRAS and J. COMBOURIEU - J. Phys. Chem. 85 (1981) 655.

4) M.J. KURYLO - Chem. Phys. Lett. 58 (1978) 233.

5) R. ATKINSON, R.A. PERRY and J.N. PITTS Jr - Chem. Phys. Lett. 54 (1978) 14.

6) R.A. COX and D. SHEPPARD - Nature 284 (1980) 330.

7) P.H. WINE, N.M. KREUTTER, C.A. GUMP and A.R. RAVISHANKARA - J. Phys. Chem. 85 (1981) 2660.

8) R. ATKINSON, R.A. PERRY and J.N. PITTS Jr - J. Chem. Phys. 66 (1977) 1578.

9) J.H. LEE and I.N. TANG - J. Chem. Phys. 78 (1983) 6646.

10) J.H. LEE and I.N. TANG - J. Chem. Phys. 77 (1982) 4459.

11) R. ATKINSON, S.M. ASCHMANN and W.P.L. CARTER - Int. J. Chem. Kinetics 15 (1983) 51.

12) R. ATKINSON, K.R. DARNALL, A.C. LLOYD, A.M. WINER and J.N. PITTS Jr - Adv. Photochem. 11 (1979) 375.

13) D.M. GRAHAM, R.L. MIEVILLE, R.H. PALLEN and C. SIVERTZ - Can. J. Chem. 42 (1964) 2250.

14) R.P. STEER and A.R. KNIGHT - J. Phys. Chem. 72 (1968) 2145.

15) S. HATAKEYAMA and AKIMOTO - J. Phys. Chem. 87 (1983) 2387.

16) I. BARNES, K.H. BECKER and E.H FINK - CAGCP Symposium on Tropospheric Chemistry - 28 Aug. - 3 Sept. Oxford (G.B.)

REACTIONS OF OH RADICALS WITH REDUCED SULFUR COMPOUNDS UNDER ATMOSPHERIC CONDITIONS

I. BARNES, V. BASTIAN, K.H. BECKER and E.H. FINK
Physikalische Chemie/FB 9, Bergische Universität-GH Wuppertal
D-56 Wuppertal 1, W. Germany

Summary

Rate constants for the reactions of OH radicals with H_2S, CH_3SH, CH_3SCH_3 and thiophene in the presence of various partial pressures of O_2 at 700 Torr total pressure in N_2 diluent have been measured in 38 1 and 420 1 reaction chambers using competitive kinetic techniques. For CH_3SH and CH_3SCH_3 the measured rate constants were found to increase with increasing pressure of O_2. In the case of H_2S and thiophene the rate constants showed no dependence on the pressure of O_2 present in the system. The O_2-dependence of the reaction of OH with CH_3SH and CH_3SCH_3 is discussed in terms of a mechanism involving the addition of OH to form a complex which can react with O_2 or decompose back to the reactants. Possible reactions pathways for $OH + H_2S$ and thiophene are also discussed in terms of the present results and data available from the literature.

1. INTRODUCTION

In recent years extensive effort has gone into investigating the tropospheric oxidation mechanisms of reduced sulfur compounds such as H_2S, CH_3SH, CH_3SCH_3 etc. Most of these sulfur compounds have both, natural and anthropogenic sources in the troposphere (Graedel, 1977). It is important to understand their role in the atmospheric sulfur cycle since their degradation products which include SO_2 and sulfonic acids could significantly contribute to aerosol formation and acid rain. Kinetic experiments have shown that the degradation of reduced sulfur compounds in the atmosphere is in most cases initiated by their reaction with OH radicals. Although rate constants for the reaction of these compounds with OH radicals are now becoming available there are often discrepancies between k_{OH}-values determined using absolute methods at low pressure in inert gas diluent and relative methods in one atmosphere of air. Further, the data on the pressure and temperature dependence of these reactions are also insufficient and often do not allow any conclusions to be made concerning the nature of their reaction mechanisms in the atmosphere.

Evidence has been emerging over the last few years that the rate constants for the reaction of OH radicals with some sulfur compounds depend on the pressure of O_2 present in the system. For example, studies on the reaction of OH with CS_2 carried out at one atmosphere total pressure in $N_2 + O_2$ diluent showed that the measured rate constant increased as the partial pressure of O_2 in the system increased (Jones et al., 1982; Barnes et al., 1983). The experimental results were interpretated in terms of a complex mechanism involving the addition of OH radicals to CS_2 to form a CS_2-OH adduct which either decomposes back to the reactants or reacts further with O_2 to form products. It seems likely that under atmospheric conditions molecular oxygen may be actively involved in the OH initiated oxidation of other reduced sulfur compounds too. If the rate constants for the reaction of OH with other sulfur compounds show an O_2 dependence then some of the decrepancies between absolute and relative rate measurements could be explained.

Reported here are competitive kinetic studies performed at one atmosphere total pressure and 300 K on the dependence of the rate constants for the reaction of OH radicals with H_2S, CH_3SH, CH_3SCH_3 and thiophene on the partial pressure of O_2. The aim of the present work was to determine k_{OH} rate constants for the above compounds under atmospheric conditions.

2. EXPERIMENTAL

The kinetic experiments were carried out in a 38 l cylindrical Duran-glass reactor (100 cm long, 20 cm diameter) capped with glass end-flanges and surrounded by 19 fluorescent lamps (Philips TLA 40 W/05). A competitive rate technique was used to measure the rate constants at a total pressure of 740 Torr in N_2 diluent which contained partial pressures of O_2 in the range 2 - 700 Torr. Ethene and propene were used as reference hydrocarbons in the experiments with rate constants for their OH reaction of 8×10^{-12} and 2.6×10^{-11} $cm^3 s^{-1}$ (Atkinson et al., 1979; Klein et al., 1984), respectively. Studies in this laboratory have shown that these reference rate constants are not influenced by O_2 (Klein et al., 1984).

In the experiments the effective decay of the sulfur compound under study was measured relative to that of one of the reference hydrocarbons. The relative rate constant for that particular compound was then derived from the integrated relative rate equation (Atkinson et al., 1979; Barnes et al. 1982),

$$\frac{k_S}{k_R} = \frac{\ln ([S]_t / [S]_o)}{\ln ([R]_t / [R]_o)} \tag{1}$$

where the subscripts o and t denote the concentrations of the sulfur compound and reference hydrocarbon at the reaction time o and t. The rate ratio k_S/k_R was determined from plots of $\ln ([S]_t / [S]_o)$ against $\ln ([R]_t / [R]_o)$ for a series of t values in steps of 5 min. over a period of ~ 60 min.

The photolysis of methyl nitrite, ethyl nitrite (Niki et al., 1981) and the NO_x/hydrocarbon smog system (Doyle et al., 1975) were used as OH sources in the experiments. Ethene, propene, CH_3SH, CH_3SCH_3 and thiophene concentrations in the ppm range were measured by gas chromatography (Hewlett Packard 5730 A GC) using a 2 m Teflon column packed with Porapak QS and a flame ionisation detector. H_2S was measured using a flame photometric detector (Tracor Inc.) after separation on a 1.5 m Teflon column packed with acetone washed Porapak QS operated at $160^{\circ}C$ and mounted in an Intersmat IGC 120 FL gas chromatograph.

Some kinetic experiments were also performed in a 420 l reaction chamber where the concentration-time behaviour of the sulfur compound under study and the reference hydrocarbon was monitored using in-situ infrared spectroscopic analysis. For details of this experimental system see a previous publication (Barnes et al., 1983). Methyl and ethyl nitrite were prepared by the dropwise addition of 50% H_2SO_4 to a saturated solution of $NaNO_2$ in the appropriate alcohol (Taylor et al., 1980). All other gases were used as supplied commercially by Messer Griesheim.

3. RESULTS

The concentration-time data from all of the OH + sulfur compound reactions investigated gave good linear relationships when plotted according to equation 1. Collected in Table I are the measured rate constants for the reaction of OH radicals with H_2S, CH_3SH, CH_3SCH_3 and thiophene, respectively, as determined in the 420 and 38 l reactors at 300 K for various proportions of N_2/O_2 diluent at a total pressure of 740 Torr. The reference hydrocarbon, either ethene or propene, and the OH source used for the respective determinations are also listed. Each quoted rate constant represents a minimum of four experiments and the errors refer to precision only (2σ). Fig. I shows the measured rate constants for OH with H_2S, CH_3SH and CH_3SCH_3 which were determined with CH_3ONO as the OH source in the 38 l reactor and plotted as a function of the O_2 partial pressure. The O_2-dependence of the k values for the OH + CS_2 reaction (Barnes et al. (1983)) are also shown for comparison.

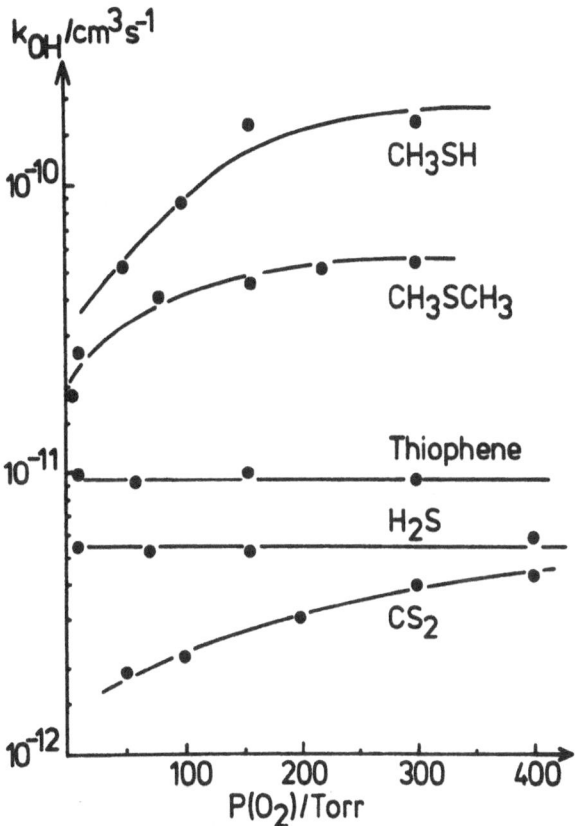

Figure I

Plots of the rate constants as a function of the O_2 partial pressure for the reactions of OH with CH_3SH, CH_3SCH_3, H_2S, CS_2 and thiophene as determined at 740 Torr total pressure (N_2 diluent) and 300 K.

4. DISCUSSION

1) Reaction OH + H_2S

From the results in Table I a value of $(5.2 \pm 0.8) \times 10^{-12}$ $cm^3 s^{-1}$ was obtained as rate constant for the reaction of OH with H_2S in 740 Torr synthetic air at 300 K. Within the error limits and the range of partial pressures of O_2 investigated the measured rate constant was found to be independent of O_2. Room temperature rate constants measured by different investigators using a variety of techniques range from 3.1 to 5.5 x 10^{-12} $cm^3 s^{-1}$. Reviews of these determinations can be found in the recent publications of Leu and Smith (1982), Michael et al. (1982) and Lin (1982).

The present results do not allow any conclusions to be drawn about the mechanism of the reaction of OH with H_2S. In the literature two types of temperature dependence have been reported for OH + H_2S, Arrhenius behaviour (Westenberg and De Haas, 1973; Parry et al., 1976; Wine et al., 1981a) and non-Arrhenius behaviour (Leu and Smith, 1982; Michael et al., 1982) where k(T) was found to go through a minimum near room temperature. Michael et al. (1982) have suggested that the non-Arrhenius behaviour observed in their work could be explained by a mechanism where H-abstraction dominates at higher temperature while complex formation becomes increasingly more important at lower temperatures. However, the measured rate constants of Lin (1982) and Lee and Smith (1982) show no variation with pressure at low temperatures. Further pressure and temperature dependence studies are under way in order to gain more information on the nature of the reaction of OH with H_2S.

2) Reaction OH + CH_3SH

It is evident from the list of measured rate constants for the reaction of OH with CH_3SH in Table I and Fig. I that the rate increases with increasing O_2 partial pressure. From a consideration of the data in Table I a value of $(12.6 \pm 3.2) \times 10^{-11}$ $cm^3 s^{-1}$ is obtained as rate constant for the reaction in 740 Torr synthetic air at 300 K. This value lies between the values of 14 x 10^{-11} and $(9.04 \pm 0.85) \times 10^{-11}$ $cm^3 s^{-1}$ determined by Cox (1975) and Cox and Sheppard (1980), respectively, using continuous photolysis methods at 1 atm total pressure of air. The value is a factor of ∼3-4 higher than those reported by workers using flash photolysis-resonance fluorescence (FP-RF) (Atkinson et al., 1977; Wine et al., 1981b) and discharge flow (DF) (Mac Leod et al., 1983; Lee and Tang, 1982) techniques at low pressure in inert diluent gas. However, for conditions of 740 Torr total pressure and 2 Torr O_2 partial pressure a value of $(3.4 \pm 0.4) \times 10^{-11}$ is obtained for OH + CH_3SH (see Table I), and this value is in reasonable agreement with the room temperature values obtained using the FP-RF and DF techniques.

Two reaction channels are possible for the reaction of OH with CH_3OH, either abstraction of an H atom from the S atom or addition to the S atom to form an adduct. A negative temperature dependence of the rate constant has been observed by Atkinson et al. (1977) and Wine et al. (1981a) and would be in favour of an addition mechanism. Wine et al. (1981 b), however, found the rate constant to be independent of pressure over the range 25-200 Torr SF_6. The results of the present study demonstrate that the rate of reaction of OH with CH_3SH is accelerated if O_2 is present in the system. This observation indicates that under atmospheric conditions H-atom abstraction by OH can not be the main reaction pathway. A mechanism which could possibly explain the observed O_2 effect involves the formation of an adduct which as in the case of OH + CS_2 can decompose back to the reactants or react further with O_2.

$$OH + CH_3SH + M \overset{k_a}{\underset{k_b}{\rightleftharpoons}} \overset{H}{CH_3-S} \ldots OH + M$$

$$\overset{H}{O_2} + CH_3S \ldots OH \overset{k_c}{\longrightarrow} products$$

If the above reaction mechanism is operative in the reaction system then a plot of the reciprocal of the effective rate constant against the recipro- cal of the O_2 concentration should be linear. The results from Table I for CH_3SH when plotted in this manner show reasonable linearity and suggest that under atmospheric conditions addition of the OH radical to CH_3SH is the major reaction step.

More direct and indirect measurements on the rate constant for the OH reaction with CH_3SH under different conditions of temperature, total pressure and gas composition are necessary to gain more information on the mechanism of the reaction.

3) Reaction OH + CH_3SCH_3

As for the reaction of OH with CH_3SH the rate constant for the rate of reaction of OH with CH_3SCH_3 was found to be dependent on the partial pressure of O_2 present in the system (Table I). For room temperature and 1 atm air a rate constant of $(4.3 \pm 0.9) \times 10^{-11}$ $cm^3 s^{-1}$ was found for the reaction. This value is approximately a factor of 5 higher than other room temperature values obtained using experimental techniques as diverse as FP-RF (Kurylo, 1978; Atkinson et al., 1978; Wine et al., 1981a), DF (Mac Leod et al., 1983; Lee and Tang, 1982) and continuous photolysis (Cox and Sheppard, 1980).

Again, as for the case of OH + CH_3SH the various reported investiga- tions, including pressure and temperature effects, as listed above would not allow any conclusive inferences to be made about the nature of the reaction mechanism (see Mac Leod et al., 1983, for review). The observa- tion of a dependence of the measured rate constant on the pressure of O_2 present in the system, again, supports a mechanism involving the addition of OH to CH_3SCH_3 to form an adduct which decomposes back to the reactants or reacts with O_2. A plot of the reciprocal of the measured effective rate constants in Table I against the reciprocal of the oxygen concentration shows reasonable linearity.
The large difference between the rate constant determined in this study and those determined by other workers for the reaction of OH with DMS

could possibly be explained by the effect of O_2 on the reaction rate. However, the present results need to be checked over a wide range of experimental conditions to completely exclude interferences from species, such as, $O(^3P)$, O_3 and possibly NO_3 and from wall reactions. The present results, therefore should be taken as very preliminary.

4) Reaction OH + thiophene

From the results in Table I a rate constant of $(9.6 \pm 1.5) \times 10^{-12}$ $cm^3 s^{-1}$ was obtained for the reaction of OH with thiophene in 740 Torr synthetic air at 300 K. As in the case of H_2S no dependence of the measured rate constant on O_2 could be established within the experimental error limits and the range of O_2 partial pressures investigated. The value is in excellent agreement with the value of $(9.58 \pm 0.38) \times 10^{-12}$ determined by Atkinson et al. (1983) using a continuous photolysis method. However, Lee and Tang (1982) and Mac Leod et al. (1983), who both used a discharge flow method, found values of $(5 \pm 0.4) \times 10^{-11}$ and (4.77 ± 0.63) $\times 10^{-11}$ $cm^3 s^{-1}$, respectively, for the rate constant at room temperature. This gives a factor of ~5 difference between the value reported by Aktinson et al. (1983) and that found in this study as compared to the results of Lee and Tang (1982) and Mac Leod et al. (1983).

At present there is no obvious explanation for the discrepancy found between the rate constants determined using the two experimental methods. The discrepancy cannot be explained by a pressure effect since the value obtained by the relative methods in 1 atm of air (Atkinson et al., 1983, and this work) would be expected to be higher than those determined using absolute methods at low pressure.

Mac Leod et al. (1983) have found a negative temperature dependence for the rate constant which supports an addition mechanism for the reaction of OH with thiophene. They were not able to differentiate between an addition to the S atom or the carbon–carbon double bond in their work. Lee and Tang (1982) measured the rate constant for the reaction of OH with furan as well as with thiophene. From a comparison of the k-values and a consideration of structural differences they suggested that the OH adds onto the S atom in thiophene. However, Atkinson et al. (1983) in a recent study on the reaction of OH with furan using a continuous photolysis method find a rate constant a factor ~2.5 lower than the value reported by Lee and Tang (1982). It is perhaps interesting to note that as in the case for OH + thiophene the continuous photolysis method yields a value for the rate constant for OH + furan much lower than that obtained using the discharge flow technique. Whether the disagreement between the rate constants determined for these reactions using the different methods is due to experimental artifacts in one of the techniques or a change in the mechanism on going from low to high pressure remains to be resolved. Until this issue is clarified further speculation on the mechanism of OH + thiophene seems unwarranted.

In all of the reactions reported in this study SO_2 has been found as a major product. In the case of OH + CH_3SH and CH_3SCH_3, methanesulfonic acid $(CH_3(SO_2)OH)$ also is formed in a large proportion. Quantitative product analyses of all of the reactions are currently in progress in this laboratory.

The reaction of OH radicals with SO_2 has also been studied. Initial experiments using GC-analysis for the detection of SO_2 led to erroneously high rate constants as well as on O_2 dependence for the reaction since the columns used were not inert towards the chemical systems investigated. However, more recent measurements on SO_2 / n-butane / CH_3ONO / O_2 / N_2 photolysis systems using a Chromosil column for SO_2 detection gave good reproducible results. Within the error limits and range of O_2 partial pressures investigated the measured rate constant was found to be independent of O_2. A value of (1.1 ± 0.2) x 10^{12} $cm^3 s^{-1}$ was obtained as rate constant for the reaction in 740 Torr synthetic air and 300 K. This result compares well with the value of (0.99 ± 0.17) x 10^{12} $cm^3 s^{-1}$ also determined in this laboratory using similar reaction mixtures as above but with mass spectrometric monitoring of the SO_2 and the reference benzene. Both of these values agree well with other measurements (Stockwell and Calvert, (1983)) and with the recent determination of Izumi et al. (1984) who obtained (1.22 ± 0.13) x 10^{12} $cm^3 s^{-1}$ for the reaction at 780 Torr air and 303 K using a competitive rate technique.

ACKNOWLEDGEMENTS

The financial support by the "Bundesminister für Forschung und Technologie" (BMFT) is gratefully acknowledged.

5. REFERENCES

Atkinson R., Perry R.A. and Pitts J.N., Jr. (1977)
J. Chem. Phys. 66, 1578-1581.

Atkinson R., Darnall K.R., Lloyd A.C., Winer A.M. and Pitts J.N., Jr. (1979) Adv. Photochem. 11, 375-538.

Atkinson R., Aschmann S.M. and Carter W.P.L. (1983)
Int. J. Chem. Kinetics 15, 51-61.

Barnes I., Bastian V., Becker K.H., Fink E.H. and Zabel F. (1982)
Atm. Environ. 16, 545-550.

Barnes I., Becker K.H., Fink E.H., Reimer A. and Zabel F. (1983)
Int. J. Chem. Kinetics 15, 631-645.

Cox. R.A. (1975) Atmospheric Photooxidation Reactions. The Gas Phase Reaction of OH Radicals with Some Sulphur Compounds. United Kingdom Atomic Energy Authority Report AERE - R 8132.

Cox. R.A. and Sheppard D. (1980) Nature 284, 330 - 331.

Doyle G.J., Lloyd A.C., Darnall K.R., Winer A.M. and Pitts J.N., Jr. (1975) Environ. Sci. Technol. 9, 237-240.

Graedel T.E. (1977) Rev. Geophys. Space Phys. 15, 421 - 428.

Izumi K., Motoyuki M., Yoshloko M., Murano K. and Fukuyomo T. (1984) Environ. Sci. Technol. 18, 116 118

Jones B.M.R., Burrows J.P., Cox R.A. and Penkett S.A. (1982)
Chem. Phys. Lett. 88, 372-376.

Kurylo M.J. (1978) Chem. Phys. Lett., 58, 233-238.

Klein T., Barnes I., Becker K.H., Fink E.H. and Zabel F. (1984)
J. Phys. Chem. (in press)

Lee J.H. and Tang I.N. (1982) J. Chem. Phys. 77, 4459-4463.

Leu M.T. and Smith R.H. (1982) J. Phys. Chem., 86, 73-81.

Lin C.L. (1982) Int. J. Chem. Kinetics 14, 593-598

Mac Leod H., Poulet G., and Le Bras G. (1983)
J. Chimie Physique 80, 287-292.

Mac Leod H., Jourdain J.L., Poulet G., and Le Bras G. (1983) Paper presented
at CACGP Symposium on Tropospheric Chemistry, Christ Church, Oxford, England,
28. Aug. - 3. Sept., Abstract VI-2.

Michael J.V., Nova D.F., Brobst W.D., Borkowski R.P. and Stief L.J. (1982)
J. Phys. Chem. 86, 81-84.

Niki H., Maker P.D., Savage C.M. and Breitenbach L.P. (1981)
Chem. Phys. Lett. 80, 499-503.

Perry R.A., Atkinson R. and Pitts J.N., Jr. (1976)
J. Chem. Phys. 64, 3237-3239.

Stockwell R.S. and Calvert J.G. (1983) Atmos. Environ. 17, 2231 - 2235

Taylor W.D., Allston T.D., Moscato M.J., Fazekes G.B., Kozlowski R. and Takacs
G.A. (1980) Int. J. Chem. Kinet. 12, 231-240.

Westenberg A.A. and DeHaas N. (1973)
J. Chem. Phys. 59, 6685 - 6686.

Wine P.H., Kreutter N.M., Gump C.A. and Ravishankara A.R. (1981a)
J. Phys. Chem. 81, 2660-2665.

Wine P.H., Kreutter N.M., Gump C.A. and Ravishankara A.R. (1981b)
J. Phys. Chem. 85, 2660-2665.

Table I

Rate constants for the reaction of OH radicals with H_2S, CH_3SH, CH_3SCH_3 and thiophene in $O_2 + N_2$ diluent at 740 Torr total pressure and 300 K as determined relative to OH + C_3H_6

OH-Source/ Reactor	O_2 Pressure (Torr)	$k_{OH + H_2S}$ (10^{-12} cm³/s)	$k_{OH + CH_3SH}$ (10^{-11} cm³/s)	$k_{OH + CH_3SCH_3}$ (10^{-11} cm³/s)	$k_{OH + Thiophene}$ (10^{-12} cm³/s)
NO_2–RH/38 1	2	–	3.3 ± 0.3	–	–
	10	–	4.7 ± 0.7	2.2 ± 0.2	–
	50	–	6.5 ± 1.7	–	–
	155	–	12.4 ± 2.8	4.2 ± 0.4	–
	220	–	–	4.6 ± 0.4	–
	300	–	11.0 ± 3.0	5.3 ± 0.6	–
CH_3ONO/38 1	2	–	–	–	–
	10	5.4 ± 1.0	–	1.7 ± 0.3	–
	50	–	–	2.6 ± 0.5	9.7 ± 1.4(9.5±1.4)*
	70	5.2 ± 0.6	5.2 ± 1.5	–	–
	80	–	–	4.1 ± 0.6	9.4 ± 1.8
	100	–	8.7 ± 1.7	4.5 ± 0.6	–
	155	5.2 ± 0.8	14.0 ± 3.0	5.2 ± 0.7	9.9 ± 1.5(9.2±1.3)*
	220	–	–	5.4 ± 0.8	–
	300	–	13.0 ± 3.0	–	9.5 ± 1.7
	400	5.9 ± 0.9	–	–	–
CH_3ONO/420 1	2	–	–	1.8 ± 0.4	–
	10	–	–	2.5 ± 0.4	–
	50	–	6 ± 2	–	–
	80	–	–	4.2 ± 0.6	–
	155	–	–	4.7 ± 0.7	–
	300	–	–	5.1 ± 0.7	–
	700	–	12 ± 3	–	–
C_2H_5ONO/38 1	2	–	3.4 ± 0.4	–	–
	10	–	4.9 ± 0.5	–	–
	50	–	8.6 ± 1.0	–	–
	155	–	13.0 ± 2.6	–	–

* measured relative to OH + C_2H_4

- 157 -

LIF STUDIES OF FORMATION AND KINETICS OF PRIMARY RADICAL PRODUCTS IN OH-OXYGENATED HYDROCARBON REACTIONS

K. LORENZ, D. RHÄSA, and R. ZELLNER
Institut für Physikalische Chemie, Universität Göttingen
3400 Göttingen, FRG

Summary

The branching ratio (ρ) of formation of CH_3O (Methoxi) and CH_2CHO (Vinoxy) in reactions of the OH radical with CH_3OH ((1a) OH + $CH_3OH \rightarrow CH_3O + H_2O$) and ethylene oxide ((2a) OH + $C_2H_4O \rightarrow CH_2CHO + H_2O$), respectively, has been determined using time resolved laser induced fluorescence. Whereas (1a) is a minor channel of the total reaction at 298 K ($\rho_{1a} = 0.11$), the yield of (2a) is dependent on pressure and is predicted to be dominant under tropospheric conditions. The subsequent reactions of CH_3O with NO (3) $CH_3O + NO \rightarrow$ products and of vinoxy with O_2 (4) $CH_2CHO + O_2 \rightarrow$ products are found to be pressure dependent, indicating a dominant recombination process. At pressures above 250 mbar and at 298 K we obtain $k_3 = (2.5\pm0.5)10^{-11}$ and $k_4 = (2.5 \pm 0.5)10^{-13}$ cm^3/molecule·s.

1. INTRODUCTION

Elementary reactions controlling the oxidation of partially oxygenated hydrocarbons are of great practical importance in modelling their chemistry in the atmospheric environment. The relevant information needed focusses on two points: i) the overall rate coefficient k(T) and ii) the distribution of primary radical products. The former determines the atmospheric residence time (or at least an upper limit) of a chemical and the latter is responsible for its degradation pathway. Whereas there is now a wealth of kinetic data on OH reactions with atmospheric trace constituents (k_{OH}-values) - well exemplified by various reviews and reports, including these within this Symposium series -, corresponding studies of the primary products of these reactions and their consecutive reactions with the dominant atmospheric reagents (O_2, NO) are still scarce.

In the present work we report on

i) direct studies of the branching ratio (specific product channel) of formation of CH_3O in the reaction

(1a)   OH + $CH_3OH \rightarrow \overline{CH_3O} + H_2O$ + 77 kJ/mol

(1b) $\rightarrow \overline{CH_2OH} + H_2O$ + 106 kJ/mol

and of CH_2CHO in the reaction

(2a)   OH + $C_2H_4O \rightarrow CH_2CHO + H_2O$ + 203 kJ/mol

(2b) $\rightarrow \overline{CH_3CO} + H_2O$ + 245 kJ/mol

Both reactions are exothermic. The yield into the different channels is therefore in general expected to be governed by the relativ exothermicities. However, as will be shown, the products of reaction (2) originate via the decomposition of a common primary cyclic C_2H_3O radical. They are therefore formed vibrationally excited and their relativ yield becomes pressure dependent.

In addition to (i) we have also studied
(ii) the subsequent atmospheric reactions

(3) $CH_3O + NO \rightarrow$ products

and

(4) $CH_2CHO + O_2 \rightarrow$ products

for which we report rate constants at 298 K and at different total pressures. The observed kinetic behaviour indicates that both reactions proceed predominantly via addition, forming CH_3ONO and the radical O_2CH_2CHO, respectively, as products.

Product identification and kinetic studies have been performed using a combination of excimer laser photolysis to generate either OH or CH_3O/CH_2CHO and time resolved laser induced fluorescence for the detection of methoxi and vinoxy radicals.

2. EXPERIMENTAL

A schematic representation of the combined laser photolysis / laser induced fluorescence (LPLIF) system is provided in Fig. 1. For studies of the product yields in reactions (1) and (2) OH radicals are generated by excimer laser photolysis

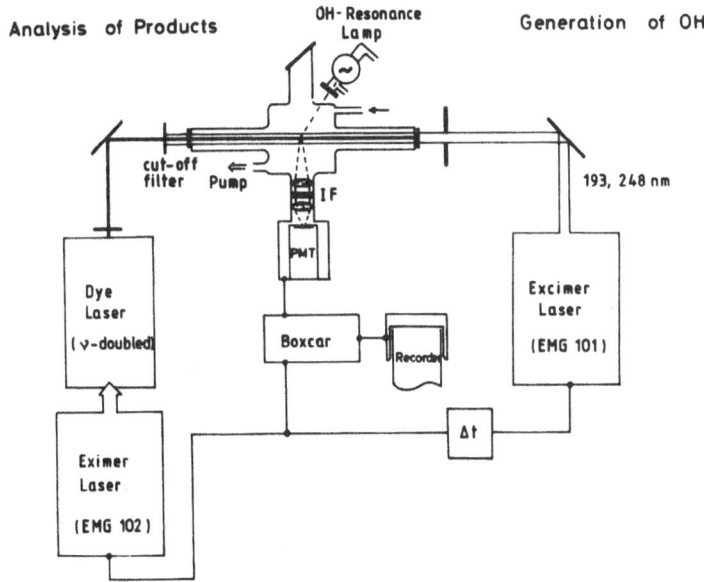

Fig. 1: Schematic representation of laser flash photolysis system with combined detection facilities for OH (by conventional resonance fluorescence) and for CH_3O and CH_2CHO by LIF

of HNO_3 at 248 nm (KrF). With HNO_3 partial pressures of ≤ 0.1 mbar typical OH concentrations are $\sim 4 \times 10^{11}$ cm^{-3}. In the presence of a reactant (CH_3OH, C_2H_4O) the decay of OH is monitored using conventional A$^-$X resonance fluorescence. A second pulsed frequency doubled dye laser system serves as a tunable analysis light source. Its output is directed through the fluorescence cell from the opposite end, but coaxially with the photolysis laser beam. Fluorescence excited with this laser is collected via cut-off filters and measured in analog mode by means of a gated boxcar integrator. In order to obtain time dependent profiles of the radical products (CH_3O, CH_2CHO) the delay time Δt between photolysis and analysis laser beam is varied. Further details of the experimental arrangement can be found elsewhere [1].

CH_3O and CH_2CHO as reaction products are identified by means of comparative fluorescence spectroscopy. Fig. 2 shows part of the fluorescence excitation spectrum of CH_3O which has previously been assigned by Inoue et al. [2]. A similar spectrum, however, shifted to larger wavelengths (≥ 331 nm) holds for the vinoxy radical [3]. It can be obtained in the photolysis of methyl-vinyl-ether (MVE).

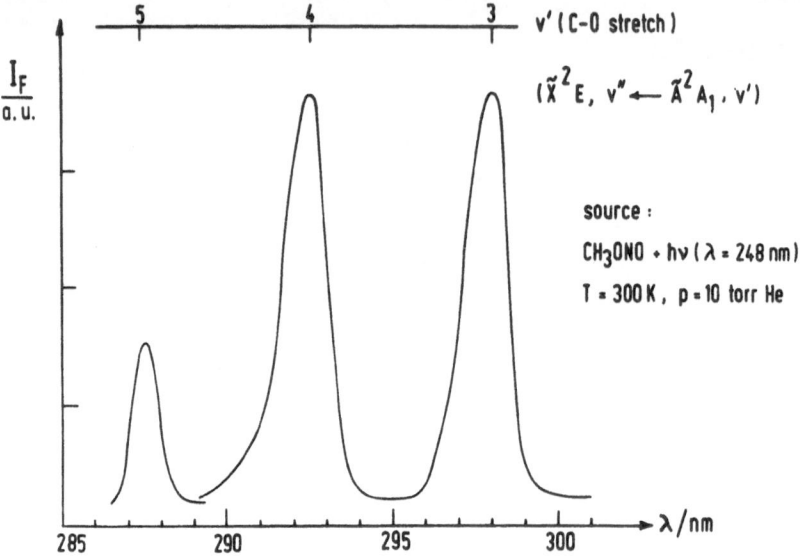

Fig. 2: Fluorescence excitation spectrum of CH_3O as obtained in the laser photolysis of CH_3ONO

In order to quantitatively determine the yield of the primary radical products, an absolute calibration of the fluorescence intensity has to be made. This is done by comparing the fluorescence intensity observed in the reaction (I_F^R) to the one observed in the photolysis of the parent molecule alone (I_F^0), allowing for quenching effects by all components of the reaction mixture. Details of this procedure are fully described in ref. [1].

For kinetic studies on reactions (3) and (4), CH_3O and CH_2CHO are again generated via the excimer laser photolysis of

CH_3ONO (λ = 248 nm) and MVE (λ = 193 nm), respectively. Their decay in the presence of NO/O_2 is then followed by LIF using the bands at 292 nm and 337 nm.

3. RESULTS AND DISCUSSION

3.1. Formation of Primary Radical Products

3.1.1. $OH + CH_3OH \rightarrow CH_3O + H_2O$

The formation of CH_3O in reaction (1) has been studied at 298 and 393 K. Typical results are presented in Fig. 3 where we have plotted (on a logarithmic scale) the time dependence of the CH_3O fluorescence intensity together with the corresponding result for OH. As can be seen CH_3O is formed as OH decays. However, it does not attain a time independent value, but rather passes through a maximum. This is due to conse-

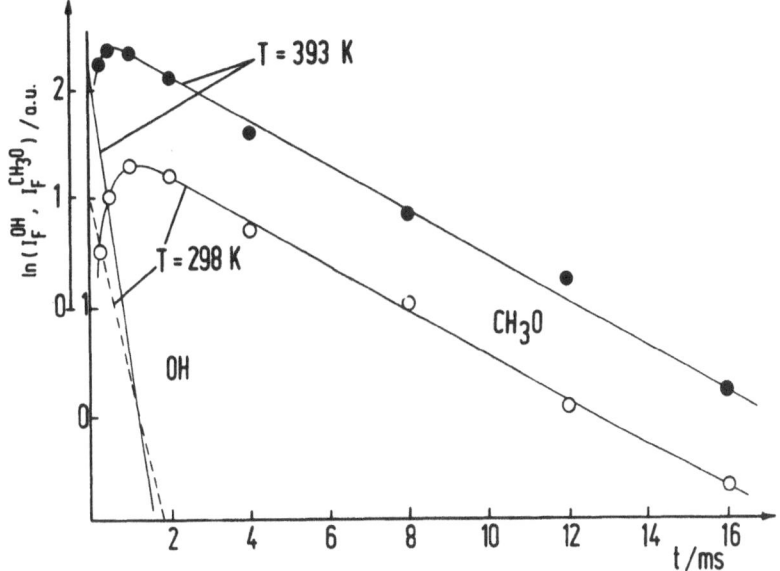

Fig. 3: Time dependence of CH_3O and OH fluorescence intensities observed in the reaction $OH+CH_3OH \rightarrow$ products

cutive loss of CH_3O by diffusion (out of the excitation volume) and possibly reaction with other components of the mixture. However, by applying to the CH_3O profile the time dependent solution of a differential equation pertaining to consecutive reactions of first order we can easily compute the CH_3O concentration (without additional loss) from the one observed at maximum [1]. We obtain $[CH_3O]_{\infty} = 1.33 [CH_3O]_{max}$. From this, and by including the quenching corrections, we find that the relative yield (branching ratio) of the CH_3O channel

(1a) $OH + CH_3OH \rightarrow CH_3O + H_2O$

(1b) $\rightarrow CH_2OH + H_2O$

is $\rho_{1a} = k_{1a}/k_1 = 0.11$ at 298 K and 0.22 at 398 K. Hence, formation of CH_3O is a minor channel at room temperature and below. Together with the overall rate coefficient ($k_1 = (1.2 \pm 0.3)10^{-11}$ $\exp[-(810 \pm 50)K/T]$ [1]) we obtain for the channel specific rate coefficient at 298 K

$$k_{1a} = 8.7 \times 10^{-14} \text{ cm}^3/\text{molecule·s}$$

The corresponding activation energy is
$$E_{1a} = (13.6 \pm 3) \text{ kJ/mol.}$$

Hence, compared to the overall reaction (with an activation energy of $E_1 = 6.7$ kJ/mol) the height of the activation barrier is roughly doubled for a decrease in exothermicity from 106 kJ/mol (channel (1b)) to 77 kJ/mol (channel (1a)).

3.1.2. $OH + C_2H_4O \rightarrow CH_2CHO + H_2O$

The yield of vinyloxy in the reaction of OH radicals with ethylene oxide has been studied at 298 K and at pressures of 13 and 80 mbar (He). Typical results are presented in Fig. 4, where we again show the CH_2CHO fluorescence intensity together with the corresponding intensity for OH.

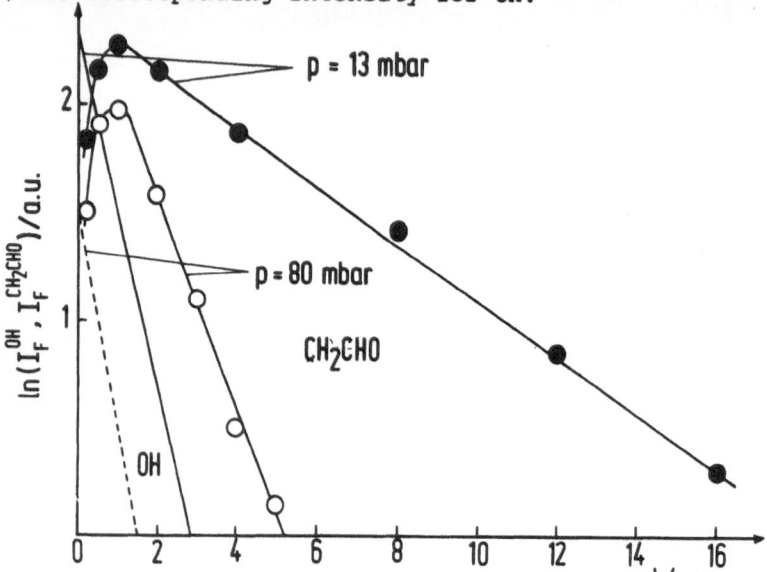

Fig. 4: Time dependence of CH_2CHO and OH fluorescence intensity observed in the reaction OH $+C_2H_4O \rightarrow$ products

Clearly, the vinoxy radical is a primary product of the reaction, the yield of which increases with total pressure. Using the same kind of fluorescence intensity calibration and analysis of time dependent intensities as described above, we obtain for the branching ratio of the CH_2CHO channel

(2a) $OH + C_2H_4O \rightarrow CH_2CHO + H_2O$

(2b) $\rightarrow CH_3CO + H_2O$

$$\rho_{2a} = k_{2a}/k_2 = 0.08 \text{ (13 mbar)}$$
$$= 0.23 \text{ (80 mbar)}$$

In here, k_2 is the overall rate coefficient for the decay of OH, which we have previously determined to be $k_2 = (1.1 \pm 0.4)10^{-11} \exp(-1460 \text{ K/T})$, corresponding to $(8.1 \pm 1.6)10^{-14} \text{ cm}^3/\text{molecule·s}$ at 298 K [4].

Ethylene oxide is a cyclic ether. The above results, therefore, suggest that the attack of OH on C_2H_4O must be accompanied by essentially simultaneous ring opening. This provides additional exothermicity to the reaction and therefore the linear products must be formed highly vibrationally excited. Fig. 5 shows an energy diagram of the overall reaction. Although there are a number of unknowns in this diagram (i.e. the heat of formation of cyclic-C_2H_3O and the barrier heights to isomerization: cyclic $C_2H_3O \rightarrow CH_2CHO$ and $CH_2CHO \rightarrow CH_3CO$) the following picture seems to emerge: OH primarily abstracts an H-atom from C_2H_4O to form cyclic-C_2H_3O.

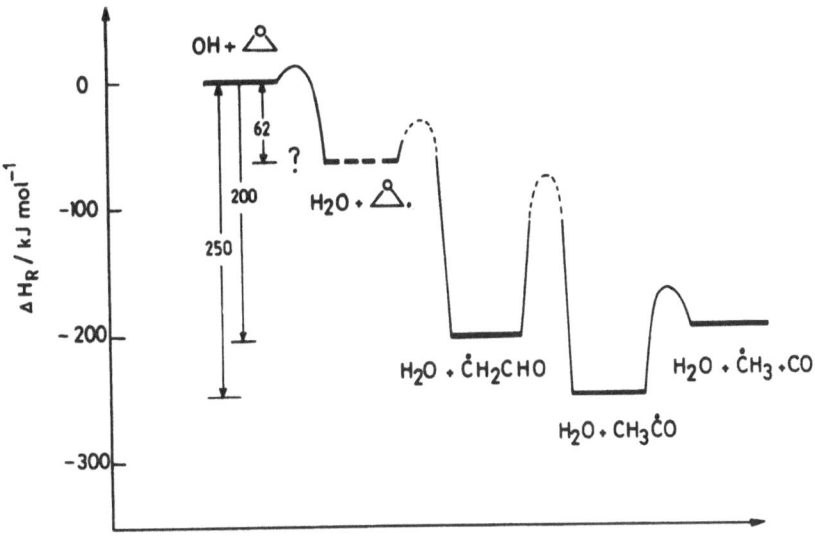

Fig. 5: Energy diagram for OH + ethylene oxide

This is thermally unstable and splits the C-O bond yielding CH_2CHO. The total energy of CH_2CHO is above the isomerization threshold and hence the reaction may continue in the exothermic direction leading to acetyl (CH_3CO) and CH_3 + CO. The amount of CH_2CHO that we observe in our experiments then corresponds to the fraction which is trapped in the CH_2CHO potential well by collisional deactivation. Hence, its amount increases with pressure. The extrapolated CH_2CHO yield for atmospheric pressure is expected to be close to unity. Hence, under these conditions the overall reaction may be represented by (2a) only.

3.2. Reactions of Primary Radical Products

3.2.1. $CH_3O + NO \rightarrow$ Products

The kinetics of this reaction have been studied at 298 K and at pressures of 13 and 254 mbar. Fig. 6 shows a plot of the first order decay constant for CH_3O as a function of added NO. The slope of the lines corresponds to the second

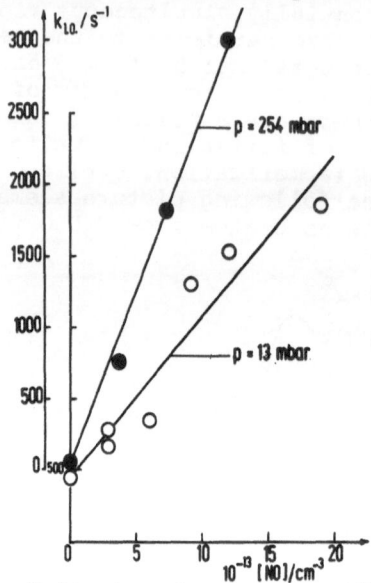

Fig. 6: Dependence of first order rate coefficients for $CH_3O + NO \rightarrow$ products on the NO concentration. For clarity, the abscissa to measurements at 254 mbar has been displaced by 500 s^{-1}

order rate coefficient k_3, which is seen to increase with pressure. We obtain

$$k_3 \text{ (13 mbar)} = (1.1 \pm 0.2)10^{-11} \text{ cm}^3/\text{molecule's}$$
$$k_3 \text{ (254 mbar)} = (2.5 \pm 0.5)10^{-11} \text{ "} \qquad \text{"}$$

The value at 254 mbar agrees within the error limit with the one recommended by the CODATA group [5] ($k_3 = 2 \times 10^{-11}$ cm^3/molecule's) and which is based on a previous direct measurement by Sanders et al. [6] as well as on a measurement of the reverse dissociation by Batt et al. [7]. However, the present work provides the first indication of a pressure dependence of k_3 as one would expect for a dominant recombination process, viz.

(3a) $CH_3O + NO \text{ (+M)} \rightarrow CH_3ONO \text{ (+M)}$

The possibility of an alternative bimolecular channel

(3b) $CH_3O + NO \rightarrow CH_2O + HNO$

has to be further investigated by direct observation of the HNO product. However, from both the pressure dependence observed here and preliminary data on the temperature dependence of k_3 [8] we conclude that $k_{3b} \simeq 4 \times 10^{-12}$ cm^3/molecule·s, in agreement with a previous estimate of $k_{3a}/k_{3b} \simeq 0.17$ by Batt et al. [7].

3.2.2. $CH_2CHO + O_2 \rightarrow$ products

Rate constants for this reaction have been measured at 298 K and at pressures of 27, 135, and 276 mbar (M = He). Fig. 7 represents the first order rate constants for the decay of vinoxy as a function of added O_2. As can be seen, $k_{1.0.}$ and

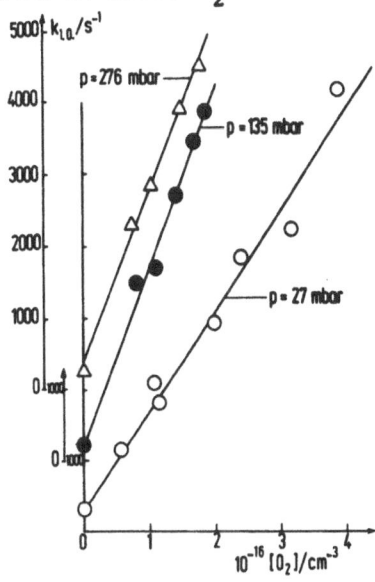

Fig. 7: Dependence of first order decay constant for $CH_2CHO + O_2 \rightarrow$ products on the O_2 concentration for different total pressures. For clarity, the abscissa are displaced.

hence k_4 increase with pressure, but are essentially pressure independent above 100 mbar (Fig. 8). In the high pressure limit we obtain

$$k_4^\infty = (2.6 \pm 0.5) 10^{-13} \text{ cm}^3/\text{molecule·s}$$

This is in excellent agreement to a result obtained recently by Gutman and Nelson [9] using a similar technique.

The size of the rate coefficient and its dependence on pressure suggests that the reaction proceeds as a recombination process, viz.

$$(4) \quad CH_2CHO + O_2 \rightarrow O_2CH_2CHO$$

For the alternative bimolecular abstraction route

$$(4') \quad CH_2CHO + O_2 \rightarrow CH_2CO + HO_2$$

- which is exothermic by ~ 62 kJ/mol - one would expect a much lower rate coefficient. This conclusion is based on a comparison with other O_2 abstraction reactions and their respective exothermicities (i.e. $CH_3O + O_2 \rightarrow CH_2O + HO_2 + 96$ kJ/mol,

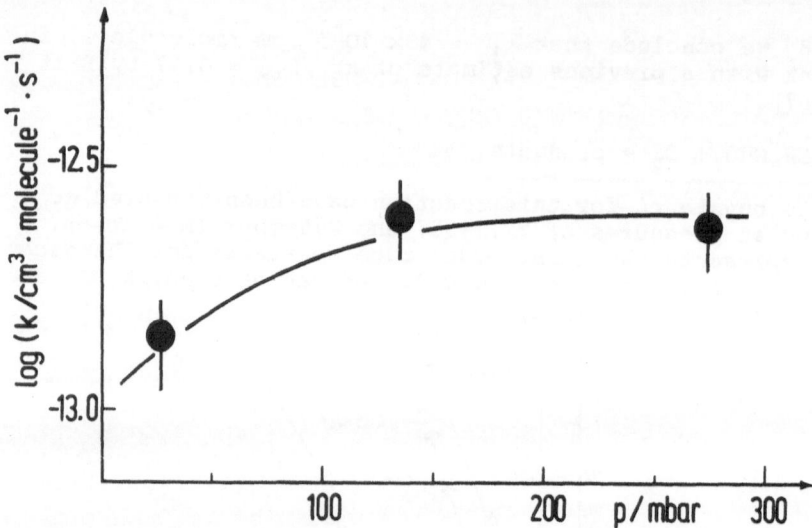

Fig. 8:   Pressure dependence of rate coefficient k_4.
T = 298 K, M = He

$k = 6 \times 10^{-16}$ cm^3/molecule·s; CHO + O_2 → CO + HO_2+134 kJ/mol;
$k = 5 \times 10^{-12}$ cm^3/molecule·s).

The product of reaction (4), O_2CH_2CHO, differs from
normal alkylperoxi radicals through the weakly bound (aldehyde) H-atom. This allows intramolecular rearrangement via
1,4-hydrogen shift, possibly followed by subsequent unimolecular decomposition, viz.

(5) O_2CH_2CHO → $HOOCH_2CO$ → CH_2O + CO + OH

This mechanism is at present speculative and needs to be
tested, for example by direct detection of OH or CH_2O. It
offers though the extremely interesting and unusual perspec-
tive that the overall reaction between OH and ethylene oxide
under atmospheric conditions proceeds as an OH catalyzed
oxidation by O_2 of net form:

(6) C_2H_4O + O_2 → CH_2O + CO + H_2O.

ACKNOWLEDGEMENT

Support of this work by Bundesministerium für Forschung
und Technologie (BMFT) is gratefully acknowledged. R.Z. thanks
the Deutsche Forschungsgemeinschaft (DFG) for a Heisenberg-
Stipendium.

REFERENCES

[1] J. Hägele, K. Lorenz, D. Rhäsa, R. Zellner, Ber. Bunsen-
ges. phys. Chem. 87, 1023 (1983)
[2] G. Inoue, H. Akimoto, M Okuda, J. Chem. Phys. 72, 1769
(1980)

[3] G. Inoue, H. Akimoto, J. Chem. Phys. $\underline{74}$, 425 (1981)
[4] B. Fritz, K. Lorenz, W. Steinert, R. Zellner, in "Physico-chemical Behaviour of Atmospheric Pollutants", B.Versino, H. Ott (Eds.) Reidel, 1982, p. 192
[5] D.L. Baulch, R.A. Cox, P.J. Crutzen, R.F. Hampson, J.A. Kerr, J. Troe, R.T. Watson, J. Phys. Chem. Ref. Data $\underline{11}$, 327 (1982)
[6] N. Sanders, J.E. Butler, L.R. Pasternak, J.R. McDonald, Chem. Phys. Lett. $\underline{48}$, 203 (1980)
[7] L. Batt, R.T. Milue, R.D. McCulloch, Int. J. Chem. Kin. $\underline{9}$, 567 (1977)
[8] B. Fritz, V. Handwerk, K. Lorenz, R. Zellner, to be published
[9] D. Gutman, H.H. Nelson, J. Phys. Chem., to be published

THE TEMPERATURE DEPENDENCE OF THE FORWARD-BACKWARD REACTIONS OF THE ADDITION OF OH TO BENZENE, ANILINE AND NITROBENZENE

F. WITTE and C. ZETZSCH

Lehrstuhl für Physikalische Chemie I, Ruhr-Universität Bochum
Postfach 102148, D-4630 Bochum

and

Fraunhofer-Institut für Toxikologie und Aerosolforschung, Stadtfelddamm 35,
D-3000 Hannover 61, West Germany

Summary

The temperature dependence of the reaction of OH with benzene, aniline and nitrobenzene has been investigated using the flash photolysis/ resonance fluorescence technique. The solution of the according system of differential equations allows us to calculate the rate constants for the addition of OH to benzene ($k(T) = (2.3+/-0.4) \cdot 10^{-12} exp(-(190+/-60)/T)$ $cm^3 s^{-1}$), aniline ($k(T) = 1.7_3 10^{-11} exp((440+/-40)/T) cm^3 s^{-1}$) and nitro-benzene ($k(T) = (6+/-2) \cdot 10^{-13} exp(-(560+/-30)/T) cm^3 s^{-1}$), the unimolecular decay of the adducts benzene-OH ($k(T) = 3.4 \cdot 10^{12} exp(-8200+/-700)/T) s^{-1}$) and aniline-OH ($k(T) = 6 \cdot 10^{10} exp(-(8400+/-1100)/T) s^{-1}$) back to the reactants and a further reaction path of the adduct aniline-OH not leading back to the reactants ($k(T) = 8 \cdot 10^{10} exp(-(7400+/-1100)/T) s^{-1}$). The bond dissociation energies of OH to benzene and aniline in the adducts have been determined.

Introduction

For most anthropogenic air pollutants the reaction with OH radicals is a major sink. The knowledge of the half-lives of organic substances in the atmosphere is very important and rate constants for the reactions of OH with various compounds have been determined in order to be able to estimate their half-lives. The rate constants for the reactions of OH with aromatics - especially with benzene as their parent compound - have been determined by various authors[1-12]. Some of these publications deal with the temperature and pressure dependence[1,7-10]. In the present study the temperature dependence for the reaction of OH with benzene, aniline and nitrobenzene using the flash photolysis/ resonance fluorescence technique is investigated.

Experimental

Figure 1 gives a schematic view of the apparatus. A detailed description may be found elsewhere[13]. The reaction cell is made of black anodized aluminium. The temperature of the cell can be adjusted using circulating fluids (e.g. methanol at low temperatures and water or oil at high temperatures). Photolysis of H_2O as a source of OH is performed by a flash lamp with a flash energy of 2 J at wavelengths $\lambda > 160$ nm using a quartz window for experiments with benzene and aniline, or $\lambda > 125$ nm using a CaF_2 window in the case of nitrobenzene. The concentrations of water and the reactant are controlled using the gas saturation technique[11-15]. The resonance fluorescence of OH is excited using a microwave discharge in a slowly flowing mixture of H_2O and Ar at 1 Torr pressure as a light source. The resonance fluorescence signal is detected by a photomultiplier (EMI 9789QB) using an interference filter at 308 nm. The photon pulses are amplified and converted to TTL-signals by an amplifier/discriminator and are transmitted to a multichannel scaler (Tracor Northern TN 1710). The multichannel scaler contains a microprocessor (DEC LSI 11/2). Using a 16-channel

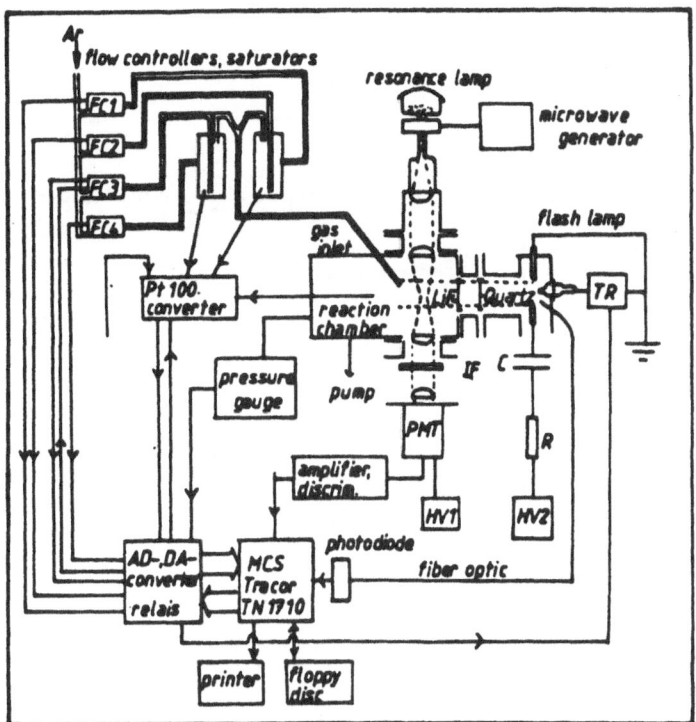

Figure 1:   Schematic view of the apparatus
FC 1-4: Flow controller; TR: Trigger equipment for the flash lamp;
PMT: Photomultiplier; MCS: Multichannel scaler; IF: Interference
filter; HV 1: High voltage supply for the photomultiplier;
HV 2: High voltage supply for the flash lamp

16-bit A/D-converter and an 8-bit D/A-converter, parameters like pressures,
temperatures and mass flows are transferred to the computer, and the concentration of the reactant is adjusted software controlled.
      Special algorithms allow to evaluate the parameters of the observed decay
curves. Using concentrations of H_2O of approximately 0.05 Torr in 100 Torr
inert gas (Argon) initial concentrations of OH of $2 \cdot 10^{10} cm^{-3}$ are obtained.
The decay of OH can be observed by about two orders of magnitude of the concentration, down to $10^8 cm^{-3}$.

Results
      The decay curves of OH are observed to be exponential in the presence of
benzene at T < 298 K and in the presence of aniline at T < 336 K. Above these
temperatures biexponential behaviour of the decay curves can be observed. The
decay curves observed in the presence of nitrobenzene in the whole temperature
region (259 K < T < 342 K) cannot be described as a sum of two exponential
decays.
      Figure 2 shows biexponential decay curves of the resonance fluorescence
signal of OH in the presence of benzene at two different concentrations. The
extrapolated initial intensity of the second, slower exponential function, I_{02},
decreases with increasing concentration, while the decay rate τ_2^{-1} is remaining

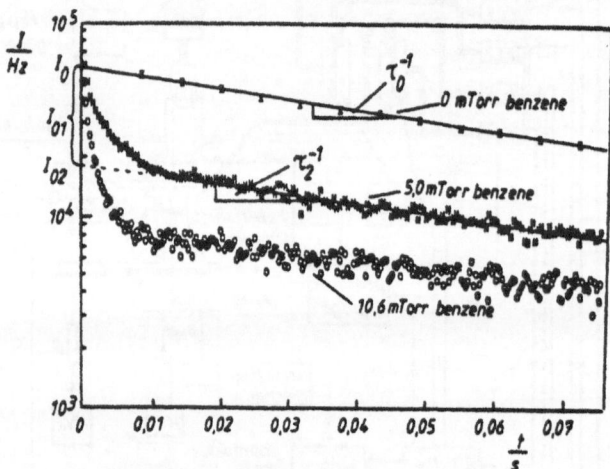

Figure 2: The effect of the concentration of the aromatic on the course of the biexponential decay curves (benzene as aromatic at 342 K, at a total pressure of 150 Torr and a partial pressure of H_2O of 0.05 Torr; explanation of the abbreviations see text)

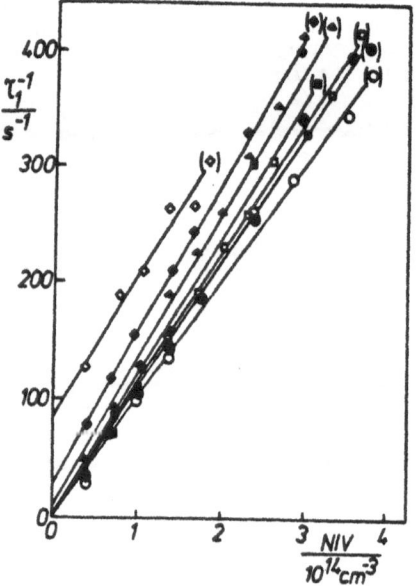

Figure 3:

Some plots of τ_1^{-1} against the benzene concentration at 150 Torr total pressure and 0.05 Torr H_2O,
(o) 239 K; (⊡) 277 K;
(•) 299 K; (■) 312 K;
(◆) 328 K; (◇) 342 K

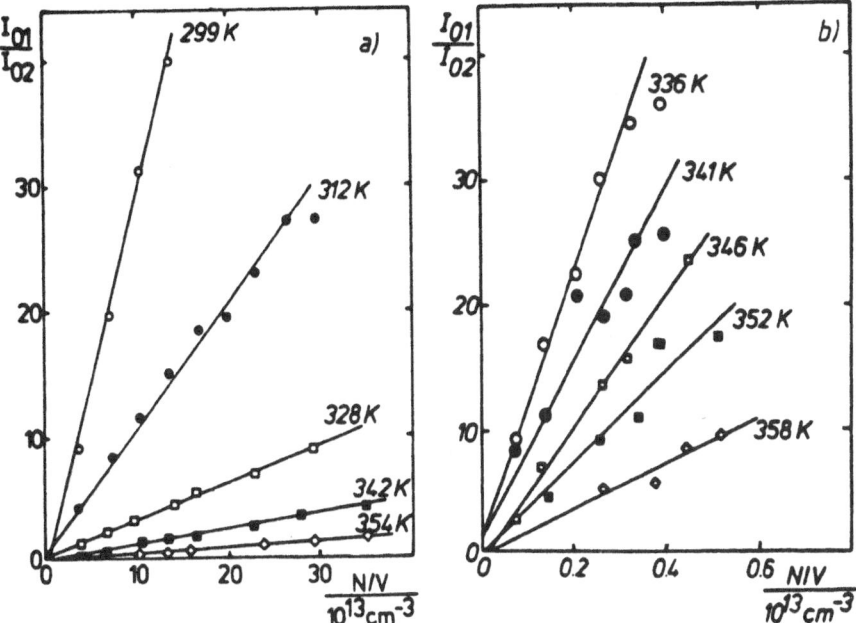

Figure 4:   Plots of the ratio I_{01}/I_{02} vs. the concentration of the aromatic
a) Benzene in the presence of 150 Torr of Ar and 0.05 Torr of H_2O
b) Aniline in the presence of 100 Torr of Ar and 0.05 Torr of H_2O

constant and approximately equal to τ_0^{-1}, the decay rate in the absence of
reactant. Using a computer program the exponential functions are separated
by calculating first the parameters I_{02} and τ_2^{-1} of the slower, second exponential decay from the final part of the decay curve. Subtraction of this
exponential function from the original curve yields the first, faster exponential decay with the parameters I_{01} and τ_1^{-1}.
    Figure 3 shows some plots of the decay rate τ_1^{-1} of the first, faster expo-
nential decay vs. the concentration of benzene at different temperatures. All
these plots are straight lines. The slope is not very much dependent on temperature, but the intercept increases with increasing temperature.
    Figures 4a and 4b show plots of the ratio I_{01}/I_{02} vs. the concentrations of
benzene and aniline. These plots are straight lines to a good approximation.
The slope decreases significantly with increasing temperature. For benzene the
decay rate τ_2^{-1} of the second exponential function is equal to the decay rate τ_0^{-1},
obtained in the absence of reactant. For aniline τ_2^{-1} depends on temperature and
on the concentration of aniline as shown in figure 5.

Discussion

    The observed temperature and pressure dependence of the reaction of OH with
aromatics agrees with the following reaction mechanism[1,7-16]:

OH + aromatic -> H_2O + products	(1)
OH + aromatic -> adduct	(2)
adduct -> OH + aromatic	(-2)

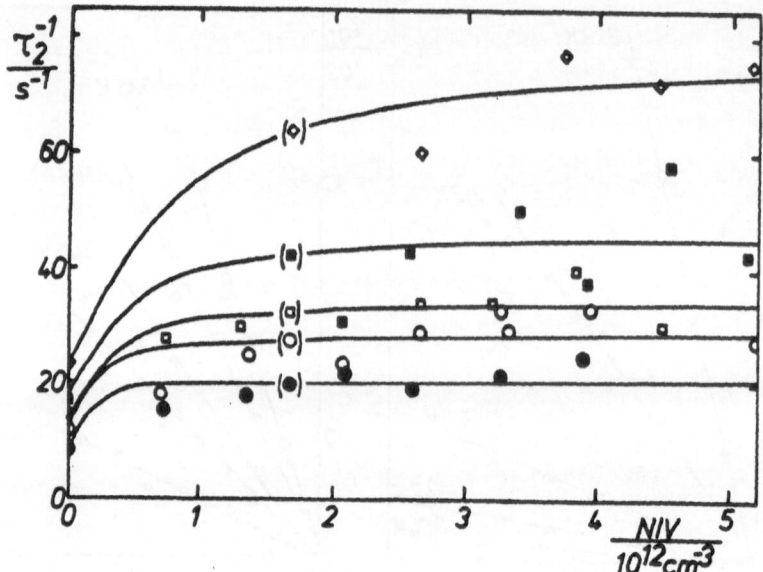

Figure 5: The effect of temperature and of the concentration of aniline on τ_2^{-1}.
The decay rate, τ_2^{-1}, increases with the concentration of the reactant leading to a constant value at higher concentrations. In addition τ_2^{-1} increases with temperature at a fixed concentration.
(●) 336 K; (○) 342 K; (□) 346 K; (■) 352 K; (◇) 359 K

In addition the diffusion of OH and the diffusion of the adduct (or other reaction paths not leading back to the reactants) have to be taken into account:

OH   -> diffusion   (3)
adduct   -> diffusion   (4)

The resulting system of differential equations has been solved by Wahner and Zetzsch[11]. The solution confirms that the decay of the concentration of OH can be described by the sum of two exponential functions:

$$[OH] = [OH]_{01} \exp(-t/\tau_1) + [OH]_{02} \exp(-t/\tau_2) \qquad \text{(eq. I)}$$

Since the resonance fluorescence intensity is proportional to the concentration of OH, we obtain for the ratio I_{01}/I_{02}:

$$\frac{[OH]_{01}}{[OH]_{02}} = \frac{I_{01}}{I_{02}} = \frac{(K_1 + \tau_2^{-1})(K_4 + \tau_1^{-1})}{K_2 K_3} \qquad \text{(eq. II)}$$

$$K_1 = -(k_1 + k_2) [\text{aromatic}] - k_3 \qquad \text{(eq. III)}$$

$$K_2 = k_{-2} \qquad \text{(eq. IV)}$$

$$K_3 = k_2 [\text{aromatic}] \qquad \text{(eq. V)}$$

$$K_4 = - k_{-2} - k_4 \qquad \text{(eq. VI)}$$

$$\tau_1^{-1} = - \frac{(K_1 + K_4)}{2} + \sqrt{\frac{(K_1 - K_4)^2}{4} + K_2 K_3} \qquad \text{(eq. VII)}$$

$$\tau_2^{-1} = - \frac{(K_1 + K_4)}{2} - \sqrt{\frac{(K_1 - K_4)^2}{4} + K_2 K_3} \qquad \text{(eq. VIII)}$$

Investigations by Perry et al.[1] and Tully et al.[8] show that abstraction of an H atom is negligible for the reaction with benzene at temperatures below 370 K (less than 10 % portion of reaction 1). It can be estimated that it might be negligible for the reaction with aniline as well but perhaps not negligible with nitrobenzene due to the slow rate constant k_2. The diffusion of OH, k_3, is of the order of 2-4 s^{-1}. The diffusion of the adduct, k_4, should be a quarter of this value because of the higher molecular weight. Neglecting reactions (1), (3) and (4) we obtain the approximation:

$$I_{01}/I_{02} = k_2/k_{-2} \text{ [aromatic]} \qquad \text{(eq. IX)}$$

$$\tau_1^{-1} = k_{-2} + k_2 \text{ [aromatic]} \qquad \text{(eq. X)}$$

In this simple case the decay rate of the second exponential, τ_2^{-1}, becomes zero, which is essentially equal to the decay rate, τ_0^{-1}, in the absence of reactant since the disappearance of OH by diffusion has been neglected. Using this approximation the rate constant, k_2, of the addition of OH to the aromatic can simply be determined from the slope of a plot of τ_1^{-1} vs. the concentration of the aromatic. The rate constant for the unimolecular decay of the adduct back to the reactants can be obtained from the slope of the plot of I_{01}/I_{02} vs. the reactant concentration if k_2 is known.

Figure 6 shows Arrhenius plots of the rate constants k_2 for benzene, aniline and nitrobenzene. Figure 7 shows Arrhenius plots of the rate constants, k_{-2}, of the unimolecular decay back to the reactants of the adducts benzene-OH and aniline-OH obtained using the approximation. The dependence of τ_2^{-1} on temperature and on the concentration of aniline (Fig. 5) can be explained by an additional reaction path of the adduct aniline-OH not leading back to the reactants. Although a value for the rate constant of this reaction (simulated by increasing k_4) can be calculated by the computer using an iterative method, no further information about the products of this reaction is obtained. An Arrhenius plot of these rate constants is given in figure 7.

The iterative calculation method using the exact solution can be applied to the determination of k_{-2} as well. The rate constant for the abstraction channel channel can be estimated to be 5·k_1/6, taking k_1 for benzene[1,8]. The decay rate, τ_0^{-1}, in the absence of reactant can be set equal to k_3 and the value for k_2 is obtained using the approximation (eq. X). The rate constants obtained by this method for the unimolecular decay of the aniline-OH adduct back to the reactants are also included in figure 7. They do not differ very much from the values determined using the approximation, but we prefer calculating the according Arrhenius expression with these values.

The reaction of OH with nitrobenzene is very slow compared to the reaction of OH with benzene and aniline. Hence, abstraction of H atoms might be a major portion of the total reaction. It can be estimated that below 342 K abstraction is still negligible (less than 20 %[8]). After subtracting the final part of the

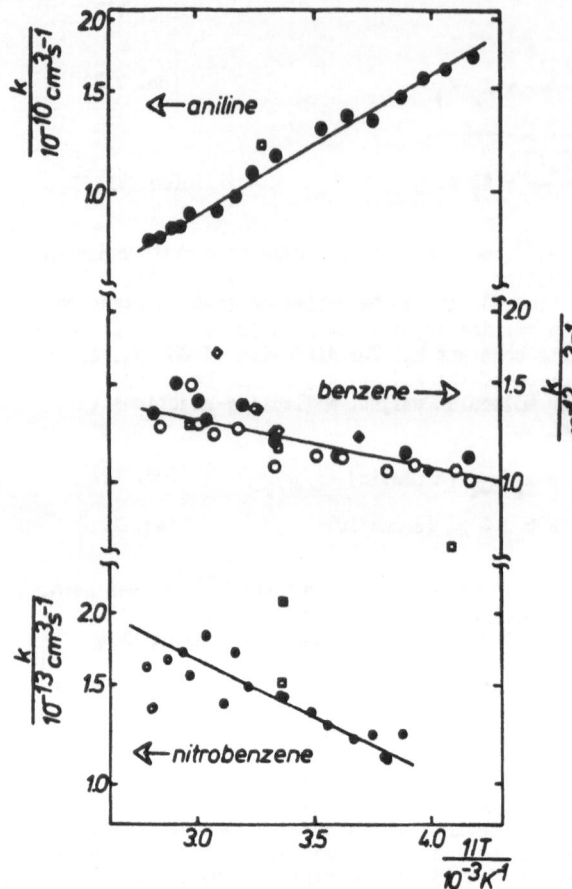

Figure 6:

Arrhenius plots of the rate constants of the addition of OH to the aromatic (k_2)

OH + aniline:
The value obtained by Rinke and Zetzsch[14] (□) at room temperature is in good agreement with our latest results (●).

OH + benzene:
(◇) Perry et al.[1]
(◆) Tully et al.[8]
(□) Lorenz and Zellner[10]
(○) this work at 100 Torr Ar
(●) this work at 150 Torr Ar

OH + nitrobenzene:
The rate constant obtained by Zetzsch[15] (■) at room temperature has been corrected (□) using the vapour pressure value of nitrobenzene determined in the present study by the technique described previously[11]. The rate constants obtained at higher temperatures (○) have not been taken into account to calculate the Arrhenius expression, since the portion of the abstraction can probably not be neglected (more than 20 %).

original decay curves there remains a nonexponential rest. This can probably be explained by a production of OH by the reaction of NO_2 (split off from nitrobenzene) with H atoms (produced by photolysis of H_2O). Further work is necessary to understand the mechanism. The preexponential factors and activation energies of the addition of OH to benzene, aniline and nitrobenzene, the unimolecular decay of the adducts back to the reactants and the unimolecular decay of the aniline-OH adduct not leading back to the reactants are listed below:

OH + aromatic -> adduct:

	$A/10^{-13} cm^3 s^{-1}$	$(E_A +/- 2\sigma)/kJ\ mol^{-1}$	
Benzene	23	1.6+/-0.5	239 K < T < 354 K
Aniline	170	-4.6+/-0.3	239 K < T < 359 K
Nitrobenzene	6	3.7+/-0.3	259 K < T < 342 K

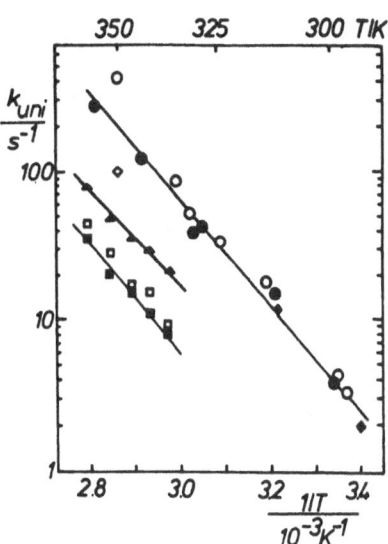

Figure 7: Arrhenius plots of the rate constants for the unimolecular decay
of the adducts benzene-OH and aniline-OH back to the reactants and
for the unimolecular decay of the adduct aniline-OH not leading back
to the reactants.
Benzene-OH -> Benzene + OH:
(◇) Tully et al. (estimated value); (◆) Wahner and Zetzsch[11];
(○) this work at 100 Torr Ar; (●) this work at 150 Torr Ar
Aniline-OH -> Aniline + OH:
(□) rate constants obtained using the approximation;
(■) rate constants obtained using the iterative method
Aniline-OH -> Products (no OH):
(▲) rate constants obtained using the iterative method

Adduct -> OH + aromatic:

	$A/10^{11} s^{-1}$	$(E_A +/- 2\sigma)/kJ\,mol^{-1}$	
Benzene	34	68+/-6	298 K < T < 354 K
Aniline	6	70+/-10	336 K < T < 359 K

Adduct -> products (no OH):

	$A/10^{10} s^{-1}$	$(E_A +/- 2\sigma)/kJ\,mol^{-1}$	
Aniline	8	62+/-10	336 K < T < 359 K

The knowledge of both, the activation energy of the addition and of the
unimolecular decay back to the reactants, allows us to calculate the bond
dissociation energies (BDE) of the OH radical to the adducts given below:

$$\text{Benzene-OH:} \quad \text{BDE} = (69+/-6)\ kJmol^{-1}$$
$$\text{Aniline-OH:} \quad \text{BDE} = (77+/-11)\ kJmol^{-1}$$

Although in agreement within the error limits of previous studies the BDE of
OH to benzene determined in the present study is slightly lower than estimated
values obtained by Perry et al. (BDE = (77+/-13) kJmol^{-1}) and Lorenz and
Zellner[10] (BDE = (77+/-6) kJmol^{-1}).
   No literature value for the BDE of OH to aniline is available for comparison.
The slightly higher BDE of aniline-OH than observed for benzene-OH is reflected
in the higher rate constant for the addition of OH to aniline in comparison to
benzene.

If the decomposition channel of the adduct aniline-OH not leading back to the reactants delivers NH_2 the activation energy, 62 kJmol^{-1}, would correspond to this bond in the adduct. An exact value for the bond energy could be determined, if the activation energy for the reaction of NH_2 with phenol were known.

References

1   R. A. Perry, R. Atkinson and J. N. Pitts Jr.
    J. Phys. Chem. 81, 296 (1977)

2   D. A. Hansen, R. Atkinson and J. N. Pitts Jr.
    J. Phys. Chem. 79, 1963 (1975)

3   G. J. Doyle, A. C. Lloyd, K. R. Darnall, A. M. Winer and J. N. Pitts Jr.
    Environ. Sci. Technol. 9, 237 (1975)

4   D. D. Davis, W. Bollinger and S. Fischer
    J. Phys. Chem. 79, 293 (1975)

5   A. R. Ravishankara, S. Wagner, S. Fischer, G. Smith, R. Schiff, R. T. Watson, G. Tesi and D. D. Davis
    Int. J. Chem. Kinet. 10, 783 (1978)

6   R. Atkinson, K. R. Darnall and J. N. Pitts Jr.
    J. Phys. Chem. 82, 2759 (1978)

7   R. A. Perry, R. Atkinson and J. N. Pitts Jr.
    J. Phys. Chem. 81, 1607 (1977)

8   F. P. Tully, A. R. Ravishankara, R. L. Thompson, J. M. Nicovich, R. C. Shah, V. M. Kreutter and P. H. Wine
    J. Phys. Chem. 85, 2262 (1981)

9   J. M. Nicovich, R. L. Thompson and A. R. Ravishankara
    J. Phys. Chem. 85, 2913 (1981)

10 K. Lorenz and R. Zellner
    Ber. Bunsenges. Phys. Chem. 87, 629 (1983)

11 A. Wahner and C. Zetzsch
    J. Phys. Chem. (in press)

12 F. Witte, A. Wahner and C. Zetzsch
    Bull. Soc. Chim. Belg. 92, 625 (1983)

13 F. Witte, Diplomarbeit, Ruhr-Universität Bochum 1983

14 M. Rinke and C. Zetzsch
    Ber. Bunsenges. Phys. Chem. (in press)

15 C. Zetzsch
    Presented at XVth Informal Conference on Photochemistry, Stanford, USA, June 1982

Acknowledgement
This work was supported by the Bundesminister für Forschung und Technologie and by the Fonds der Chemischen Industrie.
The experiments were performed at the Ruhr-Universität Bochum, and we thank Prof. H. Richtering for providing the laboratory space.

ABSOLUTE RATE CONSTANT MEASUREMENTS OF OH REACTIONS UNDER ATMOSPHERIC CONDITIONS BY LASER PHOTOLYSIS/DYE LASER FLUORESCENCE

V. SCHMIDT, Gui-Yun ZHU[*], K. H. BECKER and E. H. FINK
Physikalische Chemie – Fachbereich 9
Bergische Universität – Gesamthochschule Wuppertal
D-5600 Wuppertal 1, FRG

Summary

Absolute rate constants of OH reactions have been measured under tropospheric conditions by applying the sensitive pulsed dye laser induced fluorescence technique for monitoring OH radicals as a function of reaction time. The results of test measurements with simple alkanes and alkenes at a total pressure of 1 atm of air were found to be in good agreement with literature data. Detailed measurements were performed on the reaction of OH with acetylene at 295 K. Similar pressure dependences of the reaction rate were measured for inert buffer gases (He, Ar, N_2) between 8 and 1053 mbar with rate constants in the range $(0.9 - 8.0) \cdot 10^{-13}$ $cm^3 s^{-1}$. In the presence of O_2, the effective disappearance rate of OH was diminished at all total pressures by about a factor of four due to reproduction of OH by fast secondary processes. Besides the decay of OH, the formation and decay of vinoxy radicals as intermediates and the formation of glyoxal as a major product of the reaction in the presence of O_2 could be followed with the LIF technique.
For the reaction of SO_2 with OH at 1 atm total pressure a rate constant of $(8.6 \pm 2.0) \cdot 10^{-13}$ $cm^3 s^{-1}$ was measured in Ar as well as in synthetic air as diluent gases excluding a major effect of O_2 on the reaction rate.

1. Introduction

Rate constants of OH reactions under tropospheric conditions, i.e. in the presence of 1 atm of air, have mostly been measured by relative rate methods in smog chamber experiments /1/. Due to their restriction to low pressure conditions and/or diluent gases with low qenching probabilities towards electronically excited $OH(A^2\Sigma^+)$, most experimental techniques yielding absolute rate data such as flow tube studies or the classical flash photolysis/resonance fluorescence method can not be applied at tropospheric conditions. Methods involving OH detection by classical resonance absorption require high OH concentrations which complicates the data analysis and can induce interferences by secondary reactions.

The intention of the present work was to utilize the high sensitivity of the laser-induced fluorescence technique (LIF) for absolute rate constant measurements of OH reactions up to total pressures of 1 atm and in the presence of strongly quenching diluent gases like O_2 and H_2O. An experimental set-up is described which involves an excimer laser for the photolytic generation of OH radicals from H_2O_2 or HNO_3 parent molecules and a pulsed, frequency doubled, high energy dye laser for the measurement of relative OH concentrations by LIF using the $OH(A^2\Sigma^+ - X^2\pi)$ transition near 300 nm. Besides test measurements with alkanes and alkenes, studies of the reactions of OH with acetylene and with SO_2 are reported to illustrate the potential of the method.

[*] Guest scientist from the Department of Chemistry, University of Chandong, Jinang, People's Republic of China.

Acetylene is an atmospheric trace gas /2/. It is known to play an important role in most hydrocarbon combustion systems, especially in the formation of soot in fuel-rich flames /3/. In order to determine the rate constants and products of the elementary processes in acetylene combustion, C_2H_2 reactions with hydrogen and oxygen atoms and with OH radicals have been studied with different experimental techniques including flow tubes, crossed molecular beams, and flash pholysis/resonance fluorescence measurements. Unfortunately, the results often are inconsistent. For instance, large discrepancies still exist in the results of rate constant measurements for the reaction C_2H_2 + OH at room temperature, especially concerning the pressure dependence of the rate and the contribution of a bimolecular process (Fig. 1 and references cited therein). Moreover, there is little

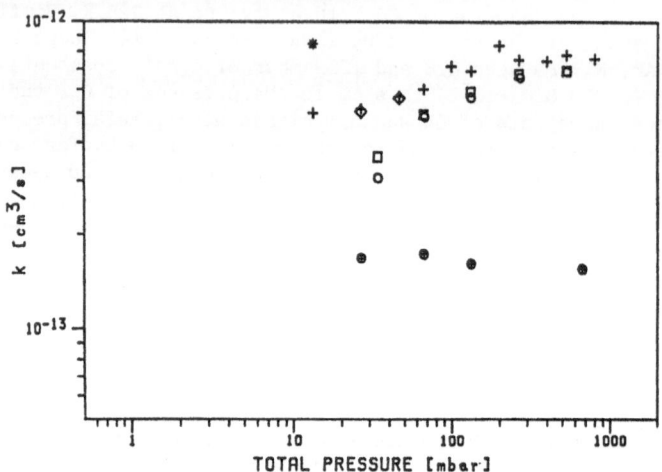

Fig. 1: Literature data on the pressure dependence of the rate constant of the reaction OH + C_2H_2. $*$, Smith & Zellner 1973 /5/; $\otimes$,M=He, Davis et al. 1975 /6/; $\square$,M=Ar, Perry et al. 1977 /7/; $\Diamond$,M=He, +,M=Ar, Michael et al. 1980 /8/; O,M=Ar, Perry & Williamson 1982 /9/.

known about the products of this reaction. Mainly three different elementary steps (1a-c) resulting from both, abstraction and addition reaction of C_2H_2 with OH, have been proposed /10/.

(1a) $\quad C_2H_2$ + OH $\longrightarrow$ C_2H + H_2O $\qquad \Delta_r H = \quad 26$ kJmol^{-1} /11/

(1b) $\qquad\qquad\qquad\quad \longrightarrow$ CH_2CO + H $\qquad \Delta_r H = - 98$ kJmol^{-1}

(1c) $\qquad\qquad\qquad\quad \longrightarrow$ CH_3 + CO $\qquad \Delta_r H = - 231$ kJmol^{-1}

Furthermore, in the presence of oxygen, three different channels (2a-c) for the reaction of the C_2H_2(OH) adduct with O_2 have been suggested /12/.

(2a) $\quad$ HC=CH(OH) + O_2 $\longrightarrow$ CHO + HCOOH

- 178 -

(2b) $HC=CH(OH) + O_2 \longrightarrow (CHO)_2 + OH$

(2c) $\longrightarrow CH_2CO + HO_2$

Support for the processes (2b) and (2c) comes from recent radiolysis studies of the OH radical induced oxidation of C_2H_2 in the presence of O_2 in
aqueous solution /13/.

In the present work we report on absolute measurements of the rate
constant of the C_2H_2 + OH reaction as a function of pressure both, in inert
gas diluent and in air, i.e. in the presence of O_2. The LIF technique was
used to follow the concentrations of OH, CH_2CHO (vinoxy), and $(CHO)_2$ as
well as that of OD in the reaction of C_2D_2 with OH. It could be proved that
vinoxy radicals are formed in the primary reaction (1d). In the presence of

(1d) $C_2H_2 + OH \xrightarrow{M} CH_2CHO$ $\Delta_rH = - 241 \text{ kJmol}^{-1}$ /14/.

O_2, the product formation of $(CHO)_2$ and fast regeneration of OH according
to the overall reaction (2b) was observed reducing the effective rate of OH
disappearance in the C_2H_2 + OH + O_2 system.

The rate constant of the reaction of SO_2 with OH was measured in 1 atm
of inert diluent (Ar) and of air, since relative rate measurements of OH
reactions with sulfur containing compounds performed in our laboratory suggested that the rate constant of this reaction might be influenced by O_2,
as has been shown for the reactions of OH with CS_2, CH_3SH, and CH_3SCH_3
/15-17/.

2. Experimental

Several experimental problems are connected with the application of
the LIF technique for measuring relative OH concentrations at 1 atm of air:
- Compared to low pressure and inert buffer gas conditions, the OH fluorescence yield is reduced by a factor of 100 to 1000 due to quenching of the
  electronically excited $OH(A^2\Sigma^+)$ state by O_2 and N_2. Moreover, the
  effective lifetime of the excited OH molecules is reduced to some nanose-
  conds. Thus the fluorescence light coincides in time with the laser pulses and can not be discriminated from scattered laser light by electronic
  techniques.
- At pressures above 0.1 atm and low OH concentration of 10^9-10^{11} OH/cm^3,
  the straylight due to Rayleigh scattering of the exciting laser light is
  much more intense than the OH fluorescence.
- At 1 atm of air, Raman scattering of the laser light by N_2 and O_2 results
  in first Stokes band intensities which also exceed that of the OH fluorescence.
These problems could be overcome by
- using a pulsed, high energy dye laser for the fluorescence excitation of
  OH,
- choosing different wavelengths for excitation and detection of the OH
  fluorescence, and
- using a double monochromator to suppress the Rayleigh-scattered light as
  well as the Raman bands of N_2 and O_2.
Figure 2 shows the experimental set-up and the energy levels of OH
involved in the LIF detection. The OH radicals were produced by pulse photolysis of H_2O_2 or HNO_3 with the unfocussed output of an excimer laser
(LAMBDA PHYSIK, Model EMG 102) at 193 nm (ArF) or 248 nm (KrF). In most

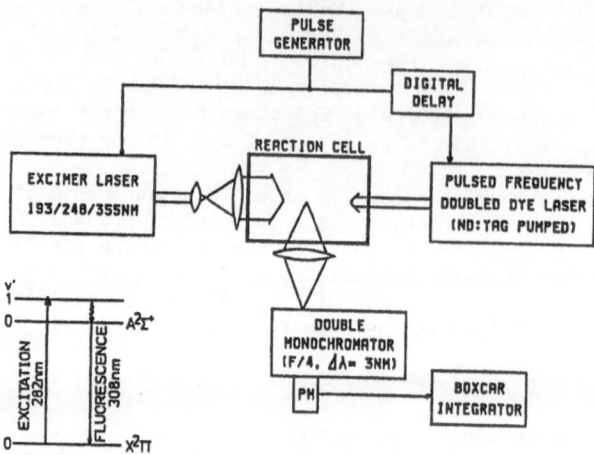

Fig. 2: Schematic view of the experimental set-up.

experiments reported here, H_2O_2 was used as parent compound for the generation of OH. With a pulse energy of ≈ 80 mJcm^{-2} of the KrF laser emission and H_2O_2 partial pressures of 0.1 - 1 μbar, the initial OH concentration is estimated to be in the range of $10^{10}-10^{11}$ OH/cm^3. The OH ground state concentration was probed as a function of reaction time by delayed fluorescence excitation on the nearly coinciding $Q_1(1)$ and $R_2(3)$ lines of the (1,0) band of the $A^2\Sigma^+ \leftarrow X^2\pi$ transition at 282 nm with the frequency doubled dye laser (QUANTEL, Model YG 481/TDL III) and monitoring the wavelength shifted emission of the (0,0) band at 308 nm. To suppress the intense background emission of Rayleigh-scattered laser light and to discriminate the OH fluorescence at 308 nm from the first Stokes Raman bands of O_2 and N_2 at 295 nm and 302 nm, respectively, a double monochromator (SPEX, Doublemate) with a bandwidth of 3-5 nm was used. The fluorescence light pulses were measured with a photomultiplier and averaged by a boxcar integrator. A typical fluorescence and straylight spectrum at 1 atm of air is shown in Fig. 3.

LIF detection of OD correspondingly was achieved by exciting the radicals via the $Q_1(1)$ or $Q_1(2)$ lines of the $A^2\Sigma^+$,v'=1 $\leftarrow X^2\pi$,v''=0 band near 288 nm and monitoring the (0,0) band fluorescence at 308 nm /18/. Vinoxy radicals were laser pumped within the (0',0',0') - (0",0",0") band at 347 nm, and the fluorescence was measured near 388 nm with a bandwidth of about 10 nm /19/. Glyoxal was excited in the (0,0) band of the $^1A_u - ^1A_g$ system at 455 nm, and the fluorescence was monitored through an interference band filter of 20 nm band width and peak transmission at 486 nm /20/.

The rates of OH reactions were measured in an excess of the reactants resulting in pseudo-first-order kinetics for the decay of OH. By properly choosing the reactant concentration, OH decay times of 10 μs to 20 ms could be attained. The OH decay curves were measured over 1-2 orders of magnitude by shifting the time delay between the excimer laser and the dye laser pulses. To reduce impurity effects and to avoid the piling-up of reaction

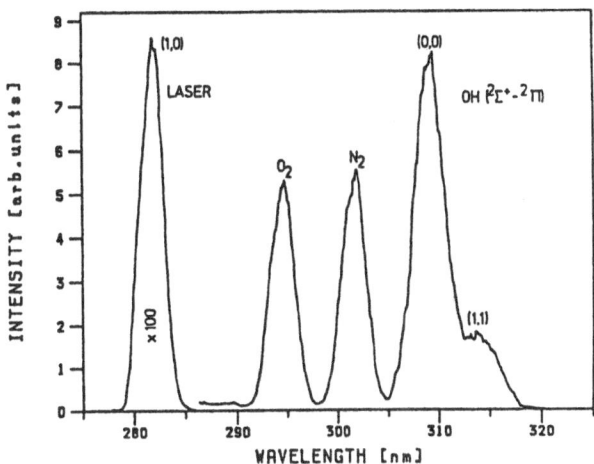

<u>Fig. 3:</u> OH fluorescence and straylight spectrum at a total pressure of
1 atm of air and a H_2O_2 pressure of ≈ 1 μbar.

products, a slow flow of carrier gas (He, Ar, N_2, or synthetic air) containing suitable amounts of the OH parent compound and the reactant under
study, was continuously passed through the stainless steel fluorescence
cell. The concentrations of the different gases were calculated from the
total pressure as measured with a BARATRON pressure transducer, and from
the total and the individual gas flow rates which were measured with calibrated flow meters. All experiments were performed at room temperature (295
$\pm$ 2 K).

Aqueous solutions of $\approx$ 90 % and 65 % concentration were used as
sources of the H_2O_2 and HNO_3 vapours, respectively, since low water vapour pressures ($\lesssim$ 10 μbar) showed no effect on any of the results. All
other gases were research grade and were used without further purification.

3. Results and discussion
    Test measurements with H_2, alkanes and alkenes at 1 atm of air in all
cases yielded exponential decays of the OH concentration which could be
followed over at least three 1/e decay times. Plots of the first-order decay constants as a function of reactant concentration gave straight lines
from the slopes of which the k_{OH} rate coefficients were deduced. The results are collected in Table I. The given error limits represent the total
uncertainty estimated from the standard deviation of the least squares line
fit to the $1/\tau_{eff}$ vs. reactant concentration data, the uncertainty in the
determination of the reactant pressure, and the possible influence of parallel OH reactions with impurities in the sample gases. Within these error
limits, the results are in agreement with literature data.

    In the reaction with acetylene, the decay rate of OH at constant C_2H_2
concentration was found to be strongly dependent on total pressure and on
the partial pressure of O_2. Figure 4 shows the pressure dependence of the
rate constant in inert buffer gases and in synthetic air. As would be expected, similar fall-off curves were found for He, Ar, and N_2 in the range

Table I: Rate constants of OH reactions at a total pressure of 1 atm of
air and T = 295 K.

Reactant	k_{OH} $(cm^3 s^{-1})$
H_2	$(5.8 \pm 1.0) \cdot 10^{-15}$
C_2H_6	$(2.2 \pm 0.3) \cdot 10^{-13}$
C_3H_8	$(1.0 \pm 0.15) \cdot 10^{-12}$
$n-C_4H_{10}$	$(2.3 \pm 0.3) \cdot 10^{-12}$
$i-C_4H_{10}$	$(1.9 \pm 0.25) \cdot 10^{-12}$
C_2H_4	$(7.3 \pm 1.0) \cdot 10^{-12}$
C_3H_6	$(2.2 \pm 0.4) \cdot 10^{-11}$

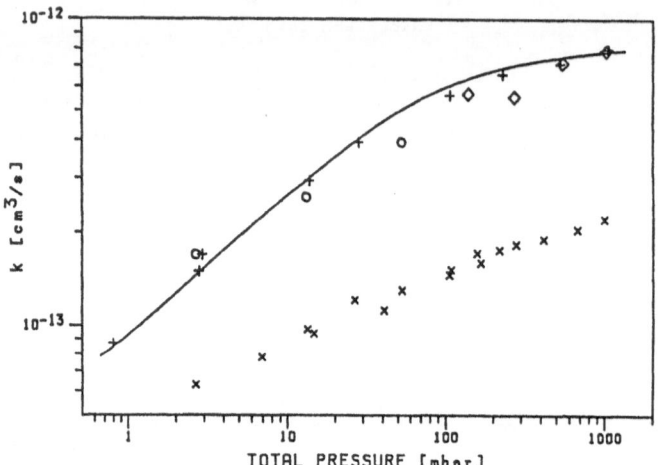

Fig. 4: Rate constants of the reaction OH + C_2H_2 as a function of total
pressure. ○, M=He; +, M=Ar; ◊, M=N_2; ×, M=synthetic air.

of 1 to 1000 mbar total pressure. The data for He are only slightly lower
than the closely coinciding values for Ar and N_2. The argon data are in
reasonable agreement with the results of Perry et al. /7,9/ but are substantially lower than the data of Michael et al. /8/, especially at pres-
sures below 40 mbar. The k_{OH} values of $\approx 8 \cdot 10^{-13}$ $cm^3 s^{-1}$ at high pressure
are in good agreement with recent results of relative rate measurements in
smog chambers /21,22/. In these relative rate measurements no difference in
the C_2H_2 consumption rate was found in Ar and in synthetic air as diluent
gases. Davis et al. /6/ have measured pressure independent rate constant
data for the reaction of OH with acetylene in He buffer gas between 25 and
650 mbar. Their results are close to our low pressure data and to the
effective rate constants for the decay of OH in the presence of O_2 (see
below), suggesting that their measurements have been falsified by O_2
impurities in the sample gases. Considering the pressure range for which

their data were obtained, their results are clearly in disagreement with both, our data and the two values measured by Michael et al. /8/.

The argon data were fitted to equation (E1) in which the first term, k_{bi}, accounts for a bimolecular abstraction reaction, and the second term is the semi-empirical expression for addition reactions developped by Troe /23/, $k_o(M)$ and k_∞ being the low and high pressure limiting values of the pressure dependent part of the rate constant, and F_c an adjustable broadening factor of the fall-off curve.

$$(E1) \quad k = k_{bi} + k_o/(1 + k_o(M)/k_\infty) \; F_c^{(1/(1 + (\log(k_o(M)/k_\infty))^2)}$$

Setting $F_c = 0.6$, the following parameters were obtained:

$$k_o = (2.5 \pm 0.3) \cdot 10^{-30} \; cm^6 s^{-1}, \qquad k = (8.3 \pm 0.8) \cdot 10^{-13} \; cm^3 s^{-1},$$

$$k_{bi} = (0.5 \pm 0.3) \cdot 10^{-13} \; cm^3 s^{-1}.$$

The curve shown in Fig. 4 is the least squares fit to the argon data from which these rate coefficients have been obtained. Although the contribution from the bimolecular term k_{bi} to the overall rate constant only becomes significant at low pressures, a much better least squares fit to the data in the intermediate pressure range was obtained by including k_{bi} in the analysis. However, considering the error limits of the data, this may be taken as an indication rather than a proof for the occurence of a bimolecular reaction channel in the C_2H_2 + OH reaction. Clearly, the bimolecular rate constants extrapolated by Michael et al. /8/ from their measurements are much too large. Careful measurements at total pressures below 10 mbar are necessary to definitely establish the existance and magnitude of a pressure independent pathway.

When small concentrations of O_2 were admixed to the sample gas at constant C_2H_2 and buffer gas pressures, the observed OH decay curves became biexponential (Fig. 5). The decay constant of the initial fast component

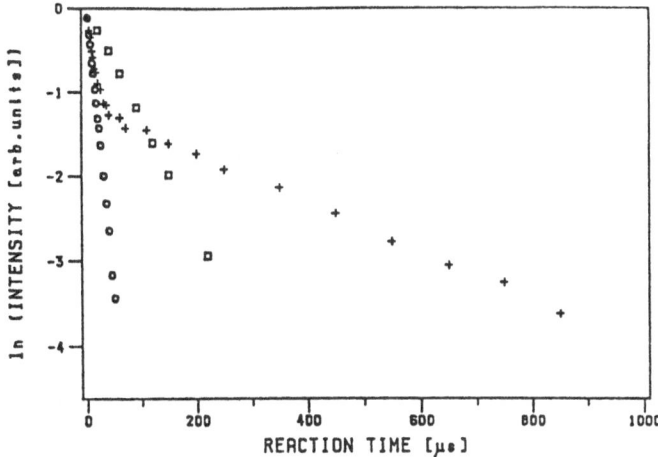

Fig. 5: Time decay of the OH concentration at a total pressure of 1 atm of Ar, a C_2H_2 pressure of 3.5 mbar, and O_2 partial pressures of O, 0 mbar; +, 0.15 mbar; □ , 36 mbar.

was nearly identical to that in the oxygen free system, whereas the time constant of the slow component was strongly dependent on the O_2 concentration approaching values about four times larger than the fast component at O_2 pressures above 25 mbar. With synthetic air as diluent gas, only the slow components of the OH decay curves could be measured which showed exponential form over more than three 1/e decay times. The effective rate constants increased from $0.6 \cdot 10^{-13}$ at 2.7 mbar to $2.2 \cdot 10^{-13}$ $cm^3 s^{-1}$ at 1055 mbar total pressure of air (Fig. 4).

The results suggest that in the oxygen containing systems OH radicals are regenerated by secondary reactions with O_2. This assumption was proved by studies of the reaction of OH with perdeuterated acetylene in the presence of O_2. In this system fast formation of OD radicals could be measured with the LIF technique, the maximum concentrations being in the same order of magnitude as those of OH. Moreover, in the C_2H_2 + OH system the formation and decay of vinoxy radicals could be directly observed (Fig. 6).

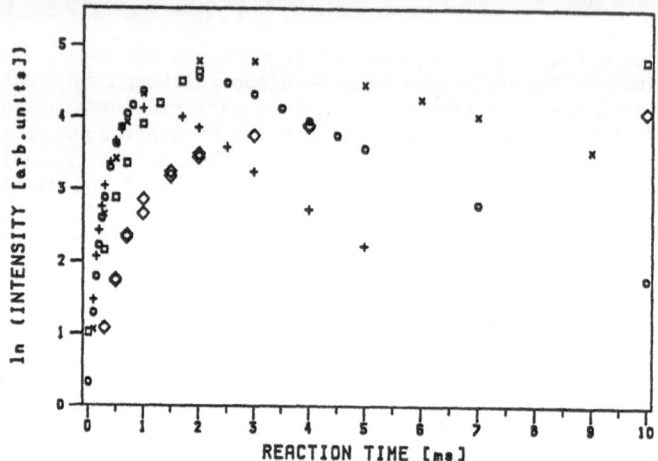

Fig. 6: CH_2CHO concentration vs. reaction time at a total pressure of 2.8 mbar (Ar), a C_2H_2 pressure of 1 mbar, and O_2 partial pressures of ◊, 0 µbar; □, 11 µbar; x, 57 µbar; O, 185 µbar; +, 742 µbar.

Addition of O_2 resulted in an increase of the CH_2CHO decay rate, and the formation of glyoxal as a stable product could be detected. The following reaction scheme is suggested to explain these results:

$$C_2H_2 + OH \underset{(1)}{\overset{M}{\rightleftharpoons}} (C_2H_2(OH)) \overset{M}{\underset{(5)}{\longrightarrow}} C_2H_3O \cdot \overset{+O_2}{\underset{(6)}{\longrightarrow}} \nearrow (CHO)_2 + OH$$
$$\searrow products$$

$k_{bi} \downarrow (9)$

$(2) \downarrow M$

$(8) \downarrow +X_i$

products

C_2H_3O

products

$(3) \downarrow M$

$CH_2CHO \overset{+O_2}{\underset{(4)}{\longrightarrow}} \nearrow (CHO)_2 + OH$
$$\searrow products$$

$(7) \downarrow +X_i$

products

- 184 -

Reactions (1)-(3) of this scheme have been proved by the detection of vinoxy radicals in the system. In separate experiments, reaction (4) has been studied and proved by generating vinoxy radicals in the presence of O_2 by ArF laser photolysis of methyl vinyl ether and measuring by LIF the decay of CH_2CHO and the formation of $(CHO)_2$ and OH. From the measured rate of CH_2CHO decay the rate constant of the reaction $CH_2CHO + O_2$, which recently has been measured by Gutman and Nelson /24/, could be confirmed.

Reaction steps (1)-(4), however, can not explain all of our results. Reactions (7)-(9) have to be included into the scheme to account for the reactions in the absence of O_2, and reactions (5) and (6) have to be postulated because the time constant of the regeneration of OH(OD) was found to be by about a factor of 5 faster than the formation rate of vinoxy which also was slower than the rate of OH consumption in the primary reaction (1). To further investigate the delayed formation of vinoxy, the reaction of oxygen atoms with ethylene was studied, which is known to yield vinoxy radicals in the primary reaction step /25/. In this reaction the formation of CH_2CHO was found to closely correspond to the primary reaction rate without any time delay. It is concluded that in the OH + C_2H_2 system two different still unknown intermediate adducts are formed. The species formed in reaction (2) could be the recently detected low lying $^2A'$ state of CH_2CHO /14,26/ or an isomeric form of C_2H_3O which is collisionally stabilized to form CH_2CHO in reaction (3). Another isomer formed in reaction (5) is assumed to undergo rapid reaction with O_2 accounting for the fast regeneration of OH(OD) observed in the system. The pressure independent branching ratio of the channels (2) and (5) must be in the order of 0.2 to account for our results. Two rough estimates of the $(CHO)_2$ yield in the C_2H_2 + OH + O_2 system have been obtained. In smog chamber experiments the total amount of $(CHO)_2$ product formed was found to be at least 50 % of the C_2H_2 consumed /27/. In the pulsed laser experiments, comparison of the initial OH and the final $(CHO)_2$ concentrations corresponded to a $(CHO)_2$ yield of > 100 % which is consistent with the assumption of a chain reaction with OH as the chain carrier and yielding $(CHO)_2$ as a major product. Clearly, more detailed studies including absolute measurements of the CH_2CHO, OD, and $(CHO)_2$ yields in the different reaction steps are necessary to fully understand the C_2H_2 + OH + O_2 reaction and the other systems in which the formation of vinoxy radicals and the regeneration of OH have been observed.

The reaction of OH with SO_2 was studied at a total pressure of 1 atm only. Figure 7 shows a Stern-Volmer plot of the effective OH decay constants as a function of SO_2 pressure in both, Ar and synthetic air as diluent gases. Within the error limits of the data points no difference is found for the two diluent gases excluding a major effect of O_2 on the reaction rate. From the slope of the least squares fit to all data points a rate coefficient of $(8.6 \pm 2.0) \cdot 10^{-13}$ $cm^3 s^{-1}$ is obtained which is in good agreement with recent literature data of Harris et al. /28/ $(1.0 \cdot 10^{-12}$ $cm^3 s^{-1}$), Paraskevopoulos et al. /29/ $(9.5 \pm 1.3) \cdot 10^{-13}$ $cm^3 s^{-1}$), and Izumi et al. /30/ $(1.22 \pm 0.13) \cdot 10^{-12}$ $cm^3 s^{-1}$).

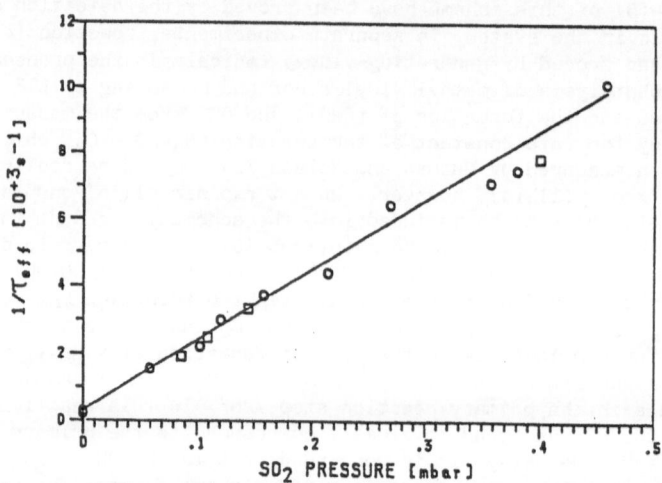

<u>Fig. 7</u>: Stern-Volmer plot for the reaction OH + SO_2 at a total pressure of 1 atm. □,M=Ar; ○,M=synthetic air.

Acknowledgement
Financial support of this work by the Bundesminister für Forschung und Technologie is gratefully acknowledged.

4. References

/1/ I. Barnes, V. Bastian, K. H. Becker, E. H. Fink and F. Zabel, Atm. Environ. 16 (1982) 545, and references cited therein.
/2/ J. Rudolph, D. H. Ehhalt and G. Gravenhorst, Proceedings of the First European Symp. "Physico-Chemical Behaviour of Atmospheric Pollutants" (B. Versino and H. Ott, Eds., ESCS-EEC-EAEC, Brussels-Luxembourg, 1980) p. 41.
/3/ J. Warnatz, H. Bockhorn, A. Möser and H. W. Wenz, Nineteenth Symp. (Intern.) on Combustion, The Combustion Institute, Pittsburgh, 1982, p. 197, and references cited therein.
/4/ J. E. Breen and G. P. Glass, Int. J. Chem. Kin. III (1970) 145.
/5/ I. W. M. Smith and R. Zellner, J. Chem. Soc. Faraday Trans. 2 69 (1973) 1617.
/6/ D. D. Davis, S. Fischer, R. Schiff, T. Watson and W. Bollinger, J. Chem. Phys. 63 (1975) 1707.
/7/ R. A. Perry, R. Atkinson and J. N. Pitts, Jr., J. Chem. Phys. 67 (1977) 5577.
/8/ J. V. Michael, D. F. Nava, R. P. Borkowski, W. A. Payne and L. J. Stief, J. Chem. Phys. 73 (1980) 6108.
/9/ R. A. Perry and D. Williamson, Chem. Phys. Letters 93 (1982) 331.
/10/ J. Vandooren and P. J. Van Tiggelen, Sixteenth Symp. (Intern.) on Combustion, The Combustion Institute, Pittsburgh, 1976, p. 1133, and references cited therein.

/11/ Most ΔH_f^O data were taken from H. Okabe: Photochemistry of Small Molecules (J. Wiley & Sons, New York, 1978) p. 375 ff.

/12/ J. M. Hay and D. Lyon, Proc. Roy. Soc. Lond. A 317 (1970) 1.

/13/ R. Anker: Hydroxylradikal-induzierte Oxidation von Azetylen in wäßriger Lösung in Gegenwart von molekularem Sauerstoff (Dissertation, Univ. Wien, 1980).

/14/ Calculated using $\Delta H_f^O(CH_2CHO) = 6$ kcalmol^{-1}, as estimated by M. Dupuis, J. J. Wendoloski and W. A. Lester, Jr., J. Chem. Phys. 76 (1982) 488.

/15/ B. M. R. Jones, J. P. Burrows, R. A. Cox and S. A. Penkett, Chem. Phys. Letters 88 (1982) 372.

/16/ I. Barnes, K. H. Becker, E. H. Fink, A. Reimer and F. Zabel, Int. J. Chem. Kin. 15 (1983) 631.

/17/ I. Barnes, V. Bastian, K. H. Becker and E. H. Fink, Third European Symp. "Physico-Chemical Behaviour of Atmospheric Pollutants", Varese (Italy), 10-12 April 1984 (submitted for publication in the Proceedings).

/18/ R. Englemann, Jr., J. Quant. Spectr. Radiat. Transfer 12 (1972) 1347.

/19/ G. Inoue and H. Akimoto, J. Chem. Phys. 74 (1981) 425.

/20/ W. Holzer and D. A. Ramsay, Can. J. Phys. 48 (1970) 1759.

/21/ I. Barnes, private communication.

/22/ W. Klöpffer, private communication.

/23/ J. Troe, J. Phys. Chem. 83 (1979) 114.

/24/ D. Gutman and H. H. Nelson, J. Phys. Chem. (1983).

/25/ K. Kleinermanns and A. C. Luntz, J. Phys. Chem. 85 (1981) 1966.

/26/ H. E. Hunziker, H. Kneppe and H. R. Wendt, J. Photochem. 17 (1981) 377.

/27/ I. Barnes, private communication.

/28/ G. W. Harris, R. Atkinson and J. N. Pitts, Jr., Chem. Phys. Letters 69 (1980) 378.

/29/ G. Paraskevopoulos, D. L. Singleton and R. S. Irwin, Chem. Phys. Letters 100 (1983) 83.

/30/ K. Izumi, M. Mozuochi, M. Yoshioka, K. Murano and T. Fukuyama, Envir. Sci. Technol. 18 (1984) 116.

PHOTOOXIDATION OF ACETALDEHYDE

G.N.BAGNALL and H.W.SIDEBOTTOM

Chemistry Department, University College, Dublin, Ireland.

Summary

The lifetime of electronically excited triplet acetaldehyde has been determined in the gas phase as a function of temperature and pressure. The results provide an estimate of the rate constant for the unimolecular decomposition of triplet acetaldehyde of $k_9 = 10^{10.9\pm0.4} \exp -(9.3 \pm0.6 kcalmol^{-1})/RT$ s^{-1}. Rate data for the quenching of the triplet state by molecular oxygen yields a value of $k_{O_2} = (3.96\pm0.02)\times10^9$ $Lmol^{-1}$ s^{-1} independent of temperature. Comparison of the present results with the available data from previous emission and photodecomposition studies indicates that the reaction of triplet acetaldehyde with oxygen occurs predominantly via an energy transfer pathway. It is suggested that the photo-induced atmospheric removal of acetaldehyde occurs predominantly via radical formation in a dissociative primary process.

1. INTRODUCTION

The photochemical oxidation of acetaldehyde has been shown to be an important process in the chemistry of photochemical smog (1). Also acetaldehyde is the major organic molecule precursor to the notorious peroxyacetyl nitrate (PAN) in the polluted troposphere (2). Thus, it is important to establish the rates of the primary photo-dissociative process which occur in acetaldehyde under atmospheric conditions. Such information is required for atmospheric models and to establish realistic emission control strategies. Despite numerous studies on the photolysis of acetaldehyde there is still some remaining uncertainty concerning the nature and extent of these primary dissociative reactions (3,4). The present work is concerned with reactions of the excited states of acetaldehyde and with the gas phase collisional quenching of triplet acetaldehyde by molecular oxygen. The results from the present work are compared with the available data from previous emission and photodecomposition studies.

2. EXPERIMENTAL

A conventional mercury-free greaseless vacuum system was used for all the experiments and pressure measurements were made with an MKS Baratron capacitance manometer. Acetaldehyde (Fluka, puriss) was degassed and purified by trap-to-trap vacuum distillation. The oxygen and carbon dioxide were both Matheson ultra high purity products and were used without further purification.

Triplet acetaldehyde molecules were produced via intersystem crossing when acetaldehyde was irradiated within the first allowed singlet band at 337.1nm using a nitrogen laser. The output energy available per pulse (~5ns) was about 2mJ and variations in the incident intensity were achieved by the use of neutral density filters. The T-shaped Pyrex reaction cell (length 10-cm, volume 516-ml) had front and side windows of diameter 5-cm and was contained in a blackened insulated housing. The temperature of the cell was varied from 0 to 80°C and was controlled to better than ±0.2°C. Phosphorescence from triplet acetaldehyde molecules was detected perpendicular to the incident laser beam with an E.M.I. 9659 QB photomultiplier fitted with a 370nm long-pass filter (Corning CS3-75). The intensity-time record of the phosphorescence decay was recorded on an oscilloscope, photographed and analysed.

3. RESULTS AND DISCUSSION

The lifetime of the first excited triplet state of acetaldehyde was determined over the temperature range 0-80°C in pure acetaldehyde and in mixtures with carbon dioxide. Phosphorescence decays were independent of laser intensity and were clearly exponential in character over at least three lifetimes. A considerable decrease in the lifetime with increasing temperature was observed and the data indicated appreciable pressure quenching of the triplet state. The decrease in lifetime with acetaldehyde concentration was linear and followed simple Stern-Volmer behaviour at all temperatures investigated. Addition of up to 1000torr of the inert diluent carbon dioxide to various concentrations of acetaldehyde did not affect the measured lifetimes within experimental error. As a check on the effect of a high number of laser pulses on a particular sample, and also on any possible effects caused by products arising from acetaldehyde decomposition, the lifetime of a particular sample was determined in the heating cycle and then remeasured as the system cooled. In each case the remeasured lifetime was in good agreement with the data obtained in the initial sequence, so that problems associated with the formation of decomposition products were not encountered.

The primary processes following excitation of acetaldehyde close to the origin of the first excited singlet state can be conveniently discussed in terms of the following mechanism (3-5):

$$A + h\nu \rightarrow {}^1A_0 \tag{1}$$
$${}^1A_0 \rightarrow A + h\nu_f \tag{2}$$
$$\rightarrow {}^3A_n \tag{3}$$
$$\rightarrow \text{Products} \tag{4}$$
$${}^3A_n \rightarrow \text{Products} \tag{5}$$
$${}^3A_n + M \rightarrow {}^3A_0 + M \tag{6}$$
$${}^3A_0 \rightarrow A + h\nu_p \tag{7}$$
$$\rightarrow A \tag{8}$$
$$\rightarrow \text{Products} \tag{9}$$
$${}^3A_0 + A \rightarrow A + A \tag{10}$$
$${}^3A_0 + M \rightarrow A + M \tag{11}$$

Excitation of acetaldehyde at 337.1nm generates the first excited singlet state with virtually no excess vibrational energy (5). At low excitation energies intersystem crossing in acetaldehyde singlets is the dominant relaxation mode (6-8). Thus, except for a small amount of fluorescence ($\phi_f \sim 2 \times 10^{-3}$(9)), excitation into low-lying levels of the first singlet state results in 100% conversion into the corresponding triplet state. Even at the lowest concentration studied in this work vibrational relaxation of the triplet can be assumed to be complete before the onset of phosphorescence. Consequently the present results are concerned with the kinetics of thermalized triplet molecules.

In terms of the above mechanism, the lifetime of thermalized triplet molecules will be given by the function:

$$1/_\tau \quad = \quad k_7 + k_8 + k_9 + k_{10}[A]$$

Since the lifetimes were found to be insensitive to the addition of high pressures of the inert diluent CO_2 collisionally induced intersystem crossing to ground state must be negligible, reaction (11). The results at 20°C give

$$1/_\tau(20^oC) = 3.3\times10^4 s^{-1} + 1.6\times10^7 [A] Lmol^{-1} s^{-1}$$

in excellent agreement with the values previously reported by Gandini and Hackett (10). The data are also consistent with the concentration dependent phosphorescence quantum yield results of Parmenter and Noyes (6) for excitation at 334nm. Since decomposition of acetaldehyde has been shown to decrease with acetaldehyde pressure (3), collisional quenching of triplets by ground state acetaldehyde, reaction(10) must be physical in nature.

The phosphorescence lifetime was found to be strongly temperature dependent, the triplet decay rate increasing with temperature. Both the unimolecular and collisional quenching triplet decay rates showed a strong temperature dependence. The latter rate constant increasing by about an order of magnitude over the temperature range 0-70°C. The rate constants for radiative decay, k_7, and for intersystem crossing from the triplet to the ground electronic state, k_8, are expected to show little temperature dependence (11). Hence, an increase in the unimolecular dissociation of triplet molecules must account for the experimentally observed increase in the first order triplet decay rate with increasing temperature. The data suggests that at temperatures below 10°C a limiting value exists for the triplet lifetime. At these temperatures the rate of unimolecular dissociation is negligible compared to the rates of radiative decay and intersystem crossing. Thus at temperatures below 10°C the observed triplet lifetime gives directly values of k_7+k_8. The lifetime data can thus be used to derive the rate constant for unimolecular decay of the acetaldehyde triplet state over the complete temperature range investigated. The data provide a value of

$$k_9 \quad = \quad 10^{10.9\pm0.4} \exp[-(9.3\pm0.6 kcalmol^{-1})/RT]s^{-1}.$$

The lack of electronic state correlation between the n,π* triplet state of acetaldehyde and the ground state photoproducts may account for the relatively low pre-exponential factor for the decomposition.

Values for the quantum yield of dissociation of the triplet state at limiting low pressures of CH_3CHO as a function of temperature may be calculated from the above data:

$$\phi_D(25^oC) = 0.39; \quad \phi_D(50^oC) = 0.46; \quad \phi_D(80^oC) = 0.79$$

At higher pressures of CH_3CHO ϕ_D will be significantly reduced due to physical self-quenching of the triplet by ground state molecules. These results are in line with the temperature dependence for the quantum yields of decomposition reported by Cundall and co-workers (8) from product analysis studies.

Three channels have been proposed for the primary photodecomposition of acetaldehyde:

$$CH_3CHO^* \rightarrow CH_3\cdot + H\dot{C}O \qquad (I)$$
$$\rightarrow CO + CH_4 \qquad (II)$$
$$\rightarrow CH_3\dot{C}O + H\cdot \qquad (III)$$

Calvert and co-workers (3,4), from product quantum yield data obtained over a range of excitation wavelengths (290-331nm), have shown that processes II and III are relatively unimportant for wavelengths greater than 290nm ($\phi_{II} < 0.07$; $\phi_{III} < 0.08$). The experimental evidence indicates that these processes arise from high vibrational levels of the excited singlet state. The quantum yields of product formation from the important decomposition pathway I were found to decrease with increasing pressure and excitation wavelength. It was suggested that process I originates from vibrationally rich triplet molecules formed via intersystem crossing from the initially formed excited singlet state. The low activation energy determined for triplet dissociation in this work indicates that the rate of decomposition will be a strong function of the triplet energy supporting the conclusions of Calvert and co-workers. It is possible to compare semi-quantitatively Calvert's low pressure decomposition quantum yield data at various wavelengths with the values of ϕ_D calculated from the temperature dependent unimolecular rate constants for triplet molecules determined in this work if the decomposition quantum yields depend only on the energy of the states formed in the absorption process. Thermalized triplet acetaldehyde can be placed about 82.2kcalmol^{-1} above the ground state (5), thus, for example, the decomposition yield resulting from excitation at 25°C and 313nm (91.2 kcalmol^{-1}), giving an energy of 11.2kcalmol^{-1} above the vibrationally relaxed triplet, should be comparable to that obtained from thermalized triplet molecules at an effective temperature of 758°K (12). Other effective temperatures for the wavelengths employed by Calvert and co-workers are:290nm, T = 1118°K; 300nm, T = 953°K; 320nm, T = 658°K; 331nm, T = 503°K The values of ϕ_I calculated from this work are compared with Calvert's low pressure values for ϕ_I in Table I.

Table I

Wavelength,nm	ϕ_I(4)	ϕ_I(Calculated)
290	0.89	0.97
300	0.93	0.94
313	0.92	0.80
320	0.47	0.62
331	0.05	0.17

The results are in reasonable agreement and provide support for the suggestion that decomposition arises from the triplet state.

The quenching rate constant determined for thermalized triplet molecules of $k_{O_2} = (3.96\pm0.02)\times10^9$Lmol^{-1}s^{-1} independent of temperature over the range 0 - 80°C is in excellent agreement with the value previously reported by Gill et al (13) at 25°C. Efficient quenching of acetaldehyde triplets by oxygen can be rationalized in terms of an exothermic energy transfer process which generates a singlet oxygen species. Intermolecular enhancement of spin-forbidden triplet state decay by molecular oxygen is thought to result from an electronic interaction between the triplet state and an oxygen molecule within a collision complex (14,15).

$$^3A_o + O_2(^3\Sigma_g^-) \rightarrow (^3A_o ---- ^3O_2) \begin{cases} \xrightarrow{k_\Sigma} A + O_2(^1\Sigma_g^+) \\ \qquad \Delta H = -43 \text{kcalmol}^{-1} \\ \xrightarrow{k_\Delta} A + O_2(^1\Delta_g) \\ \qquad \Delta H = -59 \text{kcalmol}^{-1} \end{cases}$$

It is also possible that the reaction of acetaldehyde triplets with oxygen may lead directly to products. Weaver et al. (16) have proposed that product formation via this direct reaction channel is a significant process for atmospheric photooxidation of acetaldehyde.

$$^3A + O_2 \rightarrow CH_3\cdot + HO_2\cdot + CO$$

It has been established that the phosphorescence of acetaldehyde is efficiently quenched by oxygen whereas the singlet emission is unaffected (6). Thus any interaction between oxygen and excited acetaldehyde molecules must involve only the triplet state. The quantum yield of vibrationally relaxed triplet molecules approaches unity following excitation at wavelengths greater than 320nm and at atmospheric pressure (10). Data from the present study indicates that quenching of vibrationally relaxed triplets by oxygen is sufficiently rapid that direct dissociation will be unimportant at pressures of oxygen above a few torr. From the above considerations, photolysis of acetaldehyde-oxygen mixtures at 331nm and at total pressures of about 200torr must mainly involve direct reaction of vibrational relaxed triplets with molecular oxygen. The relatively low quantum yield of photooxidation observed by Horowitz and Calvert (4) under these conditions, $\phi_{CO} = 0.03$ suggests that in the reaction of acetaldehyde triplets with molecular oxygen the energy transfer process is considerably more efficient than the direct reaction channel. Excitation of acetaldehyde at shorter wavelengths in the presence of oxygen results in significantly higher photooxidation yields (4,16); $\phi_{CO}(290nm) \sim 0.7$, $\phi_{CO}(313nm) \sim 0.2$. At high excitation energies there is competition between decomposition and collisional relaxation of the highly vibrationally excited triplet molecules formed via intersystem crossing for the singlet state.

$$^3A_n \rightarrow CH_3\cdot + H\dot{C}O$$
$$^3A_n + M \rightarrow ^3A_o + M$$

The rate parameters determined for decomposition of triplet molecules in this work suggest that dissociation is important relative to vibrational relaxation within the triplet manifold even at atmospheric pressure. Hence photooxidation of acetaldehyde at these shorter wavelengths must arise from the reaction of radicals formed by disscoiation of the initially formed vibrationally excited triplet state.

The above results indicate that the photo-induced removal of acetaldehyde under solar irradiation arises from absorption in the shorter wavelength region of the spectrum where direct decomposition into radicals is important.

4. REFERENCES

1. S.L.Kopozynski, A.P.Altshuller and F.D.Sutterfield, Environ.Sci.Technol. 8, 909 (1974).

2. K.L.Demerjian, J.A.Kerr and J.G.Calvert, Adv.Environ.Sci.Technol. , 4, 1 (1974).

3. A.Horowitz, C.J.Kershner and J.G.Calvert, J.Phys.Chem., 86, 3094 (1982).

4. A.Horowitz and J.G.Calvert, J.Phys.Chem., 86, 3105 (1982).

5. E.K.C.Lee and R.S.Lewis, Adv.Photochem., 12, 1 (1980).

6. C.S.Parmenter and W.A.Noyes,Jr., J.Am.Chem.Soc., 85, 416 (1963).

7. A.S.Archer, R.B.Cundall, G.B.Evans and T.F.Palmer, Proc.Roy.Soc.Lond., A333, 385 (1973).

8. A.S.Archer, R.B.Cundall and T.F.Palmer, Proc.Roy.Soc.Lond., A334, 411 (1973).

9. D.A.Hansen and E.K.C.Lee, J.Chem.Phys., 63, 3272 (1975).

10. A.Gandini and P.A.Hackett, Chem.Phys.Lett., 52, 107 (1977).

11. P.Avouris, W.M.Gelbart and M.A.El-Sayed, Chem.Rev., 77, 793 (1977).

12. S.W.Benson, F.R.Cruickshank, D.M.Golden, G.R.Haugen, H.E.O'Neal, A.S. Rogers, R.Shaw and R.Walsh, Chem.Rev., 69, 279 (1969); Thermodynamic properties of acetaldehyde.

13. R.J.Gill, W.D.Johnson and G.H.Atkinson, Chem.Phys., 58, 29 (1981).

14. D.R.Kearns, Chem.Rev., 71, 395 (1971).

15. O.L.J.Gijzeman, F.Kaufman and G.Porter, J.Chem.Soc.,Faraday Trans.II, 69, 727 (1973).

16. J.Weaver, J.Meagher and J.Heicklen, J.Photochem., 6, 111 (1976/1977).

A FTIR SPECTROSCOPIC STUDY OF THE PHOTOOXIDATION OF ACETALDEHYDE IN AIR.

Geert K. MOORTGAT and Robert D. McQUIGG.&
Max-Planck-Institut für Chemie, Air Chemistry Division
Saarstrasse 23, D6500 MAINZ, FGR.

Summary

Long path, Fourier transform infrared spectroscopy has
been employed to study the mechanism of acetaldehyde photooxidation in dilute gaseous mixtures of CH_3CHO and $Cl_2/$
CH_3CHO (ppm level) in air at 700 torr and 25 °C. The
concentrations of the reactants and products (CO, CO_2,
H_2CO, HCOOH, CH_3OH, CH_3OOH, CH_3COOH, CH_3COOOH, etc.) were
studied as a function of irradiation time and ratio of reactants. Black lamps (310-400 nm) and/or UVB fluorescent
lamps (280-360 nm) were used as irradiation source. The
experimental data were computer simulated based on earlier acetaldehyde quantum yield measurements and currently accepted reaction rate constants of elementary reactions. A mechanism explaining the experimentally obtained
results is presented. CH_3COOH is only generated in the
CH_3CHO photolysis, probably as product of the HO_2 + CH_3CHO
reaction. CH_3COOOH (peracetic acid) is only detected in
the Cl_2/CH_3CHO system. These results imply that CH_3CHO is
likely a precursor of CH_3COOH, which is a component of
natural precipitation in the troposphere.

1. INTRODUCTION

CH3CHO is formed in and enters the atmosphere in a variety of ways. It is generated in the oxidation of non-methane
hydrocarbons (C_2H_6, isoprene, etc.) both in the background
troposphere (1) and in photochemical smog (2). It is injected
directly into the troposphere as a product of incomplete combustion e.g. exhaust emissions from petrol, diesel or jet engines, waste burning and slash and burn agricultural methods (3).
It may also be produced as a bioproduct of fermentation and
large quantities of CH_3CHO are manufactured for use as an industrial solvent. Present atmospheric theories (4,5) ascribe
the importance of CH_3CHO to its being a precursor of $CH_3COO_2NO_2$
(peroxyacetylnitrate = PAN). PAN is thought to play an important role in the troposphere where it couples together the carbon and nitrogen cycles and, by virtue of its temperature dependent decomposition and low solubility, it can be transported
through the middle and upper troposphere, conseqently releasing

---

& Fulbright Visiting Scientist (1982-83) from Ohio Wesleyan
University, Delaware, Ohio 43015 USA.

its load of oxides of nitrogen. CH_3CHO is likely to be a precursor of CH_3COOH which is a component of natural precipitation and contributes to its acidity (6).

There have been several recent acetaldehyde photolysis
studies. Meyrahn et al.(7) measured the wavelength dependence
of CO and CH_4 quantum yield of the photodissociation of CH_3CHO
at atmospheric pressure. This work established the wavelength
dependence of the two photochemical pathways I and II:

$$CH_3CHO + h\nu \longrightarrow CH_3 + CHO \qquad (I)$$
$$\longrightarrow CH_4 + CO \qquad (II)$$
$$\longrightarrow CH_3CO + H \qquad (III)$$

From the CO_2 generated in the system they found evidence for a
small contribution from pathway III, in agreement with the low
pressure photolysis study by Calvert et al.(8,9). Using a modulated photolysis broad band apparatus, Burrows et al.(10)
performed a preliminary study of the photolysis of CH_3CHO in
the presence of O_2. Although CH_3O_2 was detected and its rate
of removal measured, any HO_2 formed in the system was lower
than the detection limit. In an earlier photolysis study of
CH_3CHO in air at 313 nm, Meagher et al. (11) observed an increase of the quantum yield of the products CO_2, CH_3COOOH (peracetic acid) and CH_3OH with increasing partial pressure of
CH_3CHO. Recent investigations by Meyrahn et al. (12) also
reveal CO_2 quantum yields larger than expected for pathway III,
which increases with CH_3CHO concentration indicating that a
chain reaction is taking place.

In the present paper, a long path FTIR spectroscopic study
of the photolysis of CH_3CHO under atmospheric conditions is reported. The aim of this study was to establish the mechanism
of the acetaldehyde photooxidation, in particular to investigate if the HO_2 radical, arising from the secondary reaction
$CHO + O_2 \longrightarrow HO_2 + CO$, reacts with CH_3CHO leading to the formation of CH_3COOH analogous to the reaction of HO_2 with HCHO
forming HCOOH (13-15). Computer simulation of the observed
photooxidation products was used to unravel the processes leading to the high CO_2 yields. In addition, the photolysis of
dilute Cl_2-CH_3CHO mixtures in air was studied, and the products
initiated by the $Cl + CH_3CHO$ reaction examined. The fate of
the expected radical product CH_3CO in air was established, and
computed simulated.

2. EXPERIMENTAL

The studies described in this work were carried out with
a BOMEM DA03.01 Fourier transform infrared spectrometer coupled to a long-path cell. A 1.1 m long, 12 cm i.d.pyrex or
quartz cell was equipped with KBr windows and internal "White"
mirrors, and surrounded with 6 near-uv fluorescent lamps (1.2 m
long) for photochemical experiments. The cell gave adequate
absorption for compounds in the ppm concentration range. A
Ge/KBr beamsplitter and liquid-helium cooled Cu-doped Germani-
um detector were used to obtain spectra from 400-4000 cm^{-1} at
1 cm^{-1} resolution.

Gas handling was accomplished with an all-glass vacuum

system with greaseless stopcocks and Datametrics absolute pressure gauges. The mixtures of pure CH_3CHO (50-100 ppm) in 700 torr synthetic air were photolysed with UVB fluorescent lamps (TL40W/12 Philips) and the Cl_2 (15-70 ppm)/CH_3CHO (30-60 ppm) mixtures with UV Blacklamps (TL40W/08,Philips). The reactor temperature was 25 ± 2 °C. CH_3CHO, CH_3OH, $HCOOH$, CH_3COOH (all Merck) were purified by vacuum distillation. CH_2O was obtained by heating paraformaldehyde under vacuum, and collecting the monomer at -78 °C. CH_3COOOH was prepared by reacting 30% H_2O_2 with acetic acid anhydride (16). 1.06 % Cl_2 gas in pure N_2 (Linde AG) and synthetic air (20 % O_2, 80% N_2, high grade, l´ Air Liquide) were used as received. The reference calibration curves for the various compounds were obtained from dilute mixtures (0-30 ppm) of the pure compound in 700 torr synthetic air at 25 °C.

3. RESULTS

Photooxidation of CH_3CHO-air mixtures gave CO, CO_2, CH_2O, $HCOOH$, CH_3OH, H_2O and CH_3COOH as readily identifiable products. Figure 1 shows the infrared absorbance spectrum of the products of the photolysis of 58 ppm CH_3CHO in 700 torr air after 50 min irradiation in the quartz cell, using 12 UVB lamps. This spectrum represents a difference spectrum since the spectrum of the reactant mixture before the photolysis has been subtracted. Characteristic absorption due to the products CO and CO_2 can readily be discerned. The upper trace of figure 2 shows the same spectrum after expansion in the range 750-1850 cm^{-1} and stripping the consumed CH_3CHO; the products H_2O, CH_3OH, $HCOOH$

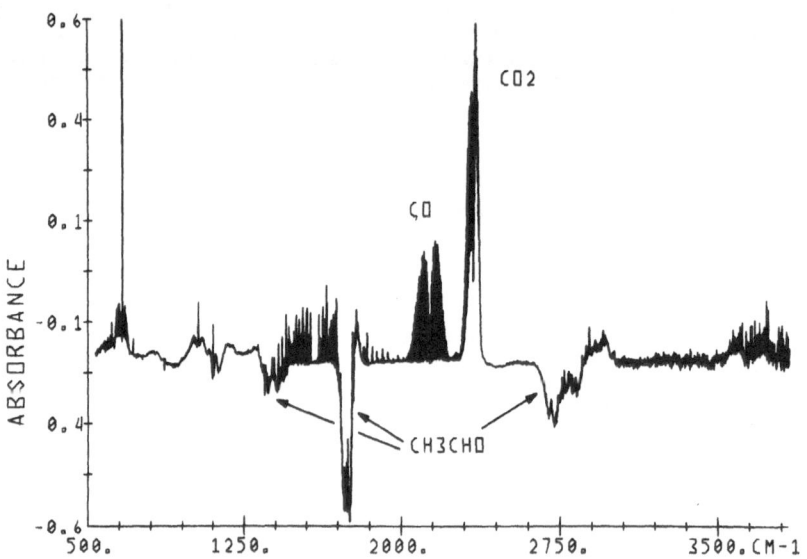

Figure 1. Spectral data from the photolysis of CH_3CHO (57.7 ppm) in air (700 torr), after 50 min.irradiation with 12 UVB lamps.

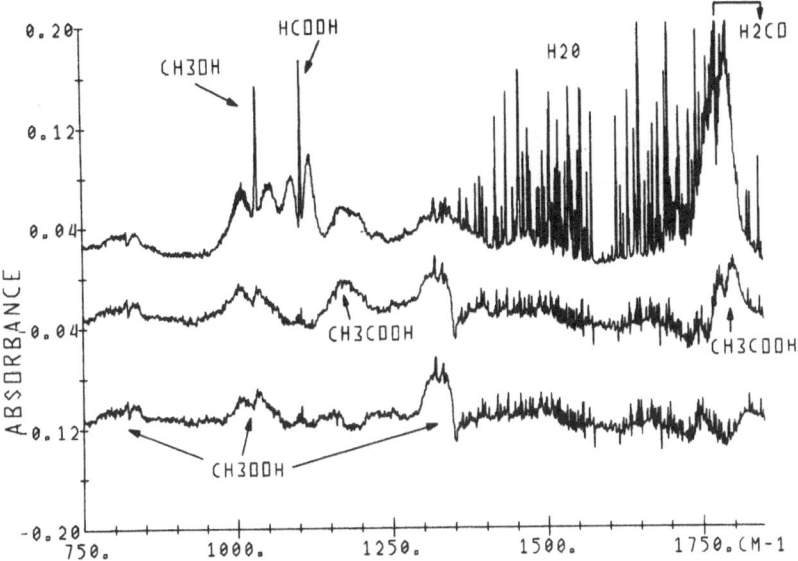

Figure 2. Spectra data from the 50 min photolysis of CH₃CHO (57.7 ppm) in air (700 torr) (see text)

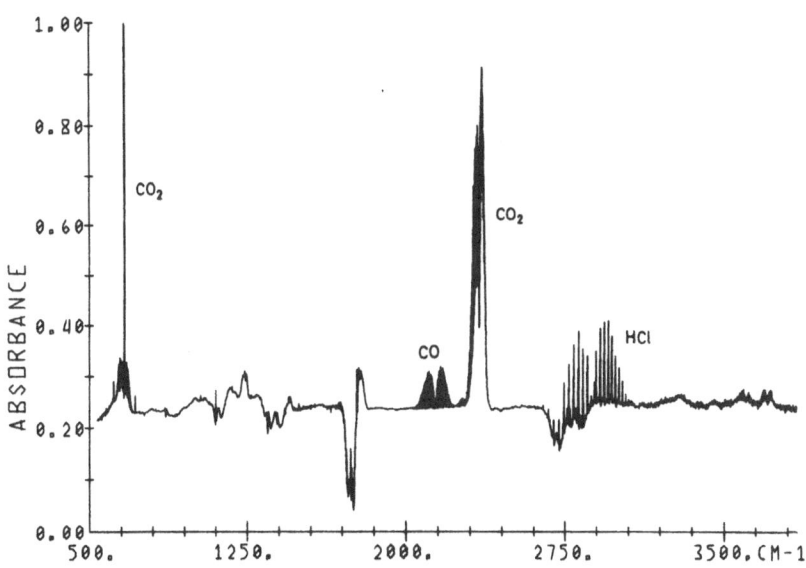

Figure 3. Spectral data from the photolysis of Cl₂ (38.0 ppm) and CH₃CHO (26.5 ppm) in air (701 torr), after 11 min photolysis.

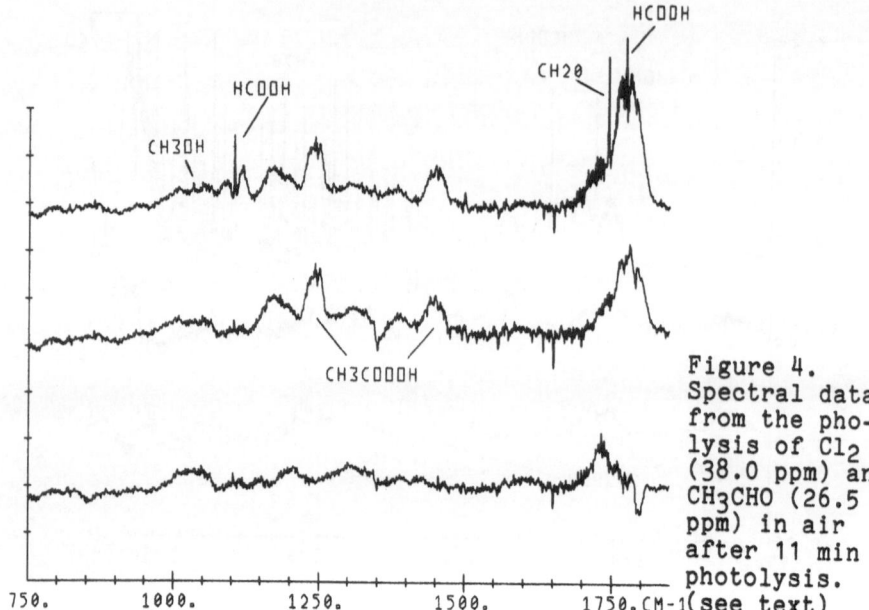

Figure 4.
Spectral data from the pho-lysis of Cl_2 (38.0 ppm) and CH_3CHO (26.5 ppm) in air after 11 min photolysis. (see text)

and H_2CO are readily identified. The second trace represents the spectrum after stripping CH_3OH, HCOOH, H_2O, and H_2CO using available reference spectra. CH_3COOH is identified by its characteristic absorption features at 1183.1 and 1790.1 cm^{-1}. After stripping CH_3COOH, the residual trace reveals distinctive features at 821.1, 1320.9, 1332.7 and 2963.6 (not shown) cm^{-1}, which are attributed to CH_3OOH, methyl hydroperoxide (17,18) Expected products, such as H_2O_2 and CH_3OOCH_3 were not identi-fied, probably due to loss on the wall.

The spectrum of the Cl-initiated photooxidation (after 11 min. irradiation) is shown in Figure 3, again after substrac-tion of the reactants at time zero. The top trace of Figure 4 represents the same spectrum after subtracting the reacted CH_3CHO in the range 750-1850 cm^{-1}. After stripping HCOOH, CH_3OH and CH_2O, distinct features of CH_3COOOH at 1244 and 1451 cm^{-1} are observed in the second trace. The bottom trace repre-sents the residual spectrum after subtracting the CH_3COOOH reference spectrum.

Figures 5 and 6 represent the concentration-time profiles of the identified products in the acetaldehyde photolysis and the Cl_2/CH_3CHO system, respectively. The experimental data were computer simulated, using a program called LARKIN (19). Table I lists the elementary reactions [1] to [39] with the appropriate reaction rate constants and photoconversion rates used to simulate the experimental results. Concentration-time profiles were obtained for a mixture containing initially 57.7 ppm in air. The product profiles are displayed in Figures 7a and 7b, where 1 x 10^{14} molec.cm^{-3} correspond to near 4.4 ppm.

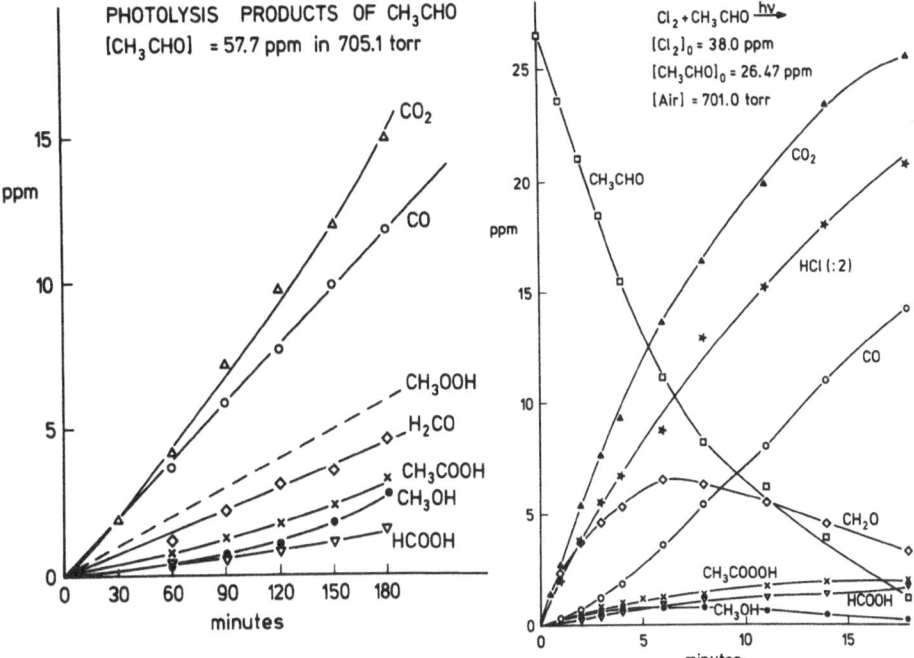

PHOTOLYSIS PRODUCTS OF CH₃CHO
[CH₃CHO] = 57.7 ppm in 705.1 torr

Figure 5.

Cl₂ + CH₃CHO →hv
[Cl₂]₀ = 38.0 ppm
[CH₃CHO]₀ = 26.47 ppm
[Air] = 701.0 torr

Figure 6.

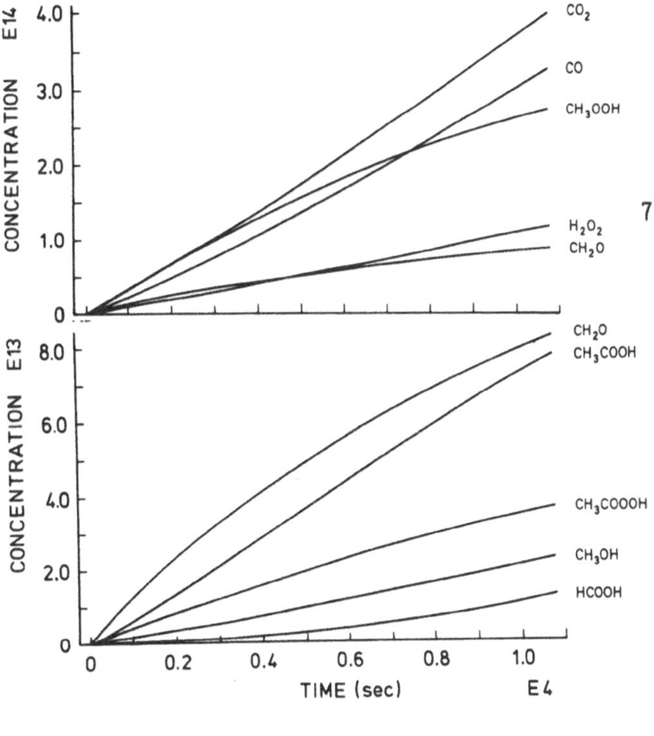

7 A

7 B

Figure 7
computer simula-
tion of the photo-
oxidation of 57.7
ppm CH₃CHO in 700
torr air, after 3h
photolysis.

- 199 -

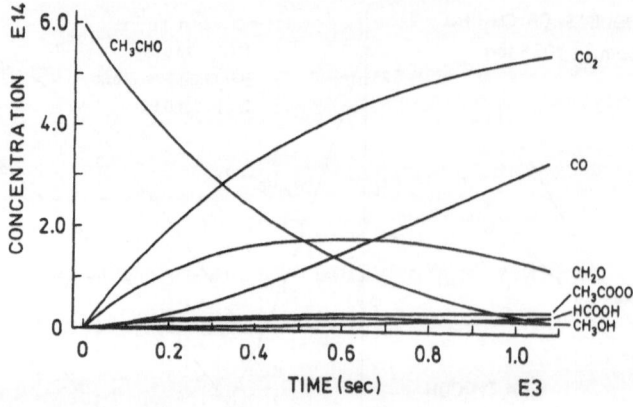

Figure 8. Computer simuila-tion of the Cl - atom initiated photooxidation of CH_3CHO (26.5 ppm in air (701 torr), after 11 min. photolysis

Additional reactions [40]-[48] were inserted to simulate the Cl-initiated photooxidation of 26.5 ppm CH_3CHO, and the obtai-ned concentration-time profiles are displayed in figure 8.

4. DISCUSSION

The mechanism leading to the formation of the products is initiated by the photolysis path I, leading to the formation of HO_2 and CH_3O_2, followed by the reaction sequence [3] to [6] producing mainly CH_3OH and CH_2O. Reaction [1] is the main source of CO, a small contribution being accounted for by the photolysis of CH_2O, which occurs with a photolysis rate about 20 times larger than that of acetaldehyde. For the distribu-tion of the product channels of the recombination of CH_3O_2 ra-dicals (3a:3b:3c), the data of Niki et al.(18) were preferred (0.60:0.32:0.08) over the data from Kan et al.(17) (0.52:0.40: 0.07), since they resulted in larger CH_3OH yields.

Recent studies of the photooxidation of CH_2O (13-15) have shown that the formation of HCOOH occurs via a chain reaction of CH_2O with HO_2, as is indicated in the reaction sequence [10] to [15]. Several values for k_{10} and the associated rever-se reaction k_{11} have been measured. Su et al. (14) obtained $k_{10} = 1 \times 10^{-14}$ cm3.molec^{-1}.s^{-1} and $k_{11} = 1.5$ s^{-1}, whereas Veyret et al. (15) deduced $k_{10} = 7.5 \times 10^{-14}$ cm3.molec^{-1}.s^{-1} and $k_{11} = 30$ s^{-1}. Two other independent measurements of the decay of HO_2 in the presence of H_2CO yielded $k_{10} = 2-4 \times 10^{-14}$ cm3.molec^{-1}.s^{-1} by Thrush and Tyndall (26), and $k_{10} = 5 \times 10^{-14}$ cm3.molec^{-1}.s^{-1} by Barnes et al. (27). Both sets of data (14-15) were inserted, the faster set of Veyret et al. (15) fitting better the observed HCOOH yield.

Using an analogous mechanism, the formation of CH_3COOH can be explained by the reaction of HO_2 radicals with CH_3CHO via an adduct complex A:

$$HO_2 + CH_3CHO \underset{17}{\overset{16}{\rightleftharpoons}} (HO_2CH_3CHO) \quad \text{complex A}$$

Table I.   * This work

	Reaction		Rate Constant	Ref
I	AcH + hv $\longrightarrow$	CH_3 + HCO	1.5 E-05	*
II	$\longrightarrow$	CH_4 + CO	7.0 E-07	*
1	$HCO + O_2 \longrightarrow$	HO_2 + CO	5.5 E-12	20
2	$CH_3 + O_2 + M$ ---	CH_3O_2 + M	3.9 E-32	20
3a	$2\ CH_3O_2 \longrightarrow$	$CH_3OH + CH_2O + O_2$	2.2 E-13	21
3b	$\longrightarrow$	$2\ CH_3O + O_2$	1.2 E-13	21
3c	$\longrightarrow$	$CH_3OOCH_3 + O_2$	3.0 E-14	21
4	$2\ HO_2 \longrightarrow$	$H_2O_2 + O_2$	2.8 E-12	20
5	$HO_2 + CH_3O_2 \longrightarrow$	$CH_3OOH + O_2$	1.3 E-12	17
6	$CH_3O + O_2 \longrightarrow$	$CH_2O + HO_2$	1.3 E-15	20
7	$CH_2O + hv \longrightarrow$	HCO + H	1.3 E-04	*
8	$\longrightarrow$	H_2 + CO	1.7 E-04	*
9	$H + O_2 + M \longrightarrow$	HO_2 + M	5.5 E-32	20
10	$HO_2 + CH_2O \longrightarrow$	$OOCH_2OH$	7.5 E-14	15
11	$OOCH_2OH \longrightarrow$	$HO_2 + CH_2O$	30	15
12	$OOCH_2OH + HO_2 \longrightarrow$	$HO_2CH_2OH + O_2$	2.0 E-12	14
13	$2\ OOCH_2OH \longrightarrow$	$2\ OCH_2OH + O_2$	2.3 E-13	15
14	$OCH_2OH + O_2 \longrightarrow$	HO_2 + HCOOH	3.5 E-14	15
15	$HO_2CH_2OH + hv,wall \longrightarrow$	H_2O + HCOOH	1.0 E-03	*
16	$HO_2 + AcH \longrightarrow$	$HAcHO_2$	1.0 E-15	*
17	$HAcHO_2 \longrightarrow$	HO_2 + AcH	1.5	*
18	$2\ HAcHO_2 \longrightarrow$	$2\ HAcHO$	2.3 E-13	*
19	$HAcHO_2 + HO_2 \longrightarrow$	$HAcHO_2H + O_2$	2.5 E-12	*
20	$HAcHO + O_2 \longrightarrow$	$CH_3COOH + HO_2$	3.5 E-14	*
21	$HAcHO_2H + hv,wall \longrightarrow$	$CH_3COOH + H_2O$	1.0 E-02	*
22	$CH_3O_2 + AcH \longrightarrow$	$CH_3OOH + CH_3CO$	3.4 E-16	*
23	$CH_3O + AcH \longrightarrow$	$CH_3CO + CH_3OH$	2.0 E-13	*
24	$CH_3CO + O_2 \longrightarrow$	CH_3CO_3	2.0 E-12	9
25	$2\ CH_3CO_3 \longrightarrow$	$2\ CH_3CO_2 + O_2$	2.5 E-12	22
26	$CH_3CO_2 \longrightarrow$	$CH_3 + CO_2$	2.2 E+10	23
27	$CH_3CO_3 + AcH \longrightarrow$	$CH_3COOOH + CH_3CO$	5.2 E-17	11
28	$CH_3CO_3 + CH_3O_2 \longrightarrow$	$CH_3CO_2 + CH_3O + O_2$	3.0 E-12	22
29	$HO_2 + CH_3CO_3 \longrightarrow$	$CH_3COOOH + O_2$	1.8 E-14	24
30	$2\ CH_3O \longrightarrow$	$CH_3OH + CH_2O$	2.5 E-11	17
31	$\longrightarrow$	CH_3OOCH_3	6.6 E-12	17
32	$CH_3O + CH_3O_2 \longrightarrow$	$CH_3OOH + CH_2O$	6.8 E-14	17
33	$CH_3O + CH_2O \longrightarrow$	$CH_3OH + HCO$	1.0 E-15	17
34	$CH_3OOH + hv \longrightarrow$	$CH_3O + OH$	3.0 E-05	24
35	$H_2O_2 + hv \longrightarrow$	2 OH	5.0 E-05	24
36	$OH + AcH \longrightarrow$	$CH_3CO + H_2O$	1.6 E-11	21
37	$OH + CO \longrightarrow$	$CO_2 + H$	2.7 E-13	20
38	$OH + CH_3OOH \longrightarrow$	$CH_2O + OH + H_2O$	4.3 E-12	25
39	$\longrightarrow$	$CH_3O_2 + H_2O$	7.5 E-14	25
40	$Cl_2 \longrightarrow$	2 Cl	6.4 E-04	*
41	$Cl + AcH \longrightarrow$	$HCl + CH_3CO$	8.0 E-11	*
42	$Cl + H_2CO \longrightarrow$	HCl + HCO	7.3 E-11	20
43	$Cl + CH_3OH \longrightarrow$	$HCl + CH_2OH$	6.3 E-11	20
44	$Cl + HCOOH \longrightarrow$	$HCl + HCO_2$	5.0 E-13	*
45	$Cl + H_2O_2 \longrightarrow$	$HCl + HO_2$	4.1 E-13	20
46	$Cl + CH_3OOH \longrightarrow$	$CH_2O + OH + HCl$	2.0 E-10	25
47	$CH_2OH + O_2 \longrightarrow$	$CH_2O + HO_2$	2.0 E-12	20
48	$HCO_2 + O_2 \longrightarrow$	$HO_2 + CO_2$	2.5 E+08	23

Complex A can rearrange intramolecularly to form a peroxy radical B:

$$CH_3 - C\begin{smallmatrix} O \\ H \quad OOH \end{smallmatrix} \quad (A) \quad \longrightarrow \quad CH_3 - C\begin{smallmatrix} OH \\ H \quad OO. \end{smallmatrix} \quad (B)$$

Two peroxy radicals B can recombine to oxy radicals (C), which can react with O_2 to form CH_3COOH

$$2 \ B \quad \longrightarrow \quad 2 \quad CH_3 - C\begin{smallmatrix} OH \\ H \quad O. \end{smallmatrix} \ (C) \quad + \ O_2 \quad [18]$$

$$C + O_2 \quad \longrightarrow \quad CH_3COOH + HO_2 \qquad [20]$$

The peroxy radical (B) can also react with HO_2 to form an acid $CH_3 - CH(OH)(OOH)$ which is photolysed or reacts on the wall to CH_3COOH and H_2O. None of the proposed intermediates have been identified so far by this FTIR work, however this mechanism provides a source of CH_3COOH in air as is observed recently (6).

Barnes et al.(27) also measured an overall decay rate of HO_2 in the presence of CH_3CHO, $k_{16} = 1 \times 10^{-17}$ $cm^3.molec^{-1}.s^{-1}$; however, since HO_2 is regenerated in reaction [20], their system may not have been sensitive to reaction [16]. On the other hand, Barnes et al.(27) obtained the decay rate for HO_2 in the presence of glyoxal, methylglyoxal and dimethylglyoxal as 1×10^{-15}, 2×10^{-15} and 4×10^{-16} $cm^3.molec^{-1}.s^{-1}$ respectively. These results would imply that the aldehydic group is attacked by HO_2 with a rate near 1×10^{-15} $cm^3.molec^{-1}.s^{-1}$, a value which was used in the computer simulation. For reactions [18] to [20], similar rate constants were taken as for the analogous reactions [12] to [14]; k_{17} and k_{21} were adapted to simulate the CH_3COOH yield.

The high CO_2 yield in the CH_3CHO photooxidation system is surprising, as was observed in earlier photolysis studies (11-12). It is generally accepted that the oxidation of CH_3CO radicals occurs via CH_3CO_3 to produce CH_3 and CO_2, (reactions 24 to 26). Although CH_3CO is produced as primary photolysis product, yield is only a few % (8,9,12) and cannot account for the high CO_2 yield. On the other hand, many radicals, such as CH_3CO, HO_2, CH_3O and OH, may abstract an H-atom from CH_3CHO to produce a CH_3CO radical. In this model calculation we set the photolysis path III to zero in order to investigate which radical is responsible for CO_2 formation. Reaction [23]

$$CH_3O + CH_3CHO \longrightarrow CH_3OH + CH_3CO, (\Delta H_r = -17.6 \text{ kcal/m}) \quad [23]$$

predicts, besides the large CO_2 yield, a large CH_3OH yield, which is not observed. The products of the reactions

$$HO_2 + CH_3CHO \longrightarrow CH_3CO + H_2O_2 , (\Delta H_r = -2.2 \pm 1 \text{ kcal/m}) \quad [16a]$$
$$CH_3O_2 + CH_3CHO \longrightarrow CH_3CO + CH_3OOH, (\Delta H_r = -1.2 \pm 2 \text{ kcal/m}) \quad [22]$$

are H_2O_2 and CH_3OOH, repectively. The relative yield of these peroxides should give information on the participation of reactions [16a] and [22]. H_2O_2 and CH_3OOH are also reaction products in the initial reaction scheme [4] to [5]. H_2O_2 was not detected, although its characteristic infrared absorption feature near 1265 cm^{-1} is known (28). It is not possible to determine why hydrogen peroxide is not observed in our system.

One feasible explanation is heterogeneous decomposition on the wall; such a process would yield H_2O, a product which was observed in our study.

In order to simulate the CO_2 yield in view of the available data, a value for $k_{22} = 3.4 \times 10^{-16}$ $cm^3.molec^{-1}.s^{-1}$ was taken, which is identical with the value Kan et al.(17) used for the H-abstraction of CH_2O by CH_3O_2. This value resulted in a CO_2 yield near identical with the experimental data. Reaction [16a] was not included in the present model, since it was assumed that HO_2 attaches to CH_3CHO via reaction [16]. As is seen in Fig 7a, the computer simulation predicts large yields of CH_3OOH. The model only considers the photodecomposition of CH_3OOH and H_2O_2 (reactions 34 and 35), with photodecomposition rates respectively 2x and 3x faster than the CH_3CHO photolysis rates, according to the estimates from Stockwell and Calvert (24). It was also observed that the lower value $k_5 = 1.4 \times 10^{-12}$, as determined by Kan et al.(17) fitted our data better than the higher value $k_5 = 6.5 \times 10^{-12}$ $cm^3.molec^{-1}.s^{-1}$ determined by Cox and Tyndall (29). Niki et al. (30) found a similar value 1.5×10^{-12} $cm^3.molec^{-1}.s^{-1}$ for the reaction between HO_2 and $C_2H_5O_2$ radicals.

Finally, the simulation of the Cl atom initiated oxidation of CH_3CHO, as shown in Figure 8 is generally in good agreement with the observed experimental product profiles, displayed in Figure 6. The rate of the reaction Cl + CH_3CHO, $k_{41} = 8.0 \times 10^{-11}$ $cm^3.molec^{-1}.s^{-1}$ was measured relative to the Cl + CH_3OH reaction, $k_{41}/k_{43} = 1.3$ in an independent study(31), in good agreement with a relative value listed by Niki et al. (30). For the rate constant k_{44} for Cl + HCOOH, a value $k_{44} = 5.0 \times^{-13}$ $cm^3.molec^{-1}.s^{-1}$ was used similar to the rate of the OH + HCOOH reaction, $k= 3.5 \times 10^{-13}.molec^{-1}.s^{-1}$ (30). Peracetic acid was detected, in agreement with recent measurements of Hanst and Gay (33), who measured peracetic acid in the Cl atom initiated oxidation of a various of hydrocarbons at ppm level in air, in the absence of nitrogen oxides. Reaction [27] is the main step leading to peracetic acid.

References.

1. J.G. Calvert, Nato Advanced Study Institute Series, Ed. H.W. Georgii and W. Jaeschke, 425 (1982).
2. K.L. Demerjian, J.A. Kerr, J.G. Calvert, Adv. Environ. Sci. Technol., 4, 1 (1974).
3. R.J. Gelinas and P.D. Skewes-Cox, J. Phys. Chem., 81, 2468 (1977).
4 P.J. Crutzen, "The major biogeochemical cycles and their interactions" Ed. B. Bolin, R.B. Cook, SCOPE, 67 (1983).
5. R.A. Cox and R.G. Derwent, Specialist Periodicals Review, 30, vol 4 (1981).
6. G.A. Dawson, J.C. Farmer and J.L. Moyers, Geophys. Res. Lett., 7, 725 (1980).
7. H. Meyrahn, G.K. Moortgat and P. Warneck, Proc. 5th Biann. Coll. of SFB-73, Verlag Fried. Viehweg & S., 1982.

8. A. Horowitz and J.G. Calvert, J. Phys. Chem., 86, 3105 (1982).
9. A. Horowitz, C.J Kershner and J.G. Calvert, J. Phys. Chem., 86, 3094 (1982).
10. J.P. Burrows, R.A. Cox, R. Patrick, private communication.
11. J. Weaver, J. Meagher and J. Heicklen, J. Photochem., 6, 111 (1976/77).
12. H. Meyrahn, G.K. Moortgat and P. Warneck, in preparation.
13. F. Su, J.G. Calvert, J.H. Shaw, H. Niki, P.D. Maker, C.M. Savage and L.P. Breitenbach, Chem. Phys. Lett., 65, 221 (1979).
14. F. Su, J.G. Calvert and J.H. Shaw, J. Phys. Chem., 83, 3185 (1979).
15. B. Veyret, J.C. Rayez et R. Lesclaux, J. Phys. Chem., 86, 3424 (1982).
16. T. Nielsen, A.M. Hansen and E.L. Thomsen, Atmosph. Environm., 16, 2447 (1982).
17. C.S. Kan, J.G. Calvert and J.H. Shaw, J. Phys. Chem., 84, 3411 (1980).
18. H. Niki, P.D. Maker, C.M. Savage and L.P. Breitenbach, J. Phys. Chem., 85, 877 (1981).
19. P. Deuflhard, G. Bader, U. Novak, Univ. Heidelberg, SFB 123, Techn. Rept. 100, (1980).
20. JPL Publication No. 5, 82-57, "Chemical Kinetics and Photochemical data for use in stratospheric modeling.", Jet Propulsion Laboratory, Pasadena, California.
21. CODATA, J. Phys. Chem. Reference Data, 11, 327 (1982).
22. M.C. Addison, J.P.Burrows, R.A. Cox and R. Patrick, Chem. Phys. Lett., 73, 282 (1980).
23. E. Hesstvedt, Ö. Hov and I.S.A. Isakson, Intern. J. Chem. Kin., 10, 971 (1978).
24. W.R. Stockwell and J.G. Calvert, Can. J. Chem., 61, 983 (1983).
25. H. Niki, P.D. Maker, C.M. Savage and L.P. Breitenbach, J. Phys. Chem., 87, 2190 (1983).
26 B.A. Thrush and G.S. Tyndall, J. Chem. Soc. Faraday Trans. 2, 78, 1469 (1982).
27. I. Barnes, K.H. Becker, E.H. Fink, F. Zabel, Private communication., Abt. Phys. Chem., 6500 Wüppertal.
28. H. Niki, P.D. Maker, C.M. Savage and L.P. Breitenbach, Chem. Phys. Lett., 73, 43 (1980).
29. R.A. Cox and G.S. Tyndall, Chem. Phys. Lett, 65, 357 (1979).
30. H. Niki, P.D. Maker, C.M. Savage and L.P. Breitenbach, J. Phys. Chem., 86, 3825 (1982).
31. G.K. Moortgat, unpublished results from this laboratory.
32. C. Zetzsch and F. Stuhl, Physico-Chemical behaviour of atmospheric pollutants, Proc.2nd Eur.Symp.Varese, 129 (1981)
33. P. Hanst and B. Gay, Atmosph. Env., 17, 2259 (1983).

ABSORPTION SPECTRUM AND KINETICS OF THE NO₃ RADICAL

R.A. Cox, R.A. Barton, E. Ljungstrom and D.W. Stocker
Environmental and Medical Sciences Division,
A.E.R.E. Harwell, Oxon, U.K.

Abstract

The photolysis of Cl_2 in the presence of $ClONO_2$ was used as a source of NO_3 radicals via the reaction $Cl + ClONO_2 \rightarrow Cl_2 + NO_3$. The absorption spectrum of NO_3 was recorded on a diode array spectrometer and the absolute absorption cross section at 662 nm determined from kinetic measurements was 1.63×10^{-17} cm² molecule⁻¹. Secondary chemistry in the system appeared to involve reactions of Cl and ClO with NO_3 and estimates of the rate constants for these reactions were obtained.

1 Introduction

It has been recognised for some time that the nitrate radical plays an important role in atmospheric NO_x chemistry. Its principle formation route is the reaction of NO_2 with ozone[1]:

$$NO_2 + O_3 \rightarrow NO_3 + O_2 \qquad (1)$$

Further reaction of NO_2 with NO_3 gives nitrogen pentoxide which reacts with water, probably heterogeously, to form nitric acid:

$$NO_3 + NO_2 + M \rightarrow N_2O_5 + M \qquad (2)$$
$$N_2O_5 + H_2O \rightarrow 2HNO_3 \qquad (3)$$

This sequence of reactions therefore provides a route for atmospheric oxidation of NO_2 to HNO_3 and moreover since photochemically derived radicals are not involved, the process can occur at night. In fact the efficiency of the process is considerably reduced in daylight due to the removal of NO_3 by photolysis and photochemically derived NO, in competition with reaction (2).

$$NO_3 + h\nu \rightarrow NO + O_2 \qquad (4a)$$
$$\rightarrow NO_2 + O \qquad (4b)$$
$$NO_3 + NO \rightarrow 2NO_2 \qquad (5)$$

NO_3 radicals have recently been identified as potential attacking species in the atmospheric oxidation of organics, particularly aromatic hydrocarbons.[2] Assessment of the importance of this process as well as the rate of oxidation of NO_2 requires definition of the steady state concentration of NO_3. Accurate kinetic and photochemical data for the reactions controlling NO_3 concentration in the atmosphere are therefore needed.

Until recently most of the photochemical and kinetic information concerning NO_3 reactions was due to H. Johnson and co-workers,[3,4] who investigated the chemistry of the NO_2-O_3-N_2O_5 system. This work provided indirect determinations of rate coefficients for a number of elementary reactions of NO_3 and also the absorption cross section, σ, and quantum yields for photodissociation of NO_3 in its broad, structured visible absorption band. Mitchell et al[5] have also used the $NO_2 + O_3$ reaction as a source of NO_3, and by observations made under conditions with complete conversion of NO_2 to NO_3, they were able to obtain a fairly direct estimate of the absorption cross-section in the 0-0 vibronic band at 622 nm. Their result was approximately 50% lower than the most recent value from Marinelli et al[4].

Very recently, Ravishankara and Wine[6] have reported measurements of the NO_3 absorption cross sections in the region 565-673 nm. NO_3 was generated in the reaction of F atoms with HNO_3 in a discharge flow system, and its concentration determined by chemical titration with NO, reaction (5). These results confirm the general agreement as to the features of the NO_3 absorption spectrum. The value of σ at the peak of the 662 nm band showed good agreement with the Marinelli et al result.[4]

Recent work on the chemistry of chlorine nitrate, $ClONO_2$, carried out in our laboratory[7] and elsewhere[8], has shown that the reaction of Cl with $ClONO_2$ is much faster than hitherto believed. This offers a potentially useful source of NO_3 radicals by the reaction:

$$Cl + ClONO_2 \rightarrow Cl_2 + NO_3 \qquad (6)$$
$$(k = 1.04 \times 10^{-11} \text{ cm}^3 \text{ molecule}^{-1}\text{s}^{-1} \text{ at } 298K)$$

In the present paper we describe experiments in which Cl_2 photolysis was employed to generate Cl atoms in the presence of $ClONO_2$. NO_3 formation was monitored in absorption using a diode array spectrometer to characterise the spectral features in the region 490 - 670 nm, and a multichannel scaler for fast time resolved absorption measurements at a single wavelength. Measurements of NO_3 formation and kinetics during illumination and in the dark enabled a determination of the absorption cross section σ_{NO_3} at 662 nm and rate coefficients for reactions of Cl and ClO with NO_3.

2   Experimental

The experiments were performed in a 120 cm long jacketed quartz reaction cell at atmospheric pressure and 296 K. The cell was surrounded by 6 fluorescent blacklights (Philips TL 40/08) any number of which could be activated by an electronically switched 250 v DC power supply for a preset period. The spectral monitoring beam from a high stability tungsten-halogen lamp passed along the cell before dispersion on an 0.75 m Spex monochromator and detection on a photomultiplier. The photo-multiplier signal was monitored on a chart recorder and also recorded on a fast digital multichannel scaler, giving time resolved absorption profiles with millisecond time resolution. The details of this device have been reported elsewhere.[9]

For multichannel spectroscopy the analysing beam was dispersed on an 0.2 m holographic grating polychromator (Jarrel Ash) before detection on a cooled 1024 channel diode array (Reticon), which covered a spectral range of approximatly 74 nm. The detector was coupled to a multichannel analyser (TRACOR Model 1710), which controlled the detector functions and allowed on-line data treatment. Wavelength calibration of the array was based on characteristic lines from a low pressure mercury lamp. The spectral slit width was 0.4 nm.

The reaction cell was filled with a mixture of Cl_2 ($\sim 10^{15}$ molecule cm^{-3}) and chlorine nitrate ($\sim 2 \times 10^{14}$ molecule cm^{-3}) in N_2 diluent, from a flow manifold at atmospheric pressure. $ClONO_2$ was introduced by passing part of the diluent through a bubbler containing a sample of liquid $ClONO_2$ at $-60°C$. The $ClONO_2$ was prepared in our laboratory using the reaction of chlorine monoxide with nitrogen pentoxide. Cl_2 was taken from a cylinder (BOC 5% Cl_2 in N_2) through a flowmeter and needle valve. Concentrations of Cl_2 and $ClONO_2$ in the reaction cell were determined by absorption at 350 nm and 220 nm respectively using cross sections of 1.85×10^{-19} cm^2 and 3.2×10^{-18} cm^2 respectively.

Absorption spectra of NO_3 were recorded with the reaction mixture flowing through the cell with all 6 photolysis lamps activated. For the kinetics experiments the flow through the vessel was stopped and photolysis performed on static gas mixtures. Time resolved absorption during a selected period of illumination followed by an equal period of dark chemistry was recorded.

The rate of photolysis of Cl_2 for any combination of the 6 lamps was determined in separate experiments in which the rate of decay of Cl_2 in the presence of excess H_2 and O_2 was monitored.[10] The rate constant determined for the photodissociation reaction:

$$Cl_2 + h\nu \rightarrow 2Cl \qquad (7)$$

was $k_a = 8.84 \times 10^{-4}$ s^{-1} per lamp.

3   Results

Fig. 1 shows the absorption spectrum of NO_3 over the range 606–674 nm. The position and shape of the principle bands in the spectrum are in excellent agreement with the previous work.[5,6] The cross-sections are based on measurements at 662 nm (q.v.)

Fig. 2 shows a typical absorption vs time recording in the photolysis of a static mixture containing initially 1.36×10^{15} molecule cm^{-3} Cl_2 and 2.7×10^{14} molecule cm^{-3} $ClONO_2$. The wavelength, 663 nm, corresponds to the band head of the 0–0 vibronic band of NO_3 and since no other species present exhibit significant absorption in this region, the profile is representative of the concentration changes of NO_3. It will be seen that NO_3 initially increases steadily on photolysis but then levels off to a steady state. On cessation of photolysis the radical decays rather slowly, taking about 30 s for 97% decay.

The absorption cross section for NO_3 was determined from measurement of the initial rate of increase in absorption in experiments covering a 20-fold range of NO_3 production rate, obtained by varying $[Cl_2]$ and the number of photolysis lamps. Fig. 3 shows a plot of these rates as a function of the quantity $2ka[Cl_2]$. Based on the assumption that each Cl atom produced photolytically yields an NO_3 radical, the slope of this plot gives the product $\sigma_{NO_3} \times$ path length ($= 120$ cm). The average of all the data give

$$\sigma_{NO_3} = (1.63 \pm 0.15) \times 10^{-17} \ cm^2 \ molecule^{-1}$$

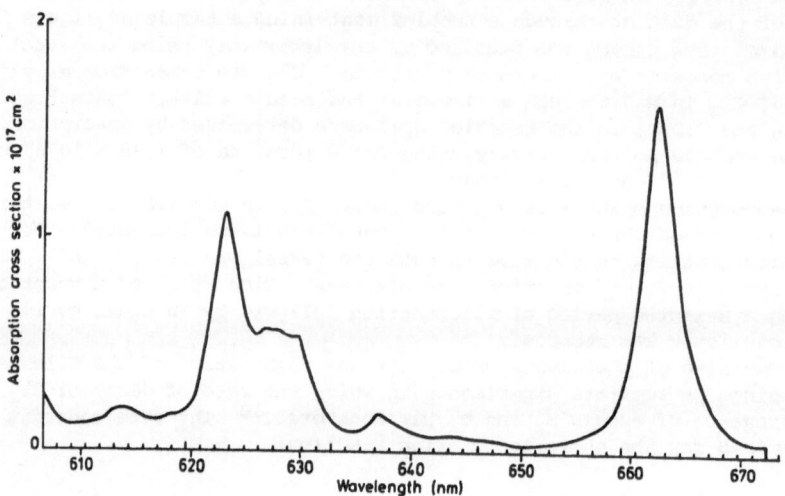

Fig. 1: Absorption Spectrum of NO₃ recorded on
diode array.   Resolution = 0,4 nm

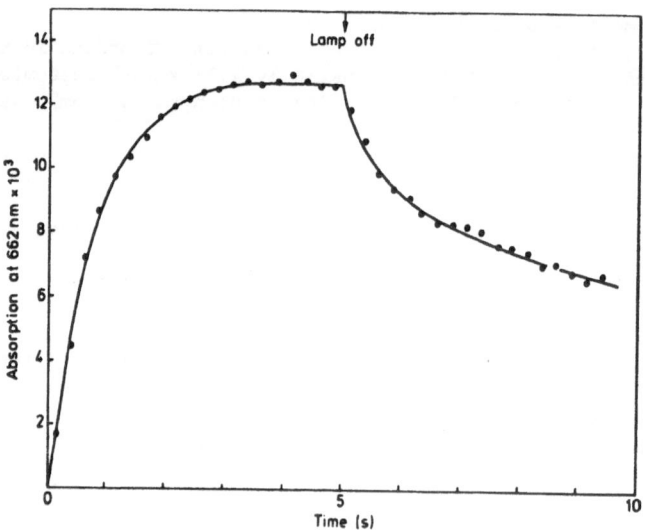

Fig. 2: Concentration behaviour of NO₃ from
absorption at 662 nm

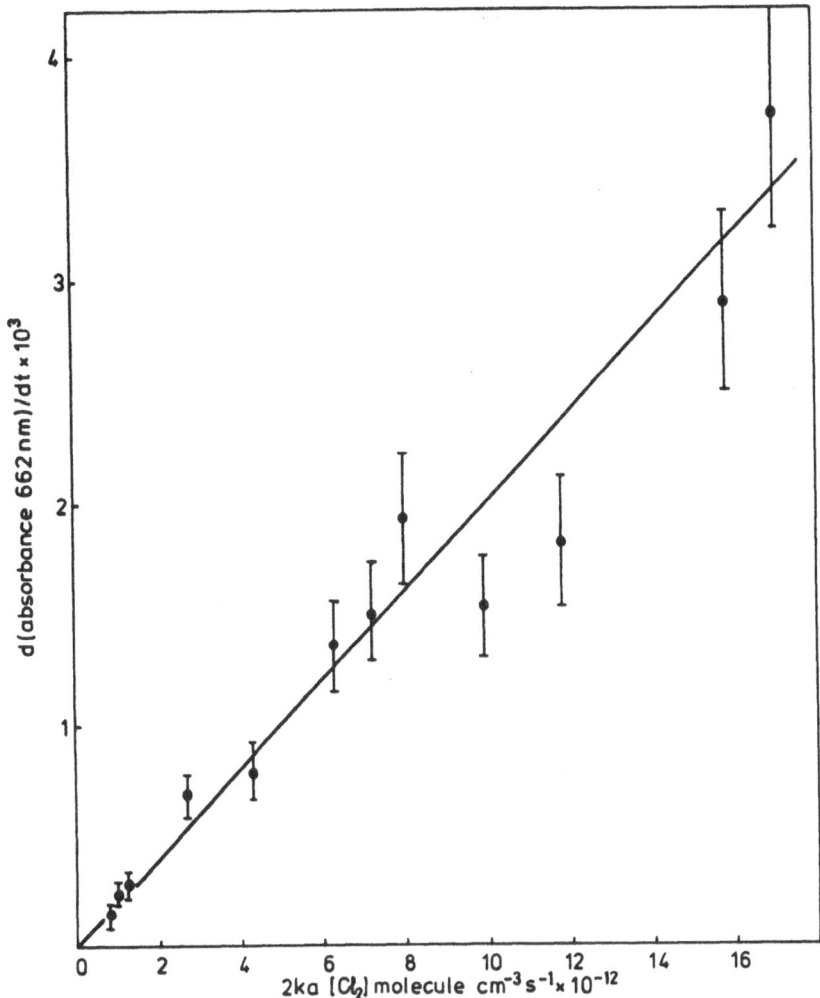

Fig. 3: NO₃ cross-section - Plot of initial rate of absorbance against photolysis rate

The kinetic behaviour of NO_3 during photolysis and in the dark was also analysed. Examination of fig. 2 showed that the rise to steady state followed first order kinetics fairly closely, i.e. NO_3 is removed by a pseudo-first order process. Assuming that NO_3 is formed at a constant rate during illumination a plot of $\ln (A_{max}/(A_{max}-A_t))$ vs time (A = absorption of NO_3) should be linear with slope equal to the first order rate constant, k^I, for the reaction: NO_3 (+X) → products, where X is a reactant present at constant (steady state or excess) concentration or an absorbed photon. Fig. 4 shows typical examples of the data plotted in this way. The plots were linear up to at least 3 half-lives giving values of k_I in the range $0.25 - 3.0$ s^{-1}.

Direct photolysis can be ruled out as a major loss process for NO_3 since it would require a photodissociation rate approximately 1000 greater

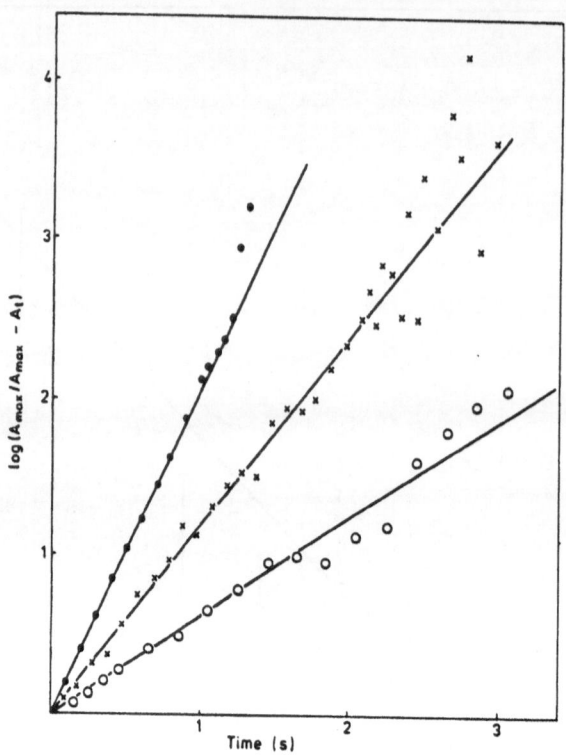

Fig. 4: *First order plots of rise to steady state*

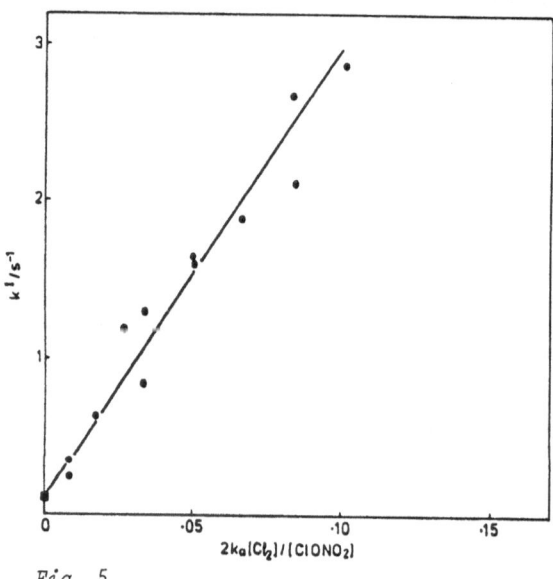

Fig. 5

than that for Cl_2 in the 350 nm region to give the required value of k_I. Although NO_3 may absorb in this region a cross section of the order of 10^{-16} cm^2 molecule^{-1} is unreasonably large. However k_I did seem to vary systematically with both light intensity and $ClONO_2$ concentration. Fig. 5 shows a plot of k^I against the ratio $2k_a[Cl_2]/[ClONO_2]$ for initial conditions. The plot was reasonably linear indicating that the concentration of X was proportional to the rate of Cl atom production and inversely proportional to chlorine nitrate concentration. This observation rules out reaction of NO_3 with $ClONO_2$ which would show the opposite dependence on $[ClONO_3]$ to that observed. The most likely candidate for X appears to be Cl atoms, since a steady state involving reactions (6) and (7) would give:

$$[Cl] = \frac{2k_a[Cl_2]}{k_6[ClONO_2]}$$

If $k_I = nk_8$ [Cl] where n is the number of NO_3 radicals removed following reaction (8),

$$Cl + NO_3 \rightarrow products \qquad\qquad (8)$$

then the slope of the plot in Fig. 5 (=27.2) gives the rate constant ratio nk_8/k_6. The value for k_6 quoted above leads to $nk_8 = 30.5 \times 10^{-11}$ cm^3 molecule^{-1} s^{-1}.

The observed first order kinetics and the above analysis require that the occurrence of reaction (8) does not lead to net loss of Cl atoms. If reaction (8) was a sink for Cl both the rate of production of NO_3 and the pseudo first order removal rate coefficient k_I would decrease with time, giving rise to complex kinetic behaviour. A plausible mechanism to explain the observed behaviour involves the following elementary processes.

$$Cl + NO_3 \rightarrow ClO + NO_2 \qquad \Delta H^\circ = -57 \text{ kJ mol}^{-1} \qquad (8)$$

$$ClO + NO_3 \rightarrow ClOO + NO_2 \qquad \Delta H^\circ = -46 \text{ kJ mol}_{-1} \qquad (9)$$

$$ClOO + M \rightarrow Cl + O_2 + M \qquad\qquad (10)$$

$$NO_2 + NO_3 + M \rightleftharpoons N_2O_5 + M \qquad\qquad (11)$$

Of these reactions, (10) and (11) are reasonably well established and the rate coefficients at 1 atm pressure can be calculated from the pressure dependencies given in the recent NASA evaluation.[11] Reactions (8) and (9) do not appear to have been characterised previously but both are reactions between two radical/atomic species and therefore likely to be relatively rapid at ambient temperature.

In the above scheme, 4 NO_3 molecules are removed in the sequence (8) + (9) + 2x(11), i.e. n = 4. This gives a value of

$$k_8 = (7.6 \pm 1.1) \times 10^{-11} \text{ cm}^3 \text{ molecule}^{-1} \text{ s}^{-1} \text{ at } 296 \text{ K.}$$

The steady state concentration of NO_3 reached during illumination and controlled by reactions (6) - (11) is given by

$$[NO_3]_s = \frac{k_6[Cl][ClONO_2]}{k_8[Cl] + k_9[ClO] + 2k_{11}[NO_2]} \qquad \text{(ii)}$$

Furthermore if Cl, ClO and NO_2 are also in steady state under these conditions, which is a reasonable assumption, the rates of reactions (8), (9) and (11) must be equal, provided these are the only removal reactions for ClO, NO_2 and NO_3, i.e.

$$k_8[Cl] = k_9[ClO] = k_{11}[NO_2] \qquad \text{(iii)}$$

$$\text{and} \quad [NO_3]_s = \frac{k_6}{4k_8}[ClONO_2] \qquad \text{(iv)}$$

Although the steady state absorption due to NO_3 increased with $[ClONO_3]$ the $[NO_3]_s$ values also exhibited a dependence on light intensity, which is not predicted by equation (iv). This would arise if a [Cl] independent loss route for NO_3 was present. The occurrence of such a loss route is indicated by the intercept on the plot in Fig. 6 and the slow dark decay of NO_3 (q.v.). An approximate value of $4k_8/k_6 = 40 \pm 20$ could be obtained from the data, which is consistent with the value derived from Fig. 5.

Information on the kinetics of the rate determining step on the reaction sequence (8) + (9) + (11) can be obtained fom the observed decay of NO_3, on cessation of photolysis. Close examination of Fig. 2 shows that on cessation of photolysis NO_3 decays rapidly at first and then more slowly. Analysis of the decay from a number of experiments showed that the slow decay was first order with a rate coefficient of 0.12 ± 0.03 s^{-1} independent of $[NO_3]_s$ i.e. similar to the intercept on Fig. 5. This may be a wall removal process or it may arise from decomposition of product N_2O_5 either thermally or photolytically, due to the unfiltered monitoring lamp beam.

In order to analyse the initial decay of NO_3, the log-linear portion of the NO_3 absorption decay curve was extrapolated back to t_o corresponding to termination of photolysis and the 'excess' NO_3 removal was obtained by difference, as illustrated in Fig. 6a. Fig. 6b shows a logarithmic plot of the 'excess' NO_3 decay, which occurs with a half life of 0.35s, and represents the time constant of the slowest, rate determining step in the sequence (8) + (9) + (11)

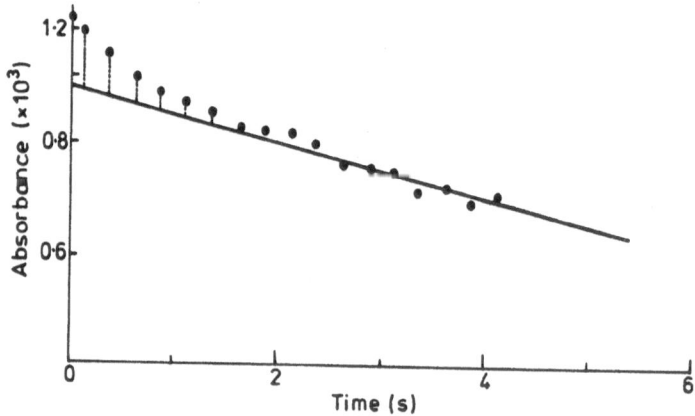

Fig. 6a: Decay of NO_3

- 212 -

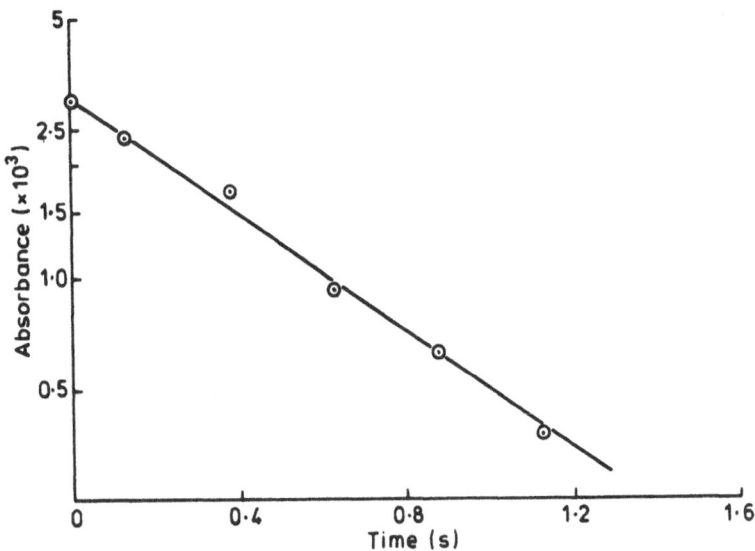

Fig. 6b: Decay of "Excess NO_3" (from fig. 6a)

The lifetime of Cl is determined by reaction (6) which for the conditions in this experiment, gives $\tau_{Cl} = (k_6[ClONO_2])^{-1} = 0.43$ mS. Since $k_{11} = 1.2 \times 10^{-12}$ cm^3 molecule^{-1} s^{-1} at 1 atm pressure and 296K, $\tau_{NO_2} = (k_{11}[NO_3])^{-1} = 0.09s$. Therefore both Cl and NO_2 decay lifetimes are considerably shorter than the observed decay time for 'excess' NO_3. It is concluded therefore that reaction (9) is the slowest step and it follows that ClO is the radical present at highest concentration. Assuming that the observed decay half-life of 0.35 s reflects the rate of reaction of ClO with NO_3 at an average concentration of 5.6×10^{12} molecule cm^{-3} and the NO_2 formed in (10) also removes an NO_3 molecule, the observed rate gives $k_9 = 3.5 \times 10^{-13}$ cm^3 molecule^{-1}.

Neglecting the contribution due to Cl which is only present at very low steady state (~ 3×10^9 molecule^{-1} cm^{-3}), the 'excess' NO_3 is related stoichiometrically to the steady states of ClO and NO_2 by:

$$[NO_3]_e = 2[ClO]_s + [NO_2]_s \qquad (v)$$

The steady state relationships for $[ClO]_s$ and $[NO_2]_s$, eq iii, then give

$$k_9 = 2k_8[Cl]_s/([NO_3]_e - [NO_2]_s)$$

Using the value of k_8 derived above and the experimental data in Fig. 6 we obtain $k_9 = 3.6 \times 10^{-13}$ cm^3 molecule^{-1} s^{-1}, in excellent aagreement with the absolute measurement from the 'excess' NO_3 decay. Treatment of several sets in this manner gave an average value of

$$k_9 = (4.0 \pm 1.7) \times 10^{-13} \text{ cm}^3 \text{ molecule}^{-1}\text{s}^{-1} \text{ at 296K.}$$

4   Discusson

The observation in the present work of the characteristic spectrum of NO_3 in the visible region confirms that this radical is a major product of reaction (6).   No fine structure was visible in the vibronic bands, confirming the conclusions of the high resolution studies of Ramsay[12] and Marinelli et al[4]. Table 1 shows a comparison of the absorption cross section at 662 nm and also the integrated cross section over the 0—0 band from 14910 – 15290 cm^{-1} obtained in the present work, with those reported by others. Our data supports the higher value of σ_{662nm} and confirms the integrated band intensities, for which there appears to be a concensus.

Kinetic data for reactions (8) and (9) have not been reported before. The value for k_8 is similar to the rate coefficient for the reactions of Cl with other radical species such as HO_2, OClO, ClOO, which are all $> 10^{-11}$ cm^3 molecule$^{-1}s^{-1}$. The reaction of ClO with NO_3 is considerably slower, which is in accord with the lower reactivity of ClO compared to Cl, particularly with polyatomic species. An interesting analogy to reaction (9) is O transfer in the self reaction of ClO:

$$ClO + ClO \rightarrow ClOO + Cl \qquad\qquad (13)$$

This reaction is also slow, $k_{13} \approx 3.5 \times 10^{-15}$ cm^3 molecule^{-1} s^{-1} at 298K,[13] the slow rate resulting from the endothermicity of the reaction (11.5 kJ mol^{-1}) and also the low A factor ($A_{13} \approx 3 \times 10^{-13}$ cm^3 molecule$^{-1}s^{-1}$). Since reaction (9) is exothermic and $k_9 \approx A_{13}$, reaction (9) may have near zero activation energy and hence a very small temperature dependence.

The reactions of Cl and ClO with NO_3 comprise yet further coupling between the ClO_x and NO_x cycles in the chemistry of the stratosphere, and provide a sink for ozone through the reaction sequence

$$Cl + NO_3 \rightarrow ClO + NO_2$$

$$ClO + NO_3 \rightarrow Cl + O_2 + NO_2$$

$$2NO_2 + 2O_3 \rightarrow 2NO_3 + 2NO_2$$

$$\text{net } 2O_3 \rightarrow 3NO_2$$

This reaction scheme is unlikely to be a major influence on the O_3 budget but it is of interest because it provides a route for converting the less reactive ClO species into Cl atoms during nightime, when usual reaction partners O and NO, which result form photodissociation, are absent. Calculations using a 1 dimensional atmospheric model with diurnal variations of photochemistry are needed for a quantitative assessment of the role of reaction (8) and (9) in the stratosphere.

Acknowledgement

This work was partly supported by the UK Department of the Environment.

References

1. R.A. Cox and G.B. Coker, J. Atmosph. Chem. $\underline{1}$ 000 (1983)

2. W.P.L. Carter, A.M. Winer, and J.N. Pitts, Env. Sci Technol $\underline{15}$, 829 (1981)

3. R.A. Graham and H.S. Johnson, J. Phys. Chem., $\underline{82}$, 254 (1978)

4. W.J. Marinelli, D.W. Swanson and H.S. Johnson, J. Chem. Phys., $\underline{76}$, 2864 (1982)

5. D.N. Mitchell, R.P. Wayne, P.J. Allen, R.P. Harrison and R.J. Twin, J. Chem. Soc. Faraday II, $\underline{76}$, 785 (1980)

6. A.R. Ravishankara and P.H. Wine, Chem. Phys. Lett. $\underline{101}$, 73 (1983)

7. R.A. Cox, J.P. Burrows and G.B. Coker, J. Chem. Kinet. in press (1983)

8. J.J. Margitan, J. Chem. Phys, $\underline{87}$, 674 (1983)

9. R.A. Cox and D.W. Sheppard, J. Photochem. $\underline{19}$, 189 (1982)

10. R.A. Cox and J.P. Burrows, J. Phys. Chem. $\underline{83}$, 2560 (1979)

11. 'Chemical Kinetics and Photochemical Data for use in Stratospheric Modelling', Eval. 6 NASA Panel for evaluation, W.B. DeMore et al. J.P.L. Publication 383-62 (1982)

12. D.A. Ramsay, Proc. Colloq. Spectrosc. Int. $\underline{10}$ (1962)

13. R.A. Cox, R.G. Derwent, A.E.J. Eggleton and H.J. Reid, J. Chem. Soc. Faraday II, $\underline{75}$ 1648 (1979)

Table 1   Absorption cross section data for NO_3 in the O–O band

	σ_{max} (622nm)	$\int\sigma$
	$10^{17} cm^2 molecule^{-1}$	$10^{15} cm\ molecule^{-1}$
Ravishankara and Wine (1983)	1.78	1.88
Marinelli et al (1982)	1.90	2.02
Mitchell et al (1980)	1.21	2.06
Graham and Johnson (1978)	1.7	1.99
This work	1.63	1.85

HYDROXYL RADICAL CONCENTRATION IN AMBIENT AIR AT A
SEMIRURAL SITE ESTIMATED FROM $C^{13}O$ OXIDATION

J. Hjorth, G. Ottobrini, F. Cappellani, G. Restelli, H. Stangl
Commission of the European Communities
Joint Research Centre - Ispra Establishment
21020 Ispra (Va) - Italy

C. Lohse
University of Odense, Chemistry Dept., Odense, Denmark

Summary

An estimate of the average OH radical concentration on sunny summer days at a semirural site was made by following $C^{13}O$ oxidation. Teflon bags filled with ambient air with added $C^{13}O$ were exposed to sunlight and the $C^{13}O$ decrease measured by long path IR absorption spectroscopy. From the results of thirteen experiments (May - July) the average OH concentration was estimated as being equal to $2 \cdot 10^6$ cm^{-3} with an upper limit of $4 \cdot 10^6$ cm^{-3}.

1. INTRODUCTION

The problem of the formation and increase of tropospheric ozone is a subject of continuing interest. In this regard the related tropospheric chemistry with particular attention to non-urban sites merits special attention. In connection with previous studies /1/ on the relationship between ambient carbon monoxide and the formation of tropospheric ozone in a semirural site, a series of experiments have been designed and performed.

The oxidation of carbon monoxide by OH-radicals was studied in samples of ambient air irradiated in Teflon bags on sunny summer days. From the results the average OH radical concentration was estimated, as originally suggested by Campbell /2/. The experiments were made in the Ispra area, a semirural site at circa 45°N latitude.

The isotopic species $C^{13}O^{16}$ (indicated in the following as $C^{13}O$) was used as a tracer and quantitatively detected independently from natural $C^{12}O^{16}$ (indicated in the following as $C^{12}O$) by long path infrared absorption spectroscopy.

The concept of the OH radical as the principal driving force of atmospheric chemistry has created much interest in methods to be used for the measurement of OH concentrations in ambient air; however there is still little and ambiguous quantitative knowledge about the prevailing hydroxyl distribution in the troposphere /3/.

Direct measurements have been performed using laser induced fluorescence /4,5/ or long path laser absorption (e.g. by G. Hübles et al. /6/). Indirect determinations have been made e.g. by Watanabe et al. /7/ who used a spin trapping method, and by Campbell et al. /2/ who measured the $C^{14}O$ oxidation rate.

Estimates have also been made from atmospheric models and in conjunction with the evaluation of budgets of atmospheric species /3/.

The calculation of the OH radical concentration at specific sites in the continental boundary layer, using models requires detailed data about the local air chemistry; this involves large uncertainties. On the other hand, direct measurements up to now exhibit very low precision and might be affected by systematic errors.

The experiments described here were performed in the period from May 20th to July 26th; from the results an OH radical concentration averaged over the daily irradiation time ($\sim$ 8 hours around noon) and over thirteen experiments was calculated. In the following, the method which makes use of $C^{13}O$ is described. The experimental details will be given and the results obtained will be presented. Finally the limits of the experiment will be discussed together with new procedures that might improve the sensitivity and reduce the uncertainty of the results.

2. EXPERIMENTALS

2.1 Method

Daytime averages of OH radical concentrations are calculated from the consumption of $C^{13}O$, added in small amounts ($\sim$ 1 ppm) to ambient air irradiated in large Teflon bags. $C^{13}O$ introduced at this concentration level should not severely modify the photochemical characteristics of ambient Ispra air as could happen with the addition of other reactants. On the other hand, the consumption of $C^{13}O$ in the bags can be determined, even though some CO is released from the bag walls, because $C^{13}O$ constitutes only a small and known fraction of the $C^{12}O$.

The concentration of $C^{13}O$ in air is measured using long path infrared absorption spectroscopy which allows the simultaneous measurement of $C^{12}O$ and of CF_2Cl_2 added at < 1 ppm level as an inert internal standard.

The oxidation of carbon monoxide ($C^{13}O$) in ambient air is believed to proceed almost exclusively with the OH radical according to reaction (a) followed by reaction (b)

$$CO + OH \rightarrow CO_2 + H \qquad \text{(a)}$$

$$H + O_2 \overset{+M}{\rightarrow} HO_2 \qquad \text{(b)}$$

The calculations are then based on the assumption that (a) is the only reaction of $C^{13}O$ in ambient air; for the rate constant the value of $2.66 \cdot 10^{-13}$ cm^3 mol^{-1}s^{-1} measured for $C^{12}O$ is used, disregarding any isotopic effect. This represents an approximation which is expected to introduce a negligible error with respect to the overall uncertainty of the measurement. Reaction (b) provides a way of regenerating the hydroxyl radicals via $HO_2 + NO \rightarrow NO_2 + OH$. The efficient regeneration of the OH radicals is crucial for the success of the method.

Modelling /8/ shows that the recycling can be very efficient if sufficient NO is present. Under the conditions of the experiments where NO_x-concentrations of 3-10 ppb are involved, the OH concentration should not be depressed due to the reaction with CO. In addition the carbon monoxide is most probably responsible only for a small part of the OH reactions in the bag. This was confirmed by the results of the experiments which did not show a negative correlation of the estimated OH concentration with the carbon monoxide concentration.

2.2 Teflon bags

Cylindrical Teflon (E.I. du Pont Inc.) bags with a volume of 2 m^3 and a diameter of 1 m were used. The bags were prepared by carefully heat sealing 2 mil. (0.05 mm) thick film followed by an outside mechanical support of the sealing. The support consisted of two strips 2 mm thick of Teflon fastened by staplers. Prepared in this way, the bags can be used for outdoor experiments but they are fragile and must be protected against the wind.

Bags prepared from new Teflon films release under sunlight irradiation large amounts of carbon monoxide (up to a few ppm in a 2 m^3 bag after one day of irradiation). The phenomenon is correlated to UV irradiation and it is not clear if carbon monoxide is released from the walls as such or as an end product of reactions of other compounds. Following 20 - 40 hour irradiation in intense sunlight, the release is reduced to levels equivalent to the build up of $\sim$ 100 ppb of CO in the 2 m^3 bag after one day of irradiation; this was considered an acceptable value.

Before the experiments, bags were thus filled with clean air and conditioned as described above using sunlight or UV lamps. The capability of well conditioned Teflon bags to form ozone when filled with purified air (AADCO 737 clean air generator) or with purified air plus NO_2 and exposed to intense sunlight was tested. The results (Table I) indicate that the concentration of the radicals (OH?) responsible for the O_3 production and most probably provided by the Teflon film is much reduced under these conditions by ambient air levels of NO_x.

TABLE I - Ozone production in air samples irradiated for 8 hours in sunlight in conditioned Teflon bags

Air sample	O_3 production (ppb)
Purified air	5 - 15
Purified air + NO_2 (5-8 ppb)	3 - 4

On days with little or no wind, the buildup of ozone in outside air and in sun-irradiated ambient air in the bags are similar within $\sim$ 20%. This also suggests that the radical source provided by the bags could only be of moderate size.

The lifetime of ozone in unirradiated bags is large: the halflife is 20 hours or more. Experiments using high (ppm) concentrations of ozone indicate, however, that the lifetime might be reduced when the bags are irradiated.

The light transmission of the Teflon film is excellent in the solar spectrum. For the UV-range as a whole it is 95%; at the short wavelengths around 300 nm - relevant to OH-formation from O_3 - the transmission is reduced to 90 - 95% /9-10/. Old, used bags however show increasing UV absorption.

The emission of organic compounds from irradiated and unirradiated Teflon bags filled with purified air has been investigated. No significant pollutants have yet been found, but further investigations of the irradiated bags seem to be necessary.

The wall-scavenging of shortlived radicals such as OH and HO_2 in experiments similar to this has been discussed by some authors /9,2/ and has been demonstrated to be of little importance at NO_x-levels like those in our air samples. However, products from wall-reactions of OH-radicals might account for some excess reactivity in the bags, especially at low NO_x-concentrations /8/.

2.3 Spectroscopic measurement

The advantages and limits of infrared absorption spectroscopy as an analytical technique for the laboratory study of atmospheric chemistry have been discussed in many papers (see e.g. Refs.11 and 12). In this experiment a Bruker IFS 113 V Fourier Transform Spectrometer in conjunction with a 70 m fixed path 25 litres volume multiple reflection cell was used. The cell was fitted to the FTS sample compartment to preserve "in vacuo" operation of the instrument. Interferograms were collected for each sample by filling the gas cell at known pressure and temperature throughout the day and storing and computing overnight. Spectra were obtained at 0.06 cm^{-1} instrumental resolution using Happ-Genzel apodization and coadding 100 scans for each spectrum. The reference file used to calculate the absorbance spectrum (cell empty) was recorded with 400 scans coadded.

The concentrations of $C^{13}O$, $C^{12}O$ and CF_2Cl_2 were determined from the absorbance spectra. All the gases are characterized by sharp spectral features located in atmospheric windows which avoids at least for a suitable number of transitions the most disturbing interference of water vapour absorption. Many lines are available for the evaluation of $C^{12}O$; four lines of the most intense ones located between 2168 and 2180 cm^{-1}, R(6) through R(9) transitions, were used.

The concentration of CF_2Cl_2 was evaluated from the sharp Q-branch of the $\nu6$ band at 1160 cm^{-1}. The isotopic species $C^{13}O$ shows a spectrum similar to that of $C^{12}O$, displaced 47 cm^{-1} towards longer wavelengths. Many of the most intense transitions interfere with $C^{12}O$ or with H_2O lines. Four transitions located in the region 2130 - 2145 cm^{-1} (see Fig.1) were used. The concentrations of carbon monoxide were calculated from the peak absorbances using experimentally determined coefficients to convert absorbance units into ppbv.

Problems were encountered in the definition of the peak height because of baseline fluctuations due to the combination of the instrumental noise with many weak lines due to N_2O, CO_2 and H_2O. Since only the relative variations were needed, the baseline of each peak was defined using ad hoc criteria to be followed in all the spectra. The use of routines for calculating peak areas did not result in better precision. For each sample three spectra were recorded so that the $C^{13}O$ and $C^{12}O$ concentrations were obtained from an average over twelve data and for the CF_2Cl_2 value from three data. In the case of the carbon monoxide lines, the instrumental resolution is too close to the width (0.12 - 0.14 cm^{-1}) of atmospheric pressure broadened lines and fluctuations may be expected because of imperfect description of the absorption line shape by the instrument.

Following the procedure indicated above, a precision of about 1% (one standard deviation) for all the three species and for the concentration range used in the experiments was observed.

2.4 Experimental procedure

The bags were filled with ambient air in the morning to one half of their volume, then suitable amounts of $C^{13}O$ (Stohler Isotopes Co.) and CF_2Cl_2 (Matheson Inc.) were added; they were then completely filled with air. This procedure was demonstrated to assure a complete mixing. In some experiments two bags were run in parallel.

The bags were exposed to sunlight on the top of a building situated on a small hill outside the research centre of Ispra. The albedo of the roof where the experiments were performed is approximately 10%.

During exposure to sunlight, O_3, NO, NO_2 concentration, temperature and relative humidity of the air inside the bag were measured at one hour

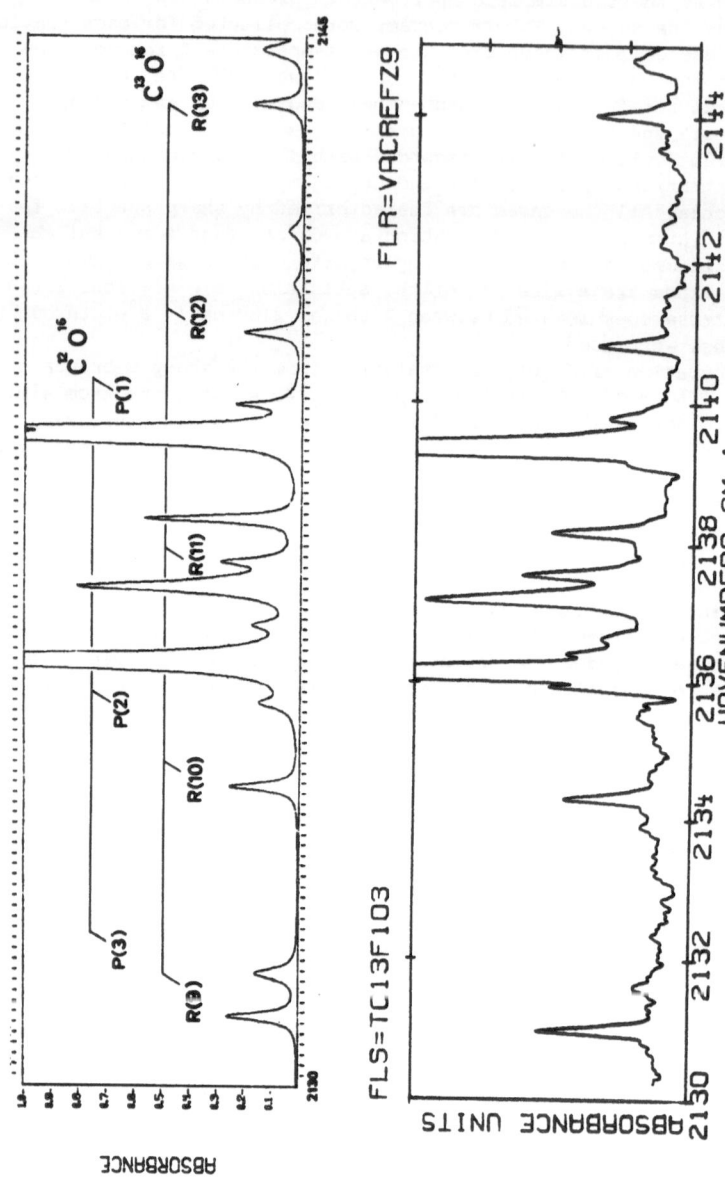

Fig.1 – Infrared absorption spectrum recorded in the region of C¹³O lines used for the experiments. FTS at 0.06 cm⁻¹ resolution, 70 m multiple reflection cell (lower spectrum). The spectrum of ambient air with added C¹³O (the amounts of carbon monoxide and H₂O are not equal in the two spectra) generated from the AFGL Tape /13/ using a modified program for isotopic conditions different from natural conditions, is shown in the upper part of the figure.

intervals. The same parameters were monitored in outside air. UV radiation was continuously recorded by an Eppley ultraviolet radiometer with a spectral response in the region 290 - 385 nm.

Ozone was measured by a Monitor Lab 8410 analyzer calibrated using a Dasibi ultraviolet photometer. NO and NO_2 were measured by a Monitor Lab 8440 chemiluminescence NO_x analyzer, calibrated using NO standards prepared by titration with O_3. Relative humidity and temperature were measured by an Ultrakust wet/dry temperature hygrometer.

Samples for the FTIR-determination of $C^{13}O$, $C^{12}O$ and CF_2Cl_2 were taken at the beginning and at the end of the irradiation period by filling 50 l Tedlar (du Pont) bags with air from the Teflon bags. These samples were transported to the laboratory and transferred to the FTIR measuring cell.

The determination of the concentration of $C^{13}O$ before and after sun irradiation can be affected by various parameters in the spectroscopic measurement. Most of them, such as filling temperature and pressure in the measurement cell and instrumental response different from one spectrum to the other, can be either measured (T,P), or controlled in the recorded spectra (linewidth).

The possible occurrence of dilution of the sample with ambient air during the operation of extraction from the irradiation bag and transfer into the measuring cell could not be checked up from the variation of species already contained in ambient air. For this reason CF_2Cl_2 gas at a concentration around 1 ppm was added. This compound was selected because, from a chemical point of view, it does not perturb the system. Absorbance phenomena in the bag at this concentration level were not expected to be important.

The data for $C^{13}O$ were then normalized to those of CF_2Cl_2: this correction was essential in order to obtain consistent data for the $C^{13}O$ decrease.

3. RESULTS

The data concerning the experiments performed are given in Table II. As previously indicated the $C^{12}O$ and $C^{13}O$ concentrations as well as the CF_2Cl_2 absorption were calculated from the results of three spectra recorded for each sample; in a few cases two or four spectra were recorded. The uncertainty (one standard deviation) of the single $C^{13}O$ determination was equal to 1.7%, reduced to 1.0% using the average over three spectra (four lines used for each spectrum). For CF_2Cl_2 these figures were 1.1 and 0.6%, respectively.

Within the concentration range used the average relative standard deviations appeared to be nearly constant.

In the calculation of $C^{13}O$, the $C^{12}O$ and $C^{13}O$ values were normalised according to the CF_2Cl_2 values. The $C^{13}O$ final value was also corrected for an increase corresponding to carbon monoxide release during the irradiation, taking into account the C^{13} natural abundance (1.1%).

The NO concentrations were too small to be measured (< 2 ppb). The NO_2 values given in Table II might be incorrect because the instrument uses a molybdenum oxide converter at 195°C. This means that other NO_y species (e.g. HNO_3 and PAN) may contribute to the "NO_2" signal.

In three experiments, metal foils were used as reflectors, placed beneath parallel bags on June 28th (a day of low photochemical activity) and in the two last experiments they were used to enhance the irradiation of the most used bag. This bag was proved to give lower UV transmission and consequently low ozone formation in the ambient air sample.

TABLE II - Data of ambient air Teflon bag irradiation experiments. All the values refer to the contents of the bags except "ozone - outside air"; $\Delta c^{13}O$ is the percentual change of the $c^{13}O$-concentration after sun irradiation, corrected for photochemical $c^{13}O$-production.

Date	Bag # *	Time	Ozone ppb Ambient Air	Bag	NO$_2$ ppb	C^{12}O ppb	C^{13}O ppb	Max temp. °C	Rel.hum.% at T max.	UV max. arbitrary units	integr.UV units	$\Delta c^{13}O$ %
May 20	832	7.30	15	15	5–6	392	216	27	42	3.9	17.5	−1.5
		15.00	35	110	5–8	469	214					
May 31	832	7.20	25	25	6	315	173	29	31	3.3		−0.7
		16.20	65	76	2–3	432	173					
June 9	832	7.55	35	28	9–10	400	377	33	32	3.1	20.3	−1.5
		16.45	90–95		2–5	494	373					
June 9	834	7.55	35	28	8–10	391	304	33	31	3.1	21.5	−0.6
		17.45	90–95	107	3–6	500	303					
June 15	832	8.30	43	40	8–10	295	1110	35	< 20	3.6	23.3	−3.0
		16.30	60	122	5	361	1077					
June 15	834	8.30	43	43	8–10	322	2718	35	< 20	3.6	26.2	−5.0
		18.30	62	135	3–4	388	2583					
June 23	834	8.30	30	35	8–10		916	28	44	3.5	12.1	+0.1
		14.45	80	103	6–10		918					
June 28	832	8.45	30	20	4–6	285	1043	29	42			+0.5
		16.00	50	75	6	362	1049					
June 28	834	8.45	30	20	4–6	292	1097	28	51			−1.3
		18.00	50	87	5	397	1083					
July 22	832	8.45	40	40	4	242	728	34	44	2.8	18.5	−4.6
		17.45	120	98	2–3	342	695					
July 22	834	8.45	40	37	4	248	759	34	39	2.8	15.9	−2.9
		15.45	123	100	3	335	738					
July 26	832	8.20	30	30	9–10	431	1061	36	37	3.0	18.3	−1.3
		16.25	95	115	8–10	553	1049					
July 26	834	8.20	30	30	8–9	387	1130	29^{xx}	54	3.0	19.6	−0.9
		18.30	95	108	8–10	495	1121					

* in this case a reflector was used (see text).

** this value is not the maximum temperature but the temperature corresponding to the indicated relative humidity.

By use of the reflector quite good agreement was obtained for the ozone formation in the two bags. Large differences between ambient air and bag ozone levels occurred only when relatively strong winds were prevailing; under quiet weather conditions the difference was 20% or less.

As can be seen from the data in Table II, generally the percentage $c^{13}O$ decrease was too small to permit meaningful calculations of the average OH radical concentrations in the bags for each experiment because of the uncertainty affecting the $c^{13}O$ values. However, by using all the data an estimate of the average OH concentration level for these summer days can be obtained. For this purpose it was assumed that the duration of each experiment was 8 hours and that all experiments were performed during the same hours of the day, so that the results of the individual experiments were compatible. This, of course, introduces an error, but not a big one because all the experiments include the hours 9 a.m. - 3 p.m., where the major photochemical activity takes place.

The average $\Delta c^{13}O$ is -1.75%. This corresponds to an average OH concentration of $2.3 \cdot 10^6$ molecules/cm^3 for the 8 hour period considered.

Assuming a normal distribution of the data, the upper limit can be derived from the following expression

$$\Delta c^{13}O \text{ mean } - ts/\sqrt{n} = \Delta c^{13}O \text{ minimum,}$$

where t is the student t value (2.18 at a 95% level of significance), s is the standard deviation (1.68) of the 13 $c^{13}O$-determinations and n is the number of measurements. This upper limit (lower limit of the $\Delta c^{13}O$) becomes -2.8%, corresponding to an OH-concentration of $3.7 \cdot 10^6$ molecules/cm^3.

To avoid making assumptions about the distribution of the values, non-parametric statistics must be used. The values are ranked and the median is found to be -1.3%, corresponding to an OH-concentration of $1.7 \cdot 10^6$. The upper limit is -3.0% at a 95% level of significance, corresponding to an OH concentration of $4.0 \cdot 10^6$ molecules/cm^3. The lower limit found by assuming a normal distribution is $1.0 \cdot 10^6$ molecules/cm^3. From the non-parametric statistics a limit of $0.8 \cdot 10^6$ molecules/cm^3 is obtained.

The ratio between the maximum and average UV-irradiation for the ten experiments where these data were recorded was on average 1.4. If this factor only is taken into consideration the OH-concentrations at noon thus might be estimated to be on average 40% higher than the 8 hour average.

4. DISCUSSION

The main problem of the experiment is the overall uncertainty in the determination of the $c^{13}O$ decrease which is too close to the magnitude of the effect we are measuring. This not only makes a one day determination generally meaningless but gives rise to some doubts about the effects of unidentified artefacts or systematic errors. To obtain more confidence in the method an experiment was performed with the intention of reaching a more pronounced $c^{13}O$ decrease. To this end $c^{13}O$, O_3, water vapour and small amounts of NO were added to purified air in bags like those used for the previous ambient air experiments and submitted to sun-irradiation. OH should be formed in this system via the photolysis of O_3 /13/.

$$O_3 + h\nu \rightarrow O('D) + O_2 \quad (285 \text{ nm} < \nu < 315 \text{ nm})$$

followed by $O('D) + H_2O \rightarrow 2HO$.

Experiments were run on different days in duplicate. The expected (nonlinear) correlation between the relative $c^{13}O$ decrease and the product of integrated UV-radiation, absolute humidity and average O_3-concentration was observed for each day (see Table III). The difference observed between experiments performed on different days might be tentatively attributed to a different spectral distribution of the incoming UV radiation. From a series of eight experiments of this type, the photolysis rate of ozone was estimated using a rough model. This value was found to exceed by a factor 2 the photolysis rate calculated from the model of Demejian and Schere /13/ for conditions similar to those of the experiments; i.e. the method led to an estimated OH radical concentration higher than expected. The calculations involve many approximations and the results are only tentative; however, it can be argued that the method very probably leads to some overestimation of the OH radical concentration.

TABLE III – Data of sun irradiations of mixtures of $C^{13}O$, O_3, H_2O and NO in purified air. The NO_x initial concentration in all bags was $\sim$ 19 ppb.

Date	Bag #	$\Delta C^{13}O$ %	Integrated UV(1) Arbitrary units	H_2O 10^{18} molec.cm^{-3} (2)	Average O_3 ppb (3)	Product of (1),(2)and (3) (arbitrary units)
20/8	1	−2.5	103	0.41	435	184
	2	−3.8	103	0.42	955	413
30/8	1	−4.7	85	0.32	670	182
	2	−7.9	92	0.32	1473	434

It is generally found that smog chambers do have a radical initiating effect. Though we do not find evidence for a strong effect of this type in the Teflon bags, we find it reasonable to suppose that the upper limit of the OH-concentration given in the preceding paragraph is not exceeded in the real atmosphere (outside the bags).

The $C^{13}O$-decreases observed in the ambient air experiments (Table II) were also compared to the integrated UV-light and to the product of integrated UV-light, ozone and water-vapour concentrations in the cases where all the data were available. When experiments with similar concentrations of NO_x in the bags were compared (see Table IV) a correlation with the product was obvious, while there was no correlation with UV-light alone. This supports our view that the measured $C^{13}O$ decreases really reflect OH-concentrations. In most of the experiments performed, the CF_2Cl_2 con-

TABLE IV – Analysis of results from Table II. "Low" initial NO_2 means $\sim$ 5 ppb, "high" means $\sim$ 9 ppb. "Product" as in Table III is the product of integrated UV-light, absolute humidity and "average" ozone (= the average of morning and evening ozone concentrations in the bags).

Date	Bag #	$\Delta C^{13}O$ %	Initial	Integrated UV (arbitrary units)	Product as for Table III (arbitrary units)
20/5	382	−1.5	low	17.5	1.2
22/7	834	−2.9	low	15.9	1.7
22/7	832	−4.6	low	18.5	2.3
23/6	834	+0.1	high	12.1	1.0
1/6	834	−0.6	high	21.5	1.7
26/7	834	−0.9	high	19.6	2.2
26/7	832	−1.3	high	18.3	2.2

centration was found to be slightly higher in the afternoon sample. This unexpected behaviour could be explained by a different dilution of the sample during filling of the transfer Teflar bags. Tests are, however, in progress to clarify this problem.

A remarkable improvement could probably be obtained in the future by using an InSb detector in place of the HgCdTe detector used. The higher detectivity of InSb would in fact allow a better signal to noise ratio in the spectra or for the same signal to noise ratio, the use of a measuring cell with a longer path length; both can be expected to lead to an increased sensitivity and to better precision. However, this is bound to the use of another inert tracer with spectral features in the range of the InSb detector (cut off at 1700 cm^{-1}); this can be tentatively identified in one isotopic species of nitrous oxide.

From a general point of view a possible improvement would be better knowledge of the air chemistry inside the bag. While at the moment the Teflon bags and sunlight represent the best approach to simulating ambient air conditions, some comments are mandatory. The air kept in the Teflon bag is not quite the same over the irradiation time as the ambient air outside with its continuous inputs of pollutants during the day and dilution effects. Some species can be adsorbed by the walls or species previously adsorbed can be released during the experiment /14/ though several washings of the bag with air should minimize this effect.

The impact of UV irradiation on the Teflon film is not well known; at the moment we only know that CO is released and ozone might be consumed.

In conclusion, keeping in mind the above limitations, some considerations can be made about the results obtained. The average OH radical concentration detected in this semirural site (moderately polluted) on sunny summer days is in fairly good agreement with values reported in the literature /6/.

Concerning the importance of carbon monoxide as an ozone precursor at Ispra, the estimate of the OH radical concentration allows it only a relatively unimportant role. The average concentration of carbon monoxide in ambient air during the period considered was equal to 333 ppb and the maximum amount of CO oxidized by OH ($4 \cdot 10^6$ molecules/cm^3) during 8 hours would be 10 ppb. This upper limit is to be compared with the net ozone production, observed from monitoring on days with little influence of transport, of about 70 - 80 ppb. This last result is in agreement with other experiments already performed which showed no significant influence of CO added to ambient air on ozone production.

REFERENCES

1. C.LOHSE et al., Proc. of 2nd European Symp. on Physico-Chemical Behaviour of Atmospheric Pollutants, Ispra 1981, September 29 - October 1, B.Versino and H.Ott, Eds. Reidel Publ. Co., 1982, p.212.
2. J. CAMPBELL, J.C. SHEPPARD, B.F.AU; Geophys. Res. lett., Vol.6, No.3 (1979) P.175.
3. P.J. CRUTZEN, Report of the Dahlem Workshop on Atmospheric Chemistry, Berlin 1982, May 2-7, E.D.Goldberg, Ed. Springer-Verlag 1982, p.313.
4. D.D. DAVIS, W. HEAPS and T.McGEE, Geophys. Res. Letters 3 (1979) p.331.
5. C.C. WANG, L.I. DAVIS Jr., P.M.SELZER and R.MUNOZ, J.Geophys.Res. 86, C2 (1981) p.1181.
6. G.HUEBLER et al., Proc. of 2nd Eur.Symp. on Physico-Chemical Behaviour of Atmospheric Pollutants, Ispra, 1981, September 29 - October

1, B.Versino and H.Ott, Eds. Reidel Publ. Co. 1982, p.2.

7. T. WATANABE, M. YOSHIDA and S. FUJIWORA, Anal.Chem., 54 (1982) p.2470.

8. B. WEINSTOCK, H. NIKI and T.Y. CHANG, Adv.Environ.Sci.Technol. Vol.10 (1980) p.221.

9. N.A. KELLY, Env.Sci. and Technol., Vol.16 (1982) p.763.

10. B. DIMITRAIADES, Journ. of Air Poll. Contr. Assoc., Vol.17 (1967) No.7 p.460.

11. P.L. HANST, Adv.Environ.Sci. Technol., Vol.2, (1971) p.91.

12. H. NIKI, P.D. MAKER, C.M. SAVAGE and L.P. BREITENBACH, Spectroscopy in Chemistry and Physics: Modern Trend; Elsevier, Amsterdam 1980 p.1.

13. L. DEMERJIAN and L. SCHERE, Adv.Environ.Sci.Technol., Vol.10 (1980) p.369.

14. N.A. KELLY, General Motors Research Publication 3989 (1982).

15. L.S. ROTHMAN, Applied Optics 20 (1981) p.791.

TEMPERATURE DEPENDENCE OF THE REACTIONS
SO + O$_3$ (1) AND SO + O$_2$ (2)

U. SCHURATH and H.-J. GOEDE
Institut für Physikalische Chemie der Universität Bonn,
Wegelerstr. 12, D-53 Bonn 1

Summary

Two aeronomically important reactions of the SO radical,
(1) SO + O$_3$ → SO$_2$ + O$_2$, and (2) SO + O$_2$ → SO$_2$ + O, have
been studied in the temperature ranges 265 - 364 K, and
220 - 364 K, respectively. The decay of the chemiluminescence due to reaction (1) was measured in a large excess
of O$_3$/O$_2$ under stopped-flow conditions in a glass sphere
of 113 liter volume, at total pressures of less than 200
mtorr. The first order decay occurred on a time scale of
several seconds up to about one minute. The results can
be expressed in Arrhenius form:

$$k_1 = (2.99^{+3.77}_{-1.67}) \times 10^{-12} \exp(-(1070\pm242)/T) \; ;$$

$$k_2 = (1.00^{+0.46}_{-0.32}) \times 10^{-13} \exp(-(2180\pm117)/T) \; .$$

The units are cm^3 molecule^{-1} s^{-1}, and the error limits represent two standard deviations. The results are compared
with available literature data.

1. INTRODUCTION

The SO radical is a short lived intermediate in the combustion of reduced sulphur compounds (1,2), and earlier studies
of its reactions have therefore been confined to high temperatures (3). Relatively little is known about the role of sulphur
monoxide in the photooxidation of H2S and the organic sulphides
in the troposphere. For example, H2S is quantitatively converted to SO2 under atmospheric conditions (4), although the reaction of SH with O2 seems to be slow (5) and does not necessarily release SO as an intermediate. The importance of SO in
the stratospheric sulphur cycle is closely linked with the
source and persistence of the Junge layer (6): Direct transport
of the majority of tropospheric sulphur compounds through the
tropopause is rather inefficient, with the exception of SO2 injections in volcanic eruptions, owing to the short tropospheric
lifetimes of the compounds. The only important exception is COS
which possesses several natural and anthropogenic sources at
the earth's surface, and is probably a product of the tropospheric photooxidation of CS2 (7,8,9). The reaction of OH with
COS has been found to be too slow to measure in the laboratory
(4), and the molecule is thus sufficiently stable in the troposphere to diffuse into the stratosphere, where it is eventually
photolyzed in the 200 nm window between the Hartley band of
ozone and the onset of strong absorption by molecular oxygen
(10). The dissociation products are CO and sulphur atoms which
react nearly instantaneously with molecular oxygen to yield

sulphur monoxide (11). Two competitive pathways of the SO radical are important under stratospheric conditions:

(1) $SO + O_3 \rightarrow SO_2 + O_2$; (2) $SO + O_2 \rightarrow SO_2 + O(^3P)$.

Both reactions yield sulphur dioxide, which is ultimately converted to particulate matter, mainly sulphuric acid droplets, in the altitude range of the Junge layer.

Reaction (1) was first studied by Halstead and Thrush in a flow system (12). Black et al. have re-examined the reaction in an overlapping temperature range, and have measured the temperature dependence of k2 from above to below room temperature for the first time (13,14). They generated SO by ArF excimer laser photolysis of SO2 in a large excess of molecular oxygen and ozone, and measured the decaying chemiluminescence due to reaction (1). They found that k2 was substantially larger than anticipated by extrapolation of the high temperature data (15). However, it was not quite clear to what extent the authors had been able to kinetically isolate reaction (2) from impurity reactions, particularly at low temperatures.

We report an independent investigation of the rate constant k1 and k2 which govern the pathways of SO in the stratosphere, and are potential reference rate constants in kinetic studies of sulphur containing molecule and radical reactions. The results of this work compare well with published rate constants of reaction (1) (12,16), and are in acceptable agreement with the only available study of reaction (2) at medium and low temperatures (13,14).

2. EXPERIMENTAL

The reaction of SO with an excess of O2/O3 was studied at very low total pressures (1 ≤ p ≤ 200 mtorr) in a borosilicate glass sphere of 113 liter volume under stopped flow conditions. The reaction of SO with ozone produces electronically excited SO2 in singlet and triplet states which emit mainly in the wavelength range 280 - 500 nm (12). The reactor was thermally isolated in a light tight metal box embedded in plastic foam. Constant temperatures in the range + 92 to - 52 °C could be maintained by pumping hot water or cold methanol through a long copper tube which was soldered to the inner surface of the metal box, as shown schematically in figure 1. Cooling of the flanges and of the valve mechanism was enforced with dry cold air from a liquid nitrogen cooled heat exchanger. The air inside the box was vigorously stirred with a fan to improve temperature homogeneity. The temperature of the glass body, of the flanges, and in the center of the reactor, were measured with nine calibrated solid state sensors, as marked in figure 1, and automatically recorded in each experiment.

Two stainless steel flanges were attached to the reactor ports with silicon rubber adhesive sealant. The lower flange housed a high vacuum valve separating the reactor from a 150 liter/s turbomolecular pump which allowed the system to be evacuated to 10-7 torr. The valve was especially constructed to equalize the temperature of the pyrex plated valve lid with the reactor temperature, see figure 1. The upper flange supported an evacuated dewar with a large quartz front window facing the reaction chamber. It housed a cooled EMI 9635 QB photomulti-

plier which has an average quantum efficiency of at least 20 %
in the range of the SO2 chemiluminescence. It was used without
filter. Reactant gases were admitted via metal armoured pyrex
tubes which protruded through the quartz window of the reactor.
The total pressure was measured with a 0 - 1 torr Baratron gage
with better than 0.1 mtorr resolution.

Ozone was generated continuously in 1 atm pure oxygen in
an all glass ozonizer. The ozone concentration, which was measured optically at 253.7 nm with a solar blind photomultiplier/
interference filter combination, could be varied in the range
0.3 to 6 % by adjusting the supply voltage of the ozonizer. The
ozonized gas passed on through the straight borehole of a glass
stopcock, volume ca. 70 μl. Its contents could be injected into
the reactor by a half-rotation of the stopcock. The volume of
the borehole was determined by measuring the pressure increase
in the reactor for a large number of injections. Ozone concentrations $\geqslant 5 \times 10^{10}$ molecule/ccm could thus be made up in the
reactor with an estimated uncertainty of $\pm$ 4 %.

In order to initiate the reaction, very small amounts of
SO, of the order of 1 % of the ozone concentration already present, had to be injected into the reactor as quickly as possible. SO was generated in a flow tube by reacting oxygen atoms
from a microwave discharge in He/O2 with a slight excess of COS
at total pressures of a few torr. This reaction is a clean
source of SO radicals if care is taken to sample the product
after short reaction times. Further downstream of the mixing
zone, increasing amounts of secondary products, in particular
of the dimer OSSO and SSO, have been detected in a similar flow
system (17). SO was therefore sampled directly from the mixing
zone in the flow tube, as shown schematically in figure 2: after
closing valve 1 and opening valve 2 by hand, a computer routine
was started which activated the greaseless solenoid valve for
typically 1 - 3 s, injecting the desired amount of SO into the
reactor. The resulting increase in total pressure amounted to
less than 0.1 mtorr. Complete mixing of the reactants in the
vessel required less than 1 s. The intensity of the chemiluminescence was measured as function of time with the photomultiplier, digitized with a fast voltmeter (HP 3437A) at a rate of
400 - 2000 readings/s, depending on the duration of the experiment, and transferred to a HP 9835A desktop computer for later
analysis. Other experimental parameters, such as the background
signal of the photomultiplier immediately before SO was injected, the readings of the temperature sensors, and the total
pressure, were either stored automatically in the computer
memory, or entered via keyboard.

The rate constants k1 and k2 could be derived from the intensity profiles which decayed exponentially under the condition
(SO) $\ll$ (O3), which was safely established in the kinetic runs.
Either the ozone or the molecular oxygen concentrations were
varied in each series of constant temperature runs. Ozone was
also needed as a chemiluminescence indicator of the reaction
rate in measurements of k2, since reaction (2) is non-chemiluminescent under the experimental conditions: no light emission
occurred in the absence of ozone. SO2 chemiluminescence could
have resulted, in principle, from the three-body recombination
of SO with oxygen atoms which are produced in reaction (2). However, this chemiluminescent reaction is extremely inefficient

under the experimental conditions, due to the rapid destruction of the atoms on the reactor wall, and by other non-chemilumi-nescent reactions.

In order to establish the necessary condition $(SO) \ll (O_3)$ in the kinetic runs, the amount of SO injected into the reactor via the capillary had to be measured as function of injection time. This was done by titrating a known amount of ozone in the reactor with a continuous flow of SO through the solenoid valve. In order to keep the titration time reasonably short, the standard capillary used for the kinetic experiments was replaced by a wider and shorter one. It was not possible to find the endpoint of the titration simply by inspection of the intensity profile, figure 3, owing to the relative slowness and complexity of the reactions involved. The following titration mechanism was considered:

(1) $SO + O_3 \quad \rightarrow SO_2^* + O_2 \quad k_{1a}$
$\left. \right\}$
$\quad\quad\quad\quad\quad \rightarrow SO_2 + O_2 \quad k_{1b}$ $\quad k_1 = 8 \times 10^{-14}$

(2) $SO + O_2 \quad \rightarrow SO_2 + O \quad k_2 = 7.2 \times 10^{-17}$

(3) $SO_2^* \quad\quad \rightarrow SO_2 + h\nu \quad k_3$
$\left. \right\}$ $k_3/k_4 = 1.3 \times 10^{15}$

(4) $SO_2^* + M \quad \rightarrow SO_2 + M \quad k_4$

(5) $O + O_3 \quad \rightarrow 2 O_2 \quad k_5 = 9.0 \times 10^{-15}$

(6) O (+ wall) $\rightarrow 1/2\ O_2 \quad k_6 = 1.6 \times 10^{-1}$

(7) SO (+ wall) $\rightarrow$ loss $\quad k_7 = 1.1 \times 10^{-2}$

(8) O_3 (+ wall) $\rightarrow 3/2\ O_2 \quad k_8 = 1.2 \times 10^{-3}$

The listed room temperature rate constants (in ccm molecule^{-1} s^{-1} or 1/s) are either evaluated literature data (11), or were determined in our system. k7 was measured by injecting reproducible amounts of SO and ozone, in that order, into the evacuated reactor. The time interval Δt between the injections was varied systematically in the experiments between a few seconds and 150 seconds. The intensity peak caused by the injection of ozone (acutally obtained by back-extrapolation of the following decay to the moment of injection) is plotted in figure 4 as function of the time interval Δt. It is proportional to the amount of SO still present in the reactor. A computer fit of the data (solid line in figure 4) yielded a first order rate constant k7 = 1.1 ± 0.1 x 10^{-2} which corresponds to a collision efficiency $\gamma = 10^{-5}$ for the destruction of SO on glass. This may be compared with a collision efficiency of $\gamma = 5 \times 10^{-5}$ which was measured for the stainless steel surface of a huge spherical reactor (18). We arrive at a still larger estimate of $\gamma \cong 2 \times 10^{-4}$ for a tubular glass reactor described in reference 17. The discrepancy between the collision efficiencies is thought to result from different surface conditions. - k8 was measured in exact analogy to k7, only reversing the order in which the reactants were injected.

In evaluating the titration profiles, the pressure dependence of the chemiluminescence intensity had to be taken into account. The pressure dependence of the superimposed emissions from a relatively long living triplet (19) and a short living singlet state which can be interconverted to the triplet in collisions or quenched to the ground state (20), is presumably quite complex in the mtorr range. However, in lack of applicable experimental data, simple Stern-Volmer behaviour of the intensity was assumed in the titration scheme, using k3/k4 = 40 mtorr as an effective half-quenching pressure for the regime. This estimate is not a critical parameter in the evaluation, since the titration profile showed a maximum at 6 mtorr already.

Computer simulations of the titration were carried out by means of a suitable kinetics program (21), treating the SO injection rate as a free parameter. Three computed profiles of the chemiluminescence intensity, normalized to the intensity maximum, are compared with the experimental profile in figure 3. The best fit is obtained with a source strength of 1.1 x 10^{10} SO molecules per ccm per s. The tailing of the experimental profile, which is not reproduced by the computer simulation, must be due to oxygen atoms and/or other short living species being sampled from the flow tube. The chemiluminescence in the tail of the titration profile ceased abruptly when the SO valve was closed, and thus did not interfere with the determinations of the SO decay rate.

3. RESULTS

Essentially the same procedure was adopted to determine the rate constants k1 and k2. Each kinetic run was initiated by injecting a small quantity of SO into a large excess of ozone in molecular oxygen, already present in the reactor. The reaction was followed over a minimum of two (usually several more) lifetimes, ranging from a few seconds to about one minute, depending on temperature and reactant concentrations. According to the kinetic scheme (1) - (8), the decay is governed by three reactions:

(A) $\quad - \dfrac{d \ln(SO)}{dt} = k_1(O_3) + k_2(O_2) + k_7 = k_{(c,T)}.$

Three body recombination of SO with itself and with oxygen atoms is too slow at the low radical concentrations and pressures to contribute significantly to the consumption of SO in the system. Heterogeneous removal of ozone on the wall, reaction (8), was negligible on the experimental time scale, as was the destruction of ozone by reactions (1), (2) and (5), owing to the small amount of SO used. In conclusion, the decay of SO in the system was strictly exponential, its time constant k(c,T) being given by equation (A) to a very good approximation. At constant pressure and ozone concentration the chemiluminescence intensity of reaction (1a) varied in proportion with the SO concentration,

(B) $\quad I_t = \dfrac{k_{1a}(O_3)}{1 + (M)k_4/k_3} \times (SO)_t = \text{const.} \times (SO)_t$

and the constant k(c,T) could thus be conveniently obtained from the intensity profile.

Three examples of relatively slow, medium, and fast chemiluminescence decays, at different temperatures and relatively high O2 pressures are reproduced in figure 5. Considerably better signal-to-noise ratios were achieved at lower pressures and higher ozone concentrations. The computed semilog plots are also shown. The faster than exponential decay immediately after the injection of SO is due to excess reactant concentration in front of the photomultiplier where injection takes place. It can be seen that less than 1 s was required for complete mixing of the reactants at total pressures near 100 mtorr, while the effect of incomplete mixing in the intensity profile was not noticeable near 1 mtorr total pressure where mixing by diffusion is faster.

Reactions (1) and (2) were investigated at fixed temperatures by varying either the ozone or the oxygen concentration, while keeping all other parameters constant. The rate constants were obtained from the slopes of k(c,T)-versus-concentration plots. Some raw data pertaining to reaction (2), which were measured at 6 temperatures, are shown in fiugure 6. The irregular variation of the intercepts in figure 6 with temperature is due to variations in the amount of ozone used at the particular temperatures, which mask the smooth T-dependence of k7.

The rate constants thus obtained for reactions (1) and (2) are plotted in Arrhenius form in figures 7 and 8 (diamond shaped symbols and solid lines). The vertical diagonal of the diamonds equals the statistical errors (2σ) of the measurements, while the horizontal width indicates the actual spread of 7 temperature readings on the reactor surface and in its center. The flanges, which cover less than 2 % of the inner surface, could be up to 3 K warmer or colder than the reactor. The marked diamond in figure 7 derives from 20 measurements of k1 at 290 K in a stainless steel reactor of 220,000 liter volume (22). The experimental procedure was basically the same in both systems, but the range of ozone concentrations studied in the big sphere (0.5 to 6.5 x 10^{11} molecule per ccm) was nearly an order of magnitude below the range in the glass vessel.

The solid lines in figures 7 and 8 represent the following fitted Arrhenius expressions (error limits represent 2 standard deviations):

$$k_1 = (2.99^{+3.77}_{-1.67}) \times 10^{-12} \exp(-(1070 \pm 242)/T) \ ,$$

$$k_2 = (1.00^{+0.46}_{-0.32}) \times 10^{-13} \exp(-(2180 \pm 117)/T) \ .$$

The units are ccm per molecule per s. Available literature data on k1 and k2 are plotted in figures 7 and 8 for comparison (circles, filled squares, and dotted lines). Error bars of individual measurements are also shown, where possible.

4. DISCUSSION

The potentially largest systematic error in the present determination of k1 is believed to involve the ozone concentration which rosulted after injection of a small amount of ozonized oxygen into the reaction vessel. The possibility of heterogeneous ozone loss in this procedure cannot be entirely excluded, particularly at the highest temperature of 364 K. However, the ozone loss could be delimited to less than 5 % for the set of rate determinations in the big stainless steel reactor at 290 K (results marked by arrow in figure 7), which agree well with measurements in the glass vessel at adjacent temperatures. Optical in situ measurements of the ozone concentration were possible in the big stainless steel reactor, as described in reference 22, which yielded excellent agreement with the expansion method used in this work.

Four independently measured room temperature rate constants and three sets of Arrhenius parameters, based on different techniques and experimental conditions in partially overlapping temperature intervals, are now available for reaction (1). Table 1 summarizes the data at 298 K and 220 K which delimit an important atmospheric temperature range. The averages of k1, taken separately at both temperatures, fit the following Arrhenius law:

$$k_1 = 3.54 \times 10^{-12} \exp(-1115/T) \ cm^3 \ molecule^{-1} \ s^{-1}.$$

The experimental results at all temperatures, cf. figure 7, are scattered around this function within a band of ± 25 % width. The function digresses from the CODATA evaluation of 1982 (11) by +4 to +5 %, and should be reliable within confidence limits of $\Delta \log k = 0.1$, in the temperature range 220 - 420 K.

Unlike the measurements pertaining to reaction (1), our determination of k2 should be unaffected by systematic errors in ozone concentration. The Baratron gage which measured the oxygen pressure in the reactor was found to be accurate within ± 2 %. The instrument was compared with a conventional mercury barometer by expanding 1 atm He contained in a 100 ml flask into the reactor. The volume of the reactor (113.5 ± 0.1 liter) had been previously determined with the same barometer by pV-measurements, using a comparison volume of 20 liter. Interference by a reactive impurity in our tank oxygen (99.998 % purity) seems unlikely in view of the fact that our rate constants fall 15 - 30 % short of the Arrhenius expression given by Black et al. (14), in the overlapping temperature range 258 - 364 K, cf. figure 8. The reaction of SSO with ozone is non-chemiluminescent (23), but nothing is known about ozone reactions of other SO-byproducts, such as OSSO. The effect of these secondary products was tested as follows: The concentration ratio of SO and its secondary products was varied by changing the pressure and the distance between the mixing zone and the sampling point in the flow tube, figure 2. The rate determinations were unaffected by these variations, within experimental scatter, although a weak background of a long-living chemiluminescence could be detected under unfavourable conditions.

Table 2 summarizes the available rate constants for reaction (2) at 298 and 220 K. Clearly the CODATA evaluation of 1982 (11) has to be revised considerably beyond its estimated uncertainty limits in this important temperature range. The large estimated uncertainties in the measurements of Black et al., figure 1 in reference 14, and the fact that their Arrhenius fit was drawn slightly in favour of the lower values in the low temperature part of their study, renders the difference in activation energies, 9915+835, -1045 kJ/mol in reference 14 versus 9120±490 kJ/mol in this work, rather insignificant. The Arrhenius A factor of the reaction, which is of the order 10^{-13} ccm per molecule per s, might seem rather low for a simple atom transfer reaction. It should be noted, however, that the negative entropy change of $\Delta S = -17.6$ J per mol per K (24) reduces the A factor of reaction (2) by a factor $A_2/A_{-2} = \exp(-\Delta S/R) = 0.12$ with respect to the reverse reaction

(-2) $SO_2 + O \rightarrow SO + O_2$,

which has been studied above 3000 K (25). An equally low A

factor of 1.2 x 10^{-13} has been reported for the atom transfer reaction NO2 + O3 $\rightarrow$ NO3 + O2 which involves a negative entropy change of ΔS = -20.9 J per mol per K (24).

We conclude that the important rate constant ratio k1/k2 is now sufficiently well characterized for atmospheric chemistry applications. Reaction (1) is of minor importance (< 10 %) as a sink of SO radicals at all tropospheric and stratospheric temperatures and ozone mixing ratios. However, improved measurements of k2, particularly of its temperature dependence in the intermediate range between the more recent studies (reference 14 and this work) and the high temperature determination of Homann et al. (3), would improve our understanding of this simple atom transfer reaction.

Acknowledgement: Support of this work by the Deutsche Forschungsgemeinschaft is gratefully acknowledged.

Table 1: Rate constants k1 in ccm per molecule per s, and their averages, at 298 K and 220 K. Data in brackets are not included in the averages.

10^{14} x k1, 298 K	10^{14} x k1, 220 K	references
7.2 ± 0.5	2.04 ± 0.15	Halstead and Thrush (12)
8.7 ± 1.6		Robertshaw and Smith (16)
(10.6 ± 1.6)		Black et al. (13)
$(7.9^{+3.3}_{-2.9})$	$(2.1^{+1.3}_{-0.8})$ *)	CODATA (11)
9.5 ± 1.5	$3.25^{+0.25}_{-0.30}$ *)	Black et al. (14)
8.2 ± 1.0	$2.3^{+0.7}_{-0.6}$ *)	this work
8.4	2.23	average

Table 2: Rate constants k2 in ccm per molecule per s at 298 K and 220 K.

10^{17} x k2, 298 K	10^{18} x k2, 220 K	references
$0.9^{+1.9}_{-6}$	$0.18^{+0.14}_{-0.08}$ *)	CODATA (11)
10.7 ± 1.6		Black et al. (13)
8.4 ± 1.5	5.0 ± 1.3 *)	Black et al. (14)
7.03 ± 0.35	5.0 ± 0.5	this work

*) Extrapolated Arrhenius fits. The upper and lower bounds refer to the uncertainty in E/R only and do not include the uncertainty in the A factor.

5. REFERENCES

1   E.L. Merryman and A. Levy, J.Phys.Chem. 76,1925-1931(1972)
2   A. Levy, E.L. Merryman and W.T. Reid, Environ.Sci.Technol. 4,653-662(1970)
3   K.H. Homann, G. Krome and H.Gg. Wagner, Ber.Bunsenges. 72, 998-1004(1968)
4   R.A. Cox and D. Sheppard, Nature 284,330-331(1980)
5   M.B. McElroy, S.C. Wofsy and N. Dak Sze, Atmos.Environ. 14, 159-163(1980)
6   P.J. Crutzen, Geophys.Res.Lett. 3,73-76(1976)
7   A.R. Bandy, B.J. Maroulis, L. Shalaby and L.A. Wilner, Geophys.Res.Lett. 8,1180-1183(1981)
8   P.H. Wine, W.L. Chameides and A.R. Ravishankara, Geophys. Res.Lett. 8,543-546(1981)
9   I. Barnes, K.H. Becker, E.H. Fink, A. Reimer, F. Zabel and H. Niki, Int.J.Chem.Kinet. 15,631-645(1983)
10 L.T. Molina, J.J. Lamb and M.J. Molina, Geophys.Res.Lett. 8,1008-1011(1981)
11 D.L. Baulch, R.A. Cox, P.J. Crutzen, R.F. Hampson Jr., J.A. Kerr, J. Troe and R.T. Watson, J.Phys.Chem.Ref.Data 11,327-496(1982)
12 C.A. Halstead and B.A. Thrush, Proc.Roy.Soc. A295,380-398 (1966)
13 G. Black, R.L. Sharpless and T.G. Slanger, Chem.Phys.Lett. 90,55-56(1982)
14 G. Black, R.L. Sharpless and T.G. Slanger, Chem.Phys.Lett. 93,598-602(1982)
15 D.L. Baulch, R.A. Cox, R.F. Hampson, J.A. Kerr, J. Troe and R.T. Watson, J.Phys.Chem.Ref.Data 9,295-471(1980)
16 J.S. Robertshaw, I.W.M. Smith, Int.J.Chem.Kinet. XII,729-739(1980)
17 J.T. Herron, R.E. Huie, Chem.Phys.Lett. 76,322-324(1980)
18 Earlier unpublished measurements of k1 in a 220,000 liter stainless steel reactor, see also caption to figure 7
19 Fu Su, J.-W. Bottenheim, D.L. Thorsell, J.G. Calvert and E. K. Damon, Chem.Phys.Lett. 49,305-311(1977)
20 J. Heicklen, N. Kelley and K. Partymiller, Rev.Chem.Intermed. 3,315-404(1980)
21 E.M. Chance, A.R. Curtis, I.P. Jones and C.R. Kirby, A.E.R. E. Report R-8775, Harwell, December 1977
22 U. Schurath, H.H. Lippmann and B. Jesser, Ber.Bunsenges. Phys.Chem. 85,807-813(1981)
23 D.H. Stedman, H. Alvord and A. Baker-Blocker, J.Phys.Chem. 78,1248-1250(1974)
24 Thermocehmical Kinetics, 2nd Ed., S.W. Benson, John Wiley & Sons, 1976
25 M. Slack and A. Grillo, J.Chem.Phys. 73,987-988(1980)

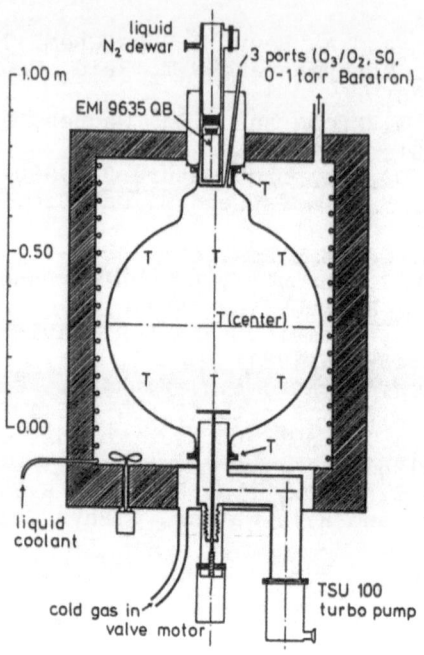

Figure 1: Schematics of the thermostated reactor.

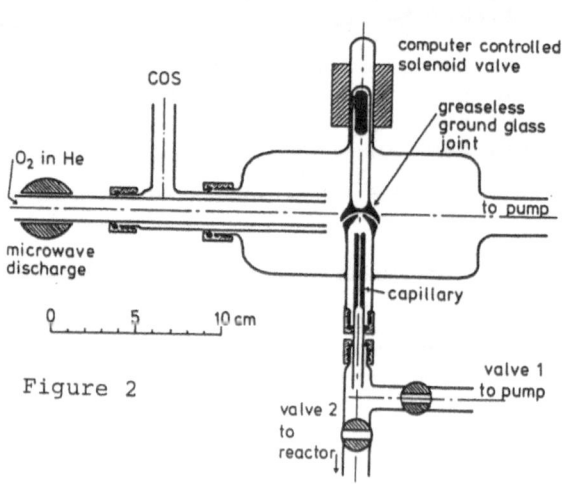

Figure 2

Figure 2: Schematics of the SO generator and valve.

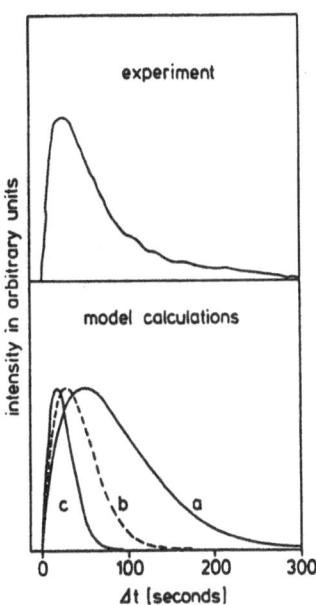

Figure 3: Titration of 2.1 x 10^{11} molecule per ccm ozone in the reactor with SO in He from the generator flow tube. Upper trace: Experimental results. Lower traces: Model calculations assuming source strengths of a) 3 x 10^9; b) 1.1 x 10^{10}; c) 3 x 10^{10} molecule per ccm per s of SO.

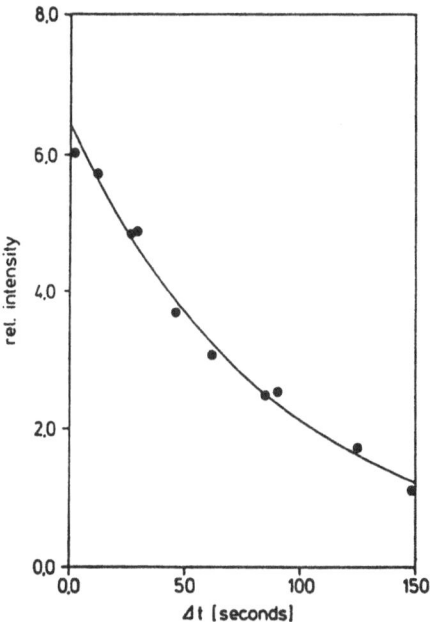

Figure 4: Determination of k7: Peak intensities are plotted for ozone injections into the reactor, Δt s after SO was injected.

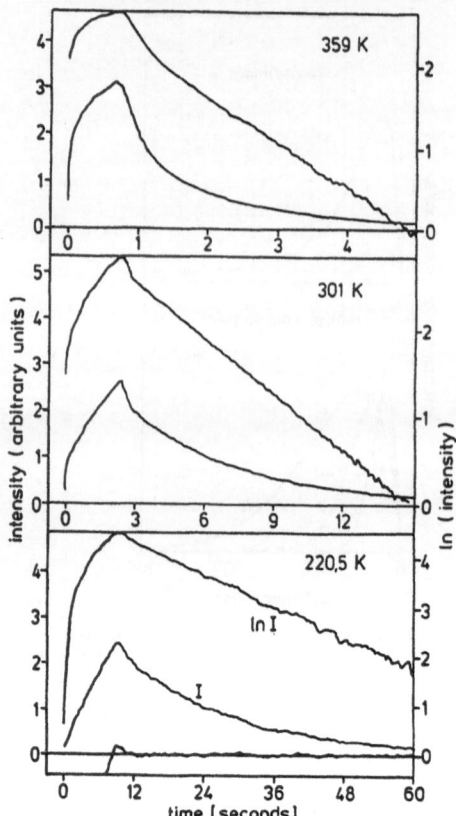

Figure 5: Experimental profiles and semilog plots of chemiluminescence intensity in measurements of k2. Initial concentrations
(molecule per ccm) of the reactants were (O3) = 2.8 x 10^{11}, (O2)
= 2.7 x 10^{45} at 359 K; (O3) = 7.1 x 10^{11}, (O2) = 2.0 x 10^{45} at
301 K; (O3) = 5.6 x 10^{11}, (O2) = 3.0 x 10^{15} at 220.5 K.

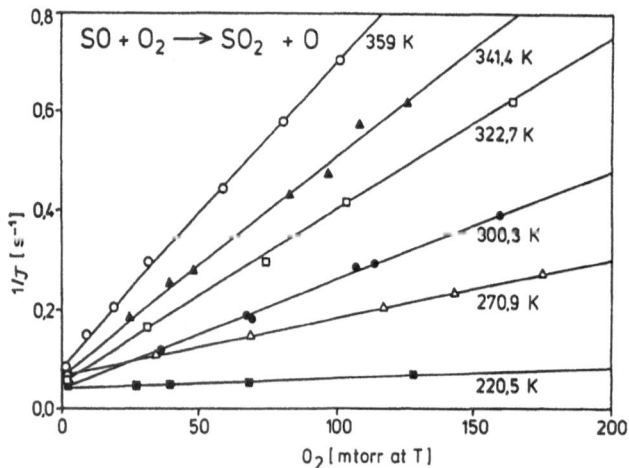

Figure 6: Decay constants k(c,T) ≡ 1/τ as function of O2 pressure. At each temperature a fixed ozone concentration in the
range 1.5 to 9 x 10^{11} molecule per ccm was used.

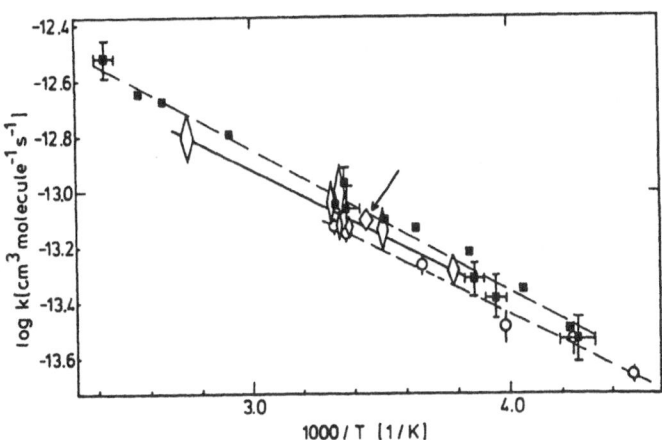

Figure 7: Arrhenius plot of k1: filled squares and dotted line, reference 14; open circles and dotted line, reference 12;, diamonds and solid line, this work. The arrow marks the rate constant which was obtained at 290 K in a stainless steel reactor of 2.2 x 10^{5} liter volume, as explained in the text.

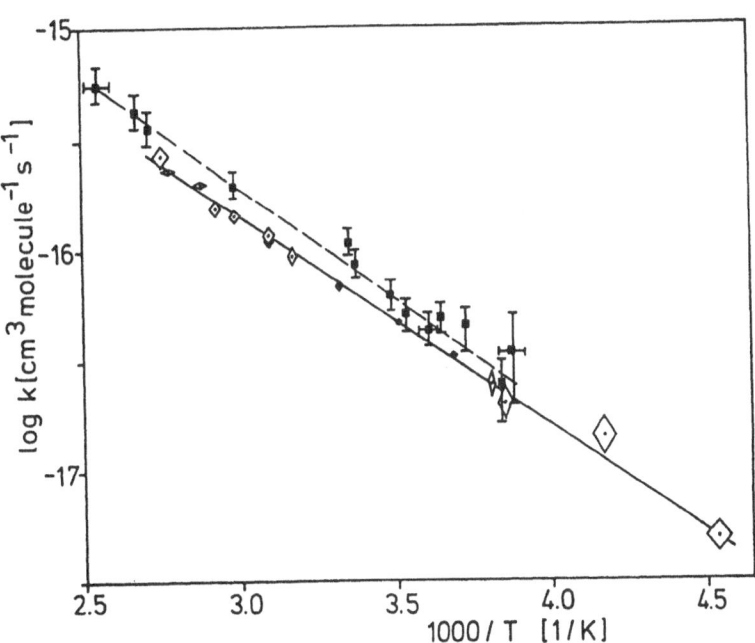

Figure 8: Arrhenius plot of k2: filled squares and dotted line, reference 14; diamonds, this work. The sizes of the diamond shaped symbols are explained in the text.

A STUDY OF N_2O_5 AND NO_3 CHEMISTRY AND THE PHOTOLYSIS OF N_2O_5 MIXTURES

J.P.Burrows, G.S.Tyndall and G.K.Moortgat
Max Planck Institut für Chemie
Postfach 3060,D-6500 Mainz, F.R.G.

Summary
    Results are presented from a recent study of the chemical behaviour of of NO_3 and N_2O_5 in flowing mixtures of NO_3 and N_2O_5. The absorption spectrum of NO_3 has been determined in the photolysis of Cl_2, $ClNO_3$ and NO_2 mixtures. NO_2, NO_3 and N_2O_5 have been observed directly by their U.V.-visible absorptions. N_2O_5 mixtures were photolysed using low pressure Hg lamps driven by a novel 'oblong' wave generator:

$$N_2O_5 + h\nu(254nm) \rightarrow NO_2 + NO_3 \qquad (1a)$$

$$\rightarrow O + NO_2 + NO_2 \qquad (1b)$$

$$\rightarrow NO + NO_2 + O_2 \qquad (1c)$$

$$\rightarrow quenching \qquad (1d)$$

In a limited set of experiments the equilibrium for reactions (2) and (-2),

$$N_2O_5 + M = NO_2 + NO_3 + M \qquad (2)$$

was investigated at 294K and 30 torr, where

$$K_2 = [NO_2] \times [NO_3] / [N_2O_5]$$

The value obtained for K_2 is in good agreement with the literature value. The recombination reaction k_{-2} has been studied directly in the photolysis of Cl_2, $ClNO_3$ and NO_2 or F_2, HNO_3 and NO_2 mixtures between 20 and 40 torr. The flowing photolysed system has been simulated using the LARKIN computer program. Analysis yields the following branching ratios:

$$k_{1a}/k_1 = 0.75 \text{ and } (k_{1b}+k_{1c})/k_1 = 0.25$$

for an N_2O_5 concentration of 4×10^{14} molecule cm^{-3}.

1. INTRODUCTION

    The role played by N_2O_5 in the chemistry of the atmosphere is not yet fully understood. However it is a potential temporary reservoir for NO_x (NO, NO_2, NO_3) as it is produced by the association of NO_2 and NO_3 and it thermally decomposes to regenerate the same radicals:

$$N_2O_5 + M = NO_2 + NO_3 + M \qquad (2)$$

N_2O_5 is known to react heterogeneously with H_2O to produce HNO_3:

$$N_2O_5 + H_2O = 2HNO_3 \qquad (3)$$

Since the precursor of N_2O_5, NO_3, has a large visible photolysis rate it is likely that N_2O_5 will be in larger concentrations only at night. Due to its thermal instability N_2O_5 is able to play a variety of roles in the atmosphere, being formed in cooler regions of the atmosphere and releasing radicals in warmer regions.

The chemistry of N_2O_5 has been investigated for many years; as early as 1875 Bertholet attempted to measure the heat of solution of N_2O_5 (1). However the instability and heterogeneous reactivity of N_2O_5 pose serious problems to the experimental investigation of its chemistry. Since the early 1950's several aspects of the gas phase chemistry of N_2O_5 have been investigated. The studies of Johnston and coworkers on the decomposition of N_2O_5 (2-6) appear to be in good agreement. Malko and Troe (7) in a theoretical study of reaction (1) determined limiting low pressure and high pressure values for k_1 from the data of Connell and Johnston (6), and Viggiano et al. (8). Schott and Davidson studied the equilibrium (1) in a shock tube from 450-550K (9). Graham and Johnston investigated equilibrium (1) by observing the decomposition of O_3 in the presence of N_2O_5 (10). Results are presented here from a study of the chemistry of NO_3 and N_2O_5. The absorption cross section of NO_3 has been determined and the rate coefficient of the reaction between NO_2 and NO_3 has been measured. Both the N_2O_5 equilibrium and the products of the photolysis of N_2O_5 at 254nm have been determined.

2. EXPERIMENTAL

The apparatus used to study N_2O_5 chemistry is shown schematically in Figure 1. The reaction vessel is at present a quartz tube 145cm long and 3.5cm in diameter. The reactants enter the vessel through 4 inlet jets simultaneously. The viscous pressure drop between the jets is designed to be insignificant. The products are pumped out of the cell through 3 outlet jets, which are situated halfway between the inlet jets, and also have negligible viscous pressure drop between them. For the photolysis experiments, low pressure Hg lamps were used and for the calibration of the NO_3 absorption cross section blacklamps were used. The output of the low pressure Hg lamps is predominantly at 254nm; however small amounts of the other Hg emission lines are observed. The output of the blacklamps is maximum at 350nm

The reaction vessel approximates to a well-mixed reactor and the residence time, t_{res}, is given by,

$$t_{res} = V/F$$

where V is the volume of the cell (1.0 litre) and F is the volume flow rate through the cell.

Figure 1.

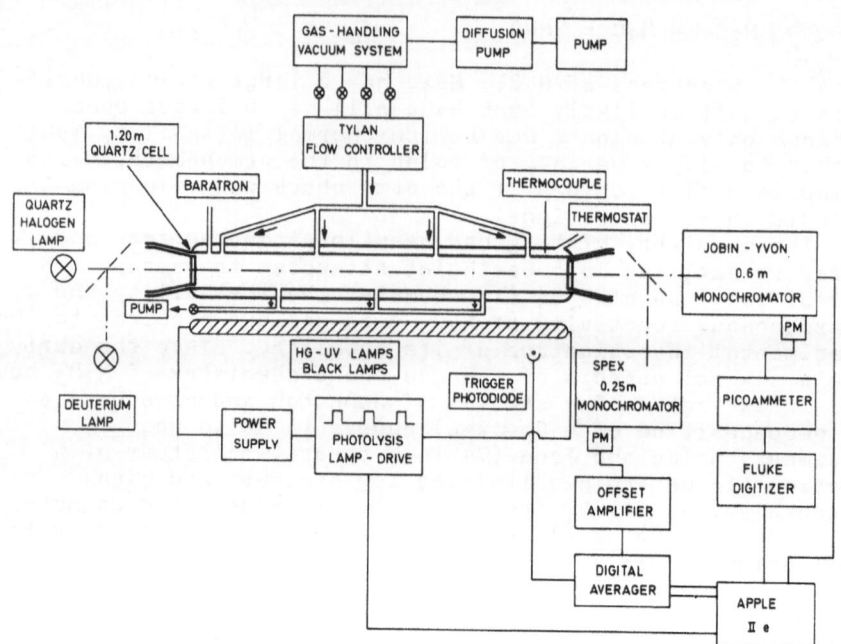

UV-VIS ABSORPTION MODULATED PHOTOLYSIS APPARATUS

N_2O_5 was prepared by repeatedly flowing an O_3-NO mixture between two traps at 195K (11). A flow of N_2O_5 was produced by passing N_2 through a trap held at 195K containing solid N_2O_5 and subsequently mixing this with a second larger flow of N_2 prior to entry into the reaction vessel.

The photolysis were driven by a novel waveform generator. The light output intensity could be varied stepwise from a square wave to an oblong wave in which the off time is up to ten times the on time. The N_2O_5 concentration was determined by its absorption at 220nm using a D_2 lamp, a 0.3m spectrometer with a resolution of 5nm and cross sections taken from a recent compilation (12). NO_2 and NO_3 were both monitored using light from a quartz halogen lamp. NO_2 concentration was determined from the absorption at 350nm recorded using a 5nm resolution and appropriate cross section. The absorption cross section of NO_2 has been measured and agrees with published values in the 300-400nm region at 298K (13). A resolution of 1nm was chosen for NO_3 detection and a 610nm glass filter was used to remove scattered light. A second 610nm cut off filter was used to investigate any photlysis of NO_3 by the monitoring lamp. This was found to be negligible,(<10%). The output from the photomultiplier was sent either to a chart recorder or a digital signal averager, triggered from the photolysis lamps. In this way time-resolved growth and decay curves of NO_2 and NO_3 could be obtained for time periods varied between 20ms and 20s. Chemical information can be obtained from both the short time behaviour and the grosser changes at longer times.

The chemical system was modelled using the LARKIN computer program developed for chemical kinetic simulation by Deuflhard et al. (14). The scheme of reactions used to model the system is listed in Table I.

Table 1. Reaction mechanism used to simulate the photolysed flowing mixture.

Reaction	$k^{(a)}$
$N_2O_5 + N_2 \rightarrow NO_2 + NO_3 + N_2$	1.2×10^{-21}
$NO_2 + NO_3 + N_2 \rightarrow N_2O_5 + N_2$	$5 \times 10^{-31} cm^6 molecule^{-2} s^{-1}$
$NO_3 + NO_3 \rightarrow NO_2 + NO_2 + O_2$	2.4×10^{-16}
$N_2O_5 + h\nu \rightarrow NO_2 + NO_3$	Varied, s^{-1}
$\rightarrow O + NO_2 + NO_2$	Varied, s^{-1}
$\rightarrow NO + NO_2 + O_2$	Varied, s^{-1}
$NO_2 + h\nu \rightarrow NO + O$	$3.4 \times 10^{-4} s^{-1}$
$O + NO_2 \rightarrow NO + O_2$	9.0×10^{-12}
$O + NO_3 \rightarrow NO + O_2$	1.0×10^{-11}
$NO + NO_3 \rightarrow NO_2 + NO_2$	2.0×10^{-11}
$N_2O_5 \rightarrow$ flow out	$k_f, s^{-1(b)}$
$NO_3 \rightarrow$ flow out	k_f, s^{-1}
$NO_2 \rightarrow$ flow out	k_f, s^{-1}
flow in $\rightarrow N_2O_5$	$F(N_2O_5)^{(c)}$
flow in $\rightarrow NO_3$	$F(NO_3)$
flow in $\rightarrow NO_2$	$F(NO_2)$

Notes:
(a) Units in cm^3 molecule$^{-1}s^{-1}$ unless otherwise stated.
(b) $k_f = 1/t_{res}$.
(c) $F(X) = k_f X$ molecule $cm^{-3}s^{-1}$.
(d) $k_{NO2+NO3}$ and $F(X)$ were adjusted to give the observed concentrations without photolysis. $k_{N2O5+N2}$ is the effective second order rate at 30 torr and 294.

3. RESULTS

The NO_3 absorption spectrum was determined in the following manner. Flows of Cl_2, $ClNO_3$ and NO_2 in N_2 were passed through the photolysis cell and the mixture photolysed using blacklamps driven by the oblong wave generator. The absorption at 623nm was observed to increase to a maximum with the lights on and to decrease to zero when the lights went off. This behaviour is attributed to the following mechanism,

$$Cl_2 + h\nu(310-400nm) \rightarrow Cl + Cl \qquad (4)$$

$$Cl + ClNO_3 \rightarrow Cl_2 + NO_3 \qquad (5)$$

$$NO_2 + NO_3 + M \rightarrow N_2O_5 + M \qquad (-2)$$

The absorption cross section of NO_3 is obtained from a knowledge of the stationary state concentration of NO_3, the rate of production of NO_3 and its decay. At the stationary state, NO_3 concentration is given by,

$$[NO_3] = 2k_4[Cl_2]/k_0$$

where k_0 is the effective first order loss rate for NO_3 in the system. First order plots for the rise and decay of NO_3 yielded values of k_0 within 10% of each other indicating that the decay processes in the dark are the same as those with the lights on. The Cl_2 photolysis rate, k_4, was determined by measuring the decay of Cl_2 concentration in the photolysis of H_2, O_2 and Cl_2 mixtures. The NO_3 absorption cross section at 623nm was determined to be $(1.17\pm0.10) \times 10^{-17} cm^2 molecule^{-1}$. Measuring the NO_3 concentration as a function of wavelength enabled the cross section of NO_3 to measured relative to the absolute value obtained at 623nm. The spectrum of NO_3 obtained here is shown in figure 2.

When the ratio $[ClNO_3]/[Cl_2]$ in the photolysis mixture was reduced, the NO_3 loss processes in the lights on time appeared to be larger than the loss processes in the dark. This behaviour is attributed to a reaction between Cl and NO_3 the most thermodynamically favourable products being ClO and NO_2:

$$Cl + NO_3 \rightarrow ClO + NO_2 \qquad (6)$$

From the steady state concentration of NO_3 the ratio and a simple analysis the ratio k_6/k_5 lies in the range 2-15, an exact figure requires a more detailed analysis.

Measuring the decay rate of NO_3 in the modulated photolysis of Cl_2, $ClNO_3$ and NO_2 mixtures and F_2, HNO_3 and NO_2 mixtures enables the rate coefficient for the reaction between NO_3 and NO_3 to be determined. Measurements were made between 20 and 40 torr. The rate coefficient k_{-2} was found to be $(3\pm1) \times 10^{-13} cm^3 molecule^{-1}s^{-1}$ at 20torr and $(5\pm1.5) \times 10^{-13} cm^3 molecule^{-1}s^{-1}$ at 40torr total pressure of N_2 and 294K.

Figure 2. Absorption spectrum of NO_3.

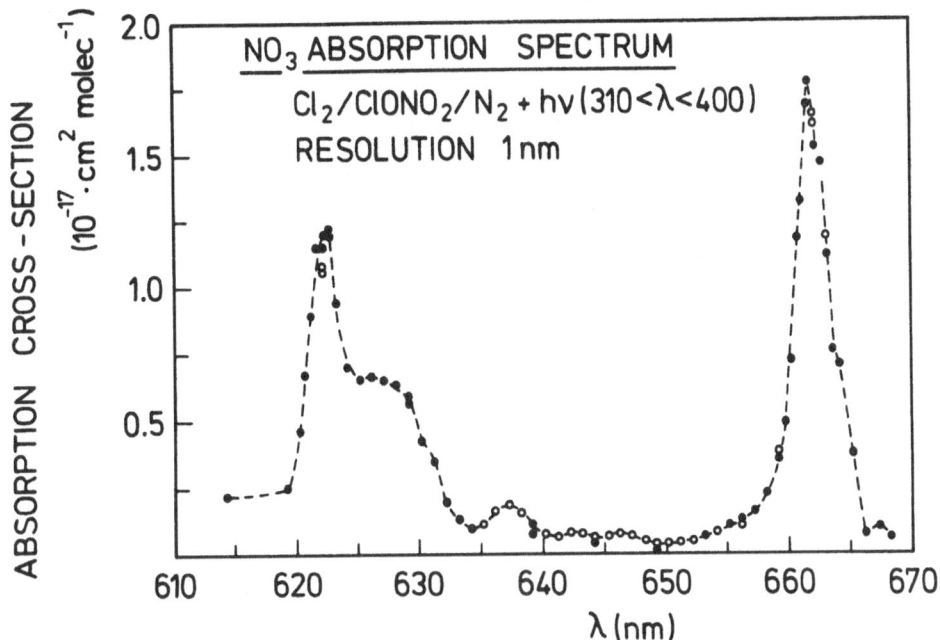

Figure 3. NO_3 concentation modulation in the photolysis of N_2O_5 at 254nm.

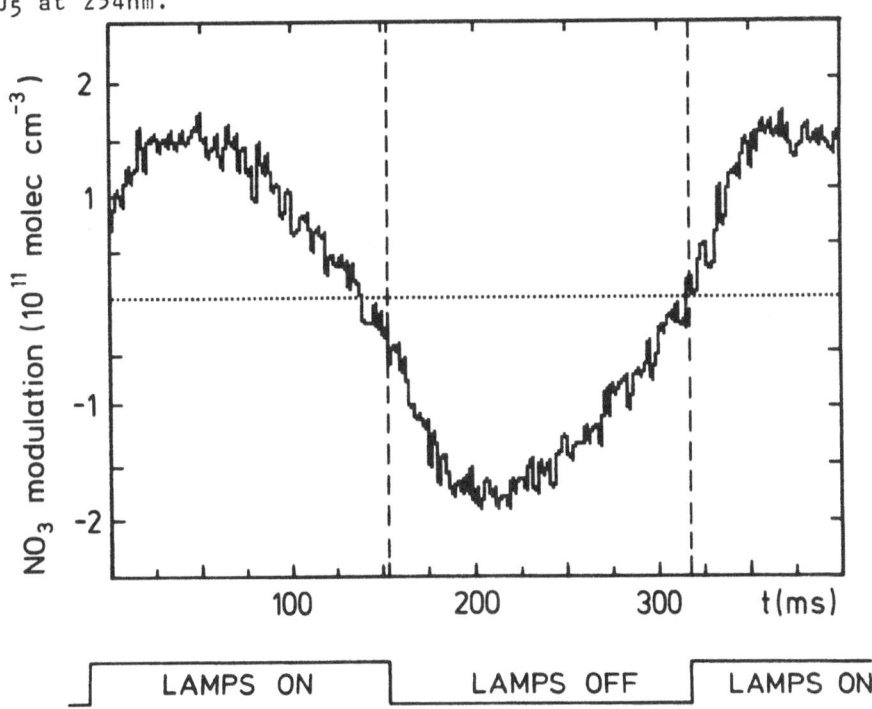

In a limited set of experiments NO_2, NO_3 and N_2O_5 concentrations were measured simultaneously in flowing mixtures of N_2O_5 in N_2. Limited sensitivity to NO_2 restricted the range of N_2O_5 concentrations usable. For N_2O_5 concentrations in the range 3-20×10^{14} molecule cm^{-3}, K_2 was determined to be $(6 \pm 2) \times 10^{10}$ molecule cm^{-3}, verifying the determination of Graham and Johnston (10).

Finally flowing mixtures of NO_2, NO_3 and N_2O_5 at equilibrium in N_2 were photolysed using low pressure Hg lamps driven by the modulated waveform generator. On a time scale of several seconds, NO_3 appeared to decay linearly and NO_2 to increase linearly. However, when short modulation times were used and the signal averaged over many cycles, it was found that the NO_3 first increased, passed through a maximum after typically 100ms and then decayed. This leads to the NO_3 apparently lagging behind the lamps as shown in figure 3. This behaviour is consistent with the following photolysis steps:

$$N_2O_5 + h\nu(\lambda=254nm) \ -> \ NO_2 + NO_3 \qquad (1a)$$

$$-> \ O + NO_2 + NO_2 \qquad (1b)$$

$$-> \ NO + NO_2 + O_2 \qquad (1c)$$

$$-> \ quenching \qquad (1d)$$

The pathways (1b) and (1c) are chemically equivalent in this system, since they both lead to the overall production of NO_2 through the rapid reactions (7), (8) and (9):

$$O + NO_2 \ -> \ O_2 + NO_2 \qquad (7)$$

$$O + NO_3 \ -> \ O_2 + NO_2 \qquad (8)$$

$$NO + NO_3 \ -> \ NO_2 + NO_2 \qquad (9)$$

Combining the reactions in the following manner, (1b) with (7) or (1b) with (8) and (1c) with (9) leads to an overall stoichiometry of

$$N_2O_5 + NO_3 + h\nu \ -> \ 3NO_2 + O_2$$

This accounts for the long time chemistry observed. The system is further complicated by the condition that N_2O_5, NO_3 and NO_2 are related through the equilibrium reactions (2) and (-2). The overall stoichiometry at any time is consequently not described simply by integer values.

A simple analysis of the system for $[N_2O_5] = 4 \times 10^{14}$ molecule cm^{-3} yields:

$$k_{1a} = 2.4 \times 10^{-3} \ s^{-1}$$

$$k_{1b} + k_{1c} = 6.0 \times 10^{-4} \ s^{-1}$$

From a comparison of the N_2O_5 absorption cross section with those of NO_2 and H_2O_2, and the measured photodissociation rates of NO_2 and H_2O_2 in the system, the overall value for k_1 is estimated to be $(3+0.5) \times 10^{-3} s^{-1}$. The quenching pathway (1d) appears therefore to be negligible for 254nm radiation and at this concentration of N_2O_5. A full analysis using both computer simulation and least squares fitting methods is planned to complete the study.

4. DISCUSSION

The results presented here represent some preliminary studies of the N_2O_5-NO_2-NO_3 system. The recombination reaction between NO_2 and NO_3 has not been studied directly previously and our measurement of the equilibrium K_2 is novel in that all three components NO_2, NO_3 and N_2O_5 are measured directly. The agreement with literature values of K_2 and the calculated values of k_{-2} validates the methods used here. The absorption spectrum of NO_3 reported here is in good agreement with recent high resolution studies.

Preliminary results of the photolysis of N_2O_5 suggest that at 254nm pathway (1a) is the dominant channel. Two reports from Johnston's group suggest that considerable quenching takes place when N_2O_5 is photolysed (15,16). In contrast we find little evidence for self quenching.

Further work is planned in this study, extending the measurement of k_{-2} to 1 atmosphere total pressure and improving the data base in the N_2O_5 photolysis.

5. ACKNOWLEDGEMENTS

This work has been supported by the Max Planck Gesell-schaft and the Deutsche Forschung Gesellschaft. The authors are indebted to G.Schuster who designed and constructed parts of the apparatus.

6. REFERENCES

(1) M.Bertholet, Ann.Chim.Phys.(5), 4 8 (1875).
(2) R.L.Mills and H.S.Johnston, J.Am.Chem.Soc., 73 938 (1951).
(3) H.S.Johnston and R.L.Perrine, J.Am.Chem.Soc., 73 4782 (1951).
(4) H.S.Johnston, J.Am.Chem.Soc., 75 1567 (1953).
(5) D.J.Wilson and H.S.Johnston, J.Am.Chem.Soc., 75 5763 (1953).
(6) P.S.Connell and H.S.Johnston, Geophys.Res.Lett., 6 533 (1979).
(7) M.W.Malko and J.Troe, Int.J.Chem.Kin., 14 399 (1982).
(8) A.A.Viggiano, J.A.Davidson, F.C.Fehsenfeld and E.E.Fergusson, J.Chem.Phys., 74 6113(1981)
(9) G.Schott and N.Davidson, J.AM.Chem.Soc.,73 2948 (1951).
(10) R.A.Graham and H.S.Johnston, J.Phys.Chem.,82 254 (1978).

(11) J.A.Davidson, A.A.Viggiano, C.J.Howard, I.Dotan, and F.
     C.Fehsenfeld, J.Chem.Phys., $\underline{68}$ 2085 (1978).
(12) D.L.Baulch, R.A.Cox, R.F.Hampson, J.A.Kerr, J.Troe and
     R.T.Watson, J.Phys.Chem.Ref.Data, $\underline{9}$ 295 (1980)
(13) W.Schneider and J.P.Burrows, unpublished results.
(14) P.Deuflhard, G.Baker and U.Nowak, LARKIN a software
     program for the simulation of large systems in chemical
     kinetics, Univ. Heidelberg, SFB 123: TECHN.REP.100
     (1980).
(15) P.S.Connell, Ph.D. Thesis, University of California
     at Berkeley, LBL Report No 9034 (1979).
(16) F.Magnotta, Ph.D. Thesis, University of California at
     Berkeley, LBL Report No 9981 (1979).

A STUDY OF THE REACTION BETWEEN ClO AND NO₂ USING MATRIX ISOLATION FTIR SPECTROSCOPY AND UV-VISIBLE SPECTROSCOPY

J.P.Burrows, G.S.Tyndall, G.K.Moortgat, D.W.T.Griffith,
Max Planck Institut fur Chemie, Postfach 3060
D-6500 Mainz, F. R. G.

Summary

The products of the reaction between ClO and NO_2 have been investigated by photolysing flowing mixtures of Cl_2, Cl_2O, NO_2 and N_2 with blacklamps (310-400nm):

$$ClO + NO_2 + M \rightarrow ClONO_2 + M \qquad (1a)$$

$$\rightarrow \text{isomers} + M \qquad (1b)$$

The yields $R_1 = [ClONO_2]_{ir}/[ClO]$, $R_2 = [ClONO_2]_{ir}/[NO_2]$, $R_3 = [NO_2]/[ClO]$ and $R_4 = [ClONO_2]_{ir}/[ClONO_2]_{uv}$ at 298K and a total pressure of 22 torr are (1.0 ± 0.2), (1.0 ± 0.3), (1.0 ± 0.3) and (0.9 ± 0.3) respectively. No kinetic or spectroscopic evidence was found for the formation of isomer by reaction (1) over the temperature range 253K-298K.

1. INTRODUCTION

The potential depletion of stratospheric ozone resulting from the tropospheric release of chlorofluorocarbon compounds (1) necessitates a detailed understanding of the elementary reactions which control the O_3 budget. The reaction between ClO and NO_2 plays an important role in models of the chemistry of the stratosphere as it couples together the $O_x(O,O_3)$ catalytic destruction cycles by $NO_x(NO,NO_2)$ and $ClO_x(Cl,ClO)$ species and generates $ClONO_2$, a temporary reservoir for NO_x and ClO_x in the stratosphere (2,3):

$$ClO + NO_2 + M \rightarrow ClONO_2 + M \qquad (1a)$$

$$\rightarrow \text{isomers} + M \qquad (1b)$$

Reaction (1) has been studied recently using a variety of detection methods and kinetic techniques (4 to 9). Several investigations monitored the pseudo first order decay of ClO in the presence of excess NO_2 and agree on a rate coefficient, k_1, in the low pressure limit and over the range 250K to 400K $k_1 = 1.7 \times 10^{-31} \times (T/300)^{-3} \times N_2$ $cm^6 molecule^{-2}s^{-1}$ (10). However Molina et al. noted that in the presence of excess OClO k_1 apparently decreases by up to a factor of 3 (8). Also Knauth et al., measuring the decomposition of $ClONO_2$, k_{-1}, and calculating the equilibrium constant K_1, estimate $k_1 = K \times k_{-1}$ to be a factor of 3 lower than the measured value (11). One possible explanation of these observations is that reaction (1) forms isomers of ClONO$_2$. Chang et al. in a theoretical study of reaction (1) concluded that the less stable isomer ClOONO could be formed. Cox et al. observed the formation of ClONO$_2$ in the gas phase by reaction (1) using an infrared diode laser (13), Margitan detected the yield of Cl atoms from the products from reaction (1) (14), and Bhatia et al. used a combination of flow tube kinetics and matrix isolation IR spectroscopy to study reaction (1) (15). Cox et al. and Margitan concluded that no isomers of ClONO$_2$ are produced in reaction (1) whereas Bhatia et al. found evidence for the existence of isomers. Here a recent investigation of the reaction between ClO and NO$_2$ is reported.

2. EXPERIMENTAL

The apparatus constructed for the investigations of ClO-NO$_2$ chemistry is shown schematically in figure 1.

Figure 1.

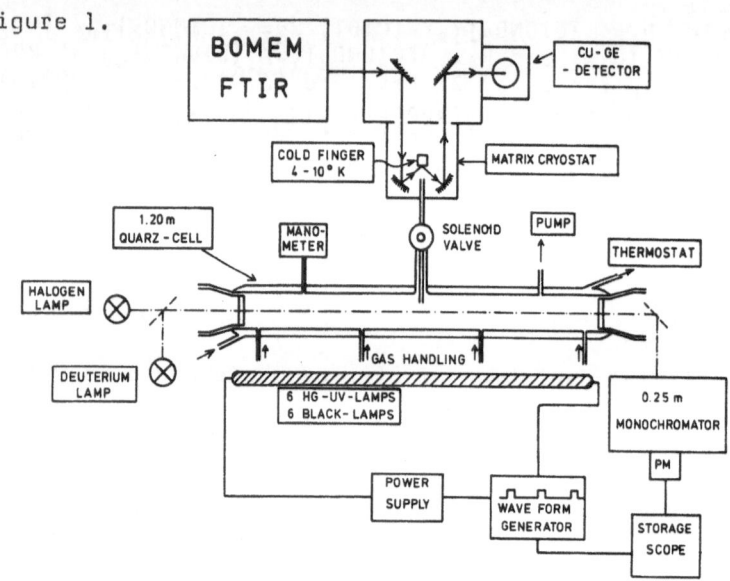

MATRIX ISOLATION FTIR, U.V.-VIS. ABSORPTION and MODULATED PHOTOLYSIS APPARATUS

The reaction vessel consists of a jacketed quartz photolysis cell, 145cm long and 3.8cm diameter. Ethanol was circulated through the jacket and the temperature of the cell regulated by a Haake 85 temperature controller. The temperature of the gas within the cell was measured using a CrAl thermocouple probe. The optical path length between the windows of the cell is 115cm. Gases are premixed before entering the vessel through 4 inlet jets simultaneously. The reaction mixture is subsequently pumped out of the cell via 3 outlet jets, which are situated halfway between the inlet jets and on the opposite side of the reaction vessel. The cell approximates to a well-mixed vessel, the residence time within the vessel, t_{res}, being given by the expression:

$$t_{res} = V/F$$

where V is the volume of the vessel (1.3 l) and F the volume flow rate of gas through the vessel. In these experiments t_{res} was varied between 3s and 60s, the total pressure in the vessel was maintained at approximately 22 torr and measured continuously by a Datametrics 1174 electronic manometer. The flow through the vessel was set using Tylan mass flow meters.

The concentrations of the reactants in the gas phase were measured by UV absorption. Light from a D_2 lamp was collimated using a quartz lens into a parallel beam and passed through the vessel, subsequently being focussed onto the entry slit of a 0.3m monochromator. The resolution of the monochromator used here was either 1nm or 5nm.

A fast deposition procedure was selected for growing matrices. This minimised the effects of aggregation in the matrix and the disturbance to the gas phase chemistry in the photolysed mixture. Matrices were grown by opening a teflon solenoid valve, typically for 0.15s, connected on one side to the centre of the photolysis cell and on the other to a cryochamber, pumped to 10^{-6} torr. Approximately 1% of the gas mixture in the cell passed through the solenoid valve subsequently striking a cold finger maintained at 10K. The amount of gas deposited on the cold finger was estimated to be approximately 5% of the total leaving the cell. Once a matrix was grown its FTIR spectrum was recorded using a BOMEM spectrometer. During these experiments a reso lution of $0.5cm^{-1}$ was used.

The bulk flow through the vessel was chosen to be a 10ppm mixture of N_2O in N_2. The N_2O was used as an internal matrix thickness calibrant by its absorption at $2235cm^{-1}$. Cl_2O was prepared continuously by passing Cl_2 over a column of HgO (16):

$$2Cl_2 + (n + 1)HgO \rightarrow Cl_2O + HgCl_2.nHgO \quad (2)$$

N_2O_5 was prepared by flowing a mixture of NO and O_3 repetitively between two traps held at 196K (17). $ClONO_2$ was prepared by reacting Cl_2O and N_2O_5 together between 196K and 273K (18):

$$Cl_2O + N_2O_5 \rightarrow 2ClONO_2 \quad (3)$$

The resulting mixture was purified by a 3-trap distillation procedure. The mix was held at 196K and connected in series to two traps, one held at 153K and the other at 78K. $ClONO_2$ forms in the trap at 153K.

The wall of the reaction vessel was coated with either teflon or silicone to minimise any wall losses e.g.

$$ClONO_2 \rightarrow products \quad (4)$$

The mixture flowing through the reaction vessel was photolysed using commercially available blacklamps (310nm-400nm). The lamps surrounded the reaction vessel and were enclosed in an aluminium coated can.

3.RESULTS

The matrix isolation FTIR spectra of the species Cl_2O, NO_2, N_2O_4, HNO_3 and $ClONO_2$ were calibrated by growing matrices from mixtures of known concentrations of the individual species in a 10ppm N_2O in N_2 mixture. The concentrations of reactant species were determined from their UV-Visble absorptions, cross sections of the species being taken from recent literature (e.g. reference (10)). The ratio of the band strength of the absorbing species, S_x, and the band strength of the v_3 vibration of N_2O , S_{N2O}, is given by

$$S_x/S_{N2O} = A_x c_{N2O}/(A_{N2O} c_x).$$

Care must be taken in interpreting the spectra of NO_2 since N_2O_4 forms readily in the matrix if NO_2 diffuses.

$$NO_2 + NO_2 = N_2O_4 \qquad\qquad (5)$$

The rate of photolysis of Cl_2 in the cell by black-lamps (310nm-400nm) was determined in the photolysis of mixtures of Cl_2, H_2, O_2 and N_2, typical concentrations being 9×10^{15}, 5×10^{17}, 9×10^{17}, and 2.2×10^{19} molecule cm^{-3} respectively. The Cl_2 decays via the mechanism:

$$Cl_2 + hv(310-400nm) \rightarrow Cl + Cl \qquad\qquad (6)$$

$$Cl + H_2 \rightarrow H + HCl \qquad\qquad (7)$$
$$k_7 = 7.1\times10^{-14} cm^3 molecule^{-1} s^{-1}$$
$$H + O_2 + M \rightarrow HO_2 + M \qquad\qquad (8)$$
$$k_8 = 1.0\times10^{-31} cm^6 molecule^{-2} s^{-1}$$
$$HO_2 + HO_2 \rightarrow H_2O_2 + O_2 \qquad\qquad (9)$$
$$k_9 = 2.4\times10^{-12} cm^3 molecule^{-1} s^{-1}$$

The decay of Cl_2 was monitored at 310nm, and is given by

$$dln[Cl_2]/dt = -k_6$$

A value of $k_6 = (1.5\pm0.2)\times10^{-3} s^{-1}$ per lamp was obtained. The photolysis rate of $N\bar{O}_2$ was measured by photolysing pure NO_2:

$$NO_2 + hv(310-400nm) \rightarrow NO + O \qquad\qquad (10)$$

$$O + NO_2 \rightarrow NO + O_2 \qquad\qquad (11)$$
$$k_{11} = 9.1\times10^{-12} cm^3 molecule^{-1} s^{-1}$$

The NO_2 decay, assuming the O atoms achieve a stationary state is given by

$$d\ln[NO_2]/dt = -2k_{10}$$

This yielded a value of $k_{10}=(3.48\pm0.3)\times10^{-3}s^{-1}$. The photolysis rate of Cl_2O was estimated from a knowledge of the output intensity of the blacklamps and the absorption spectra of Cl_2 and Cl_2O:

$$Cl_2O + h\nu(310-400nm) \rightarrow Cl + ClO \tag{12}$$

k_{12} was estimated to be approximately $0.2\times k_6$. $ClONO_2$ is decomposed slowly in the cell, the products observed being NO_2, HNO_3 and $HOCl$ indicating that the wall reaction involving the heterogeneous reaction with H_2O and a thermal decomposition are taking place. The $ClONO_2$ photolysis rate, k_{13}:

$$ClONO_2 + h\nu(310-400nm) \rightarrow products \tag{13}$$

was observed to be slow, $k_{13}= 1\times10^{-4}s^{-1}$.
    Flowing mixtures of Cl_2, Cl_2O and NO_2 were selected, so that the Cl atoms, generated by photolysis of Cl_2 would predominantly react with Cl_2O to form ClO. The ClO then reacts with NO_2:

$$Cl_2 + h\nu(310-400nm) \rightarrow Cl + Cl \tag{6}$$

$$Cl + Cl_2O \rightarrow Cl_2 + ClO \tag{14}$$
$$k_{14}=1.0\times10^{-10}cm^3molecule^{-1}s^{-1}$$
$$ClO + NO_2 + M \rightarrow ClONO_2 + M \tag{1}$$

The following secondary reactions must however be considered:

$$NO_2 + h\nu(310-400nm) \rightarrow NO + O \tag{10}$$

$$Cl_2O + h\nu(310-400nm) \rightarrow Cl + ClO \tag{12}$$

$$O + Cl_2O \rightarrow ClO + ClO \tag{15}$$
$$k_{15}=4.1\times10^{-12}cm^3molecule^{-1}s^{-1}$$
$$O + NO_2 \rightarrow NO_2 + O_2 \tag{11}$$

$$O + ClONO_2 \rightarrow products \tag{16}$$
$$k_{16}=2.0\times10^{-12}cm^3molecule^{-1}s^{-1}$$
$$Cl + ClONO_2 \rightarrow Cl_2 + NO_3 \tag{17}$$
$$k_{17}=1.0\times10^{-11}cm^3molecule^{-1}s^{-1}$$

As reaction (16) is relatively slow the fate of O atoms in the system is to react with NO_2 or Cl_2O. The fraction $k_{11}[NO_2]/(k_{11}[NO_2] + k_{14}[Cl_2O])$, depletes NO_2 and the fraction $k_{14}[Cl_2O]/(k_{11}[NO_2] + k_{14}[Cl_2O])$ generates two ClO. Mixtures of Cl_2, Cl_2O, and NO_2 in N_2 were flowed through the cell for different residence times, matrices being grown both with and without photolysis. A typical set of FTIR spectra is shown in figure 2.

Figure 2. Matix isolation FTIR spectra of an unphotolysed (a) and photolysed (b) mixture of Cl_2, Cl_2O and NO_2 and bulk synthesised $ClONO_2$ (c) in N_2: (A) $500-1200 cm^{-1}$, (B) $1200-1800 cm^{-1}$.

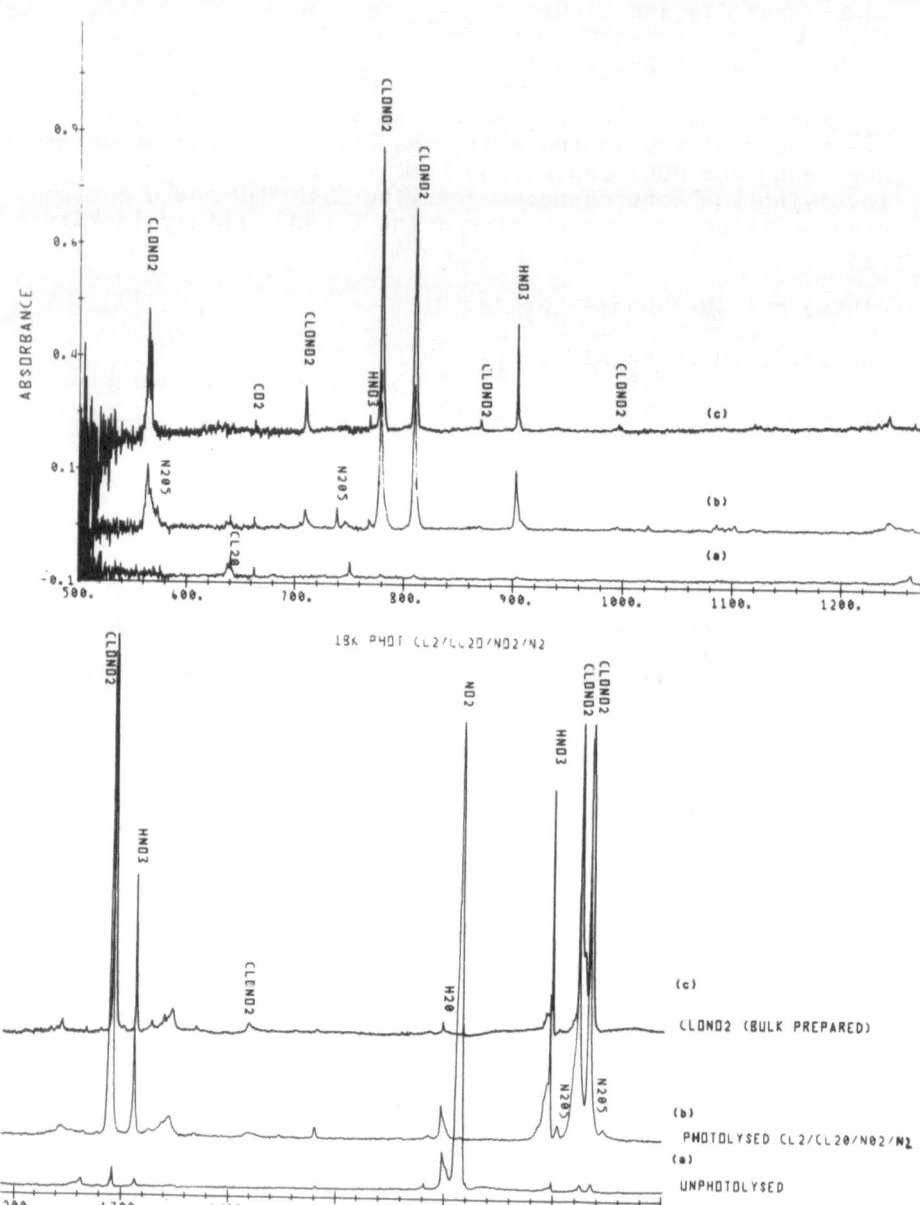

Absorptions due to $ClONO_2$ were observed to grow as the NO_2 absorptions decreased. The Cl_2O absorptions in the infrared are too weak to provide useful data. These qualitative observations were quantified by use of the calibrations. This enabled the change in concentration of the $ClONO_2$, $[ClONO_2]_{ir}$, and the change in concentration of the NO_2, $[NO_2]_{ir}$, to be determined. The change in NO_2 concentration was corrected for photolysis and O atom removal. The change in concentration in the ClO, $[ClO]$, was calculated from a knowledge of the photolysis rates and corrected for production from O atom reaction. In a few expriments the UV absorption change at 220nm was also measured. From the above the following ratios were determined:

$$R_1 = [ClONO_2]_{ir} / [ClO]$$

$$R_2 = [ClONO_2]_{ir} / [NO_2]$$

$$R_3 = [NO_2] / [ClO]$$

$$R_4 = [ClONO_2]_{ir} / [ClONO_2]_{uv}$$

At 298K and 22 torr the following values were determined: $R_1= (1.0\pm0.2)$, $R_2= (1.0\pm0.3)$, $R_3= (1.0\pm0.3)$, $R_4= (0.9\pm0.4)$. The results are presented in Table 1. Experiments were performed at temperatures down to 250K and no significant changes in R1 to R3 were observed.

Table 1. Results from the photlysis of Cl_2, Cl_2O and NO_2 mixtures

T	lamps	t_{res}	Cl_2 $\times10^{15}$	Cl_2O $\times10^{14}$	NO_2 $\times10^{14}$	N_2	R_1	R_2	R_3	R_4
K		s	cm^{-3}	cm^{-3}	cm^{-3}	torr				
298	3	25.6	1.76	17.5	6.5	22.5	0.63	1.03	0.6	
298	1	25.6	1.75	17.5	6.5	22.5	1.17	0.9	1.3	
298	2	25.6	1.6	19	6	22.0	0.91	1.03	0.9	
298	1	40.1	5.2	39	14	21.2	1.07	0.91	1.2	
298	1	7.4	1.8	7.5	4.2	22.3	1.1	1.1	1.0	
298	2	7.4	2.4	7	4.3	22.3	1.2	1.2	1.0	
298	3	7.4	2.4	7	4.0	22.3	0.97	1.15	0.8	
298	1	4.0	12.9	2	12.3	23.5	0.7			0.9
298	1	3.3	5.5	1.5	4.5	19.0	0.85			
298	1	3.3	5.5	1.5	4.5	19.0	0.96			
298	1	3.3	2.2	2.2	4.5	19.0	1.08			0.9
298	1	3.3	2.2	2.2	4.5	19.0	1.33			0.7
272	1	3.7	2.4	2.3	4.6	19.0	1.14			
266	1	3.8	2.5	2.4	4.7	19.0	0.88			
253	1	3.9	2.6	2.4	4.8	19.0	1.2			
250	1	4.0	2.6	2.4	4.8	19.0	1.1			

4. DISCUSSION

In comparing the spectra obtained from the photolysis
of the $Cl_2/Cl_2O/NO_2$ mixtures (see Figure 2, trace b) with
the bulk $ClONO_2$ spectrum (see Figure 2, trace c) it is
evident that $ClONO_2$ is the major product of reaction (1).
The trace amounts of N_2O_5 formed in the photolysis are due
to the secondary reaction $Cl + ClONO_2$, which is known to
produce N_2O_5. If any isomers other than planar $ClONO_2$ are
produced, such as the nitrites ClOONO and/or OClONO, they
must be formed in a yield of less than 10 %. In the experiments reported here radiation in the 300-400nm region is
used. The nitrite ClOONO, if it exists, is likely to absorb
strongly in this region. HONO has such an absorption feature, which overlaps the output of the blacklamps. However
its photolysis rate k_{18},

$$HONO + h\nu(310-400nm) \rightarrow OH + NO \qquad\qquad (18)$$

is estimated as $k_{18} = 1.5 \times k_6 = 2.25 \times 10^{-3} s^{-1}$ from a comparison of its absorption cross section with Cl_2 and a knowledge
of the output intensity of the blacklamps. Consequently even
at residence times of 60s the depletion of any ClOONO is, by
comparison with HONO, probably less than 15%.

The results obtained in this study of reaction (1)
imply that, when ClO reacts with NO_2, planar $ClONO_2$ is the
only significant product over the range 253-298K at 22 torr
total pressure of N_2. This result is in good agreement with
Cox et al. (13) and Margitan (14), but disagrees with the
findings of Bhatia et al. (15). In separate papers (19,20)
the discrepancies between the spectroscopic conclusions of
Bhatia et al. and the interpretations of the matrix
isolation infrared spectrum here are discussed in detail.
In summary the features attributed by Bhatia et al. to an
isomer are assigned there to either $ClONO_2$ absorptions or
impurity HNO_3 absorptions.

In the atmosphere the significance of the production
of only $ClONO_2$ by reaction (1) lies in the fact that the
photolysis rate of $ClONO_2$ in the stratosphere is relatively
slow. Consequently in the models of the depletion of
stratosphere O_3 the effect of the ClO_x destruction cycle is
diminished as compared with the production of a ClOONO
species, which is probably very rapidly photolysed in the
stratosphere.

5. ACKNOWLEDGEMENTS

The work reported here was supported by the Max Planck
Gesellschaft and the Deutsche Forschung Gesellschaft
through MAP. The authors are indebted to the work of
G. Schuster and K.H. Moebus

6. REFERENCES

(1) M.J.Molina and F.S.Rowland, Nature, _249_ 810 (1974).
(2) F.S.Rowland, J.E.Spencer and M.J.Molina, J.Phys.Chem.,
     _80_ 2711 (1976).
(3) A.E.J.Eggleton, R.A.Cox and R.G.Derwent, New
     Scientist, May 20, 402 (1976).
(4) J.W.Birks, B.Shoemaker,T.J.Leck, R.A.Borders and
     L.J.Hart, J.Chem.Phys., _66_ 4591 (1977).
(5) M.S.Zahniser, J.S.Chang and F.Kaufman, J.Chem.Phys.,
     _67_ 997 (1977).
(6) M.T.Leu, C.E.Lin and W.B.DeMore, J.Phys.Chem., _81_ 190
     (1977).
(7) R.A.Cox and R.Lewis, J.Chem.Soc.Faraday Trans. I, _75_
     2649 (1979).
(8) M.J.Molina, L.T.Molina and T.Ishiwata, J.Phys.Chem.,
     _84_ 3100 (1979).
(9) Y.P.Lee, R.M.Stimpfle, R.A.Perry, J.A.Muck,
     K.M.Evenson, D.A.Jennings and C.J.Howard, Int.J.Chem.
     Kinet., _14_ 711 (1982).
(10) D.L.Baulch, R.A.Cox, R.F.Hampson, J.A.Kerr, J.Troe and
     R.T.Watson, J.Phys.Chem.Ref.Data, _9_ 295 (1980).
(11) H.D.Knauth, G.Schonle and R.W.Schindler, J.Phys.Chem.,
     _83_ 3297 (1979).
(12) J.S.Chang, A.C.Baldwin and D.M.Golden, J.Chem.Phys.,
     _71_ 2021 (1979).
(13) R.A.Cox, J.P.Burrows and G.Coker, Int.J.Chem.Kinet.,
     in press (1983).
(14) J.J.Margitan, J.Geophys.Res., _88_ 5416 (1983).
(15) S.C.Bhatia, M.George-Taylor, C.W.Meredith and
     J.H.Hall, J.Phys.Chem., _87_ 1091 (1983).
(16) J.L.Gay-Lussac, Compt.Rend., _14_ 927 (1842).
(17) J.A.Davidson, A.A.Viggiano, C.J.Howard, I.Dotan,
     F.C.Fehsenfeld, D.L.Albritton and E.E.Ferguson,
     J.Chem.Phys., _68_ 2085 (1978).
(18) M.Schmeisser, W.Fink and K.Brendle, Angew.Chem., _70_ 97
     (1958).
(19) D.W.T.Griffith, G.S.Tyndall, J.P.Burrows and
     G.K.Moortgat, in press Chem.Phys.Lett. (1984).
(20) J.P.Burrows, D.W.T.Griffith, G.K.Moortgat and
     G.S.Tyndall, submitted to J.Phys.Chem.

OXIDATION OF METHYLCHLOROFORM

L.Nelson, J.J.Treacy and H.W.Sidebottom
Chemistry Department, University College, Dublin, Ireland

SUMMARY

The reaction mechanisms for oxidation of CH_3CCl_2 and CCl_3CH_2 radicals, formed via the atmospheric degradation of methylchloroform, have been elucidated. The primary oxidation products from these radicals are CH_3CClO and CCl_3CHO respectively. Rate constants for the reactions of hydroxyl radicals with CH_3CCl_3, CH_3CClO and CCl_3CHO have been determined. The effect of chlorine substitution on the reactivity of organic compounds towards hydroxyl radicals is discussed.

INTRODUCTION

The increased industrial use of methylchloroform as a degreasing and cleaning agent has recently given rise to some concern (1). The main tropospheric sink for methylchloroform is due to reactions with hydroxyl radicals:

$$CH_3CCl_3 + OH\cdot \longrightarrow CCl_3CH_2\cdot + H_2O$$

However, this reaction is relatively slow and, as a result, a significant fraction of released methylchloroform can reach the stratosphere where photodissociation leading to chlorine atoms can occur:

$$CH_3CCl_3 + h\nu(\lambda 170\text{-}240nm) \longrightarrow CH_3CCl_2\cdot + Cl\cdot$$

An understanding of the oxidation reactions of the chloroethyl radicals generated in the above reactions is of some importance since they may liberate further chlorine atoms via reactions in the atmosphere. The removal mechanisms and the nature and fate of the initial oxidation products for these species have not so far been unequivocally established, though it has been generally assumed that the remaining chlorine atoms will eventually be released. Thus in the stratosphere the total number of chlorine atoms in the chlorocarbon are expected to be available to enter the catalytic ozone destruction pathway, while in the troposphere the chlorine atoms will eventually be washed out as HCl. The aim of the present work was to elucidate the mechanisms for the oxidation of chloroethyl radicals.

EXPERIMENTAL

A mercury-free greaseless vacuum system was used for all the experiments and pressure measurements were made with an MKS Baratron capacitance manometer. Product analysis studies were carried out in a conventional static system using infrared analysis to determine product concentrations. Product identification was confirmed by means of gas chromatography coupled mass spectrometry. Radiation of wavelengths 360 and 253.7 nm was isolated from a Hanovia 500 watt medium pressure mercury arc by means of the appropriate Corning glass filters. The T-shaped reaction vessel was housed in the sample compartment of a Perkin-Elmer infrared spectrometer (Model 137), with sodium chloride windows in the infrared beam (pathlength 10 cm). Along the perpendicular axis (pathlength 10 cm) was the mercury arc with its associated filter system.

Kinetic experiments were carried out in an ~ 70 L reaction chamber, fabricated of FEP Teflon, surrounded by 10 blacklamps (Philips TL 20W/08) and 10 sunlamps (Philips TL 20W/05). Light intensities were varied by switching off sets of lamps and the reaction temperature maintained at $25 \pm 3°C$ by forced air circulation. Hydroxyl radicals were generated by the photolysis of methylnitrite in air:

$$CH_3ONO \;+\; hv \;\longrightarrow\; CH_3O\cdot \;+\; NO$$

$$CH_3O\cdot \;+\; O_2 \;\longrightarrow\; CH_2O \;+\; HO_2\cdot$$

$$HO_2\cdot \;+\; NO \;\longrightarrow\; NO_2 \;+\; OH\cdot$$

Nitric oxide was added to the reaction mixtures in order to minimize ozone formation during the photolyses. Measured amounts of CH_3ONO (1-10ppm), NO (1-10ppm) and the reference and reactant species (5-20ppm) were flushed from a Pyrex bulb into the reaction chamber by a stream of ultra-zero air. The bag was then filled with ultra-zero air up to a pressure of 740 torr. Prior to irradiation the reaction chamber was covered to avoid any pre-photolysis. All quantitative analyses were carried out on a Gow-Mac gas chromatograph with a flame ionization detector.

RESULTS AND DISCUSSION

(i) Oxidation of CH_3CCl_2 and CCl_3CH_2 radicals

Photolysis of chlorine/methylchloroform mixtures in the presence of oxygen proceeds by a chain reaction in which chloral and phosgene are the only chlorine-containing oxidation products. As the photolysis time increased chloral was removed and the yield of phosgene increased, suggesting that phosgene is a secondary product arising from chloral decomposition. The following mechanism is consistent with the experimental data:

$$Cl_2 \;+\; hv\,(\lambda 360nm) \;\longrightarrow\; 2Cl\cdot \qquad (1)$$

$$Cl\cdot \;+\; CH_3CCl_3 \;\longrightarrow\; CCl_3CH_2\cdot \;+\; HCl \qquad (2)$$

$$CCl_3CH_2\cdot \;+\; O_2 \;\longrightarrow\; CCl_3CH_2O_2\cdot \qquad (3)$$

$$2CCl_3CH_2O_2\cdot \;\longrightarrow\; 2CCl_3CH_2O\cdot \;+\; O_2 \qquad (4)$$

$$CCl_3CH_2O\cdot \;+\; O_2 \;\longrightarrow\; CCl_3CHO \;+\; HO_2\cdot \qquad (5)$$

$$Cl\cdot \;+\; CCl_3CHO \;\longrightarrow\; CCl_3CO\cdot \;+\; HCl \qquad (6)$$

$$CCl_3CO\cdot \;\longrightarrow\; CCl_3\cdot \;+\; CO \qquad (7)$$

$$CCl_3CO\cdot \;+\; O_2 \;\longrightarrow\; CCl_3COO_2\cdot \qquad (8)$$

$$2CCl_3COO_2\cdot \;\longrightarrow\; 2CCl_3CO_2\cdot \;+\; O_2 \qquad (9)$$

$$CCl_3CO_2\cdot \;\longrightarrow\; CCl_3\cdot \;+\; CO_2 \qquad (10)$$

$$CCl_3\cdot \;+\; O_2 \;\longrightarrow\; CCl_3O_2\cdot \qquad (11)$$

$$2CCl_3O_2\cdot \;\longrightarrow\; 2CCl_3O\cdot \;+\; O_2 \qquad (12)$$

$$CCl_3O\cdot \;\longrightarrow\; CCl_2O \;+\; Cl\cdot \qquad (13)$$

The U.V. absorption spectra of CH_3CCl_3 is continuous, showing no fine structure, and can be attributed to an $n \to s^*$ transition to a repulsive state. It is therefore expected that photolysis of CH_3CCl_3 will lead to decomposition with unit efficiency:

$$CH_3CCl_3 \quad + \quad h\nu \quad \longrightarrow \quad CH_3CCl_2\cdot \quad + \quad Cl\cdot \quad (14)$$

Photolysis of methylchloroform at 253.7nm in the presence of oxygen gave phosgene and chloral as the main chlorine-containing oxidation products with smaller amounts of acetyl chloride in substantial agreement with the work of Christiansen et al (2). As the reaction time increased CCl_2O was removed from the system via photolysis and CO was produced. The removal rate of CH_3CCl_3 was found to be considerably decreased when small amounts of Br_2 or NO were added to the system in order to trap Cl atoms. Under these conditions phosgene and chloral formation were virtually eliminated and acetyl chloride was the major product. It is suggested that CCl_2O and CCl_3CHO arise from chlorine atom sensitized oxidation of CH_3CCl_3 as previously discussed and that CH_3CClO is the major oxidation product from CCl_3CH_2 radicals. The data is consistent with the following scheme:

$$CH_3CCl_3 \quad + \quad h\nu \quad \longrightarrow \quad CH_3CCl_2\cdot \quad + \quad Cl\cdot \quad (14)$$
$$CH_3CCl_2\cdot \quad + \quad O_2 \quad \longrightarrow \quad CH_3CCl_2O_2\cdot \quad (15)$$
$$2CH_3CCl_2O_2\cdot \quad \longrightarrow \quad 2CH_3CCl_2O\cdot + \quad O_2 \quad (16)$$
$$CH_3CCl_2O\cdot \quad \longrightarrow \quad CH_3CClO \quad + \quad Cl\cdot \quad (17)$$

(ii) <u>Reaction of OH radicals with CH_3CCl_3, CCl_3CHO and CH_3CClO</u>

Photolyses of $CH_3ONO/NO/$chlorocarbon mixtures were carried out at $25\pm3°C$. Relative rate constants for attack by hydroxyl radicals were determined by comparing the decay of the reactant chlorocarbon with that for a reference compound.

$$OH\cdot \quad + \quad Reactant \quad \xrightarrow{k_{react}} \quad products$$
$$OH\cdot \quad + \quad Reference \quad \xrightarrow{k_{ref}} \quad products$$

Assuming reaction with hydroxyl radicals is the only significant loss process for the reactant and reference compounds and since dilution due to sampling is avoided by use of a collapsible reaction vessel it can be shown that:

$$\ln \frac{[Reactant]_o}{[Reactant]_t} = \frac{k_{react}}{k_{ref}} \ln \frac{[Reference]_o}{[Reference]_t} \quad (A)$$

The rate constant ratios k_{react}/k_{ref} were determined from plots of the above function and k_{react} evaluated from literature values for k_{ref}. The reference compounds used were CH_3Cl for CH_3CCl_3, $k_{ref} = 2.6\times10^7$ L mol^{-1} s^{-1} (3,4), $CH_3COOC_2H_5$ for CCl_3CHO, $k_{ref} = 1.2\times10^9$ L mol^{-1} s^{-1} (5) and $CHCl_3$ for CH_3CClO, $k_{ref} = 6.5\times10^7$ L mol^{-1} s^{-1} (4,6).

The photolysis rate of methylchloroform in pure air was insignificant under the experimental conditions used. However, photolysis of chloral/air or acetyl chloride/air mixtures resulted in a fairly rapid decay of starting material. The presence of NO was shown to decrease the decay rate appreciably, Fig. 1. In a similar manner addition of C_2H_6 reduced the rate of chloral or acetyl chloride disappearance to negligible amounts over the time scale of the hydroxyl radical kinetic experiments. It is proposed that photolysis of chloral or acetyl chloride leads to chain reactions involving chlorine atoms:

$$CCl_3CHO \ + \ h\nu \ \longrightarrow \ CCl_3\cdot \ + \ CHO\cdot \tag{18}$$

$$CCl_3\cdot \ \xrightarrow{\ O_2\ } \ CCl_2O \ + \ Cl\cdot \tag{11-13}$$

$$Cl\cdot \ + \ CCl_3CHO \ \xrightarrow{\ O_2\ } \ CCl_2O \ + \ CO + CO_2 + Cl\cdot \tag{6-13}$$

and

$$CH_3CClO \ + \ h\nu \ \longrightarrow \ CH_3\cdot \ + \ CClO\cdot \tag{19}$$

$$CClO\cdot \ \longrightarrow \ CO \ + \ Cl\cdot \tag{20}$$

$$Cl\cdot \ + \ CH_3CClO \ \longrightarrow \ \cdot CH_2CClO \ + \ HCl \tag{21}$$

$$\cdot CH_2CClO \ \xrightarrow{\ O_2\ } \ \text{oxidation products} + Cl\cdot \tag{22}$$

Addition of NO or C_2H_6 efficiently removes the chlorine atoms and the chain reaction is then unimportant.

The concentration-time profiles for the three OH + chlorocarbon systems studied gave good linear relationships when plotted in the form of equation (A), Fig. 2. Data were determined over a period of about 45 min. at 10 min. intervals and provide estimates of the rate constants for reaction of OH radicals with CH_3CCl_3, CCl_3CHO and CH_3CClO, Table I.

Table I : Rate constants for the reaction of OH radicals with various compounds at $25^\circ C$.

$$OH\cdot \ + \ RH \ \longrightarrow \ H_2O \ + \ R\cdot$$

RH	k_{OH}, L mol^{-1} s^{-1}	Reference
CH_3CCl_3	6.3×10^6	7
	6.5×10^6	8
	5.2×10^6	This work
CCl_3CHO	1.3×10^9	This work
CH_3CClO	4.3×10^7	This work
CH_3CH_3	1.6×10^8	9
CH_3CHO	9.0×10^9	10
CH_3COCH_3	4.0×10^8	11

Comparisons of the above data show that chlorine substitution decreases the reactivity of a compound with respect to attack by the hydroxyl radical. Thus in going from ethane to methylchloroform the rate constant per available hydrogen atom is reduced by about an order of magnitude. Similarly the reactivities of CCl_3CHO and CH_3CClO are reduced relative to CH_3CHO and CH_3COCH_3 respectively. Since substitution of chlorine in these compounds does not significantly affect the carbon-hydrogen bond dissociation energies the reduction in reactivity must be due to polar effects. It is proposed that chlorine substitution increases the repulsive polar forces in the transition state for the reaction of the electrophilic hydroxyl radical, that is,

$$\overset{\delta+}{HO} \text{------------} \overset{\delta+}{H} \text{---------------} \overset{\delta-}{CH_2CCl_3}$$

$$\overset{\delta+}{HO} \text{------------} \overset{\delta+}{H} \text{---------------} \overset{\delta-}{COCCl_3}$$

$$\overset{\delta+}{HO} \text{------------} \overset{\delta+}{H} \text{--------------} \overset{\delta-}{CH_2CClO}$$

A similar effect has been observed for the reaction of hydroxyl radicals with the series CH_4, CH_3Cl, CH_2Cl_2 and $CHCl_3$. In this case the C-H bond dissociation energy reduction in going from CH_4 to $CHCl_3$ is counterbalanced to some extent by increased repulsive polar forces. Hence although the activation energy for hydrogen abstraction by OH radicals decreases in going from CH_4 to CH_2Cl_2 there is an increase in going from CH_2Cl_2 to $CHCl_3$ (6).

REFERENCES
1. P.J.Crutzen, J.Geophys.Res., 83, 345(1978).
2. V.O.Christiansen, J.A.Dahlberg and H.F.Anderson, Acta Chem.Scand., 26, 3319 (1972).
3. R.A.Perry, R.Atkinson and J.N.Pitts, Jr., J.Chem.Phys., 64, 1618(1976).
4. D.D.Davis, G.Machado, B.Conaway, Y.Oh and R.Watson, J.Chem.Phys., 65, 1268(1976).
5. I.M.Campbell and P.E.Parkinson, Chem.Phys.Lett., 53, 385(1978).
6. C.J.Howard and K.M.Evenson, J.Chem.Phys., 64, 197(1976).
7. K.M.Jeong and F.Kaufman, Geophys.Res.Lett., 6, 757(1979).
8. M.J.Kurylo, P.C.Anderson and O.Klais, Geophys.Res.Lett., 6, 760(1979).
9. F.P.Tully, A.R.Ravishankara and K.Carr, Int.J.Chem.Kinet., 15, 1111(1983).
10. H.Niki, P.D.Makee, C.M.Savage and L.P.Breitenbach, J.Phys.Chem., 82, 132 (1978).
11. C.Chiorboli, C.A.Bignozzi, A.Maldotti, P.F.Giardini, A.Rossi and V.Carassiti, Int.J.Chem.Kinet., 15, 579(1983).

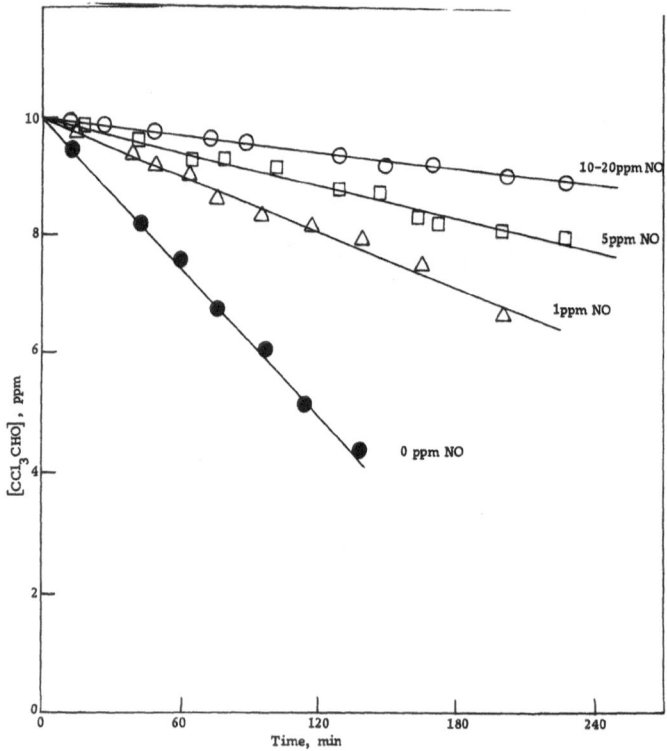

Fig. 1: Plot of CCl_3CHO concentration against photolysis time as a function of NO concentration at $25°C$ in 740torr air; $[CCl_3CHO] = 10$ ppm.

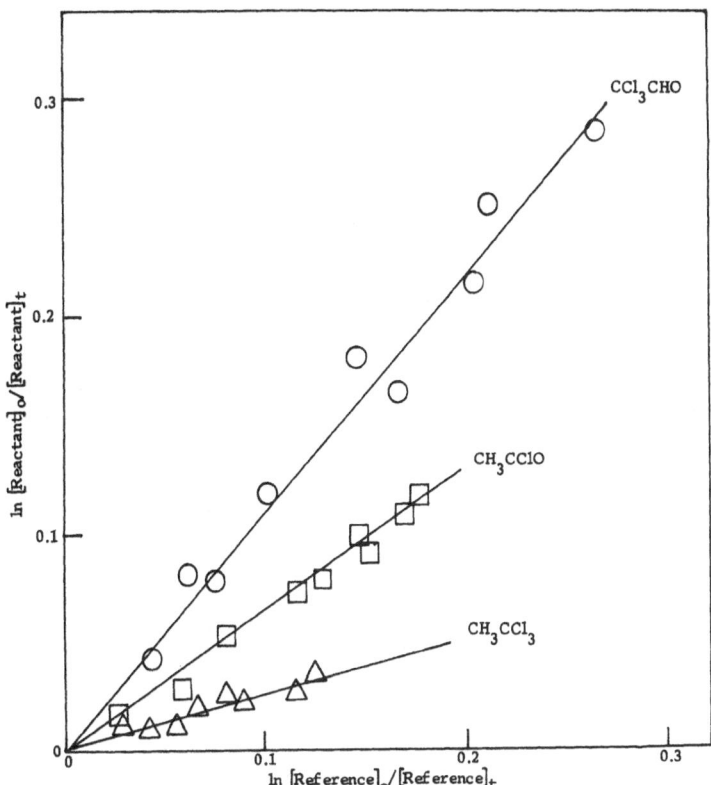

Fig. 2: Plot of ln [reactant]$_o$/[reactant]$_t$ against ln [reference]$_o$/[reference]$_t$ for the CH_3CCl_3/CH_3Cl, $CCl_3CHO/CH_3COOC_2H_5$ and $CH_3CClO/CHCl_3$ systems at 25±3°C.

TRANSFORMATION OF REACTIVE PAH ON PARTICLES BY EXPOSURE TO OXIDIZED NITROGEN COMPOUNDS

A. LINDSKOG, E. BRORSTRÖM-LUNDÉN AND A. SJÖDIN
Swedish Environmental Research Institute

Summary

In order to study the conditions for atmospheric transformation of reactive PAH on particles by exposure to
NO_2, HNO_2 and HNO_3, a set of experiments has been carried
out in a flow reactor. Soot and PAH were generated in a
smoke gas generator and collected on a glass fiber filter.
The filter was attached to the teflon coated wall in the
reactor tube and the reactive gas was added to the purified and humidified air pumped through the reactor.
After exposure, the filter was soxhlet extracted with
methylene chloride or acetone and fractionated on a silica
gel column into five fractions with increasing polarity.
        The results indicate that reactive PAH are trans-
formed when exposed to NO_2 while HNO_2 and HNO_3 have no
effect. The decrease of the concentration of PAH is dependent on the NO_2 concentration, the time of exposure
and the relative humidity. The presence of HCl or SO_3 increased the reactivity while HNO_3 seems to have no effect
in this respect. So far, only a few transformation products have been identified. With the exception of some
NO_2-PAH, quinones and ketones, most of them are eluted
in the most polar fraction and thus not easily analysed
by gas chromatography-masspectrometry.

1. INTRODUCTION

During the last few years it has been demonstrated that
reactive PAH can undergo chemical transformations when exposed
to reactive gases like NO_2, HNO_3 and SO_3 or polluted ambient
air (1-9). Benzo(a)pyrene (BaP) adsorbed on glass fiber
filters will thus form nitroderivates by exposure to NO_2 and
traces of nitric acid (2). Recently it was demonstrated that
exposure to NO_2 and HNO_3 during sampling of ambient air
particles will cause degradation of individual PAH-components
and a corresponding formation of mononitro-PAH (10). The reactivity is dependent on the kind of carrier on which the
components are adsorbed (4, 11, 12), as well as the presence
of water (2,13). In dry atmosphere BaP on soot is thus not
affected even when exposed to 10 ppm NO_x (14). As some of the
transformation products formed by these reactions are direct
mutagens in the Ames Salmonella assay as opposed to their
parent PAH, their possible formation, both in the atmosphere
and during sampling, is of great importance. In this paper the
transformation of reactive PAH on soot by exposure to NO_2,

HNO_2 and HNO_3 in a reactor tube is demonstrated.

2. EXPERIMENTAL

The experiments were carried out in a flow reactor (see figure 1). The incoming air was purified and humidified and the reactive gases were introduced to the tube by different dosage systems. NO_2 was delivered from permeation tubes or a gas cylinder. The concentration in the reactor tube was determined with a continuously recording instrument (Monitor Labs) based on the chemiluminescent reaction with ozone and varied less than 5%. Gaseous HNO_2 was prepared by adding, slowly and under intense mixing, a dilute solution of $NaNO_2$ to a dilute H_2SO_4-solution. A part flow of the purified air ventilated the reaction vessel continuously and the outlet air stream was mixed with the main air flow just before the reactor tube. The concentration was measured with the Monitor Labs instrument, (with and without a nylon filter) and was found to vary less than 10%. Gaseous HNO_3 was obtained from a small glass container (i.d. 6 mm) with concentrated HNO_3 placed in the reactor tube about 100 mm before the filter. The concentration of HNO_3 (g) in the air stream was determined by using a denuder coated with sodium carbonate according to Ferm (15).

When emitted from incomplete combustion of organic material the PAH will be attached to soot particles. In these experiments the soot and PAH were generated in a smoke generator (9) and collected on a glass fiber filter. The filter was devided and one half was attached to the Teflon® coated wall in the reactor tube while the other half was used as reference. After exposure, the filter was soxhlet extracted with methylene chloride or acetone and fractionated on a silica gel column (10% water) into five fractions with increasing polarity, according to Ahlsberg (16). After concentration the fractions were analysed on a Carlo Erba Fractovap 4130 using three different capillary columns: 50 m x 0.38 mm i.d. glass coated with SE-54, 25 m x 0.32 mm i.d. glass coated with JXR-0.4 µm and 25 m x 0.33 mm i.d. fused silica BP5-0.5 µm. Some of the fractions were also analysed on a Hewlet Packard 5990A GC-MS with a fused silica column BP5-0.5 µm. Both a peak finder mode, where the masses from 70 to 400 amu were scanned every 0.4 second, and a selected ion monitoring (SIM) technic were used. Individual components were identified by means of retention times and/or mass spectra of reference standards. The concentrations of the individual, reactive PAH were calculated relative to a stable PAH with adjacent retention time. The degradation of individual PAH was calculated from the difference in relative concentration between the exposed filter and the reference filter.

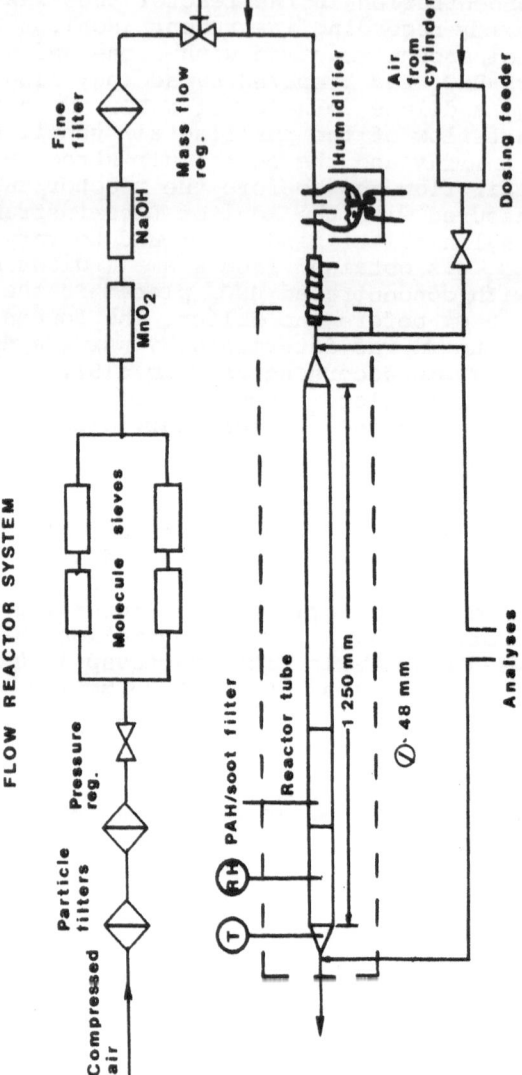

Figure 1.   Flow reactor system.

3. RESULTS AND DISCUSSION

The results from the experiments at 25°C, expressed as the percental change of concentration of individual PAH, are given in table 1 together with the experimental conditions. The results indicate that reactive PAH were transformed when exposed to NO_2 while HNO_2 and HNO_3 had no effect. The degradation is dependent on the NO_2 concentration (up to 2 ppm), the time of exposure and the relative humidity. In Figure 2 the percental change of the concentration of benzo(a)pyrene as a function of the exposure time at different concentrations of NO_2 is demonstrated.

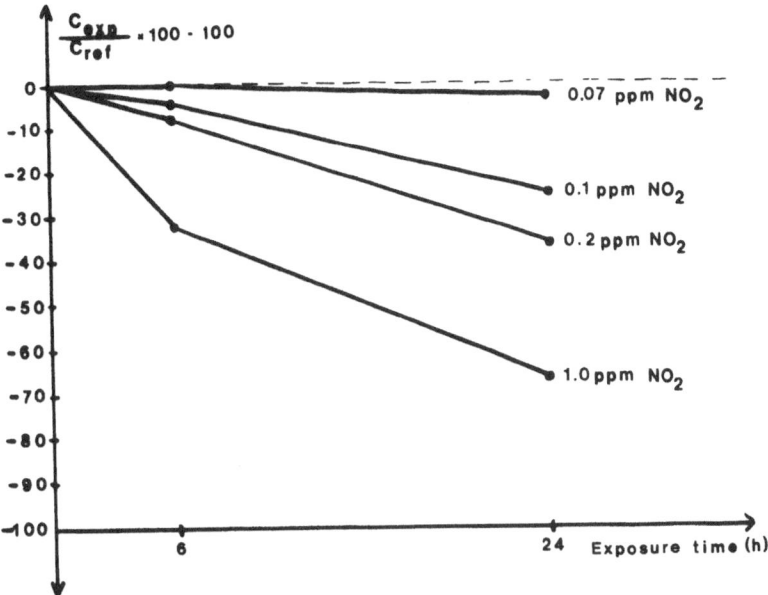

Figure 2. Percental change of the concentration as a function of the exposure time at different concentrations of NO_2.

The dependence of the relative humidity on the reactivity of BaP is demonstrated in Figure 3. In a dry system no reactions at all were obtained.

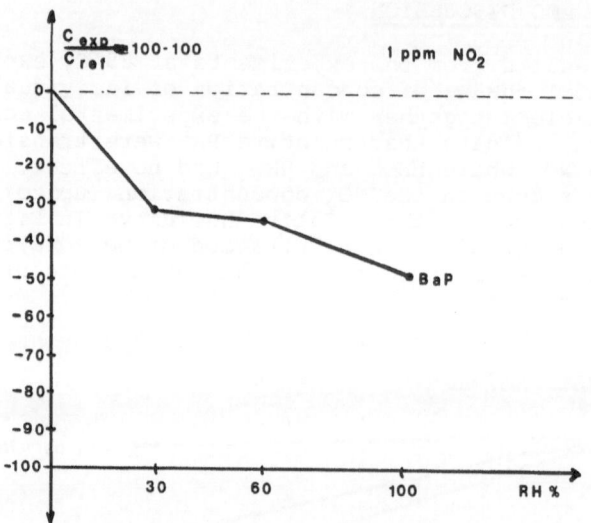

Figure 3. Percental change of the concentration of BaP as a function of relative humidity.

Three experiments (No. 26, 27 and 30-33) were carried out in order to study the stability of the transformation products formed. This was made by an initial 3 hours exposure to NO_2 or NO_2+HNO_3 followed by a 21 hours exposure to purified, humidified air only. No significant differences were obtained.

Two experiments at lower temperatures, 5 and 12°C, were also accomplished. The effect on the degradation of benzo(a)-pyrene, anthracene and cyclopenteno(c,d)pyrene is shown in Figure 4.

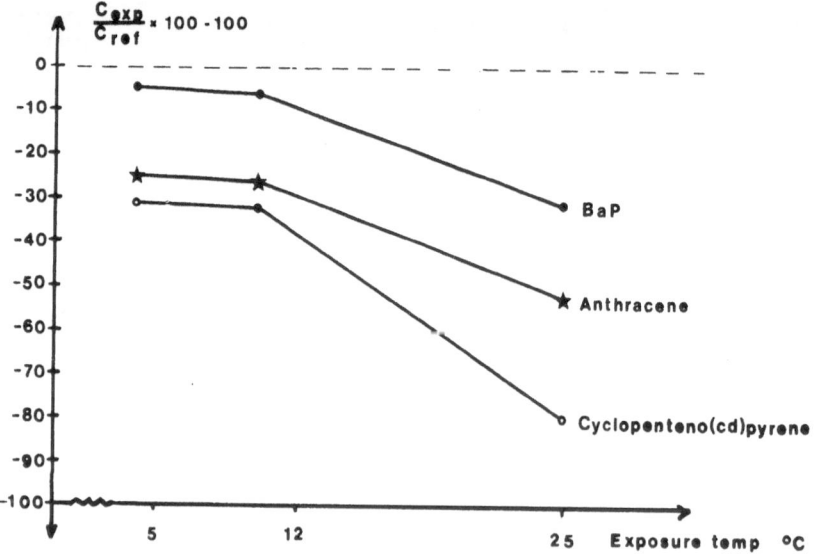

Figure 4. Percental change of the concentration of BaP, anthracene and cyclopenteno (c,d) pyrene as a function of exposure temperature.

It has earlier been demonstrated that a low pH increases the reactivity of PAH exposed to NO_2 (7,10,13). When NO_2 and HCl were introduced simultaneously the reactivity increased, probably because of the lowering of the pH on the particles, as the same effect was obtained when SO_3 had been introduced during the generation of soot (experiment No. 24). On the other hand, when NO_2 and HNO_3 were introduced simultaneously, the concentrations of reactive PAH showed no significant decrease compared to experiments with NO_2 alone. To avoid losses due to adsorbtion to the reactor walls, HNO_3 was introduced rather close to the filter. This gave rise to some uncertainty concerning the concentration of HNO_3 on the particles, and one soot-loaded filter, exposed only to HNO_3, was therefore analysed regarding strong acid and nitrate content. We found that the nitrate ions were even distributed on the filter surface while no H^+ activity was detected. It seems likely that the soot particles had a buffering effect and that the amount of HNO_3 introduced was insufficient to attain a pH below 7.

The work with the identification of transformation products is far from completed, and up to now only a few compounds have been identified. Based on earlier experiments with NO_2 exposure during sampling of ambient air, mononitro-PAH was expected to be one important group of compounds formed (10). So far, only nitro-anthracene, nitro-pyrene and nitro-BaP have been detected and only in the experiments with the highest concentration of NO_2 (No. 16 and 17). This observed difference in reaction pathways may partly be explained by diversities in the chemistry of the particles. In Sweden ambient air particles are often acidic while, as discussed above, the soot generated for our experiments were basic and had a buffering capacity. The amount and number of oxygenated PAH, eluted in the two moderate polar fractions, have increased in most of the exposed samples. The group includes ketones, aldehydes and quinonens, from simple fluorenones up to a ketone with a molecular weight of 278.

The gas chromatogram from the most polar fraction contains relatively few peaks and there is only a slight difference between the exposed sample and the reference sample. In contrast, analysis on HPLC with a fluorescence detector showed that the main part of the transformation products was present in this fraction. Due to the chemical properties of these compounds, they are not easily detected by gas chromatography-masspectrometry. Thus, most of the transformation products are still not identified.

Table 1. Experimental conditions and results. Exposure temp. $25^{\circ}C$, flow $3\ lh^{-1}$

No.	NO$_2$	HNO$_2$	HNO$_3$	Exposure time, h	RH %	Degradation of PAH in % Anthracene	Cyclopenteno-(cd)pyrene	BaP
	——— ppm ———							
1	0			24	45		1	0
2	0.07			6	44		14	0
3	0.07			24	46		44	3
4	0.1			6	43		14	4
5	0.1			24	46		80	25
6	0.2			6	40		64	9
7	0.2			6	45		50	8
8	0.2			24	45		93	31
9	0.2			24	42		92	40
10	0.5			6	40		64	27
11	1			6	0		0	0
12	1			6	29		61	32
13	1			6	58		62	35
14	1			6	100		70	49
15	1			24	40		83	66
16	2			6	45		72	69
17	4			6	40		77	68
18		0.1		24	45		0	0
19		0.1		24	45		0	0

Table 1. cont.

No.	NO$_2$	HNO$_2$	HNO$_3$	Exposure time, h	RH %	Degradation of PAH in % Anthracene	Cyclopenteno-(cd)pyrene	BaP
			ppm					
20			0.5	6	45		2	2
21		3		24	45		-7	-2
22	0.1		0.1	6	45		23	6
23(1)	0.1			6	45		20	14
24(2)	0.1		0.1	6	45		15	22
26	2		0	3		52	41	31
27	2		0	3+21		46	76	25
30	2		0.05	3		35	83	44
31	2		0.05	3+21		62	91	46
32	2		2.3	3		30	-	16
33	2		2.3	3+21		38	49	22

(1) 0.6 ppm HCl added.
(2) 10 ppm SO$_3$ added during the generation of soot/PAH.

REFERENCES

1. L. van Vaeck, G. Broddin and K.A. van Cauwenberghe,On the relevance of air pollution measurements of aliphatic and polyaromatic hydrocarbons in ambient particulate matter. Biomed.Mass.Spectr. 7 (1980) 473-483.

2. J.N. Pitts, Jr., K.A. van Cauwenberghe, D. Grosjean, J.P. Schmid, D.R. Fitz, W.L. Belser Jr., G.B. Knudson and P.M. Hynds. Atmospheric reactions of polycyclic aromatic hydrocarbons: Facile formation of mutagenic nitro derivatives. Science, 202 (1978) 515-519

3. J.N. Pitts, Jr., D.M. Lokensgard, P.S. Ripley, K.A. van Cauwenberghe, L.van Vaeck, S.D. Shafter, A.J. Thill and W.L. Belser, Jr. "Atmospheric" epoxidation of benzo(a)-pyrene by ozone: Formation of the metabolite benzo(a)-pyrene-4,5-oxide. Science, 210 (1980) 1347-1349.

4. M.M. Hughes, D.F.S. Natusch, D.R. Taylor and M.V. Zeller, Chemical transformations of particulate polycyclic organic matter. In Polynuclear Aromatic Hydrocarbons; Chemistry and biological Effects adited by A. Björseth and A.Dennis, Battelle Press, Columbus Ohio(1980) 1-8.

5. J. Peters and B. Seifert, Losses of benzo(a)pyrene under the conditions of high-volume sampling. Atmos. Environ. 14(1980) 117-119.

6. D.A.Lane and M. Katz. The photomodification of benzo(a)-pyrene, benzo(b)fluoranthene and benzo(k)fluoranthene under simulated atmospheric conditions. In Fate of Pollutants in the Air and Water Environments, Part 2, I. A. Suffet (Ed), Wiley-Interscience, New York (1977)137-154.

7. E. Brorström, P. Grennfelt and A. Lindskog, The effect of nitrogen dioxide and ozone on the decomposition of particle-associated polycyclic aromatic hydrocarbons during sampling from the atmosphere. Atmos. Environ. 17 (1983) 601-605.

8. A.H. Miguel. Atmospheric reactivity of polycyclic aromatic hydrocarbons associated with aged urban aerosols. In Polynuclear Aromatic Hydrocarbons: Formation, Metabolism and Measurement. M. Cooke and A.J. Dennis (Eds) Battelle Press, Columbus, Ohio (1983) 897-904

9. E. Brorström and A. Lindskog. Transformation of polycyclic aromatic hydrocarbons during sampling with reference to emission (submitted for publication 1983)

10. A. Lindskog. Transformation of polycyclic aromatic hydrocarbons during sampling. Environ. Health Perspectives 47 (1983) 81-84

11. J.Jäger and V. Hanus. Reaction of solid carrier-adsorbed
     polycyclic aromatic hydrocarbons with gaseous lowconcen-
     trated nitrogen dioxide. J.Hyg. Epidem. Microbiol. Immun.
     24 (1980) 1-12.

12. F.S.C. Lee, W.R. Person and J. Ezike. The problem of PAH
     degradation during filter collection of air-borne parti-
     culates. In Polynuclear Aromatic Hydrocarbons: Chemistry
     and Biological Effects, A. Björseth and A.J. Dennis (Eds)
     Battelle Press, Columbus, Ohio (1980) 543-563.

13. T. Nielsen. A study of the reactivity of polycyclic aro-
     matic hydrocarbons. Nordic PAH-project, Report No. 10.
     Published by Central Institute for Industrial Research,
     Oslo, 1981.

14. J.D. Butler and P. Crossley. Reactivity of polycyclic
     aromatic hydrocarbons adsorbed on soot particles. Atmos.
     Environ. 15(1981) 91-94.

15. M. Ferm. Method for determination of gaseous nitric acid
     and particulate nitrate in the atmosphere. Paper present-
     ed at the EMEP expert meeting on chemical matters,
     Geneva 10-12 March, 1982.

16. T. Ahlsberg, U. Stenberg, R. Westerholm, M. Strandell, U.
     Rannug, A. Sundvall, L. Romert, V. Bernson, B. Pettersson,
     R. Toftgård, B. Franzén, M. Jansson, J.Å. Gustavsson,
     K.E. Egebäck and G. Tejle. Chemical and biological
     characterization of organic material from gasoline ex-
     haust particles. Submitted for publication 1984.

THE KINETIC COEFFICIENT OF THE C_2H_4 + O REACTION

OVER EXTENDED PRESSURE AND TEMPERATURE RANGES

V. FONDERIE, D. MAES and J. PEETERS
Department of Chemistry, Katholieke Universiteit Leuven,
Belgium

Summary

A model suggested by Hunziker et al. (6) to explain the pressure dependence of the product distribution of the C_2H_4 + O reaction also implies pressure dependence of the total rate constant k_1 of the reaction. The aim of the present study was to measure the rate constant k_1 in the pressure range of 0.5 to 5 torr, at temperatures from 300 to 850 K and to compare the results with those obtained at 26-200 torr by other investigators. Using the discharge-flow technique, the rate constant was determined from the exponential decay of C_2H_4 at a large excess of O-atoms. Species concentrations were monitored by mass spectrometry using line-of-sight sampling; the sensitivity for O was determined on the basis of the NO + N reaction. Our k_1 results show no evidence for pressure dependence in the 0.5 to 5 torr range; moreover they agree very closely with the values obtained by other workers at 26-200 torr. The experimental results are at variance with the Hunziker model, which, as shown by RRKM calculations, would imply marked fall-off of k_1 in the 0.5 to 200 torr range, at least at higher temperatures.

1. INTRODUCTION

The reaction of O (^{3}P) with ethylene, which is important in combustion processes (1) and in the photochemical smog cycle (2,3), has been studied since the pioneering work of Cvetanovic (4) by several investigators using a variety of techniques. Nevertheless the nature and relative importance of primary products of this reaction is still controversial. Several investigators have recently observed the 2-oxoethyl (vinoxy) radical as a primary product of the C_2H_4 + O reaction (5,6,7,8,9) :

$$C_2H_4 + O \rightarrow CH_2CHO + H \qquad \qquad 1a$$

According to Lee et al., reaction 1a is the only pathway in crossed molecular beam conditions (5). At pressures from 40 to 760 torr however, Hunziker et al. (6) measured a 2 oxoethyl yield of only 0.36 ± 0.04, and at 80 to 760 torr a formyl yield of 0.55 ± 0.09 :

$$C_2H_4 + O \rightarrow CH_3 + CHO \qquad \qquad 1b$$

An explanation for this discrepancy was given by Hunziker et al., based on ab initio MCSCF calculations by Dupuis et al. (10). This solution takes into account that the initial triplet CH_2CH_2O diradical can be formed in two different electronic states :
i) a (π,π) state that fragments directly into CH_2CHO + H
ii) a (π,σ) state that redissociates in the absence of collisions but can

be converted by collisions into an isoenergetic singlet diradical which finally gives mainly CH_3 + CHO.

According to this model product formation by the second channel is pressure dependent. At the low densities of crossed molecular beam experiments only the first channel is accessible, whereas at higher pressures the second reaction path will also contribute to product formation. The Hunziker model thus implies fall-off behaviour of the total rate constant k_1 at decreasing pressures.

At room temperature Hunziker et al. (6) found the product distribution to be independent of pressure from 40 to 760 torr. Likewise at 300 K Davis et al. (11) found no pressure-dependence of k_1 at pressures from 1.3 to 240 torr. Thus, at 300 K the fall-off region predicted by the Hunziker model is expected at pressures below 1 torr.

One of the prime objectives of the present study was to check the validity of the model proposed by Hunziker. To that end we have extended the measurement of k_1 to lower pressures (0.5 to 5 torr). Because the fall-off region is expected to shift markedly towards higher pressures at elevated temperatures, experiments were also carried out at temperatures of about 550, 740 and 850 K. Preliminary results have been published elsewhere (12).

2. EXPERIMENTAL

Experiments were performed using a fast flow reactor in combination with molecular beam sampling and mass spectrometric detection. Figure 1 gives a schematic representation of the flow tube. The 2.8 cm - inner diameter quartz flow tube with a length of 70 cm has several fixed inlets together with a coaxial movable injector tube of 0.8 cm outer diameter.

The O atoms required were generated by passing a O_2/He mixture through a 2450 MHz microwave discharge. Ethylene, diluted by helium, was added via the central, axially movable injector tube.

The reaction time is determined by the distance z between the mixing and the sampling point (4 to 20 cm) and by the flow velocity in the kinetic region (1300 to 5000 cm s^{-1}).

Gases used in the experiments were He UHP (99.9996 %), O_2 UHP (99.995 %), N_2 UHP (99.9992 %), C_2H_4 (99.9 %) all supplied by L'Air Liquide and a UHP 1.03 % NO/He mixture from Gardner Cryogenics. The NO content of this mixture was verified by mass spectrometry using a certified 10 % NO/N_2 mixture from L'Air Liquide.

In the high temperature experiments the flow reactor was heated by an electrical furnace, mounted around the flow tube. A detailed description of this furnace has been published recently (13).

Gas sampling occurs at the exit of the flow reactor through a hollow conical quartz probe with an orifice in the tip of 0.3 mm diameter which gives access to a three-stage differentially pumped system. After being chopped in the second stage, the resulting molecular beam traverses the electron-impact ionizer of an Extranuclear quadrupole mass spectrometer mounted in the third stage. When monitoring atoms or radicals, the ionizing electron energy was chosen only 2 or 3 eV above the corresponding ionization potential. In this way interference by fragment ions is minimized. Phase-sensitive detection is applied.

3. DESCRIPTION OF THE PROCEDURE

The total rate constant of reaction 1 was determined from the exponential pseudo-first-order decay of ethylene at a large excess of O atoms

$([O]_i/[C_2H_4]_i \approx 50\text{-}100)$. Ethylene was monitored at m/e = 27 (to avoid interference by CO) at an electron energy of 50 eV. Destruction of O atoms by reaction with ethylene and by termination on the reactor wall was only 5 to 10 %. Working with excess O atoms has the advantage that destruction of C_2H_4 by secondary reactions is negligible. Reaction of C_2H_4 with H atoms :

$$C_2H_4 + H \xrightarrow{M} C_2H_5 \qquad\qquad 2$$

with H originating from reaction 1a is unimportant compared with reaction 1 because $k_2 < k_1$ and also $[H] \ll [O]$.

When determining k_1 at excess O the absolute O atom concentration has to be known accurately. The sensitivity of the quadrupole mass spectrometer with respect to O atoms was determined by generating a known O atom concentration by the reaction :

$$N + NO \rightarrow N_2 + O \qquad\qquad 3$$

A known flow of NO was allowed to react with excess N atoms. The O atoms so produced were detected as O^+ at an electron energy of 16 eV.

In the determination of k_1 effects such as the non-uniform temperature profile over the kinetic region (13) and the pressure drop associated with the flow had to be taken into account. As already mentioned in the experimental part the reactor is heated by an electrical furnace mounted around the flow tube. This leads to a pronounced temperature drop over the last few centimeters of the kinetic region (see fig. 1). The kinetic region can be divided in two zones : a longer one of constant temperature T_c and one of non-uniform temperature extending over a short distance q (≈ 4 cm) in front of the sampling probe.

For z > q, the decay of ethylene over the kinetic region can be written as :

$$[C_2H_4]_s = [C_2H_4]_i \; A \; \exp(-k_1[O] \frac{z-q}{v} f_{corr}) \qquad\qquad 3.1$$

with k_1 : the rate constant at the constant temperature T_c in the zone
    x < z-q
    z : the total length of the kinetic region (distance between the mixing point and the sampling point)
    v : the linear flow velocity
    f_{corr} : a correction function for the pressure gradient
    subscript s referring to the sampling point (x = z)
    subscript i referring to the mixing point (x = 0).
A is the additional decay factor over the zone z-q < x < z; this factor is a constant, independent of z.

Equation 3.1 can be written in logarithmic form as :

$$\ln[C_2H_4]_s = K - k_1[O] \frac{z}{v} f_{corr} \qquad\qquad 3.2$$

The correction function f_{corr} for the pressure gradient is given by :

$$f_{corr} = 1 + \frac{8\eta v_s z}{R^2 P} \qquad\qquad 3.3$$

with η being the coefficient of viscosity of the carrier gas helium, v_s the flow velocity at the end of the reactor, R the radius of the reactor (1.4 cm) and P the pressure at the end of the reactor.

According to equation 3.2 the pseudo-first-order rate constant $k_{obs} = k_1[O]$, at the temperature of the constant temperature region, can be derived from the slope of the straight line $\ln[C_2H_4]_s$ in function of z_{conv} ($z_{conv} = z\,f_{corr}$).

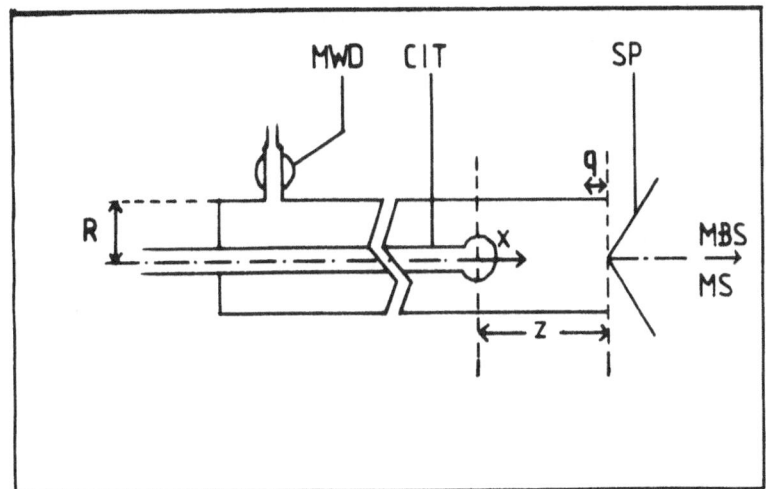

FIG.1 : Schematic drawing of the flow tube; R : internal radius of the
flow tube; CIT : central injector tube; MWD : microwave discharge; SP : sampling probe; z : distance between mixing point
and sampling point; MBS and MS : molecular beam sampling and
mass spectrometric analysis; the origin of the x-scale is
located at the mixing point

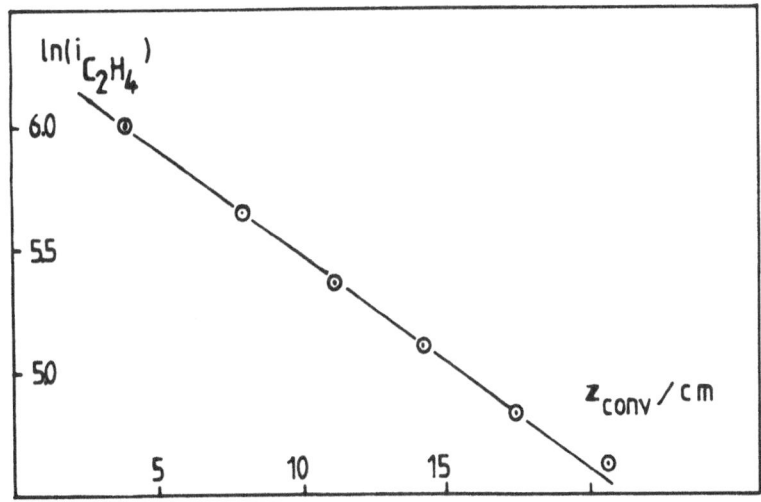

FIG. 2 : Example of determination of k_1 by plotting $\ln i(C_2H_4)$ versus
the converted distance scale z_{conv}; $T_c = 564$ K; $v = 3607$ cm s^{-1};
$[O] = 2.02 \times 10^{-10}$ mole cm^{-3}; $P = 2$ torr; $k_1 = 1.53 \times 10^{12}$
cm^3 mole^{-1} s^{-1}

Those rate coefficients k_{obs} were successively corrected for small deviations of the quasi-plug-flow (14) and for axial diffusion effects (15) by the following expressions (16) :

$$k_{corr} \simeq k_{obs} \left(1 + \frac{k_{obs} R^2}{48D}\right) \qquad\qquad 3.4$$

$$k_{corr} = k_{obs} \left(1 + \frac{k_{obs} D}{v^2}\right) \qquad\qquad 3.5$$

with D the diffusion coefficient of C_2H_4 in helium. The correction for deviation of quasi-plug-flow ranged from 0.2 to 9 % while this for axial diffusion was in the range of 0.3 to 10 %.

A plot of the corrected k_{obs}-values, measured at different [O], in function of [O] gives a straight line with slope k_1.

4. RESULTS

Figure 2 shows an example of a ln $i(C_2H_4)$ versus z_{conv} plot, and in figure 3 k_{obs} is plotted as a function of [O], yielding a straight line with slope k_1. The results, converted to temperatures of respectively 298, 552, 736 and 835 K, are listed in table I. Each value is the average of four experiments. The values of Ravishankara et al. (17) at 40 to 200 torr Ar and those of Pitts et al. (18) at 26 torr Arr are also listed. The table shows no evidence for fall-off, not even at elevated temperatures, where the fall-off region is expected to be shifted to higher pressures.

5. DISCUSSION

As already mentioned in the introduction, the pressure-dependent channel (ii) of the Hunziker-mechanism involves a competition between on the one hand redissociation of the formed (π,σ) adduct and on the other hand collision-induced intersystemcrossing of that adduct finally leading to product-formation. Calculation of the ratio $k_c[M]/(k_D + k_c[M])$
with k_c : the bimolecular rate coefficient for the collision-induced triplet-singlet conversion
[M] : the total molar concentration
k_D : the rate coefficient for the unimolecular redissociation of the (π,σ) CH_2CH_2O to C_2H_4 + O
at different total molar concentrations allows a prediction of the pressure-dependence of this second channel.

The rate coefficient k_D, which depends on the vibrational energy content of the adduct, can be calculated by means of RRKM-theory. In the calculation of $k_c[M]$ two extreme cases can be considered :
A. Direct collision-induced intersystem crossing is slow compared to collisional stabilization. In that case stabilization of the excited (π,σ) adduct to a level below the redissociation limit will be rate-determining. The "trapped" adduct can then only disappear by spontaneous intersystem crossing. In this case the ratio $k_c[M]/(k_D + k_c[M])$ can be taken equal to the probability of stabilization of the formed (π,σ) diradicals.
B. The second possibility is a fast collision-induced intersystem crossing in which long-distance-interactions between the triplet diradical and a helium atom are already sufficient to cause a triplet-singlet conversion. In the limit, the rate coefficient $k_c[M]$ could become even a few times larger than the collision frequency.

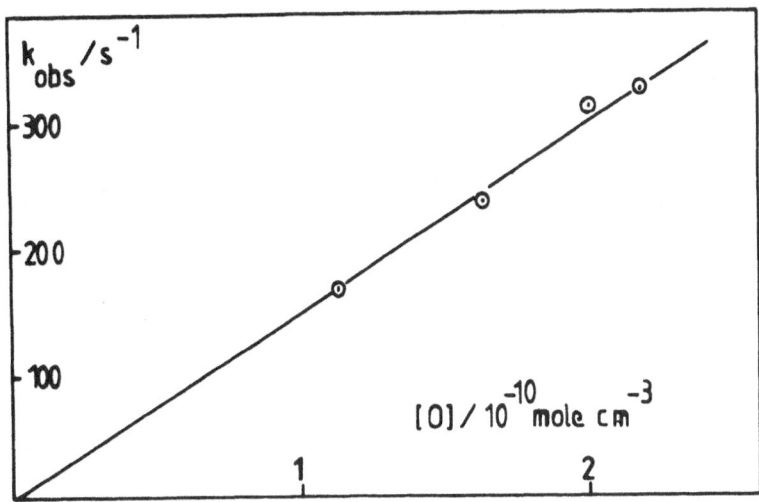

FIG. 3 : Plot of k_{obs} versus oxygen atom concentration [O];
T_c = 564 K; P = 2 torr

TABLE I

The rate constant k_1 as a function of pressure and temperature

$$(k_1 \pm \sigma)/(10^{11} \text{ cm}^3 \text{ mole}^{-1} \text{ s}^{-1})$$

T/K	THIS WORK			Pitts et al.	Ravishankara et al.
	0.5 torr	2 torr	5 torr	26 torr(Ar)	40 – 200 torr(Ar)
298	4.1 ± 0.2	4.0 ± 0.3	3.8 ± 0.1	4.6	4.3 ± 0.2
552	15.7 ± 0.3	14.5 ± 0.4	14.6 ± 0.2	14.4	16.4 ± 0.8
736	29.8 ± 1.0	27.4 ± 1.1	27.5 ± 1.3	–	26.5 ± 1.4
835	38.6 ± 1.0	36.3 ± 1.1	–	–	35.0 ± 0.9

These two possibilities and their implications regarding the pressure at which fall-off is expected in the framework of the Hunziker model will be discussed separately in the following part.

A. Slow collision-induced intersystem crossing

As already mentioned above, the ratio $k_c[M]/(k_D + k_c[M])$ can in this case be taken equal to the probability of stabilization of the formed (π,σ) adducts which can be calculated by means of a simplified model. In this model the area above the E_o-barrier (E_o being the critical energy for redissociation of the (π,σ) adducts) was divided in intervals of width $\delta E = 0.25$ kcal/mole each.

The energy distribution function $F(E)$ for formation of the (π,σ) diradicals was taken as (19) :

$$F(E) = C\ G^*(E-E_o)e^{-(E-E_o)/RT} \qquad (E > E_o) \qquad 5.1$$

with C a normalization factor and $G^*(E-E_o)$ the total amount of vibrational states between energy 0 and $E-E_o$ in the transition state. The $G^*(E-E_o)$ values were obtained by direct count.

A still excited (π,σ) adduct, with an energy $E > E_o$ in a given interval δE, can in this model loose energy by collision and go to one of the lower lying intervals or it can redissociate into $C_2H_4 + O$ with rate coefficient k_D, characteristic for species with the given energy-content.

To bring into account the possibility of negligible energy transfer on collision and also of energy gain of the adduct it was assumed that the excited (π,σ) adduct only looses energy in 50 % of the collisions. The probability for loss of a given amount of energy was assumed to be an exponentially decreasing function of the quantity of energy transferred, taken as a multiple of 0.25 kcal/mole. The width of the energy loss distribution function is determined by $<\Delta E>$, the average energy transferred per collision.

For each interval above the E_o-barrier an average rate coefficient k_D was calculated by means of the RRKM-theory. Ab initio calculated vibrational frequencies of the CH_2CH_2O diradical and also one value (the lowest) for the transition state were found in the literature (10). These values had to be reduced with 10 % because ab initio values are usually 10 % higher than experimental ones. The other vibrational frequencies of the transition state were estimated by means of ab initio values of the analogous T.S. for the reaction $C_2H_5 + C_2H_4 + H$ (20).

(π,σ) CH_2CH_2O $\quad \nu_i$: 3090, 2990, 2930, 2895, 1445, 1415, 1340, 1255, 1075, 1045, 975, 730, 640, 400 cm^{-1}, hindered internal rotation

$\qquad\qquad\qquad$ ($I_m = 1.5$ a.m.u. Å^2, $V_o = 5.17$ kcal/mole)

T.S. $\qquad\qquad \nu_i$: 2950, 2950, 2950, 2950, 1450, 1450, 1400, 1150, 1150, 850, 800, 780, 350, 260.

A crucial parameter in the RRKM-calculation of k_D is the depth of the CH_2CH_2O (π,σ) well with respect to the $C_2H_4 + O$ level. It determines the vibrational density of states of adducts formed at an energy level a given amount above the 1.4 kcal/mole barrier for the forward reaction. According to the ab initio MCSCF-calculations of Dupuis et al. the well depth is 14 kcal/mole (10), whereas a thermochemical estimate yields 32 kcal/mole (21). Accordingly, the activation energy for redissociation E_o can be taken as either 15.4 or 33.4 kcal/mole, resulting in a large uncertainty regarding k_D and hence also $P_{1/2}$, the midpoint of the fall-off region for the second channel of the $C_2H_4 + O$ reaction.

At T = 287 K, with $E_o = 15.4$ kcal/mole, the model predicts $P_{1/2} = 600$ torr for $<\Delta E> = -0.62$ kcal/mole and $P_{1/2} = 900$ torr for $<\Delta E> = -0.31$ kcal/mole. With $E_o = 33.4$ kcal/mole, $P_{1/2} = 2$ torr for $<\Delta E> =$

- 0.62 kcal/mole and $P_{1/2}$ = 3.5 torr for $\langle \Delta E \rangle$ = - 0.31 kcal/mole.

At 740 K, with E_o = 33.4 kcal/mole, the model results in a $P_{1/2}$ of 222 torr for $\langle \Delta E \rangle$ = - 0.62 kcal/mole and in a $P_{1/2}$ of 369 torr for $\langle \Delta E \rangle$ = - 0.31 kcal/mole.

B. Fast collision-induced intersystem crossing

The frequency of intersystem crossing $k_c[M]$ can in this case become a few times the collision frequency; as a limiting value $k_c[M] = 5\ Z_{LJ}[M]$ is assumed here.

The probability of triplet-singlet conversion is given by

$$P_{isc} = k_c[M]/(\bar{k}_D + k_c[M]) \qquad\qquad 5.2$$

At 287 K, with E_o = 15.4 kcal/mole, a value of 46 torr is predicted for $P_{1/2}$; taking E_o = 33.4 kcal/mole, $P_{1/2}$ would be 0.16 torr. At 740 K, with E_o = 15.4 kcal/mole, the predicted $P_{1/2}$ = 1014 torr; using E_o = 33.4 kcal/mole, $P_{1/2}$ becomes 9 torr.

6. CONCLUSIONS

It is clear from the above that, both for slow and for fast collision-induced intersystem crossing and with E_o anywhere in the range of 15 to 33 kcal/mole, the Hunziker model would imply a pronounced fall-off of k_1 in the pressure range of a few torr to atmospheric pressure, at least in part of the temperature range of 300-850 K. Since that prediction is at variance with the experimental evidence, the validity of the Hunziker model should be put into question.

7. ACKNOWLEDGMENTS

The authors wish to express their gratitude to the Belgian "Fonds voor Kollektief Fundamenteel Onderzoek" for its financial support. J.P. is indebted to the "Nationaal Fonds voor Wetenschappelijk Onderzoek" of which he is a Research Associate. V.F. and D.M. thank the "Instituut voor Aanmoediging van het Wetenschappelijk Onderzoek in Nijverheid en Landbouw" for granting doctoral fellowships.

8. REFERENCES

(1) Warnatz, J. : Eighteenth Symposium (International) on Combustion, p. 369, The Combustion Institute (1981)
(2) Herron, J.T., Huie, R.E. and Hodgeson, J.A. (ed.) : Chemical kinetic data needs for modeling the lower troposphere NBS Spec. Publ. 557, Washington (1979)
(3) Atkinson, R., Darnall, K.R., Lloyd, A.C., Winer, A.M. and Pitts, J.N. Jr. : Advances in Photochemistry, vol. 11, 375 (1979)
(4) Cvetanovic, R.J. : Can. J. Chem., 33, 1684 (1955)
(5) Buss, R.J., Baseman, R.J., Guozhong He and Lee, Y.T. : J. Photochem., 17, 398 (1981)
(6) Hunziker, H.E., Kneppe, H. and Wendt, H.R. : J. Photochem., 17, 377 (1981)
(7) Inoue, G. and Akimoto, H. : J. Chem. Phys., 74, 425 (1981)
(8) Kleinermanns, K. and Luntz, A.C. : J. Phys. Chem., 85, 1966 (1981)
(9) Sridharan, U.C. and Kaufman, F. : Chem. Phys. Lett., 102, 45 (1983)
(10) Dupuis, M., Wendoloski, J.J., Takada, T. and Lester, W.A. Jr. : J. Chem. Phys., 76, 481 (1982) and : personal communications of Dupuis et al. to Hunziker (ref. 6) and to Lee (ref. 5)

(11) Davis, D.D., Huie, R.E., Herron, J.T., Kurylo, M.J. and Braun, W. : J. Chem. Phys., $\underline{56}$, 4868 (1972)
(12) Fonderie, V., Maes, D. and Peeters, J. : Bull. Soc. Chim. Belg., $\underline{92}$, 641 (1983)
(13) Caymax, M. and Peeters, J. : Nineteenth Symposium (International) on Combustion, p. 51, The Combustion Institute (1982)
(14) Poirier, R.V. and Carr, R.W. Jr. : J. Phys. Chem., $\underline{75}$, 1593 (1971)
(15) Kaufman, F. : Prog. React. Kinet., $\underline{1}$, 3 (1961)
(16) Fonderie, V. : Ph. D. Thesis, K.U. Leuven (1981)
(17) Nicovich, J.M., Ravishankara, A.R. : Nineteenth Symposium (International) on Combustion, p. 23, The Combustion Institute (1982)
(18) Atkinson, R. and Pitts, J.N. Jr. : J. Chem. Phys., $\underline{67}$, 38 (1977)
(19) Forst, W. : Theory of unimolecular reactions, Academic Press, New York and London (1973)
(20) Hase, W.L., Schlegel, H.B. : J. Phys. chem., $\underline{86}$, 3902 (1982)
(21) Hunziker, H.E., personal communication.

Kinetics of the reaction of OH with Ethane and a series of Cl- and F-substituted Methanes at 300-400 K, studied by pulse radiolysis combined with kinetic spectroscopy

O.J. Nielsen, P. Pagsberg, A. Sillesen
Chemistry Department
Risø National Laboratory
DK-4000 Roskilde, Denmark.

Summary

Gas phase reactions of OH with ethane and a series of Cl- and F-substituted methanes were studied in the temperature range 300-400 K. In contrast to most previous experimental work our experiments were carried out at atmospheric pressure. OH was produced by pulse radiolysis of water vapour and the decay rate was studied by monitoring the transient light absorption at 3090 A. Arrhenius parameters (A, E_a) for the reaction $RH + OH \rightarrow R + H_2O$ were obtained for reactants $RH = C_2H_6$, CH_3Cl, CH_2Cl_2, $CHFCl_2$:

$k(OH+C_2H_6)$ = $8.1 \times 10^{12} \times \exp(-2160/RT)$ $cm^3 mole^{-1} s^{-1}$
$k(OH+CH_3Cl)$ = $3.2 \times 10^{12} \times \exp(-2510/RT)$ $cm^3 mole^{-1} s^{-1}$
$k(OH+CH_2Cl_2$ = $4.1 \times 10^{12} \times \exp(-2220/RT)$ $cm^3 mole^{-1} s^{-1}$
$k(OH+CHFCl_2)$= $1.1 \times 10^{12} \times \exp(-3550/RT)$ $cm^3 mole^{-1} s^{-1}$

CF_2Cl_2 which contains no C-H bonds was found to be inert toward attack by OH.

1. Introduction

Molina and Rowland (1) have pointed out the the chemical inertness and high volatility of chlorofluorocarbons (CFCs) which make them suitable for technological use also means that they enter the atmosphere upon release and since there are no obviuos sinks for them in the troposphere the CFCs will ultimately diffuse into the stratosphere. In the stratophere light absorption by CFCs gives rise to photochemical release of chlorine atoms followed by a catalytic chain reaction leading to a net destruction of ozone:

$$Cl + O_3 \rightarrow ClO + O_2$$
$$ClO + O \rightarrow Cl + O_2$$

The potential ozone depletion owing to a continued release

of CFCs may have serious consequences on the entire biosphere. Since the predictions by Molina and Rowland (1) in
1974 the CFC-problem has entered the public consciousness
and the legislative process of many governments around the
world. Consequently, atmospheric scientists have been hard
pressed to give relistict predictions about the effect of
CFC-release on stratospheric ozone. To set up a realistic
model of the atmosphere with the required predictive power
is a formidable task due to the overwhelming complexity
in chemical kinetics and transport dynamics. One of the
most important tasks is to establish a reliable data base
in terms of experimental rate constants as input for the
computer models. Experimental data from a fairly large
number of active working teams on chemical kintics are
collected and evaluated by the CODATA Task Group on
Chemical Kinetics (2). All of the chemical reactions occurring in the atmosphere can in principle be studied in the
laboratory thanks to a number of powerful experimental
techniques that have been developed within the field of
gas phase kinetics. A large amount of experimental work has
been devoted to the study of reactions between OH and halocarbons. In the present investigation we apply the technique of pulse radiolysis combined with kinetic spectroscopy to study the reactions between OH and ethane and a
number of CFCs at a pressure of 1 atm and over a range of
temperatures in order to obtain rate constants and
Arrhenius parameters for these reactions.

2. Experimental

The experimental set-up is shown schematically in
Fig. 1, and has been descibed in detail by Hansen et al.
(3). The field emission accelerator (FEBETRON 705B) provides single pulses of 2-MeV electrons with a pulse duration
of 30 nsec and a maximum current of 3000 amperes. The gas
samples are contained in a stainless steel irradiation
cell mounted directly onto the accelerator. The optical
arrangement involving an internal set of conjugate mirrors
was first described by White (4). By this device the
optical path length may be increased by multiple passages
of the analyzing light beam through the gas sample. Most of
the experiments were carried out using twelve traversals
corresponding to an optical path length of 120 cm. Gas
mixtures were prepared by admitting one component at a time
and reading the corresponding partial pressure using a MKS
Baratron model 170 absolute electronic membrane manometer
with a resolution of 10^{-5} Bar. Electric heating and temperature control provides a range of sample temperatures from
300 to 400 K. A platinum-resistance thermometer is used to
measure the temperature of the gas mixture near the central
part of the irradiation cell traversed by the analyzing
light beam. The analyzing light source is a pulsed 150-watt
high-pressure xenon lamp with short axial arc and parabolic
reflector. An optical train of suprasil lenses and front-
surface aluminized mirrors carries the analyzing light

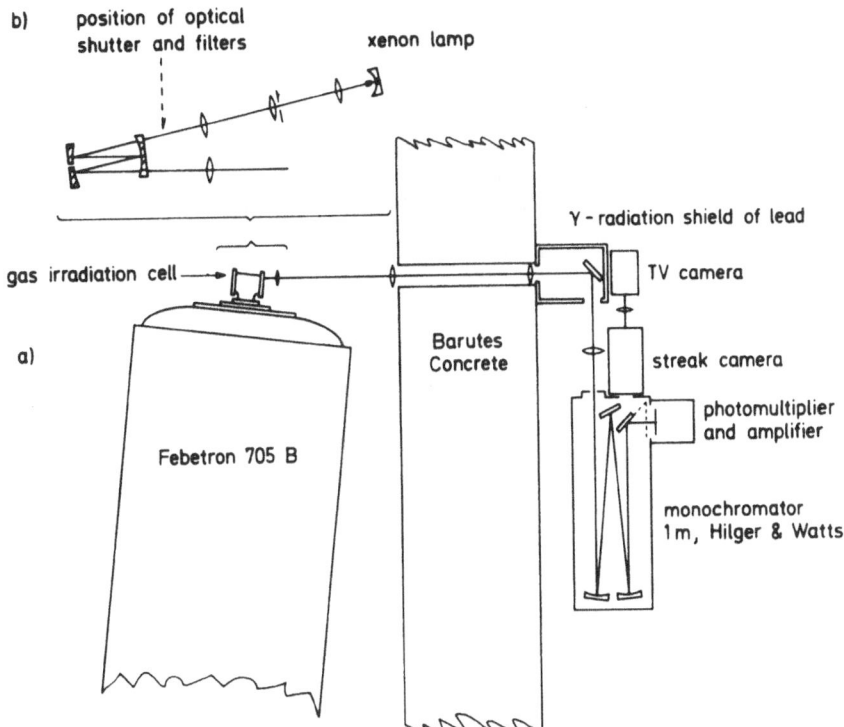

b) position of optical
   shutter and filters   xenon lamp

Y-radiation shield of lead

gas irradiation cell ⟶   TV camera

a)   Barutes
                                Concrete   streak camera

                                                   photomultiplier
                                                   and amplifier

   Febetron 705 B
                                              monochromator
                                              1m, Hilger & Watts

Figure 1. Schematic diagram of the experimental set-up.
a) Horizontal plane. b) Enlarged vertical cut showing
the analyzing light beam.

through the sample cell to the entrance slit of the Hilger
and Watts 1 meter monochromator. The xenon lamp pulser is
most essential to the performance in the ultraviolet region
where the light intensity is increased by a factor of 20-50
in light pulses stabilized to within a few percent and with
a duration of several milliseconds. The application of a
total of seven lenses implies a large loss of light intensity which makes the use of the lamp pulser indispensable.
However, a major advantage of the system is that the lenses
act as a pre-monochromator resulting in a very low level of
stray light (2% at 2000 Å). The analyzing light is dispersed
in the monochromator which has an aperture ratio of 1/10.
For the ultraviolet region we apply a ruled grating with
1200 grooves/mm and blazed at 3000 Å. The reciprocal dispersion in the plane of the exit slit is 8 Å/mm. The light
intensity over the wavelength band-pass selected by the
exit slit is monitored using a Hamamatsu R 955 photomultiplier coupled to a current input operational amplifier
with adjustable offset. The transient signals are stored in
a PDP8 minicomputer. Plots of transient absorption versus
time, as well as simple first- and second-order kinetic
plots are immediately available on a display screen and may
be displayed on an X-Y recorder or stored on magnetic tape

for further processing on a large central computer system.

We use a chemical simulation computer program developed by O. Lang Rasmussen (5). As input, this program accepts reaction schemes in the usual chemical notation, e.g., H + OH → H_2O, k=2x10^{10}, etc. The program translates the "chemical equations" into the pertinent set of differential equations which is solved by numerical integration after specification of initial concentrations, irradiation dose, etc. Computer simulations are conducted much faster than real experiments and the effect of parameter variation on the model kinetics can be studied very quickly using a graphical computer terminal. In this way we have been able to find out in which cases it is safe to apply simple analytical procedures to calculate rate constants from the experimental decay curves.

3. Analysis of pulse radiolysis data

By pulse radiolysis of water vapour backed-up with Ar to 1 atm initially Argon ions and excited Argon atoms are formed. OH is formed mainly through the reaction:

$$Ar^* + H_2O \rightarrow H + OH + Ar$$

The formation of OH is instantaneous compared to the time-scale of the OH decay. In the absence of additives the fate of H and OH is governed mainly by the reactions:

(1) $\qquad\qquad$ OH + OH + M → H_2O_2 + M
(2) $\qquad\qquad$ H + OH + M → H_2O + M
(3) $\qquad\qquad$ H + H + M → H_2 + M

Consecutive reactions of H and OH with accumulating product molecules H_2 and H_2O_2, play only a minor role in single-pulse high-dose rate experiments. Under these conditions the decay of OH and H is to a good approximation described by the differential equations:

(I) $\qquad\qquad -d[OH]/dt = 2k_1[OH]^2 + k_2[H][OH]$

(II) $\qquad\qquad -d[H]/dt = 2k_3[H]^2 + k_2[H][OH]$

The decay of hydroxyl radicals can be studied by monitoring the transient absorbance of one of the characteristic rotational lines of the $OH(A^2\Sigma^+ - X^2\Pi)$ rovibronic transition in the region 3000-3200 Å. A modified version, $A=(\varepsilon lc)^n$, of Beer's law has been used (see references (6) and (7) for further details). The observed transient absorbance is a direct measure of the concentration of hydroxyl radicals at any given time during the decay. In Figure 2 is shown an example of the transient absorbance of OH monitored at 3089.6 Å in the absence of additives. The absorbance decreases to half of the maximum value in the course of 100 μsec and this "natural half-life" must be ascribed to the combined rates of reactions (1) and (2). Another

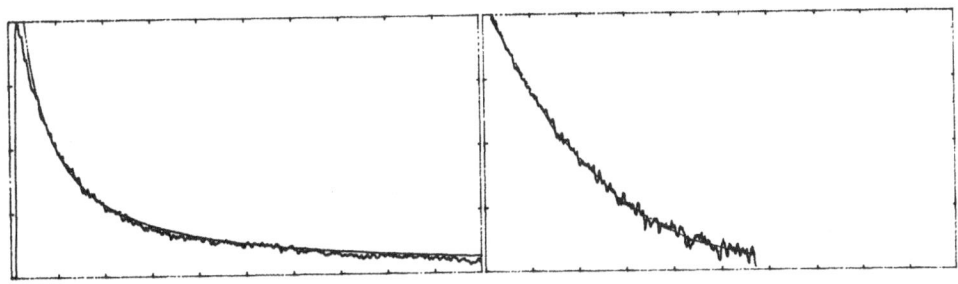

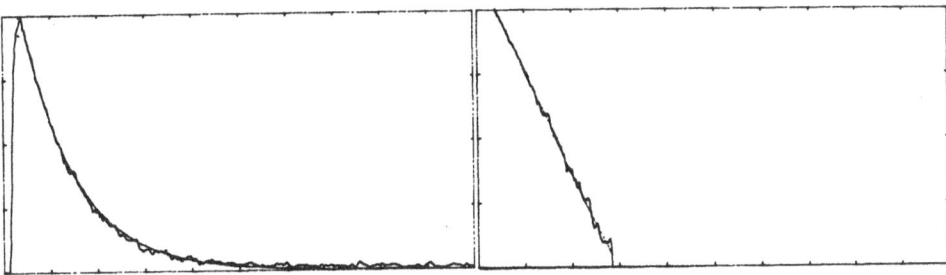

Figure 2. OH-formation and -decay at 400 K. Upper left: 200 mbar H_2O with Ar to 1 atm. Full timescale=1000 µsec. Lower left: as upper left, but with 2.7 mbar C_2H_6 added and full timescale=200 µsec. Right side: Corresponding logarithmic plots.

characteristic feature is the long "second-order tail" following the steep initial part of the decay.
  The high reactivity of OH-radicals observed in many H-abstraction reactions:

(4)   R-H + OH → R + H-OH

is due to the high bond energy, D(H-OH)=118 kcal/mole, which exceeds the bond energies D(R-H) of most other hydrides, such as simple hydrocarbons (CH_4, C_2H_6, C_3H_8, etc.), and substituted hydrocarbons like methyl chloride and chloroform. In general, the observed rate constants are higher the higher the difference in bond energies, D(H-OH)-D(R-H). Addition of a "substrate", RH reacting in accordance with eq. 4 has a characteristic effect on the observed decay of OH, as shown in Fig.2. Comparison of the two curves in Fig.2 reveals a substantial suppression of the "second-order tail" and also that the half-life is reduced relative to the "natural decay". In the presence of RH an additional term must be added to (I),

(III) $-d[OH]/dt = 2k_1[OH]^2 + k_2[H][OH] + k_4[RH][OH]$

This differential equation serves as the basis for our determinations of the rate constants k_4 for the elementary reaction between OH-radicals and selected substrate molecules, RH. The simultaneous occurrence of (1),(2),(3) and (4) implies that one (1),(2) and (3) to the observed overall decay rate. In general, this can be accomplished by studying the effect of reaction (4) over a sufficiently large range of substrate concentrations. In the limit of very high RH-concentrations, (i.e. when $k_4[RH] \gg 2\ k_1[OH]$ and $k_4[RH] \gg k_2[H]$) the differential equation (III) degenerates into:

(III*) $\qquad\qquad -d[OH]/dt = k_4[RH][OH]$

corresponding to a simple exponential decay

$[OH] = [OH]_o exp(-k_4[OH]_o t)$

with a characteristic half-life

$\tau = \ln2/k_4[RH]_o$

This expression may be used to obtain a quick estimate of the value of k_4 from a single experiment in which the RH-concentration was high enough to push the observed half-life down to a small fraction of the "natural half-life". Problems arise in studies of "slow reactions" where it may be difficult to fulfill the requirement $k_4[RH]_o > 2k_1[OH]_o$. Based on computer simulations we find that it is still possible to calculate a reliable value of k_4 based on a relatively small variation in half-lives by applying the equation:

(IV) $\quad 1/\tau = 1/\tau_o + \ln2/k_4[RH]_o$

According to this equation a plot of reciprocal half-lives versus the corresponding RH-concentrations is a linear function with a slope equal to $\ln2/k_4$. This is a simple method which allows a quick determination of the rate constant k_4 based solely on measured half-lives and the corresponding substrate concentrations. A more rigorous procedure which allows a clear distinction between first- and second-order kinetics is to analyze plots of lnA versus time, where A is the transient OH-absorbance. Such a plot becomes strictly linear when the decay degenerates into pseudo-first-order kinetics at high substrate concentrations in accordance with (III). Any significant contributions from reactions of OH with itself or other short-lived radicals will show up in terms of a more-or-less pronounced curvature. Thus, if a particular curve is found to be linear (within the experimental noise) it means that the decay is almost completely controlled by the last term in Equation (III) and hence, the slope $dlnA/dt = k_4[RH]_o$ + CORR, where "CORR" is a small correction term due to contributions from the radical-radical reactions. The magnitude of the correction term is related to the curvature which may be hard to detect due to the experimental noise level. We subtract the correction term by "double differentiation" of a series of decay curves corresponding to different substrate concentrations:

(V) $$k_4 = d(dlnA/dt)/d[RH]$$

We have applied both methods (IV) and (V) and we find that the calculated values of k_4, using either method, are the same within the experimental uncertainties. We expect of course the values calculated via (IV) to exhibit a larger scatter than those based on (V) because the former utilizes only a single point of the curve (half-life) while the latter involves all points on the decay curve.

4. Results and discussion

a. $OH + C_2H_6$

Ethane was chosen as a reference compound because the reaction is fast and the kinetics are well determined. From the two curves in Fig. 2 it is clearly seen that the OH decay in the presence of 2.7 mbar C_2H_6 has changed into pseudo-first-order kinetics, which is further clarified by the linear logarithmic plot shown in the same figure. In Table 1 the results are summarized.

Gordon and Mulac (10) gave no A and E_a values since they were measured only at two different temperatures which would give the A- and E_a-value inserted in Table 1 for the sake of completeness.

In view of the differences in techniques and experimental conditions the agreement between our result and those of Greiner (8) and Howard et al. (11) is quite good.

b. $OH + CH_3Cl$

Of the 360 kt methyl chloride, CH_3Cl, produced industrially in 1980 approximately 20 kT were emitted to the atmosphere (12). Approximately half of the CH_3Cl produced is used for tetramethyllead production. However, more than 200 times as much, 5 Mt, is released annually from other sources: combustion and the oceans (12). The only significant pathway by which CH_3Cl is removed from the atmosphere is by OH attack.

Five earlier measurements of the $CH_3Cl + OH$ rate constant have been found in the literature (11,13-16). In Table 1 rate parameters are compared with published results.

Because CH_3Cl reacts fairly slowly with OH it was necessary to apply rather high partial pressures of CH_3Cl in order to get a significant change in the OH decay. Under these circumstances there is a risk of direct fragmentation of CH_3Cl following irradiation. If a significant number of such fragments are formed together with OH it becomes very difficult to derive the OH rate constant. This is a problem especially at room temperature where the saturation pressure of water is rather low. Here one may have to consider the following additional reactions:

Table 1. Comparison of rate parameters for all reactions

Reference	T in K	k in $10^8 M^{-1}s^{-1}$	A in $M^{-1}s^{-1}$	E_a in kcal/mol
OH + C_2H_6				
Greiner (8)	300	1.83	1.1×10^{10}	2.45
Overend et al. (9)	295	1.59		
Gordon and Mulac (10)	381	4.00	3.5×10^9	1.64
Howard et al. (11)	296	1.75		
This work	300	1.95	9.7×10^9	2.33
OH + CH_3Cl				
Howard et al.(11)	296	0.22		
Perry et al.(13)	298	0.27	2.5×10^9	2.70
Davis et al.(14)	298	0.26	1.1×10^9	2.18
Paraskevopoulos et al.(15)	297	0.25		
Jeong et al.(16)	293	0.24	2.1×10^9	2.61
This work	300	0.43	3.2×10^9	2.51
OH + CH_2Cl_2				
Perry et al.(13)	299	0.87		
Davis et al.(14)	298	0.70	2.6×10^9	2.17
Howard et al.(11)	296	0.93		
Jeong et al.(16)	292	0.92	3.5×10^9	2.08
This work	300	0.88	4.1×10^9	2.22
OH + $CHFCl_2$				
Atkinson et al.(18)	297	0.029	7.3×10^8	3.25
Howard et al.(11)	296	0.020		
Watson et al.(19)	298	0.029	5.6×10^8	3.13
Chang et al.(20)	296	0.026	7.2×10^8	3.29
Handwerk et al.(21)	293	0.028	1.3×10^9	3.54
Clyne and Holt (22)	293	0.020	5.7×10^9	4.6
Paraskevopoulos et al.(15)	297	0.028		
Jeong et al.(16)	298	0.029	7.7×10^8	3.32
This work	300	0.031	1.1×10^9	3.55

(5) $OH + CH_2Cl \rightarrow products$
(6) $OH + CH_3 + M \rightarrow CH_3OH + M$
(7) $OH + Cl + M \rightarrow HOCl + M$

In the experiments at room temperature a significant reduction in the initial OH yield, OH_{max}, was observed. This is taken as evidence for direct fragmentation of CH_3Cl. In the analysis of kinetic results the $\ln[OH]$ versus time plots were most often used. Deviations from linearity of these

plots are observed when radical-radical reactions like
5-7 influences the OH decay which should be controlled by

(8) $OH + CH_3Cl \rightarrow H_2O + CH_2Cl$

Even when there is no direct fragmentation of CH_3Cl, CH_2Cl
is produced in reaction 8 and this makes reaction 5
important. However, for the initial slope at t=0 $[CH_2Cl]=0$
and again a simple expression is obtained

$$d(\ln(A)) = (2k_1+k_2)[OH]_o + k_8[CH_3Cl]_o$$

which contains no contribution from consecutive reactions.
This procedure was used for CH_3Cl.

The activation energy is in good agreement with the
result of Jeong and Kaufman (16). At higher temperatures
our rate constant values are in fair agreement with Jeong
and Kaufman. However, at 27^oC our rate constant seems to be
overestimated by a factor 1.7 as seen from Table 1. As
already explained, direct fragmentation may be the reason
for this overestimate.

c. $OH + CH_2Cl_2$

In 1980 570 kt of methylene chloride, CH_2Cl_2, was produced
(12). All methylene chloride produced is eventually released
to the atmosphere. World production is expected to increase
rapidly. Twenty per cent of the production is used in
aerosols. This application may increase and develop a
larger market percentage. No natural source of methylene
chloride has been identified. The rate-controlling step in
the tropospheric oxidation of methylene chloride is the
reaction with OH. Since this reaction is fast, CH_2Cl_2 is
not expected to contribute to the stratospheric chlorine
budget. The room temperature measurements have been left
out because of an identified systematic error in the measurement of $p(CH_2Cl_2)$ in this particular experimental
series. In Table 1. the rate parameters are compared with
published results. Our results seem to be in good agreement
with all previously published data. Only the room temperature value of Davis et al. (14) seems to be out of range.

d. $OH + CHFCl_2$

$CHFCl_2$ is an important refrigerant and has been recommended
as a substitute for $CFCl_3$, CF_2Cl_2, and CH_3CCl_3 since it
contains only one chlorine atom and reacts with tropospheric
OH. However, it reacts only slowly with OH, and $CHFCl_2$
concentrations were found to increase at an exponential rate
of 11.7% per year (17). In Table 1. our result is compared
with 8 earlier investigations of the $OH + CHFCl_2$ reaction.
The result of Clyne and Holt looks suspect compared with
the rest, which are all in fair agreement with each other.

5. Conclusion

The fate of the radicals produced by the reaction of OH with CFCs is not known with certainty. The mechanism of the subsequent oxidation of these radicals remains an obvious reseach field. Recently we have recorded the UV-spectrum of CCl3 and studies of the kinetics in the presence and absence of O_2 are in progress.

6. Acknowledgement

We are grateful to Nordic Counsil of Ministers for a grant supporting this work.

7. References

1) M.J.Molina and F.S. Rowland; Nature 249, 810 (1974)
2) D.L. Baulch, R.A. Cox, P.J. Crutzen, R.F. Hampson Jr., J.A. Kerr, J. Troe, and R.T. Watson; J.Phys.Ref.Data 11, 237 (1982)
3) K.B. Hansen, R. Wilbrandt, P. Pagsberg; Rev.Sci.Instrum. 50, 1532 (1979)
4) J.U. White; J.Opt.Soc.Am. 32, 283 (1942)
5) O.L. Rasmussen; Risø-R-395 (1984)
6) E. Bjarnov, J. Munk, O.J. Nielsen, P. Pagsberg, and A. A. Sillesen; Risø-M-2366 (1983)
7) O.J. Nielsen; Risø-R-480 (Ph.D.-thesis 1983)
8) N.R. Greiner; J.Chem.Phys. 53, 1070 (1970)
9) R.P. Overend, G. Paraskevopoulos, and R.J. Cvetanovic; Can.J.Chem. 53, 3374 (1975)
10) S. Gordon and W.A. Mulac; Int.J.Chem.Kinet.Symp. 1, 289 (1975)
11) C.J. Howard and K.M. Evenson; J.Chem.Phys. 64, 4303 (1976)
12) P.R. Edwards, I. Capbell, and G.S.Milue; Chem.Ind. (1982) p 619
13) R.A. Perry, R. Atkinson, and J.N. Pitts Jr.; J.Chem.Phys. 64, 1618 (1976)
14) D.D. Davis, G. Machado, B. Conaway, Y. Oh, and R.T. Watson; J.Chem.Phys. 65, 1268 (1976)
15) G. Paraskevopoulos, D.L. Singleton, and R.S. Irwin; J.Phys,Chem. 85, 561 (1981)
16) K. Jeong and F. Kuafman; J.Phys.Chem. 86, 1808 (1982)
17) M.A.K. Khalit and R.A. Rasmussen; Natue 292, 823 (1981)
18) R. Atkinson, D.A. Hansen, and J.N. Pitts Jr.; J.Chem.Phys. 63 1703 (1975)
19) R.T. Watson, G. Machado, B. Conaway, S. Wagner, and D.D. Davis; J.Phys.Chem. 81 256 (1977)
20) J.S. Chang and F. Kaufman; Geophys.Res.Lett. 4, 192 (1977)
21) V. Handwerk and RF. Zellner; Ber.Bunsenges.Phys.Chem. 82, 1161 (1979)
22) M.A.A. Clyne and P.M. Holt; J.Chem.Soc.Faraday.Trans.II 75, 582 (1979)

THE ROLE OF FREONS IN THE CHEMISTRY OF THE UPPER ATMOSPHERE

L. BATT

Department of Chemistry, University of Aberdeen,
Aberdeen AB9 2UE, Scotland.

Summary

In connection with the diffusion of freons into the upper atmosphere, the oxidation of halogenomethyl radicals is discussed in relation to the reactions of the trifluoromethyl peroxy and trifluoromethoxy radicals.

The importance of methyl radicals in relation to combustion and atmospheric chemistry is well established. This has recently been highlighted by the observation of an increasing amount of atmospheric carbon dioxide due to both the combustion of fossilised fuels [1] and the increase of atmospheric methane [2].

At low temperatures, the oxidation of methyl radicals proceeds via the formation of methyl peroxy radicals, which subsequently react with either a free radical (e.g. CH_3O_2, CH_3 or HO_2) or organic compounds:

$$CH_3 + O_2 + M \rightleftharpoons CH_3O_2 + M \qquad (1)$$

$$CH_3O_2 + HO_2 \longrightarrow CH_3O_2H + O_2 \qquad (2)$$

$$CH_3O_2 + CH_3 \longrightarrow (CH_3O)_2 \rightarrow 2CH_3O \qquad (3)$$

$$CH_3O_2 + CH_3O_2 \longrightarrow (CH_3O_4CH_3) \rightarrow 2CH_3O + O_2 \qquad (4)$$

$$CH_3O_2 + RH \longrightarrow CH_3O_2H + R \qquad (5)$$

In the presence of nitric oxide, the methyl peroxy radicals are converted to methoxy radicals [3].

$$CH_3O_2 + NO \longrightarrow CH_3O + NO_2 \qquad (6)$$

Molecular oxygen readily converts methoxy radicals to formaldehyde [4] which is subsequently oxidised to carbon dioxide:

$$CH_3O + O_2 \longrightarrow CH_2O + HO_2 \qquad (7)$$

Since the photolysis of halogenated methanes (CX_3Cl) dispersed in the upper atmosphere [5] produces the radicals CX_3,

$$CX_3Cl + h\nu \longrightarrow CX_3 + Cl \qquad (8)$$

it is important to consider whether similar fates await these radicals. A great deal of work has been carried out on the reactions of halogen atoms generated in the photolytic act, but much less is known about the chemistry of the fragment CX_3. Taking CX_3 to be CF_3 as an example, the most likely fate is addition to ozone [6] or oxygen:

$$CF_3 + O_3 \longrightarrow CF_3O + O_2 \tag{9}$$

$$CF_3 + O_2 \longrightarrow CF_3O_2 \tag{10}$$

The $CF_3 - O_2$ bond strength is unknown but we were able to calculate a value from the relationship [7]:

$$D(CF_3 - OOCF_3) - D(CF_3 - O_2) = D_\pi O_2$$

where $D_\pi(O_2)$ is the π bond strength in molecular oxygen. We concluded that $D(CF_3 - O_2)$ was 48.8 Kcal mol^{-1}. Thus redissociation to oxygen and trifluoromethyl radicals is very unlikely and is confirmed by the recent work of Czarnowski and Schumacher on the decomposition of bis(trifluoromethyl) trioxide [8]. Using $\Delta H^o_f(CF_3) = -112.5$ Kcal mol^{-1} [9], this makes $\Delta H^o_f(CF_3O_2) = -161.3$ Kcal mol^{-1}.

On thermochemical grouns the decomposition process (11) is very unlikely:

$$CF_3O_2 \rightarrow CF_2O + OF \tag{11}$$

since $\Delta H^o_{11} = +35.6$ Kcal mol^{-1} compared to $\Delta H^o_{12} = -18.4$ Kcal mol^{-1}:

$$CH_3O_2 \rightarrow CH_2O + OH \tag{12}$$

Despite it exothermicity, even the importance of this reaction in combustion is seriously questioned [10]. Bimolecular reactions seem the most likely processes. Although direct interaction is a viable process:

$$2CF_3O_2 \rightarrow 2CF_3O + O_2 \tag{13}$$

the rate of this reaction may well be limited by the atmospheric concentration of CF_3O_2. Other possible reactions are with HO_2, NO_2 or NO:

$$CF_3O_2 + HO_2 \longrightarrow CF_3O_2H + O_2 \tag{14}$$

$$CF_3O_2 + NO_2 \longrightarrow CF_3O_2NO_2 \tag{15}$$

$$CF_3O_2 + NO \longrightarrow CF_3O + NO_2 \tag{16}$$

Decomposition of the alkoxy radical is ruled out on thermochemical grounds since $\Delta H^o_{17} = 22.9 \pm 1.6$ Kcal mol^{-1} [11]. It is possible that reaction with nitrogen dioxide could provide a sink for these radicals:

$$CF_3O + NO_2 \longrightarrow CF_3ONO_2 \tag{17}$$

$$CF_3O + NO_2 \longrightarrow CF_2O + NO_2F \tag{18}$$

Alternatively the radical CF_3O may also react with HO_2:

$$CF_3O + HO_2 \longrightarrow CF_3OH + O_2 \tag{19}$$

Reaction of the trifluoromethoxy radical with oxygen seems remote because $\Delta H^o_{20} \sim 12$ Kcal mol^{-1} [7]

$$CF_3O + O_2 \longrightarrow CF_2O + FO_2 \tag{20}$$

Combination of the trifluoromethoxy radical and nitric oxide would probably be followed by rapid photolysis, but disproportionation may be significant [12]:

$$CF_3O + NO \longrightarrow CF_2O + NOF \qquad (21)$$

Nothing is known about the photochemistry of either trifluoromethyl hydroperoxide, nitrate or pernitrate. Both the unfluorinated hydroperoxide and the nitrate absorb at wavelengths < 320 nm [1%], and methyl nitrate has been shown to have a considerable photolytic lifetime in the atmosphere depending upon the altitude [14]. The trifluoromethyl compounds may well have their ultra-violet absorption spectra shifted to shorter wavelengths. Photolysis of these compounds remain the ultimate fate, unless they reached lower altitudes when rapid reaction with water could occur [15]. Clearly more work is required here.

The example $CX_3 = CF_3$ was taken because much more is known about the thermochemistry and kinetics of these radicals rather than of other halogeno methyl radicals. However, the radicals CF_2Cl and $CFCl_2$, which are more relevant to the chemistry of the upper atmosphere, may well behave similarly.

REFERENCES

[1] J. Gribbia, New Scientist, 90, 82 (1981).
[2] R.A. Rasmussen and M.A.K. Khalid, Atmos. Environ. (1981).
[3] J.R. Barker, S.W. Benson and D.M. Golden, Int. J. Chem. Kinet., 9, 31 (1977); L. Batt and G.N. Robinson, Int. J. Chem. Kinet., 11, 1045 (1979); R.A. Cox, R.G. Derwent, S.V. Kearsey, L. Batt and K.G. Patrick, J. Photochem., 13, 149 (1980).
[4] C.T. Pate, B.J. Finlayson and J.N. Pitts Jr., J. Am. Chem. Soc., 96, 6554 (1974).
[5] F.S. Rowland and M.J. Molina, Rex Geophys. Space Phys., 13, 1 (1975).
[6] M.S. Rossi, J.R. Barker and D.M. Golden, J. Chem. Phys., 71, 3722 (1979).
[7] L. Batt and R. Walsh, Int. J. Chem. Kinet., 15, 605 (1983).
[8] J. Czarnowski and H.J. Schumacher, Int. J. Chem. Kinet., 13, 639 (1981).
[9] D.R. Stull and H. Prophet et al., "JANAF Thermochemical Tables", 2nd ed. NSRDS - NBS 37, Washington, D.C., 1971.
[10] K.A. Bhaskaron, P. Franck and T. Just, Proc. Internat. Symp. on Shock Tubes and Waves, Jerusalem, July 16-19, 1979; A.C. Baldwin and D.M. Golden, Chem. Phys. Lett., 55, 350 (1978).
[11] L. Batt and R. Walsh, Int. J. Chem. Kinet., 14, 933 (1982).
[12] J.B. Levy, private communication.
[13] J.G. Calvert and J.N. Pitts Jr., "Photochemistry", Wiley, New York, 1966.
[14] W.D. Taylor, T.D. Allson, M.J. Mascata, G.B. Fazekas, R. Kozlowski and G.A. Takacs, Int. J. Chem. Kinet., 12, 231 (1980).
[15] G.S. Milne, private communication.

SESSION 3

AEROSOLS

Summary by the Chairman
  J.G. MADELAINE

Mesure de particules fines dans une zone polluée

Design and performance of an aerosol reactor for photochemical studies

Physical and chemical characteristics of suspended particulates during smog conditions in Berlin(W)

A study of the concentration of sulfates in the particulate matter

Aerosol neutralization by atmospheric ammonia

Elemental composition and size distribution of atmospheric aerosols during long range transport

Characterization of suspended particulate matter in a
lead smeltery area

The concentration of sulfate in broken cloud layers

Combined photolytic and radiolytic aerosol formation in
a SO_2-NO_2-air

Measurement of gaseous halogenated hydrocarbons in
ambient air

AEROSOLS

Résumé par le Président de Session
J. Guy MADELAINE
L.P.M.A./SPIN CEN-FAR - Fontenay-aux-Roses

Le groupe de travail n° 3 du COST 61a intitulé "Aérosols" est concerné par les problèmes de physique des aérosols incluant en théorie la formation et la dynamique des particules atmosphériques ainsi que la chimie associée à ces particules.

Dans la communauté scientifique, on a pris également l'habitude de traiter dans cette partie tous les problèmes inhérents à l'échantillonnage et la métrologie des particules, étant donné la spécificité des techniques utilisées qui en fait un domaine très spécialisé, contrairement à l'analyse des polluants gazeux qui est en général traitée dans des laboratoires de chimie analytique plus ou moins spécialisés dans les sciences de l'environnement. Cet état de fait a conduit, par exemple, a une réunion commune des groupes de travail 1 et 3 qui a été très profitable aux chercheurs et laboratoires de ces deux groupes.

La composition du COST 61a est telle que des recoupements sont possibles entre les différentes structures et que l'analyse des résultats obtenus concernant plus particulièrement les aérosols peut parfois concerner des travaux présentés dans les autres sessions de ce symposium.

En ce qui concerne les travaux effectués ces deux dernières années, dont une partie est présentée à ce congrès, on peut voir que les principaux domaines de la science des aérosols sont abordés :

- Expériences en laboratoire destinées à comprendre et préciser la formation et la dynamique des aérosols présents dans l'atmosphère.

- Mesures in situ servant à caractériser un état de pollution de différents sites spécifiques ou non.

- Métrologie des aérosols, non limitatif du Groupe 3.

EXPERIENCES EN LABORATOIRE

Deux études particulièrement intéressantes ont été menées.

L'une concerne le processus de formation des noyaux d'Aitken, sous forme d'H_2SO_4 et sulfates. Les résultats obtenus apportent des informations intéressantes sur les actions réciproques et combinées des rayonnements UV et nucléaires dans les processus de nucléation hétéromoléculaire des aérosols. Ce domaine ayant fait l'objet d'études nombreuses et controversées. On notera que dans certaines conditions, l'action des radiations gamma à faible dose peut contribuer à la formation d'aérosols.

La seconde concerne la description d'un réacteur de simulation destiné à l'étude de la dynamique des aérosols. Des résultats importants sont obtenus et permettent d'obtenir des données sur le temps de résidence des aérosols atmosphériques en fonction de leur dimension et leur nature et sur les paramètres (thermophorèse et photophorèse) susceptibles d'influencer cette grandeur.

EXPERIENCES IN SITU

Six études concernent le difficile problème de la mesure "sur le terrain" de la composante particulaire caractérisant un état de la pollution spécifique à différents sites.

Les résultats obtenus concernent les caractéristiques physiques : concentration, granulométrie, et la nature chimique des particules déterminée en fonction de leur dimension.

Les différentes études menées couvrent plusieurs domaines : soit la caractérisation de sites spécifiques et caractéristiques d'une pollution donnée, soit des mesures destinées à évaluer les contributions anthropogènes et naturelles et/ou étudier les mécanismes de conversion gaz-particules.

On notera que dans toutes ces études les techniques de mesures deviennent de plus en plus sophistiquées et précises. La tendance déjà constatée de la détermination de la nature chimique des particules en fonction de leur dimension se développe de plus en plus et remplace peu à peu les prélèvements globaux et intégraux : utilisation d'impacteurs en cascade, dichotomiques, d'analyseurs électriques d'aérosols, de compteurs optiques, etc.

L'utilisation de la néphélométrie servant à la mesure de la masse des aérosols submicroniques de sulfate confirme les résultats nombreux publiés aux USA et devrait se développer dans les prochaines années.

Il est intéressant de signaler, bien que ces papiers ne soient pas présentés dans cette session, l'adoption des techniques des batteries de diffusion utilisées depuis de nombreuses années en physique des aérosols à la séparation gaz-particules (SO_2-SO_4^{--}).

En conclusion, on peut noter que de nombreux progrès ont été réalisés ces deux dernières années, les résultats quant aux mesures des concentrations, nature et dimension des aérosols sont de plus en plus précis, découlant de l'utilisation de techniques de mesures plus sophistiquées. Ceci pourrait conduire, peut-être en profitant de cette action concertée, à établir une collaboration plus étroite entre les différents laboratoires européens et à une comparaison des résultats obtenus avec les différentes techniques de mesure utilisées.

MESURE DE PARTICULES FINES DANS UNE ZONE POLLUEE

M.L. PERRIN, G. MADELAINE, C. FRAMBOURT
Laboratoire de Physique et Métrologie des Aérosols
Commissariat à l'Energie Atomique
BP n°6 - 92260 FONTENAY AUX ROSES   France

Summary

We took part in the VIth European Fos remote sensing campaign 1983 to
make aerosols fields measurements around the industrial area. We use a
condensation nuclei counter, an electrical aerosol analyser and an
optical counter to characterize the in situ formation of fine particles
and compare their evolution to the one of the emitted gaseous
pollutants, for different meteorological conditions. Results are
obtained in different sites of the area and to make them complete, we
also made aircraft measurements with an electrical aerosol analyser.

1. INTRODUCTION

Une campagne Européenne de Télédétection de la Pollution Atmosphérique
s'est déroulée du 06.06 au 14.06.83 sur le site industriel de Fos-1'Etang
de Berre, au sud de la France. Les objectifs de cette campagne étaient les
suivants :

- caractériser les situations météorologiques de la région,
- déterminer le bilan total du SO_2 sur le site,
- caractériser les épisodes de pollution,
- évaluer les taux de transformation SO_2 → sulfates,
- collecter un ensemble de données validées aux fins d'analyse et
  comparaison avec les prévisions des modèles.

Notre participation a consisté à mesurer les aérosols, et plus
spécialement la partie fine du spectre, inférieure à 1 µm en diamètre, pour
caractériser les épisodes de pollution. Cette série de mesures a fait suite
à de précédentes campagnes que nous avons effectuées sur le site de
Fos-1'Etang de Berre en décembre 1981 et juin 1982 (1) (2).

2. MATERIEL ET METHODES DE MESURE

La fig. 1 est une carte de la région sur laquelle on a indiqué d'une
part les sources d'émissions et d'autre part les points de mesures. On a
disposé pendant toute la campagne de deux stations : l'une fixe implantée à
Port de Bouc et l'autre installée dans un camion laboratoire déplaçable
suivant les situations météorologiques.

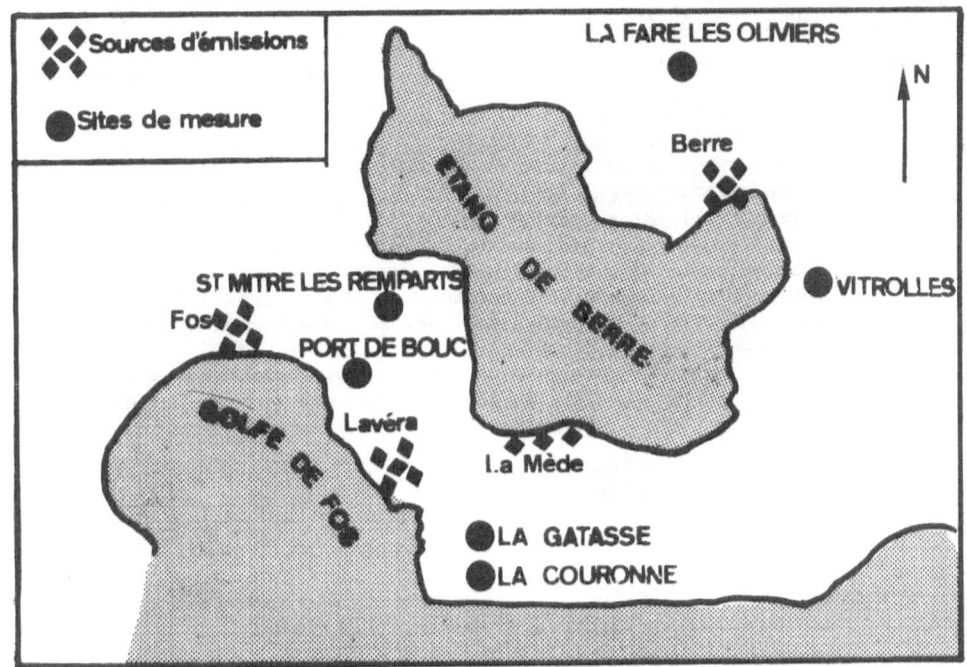

Fig.1: Zone industrielle de Fos-l'Etang de Berre

Les deux stations étaient équipées des matériels suivants :

- un compteur de noyaux de condensation qui mesure la
  concentration des particules fines
- une batterie de diffusion qui fournit le granulométrie des
  particules dont le diamètre est compris entre 0,003 µm et
  0,2 µm.
- un analyseur de mobilité électrique qui indique la granulométrie
  des particules dont le diamètre est compris entre 0,01 µm et
  1 µm.
- un compteur optique qui détecte les particules dont le diamètre
  est compris entre 0,5 µm et 10 µm.

Dans le même temps, on dispose des données météorologiques (direction
et vitesse du vent, température, pression et hygrométrie) et des émissions
gazeuses (en particulier SO_2) fournies soit par le réseau de surveillance
de la pollution atmosphérique implanté sur le site, soit par les équipes
participant à la campagne.

En plus de ces mesures au sol, on a participé à des mesures aéroportées
en installant un analyseur de mobilité électrique d'aérosols à bord d'un
avion.

3. RESULTATS DES MESURES AU SOL

Pour chaque journée de la campagne, et pour chacune des deux stations de mesures, on a pu tracer :

- une courbe d'évolution des paramètres caractéristiques de l'aérosol mesuré :
  - concentration particulaire mesurée au compteur de noyaux de condensation, N_{CNC} (cm^{-3})
  - concentrations particulaire N_{AME} (cm^{-3}), volumique V_{AME} $(\mu m^3/cm^3)$, et surfacique S_{AME} ($\mu m^2/cm^3$) mesurées avec l'analyseur de mobilité électrique
  - diamètre géométrique moyen de l'aérosol mesuré avec l'analyseur de mobilité électrique (μm) ;
- une courbe d'évolution des concentrations en polluants gazeux (dioxyde de soufre, oxydes d'azote, hydrocarbures) ou de l'acidité forte suivant le point de mesures.

Les figures 2 et 3 représentent par exemple les résultats obtenus le 08.06.83 sur la station fixe de Port de Bouc.

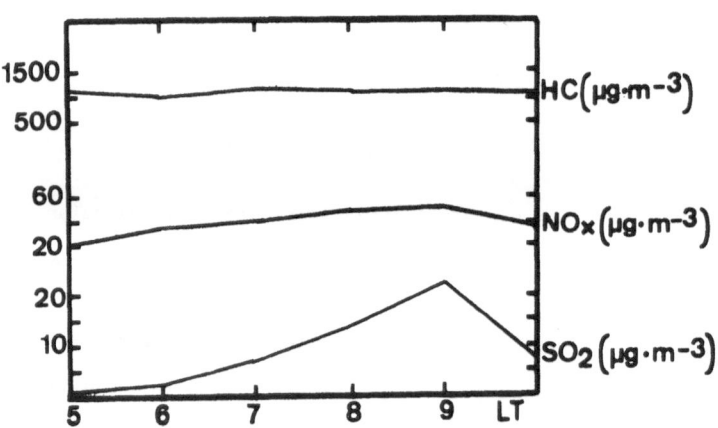

Fig.2: Concentrations en polluants gazeux
mesurées le 08-06-1983 à Port de Bouc

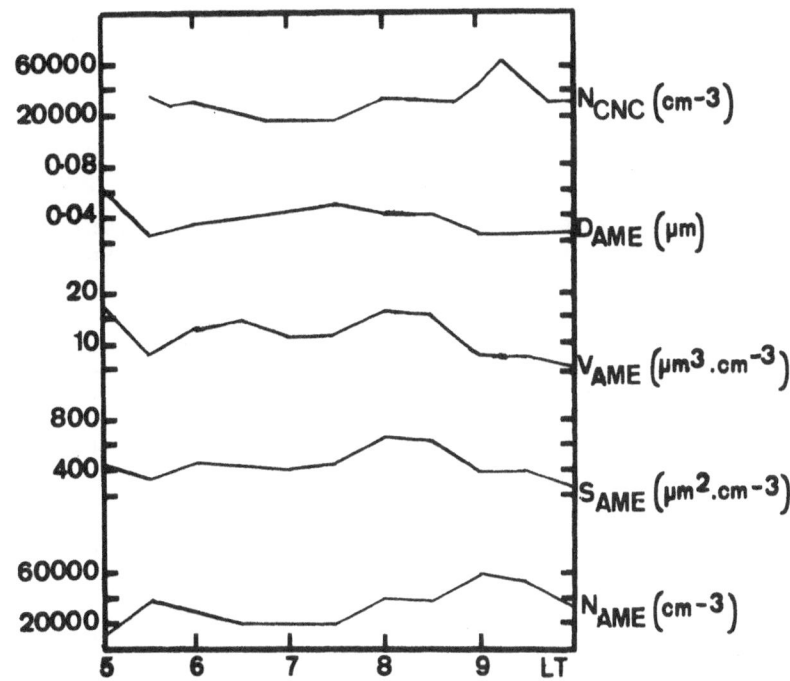

N_{CNC} (cm-3)

D_{AME} (µm)

V_{AME} (µm3.cm-3)

S_{AME} (µm2.cm-3)

N_{AME} (cm-3)

Fig.3: Résultats des mesures effectuées le 08-06-1983 à Port de Bouc

On peut distinguer deux périodes de mesures sur ces courbes :

- entre 5H et 7H30, on note :

 . une faible concentration particulaire comprise entre 20.000 et 40.000 p/cm³
 . une faible concentration en SO_2, comprise entre 2 et 6 µg/m³.
 . une concentration en NOx inférieure à 40 µg/m³.
 . le vent est pendant cette période assimilé à une brise d'étang, avec une direction comprise entre 20° et 110°.

- entre 8H et 10H, on note :

 . une augmentation du nombre des particules jusqu'à 60.000/cm³.
 . une augmentation de la quantité de SO_2 jusqu'à 25 µg/m³.
 . un vent tournant à 120° - 240°.

Cette augmentation de tous les paramètres est liée au balayage par le vent des zones de La Mède et Lavéra, rabattant ainsi la pollution sur Port de Bouc. Cette situation est caractéristique de l'établissement de la brise de mer. Les courbes des fig. 2 et 3 permettent de confirmer ce que nous avions déjà noté (1) (2) : les variations des concentrations en particules sont fortement liées à celles du SO_2. Une augmentation de ce polluant gazeux est rapidement suivie d'une augmentation de la concentration d'aérosols. Et cette remarque reste valable pour tous les résultats que nous avons obtenus pendant la Campagne.

Par ailleurs, sur les fig. 2 et 3, on s'aperçoit que les concentrations, aussi bien en SO_2 qu'en particules, sont influencées par les conditions météorologiques et plus particulièrement la direction du vent.

Pendant la durée de la Campagne, on a rencontré les régimes de vent suivants : brise d'étang, brise de mer et mistral.

L'ensemble des mesures dont on dispose permet de caractériser, pour chacune de ces situation de vent, l'aérosol prélevé sur les différents sites. Les tableaux 1, 2 et 3 résument ces résultats.

Tableau 1 : BRISE DE MER ETABLIE

	SUD/SUD-OUEST			SUD		SUD/SUD-EST
	Port de Bouc	La Fare les Oliviers	Vitrolles	Port de Bouc	La Fare les Oliviers	Port de Bouc
conc. SO_2 ($\mu g/m^3$)	négligeable	*	150	300	*	négligeable
conc. aérosols (cm^{-3})	20.000	35.000	75.000	300.000	*	20.000
surface aérosols ($\mu m^2/cm^3$)	250	*	750	500 à 1000	*	250
volume aérosols ($\mu m^3/cm^3$)	5 à 10	*	10	10 à 30	*	5 à 10
diamètre moyen (μm)	$3-4 \times 10^{-2}$	*	3×10^{-2}	2×10^{-2}	*	$3-4 \times 10^{-2}$

* paramètres non mesurés

Tableau 2 : BRISE D'ETANG

	Port de Bouc	La Fare les Oliviers	St Mitre les Remparts
concentration en SO_2 ($\mu g/m^3$)	2 à 10	*	*
concentration aérosols (cm^{-3})	20.000	20.000	20.000
surface aérosols ($\mu m^2/cm^3$)	400	*	*
volume aérosols ($\mu m^3/cm^3$)	15	*	*
diamètre moyen (μm)	4×10^{-2}	*	*

* paramètres non mesurés

Tableau 3 : MISTRAL

	Port de Bouc	La Couronne	La Gatasse	Vitrolles
concentration en SO_2 ($\mu g/m^3$)	10	*	31	20
concentration aérosols (cm^{-3})	6.000	15.000	20.000	30.000
surface aérosols ($\mu m^2/cm^3$)	125	*	*	150
volume aérosols ($\mu m^3/cm^3$)	4	*	*	2
diamètre moyen (μm)	3×10^{-2}	*	*	3×10^{-2}

* paramètres non mesurés

On a noté également que le phénomène de la reverse de la brise
(transition matinale brise de terre - brise de mer) se traduit à Port de
Bouc par une forte augmentation de la concentraiton en particules fines,
que la transition s'effectue :

- soit par le nord, et alors la zone de Fos-sur-Mer est balayée,
- soit par le sud, rabattant la pollution de Lavéra.

La lecture des tableaux 1, 2 et 3 montre que :
- même dans le cas où les points de mesure se situent en amont des
  sources d'émissions, les concentrations particulaires restent élevées
  (20.000p/cm^3), ne correspondant pas à des concentrations de bruit
  de fonds (N≈1000p/cm^3)
- lorsque le point de mesure se trouve sous le vent d'une émission, on
  trouve un grand nombre de particules très fines de diamètre moyen
  0,02 µm
- par mistral, les concentrations mesurées diminuent (<10.000p/cm^3) du
  fait d'une dilution des émissions et d'un temps de séjour des particules moins long.

Ces constatations tendraient à montrer que même en dehors d'un épisode
de pollution, on mesure un nombre important de particules fines dont la gé-
nération par photolyse pourrait être induite et favorisée par le grand ensoleillement de la région.
Tous ces résultats sont rassemblés sur la fig.4 sur laquelle on a tracé
une rose de pollution du site de Fos-l'Etang de Berre.

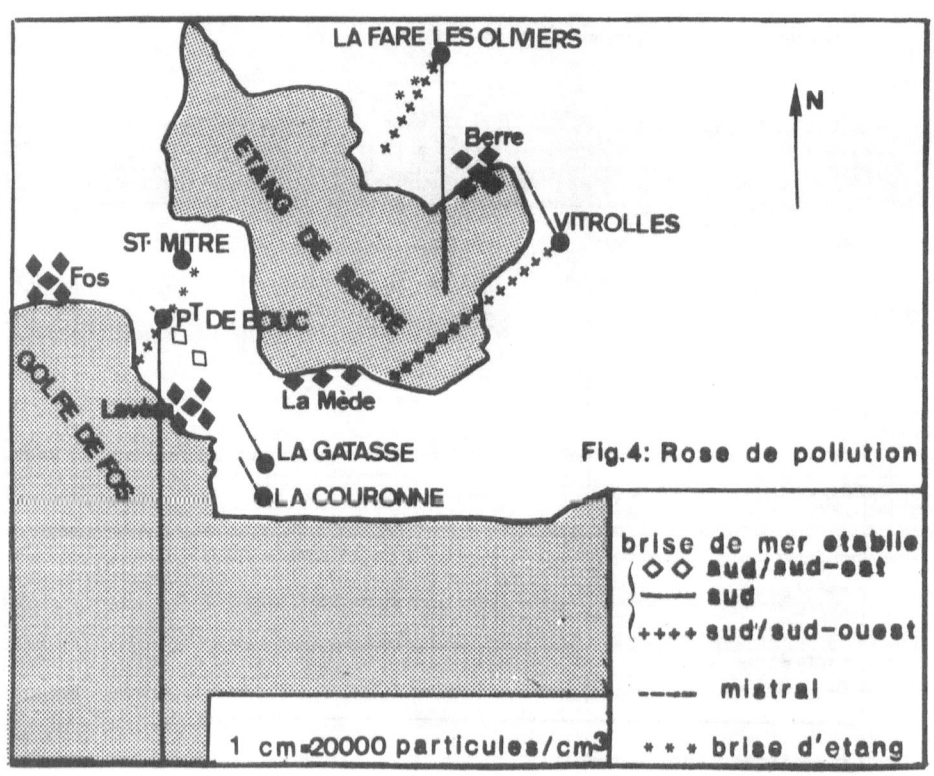

Fig.4: Rose de pollution

brise de mer etablie
◇ ◇ sud/sud-est
——— sud
++++ sud/sud-ouest

----- mistral

* * * brise d'etang

1 cm=20000 particules/cm^3

4. RESULTATS DES MESURES AEROPORTEES

Les mesures aéroportées permettent d'effectuer une cartographie de la pollution particulaire du site. Pour chaque vol, on a reporté sur une carte de la région les concentrations d'aérosols en fonction de la direction du vent fournie par les différents réseaux de météorologie installés pour la Campagne.

La figure 5 indique les mesures obtenues le 09.06 entre 11H30 et 15H. On a également signalé les sources d'émissions sur cette carte.

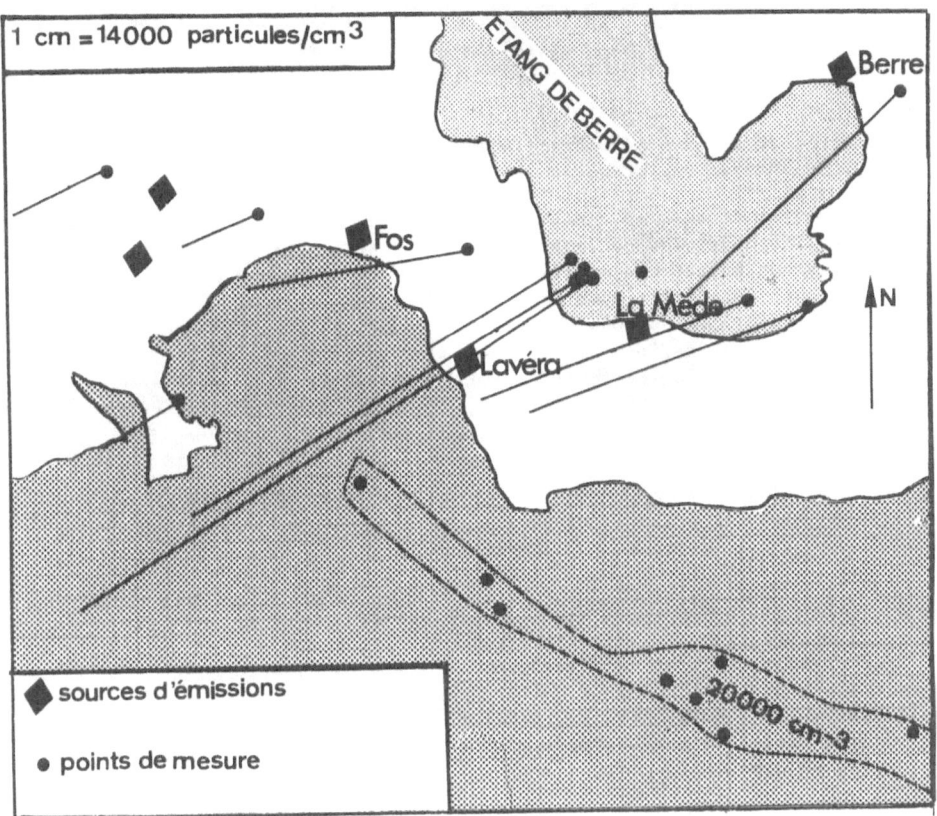

Fig.5: Mesures aeroportees effectuees le 09-06-1983

La période de mesure, correspondant à une situation de brise de mer établie au sud-ouest, est caractérisée par :

- . une concentration particulaire de l'ordre de 20.000p/cm^3 pour les mesures faites au-dessus de la mer,
- . des concentrations particulaires de l'ordre de 70.000p/cm^3 lorsqu'on se trouve sous le vent d'une émission et pouvant atteindre des valeurs plus élevées encore.

L'ensemble des mesures aéroportées effectuées pendant la campagne confirme donc le fait que les concentrations particulaires mesurées en un point donné sont largement influencées par les conditions météorologiques, et particulièrement le direction du vent :

- on mesure de l'ordre de 20.000p/cm^3, soit en l'absence de source d'émissions, soit lorsque le point de mesure se situe en amont de la source d'émissions,
- les concentrations peuvent être supérieures à 100.000p/cm^3 lorsqu'on se trouve en aval,
- le mistral provoque rapidement une dilution du panache, conduisant à des concentrations faibles inférieures à 10.000p/cm^3.

5. CONCLUSION

Du point de vue des aérosols, la VI° Campagne Européenne de Télédétection de la Pollution Atmosphérique a permis de confirmer les points suivants :

- les variations des concentrations d'aérosols sont fortement liées à celles du dioxyde de soufre,
- les concentrations, aussi bien en SO_2 qu'en particules, sur un site donné, sont influencées par les conditions météorologiques.

On a ainsi pu définir, sur chaque site de mesure et pour différentes situations météorologiques, les caractéristiques de l'aérosol présent.

Les mesures aéroportées ont conforté ces résultats et permis d'élargir le champ de mesures.

REFERENCES

(1) PERRIN M.L., MADELAINE G., NADAL R. (1982)
Fine particles measurements around Fos-Berre Area
Proceedings of the 10th conference of the GAeF, Bologne (Italie)

(2) PERRIN M.L., BOURBIGOT Y., MADELAINE G. (1982)
Fine particles measurements around an Industrial Area - Aer.Sci. and Technology 2, 165

DESIGN AND PERFORMANCE OF AN AEROSOL REACTOR
FOR PHOTOCHEMICAL STUDIES

W. HOLLÄNDER, W. BEHNKE, W. KOCH and G. POHLMANN
Fraunhofer-Institute for Toxicology and Aerosol Research
Nottulner Landweg 102, D 4400 Münster-Roxel

Abstract

The aim of the present study was to optimize a cylindrical reaction vessel of 2,54 m^3 volume for long aerosol life time under illumination. For analytical reasons, the vessel had to be made from glass and we chose 3 reaction columns of the maximum commercially available diameter of 1 m. This volume was considered enough for a time series measurement of the degradation of the chemical compounds to be studied. The design goal was a one day aerosol mass residence time under actual operating conditions (irradiation of polydisperse aerosol at a concentration of several mg/m^3). In order to assess the influence of thermophoresis and photophoresis on particle deposition, we performed model experiments with three chambers (scale 1/20, 1/10, 1/5). We found a strong increase of particle deposition with temperature difference. Photophoresis was significant only for soot particles.
With these results of the model experiments, the reactor was designed with climatization of the transparent top and of the black anodized aluminum bottom. The 1/e mass residence times for different aerosols ranged under irradiation, depending on the size, up to 40 h. The observed temperature difference dependence of the decay constant is partly due to turbulent agglomeration. The size dependence of the deposition was successfully described by the theory of Crump and Seinfeld (1981). Without irradiation longer residence times were observed.

1. INTRODUCTION

In a study on the photochemical degradation of environmental chemicals with special emphasis on the effect of airborne particles we face the problem of aerosol decay in the reaction chamber. The atmospheric aerosol residence time is - depending on particle size, emission height etc. - of the order of a few days. Hence, one cannot expect to correctly measure slow photochemical aerosol reactions of tropospheric relevance, if the aerosol chamber residence time is significantly shorter than the reaction time. On the other hand, rapidly with size increasing costs are the prohibiting factors against enlarging the chamber, which would increase the aerosol residence time.

The intensive illumination causes additional problems, because it heats the reactor volume and consequently enhances the turbulent deposition of particles as well as their thermophoretic motion to the walls. In addition, depending on the optical properties of the paricles, photophoretic motion may be of importance.

The aerosol mass concentrations required for chemical analysis of the degradation products are of the order of mg/m^3, which means that even if the mass median aerodynamic diameter before the experiment is less than 0,1 μm, due to coagulation a particle size > 0.3 μm is reached soon. For this size in the deposition minimum diffusion cannot be neglected. Of course, in addition, gravitational, thermophoretic and photophoretic deposition will be of importance, too.

2. BASIC MECHANISMS OF AEROSOL DYNAMICS IN A CONTAINMENT

The aerosol in a containment may be stagnant or in ordered or in turbulent motion. The simplest behavior should be found under the stagnant condition, because stochastic (Brownian) and directed motion (sedimentation, thermophoresis, diffusiophoresis) can be easily described for deposition from a stagnant bulk. Also, deposition under ordered motion like in rotating drums (Goldberg (1958), Frostling (1973)) or laminar convection is conceptually attractive for long residence times, since the particles will be captured in closed trajectories from which they can escape through diffusion only. Even then they will have to migrate distances of the order of the chamber dimensions before they can be deposited. The above-mentioned regimes will be called "non-stirred", because there exist concentration gradients throughout the volume.

On the other hand, in turbulent motion (with a fan or under natural convection), there is no bulk concentration variation except for a concentration boundary layer which is very thin compared to the chamber dimensions. This regime will be called "stirred". If deposition occurs through gradient forces, the latter is less desirable than the non-stirred regime, since the aerosol residence time should be reduced because of the enhanced transport through the thin concentration boundary layer.

Now the question arises, if it is possible to achieve nonstirred conditions in a photoreactor vessel and, if not, how to minimize particle losses to the wall.

Let us assume, that the transition from nonstirred to stirred regime is caused by natural temperature differences in the chamber (i.e. without mechanical stirring). The type of motion inside the chamber is then determined by the dimensionless Rayleigh number

$$Ra = h^3 \cdot \Delta T \cdot \frac{\gamma g}{\nu a} \tag{1}$$

which is directly related to the distance, a volume h^3 with a temperature difference ΔT from the surrounding can move under the action of the buoyancy force with counteracting viscous and thermal dissipation. Here

γ is the coefficient of expansion
g is the gravity acceleration
ν is the kinematic viscosity (diffusivity of momentum)
a is the diffusivity of heat.

Therefore, the last factor in (1) is a material proper-
ty. If for h the chamber height is substituted, theoretical
and experimental evidence (for large parallel planes) indicate, that for

	$Ra \lessapprox 1700$	stagnant conduction regime
$1700 \lessapprox$	$Ra \lessapprox 5 \cdot 10^4$	laminar convection
	$Ra \gtrapprox 5 \cdot 10^4$	turbulent convection

prevail (Jaluria, 1980).

Since the material factor $\gamma g / \nu a = 100 \; K^{-1} cm^{-3}$ for air at
room temperature, in a photoreactor of 300 cm height a temperature difference of $1.85 \cdot 10^{-5}$ K would be sufficient to set
up turbulent convection according to (1). Clearly such a temperature uniformity is not feasible and turbulent convection
cannot be avoided under real operating conditions.

In a recent theoretical and experimental work Crump and
Seinfeld (1981, 1983) obtained an analytical solution to the
equation

$$\nabla \cdot [(D+D_e) \nabla c] - v_s \vec{k} \cdot \nabla c = 0 \tag{2}$$

for a containment of arbitrary shape with the boundary condition

c = 0 on the containment surface
c = c_0 in the bulk outside the concentration boundary
layer. D is the particle diffusion coefficient and D_e the
eddy diffusion coefficient, which depends on position and
possibly particle size, k is the unit vector in the vertical
direction and v_s is the sedimentation velocity.

The adaptation of their result to our case of a vertical
cylinder with height L and diameter Δ yield for the aerosol
decay constant β_{cs}

$$\beta_{cs} = \frac{8}{\pi \Delta} \cdot \sqrt{k_e D} + \frac{v_s}{L} \cdot \coth \left(\frac{\pi v_s}{4 \sqrt{k_e D}} \right) \tag{3}$$

The critical parameter k_e is related to the eddy diffusivity according to the relation

$$D_e = k_e \cdot x^2 \tag{4}$$

where x denotes the distance from the wall.

As Crump and Seinfeld, we used Okuyama's (1977) expression for

$$k_e = 0,4 \left(\frac{2\varepsilon}{16\nu} \right)^{1/2} \tag{5}$$

where ε is the turbulent dissipation rate which we took as
free parameter to fit our experimental results. The important
point is that in relation (3) the second term on the right
side shows intimate coupling between diffusion and sedimentation at the floor, whereas the uncoupled linear superposition
of diffusion and sedimentation, as it was used by Van de Vate
(1980)

$$\beta_{VdV} = 0,87 \cdot D^{0,735} \cdot [1/(2L)+1/\Delta] + v_s/L \tag{6}$$

yields a much deeper minimum of the depositon constant at a

smaller particle size.

Of course, the solution of Crump and Seinfeld also applies to directed motions other than sedimentation as well. Thermophoresis and photophoresis (combined with diffusion) cannot a priori be neglected.

Since photophoresis is by far too little understood to make reliable predictions (Preining, 1966) and therefore has to be determined experimentally, the possible use of thermophoresis for reducing the overall deposition will be discussed for the case of negligible diffusion $[\pi v_s/(4\sqrt{k_e}D)$ $<<1]$: Sedimentation and thermophoresis on the lower and upper walls have opposite directions and tend to cancel each other. For an increasing bottom temperature gradient the net deposition velocity of sedimentation minus thermophoretic velocity is reduced until both equal. For larger gradients particles will be repelled from the surface. For the upper wall the situation is just reverse in that for small gradients there is no deposition. On the vertical side walls only thermophoresis is efficient for particle deposition.

3. SETUP FOR THE MODEL EXPERIMENTS

A scheme of the setup is shown in Fig. 1. As mockups of our planned 1 m diameter, 3 m height cylindrical photoreactor we used three Duran glass cylinders of the same height to diameter ratio. They were cut along the axis and equipped with flat optical windows for measuring particle concentrations with laser Doppler anemometry. The upper and lower cylinder surfaces were in good thermal contact with a thermostatized fluid, which facilitated to control the temperature difference. For illumination we used Osram HMI arc lamps, because they were chosen for the large reactor as well.

Using forward scattering, a frequency shifter (TSI Model 3180) allowed us together with an Iwatsu two channel Fast Fourier Analyzer (Model SM 2100) to determine velocity and direction of the particles. Data are transmitted via IEEE 488 interface bus to a HP 9845 desktop calculator. From the geometrical dimensions of the scattering volume, the particle velocity and the data rate we obtained the particle concentration as a function of time. The measuring point in all experiments was on the cylinder axis, 6,7 cm above the bottom in the small chamber and 9,5 cm in the medium and large chambers.

4. RESULTS OF THE MODEL EXPERIMENTS

The first series of the model experiments was performed with 0.8 μm Latex particles. As predicted by any stirred particle deposition if coagulation and electrostatic effects are negligible concentration decay was exponential ($c = c_0 \cdot \exp(-\beta t)$). From the slope of the curves for several temperature differences $\Delta T = T_1 - T_u$, where T_1 (T_u) is the temperature of the lower (upper) cylinder plane, the decay constant β can be determined.

Fig. 2 shows the relationship between β and ΔT for dark experiments with 0,8 μ Latex in the 15 cm height chamber. The data scatter is considerable due to changing ambient temperature influence on the side walls. There is a shallow minimum of the decay constant around $\Delta T \approx 16K$ which probably can be explained by thermophoretic inhibition.

For small temperature differences, the deposition constant β approaches Van de Vate's value, whereas there is a linear increase in deposition constant for large temperature differences.

A similar behavior was found for the 30 cm height chamber. Here the scatter was generally larger and some data with illumination and elevated chamber temperatures ($T_1 > T_u > T_{Room}$) are included. Again, there is a steep increase in deposition velocity with temperature difference. No clear indication for photophoretic deposition of Latex particles can be drawn from these data.

With the largest model chamber (60 cm height) we also performed experiments with (polydisperse) aerosol of interest for the photochemical studies: soot (Printex 90 furnace black, Degussa) and TiO_2 (140 m^2/g BET surface, MMAD $\approx$ 0,3 μm, primary particle size $\sim$ 0,03 μm). The aerosols were produced by atomizing, drying and neutralizing water suspensions of the powder.

Soot obviously shows a strong photophoretic effect, which is dominated, however, by thermal deposition for large temperature differences.

TiO_2 aerosol shows no significant difference between dark and illumination experiments for temperature differences larger 10K. At small temperature differences, however, irradiation helps to increase the particle residence time.

The residence times observed for Latex particles were higher for black painted bottom as compared to untreated aluminum, which is probably due to increased thermophoretic repulsion. Without climatization chamber temperature reached 110°C with rapid deposition resulting.

The problem which we have not solved yet is to calculate for given geometry and temperature difference the turbulent dissipatin rate ε, which is needed in (5) to yield the critical parameter k_e.

5. DESIGN OF THE PHOTOREACTOR

The design goal for the chamber was a residence time of one day under actual operating conditions in order to facilitate results of tropospheric relevance.

As the residence time of particles increases with increasing chamber dimensions, a large chamber is favorable, but financial constraints forced us to use commercially available 1 m diameter Pyrex glass rings. The volume should not be smaller than approximately 2 m^3 because of the need for enough material for subsequent chemical analysis. So, three of the Pyrex rings were considered sufficient. As can easily be seen from (3) or (6), a vertical cylinder position has lower sedimentation losses as compared to a cylinder with horizontal axis.

As there are indications that the irradiated chamber wall

might be an OH radical source (Carter et al, 1981), we decided to have parallel axial illumination from point sources rather than divergent radial illumination from line sources.

Furthermore, the spatial homogeneity of the illumination inside the reaction chamber should be as good as possible.

This was achieved by a layout of seven HMI lamps, which is placed about 1,5 m above the top of the reaction chamber. The outer lamps are inclined by 2 degrees. Each lamp consumes an electric power of 1,2 kW. Inside the reaction chamber this leads to a mean light intensity of about 2,5 solar constants. The intensity is continuously monitored by a photodiode (GAsP) which is calibrated by means of photochemical actinometry.

All the model experiments showed that high temperature differences are detrimental to residence time and hence, climatization of the top, where the radiation enters, and of the bottom, where most of it is absorbed, is essential.

Therefore, both the bottom plate and the top plate of the reaction chamber are water cooled. The top plate is a Pyrex glass plate of 3 mm thickness which is inserted in a slotted aluminum ring. This configuration is mounted on the topmost of the three glass rings. A laminar and vortex free stream of water is flowing over the glass plate absorbing a part of the infrared radiation of the lamps and carrying away the heat of absorption produced in the glass plate. The flow rate is of the order of 1 m^3/h so that the temperature gradient of the water in flow direction is less than 1 K/m.

The bottom of the chamber is a hollow aluminum plate with a black oxide layer on its outer surface. This reduces diffuse reflection of the light which would lead to an undesirable heating of the reactor walls. The inner surface of the plate which is in contact with the water is riffled. In combination with the high flow rate of the water (3,6 m^3/h) a good exchange of the large amount of radiation energy (7 kW) absorbed at the black outer surface is achieved. Fig. 3 shows a total view of the reaction chamber and the lamp system. UV-absorbing curtains all around the system are protecting the operating personnel against hazardous radiation.

As described in section 4, optimum aerosol life times are obtained by applying a small but finite temperature gradient in vertical direction (thermophoretic inhibition of the sedimentation) but keeping the reactor walls essentially at room temperature. This is achieved by a special climatization technique. We use two separate water circuits, one for the top plate and another for the bottom plate. One top circuit is connected to a heat exchanger with a large exchange area which is ventilated with room air. This leads to a temperature of the top glass plate which is slightly (3^{0}C) above room temperature. The desired temperature difference between the top and the bottom plate is adjusted by controlling the temperature in the bottom water circuit by means of a computer aided control circuit. The design is shown in Fig. 4.

6. PERFORMANCE OF THE PHOTOREACTOR

The decay constant β for different monodisperse Latex

particle sizes and a temperature difference of 5 K is shown in Fig. 5. The evaluation was done by electron microscopy and/or an optical particle counter (Royco 226). The best fit constants for the dark experiment are

k_e (dark) $= 0,018$ sec^{-1} and
k_e (illumination) $= 0,165$ sec^{-1}.

It is obvious, that Crump's and Seinfeld's theory can correctly describe the general size dependence of the deposition factor, because it has an adjustable parameter (contrary to Van de Vate's theory). With the limited data available so far, it cannot be decided yet, whether the larger k_e value in the illumination experiment should be attributed to a photophoretic effect or simply an increased turbulent dissipation caused by the illumination. It must be kept in mind that the actual bottom surface temperature under irradiation is higher than in the dark experiment, because the temperature difference (which is kept constant) is measured <u>in</u> the bottom heat exchanger and not at its surface (where the radiation balance would complicate a measurement even more).

The influence of temperature difference on the residence time is shown in Fig. 6 for concentrated, polydisperse aerosols under investigation (TiO_2, SiO_2, SiO_2 coated with 5 % DOP-Dioctylphthalate). The large scatter is probably caused by the different mass concentrations which result in different sizes due to coagulation.

Nevertheless, a decrease of residence time of a factor of two is observed, if the temperature difference increases from 5 K to 15 K. As electronmicrographs have revealed, part of this residence time decrease is caused by turbulent coagulation growth up to 2 μm with consequential fast deposition. A mass average residence time of approximately one day results at $\Delta T = 5$ K under irradiation for the aerosols under investigation, i.e. the design goal has been met.

The photochemical results are not presented here; they will be published in a subsequent paper.

7. ACKNOWLEDGEMENT

The present study is supported by a joint grant of the Union of Chemical Industries (Verband der Chemischen Industrie) funded by BASF, Bayer, Chemische Werke Hüls, Henkel, Hoechst and the Federal Ministery for Research and Technology (BMFT). This support is gratefully acknowledged.

8. REFERENCES

Carter, W.B.L. et al, Evidence for chamber dependent radical sources: impact on kinetic computer models for air pollution Int. J. Kinetics <u>13</u> (1981), 735-740

Crump, J.G. and J.H. Seinfeld, Turbulent deposition and gravitational sedimentation of an aerosol in a vessel of arbitrary shape J. Aer. Sci <u>12</u> (1981), 405-415

Crump, J.G., R.C. Flagan and J.M. Seinfeld, Particle wall loss rates in vessels Aerosol Sci Technol. <u>2</u> (1983), 303-309

Frostling, A rotating drum for the study of toxic substances in aerosol form. J. Aerosol Sci. 4 (1973), 411-419

Goldberg, L.J., The use of a rotating drum for the study of aerosols over extended periods of time. Am. J. Hyg. 68 (1978), 85-93

Jaluria, Y. Natural convection heat and mass transfer HMT vol. 5, Pergamon 1980

Okuyama, K. et al, J. Chem. Eng. Jpn 10 (1977), 142-147

Preining, O., Photophoresis pp. 111-135 Ch. V in: Aerosol Science Ed. C.N. Davies, Academic Press, London, 1966

van de Vate, J.F., Investigations into the dynamics of aerosols in enclosures as used for air pollution studies. Netherlands Energy Research Foundation Report ECN-86 July 1980

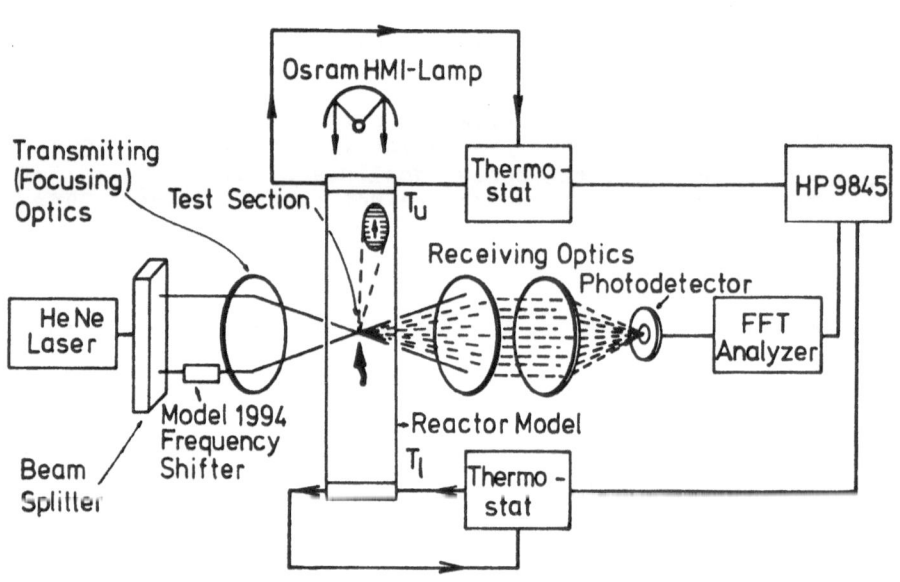

Fig. 1: Scheme of the experimental setup for the model
        experiments

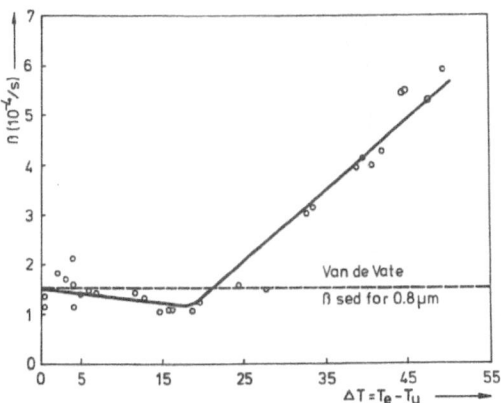

Fig. 2: Decay constant β as a function of temperature
difference for 0,8 μm Latex particles in the small
model reactor

Fig. 3: View of the photoreactor

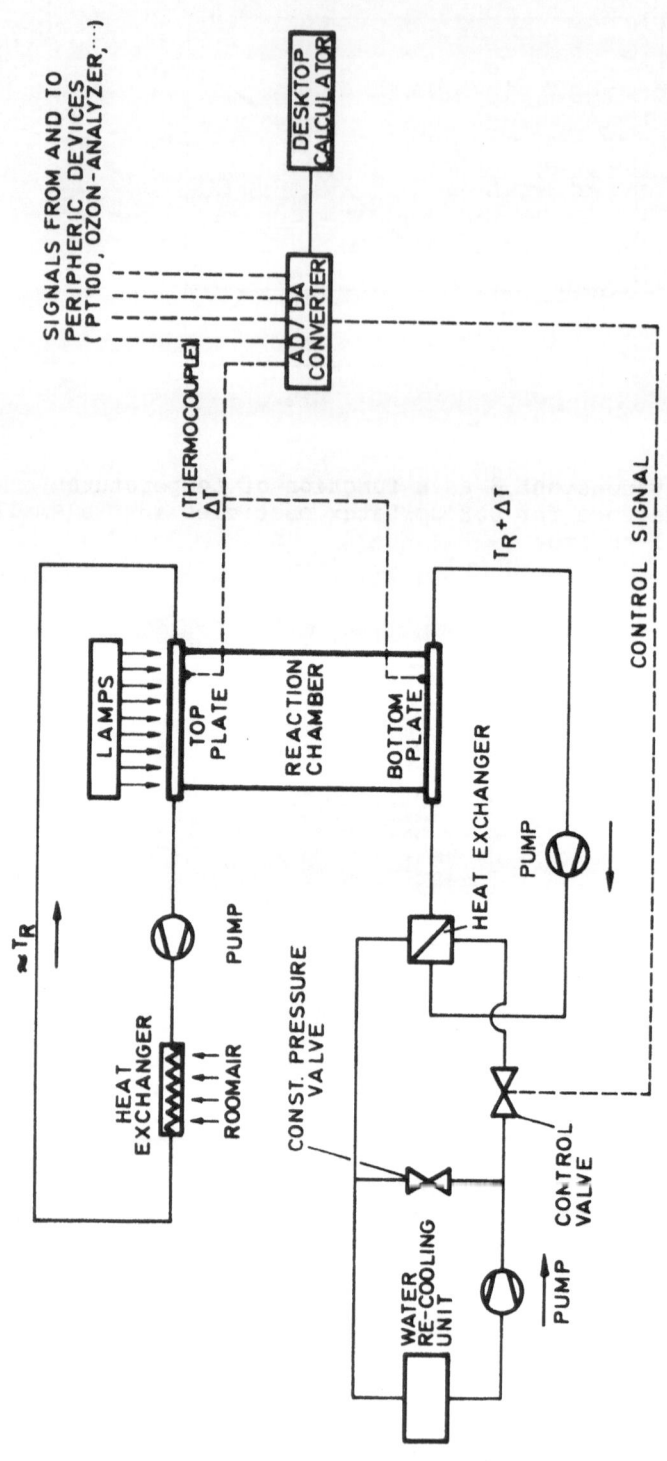

Fig. 4: Scheme of the reactor temperature control circuit

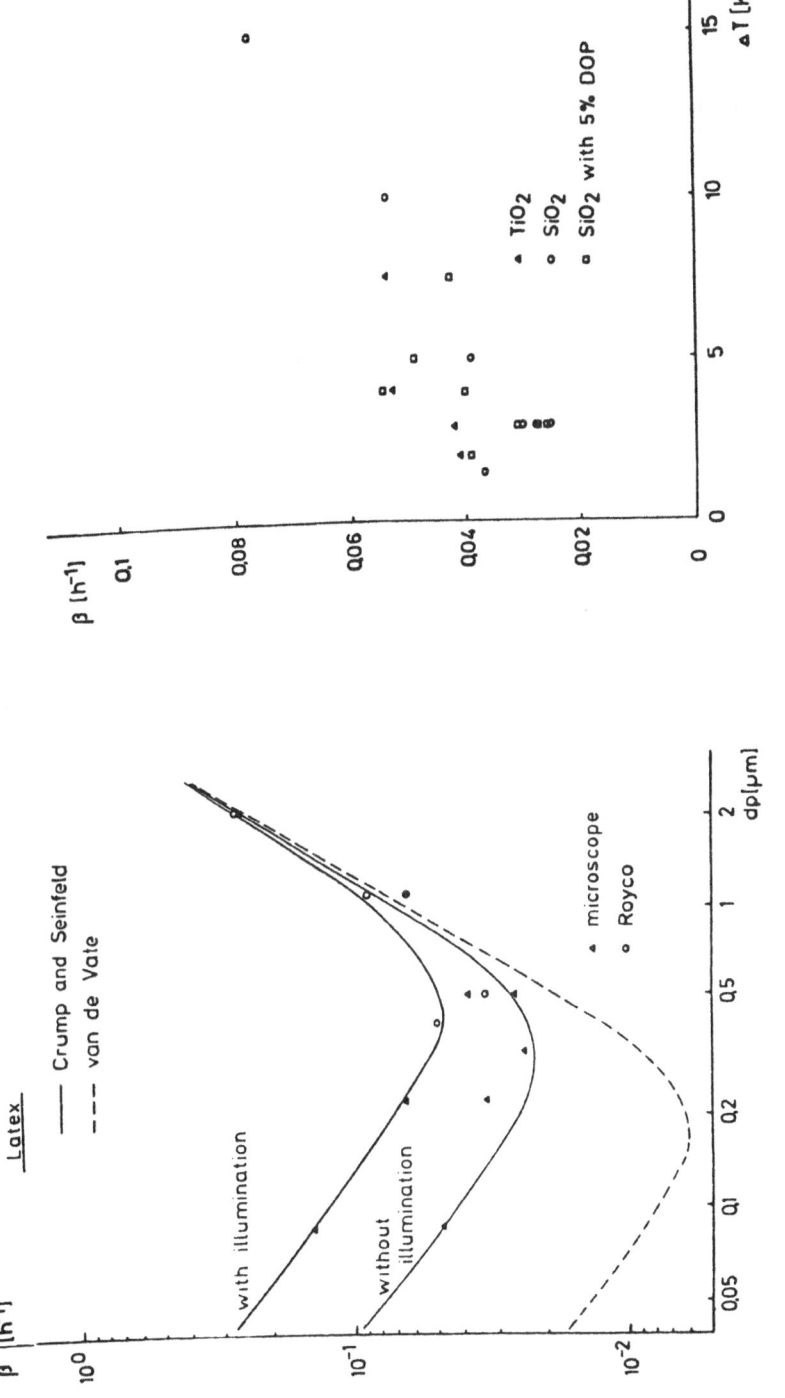

Fig. 5: Decay constant for Latex particles of various sizes (ΔT=5K)

Fig. 6: Decay constant for different aerosols (c ≈ 2 mg/m3) as a function of temperature difference ΔT

PHYSICAL AND CHEMICAL CHARACTERISTICS OF SUSPENDED PARTICULATES DURING SMOG CONDITIONS IN BERLIN(W)

G. W. Israël, H.-W. Bauer and K. Wengenroth
Fachgebiet Luftreinhaltung im Institut für Technischen Umweltschutz,
Technische Universität Berlin,
Sekr. KF 2, Straße des 17. Juni 135, 1000 Berlin 12

Summary

Major gaseous pollutants, total suspended particulates, their size
distribution, their composition and the size distribution of their
constituents have been investigated for 14 smog periods, i.e. periods
whereby the mean SO_2 and/or mean TSP levels exceed 400 $\mu g/m^3$. These
smog conditions lasted from 6 up to 100 hours. They were caused
either by the emissions of the city of Berlin proper or by advection
of pollutants from other source regions, or a mix of both. They
occured during anticyclonic weather conditions with restricted vertical mixing and low to moderate wind velocities.

Total suspended particulates were sampled by high-volume
(Staplex, Sierra) and low-volume (Kleinfiltergerät) filter samplers
and recorded with a β-mass monitor (FH62I). The results of the
samplers differed significantly from one method to the other. The
low volume sampler yielded only 66 - 83 % of the high-volume sampler
results. At high TSP levels, i.e. smog conditions, the FH62I underestimated the "Kleinfiltergerät" measured particulate mass concentration by about 50 %.

The aerosol size distribution (0,01$< d_p <$10 μm) was measured
by an electrical and optical analyser with a time resolution of
about 10 min.

A summary of the observed concentration ranges during the observed smog periods is given in table 1.

The levels of all pollutants were found to increase from summer
to winter and from normal winter conditions to smog periods. However,
the relative changes were quite different for the various species
and from one smog period to another. Ambient air quality standards
were exceeded by TSP, SO_4^{-2}, SO_2 and NO_x. Benzo-a-pyrene levels of
up to 92 $\mu g/m^3$ were observed.

The total sulfate level increases with the SO_2 concentration
and tends to exceed 100$\mu g/m^3$ for SO_2 concentrations above 600 $\mu g/m^3$.
The molar $[SO_4^{2-}]/([NH_4^+] - [NO_3^-])$ -ratio varies between 0,5 and 4
with no apparent relation to the total sulfate level. The ratio
exceeds 1 in 40 % of all cases indicating that frequently a considerable proportion of the sulfate may possibly consist of free sulfuric acid.

The fraction of secondary particulates formed in the atmosphere
by chemcial conversion, i.e. the sum of SO_4^{-}, NH_4^+ and NO_2^- ranges
from 23 up to 65 % during smog conditions and is a measure for the
age of the pollutants reaching the station. For locally produced
pollutants this fraction is less than 35 %. Larger secondary aerosol
fractions indicate a major contribution from source regions
100 - 200 km away.

Table 1

Observed concentration ranges during smog periods (N_T, S_T, V_T = total number, surface and volume concentration of particulates).

Compound		Concentration ranges	Compound		Concentration ranges
TSP-STA	$\mu g/m^3$	230 - 930	SO_4^{2-}	$\mu g/m^3$	50 - 190
Pb	ng/m^3	200 - 2200	NO_3^-	$\mu g/m^3$	3 - 50
Cd	ng/m^3	2 - 25	NO_2^-	$\mu g/m^3$	1 - 1,5
Mn	ng/m^3	90 - 470	NH_4^+	$\mu g/m^3$	12 - 55
Ni	ng/m^3				
Flu	ng/m^3	40 - 580	$N_T \times 10^3\ cm^{-3}$		40 - 107
Py	ng/m^3	30 - 460	$S_T\ \mu cm^2/cm^3$		2000 - 6500
Per	ng/m^3	1 - 13	$V_T\ \mu cm^3/cm^3$		400 - 1100
BaP	ng/m^3	8 - 92			
BghiP	ng/m^3	6 - 75	SO_2	$\mu g/m^3$	140 - 760
Cor	ng/m^3	2 - 32	NO_x als		
			NO_2	$\mu g/m^3$	90 - 320
			CO	mg/m^3	1 - 6

In general one finds that the Pb fraction of the primary particulates is markedly reduced during most smog periods in comparison to normal winter conditions while the opposite is found for the BaP fraction.

The mass, surface and volume distributions and the distributions of chemical constituents vary according to the emission pattern and the age of the particulates. 65-78 % of the particulate mass is associated with aerodynamic particle diameters smaller than 2.1 µm during smog conditions. However, more than 90 % of the mass of the PAH´s is accumulated in this small particle fraction.

The composition of the pollutants reflects the differences in emission patterns of the respective source regions and the chemical conversion during the transport to the point of observation. Thus, the observed pollutant profiles provide a means to apportion the pollutants among the contributing source regions.

A presentation of the data and their discussion can be found in the article "The Berlin Smog Project - Description and Summary of Results" that has been submitted for publication to "Atmospheric Environment" and in the two research reports of the Technical University Berlin by Israël et al:"Staubbelastung und Staubeigenschaften während Smogsituationen" and "Schadstoffzusammensetzung der Luft während Smogsituationen in Berlin (West)".

A STUDY OF
THE CONCENTRATION OF SULFATES IN THE PARTICULATE MATTER

J. DE LA SERNA, R. FERNANDEZ PATIER and F. PEREZ CARLES
Departamento de Sanidad Ambiental. Escuela Nacional de Sanidad.
Ciudad Universitaria. Madrid - 3. Spain.

Summary

A study of the concentration of sulfates in SPM, and its --
distribution, has been carried out in two different towns -
in Spain. From GF filters, sulfates were extracted by a HCl
solution in a ultrasonic bath. The range of concentrations
of sulfates was within 0-11 ug/m3. The average value of sul
fates was higher in winter than in summer, but the highest
daily concentration was found in the summer. On this study,
three variables were taken up: the two mentioned SPM and --
sulfates, and the SO2 concentrations in the air in both s--
pots. The relations SPM-sulfate and SO2-sulfate were stue--
died by bivariant distributions. The sulfate concentrations
for fixed intervals do not show gaussian distribution, and
they could be log-normal one. The same study for SO2-sulfa-
te shows neither gaussian distribution nor log-normal one.
The average concentrations of sulfates in intervals of SPM
and SO2 taken from these distributions, showed good correla
tions between the variables.

1. INTRODUCTION

The present study is a part of a project started in our Department some years ago, whose aims were the field studies of -
the principal components of SPM. Our present study is focussed
on the relationship between the concentration of sulfates and -
that of particulate matter, in two different places, having different sources of emission. Sampling point Nº1, located at the
University Campus in Madrid (situation: 40º27'N; 3º43'W; altitu
de 661m) is a residential area, where air pollution is not heavy, but noticeable (yearly average value of SO2, 22 ug/m3, du--
ring the period 1978-1982). Sampling point Nº2, is situated in
Puertollano (Ciudad Real) (situation: 38º40'N; 4º5'W; altitude
660m), an industrial town, with petrochemical industry and deri
vatives one. On this spot, the air pollution is higher than on
the first one (yearly average value of SO2, 92 ug/m3, during --
the period 1980-81, existing also, a non-frequent presence of -
NH3,as a permanent pollutant (yearly average value of NH3, 42 -

ug/m3). The occurrence of this compound led us to consider inte_
resting this study on the lower atmosphere, since many authors
had pointed out that a large portion of sulfates in the atmos--
phere is in the form of ammonium salts (1, 2, 3).

Earlier, our Department had carried out related studies --
(COST-61-A programme) on the relationship SO2-sulfate (4), stu_
dying the concentration of sulfuric acid and sulfates on the --
SPM collected on membrane filters (2, 5, 6), finding out that -
on the filter the sulfuric acid gave rise to a sulfuric-bisulfa_
te-sulfate equilibrium, which decreased the initial amounts of
the acid collected in favour of higher concentrations of bisul-
fate-sulfate.

This finding led us to view in our analytical determina---
tions the acidic forms and the neutral sulfate together for total sulfate. New experimental procedures already used (7, 8) --
will allow us soon to determine the original sulfuric acid concentrations existing in the atmospheric aerosols collected.

As well, in order to complete the main objetive of the stu_
dy, a relation between SO2 in the air, as a precursor; and sulfates in SPM, as final product, was studied in both spots.

2. MATERIAL AND METHODS

The SPM was sampled on a glass fiber filter (Whatman GF/A)
by a Hi-Vol sampler (9) at a flow rate of 40 m3/h, in periods -
of 22 hours, every day. The all-particle mass was determined --
gravimetrically. Aliquots of the filters were cut off, and some
of them, put in a hydrochloric solution, where sulfates were ex_
tracted by a ultrasonic method. The watery solution was diluted,
in order to get a pH above 3.0, to avoid interferences from the
high concentration of HCl. Total sulfates were determined spec-
trophotometrically by barium perchlorate-thorin method (10), --
previously tested and standardized in our Laboratory (11). This
method has been carried out for determining the SO2-concentra--
tions in the atmosphere, too.

A suffécient number of aliquots of the filter, was put into a flask, with 5 ml of 0.3M hydrochloric solution + 25 ml water, and extracted during 15 min. in an ultrasonic bath. The wa_
tery extract was diluted to 100 ml and total sulfate determined
as it is mentioned above. The lower limit of detection for this
determination was 0.54 ug/ml of sulfates, equivalent to 0.2 ug/
m3 in the conditions of sampling and analysis. The precision --
was 9.8 percent (95% confidence level), and the accuracy, 4.1%
(for the same level of confidence). The efficiency of extrac---
tions of sulfates from the filters was also previously tested -
by different methods. The ultrasonic one, selected, gave a 96%
extraction efficiency.

3. RESULTS

Atmospheric concentrations of sulfates

The sulfate concentrations of the two sampling points, are summarized in table 1. They indicate higher amounts of this pollutant in point 2 (Puertollano) (mean value 3.0 ug/m3) than in point 1 (Madrid) (mean value 2.3 ug/m3), as it was expected.

TABLE 1
Sulfate concentrations in the period 1978-1982

Sampling point	Period	Number of data	Mean	S.D.	X(50%)
1	1981	156	2.1	1.4	1.8
	1982	122	2.6	1.8	2.2
	Total	473	2.3	1.5	2.0
	Summer	292	2.1	1.6	1.7
	Winter	181	2.8	1.5	2.5
2	Total	101	3.0	2.4	2.6

On this table, the data of point 1, are treated by two different ways: Data of the five years of study (1978-1982) were studied as a whole, and also divided into its seasonal patterns. In addition, having a sufficient number of data of the years -- 1981 and 1982, we made a broader statistical study of these two years separately. However, due to the low number of data on sampling point 2, we could not produce but the simplest study.

The relative cumulative frequency of the sulfate concentrations in Madrid form the total and seasonal pattern periods, -- are plotted on fig. 1-a. It shows that the average concentrations of sulfates are higher in winter, than in the summer season. It is remarkable, that in the latter period, several situations of daily high sulfate concentrations are found, as it can be observed on fig. 1-a and fig. 2-a.

The fig. 1-b shows the comparative study of the relative - cumulative frequencies for points 1 and 2, in which the before mentioned higher amounts of sulfates in point 2 over point 1 -- are observed.

On fig. 2-a and 2-b are log-normal plotted the absolute -- frequencies of sulfate concentrations in Madrid and Puertollano, respectively.

Relationships between SPM-Sulfates and SO2-Sulfates

The study is carried out by means of the bivariant distributions of SPM-Sulfates and SO2-Sulfates.

TABLE 2

SPM-Sulfates bivariant distribution in Madrid(Total period78-82)

SPM intervals ug/m3	Sulfate intervals ug/m3									
	0-1	1-2	2-3	3-4	4-5	5-6	6-7	7-8	8-9	9-10
0- 30	10	30	11	11	1	0	0	0	0	0
30- 60	47	75	53	33	12	9	6	0	0	0
60- 90	10	38	23	16	7	6	2	2	0	0
90-120	2	11	9	3	3	3	0	0	0	1
120-150	1	4	6	4	1	1	3	0	0	0
150-180	2	2	1	0	1	3	1	0	0	0
180-210	0	1	3	0	0	0	0	0	0	0
210-500	0	1	0	1	1	1	0	0	0	0

The distributions of sampling point 1 are shown on table 2, that correspond to the total data analyzed. The concentration - of sulfates varies on a range between 0 to 10 ug/m3, distribu-- ted in intervals of 1 ug/m3. As the concentrations of SPM vary at a different range, between 0 to 500 ug/m3, the distributions of intervals chosen were: seven from 0 to 210, in intervals of 30 ug/m3, and the last ones, from 210 to 500 ug/m3. Separately, they were studied the same distributions on winter and summer - periods, whose results will be commented.

The data of the spot nº2 were studied in similar way for - the total SPM and sulfates, as it is shown on table 3.

TABLE 3

SPM-Sulfates bivariant distribution in Puertollano (1980-1982).

SPM intervals ug/m3	Sulfate intervals ug/m3									
	0-1	1-2	2-3	3-4	4-5	5-6	6-7	7-8	8-9	9-10
0- 30	8	3	5	1	2	0	1	0	0	0
30- 60	10	8	11	3	4	2	2	1	0	0
60- 90	3	2	3	2	1	5	3	0	0	0
90-120	0	1	3	1	1	0	0	0	0	0
120-150	1	1	1	1	1	0	2	1	0	0
150-180	0	0	0	0	1	0	0	0	0	0
180-210	0	0	0	0	1	0	0	0	0	0
210-300	0	0	0	0	0	0	1	0	1	1

The SO2-sulfates bivariant distribution in point 1 is ----

showed on table 4. The range of the SO2 concentrations was divided into intervals of 10 ug/m3, maintaining the same intervals for sulfate concentrations, as before.

The average concentration of sulfate for each interval has been calculated from the bivariant distribution expressed on the precedent tables, whose results are summarized on table 5, classified by sampling points and seasonal periods.

Finally, the results from table 5, were fitted to linear - functions, by least squares, recording the correlation coeffi-- cients for SPM-sulfates and SO2-sulfates, on table 6.

TABLE 4
SO2-Sulfates bivariant distribution in Madrid (1978-1982).

SO2 intervals ug/m3	Sulfate intervals ug/m3									
	0-1	1-2	2-3	3-4	4-5	5-6	6-7	7-8	8-9	9-10
0- 10	21	58	35	22	5	4	0	1	0	0
10- 20	12	36	19	18	4	1	3	0	0	0
20- 30	8	18	19	11	4	5	1	0	0	1
30- 40	9	14	13	5	2	6	2	0	0	0
40- 50	2	0	5	5	1	1	1	0	0	0
50- 60	0	1	3	3	1	2	0	0	0	0
60- 70	1	1	0	2	1	2	0	0	0	0
70- 80	0	1	1	0	2	0	0	0	0	0
80- 90	0	0	0	0	1	0	0	0	0	0
90-100	0	1	0	0	1	0	0	0	0	0

4. DISCUSSION

The occurrence of higher concentrations of sulfates in the industrial town (Puertollano) related to the residential area - (Madrid) is evident (table 1, fig. 1-b).

For the seasonal periods of study, winter periods were considered to be the ones, from November til March, knowing that - the central heatings in Spain work during this time. The levels of sulfates in the winter periods in Madrid, are higher than -- those in the summer ones, as it can be observed by the mean values of table 1 and the distributions on fig. 1-a. But the maximum daily average concentrations are found in the summer periods (fig. 1-a and 2-a).

These results agree with the ones published by Hickock -- (12) and Hidy (13).

Also, it is observed that, the range of sulfate concentrations found in the SPM, on the two spots studied, are lower ---

than the ones published in survey studies of sulfates, without
determining a possible cause.

TABLE 5
Average concentrations of Sulfates in the intervals of SPM and
SO2.

Pollutant Intervals		Sampling point N° 1						N° 2	
		Period						Period	
		Total		Winter		Summer		Total	
		n	Aver	n	Aver	n	Aver	n	Aver
S	0- 30	63	2.4	28	2.9	35	2.0	20	2.5
	30- 60	235	2.7	69	3.4	166	2.6	41	3.0
	60- 90	104	3.1	37	3.5	67	2.9	19	4.3
P	90-120	32	3.3	21	3.5	11	2.9	14	4.1
M M	120-150	20	3.7	13	3.8	10	3.5		
	150-180	10	3.9	7	4.1	-	-	-	-
S	0- 10	146	2.6	27	3.4	119	2.5	-	-
	10- 20	93	2.8	33	2.9	60	2.7	-	-
	20- 30	67	3.2	40	2.8	27	3.5	-	-
O	30- 40	51	3.1	42	3.1	9	2.9	-	-
	40- 50	15	3.7	10	4.2	5	2.6	-	-
2	50- 60	10	4.0	10	4.0	-	-	-	-
	60- 70	7	4.0	7	4.0	-	-	-	-
	70-100	7	3.9	7	3.9	-	-	-	-

TABLE 6
Correlation coefficients between SPM and Sulfates and SO2 and -
Sulfates.

Sampling point	Period	Relation			
		SPM-Sulfates		SO2-Sulfates	
		n	r	n	r
1	Total	6	0.996	8	0.878
	Winter	6	0.954	8	0.667
	Summer	5	0.959	5	0.159
2	Total	4	0.767	-	-

On the bivariant studies, the distribution of sulfates in
each interval of SPM, for every period studied on the two sam--
pling points, does not seem a gaussian one; may be it is a log-

normal one, but there is no certainty about it (table 2,3).

According to the distributions studied for SO2 and sulfa--
tes, it can be deduced that the distribution of sulfates in ---
each SO2 interval is neither gaussian nor seems a log-normal --
one, as it is shown on table 4. Perhaps, this conclussion would
vary if the SO2 intervals were shorter.

Table 5, shows that the average concentrations of sulfates
increase with SPM increase in all period studied and in both --
sampling points. Similar results are obtained on the total pe--
riod of SO2-sulfates relation;instead of it, it is remarkable -
that in stational periods there are not found a reliable rela--
tion.

These relations stated, we have considered of interest the
study of correlation coefficients between SPM and sulfates and
SO2 and sulfates from the results of table 5. The total period
is divided into six steps, representing 473 samples for point 1,
and 101 for point 2. The correlation coefficients confirm the -
comments on the table 5. For SPM and sulfates, "r" is near the
unit (significance level higher 99%) in Madrid; in Puertollano
the correlation coefficient is lower but inside 90% of significance level.

With reference to the relation SO2 and sulfates, the total
period shows a strong correlation (significance level 99%). In
other part, seasonal correlations show, in winter time an accep
table coefficient (95%). Instead of it, summer period presents
the lowest coefficient, and shows that the relation between variables does not agree.

From our experience, it would be advisable a broader use -
of bivariant distributions on studies, that would provide in de
termined spots, better discernible knowledge of the situation *
of each pollutant separately.

REFERENCES

(1) Altschuller, A.P. (1976), J. Air Pollut. Control Ass., 26,
318.

(2) Barton, S.C. and McCadie, W.G., (1970), Proc. 2nd. Intern.
Clean Air Congr., Washington, 379-382.

(3) Eggleton, A.E.J., (1969), Atm. Env., 3, 355-372.

(4) De la Serna, J. and Gómez, A.M., (1976), Final Symposium of
Cost-61-A Project, Ispra.

(5) Leahy, D., Siegel, R., Klotz, P. And Newman, L., (1975), --
Atm. Env., 9, 219-229.

(6) Elshout, A., (1973), Cost-61-A Project, Technical note.

(7) Mason, D.W. and Miller, H.C., (1978), Ion Chromatographic -

Analysis of Pollutants, 193-201.

(8) Slanina, J. (1982), 3rd. Meeting Working Party 1 Cost-61-A Bis Project, Rome.

(9) De la Serna, J. and Blond, L., (1975), Cost-61-A Project, - Ispra.

(10) Persson, G.A., (1966), Int. J. Air Wat. Pollut., 10, 845-- 852.

(11) Anechina, P., Alvarez, S., Merchan, M.L. and De la Serna, + J., (1974), Rev. San. Hig. Pub., 48, 115-131.

(12) Hitchcock, D., (1976), J. Air Pollut. Control Ass., 26, 210 -215.

(13) Hidy, G.M., Mueller, P.K. and Tong, E.Y. (1978), Atm., Env., 12, 735-752.

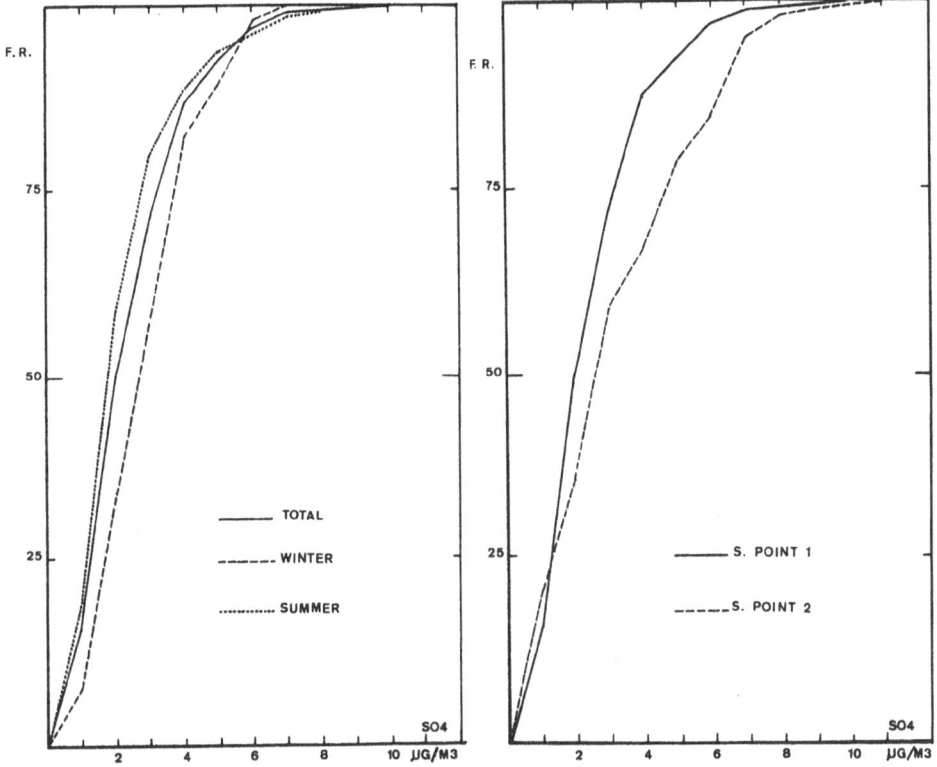

Fig. 1A: Relative cummulative frequencies of sulfate concentrations in Madrid (Point, 1)

Fig. 1B: Relative cummulative frequencies of sulfate concentrations in Madrid (Point, 1) and Puertollano (Point, 2)

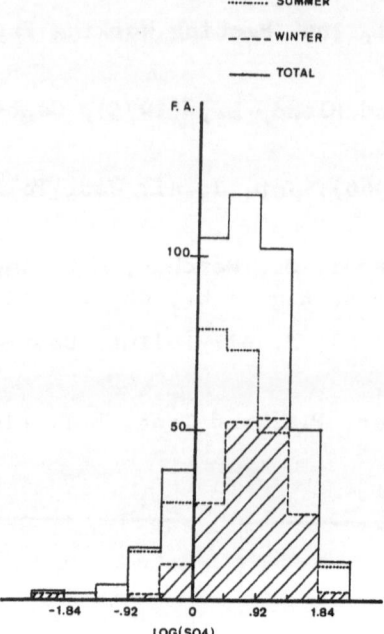

Fig. 2A: Absolute frequencies of the logarithms of sulfate concentrations in Madrid

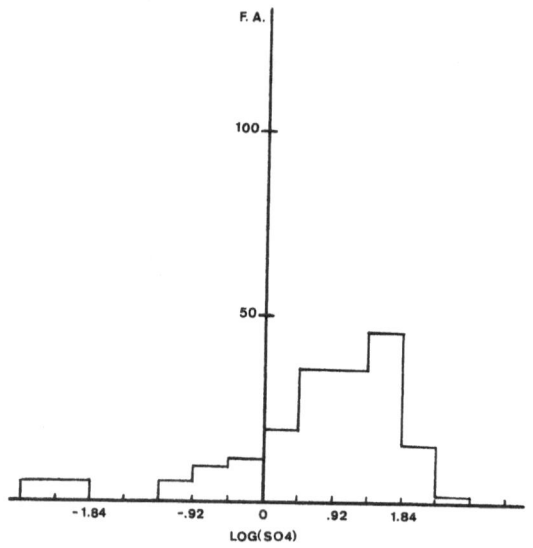

Fig. 2B: Absolute frequencies of the logarithms of sulfate concentrations in Puertollano

AEROSOL NEUTRALIZATION BY ATMOSPHERIC AMMONIA

A.G. CLARKE, M.J. WILLISON and E.M. ZEKI
Department of Fuel and Energy, Leeds University, Leeds, U.K.

Summary

The daily aerosol compositions at two sites in N. England have
been measured using dichotomous samplers for a period of 18 months.
A detailed analysis of the aerosol for one month established that
within the fine fraction the ammonium ion accounts for $\geq$ 80% of
the cations which neutralise the major anions SO_4^{2-}, NO_3^- and Cl^-.
The whole year's data was analysed for seasonal trends in the
ammonium levels and the degree of neutralisation of the aerosol.
Although there is little variation of the average monthly ammonium
levels there is a clear deficiency of ammonium relative to the sum
of the anions during winter. It is shown that this deficiency
occurs mainly with fairly clean west to north westerly air masses
of Atlantic origin having short overland traverses before reaching
the sampling site. Conversely, air masses originating from southern
U.K. and from Europe which generally have the highest aerosol
concentrations also have high ammonium levels throughout the year.

1. INTRODUCTION

Ammonia forms one of the main neutralising species present in the
atmosphere which is able to counterbalance the effects of the strong
acids H_2SO_4, HNO_3 and HCl arising from pollutant emissions. Relatively
few measurements of gaseous ammonia have been made and the emission levels
(natural and man-made) and spatial distribution are very uncertain. On
the other hand there have been a number of measurements of particulate
ammonium within the lower atmosphere from which it is possible to infer
useful information. This paper is concerned with the detailed analysis of
one such set of data.

Aerosol composition at a site in N.W. England has been the subject
of several papers by Harrison and Pio (1,2). They performed detailed ionic
balances of total suspended particulates. However, this is a time consum-
ing procedure which can only be performed on a limited number of samples.
There is also a disadvantage in using the total aerosol in that chemical
differences between coarse and fine particles are not revealed. We have
used dichotomous samplers collecting coarse and fine particle fractions
which have been separately analysed for NH_4^+, SO_4^{2-}, NO_3^- and Cl^-. More than
90% by mass of the ammonium is found in the fine fraction. Conversely,
Ca^{2+} and Mg^{2+} are generally found in the coarse fraction as is the Na^+
associated with sea salt. Thus the fine fraction is chemically fairly
simple with the predominant anions SO_4^{2-}, NO_3^-, and Cl^- being balanced
mainly by NH_4^+ if the aerosol is well neutralised or by $NH_4^+ + H^+$ if the
aerosol is acidic.

The chemical equilibria between NH_3 gas and H_2SO_4 or its ammonium
salts is such that significant levels of NH_3 in the atmosphere would only

be expected if the sulphate is completely neutralised. Conversely if an acidic aerosol is present the NH_3 levels are expected to be very low (<0.1ppb). This simple picture is complicated by the equilibria for NH_4Cl and NH_4NO_3 due to the greater volatility of HCl and HNO_3. However it will be shown that the examination of the fine particle composition can be used to deduce whether ammonia is present in excess or whether there is a deficiency relative to the other anions.

Within this paper we shall concentrate the discussion on the fine particle ammonium and a fine particle ammonium neutralisation factor which we define as the ratio of the molar ammonium concentration to the theoretical ammonium concentration assuming complete conversion of sulphate, nitrate and chloride to $(NH_4)_2SO_4$, NH_4NO_3 and NH_4Cl. i.e. expressing molar concentrations by square brackets

$$NH_4^+ \text{ Neutralisation factor (\%)} = \frac{[NH_4^+]}{2[SO_4^{2-}] + [NO_3^-] + [Cl^-]} \times 100$$

(1)

Daily sampling has been undertaken for a period of 18 months at two sites, one urban, one rural near Leeds, W. Yorkshire. An intersite comparison of the data for summer 1982, has been presented earlier (3). In this paper the data for the year October 1982-September 1983 will be analysed in terms of average aerosol composition, seasonal trends and the influence of meteorological variables.

2. EXPERIMENTAL

The geographical location of the sampling sites is shown in Fig. 1. The urban site is situated on a roof-top within the University of Leeds while the rural site (Haverah Park) is situated 20 km north of the city on open moorland. At each site Sierra Model 245 automatic dichotomous samplers were used to take 24 hour samples of fine (<2.5μm) and coarse (2.5-15μm) aerosol fractions on polypropylene backed Teflon filters. Sulphur dioxide and smoke measurements were made simultaneously using British Standard methods.

Filters were weighed before and after exposure on a microbalance and analysed for SO_4^{2-}, NO_3^- and Cl^- using ion chromatography. NH_4^+ was estimated using the colorimetric phenol-hypochlorite method. For one month only filter samples were also analysed for Ca^{2+} and Mg^{2+} using atomic absorption spectrophotometry and for Na^+ and K^+ using atomic emission.

Meteorological data was obtained from the Meteorological Office station at Wilsden 20 km west of the University and 23 km S.W. of the Haverah Park site. Three hourly information for 3 month periods was supplied on magnetic tape and transferred to the University computer for editing and subsequent statistical analysis. Automated data handling was essential since there were, excluding meteorological data, 12 parameters measured per day at each site.

For wind direction analysis the eight sectors shown in Fig. 1 were used. To avoid taking daily average wind directions each 3 hour reading was assigned to the appropriate sector and associated with the 24 hour average concentration reading together with a weighting factor of $\frac{1}{8}$. Summing the product of the concentration and the weighting factor within each sector allowed the sector average concentration to be determined for each species. Wind frequency, average wind speed and pollutant fluxes (concentration x wind speed) for each sector were also estimated.

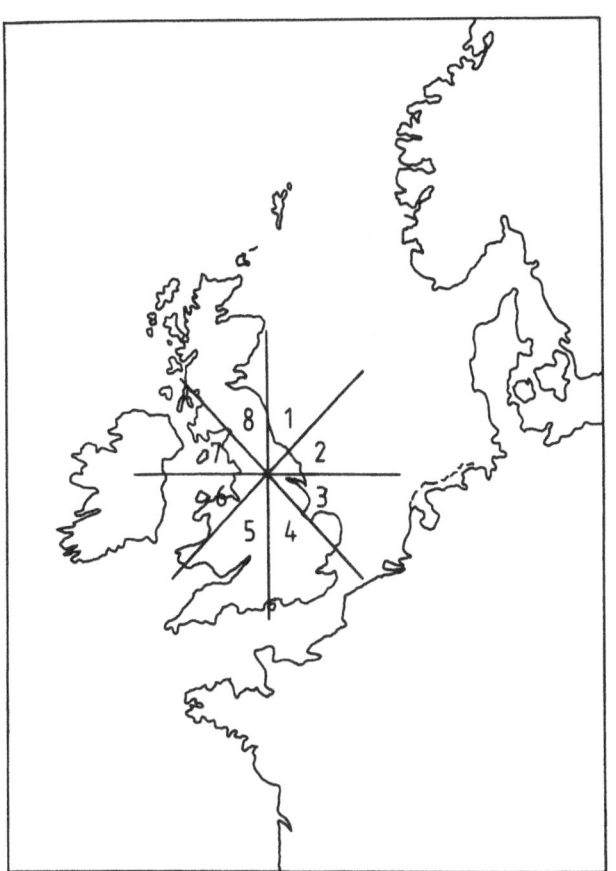

Fig. 1.   Geographical location of Leeds and definition of the sectors
           used in wind direction analysis.

3. RESULTS

Seasonal Average Aerosol Composition

Table I shows the winter (October 1982-March 1983) and summer (April
-September 1983) average concentrations for both sites. The total mass
concentrations (fine + coarse) are all below 50 µg m^{-3} indicating that the
sites are not highly polluted. The urban site has significnatly higher
mass and SO_2 concentrations than the rural site especially in winter. The
ionic compositions are very similar at the two sites providing confidence
in the measurements and indicating the uniformity of the distribution of
these secondary pollutants. In particular no significantly different
levels of NH_4^+ or trends were revealed and it is thought that neither site
is affected by major local sources of NH_3.

One feature that is clearly brought out in Table I is the virtual
disappearance of fine chloride and the marked reduction in fine nitrate
during summertime. This is presumed to be due to the volatility of
NH_4Cl and NH_4NO_3 at the higher prevailing temperatures.

Table I. Seasonal average values of aerosol and SO_2 concentrations ($\mu g\ m^{-3}$
F = Fine, C = Coarse particle fraction, n = no. of daily readings

Site	Mass		Cl^-		NO_3^-		SO_4^{2-}		NH_4^+		SO_2
	F	C	F	C	F	C	F	C	F	C	
Winter											
Hav. Park n = 140	20.7	8.0	1.1	1.4	2.1	0.8	4.4	0.8	2.3	0.2	32
Leeds n = 150	28.8	14.9	1.6	1.9	1.9	0.9	5.4	1.1	2.7	0.3	60
Summer											
Hav. Park n = 163	23.9	9.4	0.1	0.5	1.2	0.8	8.0	0.6	3.2	0.2	28
Leeds n = 180	24.3	15.6	0.1	0.6	0.6	0.9	7.7	1.0	2.8	0.2	48

Ionic Balances

For the period 18th August-17th September, 1983, a detailed chemical analysis was undertaken to attempt a full ion balance for the water-soluble aerosols. No measurements of free hydrogen ion were made and this must be roughly deduced by difference. The results are given as overall averages for the month in Table II.

Table II. Anion-cation balance for aerosols sampled at Leeds University during August-September 1983.

	ANIONS					CATIONS			
	Fine		Coarse			Fine		Coarse	
	$neq.m^{-3}$	%	$neq.m^{-3}$	%		$neq.m^{-3}$	%	$neq.m^{-3}$	%
Cl^-	6	3.3	29	42.7	Mg^{2+}	2	1.1	11	15.7
NO_3^-	24	13.0	19	27.9	Ca^{2+}	2	1.1	18	25.7
SO_4^{2-}	154	83.7	20	29.4	Na^+	9	5.1	27	38.6
					K^+	16	9.0	2	2.9
					NH_4^+	149	83.7	12	17.1
Σ anions	184	100	68	100	Σ cat-ions	178	100	70	100

In the coarse particle fraction the anions and cations balance to within 2 $neq.m^{-3}$ and the free H^+ must be zero. Chloride, nitrate and

sulphate are fairly evenly distributed in terms of the numbers of equivalents. Calcium and sodium dominate the cations with Mg^{2+} and NH_4^+ being
significant and K^+ relatively unimportant. No measurements have been made
of other ions which are possibly present such as CO_3^{2-}, Fe^{2+}, Pb^{2+}, Zn^{2+}
but these are unlikely to affect the overall picture.

In the fine particle fraction the cations sum to slightly less than
the anions leaving, on average, a possible $\leq 6neq.m^{-3}$ for free H^+. However
this would only be 3% of the cations and is subject to very large uncertainty. Sulphate dominates the anions while ammonium dominates the
cations. In both cases these ions contribute over 80% in terms of
numbers of equivalents. The fine particle neutralisation factor for
ammonium, defined previously, is 81%. Assuming that the other cations
form a roughly constant proportion of the fine particle fraction at all
times of the year, a value of the neutralisation factor well below 80%
can be taken as an indicator of an acidic aerosol. Conversely values in
the range 80-100% are probably indicative of a completely neutralised
aerosol.

Monthly Variations

In Figure 2 are presented the monthly trends of temperature,
ammonium ion concentration, ammonium neutralisation factor and the molar
ratio (fine sulphate)/(fine sulphate + SO_2 gas). Data for December are
omitted as information was only available for about half the month. In
all cases except NH_4^+ there are clear seasonal trends.

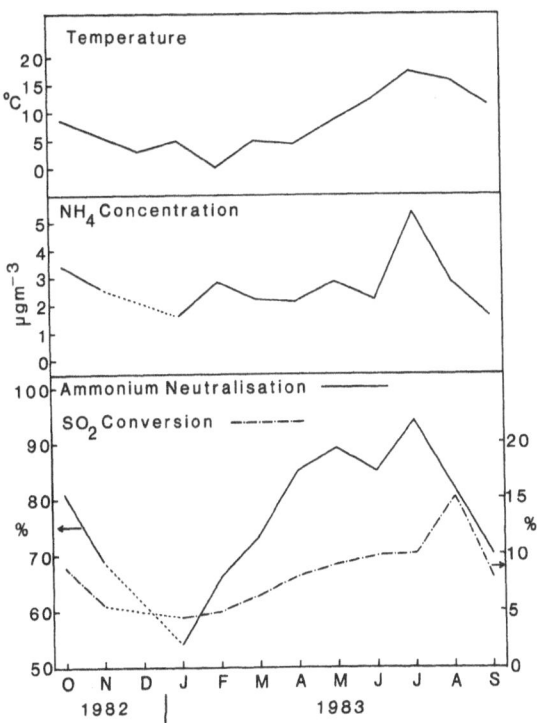

Fig. 2. Monthly Trends.

The proportion of SO_2 converted to sulphate is as low as 4-5% in winter but 10% in summer. [2]It is even higher during fine weather with anti-cyclonic conditions such as occurred during July-August, 1983. This relates to higher levels of locally produced SO_2 in winter and higher rates of conversion of SO_2 to H_2SO_4 during the summer. These two factors result in a fairly small month to month variation of sulphate ion excluding the occasional summertime episodes.

The trends in the ammonium ion concentration are also not very marked as seen in Fig. 2. However the ammonium neutralisation factor is over 80% during summer but falls to 50-70% during winter. A possible explanation of these observations is as follows. During winter in the presence of acidic aerosols the free ammonia gas concentration is expected to be very low and the measured NH_4^+ probably represents the whole of the available ammonia. In summer the aerosol is completely neutralised and there is excess free ammonia. NH_4^+ in summer is therefore limited by the available sulphate and by the positions of the NH_4Cl and NH_4NO_3 dissociation equilibria. A more marked seasonal dependence would be expected for $(NH_4^+ + NH_3$ gas) due to the increased natural NH_3 releases at higher temperatures but unfortunately this information is not available.

A similar conclusion can be drawn from the relationship of the neutralisation factor to the observed fine sulphate concentrations (summer data shown in Fig. 3). Conditions which lead to high sulphate levels are generally warm summertime situations with light winds which also lead to high levels of available NH_3 . Thus neutralisation is essentially complete (>80% in Fig. 2) and NH_4^+ and SO_4^{2-} are very highly correlated ($r = 0.99$ for 181 days). There are no days with high sulphate levels and low degree of neutralisation. Conversely during conditions which lead to low sulphate concentrations the aerosol is sometimes incompletely neutralised. These conditions occur most frequently during winter when the NH_4^+/SO_4^{2-} correlation is somewhat lower ($r = 0.93$ for 149 days). An additional factor contributing to this lower degree of correlation in winter is the presence of more fine nitrate and chloride which compete with sulphate for the available NH_3.

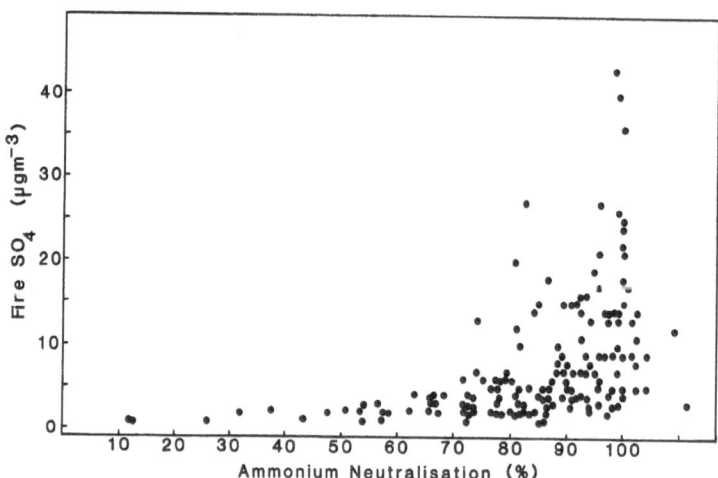

Fig. 3. The fine particle ammonium neutralisation factor in relation to the daily fine sulphate levels for summer 1983.

Wind Direction Analysis

The above analysis of seasonal and monthly trends neglects the effect of the wind direction on the observations. Sectoral averages of some of the parameters of interest are shown in Table III. All the aerosol data relate only to the fine particle fraction.

For approximately half the time the wind direction at the sampling site lies between west and south (sectors 5 and 6). These sectors are also associated with the highest wind speeds (winter averages 5.7 and 8.0 ms^{-1} respectively). Sectors 2 and 3, representing generally easterly air streams are associated with the lightest wind speeds (winter averages 3.2 and 3.1 ms^{-1} respectively). These differences have to be borne in mind in the interpretation of the concentration data.

Table III. Average meteorological and aerosol data (particles $\leqslant$ 2.5µm) for different wind directions at Leeds.

Parameter (units)	Winter/ Summer	Sector							
		1	2	3	4	5	6	7	8
Wind Freq. (%)	W	4	7	9	9	18	34	12	7
	S	11	16	11	7	11	30	8	6
Wind Speed (ms^{-1})	W	4.6	3.2	3.1	4.8	5.7	8.0	5.8	5.6
	S	3.8	3.5	2.9	3.5	4.8	5.3	3.8	4.2
Aerosol Mass (µg m^{-3})	W	26	38	53	44	35	18	17	19
	S	22	30	37	23	23	18	20	17
SO$_2$ (µg m^{-3})	W	39	94	127	90	69	44	35	35
	S	40	59	78	50	46	35	35	33
SO$_4^{2-}$ (µg m^{-3})	W	5.3	8.0	10.0	7.8	6.0	3.6	3.3	3.4
	S	8.0	10.8	11.8	6.7	6.4	5.1	6.0	5.6
NH$_4^+$ (µg m^{-3})	W	2.5	3.6	5.7	4.3	3.2	1.6	1.5	1.6
	S	2.9	4.0	4.4	2.3	2.3	1.8	2.2	2.0
Neutr. Fact. (% see eq.1)	W	79	82	89	82	74	62	63	64
	S	80	90	91	79	78	79	84	78

The highest aerosol concentrations are associated with winds in sectors 2-5 throughout the year. These sectors include air stream traversing the southern U.K. and, in many cases, parts of Western Europe as well. Table III shows that the ammonium concentrations follow this general pattern. However in winter there are significantly higher concentrations in sectors 3-5 than in summer whereas the reverse is the case for sectors 6-8.

The sector analysis of the ammonium neutralisation factor is also shown in Table III. For winds in sectors 1-5 this factor is near 80% or above in both summer and winter. From our previous discussion we may assume that these aerosols are well neutralised. However in sectors 6-8 there are very low neutralisation factors in winter (< 65%) compared to the summer values which are again near 80%. It appears that the previously noted deficiency of ammonia in winter is restricted to these wind directions.

The sectors for which this deficiency applies are those with relatively clean incoming Atlantic air with fairly short overland transit times before arrival at the sampling site. Although these sectors also have the lowest sulphate levels the net effect is that the proportion of sulphate, nitrate and chloride which can be neutralised is significantly less than for the other sectors.

4. DISCUSSION

It is of interest to compare the results obtained at Leeds with those obtained at other European sites. The situation in Berlin described by Israel and Heits (4) which relates to a considerably more polluted atmosphere than Leeds, is rather different. When the sulphate was a small fraction of the total suspended particles the aerosol was well neutralised, with NH_4^+/SO_4^{2-} ratios greater than or equal to that expected for $(NH_4)_2SO_4$. During high pollution episodes with high SO_4^{2-}/TSP ratios the amount of NH_4^+ present was less than would be expected even for NH_4HSO_4. They concluded that significant quantities of sulphuric acid were present under such conditions.

Musold and Lindqvist (5) present data for size fractionated aerosols on the south-west coast of Sweden where the mass concentrations are lower than in Leeds. The fine fraction $\leqslant 2.5\mu m$ was generally well neutralised with free H^+ being less than 10% of the cation equivalents. However the ammonium formed a somewhat smaller proportion of the fine cations than we have found, the neutralisation factors (using our definition) being in the range 55-70%.

Thus it is clear that the balance between ammonium and the anions derived from acidic air pollutants varies considerably from site to site and with the overall level of pollution. Much more information on the location and strength of ammonia emissions and their dependance on meteoroligal conditions is needed before any detailed interpretation of the data is attempted.

ACKNOWLEDGEMENT

The financial support of the Science and Engineering Research Council is gratefully acknowledged.

REFERENCES

1. R.M. Harrison and C.A. Pio. Environ. Sci. Technol. 17, 169 (1983).

2. R.M. Harrison and C.A. Pio. Atmos. Env. 17, 2539 (1983).

3. A.G. Clarke, M.J. Willison and E.M. Zeki, C.A.C.G.P. Symposium on Tropospheric Chemistry, Oxford 1983. To appear in Atmos. Env.

4. G.W. Israel and B. Heits. Second European Symposium on Physico-Chemical Behaviour of Atmospheric Pollutants. P.257, D. Reidel Pub. Co. (1982).

5. G. Musold and O. Lindqvist. Atmos. Env. 17, 1253 (1983).

ELEMENTAL COMPOSITION AND SIZE DISTRIBUTION OF ATMOSPHERIC AEROSOLS DURING LONG RANGE TRANSPORT

PETRA METTERNICH[1,2], H.-W. GEORGII[1], K.O. GROENEVELD[2]

Johann Wolfgang Goethe-Universität
6000 Frankfurt/Main
[1]Institut für Meteorologie und Geophysik
[2]Institut für Kernphysik

SUMMARY

Long range transport of air pollution is ideally investigated by measurements made over the ocean, since local pollution sources are virtually absent. The detection of pollution aerosol requires to distinguish between the fraction which has been carried from a terrestrial source and that which derived from the sea surface. Information on aerosol source types and dominating source areas is gained by a combination of size fractionating sampler and a multi-elemental technique. PIXE technique allows small sample requirements and a time resolution sufficient for meteorological interpretation.

We have examined the elemental composition of aerosol samples taken in different environments by ground and aircraft measurements. Aerosol properties evaluated by these measurements will help to understand long range transport of individual elemental constitutents.

EXPERIMENTAL

Size-fractionated aerosol samples were collected using a 1 liter/min Batelle-type cascade impactor, separating the aerosol in 8 different size fractions with aerodynamic diameters (ad) of $\leq$ 0.25 μm, > 0.25 μm, >0.5 μm, >1. μm, >2.μm, >4.μm, > 8. μm, and >16.μm for stages 0 to 7 respectively (1).

Collection was performed in the subsequent environments:

   a.   over the open Atlantic (60°N→ Antarctica), representing the unpolluted marine atmosphere (2),

b.   at Malta, a semi-remote environment in the Medi-
     terranean influenced both by continental and anthro-
     pogenic air masses (3),

c.   at St. Moritz/Switzerland and Bariloche/Argentinia,
     two remote continental environments of the Northern
     and Southern Hemisphere, respectively,

d.   at Frankfurt/Main, Germany, an urban location in a
     highly industrialized region.

In the course of ground measurements the sampling device was
mounted on a mast, about 2 m above ground level. Aerosol
sampling over the open Atlantic was performed on board the
German research vessels FS 'METEOR' during cruise Nr. 57 in
1981 and DPFVS 'POLARSTERN' during cruise Nr. 1 in 1982/83.
On board the sampler was positioned in the forward part of
the ship to avoid contamination by the ship's exhaust. Sampling was controlled by a wind vane and a General Electric
condensation nuclei counter:

- the wind vane stopped sampling whenever the wind was
  blowing from astern. Sampling was interrupted when the
  relative wind was lower than 4m/sec
- the condensation nuclei counter turned off the sampler
  when a preset concentration level was exceeded; after
  each switch off a 6 minute delay was added to clean the
  whole sampling system.

For aircraft instrumentation with a DO28 Aircraft an intake
nozzle - designed to preserve isokinetic conditions - was
mounted below the fuselage, while the sampler and the pumping system were placed inside the cabin (4).

Depending on the prevailing particle concentrations in different environments (5) sampling periods ranged between 2
days in remote regions and about 10 minutes in the polluted
atmosphere. Each size fraction was individually analyzed by
Particle Induced X-ray Emission (PIXE) at the 2.5 MV Van de
Graaff accelerator at the Institut für Kernphysik, Universität Frankfurt/Main (6).

RESULTS AND DISCUSSION

Air mass classification for each sampling period was performed by calculation of the air mass trajectories at the 1000,
850, 700 and 500 mb level by hindcasting up to 48 hours each.
Besides air mass trajectories atmospheric 222-Rn-concentra-
tions (7) gave evidence for transport of individual elements
as well as different aerosol characteristics determined for
each element in the several environments.

a.   concentration levels
b.   amount of crustal and marine fractions (8)
c.   enrichment factors

d. mass size functions
e. mass median aerodynamic diameters
f. 'excess' or 'unexplained component' (8).

The excess of an element, i.e. the amount of material which neither is soil-derived nor seasalt-derived, is a main parameter for detection of pollution aerosol. This amount of the aerosol is calculated by a bivariate regression analysis; the influence of soil and seasalt contribution to the excess of an element can be resolved with respect to particle size, using Si and Cl as reference elements. Evidence for long range transport of an element is given by the presence of a residual (after subtracting the marine and soil contribution) in the fine fraction (9), i.e. < 1 μm while an enrichment in coarse particles (> 1 μm ad) can be explained by local sources only.

SULPHUR

The difference between baseline and contaminated conditions is reflected by the fine particle sulphur in a significant way (9). Bulk concentrations of S determined over the open Atlantic didn't show any pronounced variations within experimental precision, while aerosol characteristics gave evidence for transport of anthropogenic S-particles over the ocean in the Northern Hemisphere: crustal and marine proportions were reported to be a factor of 3 lower than observed for the southern part; the 'unexplained components' accounted for more than 95% of the total mass of S, and for all samples taken in the Northern Hemisphere S containing particles were found in the ultrafine mode (< 0.25 μm ad). The amount of S in the submicron fraction (fig. 1) for northern hemispheric air masses was in good agreement with values reported by Maenhaut et al. (10): Concentrations between 30-40 ng/m^3 compares to a mean S-content for the same size range of 35 ng/m^3 over the northern tropical Atlantic and remote continental environments in Southern America (10, 11). Continental and anthropogenic influence was indicated by an increase of fine particle S in the region of NE-Atlantic (about 3000 km west of Spain and 5000 km east of United States): high S concentrations were most likely the result of influx of pollution sulfate from the Iberian Peninsula and/or from northern Africa during easterly airflow (10). Fine particle S concentrations of this order, as also observed by Maenhaut et al. (10), have already been measured over the western N-Atlantic and at Bermuda, where they were attributed to the transport of polluted air from the northeastern U.S.. High amounts of pollution sulfate were also found over the NE-Atlantic during periods of easterly winds as shown by our data. Indirect corroboration for this explanation was found in the lower concentrations obtained for samples about 1000 km west of Spain during a period of westerly winds. Besides the Iberian Peninsula (12) northern Africa, and the U.S., Great Britain seems to be another main source region for pollution aerosol. During north-easterly

winds when air masses passed over anthropogenic source areas
fine particle-S concentrations measured over the NE-Atlantic
were about 60% higher than those under purely marine conditions (13).

SOIL COMPONENTS

Of the various elements which seem to be reliably crust-de-
rived in the aerosol Al, Si, and Fe are generally considered
to be the most suitable crustal reference elements. Of
these, Fe has markedly stronger pollution sources than does
Al, and is therefore less suited for use in urban or rural
locations. Si is probably the most unambiguous elemental
indicator of crustal material. Depending on the strength
of continental influence, the Si concentration for the
various parts of the Atlantic varied in a significant way.
Increase of Si containing particles was affiliated with the
transport of mineral dust from the desert and arid regions
of Africa. The Saharan desert is known to be a source of
dust particles of great strength. Concentrations of Si
found in the region of the NE-tradewinds were of one magnitude and more higher than those observed in marine air
masses; the absolute amounts even exceeded the values for
remote continental areas (11,14). Aerosol characteristics
represented the presence of soil particles by mass median
aerodynamic diameters between 6-10 μm ad, high enrichment
factors, and an enhancement of Si mass in the particle range
2-8 μm ad mainly (fig. 2). As well as for Si the aerosol
characteristics evaluated for Al indicated that total suspended particulate concentrations may be strongly affected
by dust generating activities and that the chemical composition of terrestrial dust in the atmosphere is governed by
that of the earth crustal material from which it is derived
(15, 16). In remote marine areas of the Northern and
Southern Hemisphere Fe was found in the size fraction 0.5 -
4 μm ad only. Nearer the African continent concentrations
tended to be up to 2 orders of magnitude higher suggesting
continental effects especially from the Saharan desert.
The concentration levels over the NE-Atlantic may be at
least in part anthropogenic (16).

MARINE COMPONENTS

Na and Cl, the main cationic components of seasalt - derived
aerosols, showed highest concentrations in marine regions
known for high amounts of seasalt (16). Aerosol characteristics such as large mass median aerodynamic diameters and
marine components accounting for more than 90% of the elemental concentration, gave evidence for marine origin of
these substances due to seaspray production at the ocean surface, so highest values reported were found mainly over parts
of the southern Atlantic (Weddell Sea, 'roaring forties')
and the region around Cape Good Hope, characterized by permanent high wind speed each. Detectable Cl loss in the fraction above 0.5 μm ad was observed during episodes of conta-

minated air for Northern Hemisphere mainly (fig. 3). This
Cl loss is usually attributed to reactions of NaCl with oxidants or with sulfuric acid, both of which are of importance
under polluted conditions (17).

Otherwise under marine conditions the ratio of Cl/Na (after
subtracting the soil contribution, Cl soil corrected/Na) the
size fraction above 1 µm ad showed a significant enrichment
compared to the seawater ratio, probably the result of high
production of seasalt aerosol and the absence of reactive
pollutants. Besides Na and Cl the elements Mg, K, and Ca
are most abundant cationic components of seasalt-derived
particles, but these substances can also be soil-derived.
To determine the amount being transported over the Atlantic
the influence of the soil and seasalt contribution was resolved in the way described above. Within these elements K
showed the highest excess in the fine particle range (fig.4).
Depending on the distribution of natural and anthropogenic
sources the different parts of the Atlantic are characterized by advected K containing particles, originating from
biomass burning in the Southern Hemisphere mainly, and industrial activity in the Northern Hemisphere, respectively.
Similar to the observations made for S high concentrations
were measured over the NE-Atlantic during westerly airflow;
this transport of fine particulate K as well as S has also
been reported by Winchester (11). Comparing our data obtained for Ca to K the difference in the elemental composition of the desert soil of the various arid regions of
Africa was indicated (18). In addition high Ca-excess concentrations in the coarse fraction occur in these regions
with strong seaspray production as already mentioned. Mg
showed the smallest amount in the aerosol attributed to long
range transport, except the dust loaden region of the NE-
tradewinds. Here the amount in the fine fraction accounted
for more than 40%. This different behaviour of the three
elements over the northern tropical Atlantic can possibly be
attributed to disparities in fractionation processes at the
air/sea interface and/or the earth crust; besides this, selective removal of elements from the atmosphere has to be
taken into account as indicated by the results of our measurements at Malta: the scavenging of particulate K, Mg, Ca,
and Al with the passage of a cold front wasn't uniform.

HEAVY METALS

V, Cr, and Mn were detected in the aerosol of the marine environment during transport of mineral dust from the Saharan
desert only. Ti was found in all samples taken under continental influence, with a maximum over the northern tropical Atlantic. In contrast, Cu and Zn didn't vary in accordance with the soil element abundance; so they didn't appear
to be related to resuspension of deposited material as V,Cr,
and Mn. These two elements only found in the northern hemispheric marine environment support the presence of anthropogenic material over the open Atlantic some thousand kilo-
meters from the source area.

CONCLUSIONS

Elemental size distributions are useful in the evaluation of aerosol long range transport. The latter results could be achieved:

Fine particles of S, K, Al, Ca, and heavy metals apparently contaminate a large area of the northern hemispheric part of the Atlantic ocean and therefore probably represent long range transport. The anthropogenic source areas for this pollution aerosol-identified in this first quantitative approach - are the highly industrialized regions of the United States and the European continent; these plumes are also sufficiently reactive chemically to lead to the release of seasalt Cl in the natural marine atmosphere. While mainly anthropogenic substances pollute the remote marine environment of the Northern Hemisphere, natural constituents of crustal material dominate in the aerosol attributed to long range transport over the ocean in the Southern Hemisphere.

ACKNOWLEDGEMENT

This work is supported by Deutsche Forschungsgemeinschaft/ Bonn, Germany through Sonderforschungsbereich 73/Atmospheric Trace Substances. The crew of FS 'Meteor', DPFVS 'Polarstern', and DFVLR/Oberpfaffenhofen, Germany is gratefully thanked for their assistence; moreover the government of Malta for providing the facilities for the measurements.

REFERENCES

1. S. Baumann, P.O. Houmere, J.W. Nelson
    Nucl. Instr. and Meth. 181 (1981) 499

2. P. Metternich, H.-W. Georgii, K.O. Groeneveld
    paper presented at 3rd International Conference on
    Particle Induced X-ray Emission and its Analytical
    Applications (1983), to be published in Nucl.Instr.
    and Meth.

3. P. Metternich, H.-W. Georgii, K.O. Groeneveld
    IEEE Transactions 30 (1983) 1282

4. P. Metternich, H.-W. Georgii, K.O. Groeneveld
    Proceedings of the 5th Two-Annual Colloquium of the
    Sonderforschungsbereich 73 of the Universities Frank-
    furt and Mainz and the Max-Planck-Institut Mainz (1981)
    129

5. H. Lannefors
    Ph. D. thesis Univ. Lund/Sweden (1982)

6. P. Metternich, H.-W. Georgii, K.O. Groeneveld
    Bericht Nr. 42 des Instituts für Meteorologie und
    Geophysik, Univ.
    Frankfurt (1980)

7. B. Bonsang, B.C. Nguyen, A. Gaudry, G. Lambert
    J. Geophys. Res. $\underline{85}$ (1980) 7410

8. K.A. Rahn
    Technical Report, Graduate School of Oceanography,
    Univ. of Rhode Island, Kingston (1976)

9. M.O. Andreae
    J. Geophys. Res. $\underline{87}$ (1982) 8875

10. W. Maenhaut, A. Selen, P. Van Espen, R. Van Grieken,
    J.W. Winchester
    Proceedings of a workshop in Lund: Long range transport
    of metals and sulfur (1980) 37

11. J. W. Winchester
    Nucl. Instr. and Meth. $\underline{181}$ (1981) 367

12. J. W. Winchester, M. Darzi
    Proceedings of a workshop in Lund: Long range transport
    of metals and sulfur (1980) 32

13. H.-C. Hansson
    Ph. D. thesis Univ. Lund/Sweden (1983)

14. M. Grosch
    Bericht Nr. 36 des Instituts für Meteorologie und
    Geophysik, Univ. Frankfurt (1978)

15. B.R. Meyer, E. Le Roux, M.J. Renan, M. Peisach
    paper presented at 3rd International Conference on
    Particle Induced x-ray Emission and its Analytical
    Applications (1983), to be published in Nucl.Instr.
    and Meth.

16. J.M. Prospero
    J. Geophys. Res. $\underline{84}$ (1979) 725

17. J. Heintzenberg, H.-C. Hansson, H. Lannefors
    Tellus $\underline{33}$ (1981) 162

18. M. Sarntheim
    private communication (1983)

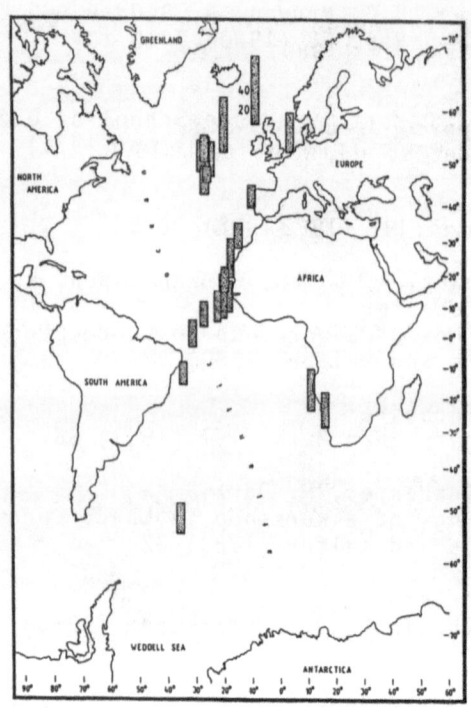

Fig. 1   S-concentrations in the fine particle range
measured over the Atlantic (ng/m³).

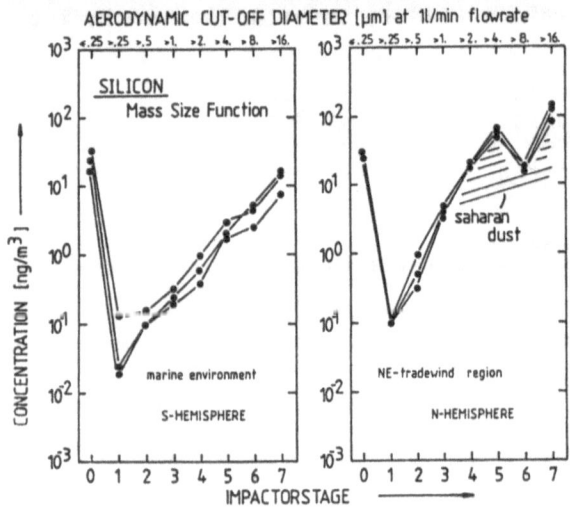

Fig. 2   Silicon mass size functions for parts of the
Atlantic in Northern and Southern Hemisphere;
the different curves represent several samples.

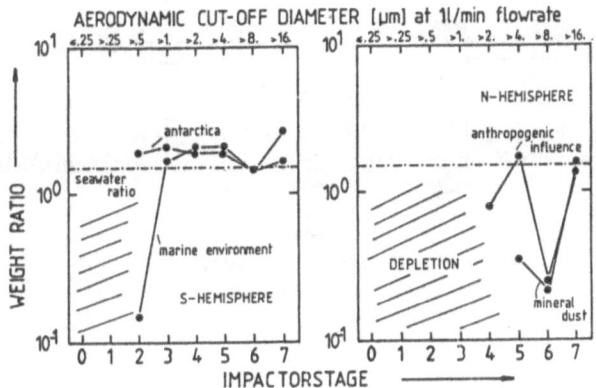

Fig. 3 (Cl soil corrected/Na) vs. particle size, given for different sampling periods in the Southern and Northern Hemisphere, respectively.

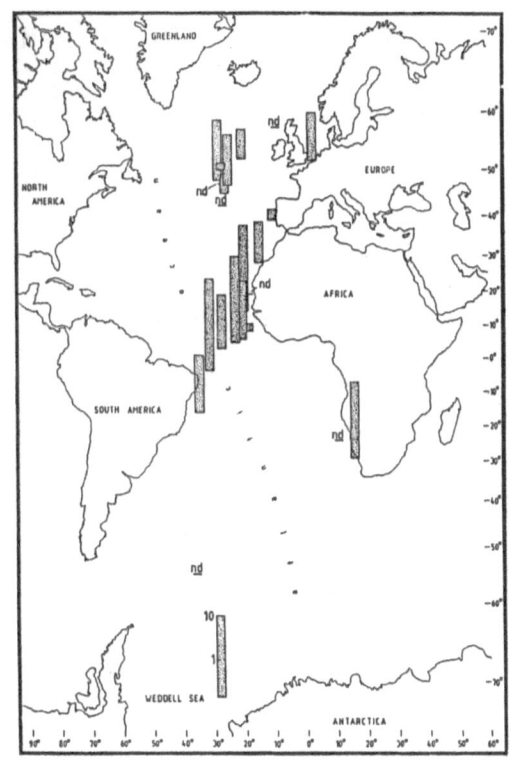

Fig. 4 Amount of fine particle K transported over the ocean relative to total mass <1 μm ad (%) nd= no advected amount estimated.

CHARACTERIZATION OF SUSPENDED PARTICULATE MATTER IN A LEAD SMELTERY AREA

M. Fugaš, J. Hršak and K. Šega
Institute for Medical Research and Occupational Health,
Zagreb, Yugoslavia

P. Souvent
Lead Mine and Smeltery
Mežica, Yugoslavia

Summary

The relationship between total suspended particulate matter (SPM) concentration and its content of Pb, Cd and SO_4^{2-} was studied in Hi-Vol samples divided by particle size and relative solubility. The samples were collected during the winter season at two sites in a lead smeltery area and at a control urban site. The SPM in the polluted area was characterized in addition to a higher lead concentration, by a lower portion of water soluble Pb and Cd, and EDTA soluble Pb and by a different particle size distribution.

1. INTRODUCTION

In the attempt to learn more about the characteristics and fate of a lead smeltery stack emissions, samples of suspended particulate matter (SPM) and depositions collected around the smeltery were analysed for lead, zinc and cadmium in respect to the distance from the stack, selective solubility and the mutual relationships. The results were related to the relevant data in ore concentrate, filter dust and flue gases (1). Further investigation was directed to correlate these findings with the concentration of the three metals in the soil (2). In this paper the relations were analysed in samples divided by particle size collected at two distances from the stack and compared with the results obtained in a large urban area with no industrial sources of Pb. Since in the smeltery area summers are windy and rainy, winter was selected for our observations.

2. EXPERIMENTAL

2.1 Collection and analysis of samples

Size selected samples were collected for 22 days during two winter months (1982/83) in a rural area close to the lead smeltery (sampling site 1) and in a small town at a distance

of about 1.5 km (sampling site 2). Sampling was done by two
Andersen Hi-Vol Fractionating Samplers with a cascade impactor
head dividing aerosol sample in four fractions with > 7.0,
7.0-3.3, 3.3-2.0 and 2.0-1.1 μm equivalent aerodynamic diameters and the back up filter retaining particles smaller than
1.1 μm. In the control area samples were collected during the
winter 1981/82 (sampling site 4).

Impaction surfaces, perforated discs 300 mm in diameter
and 200 x 250 mm back up filters were made of GF/A glass fibre
filter medium. All sampling surfaces were preconditioned to
less than 8% relative humidity by being kept in a desiccator
containing $CaCl_2$ during 24 hours, and weighed on a semimicro
balance Sartorius Model 2474. After that, samples were submitted to analysis. In size selected samples Zn was not measured
since GF filters had a rather high background Zn values.

2.2 Results

2.2.1 Total SPM

The results are presented in Table I as arithmetic means
and standard deviations for SPM, SO_4^{2-}, Pb and Cd soluble in
HNO_3 (total), EDTA, and water respectively. The relative content (%) of each measured component in SPM is shown in Table
II. The relative portions (%) of EDTA and water soluble Pb and
Cd in total Pb and Cd are shown in Table III. Correlation
coefficients for the relation between SPM and each SO_4, total
Pb and Cd, between total Cd and Pb, and between SO_4 and both
total and water soluble Pb and Cd are shown in Table IV.

2.2.2 Mass-size distribution of particles

The mass-size distribution of particles plotted as
$\Delta M/\Delta \log D_p$ vs. log D_p is shown in Fig. 1 for SPM and SO_4^{2-}, in
Fig. 2 for total, EDTA soluble and water soluble Pb and in Fig.
3 for total, EDTA soluble and water soluble Cd, for all the
three sites. The curves are rather qualitative since there are
no data on particle size distribution below 1.1 μm. Further,
the upper size limit was set at 20 μm because particles of
that size and larger are extremely rare and because of their
high settling velocity they do not enter through the sampling
inlet. A particle size of 0.1 μm was chosen as the lower limit
because mass contribution of particles of that size and smaller is negligible.

3. DISCUSSION

3.1 Total samples

SPM concentrations were higher in the polluted urban area
(site 2) than in the rural vicinity of the smeltery (site 1).
Pb concentrations were also higher at site 2 than at site 1.
The reason for this should be sought in the fact that site 2
is the point of impingement for the smeltery stack emissions.
The percentage of Pb in SPM was, however, practically the same
at both sites indicating that the pollution came from the same

source. At site 4 the SPM concentrations were highest and both the concentration of Pb and its relative content in SPM were significantly lower.

Cd concentrations were low and very similar at both polluted and the control sites when expressed in $\mu g \ m^{-3}$, which indicates that a considerable contribution of Cd comes from sources other than the smeltery. The Cd/Pb ratio was therefore significantly higher at site 4 than at the two other sites and the relative content of Cd in SPM decreased markedly from site 1 to site 4.

The selective solubility of Pb both in EDTA and in water showed a steady increase from sampling site 1 to sampling site 4 (P <0.01). The same was true for the solubility of Cd in water but not in EDTA.

The mean concentrations of sulphate in the air were not much different at all the three sites but the relative content in SPM decreased in the order: site 1 > site 2 > site 4.

Analysis of the relationship between all components showed at site 1 a significant correlation (P <0.05) between all components except between SO_4 and water soluble Cd while at site 2 significant correlations existed only between SPM and Pb, SPM and Cd, and Pb and Cd. At the control site a correlation was found only between Pb and SPM. The loss of correlation with the distance from the smeltery is obviously caused by an increasing and variable contribution of SPM from different sources.

3.2 Mass-size distribution of particles

The mass-size distribution of particles plotted as $\Delta M/\Delta \log D_p$ vs. $\log D_p$, with all the limitations mentioned earlier, is bimodal for SPM and water soluble Pb and Cd at both sites in the polluted area, with one mode at about 2-3 μm and the other above 10 μm.

Sulphates, total and EDTA soluble Pb and Cd show only one mode at about 2-3 μm.

All the curves are similar for the data of sites 1 and 2 indicating a common source of air pollution.

In the control area the mass-dize distribution curve for SPM has a similar shape, but with a shift in the lower mode towards smaller particles. For total and EDTA soluble Cd the curve is also bimodal but for the latter the second mode is in the region of <1 μm. Water soluble Cd and all fractions of lead have a completely different mass-size distribution curve, steadily descending with small particles predominating.

4. CONCLUSIONS

The analysis of concentrations, mutual relationships, selective solubility and mass-size distribution of particles in a lead smeltery area as compared to an urban control area has shown that the suspended particulate matter in the lead smeltery area is characterized in addition to a higher lead concentration and a lower Cd/Pb ratio, by a lower portions of EDTA and water soluble Pb, and water soluble Cd which increase with the increasing distance from the lead smeltery.

There is, further, a marked difference in the mass-size particle distribution curve for lead in all solubility fractions between the polluted and the control area: the former showing unimodal curves with a mode at 2-3 μm and the latter a descending curve with predominating small particles. The mass--size distributions of SPM are bimodal and very similar in both areas with one mode in the >10 μm region and the lower mode shifted from 2-3 μm in the polluted area toward smaller particles in the control area.

REFERENCES

1. J. Hršak and M. Fugaš, Proc. of 2nd European Symp. on Physico-Chemical Behaviour of Atmospheric Pollutants, Ispra 1981, B. Versino and H. Ott, Eds., Reidel, Publ. Co., 1982, pp. 287-291.

2. J. Hršak and M. Fugaš, unpublished data.

TABLE I - Mean concentration ($\bar{X}$) and standard deviation (s) of each measured component ($\mu g\ m^{-3}$)

| Component | Lead smeltery area (N=22) | | | | Control area (N=15) | |
| | Sampling site 1 | | Sampling site 2 | | Sampling site 4 | |
	$\bar{X}$	s	$\bar{X}$	s	$\bar{X}$	s
SPM	110	39.7	189	95.6	315	129
SO_4^{2-}	36.7	15.0	39.3	16.6	42.6	9.3
Pb_{tot}	5.60	4.35	8.05	5.90	1.16	0.44
Pb_{EDTA}	3.50	3.38	4.97	3.38	1.04	0.42
Pb_{H_2O}	0.097	0.077	0.190	0.184	0.170	0.098
Cd_{tot}	0.046	0.018	0.053	0.033	0.042	0.017
Cd_{EDTA}	0.036	0.014	0.045	0.031	0.027	0.006
Cd_{H_2O}	0.0034	0.0015	0.0058	0.0036	0.0113	0.0036

TABLE II - Relative content of SO_4^{2-}, Pb and Cd in SPM (%) and Cd/Pb ratio x 10^2

Component	Sampling site 1 X	s	Sampling site 2 $\bar{X}$	s	Sampling site 4 $\bar{X}$	s
SO_4^{2-}	34.3	9.86	23.6	11.5	18.5	5.63
Pb	4.6	2.28	4.37	2.66	0.38	0.007
Cd	0.045	0.018	0.029	0.015	0.016	0.009
Cd/Pb	1.13	0.712	0.83	0.519	4.10	2.13

TABLE III - Relative portion of EDTA and water soluble Pb and Cd by sampling site

Sampling site	EDTA soluble portion in % of total Pb X	s	Cd X	s	H_2O soluble portion in % of total Pb $\bar{X}$	s	Cd $\bar{X}$	s
1	58	12.72	80	8.53	1.82	0.99	8.3	4.94
2	70	17.14	84	8.92	2.67	1.52	11.1	3.66
4	90	6.13	71	16.3	14.6	5.97	30.0	10.52

TABLE IV - Correlation between measured components (N=22)

Correlated variables		Sampling site 1		Sampling site 2	
x	y	r	P	r	P
SPM	SO_4^{2-}	0.736	✗✗	0.384	
SPM	Pb	0.859	✗✗	0.438	✗
SPM	Cd	0.654	✗✗	0.612	✗✗
Pb	Cd	0.505	✗	0.852	✗✗
SO_4^{2-}	Pb	0.568	✗✗	0.160	
SO_4^{2-}	Cd	0.427	✗	0.189	
SO_4^{2-}	Pb_{H_2O}	0.689	✗✗	0.174	
SO_4^{2-}	Cd_{H_2O}	-0.078		0.173	

r= Spearman correlation coefficient
P= probability ✗ <0.05   ✗✗ <0.01

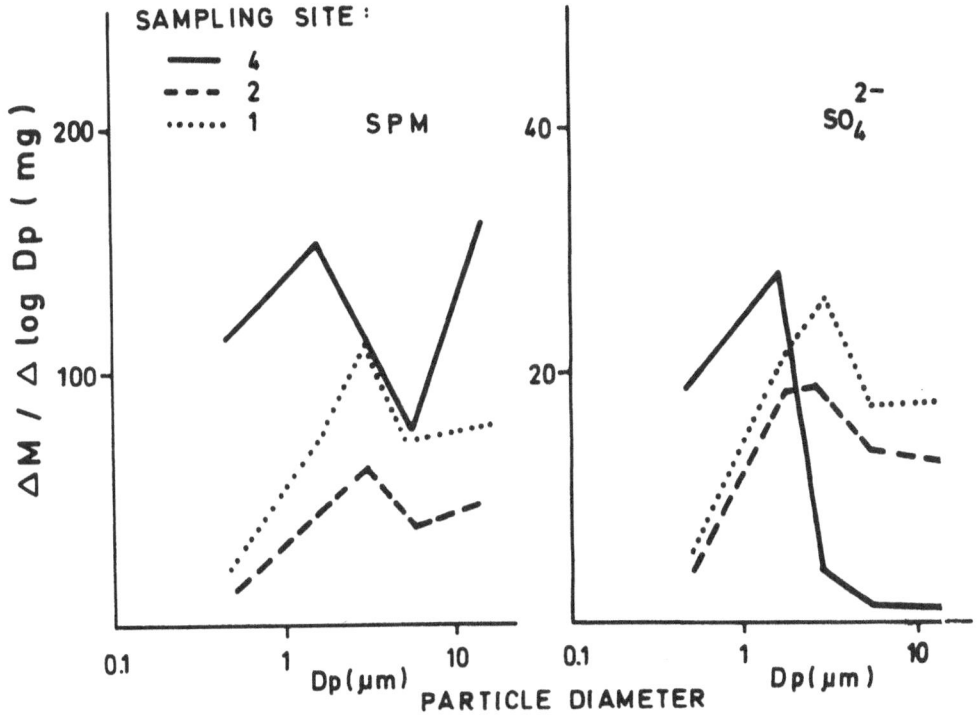

Fig. 1   MASS-SIZE DISTRIBUTION OF SPM AND SULPHATES

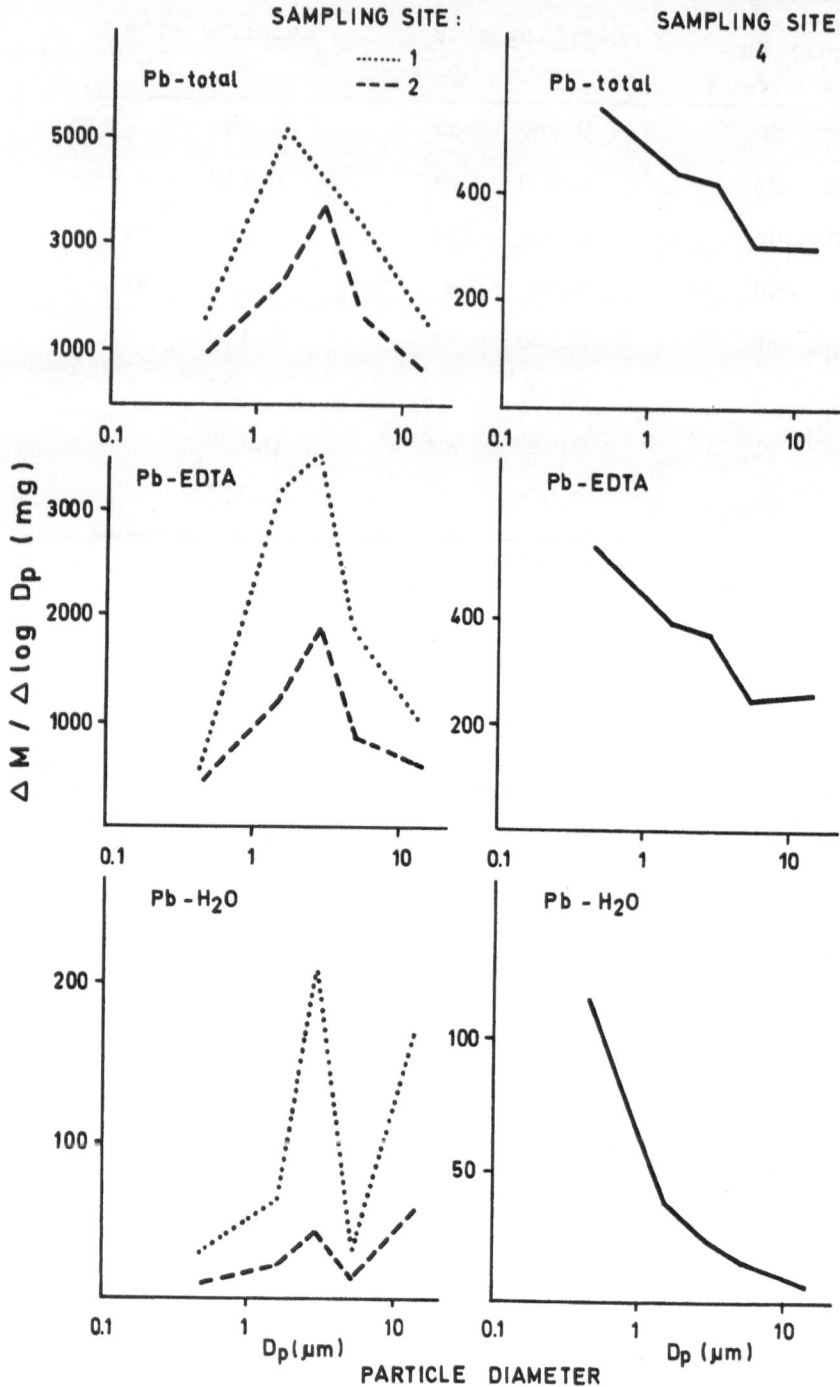

Fig. 2 MASS-SIZE DISTRIBUTION OF LEAD IN SPM

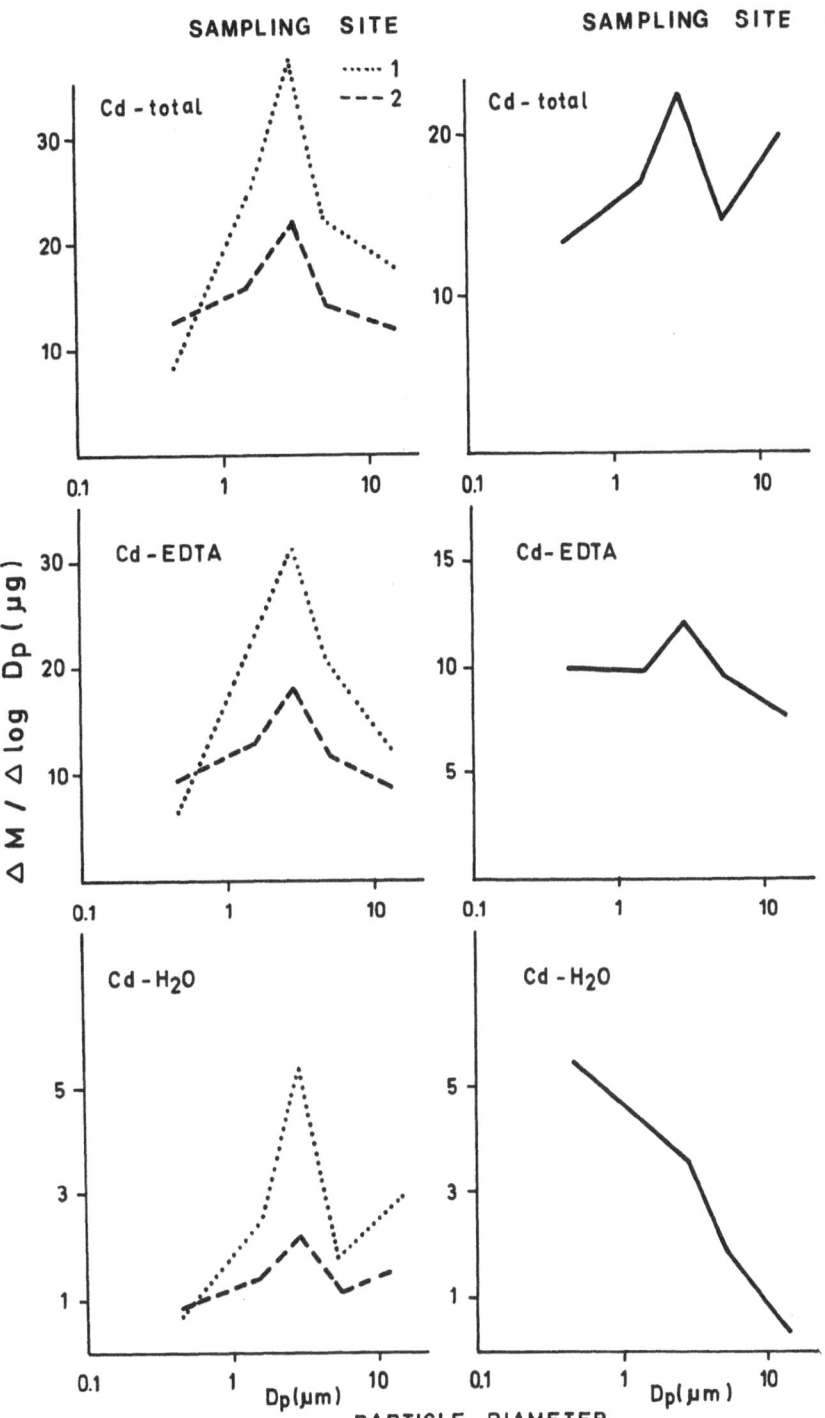

Fig. 8 MASS-SIZE DISTRIBUTION OF CADMIUM IN SPM

THE CONCENTRATION OF SULFATE IN BROKEN CLOUD LAYERS

H.M. TEN BRINK*
Chemistry Department
Netherlands Energy Research Center ECN;
1755 ZG PETTEN, The Netherlands

P.H. DAUM and S.E. SCHWARTZ
Environmental Chemistry Division, Department of Applied Science
Brookhaven National Laboratory; Upton, New York 11973, USA

Summary

The concentration of sulfate aerosol in the supersaturated
and cloud free regions of broken cloud layers was determined
via aircraft flights. Measurements were performed using two
real-time monitors, a sensitivity enhanced flame photometer
detector for sulfur and an integrating nephelometer. It is
shown that the light scattering coefficient as measured by
the nephelometer can be translated to a sulfate concentration. For the speciation of the aerosol composition filter
packs were used.
Typically the concentration of sulfate present as interstitial aerosol in the supersaturated cloud regions was found to
be less than 10% of aerosol sulfate in the adjacent cloud-
free zones. When it is assumed that the origin of the supersaturated air in the cloud layer and the cloud-free air are
of identical origin the observed difference in the sulfate
concentration is a direct measure of the scavenging efficiency for sulfate aerosol. The present results thus indicate
that over 90% of the sulfate aerosol is scavenged (incorporated in the cloud droplets). From a comparison of the concentration of sulfate in the collected cloud water and the
concentration of sulfate in the cloud interstitial air indications were found that in-cloud oxidation of sulfur dioxide
occurred. It was furthermore observed that the size of the
hygroscopic sulfate particles in the cloud zones was substantially larger than the size of these particles after
drying. This finding shows that the size of particles in the
cloud interstitial air and in the relatively dry in-flow air
of a cloud cannot be directly compared and it is recommended
that aerosol should be dried in order to deduce scavenging
effiencies from this type of measurement.

1. INTRODUCTION

In recent air pollution studies by aircraft [1] attention has been
focused on the acidification of cloud water by the incorporation of

---

* The present investigation was performed while on leave at Brookhaven
National Laboratory.

gaseous and aerosol sulfur oxides aids and salts, since ultimately the acidic cloud water constituents are delivered to the earth surface in precipitation. A major question in the acidification process is the extent to which acidic sulfate particles present prior to cloud formation are incorporated in the cloud droplets. It has been hypothesized that the efficiency of this uptake should be high [2,3] but in recent measurements, comparising sulfate concentration in the in-flow air and the air inside isolated cumulus clouds [4,5], it was reported that often only half of the total sulfate aerosol is scavenged (incorporated in cloud droplets).

In the case of frontal clouds the in-flow region of the air is ill-defined, but often breaks are observed in the cloud layer, i.e. zones of supersaturated air alternate with subsaturated regions of hundreds of meters in width. In the case of a well-mixed boundary layer it might be expected that the origin of the air in the cloud and the adjacent cloud-free breaks is identical. In the present investigation we have measured sulfate concentrations in adjacent regions of cloud and cloud-free air in broken clouds, making use of rapid-response instrumentation to measure sulfur-containing aerosol. In addition to measurements of aerosol sultate the concentration of sulfate in cloud water was determined. A comparison with the clear air concentrations of this species would reveal if production of sulfate occurs in the type of clouds under consideration here.

2. EXPERIMENTAL

Aircraft measurements in broken cloud layers were performed over Long Island, New York USA, in the summer of 1982 and 1983 and in the vicinity of Birmingham, Alabama USA, in November 1983. The aircraft used in the present flights was a Britten Norman Islander BN 2A which flies at a cruising speed of 180 km.h^{-1} (50 m.s^{-1}). The instruments available for the real-time measurement of $SO_4^=$ were a Flame Photometric Detector (FPD) for Sulfur, Meloy 285, which is specially modified for operation in aircraft [6] and an integrating nephelometer, MRI 1550.

In clouds the unactivated aerosol particles and the larger sized cloud droplets are separated by centrifugal impaction via a static vane cyclone (the CERL cloud water collector [7]). Wind tunnel calibrations showed that over 98% of the cloud droplets are removed via this separator.

With the sulfur monitor $SO_4^=$ is measured since gaseous sulfur compounds are removed upstream of the $SO_4^=$-channel by a diffusion denuder. Because of a slight drift in the base line, the FPD is periodically zeroed with sulfur-free air. The LOD (limit of detection) of the instrument was found to be as low as 0.1 ppb (fig. 1) which is almost one order of magnitude less than the LOD observed in previous flights [1a]. Under very turbulent conditions, however, the LOD increased to 0.3 ppb.

An analysis of light scattering data obtained in the previous years in clear air flights over Maryland showed that the light-sccattering coefficient, b_{scat}, and the $SO_4^=$-concentration are well-correlated, see fig 2. The relation between b_{scat} and the $SO_4^=$ concentration is:

$$b_{scat} = (0.120 \pm 0.014) \, [SO_4^=] \qquad (1)$$

with $[SO_4^=]$ in $\mu g.m^{-3}$ and b_{scat} in units $10^{-4} m^{-1}$.

This value is in excellent agreement with the value obtained by Pierson et al. [8] for the same relation in ground based measurements in the same region. The LOD for SO_4^- when measured via light scattering is 0.2 $\mu g.m^{-3}$ (0.05 ppb) and the nephelometer is insensitive to turbulence. Therefore the LOD for SO_4^- as determined via light scattering is almost an order of magnitude lower than the LOD of the sulfur monitor under turbulent conditions. The time response of the instrument is five seconds. The compatibility of the various measurement methods for SO_4^- was checked under clear air conditions at a humidity of ca. 50% and the results of the comparison of b_{scat} and the SO_4^--concentration are shown in fig. 2 and they are in agreement with equation (1). In recent flights a heated duct was installed upstream of the optical unit to heat and dry the incoming air to a humidity of 50% or less; in this way the aerosol is freed from attached water and brought to a reference dry state irrespective of the ambient humidity.

Aerosol particles were collected on filters using a quartz filter pack for the speciation of the sulfur (and nitrogen) compounds [10]. A sampling time of thirty minutes is used; SO_4^-, NH_4^+, H^+ and NO_3^- and also gaseous HNO_3 and SO_2 are determined. The LOD of the system for SO_4 is 1 $\mu g.m^{-3}$ (0.25 ppb). Further details of the system will be given in a forthcoming paper [10].

The presence of clouds is determined and the liquid-water content is recorded by means of a fast responding liquid-cloud water indicator, LWCI (Johnson-Williams hot wire detector) with an estimated time resolution of two seconds. As shown in fig. 2 and 3 the aircraft was alternatingly inside and outside of clouds during the flights.

Cloud water was collected via two slotted-rod impactors after a design by Winters et al. [1,11]. The composition of the cloud water was measured and compared to that of the cloud-free air in the breaks, both in a relative and absolute sense, using the concept of the potential partial pressure of cloud water constituents [12].

3. RESULTS

In the twenty two cloud flights conducted in the present study concentrations of sulfate as high as 2 ppb were observed in the cloud-free breaks in the cloud layer. In the adjacent cloud zones the sulfate concentration was without exception below the limit of detection (o.1 ppb, or 0.3 ppb under turbulent conditions). It was furthermore found (cf. fig. 1) that the concentration of sulfate in the successive cloud-free regions in a flight was almost a constant. The light-scattering coefficient, b_{scat}, showed a pattern which was identical to that of the sulfate concentration and b_{scat} was virtually equal to zero (less than $0.025 \times 10^{-4} m^{-1}$) inside the clouds, see fig. 3.

The average concentration of SO_4^- as derived from the filter measurements were compared to the average SO_4^--concentration as measured by the real-time monitors. The two values were equal within the error limits of the methods.

For those flights where no precipitation was encountered the composition of the cloud droplets and the surrounding cloud-free air with repect to the mentioned constituents were compared via the following procedure.

The concentration of the cloud water constituents were translated to mole fractions in air by the concept of potential partial pressure, which is defined as:

$$\hat{p} = LCRT \qquad (2)$$

where $\hat{p}$ is the potential partial pressure (atm.), L is the liquid water volume fraction, C is the aqueous concentration (molar), R is the universal gas constant, T is absolute temperature. The results are shown in table I.

4. DISCUSSION

The results available to-day show that the concentration of $SO_4^=$ in the cloud regions in a broken cloud layer is less than 5% of the concentration in the cloud-free zones. This result can be translated to a scavenging ratio of sulfate aerosol in the following sense.

From the result that the concentration of sulfate successive cloud-free zones is constant during a flight path it is concluded that these air parcels have the same origin and that the air is well-mixed over a large area. It is therefore expected that the total cloud layer has the same origin and is also well-mixed before cloud formation.

With this assumption the present results directly show that the scavenging efficiency for sulfate is better than 90% in the present flights.

In table I the concentrations of $SO_4^=$ in the clouds and in the cloud-free zones are compared. Tabulated are those flights in which precipitation was absent and also no plumes were encountered, which would be evidenced by increased NO_x and SO_2 in the clouds [8]. The conclusion that the concentration of $SO_4^=$ inside the clouds is higher is tentative, since the potential partial pressures are only known with an uncertainty of ca. 50% because of uncertainty in liquid water concentration [10]. Because of the uncertainty in liquid water content the concentration of $SO_4^=$ in the cloud zones and the cloud-free air was compared via a procedure in which NH_4^+ is used as a conservative tracer (any gaseous NH_3 would strengthen the following argument). The ratio of the $SO_4^=$ and NH_4^+ concentration (table I, underlined figures) is higher in the cloud regions than in the subsaturated air. The higher $SO_4^=$ to NH_4^+ ratio in the cloud zones is indicative of additional SO_4 which is caused by the oxidation of SO_2 to H_2SO_4. Evidently the set of data presented here is too limited for any definitive conclusions however it is suggested that the procedures outlined in the present work should be used in future investigations concerning the in-cloud oxidations in broken clouds.

Finally the results of measurements will be discussed in which an unheated nephelometer was used to measure b_{scat}. It was found that even with an unheated nephelometer b_{scat} was substantially less in the clouds than in the cloud-free domains. The correlation between b_{scat} and the $SO_4^=$-concentration is sustained, however the relation between b_{scat} under saturation humidities and the $SO_4^=$-concentration is found to be:

$$b_{scat}^{sat} = (0.25 \pm 0.05) \times [SO_4^=] \qquad (3)$$

with $(SO_4^=)$ in $\mu g \cdot m^{-3}$ and b_{scat}^{sat} in units $10^{-4} m^{-1}$. For a dried aerosol the analogeous relation is given in equation (1). The increased value of b_{scat} for the same concentration of sulfate, compare equation (1) and (3), at saturation humidities is caused by the substantial growth of the hygroscopic aerosol particles by water vapor accretion, a fact well-documented in literature [13,14]. This finding shows that the size of particles in the cloud-interstitial air and in the relatively dry in-flow air of a cloud shoud not be directly compared. It is recommended here that aerosol should be dried in order to deduce scavenging efficiencies from this type of measurement.

5. CONCLUSIONS

Pending a better knowledge of the meteorology of broken cloud layers the following conclusion is drawn concerning the present study. The scavenging of $SO_4^=$ by clouds is 90% or more efficient.

For a direct comparison of the size of particles in cloud interstitial air and the inflow air of a cloud the aerosol stream should be dried.

ACKNOWLEDGEMENTS

We thank Daniel Leahy, Robert Brown and Seymour Fink for their efforts in the aircraft measurements and Joyce Tichler for her assistance in the data handling. One of the author (HtB) expresses his special thanks to Daniel Leahy for the way he was introduced into the complicated field of cloud measurements.

This research was performed under the auspices of the United States Department of Energy under Contract DE-AC02-76CH00016.

REFERENCES

[1] a. Daum, P.H; S.E. Schwartz and L. Newman in "Precipitation Scavenging, Dry Deposition, and Resuspension. Volume 1 Precipitation Scavenging". H.R. Pruppacher et al. Eds, Elsevier New York/Amsterdam (1983), 31-52.
       b. Gervat, G.P.; A.D. Kallend and A.R.W. Marsh, this issue.
       c. Römer, F.G. and H.F.R. Reynders, this issue.
[2] Slinn, W.G.N. in "Air-Sea Exchange of Gases and Particles. P.S. Liss and W.G.N. Slinn eds, Reidel Boston/Dordrecht (1983), 299-396.
[3] Pruppacher H.R. and J.D. Klett "Microphysics of Clouds and Precipitation". Reidel Boston/Dordrecht (1978).
[4] Leaitch W.R., J.W. Strapp, H.A. Wiebe and L.A. Isaac in "Precipitation Scavenging, Dry Deposition and Resuspension. Volume 1 Precipitation Scavenging". H.R. Pruppacher et al. Eds, Elsevier New York/Amsterdam (1983), 53-70.
[5] Hegg, D.A. and P.V. Hobbs, ibid. pg 79-90.
[6] Garber, R.W.; P.H. Daum, R.F. Doering, T.D'Ottavio and R.Z. Tanner, Atmos. Environ. 17(1983), 1381-1385.
[7] Walters P.T.; M.J. Moore and A.H. Webb, Atmos. Environ. 17 (1983), 1083-1092.
[8] Pierson, W.R.; W.W. Brachaczek; T.J. Truex; J.W. Butler; and T.J. Korniski; Annals of the New York Acad. Sciences 338 (1980), 145-163.
[9] Tanner, R.L.; P.H. Daum and T.J. Kelly, Intern. J. Environ. Anal. Chem. 13 (1983) 323-335.
[10] Daum, P.H.; S.E. Schwartz and L. Newman, J. Geophys. Res. in press.

[11] Winters W.; A. Hogan; V. Mohnen and S. Banard, State University of New York at Albany, ASRC-SUNYA Publication 728, (1979).
[12] Schwartz, S.E.; P.H. Daum; M.R. Hjelmfelt and L. Newman, see ref [1a], pg 15-30.
[13] Waggoner, A.P.; R.E. Weiss; N.C. Ahlquist; D.S. Covert; W. Will and R.J. Charlson, Atmos. Environ. 15 (1981), 1891-1909.
[14] Tang, I.N.; W.T. Wong and H.R. Munkelwitz, Atmos. Environ. 15 (1981), 2463-2471.

Day . Time (a) / Location . Elevation (b)	Cloud water potential partial pressure, ppb			Cloud-free air* mole fraction, ppb		
	$SO_4^=$	H^+	$SO_4^=/NH_4^+$	$SO_4^=$	H^+ (c)	$SO_4^=/NH_4^+$
11/18,82 . 19:00 / Alabama . 750	0.9	3.1	4.5	0.5	0.1	0.8
4/25,83 . 16:30 / Lg Island . 480	1.1	2.8	2.8	0.6	0.4	0.5
6/29,83 . 15:00 / Delaware . 600	1.7	3.4	1.9	0.4	0.4	0.5
8/12,83 . 18:00 / Lg Island . 1500	0.7	1.2	1.8	0.6	<0.2	1.3
8/12,83 . 19:00 / Lg Island . 1500	1.8	4.5	2.3	2.2	<1.2	1.2

Notes:

(a) GMT
(b) in meters
(c) including H^+ from gaseous HNO_3

* calculated from half hour filter samples, assuming complete scavenging of aerosol and gas consituents in the supersaturated cloud zones.

Table I. The potential partial pressures of cloud water constituents and the mole fraction of the same constituents in the surrounding cloud free air; see text, equation (2), for the definition of the potential partial pressure.

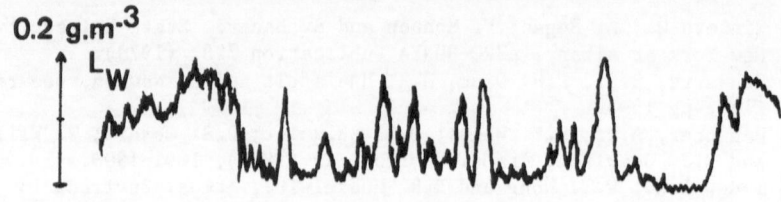

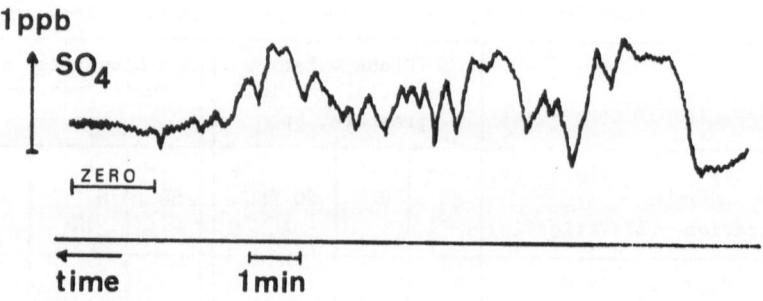

Fig. 1. The concentration of $SO_4^=$-aerosol and the liquid cloud water content, L_w, as a function of time, on November 17, 1982, Alabama, elevation 900 m, 14 : 45 GMT. ZERO indicates zeroing of the sulfur monitor, see text.

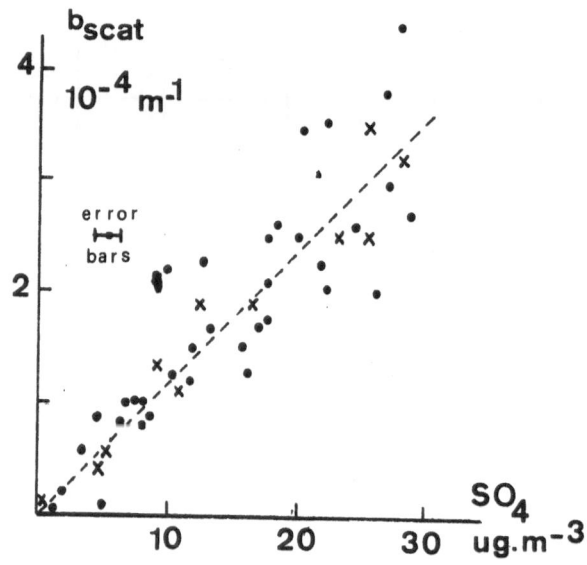

Fig. 2. The relation between light scattering coefficient, b_{scat}, and the $SO_4^=$-concentration in clear air flights. Dots indicate flights over Maryland (1980). Crosses indicate recent flights over Long Island (1982,1983).

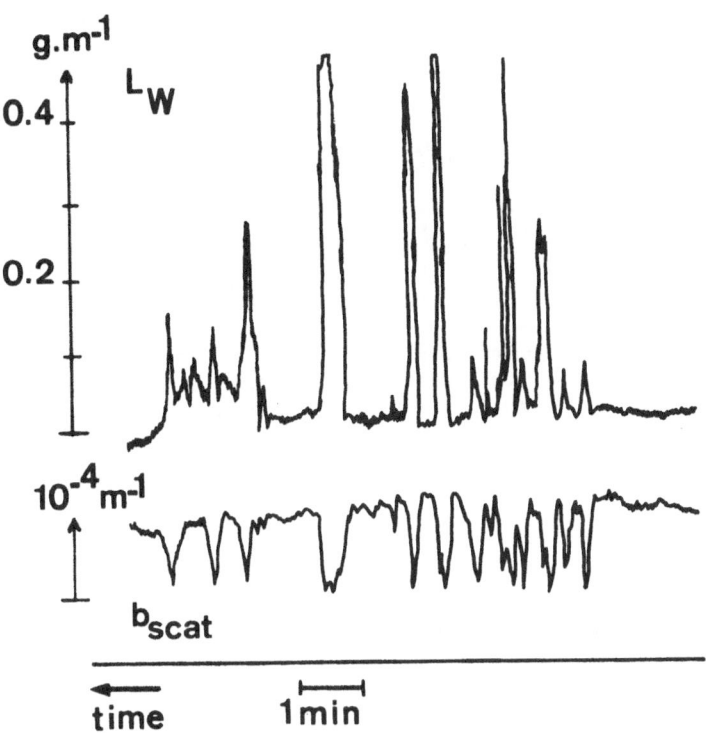

Fig. 3. Time profiles of the light scattering coefficient, b_{scat}, and the liquid cloud water content, L_W, during a flight on August 12, 1983, Long Island, elevation 1800 m, 19 : 00 GMT.

COMBINED PHOTOLYTIC AND RADIOLYTIC AEROSOL FORMATION IN A SO_2-NO_2-AIR MIXTURE (*)

F. RAES and A. JANSSENS
Rijksuniversiteit Gent
Nuclear Physics Laboratory
Proeftuinstraat 86, 9000 GENT (Belgium)

Summary

An SO_2-NO_2-air mixture is U.V. and U.V.+γ irradiated in a continously stirred tank reactor. Aerosol number and volume concentrations were measured with a continuous condensation nucleus counter coupled to a newly designed diffusion battery. Experiments at a constant mean residence time and a constant U.V. irradiation but at different dose rates of γ radiation show and increase in aerosol number and volume concentration relative to the case of U.V. radiation alone. In dry conditions the increase in aerosol volume ranges from 20 % at 4 nGy.s-1 to a factor 9 at 40 µGy.s-1. At 75 % R.H. no increase could be observed below 400 nGy.s-1; it is 60 % at 40 µGy.s-1. Experiments at different mean residence times (i.e. different concentrations of condensable products) did not allow to decide whether an increase in SO_2 transformation rate contributes to the measured effects.

1. INTRODUCTION

It is known that (H_2SO_4-) aerosols are formed when a SO_2-NO_2-air mixture is irradiated with ultra violet light. It has also been shown by many investigators that ionizing radiation is capable of forming aerosol particles in an air mixture containing SO_2 . In both situations the measurement of aerosol particles is a sensitive method to indicate that SO_2 has been transformed into H_2SO_4 respectively by photolytical or radiolytical reactions. The question of the possible contribution of radiolytical reactions in the transformation of SO_2 into aerosols when the gas mixture is irradiated simultaneously with U.V. and ionizing radiation was raised in the Seventies through the work of Vohra (1) who claimed that a synergism between U.V. and ionizing radiation at background dose rates enhances the transformation of SO_2; this topic is still of interest today in the light of the expected increase of the [85]Kr concentration in the global atmosphere due to the reprocessing of nuclear fuel (2).

The study of this problem was undertaken only occasionally and led to contradictory experimental results (1,3). The disagreement between the investigators may originate from the fact that they have been looking only at the effect of ionizing radiation on the aerosol number concentration (N): however an increase in N does not necessarily mean that more SO_2 has been transformed into H_2SO_4 since (in some conditions)

(*) This work is supported by the C.E.C., contract nr BIO-B-327-B

the gas to particle conversion of H_2SO_4 itself may be enhanced by the presence of ions. This effect will be discussed in chapter 1. On the other hand, an enhanced H_2SO_4 formation by radiolytic reactions will only lead to a higher aerosol number concentration when nucleation is the main gas to particle conversion mechanism; no increase may be observed when condensation on the preexisting aerosol is the principal mechanism, in which case an aerosol volume measurement would be more appropriate.

In this paper we present experiments in which we measured the increase in aerosol number and volume concentration in the the the case of simultaneous U.V. and γ radiation relative to the case of U.V. radiation alone. This was done at different dose rates of γ radiation and different environmental conditions.

The experiments are essentially similar to those described earlier (4) in which the effect of ionizing radiation on the aerosol volume was determined using an Electrostatic Classifier (TSI model 3071) coupled to a Continuous Condensation Nucleus Counter (TSI model 3020). Since the size of the aerosol particles formed in our experiments was near the lower bound of the size range scanned by the E.C. (0.012 - 0.85 μm), it was decided to look for another device for the determination of the particle size. In order to improve the reproducibility of the former experiments special attention was also given to the purity of the gases and of the experimental set up.

2. THEORY OF HOMOGENEOUS AND HETEROGENEOUS NUCLEATION

Nucleation in a H_2O-H_2SO_4 binary mixture is commonly described by the classical thermodynamic theory of nucleation (5)(see also reference 6 for a review of different nucleation mechanisms), where nucleation is calculated as a function of the concentration of H_2O and H_2SO_4 in the environment. The homogeneous nucleation rate is given by

$$J = 4\pi r^{*2}\beta_a N_b \exp(-\Delta G^*/kT), \qquad [1]$$

were ΔG^* and r^* are the free energy of formation and the radius of the critical cluster (i.e. the only cluster that, for given environmental conditions, is in unstable equilibrium with both H_2O and H_2SO_4 in the gas phase), k is Boltzmann's constant, T the absolute temperature, N_b is the number of H_2O molecules in the gas phase and β_a is the collision rate per unit area of H_2SO_4 molecules. This collision rate is written as

$$\beta_a = N_a(kT/2\pi m_a)^{1/2} \qquad [2]$$

where N_a is the number of H_2SO_4 molecules in the gas phase and m_a is the mass of one H_2SO_4 molecule. The rate of heterogeneous nucleation around ions is given by

$$J_i = 4\pi r^{*2} \beta_a N_i(\alpha, Q, N_a)\exp(-\delta\Delta G_i^*/kT) \qquad [3]$$

where $\delta\Delta G_i^*$ is the difference between the free energy of formation of the only ion cluster that is in stable equilibrium with the gas phase and the only ion cluster that is in unstable equilibrium with the gas phase and where $N_i(\alpha, Q, N_a)$ is the ion concentration depending on the ionisation rate Q, the recombination coefficient of ions α and N_a. According to the classical theory, nucleation rates can only be calculated in environmental conditions that allow an equilibrium between the gas phase and the droplet. In the presence of ions however, it is possible that this situation does not occur, in which case nucleation becomes kinetically rather than thermodynamically controlled and equation 2 can no longer be used. At present there is no appropriate theory describing ion induced nucleation in such conditions. We can only rely on the fact that the rate of this nucleation will be smaller or, when enough condensable products are available, will become equal to the ion formation rate.

Both nucleation rates were calculated as a function of the H_2SO_4 concentration for a fixed relative humidity of 75 % and as far as ion induced nucleation is concerned for an ionization rate of $5 \cdot 10^6$ ion pairs per cm^3 and per second. The result is plotted in Fig. 1.

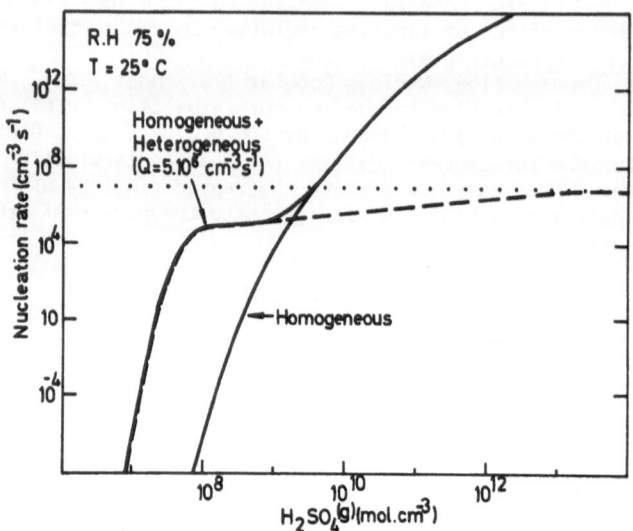

Fig. 1. Homogeneous nucleation rate and heterogeneous nucleation rate around ions as a function of the H_2SO_4 concentration in the gas phase for T= 25 °C and R.H.= 75 % according to the classical theory of nucleation.

In the case of an experiment with U.V. radiation alone, only homogeneous nucleation must be considered, in the case of simultaneous U.V. and γ radiation, the sum of homogeneous and heterogeneous nucleation must be taken as the total nucleation rate. It can be seen that for $H_2SO_4^{(g)}$ concentrations $> 2 \cdot 10^9$ molecules per cm^3 homogeneous nucleation is predominant over ion induced nucleation. In that case the presence of ionizing radiation will result in a higher number and volume concentration only if it contributes to the $H_2SO_4^{(g)}$ concentration, which is then subject to homogeneous nucleation. Aerosol number concentration measurements (or volume concentration measurements when condensation is the main gas to particle mechanism) performed under these conditions may then allow to distinguish between radiolytical formation of H_2SO_4 and ion induced nucleation. At $H_2SO_4^{(g)}$ concentrations $< 2 \cdot 10^9$ molecules per cm^3, ion induced nucleation is predominant, and its rate may become up to 10^8 times higher than the homogeneous nucleation rate without any change in H_2SO_4 concentration; in this region an increase in the aerosol number and volume concentration can not be attributed unambiguously to just aerosol physical or both radiochemical and physical mechanisms.

3. EXPERIMENTAL SET UP

The gas mixtures were irradiated in a flow reactor of the completely stirred type (C.S.T. Reactor). The set up is shown in Fig. 2 and its details have been described already in reference 4.

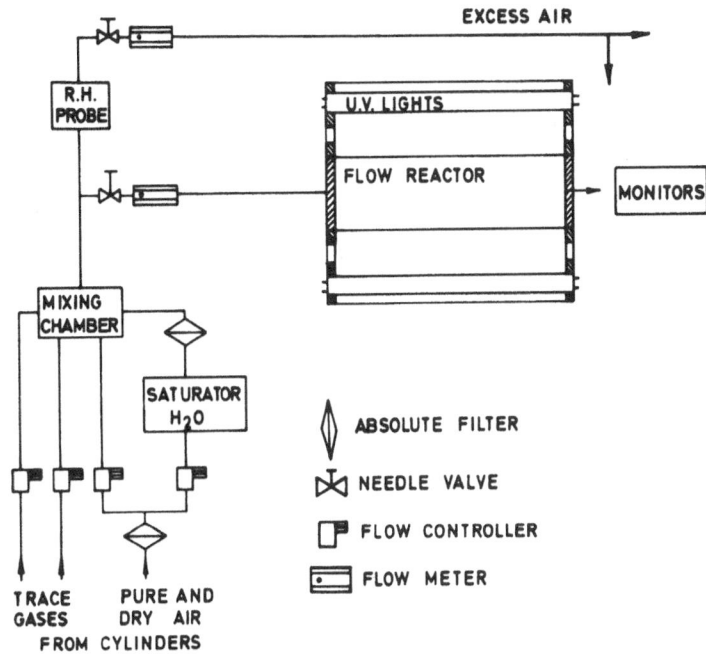

Fig. 2. The experimental set up

The gases are supplied from gas cylinders (L'Air Liquide: N2 > 99.9992 %, O2 > 99.995 %, SO_2 and NO_2 > 99%) and to keep their purity inside the reactor, the whole experimental set up was constructed using only inert materials like glass, teflon, stainless steel and aluminum. In all the experiments the U.V. (300 -450nm) intensity was held at 17 Wm^{-2}.

We choose for a C.S.T. Reactor because the steady state concentration of any product measured at its exit equals the concentration of the product throughout the reactor, which facilitates the interpretation of the results. All experimental values presented in this paper are steady state concentrations. If ratios of concentrations are presented to show the effect of a perturbation (in this case the irradiation with ionizing radiation) these are always the ratio between the steady state concentration of the perturbated aerosol to the mean of the steady state concentrations before and after the perturbation.

The aerosol number concentration was measured with the CCNC (TSI model 3020) and the mean diameter of the aerosol was determined by a newly designed diffusion battery which we call the 'diffusion caroussel' (D.C.). With this device eight sets of diffusion screens can be positioned in front of the CCNC inlet, the sets containing 0, 1, 2, 3, 4, 6, 8, and 10 screens respectively. The penetration P of an aerosol with diameter Dp through a set of n screens is given by the relationship

$$P = \exp(-2Mdn\eta_s)$$ [3]

where M is the number of fibers in the screen per unit length, d the fiber diameter and η_s the single fiber efficiency. Filtration theory shows that

$$\eta_s = bPe^{-2/3}$$ [4]

where b is a calibration coefficient and Pe is the Peclet number which itself is a function of the particle diameter Dp and the configuration of the battery (7). The calibration coefficient b can be obtained by measuring the penetration P of monodisperse aerosol particles through the screens: from P, η_s can be obtained using equation 3 and from the particle diameter Pe can be calculated. Such measurements were performed at the "Isotopenlaboratorium der Universitat Gottingen" with particles in the diameter range 4.4 -15.6 nm which were generated with the ultra fine monodisperse aerosol generator described by Scheibel et al. (8). The experimental points fit the theoretical relationship with b = 3.65. and with this value the diffusion caroussel was used in the range 2 -30 nm.

4. EXPERIMENTAL RESULTS

The first set of experiments was performed at a fixed mean residence time ($\bar{t}$) of 20 minutes. First some experiments in which the gas mixture was only U.V. irradiated were performed to examine whether the aerosol production in the SO_2 -NO_2 -air mixture is consistent with some elementary facts and with the predictions made by the photochemical model. The air mixture consisted of synthetic air, 0.5 ppmV SO_2, 0 - 0.1 ppmV NO_2 and < 2 ppmV H_2O vapor.

It can easily be shown that all products formed in a C.S.T.R. should reach the steady state at most $3*\bar{t}$ after the beginning of the irradiation. In the conditions described above, it was found that it actually takes about $50*\bar{t}$ to $100*\bar{t}$ for the aerosol concentration to reach a steady state value! Influences from the wall reactor are thought to be responsible for this effect although the right mechanism is not yet clear.

It was also observed that for a fixed SO_2 concentration, no aerosol was formed when $[NO_2]$ = 0 and when $[NO_2]$ > 0.5 ppmV. This behaviour may be explained by the photolytical model for a SO_2 -NO_2 -air mixture (9), which is one evidence that the SO_2 -NO_2 -air mixture in our reactor is not disturbed by impurities from walls or tubings, albeit not a sufficient proof.

a) experiments in dry conditions (H_2O < 2 ppmV).

For the following experiments, the NO_2 concentration was fixed at 0.015 ppmV, SO_2 = 0.5 ppmV and H_2O < 2 ppmV. The mean aerosol number concentration produced by U.V. radiation alone and detected by the CCNC was $1.93 \ 10^3 \ cm^{-3}$ +/- $1.07 \ 10^3$ (S.D.). Because at the moment of these measurements the calibration of the photometric mode of the CCNC (i.e. the counting mode for concentrations > $10^3 \ cm^{-3}$) was deficient (10), Diffusion Caroussel measurements were only performed when the total concentration was < $10^3 \ cm^{-3}$ in which region the CCNC counts the particles

in an absolute way. From these few measurements it was found by using a
simple curve stripping technique (11) that 85 % of the particles produced
by U.V. radiation alone and detected by the CCNC are peaked around 15
nm. The relative increase of the aerosol number concentration when the
gas mixture is also γ-irradiated is given in Fig. 3a as a function of

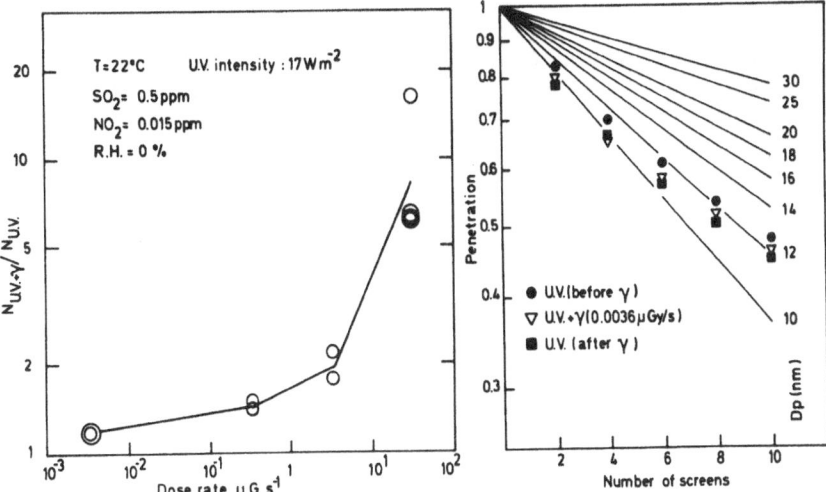

Fig. 3a. The ratio of the aerosol number concentration under simultaneous U.V. and γ radiation to the number concentration under U.V.
radiation alone, R.H. = 0 %.

Fig. 3b. Example of a D.C. measurement in the conditions mentioned
in the figure.

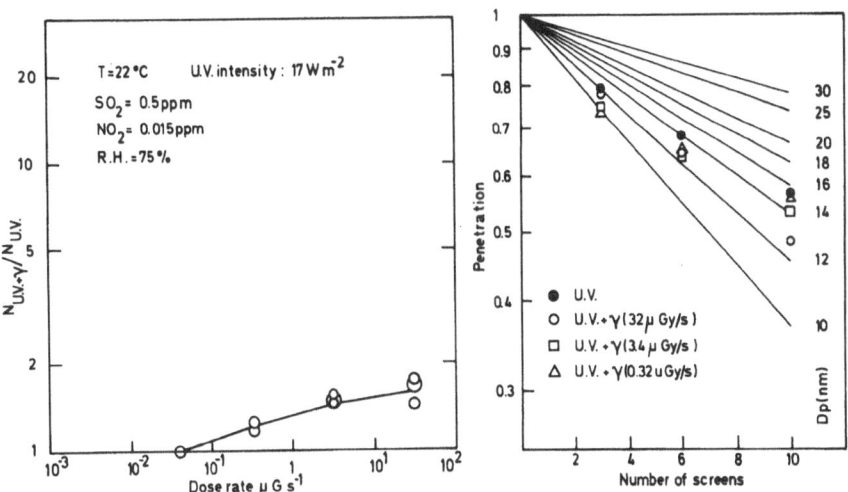

Fig. 4a. The ratio of the aerosol number concentration under simultaneous U.V. and γ radiation to the number concentration under U.V.
radiation alone, R.H. = 75 %.

Fig. 4b. The means of the D.C. measurements at the different
irradiation conditions.

the dose rate. Fig. 3b shows the result of a D.C. measurement performed before, during and after a γ -irradiation. As this result is representative for all the D.C. measurements, it is concluded that the size distribution of the aerosol did not change by the presence of γ -radiation. This means that the results of Figure 4a are also indicative of the effect of ionizing radiation on the total aerosol volume.

b)   experiments in humid conditions.

The air mixture consisted of synthetic air, 0.015 ppmV NO_2 and 0.5 ppmV SO_2 at 75 % relative humidity. In these humid conditions the aerosol concentration reaches its steady state concentration within 3 times the mean residence time as is expected. The mean aerosol number concentration produced by U.V. radiation alone and detected by the CCNC was 2.3 10^5 cm^{-3} +/- 7.5 10^4 (S.D) and 92 % of these particles peaked around 15 nm. The relative increase of the number concentration when the gas mixture is also irradiated is given in Fig. 4a as a function of the dose rate. In these experiments D.C. measurements could be performed in all cases, after recalibration of the CCNC. The means of the penetration measurements in the case of U.V. alone, and in the case of U.V. + γ (32 $\mu Gy.s^{-1}$), U.V. + γ (3.4 $\mu Gy.s^{-1}$) and U.V. + γ (0.32 $\mu Gy.s^{-1}$) are plotted in Fig. 4b. This plot suggests that the fraction of large particles becomes smaller when the gas mixture is irradiated (88 % in stead of 92 %). This would mean that within a factor 88/96 the results of Fig. 4a may again be applied to the aerosol volume.

A second set of experiments was performed in which the effect of γ radiation was studied as a function of $\bar{t}$. This showed how representative the results at $\bar{t}$ = 20 min may be for the evaluation of the effect but this also allowed us to vary the concentration of the condensable product in the reactor. The result of such a measurement performed at a dose rate of 32 $\mu Gy.s^{-1}$, at 75 % relative humidity and SO_2 and NO_2 concentrations of 0.5 and 0.015 ppmV respectively is plotted in Fig. 5a. One can see that at $\bar{t}$ = 20 the effect of ionizing radiation on the aerosol number concentration is independent of t. This was found to be the case in almost any other humidity and radiation conditions examined in this paper. In the particular conditions of Fig. 5 an extremely strong dependence on $\bar{t}$ is found below 5 min. When looking at the $\bar{t}$ dependence of the aerosol number concentration produced by U.V. in these conditions and at the corresponding particle sizes and taking into account the CCNC counting efficiency (12), a similar pattern is found for the aerosol volume concentration: it is independent of $\bar{t}$ above $\bar{t}$ = 5 min and decreases rapidly below $\bar{t}$ = 5 min.

5. DISCUSSION

The observations of Fig. 5 and of similar experiments performed in other conditions, may be explained by the fact that in the case of U.V. irradiation as well as in the case of simultaneous U.V. and γ irradiation, the loss rate of particles and the loss rate of $H_2SO_4^{(g)}$ is larger than 0.2 min-1, and the characteristic time necessary for the chemical system to reach equilibrium is smaller than 5 min. For $\bar{t}$ < 5 min one of these equilibria does not hold any more, which is reflected by the strong $\bar{t}$ dependence of the parameters. This also means that the results obtained in the flow reactor at $\bar{t}$ = 20 min are similar to those that would be obtained in a batch reactor of identical dimensions when starting from the same SO_2 and NO_2 concentrations as long as the depletion of trace gases is not important.

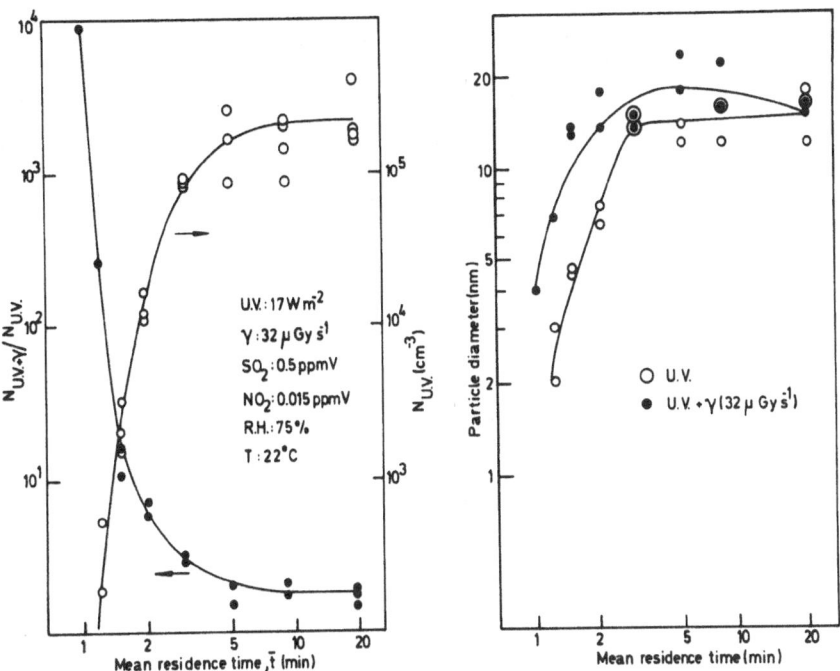

Fig. 5a. The ratio of the aerosol number concentration under simultaneous U.V. and γ radiation to the number concentration under U.V. radiation alone and the aerosol number concentration under U.V. radiation alone as a function of the mean residence time in the reactor, in the conditions mentioned in the figure.
Fig. 5b. The diameter of the aerosol particles as a function of the mean residence time in the two radiation conditions mentioned in the figure.

Since the ionizing radiation was observed to have parallel effects on the aerosol number and aerosol volume concentration (Fig. 3 and 4) the conditions in the reactor during the experiments were such that nucleation was the principal gas to particle conversion mechanism; we can thus explain the effects in terms of the theory described in chapter 1. The further discussion will concern only the experimental conditions of Fig. 5. We have already explained that for an interpretation of the results it is necessary to know what kind of nucleation prevails in the reactor. It is clear that the enormous effect of ionizing radiation on $N_{U.V.}$ at $\bar{t} < 5$ min can be explained by the fact that the $H_2SO_4^{(g)}$ concentration is smaller than $2\ 10^9$ molecules per cm^3 for small $\bar{t}$, such that ion induced nucleation prevails (see Fig. 1) The observation that the relative effect reaches a constant value for $\bar{t} > 5$ min could be attributed to the fact that the $H_2SO_4^{(g)}$ concentration has grown above the critical level, so that homogeneous nucleation is dominant and that the radiation effect would be due to an enhanced $H_2SO_4^{(g)}$ yield. However, in that case one would also expect a further increase in the aerosol volume after $\bar{t}$ = 5 min and this is in contradiction with the experimental results in Fig. 5. It is therefore more justified to conclude that the $H_2SO_4(g)$ concentration reaches its equilibrium below the critical level so that the small effect observed for long t may be caused by ion induced nucleation alone. The question can be raised if it is possible at all to create these high $H_2SO_4(g)$ concentrations ,since nucleation itself and condensation will deplete the $H_2SO_4(g)$ concentration anyhow. This will be examined by carefully modelling the interactions between $H_2SO_4(g)$ and the aerosol.

6. CONCLUSION

We have presented experimental evidence that ionizing radiation enhances the formation of aerosol particles in U.V. irradiated gas mixtures. The magnitude of the effect is very much dependent on the environmental conditions:in dry conditions an effect can already be observed at a dose rate of 0.0036 $\mu Gy.s^{-1}$ which is a factor of 50 lower than the dose rate corresponding with the maximum permissible concentration of [85]Kr in air, and a factor 10 lower than the [85]Kr dose rate that may be encountered in the plume of a reprocessing plant 1 km downwind the emission point. In humid conditions the onset of the effect shifts to about 0.3 $\mu Gy.s^{-1}$.

We have also discussed different mechanisms that may be responsible for the observed effect but we were not able to settle whether radiolytical formation of condensable products contributes to the increase in aerosol concentration or not. The question was raised if for the problem discussed in this paper aerosol measurements alone can lead to clear cut decisions about the chemical processes that might occur.

REFERENCES

(1) Vohra, K.G. (1975) IAEA-SM-197/3, 209
(2) Janssens, A. et al. (1982) Physicalia Magazine 4, 86 (in Dutch)
(3) Metayer, Y. et al. (1980) J. Aerosol Sci. 12, 198
(4) Raes, F. et al. (1983) J. Aerosol Sci. 14, 302
(5) Kiang, C.S. et al. (1973) Faraday Symp. Chem. Soc. 7, 26
(6) Hamill, P. et al. (1982) J. Aerosol Sci. 13, 561
(7) Friedlander, S (1977) Smoke Dust and Haze, 66
(8) Scheibel, H.G. et al (1983) J.Aerosol Sci. 14, 113
(9) Boulaud, D. (1977) These, Universite de Paris (in French)
(10) Raes, F. et al (1983) J. Aerosol Sci. 14, 394
(11) Sinclair, D. (1972) Am. Ind. Hyg. Ass. Journal, 729
(12) Agarwall, J.K. (1980) J. Aerosol Sci. 11,343

MEASUREMENT OF GASEOUS HALOGENATED HYDROCARBONS

IN AMBIENT AIR

Müller J., F. Riedel

Umweltbundesamt - Pilotstation Frankfurt

Feldbergstr. 45, 6000 Frankfurt (M)

Summary

Halogenated hydrocarbons (halocarbons) were measured in ambient air. For sampling, the air was sucked through little vessels partly filled with Tenax in order to absorb the halocarbons. The vessels with the collected halocarbons also were used for analysis with a headspace-gaschromatograph. In the inlet oven the samples were heated and the halocarbons desorbed and were enriched in the head space of the vessel from where a certain quantity was taken by the injection needle. The halocarbon compounds were separated in a carbopack-column and were detected in an ECD-detector.
The concentration ranges measured in an urban residential area were: chloroform $0,2 - 2,5 \, \mu g/m^3$; trichloroethylene $1 - 10$ and tetrachloroethylene $1 - 18 \, \mu g/m^3$; carbon tetrachloride $0,2 - 4,5$; trichloroethane $2 - 50 \, \mu g/m^3$.

1. INTRODUCTION

Halocarbons used as solvents, dry cleaners and propellants are considered as persistent compounds in the environment. Only a few compounds of this group are partly produced by natural sources like carbon tetrachloride (1). Most of this new atmospheric burden is man-made. Several compounds are toxic and some of them play an important role in the photochemistry of the atmosphere (2).

Therefore, it is indispensable to measure and to survey halocarbons and to attain knowledge about the seasonal variations.

By gaschromatography with ECD-detection, nowadays, it is possible to measure halocarbons also in remote areas. In this paper a measuring method with headspace-gaschromatography is described.

2. SAMPLING AND ANALYSIS

Three sampling vessels of 20 ml volume are connected in series by stainless steel tubes and all together are cooled in ice-water (Fig. 1). By addition of NaCl the aqueous solution is kept at about -10°C, a temperature sufficiently low in order to absorb halocarbons on the surface of Tenax powder.

The first vessel is empty and is used as a cooling trap in order to retain the water vapour of the air. The second and the third vessel are filled with 750 mg Tenax to absorb the halocarbons of the passing air. Two Tenax vessels are sufficient for a total absorption of the halocarbons in the air.

Behind the pumps a gasmeter is installed. The sucked air volume varies between 20 and 40 l which corresponds to a sampling time of 1 - 2 hours.

After sampling the tubes are pulled off and 5 µl cyclohexane are added to the Tenax powder. New caps consisting of butylrubber coated with Teflon and Al-frames are fixed on the vessels. The samples are stored in a refrigerator or, during winter-time, can be kept in open air.

The analysis should be carried out within some days in order to prevent a loss of the lighter halocarbons. Normally, the samples are analysed twenty four hours after sampling, a time period sufficiently long to obtain equilibrium between the halocarbons and the added cyclohexane.

The sampled vessels are set into the inlet oven of the headspace-gas-chromatograph (Perkin Elmer F 45) and for several hours are kept at a temperature of 120°C. After this time equilibrium between the solid phase and the gas phase in the head space of the vessel is established. The halocarbons which are fluid compounds are enriched in the head space.

A needle hitting through the vessel's cap sucks a certain quantity of the head space volume which at a temperature of 150°C is injected into the column system. In the column a gradient program is run increasing the temperature at a rate of 2°C/minute. The program lasting 33 min. rises the temperature from 100°C upto 166°C. The largest detected halocarbon compound tetrachloroethylene is registered after a residence time of about 22 minutes.

The separation of the compounds takes place in a 2 x 6' x 1/8'' stainless steel column filled with 0,1 % SP 1000 on carbopack C 80/100. The ECD-detector is heated up to 300°C and is run with argon/methane at a rate of 26 ml/minute.

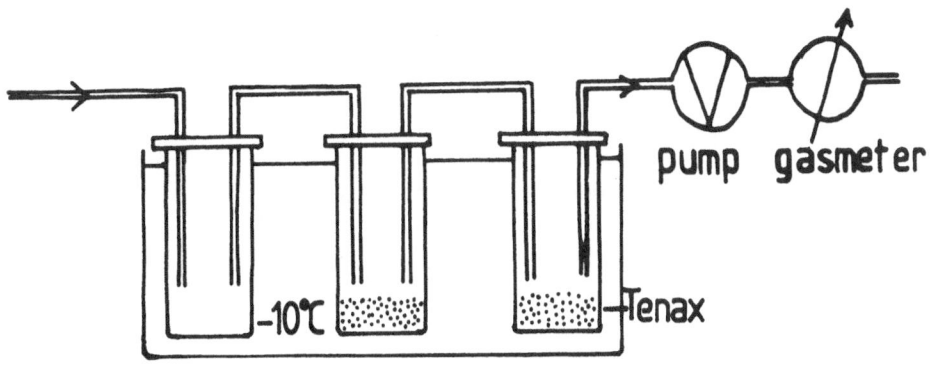

Fig. 1: Sampling device for halocarbons in ambient air

At the end of the gradient program the column for 15 minutes is back-flushed.

The compounds with the following retention times clearly can be identified: chloroform 4,95; 1,1,1 trichloroethane 6,98; carbon tetrachloride 7,56; trichloroethylene 10,70 and tetrachloroethylene 21,64. At 5,79 a peak is seen not yet surely identified (Fig. 2). This peak seems to be a mixture of methylene chloride, 1,1 dichloroethylene and trichlorofluoroethane.

The calibration is carried out with the compounds dissolved in cyclohexane. 5 µl of this solution is injected on 750 mg unloaden Tenax in the vessel. The calibration vessel is handled like the collected sample and analysed after 24 hours of equilibration time. In Fig. 3 the calibration spectrum is represented which is similar to the ambient air spectrum.

3. RESULTS AND DISCUSSION

The first vessel of the sampling device used as a cooling trap collected the water vapour of the air. The condensed vapour was also analysed in order to look for watersoluble halocarbons. In the oven of the chromatograph the water solution was heated up to 80°C but in the headspace above the liquid no halocarbons were detected. This confirms the low

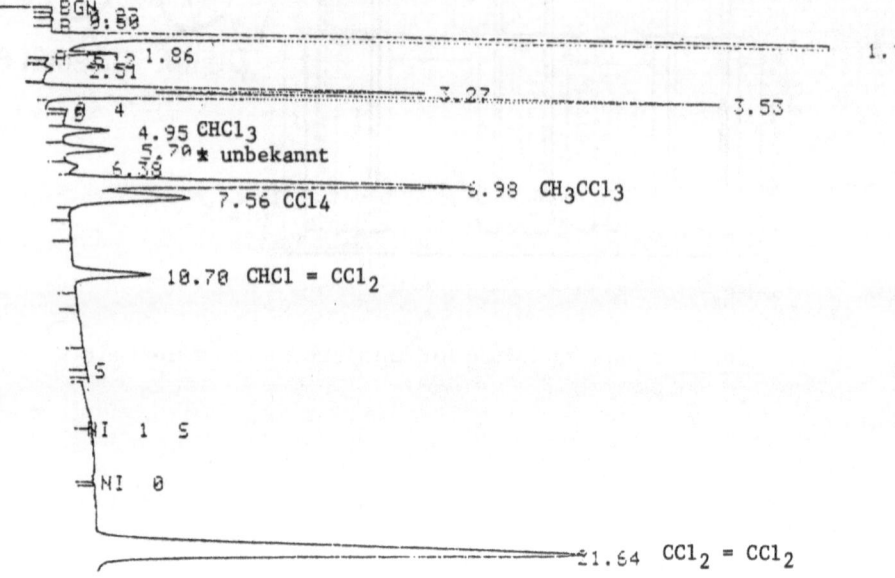

Fig. 2 : Ambient air spectrum of halocarbons

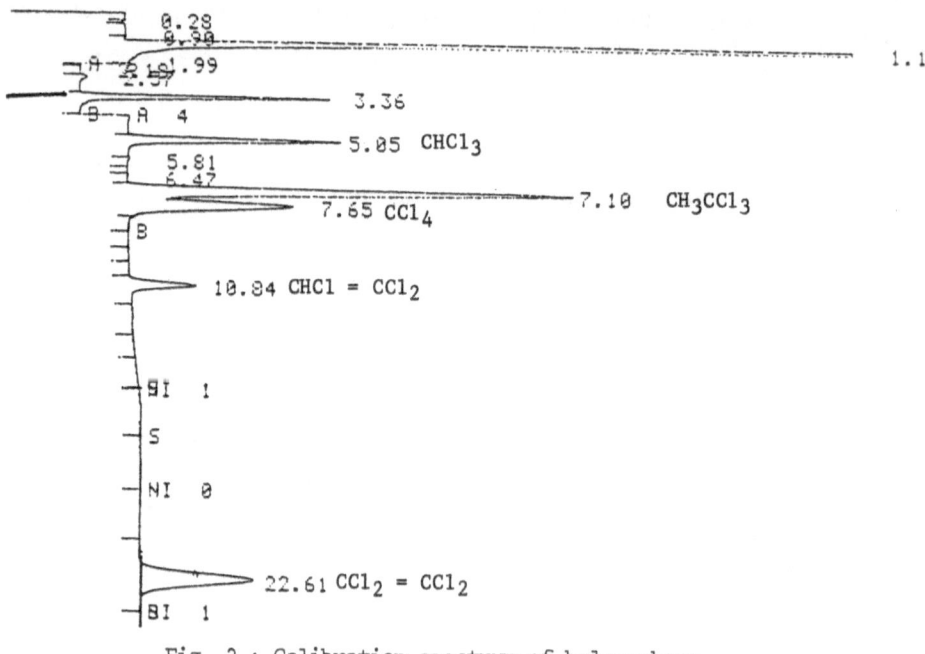

Fig. 3 : Calibration spectrum of halocarbons

water-solubility of the halocarbons. Therefore, it can be assumed that rainout and washout processes play only an unimportant role as sinks in the atmosphere.

The larger compounds trichloroethylene and tetrachloroethylene, during the sampling process, totally were absorbed in the first Tenax filled vessel. For the lighter halocarbon compounds about 70 - 90 % of the mass was found in the first Tenax vessel.

Parallel measurements with two sampling sets gave results within differences of $\pm$10 % for the total mass found in the two Tenax vessels.

The sampling of the halocarbons was carried out in a residential area at the periphery of the city of Frankfurt (M), on working days between 8 and 12 o'clock.

In the following table the monthly mean values of about 15 single measurements are listed:

Month	$CHCl_3$	CH_3CCl_3	CCl_4	$CHCl=CCl_2$	$CCl_2=CCl_2$ $\mu g/m^3$
Sep 83	0,5	8,8	0,7	3,0	11,6
Oct 83	0,6	4,3	1,0	2,6	12,8
Nov 83	1,7	13,0	1,1	5,6	9,0
Dec 83	0,4	19,4	1,0	6,3	18,0
Jan 84	0,4	10,5	0,7	1,8	8,7
Feb 84	0,5	9,9	4,5	3,1	8,8

The measurements up to now are not numerous enough to find out a seasonal or diurnal trend. But the compounds persistently can be detected in the atmosphere. By man-made activities sources exist to keep a certain amount of halocarbons steadily in urban air.

Lit:

1. Lovelock, J.E. (1977) Nature 267: 32
2. Crutzen, P.J. (1979) Ann. Rev. Earth Planet Sci. 7: 443

SESSION 4

POLLUTANT CYCLES

Summary by the Chairman
   S. BEILKE

NO_x background mixing ratios in surface air over Europe
and the Atlantic Ocean

Regional background concentrations of NO_2 in Sweden

Nitrous acid in polluted air masses - Sources and formation pathways

Study of the chemical characteristics of wet and dry
deposition in Switzerland

Vertical and horizontal profiles of hydrogen chloride in
the Mediterranean region

Bilan ionique et acidité de la précipitation antarctique

Etude de l'influence d'une source locale naturelle
intense de composés organosoufrés sur la chimie de la
troposphère en milieu non pollué

Fluctuations temporelles fines de la composition chimique de l'aérosol côtier

Toxic metals and metalloids in high altitude alpine
glaciers snow and ice

Results of many years' analyses of precipitation chemistry on samples obtained simultaneously at 30 km, 1.8 km
and 0.7 km ASL

Measurements of the latitudinal distribution of light
hydrocarbons and halocarbons over the Atlantic

POLLUTANT CYCLES

Summary of the chairman
S. BEILKE
Umweltbundesamt - Frankfurt

1. Introduction

In the follow-up programme of project COST 61a bis(1984-1985)
more emphasis will be given to problems of acid deposition.
This is done out of concern for damage to forest and freshwater
ecosystems as well as to materials in various European regions
attributed at least in part to acid deposition originating from
air pollution.
In the industrialized regions of the northern hemisphere a large
share of acid deposition arises from man-made primary emissions
largely of SO_2 and NO_x.
In the past investigations of pollutant gas cycles were mainly
concentrated on sulfur dioxide($SO2$) because $SO2$ was emitted in
much larger quantities than other gases including NO_x in most
countries of Europe.In spite of some important knowledge gaps
in the field of atmospheric $SO2$,our knowledge about its cycle
is relatively well advanced.
As opposed to $SO2$ we know a lot less about the cycles of NOx in
the troposphere.Special attention must be paid now on the investigation of the cycles of nitrogen oxides for the following
reasons:

- the anthropogenic NO_x emissions in the industrialized
  regions of the northern hemisphere have increased considerably during the last decades

- today NO_x contributes to an appreciable extent to acid
  deposition

- in contrast to $SO2$,NO_x exerts a strong influence on many
  other pollutant gas cycles.

In the following pages some aspects of the NO_x cycle will be
highlighted under special consideration of the contributions
within project COST 61a bis.I would like to emphasize that an
investigation of the budgets of pollutant gases such as NO_x is
a local or a regional problem rather than a global one not only
from the point of view of environmental effects.There are two
other points which emphasize the need for taking a regional
approach:

- the anthropogenic sources of NO_x are distributed very unevenly over the earth's surface.Most emissions occur in
  the northern hemisphere and there,too,the emissions are
  very unevenly distributed.Nitrogen oxides are mainly
  produced in the Northeast of USA and bordering areas of
  Canada,in Japan and China,and in the North and Center of
  Europe

- the mean atmospheric life-time of NO_x including its con-

version products is relatively short i.e.NO_x which is
emitted in Central Europe is generally deposited somewhere
in Europe.

In the first section some of the most important results on the
single elements of the regional NO_x cycle on the European scale
will be presented.In the second section I shall try to establish
a synthesis of these single elements to a more or less closed
cycle.Special attention is paid on indicating the most important
knowledge gaps in the field of the regional NO_x cycle.

2. Single elements of the regional NO_x cycle

The basis for the establishment of such a regional cycle is a
sound quantitative knowledge about its elements such as emissions,
concentrations,transport,conversion,and deposition of NO_x in-
cluding its conversion products.
As for most areas in the Northeast of the United States,anthro-
pogenic emissions of NO_x are overwhelmingly dominant in most of
Europe(OTTAR,1983;VAN DOP,1983;BONIS et al.,1980;REPORT of the
National Academy of Science Panel on Atmospheric Transport and
Chemical Transformation in Acid Deposition,1983).
Unlike SO_2,NO_x emissions seem still to be on a rising trend and
believed to have increased by 40-50 % over the last 10-15 years
(CEC REPORT on ACID RAIN,1983).
As far as the man-made sources are concerned,ca. 50 % of the
primary NO_x emissions are produced by vehicles in the countries
of the European Communities(ECE REPORT on ACID RAIN,1983).
As an example table 1 shows the updated NO_x emission inventory
for the Federal Republic of Germany.For the purpose of comparison,the corresponding updated SO_2 emission inventory is given
in table 2.As seen in these tables,NO_x emissions have increased
by ca. 50 % between 1966 and 1980 mainly due to an increase of
mobile sources which contributed ca. 55 % to total anthropogenic
NO_x emissions in 1982.A comparision with the SO_2 emissions shows
that NO_x emissions in 1982 were slightly higher than SO_2 emissions.

Despite the importance of the nitrogen oxides NO and NO_2 in
atmospheric chemistry there are surprisingly few measurements
of their mixing ratios outside the polluted areas.As a consequence our knowledge of NO_x background levels even on the
European continent is rather limited.
The NO_x mixing ratios can vary widely by orders of magnitude.
The concentrations in Europe can be as high as some 100 ppb
(mainly NO) in urban polluted areas and below 1 ppb(mainly NO_2)

in rural areas far from the main sources(SJÖDIN and GRENNFELT,
1984).In the tropical and polar regions over the Atlantic ocean
the NO_2 mixing ratios(NO is virtually absent) show minimum
values of about 0.03 ppb(BROLL et al.,1984).
The large variations in time and space of NO_x mixing ratios
along with its strong decrease with hight over polluted continental European areas(GEORGII and JOST,1964) are indications
for a rather short mean tropospheric residence time for NO_x
caused by rapid conversion and/or deposition processes.

These conversion and deposition processes are not well under-

	1966	1970	1974	1978	1982
% NO_x — Mt/a	2.0	2.4	2.7	3.1	3.1
Utilities,district heating	23.6	26.3	30.0	27.8	27.7
Industrial processes	30.6	25.5	21.0	16.7	14.0
Resicential/Commercial	5.8	6.0	5.0	4.5	3.7
Transportation	40.0	42.0	44.0	51.0	54.6

TAB.1 : Updated emission inventory of total NO_x-emissions(as NO_2) for the Federal Republic of Germany. (Umweltbundesamt,February 1984)

	1966	1970	1974	1978	1982
% SO_2 — Mt/a	3.2	3.6	3.6	3.4	3.0
Utilities,district heating	41.3	45.9	51.3	55.1	62.1
Industrial processes	35.7	32.3	30.0	27.8	25.2
Residential/Commercial	19.9	18.6	15.3	13.4	9.3
Transportation	3.1	3.2	3.4	3.7	3.4

TAB.2: Updated emission inventory of total SO_2-emissions for the Federal Republic of Germany(Umweltbundesamt,1984)

stood.The oxidation of the primarily emitted NO to NO_2 by O_2
and by O_3 is relatively well known.Some problems still remain
to be resolved concerning the NO-oxidation in power plant
plumes(ELSHOUT and BEILKE,1984).
The overall conversion of NO_2 under atmospheric conditions is,
however,difficult to quantify.Nitrogen dioxide is converted to
nitrous acid(HNO_2) and/or to nitric acid(HNO_3) by a series of
homogeneous gas phase reactions and by heterogeneous ones.
The conversion of NO_2 to HNO_3 by reactions with OH radicals
during daytime and the NO_2 transformation via reaction mechanisms involving O_3,NO_3,N_2O_5,H_2O,and aldehydes at night seem to
proceed rapidly(PLATT et al.,1981;CALVERT and STOCKWELL,1983).
The NO_3 radical formed by reaction of NO_2 with O_3 is in equilibrium with N_2O_5 which is most likely very effectively removed
by reactions with atmospheric droplets and aerosol particles
(PLATT et al.,1981).Another possible reaction path for HNO_3
generation is a gas phase reaction of N_2O_5 with H_2O and of
NO_3 with aldehydes(CALVERT and STOCKWELL,1983).
Model calculations of these authors have shown that even in
the absence of the N_2O_5 - H_2O gas phase reaction(most atmospheric chemists favor a heterogeneous reaction of N_2O_5),
relatively large rates of HNO_3 generation may occur during the
nighttime due to the NO_3-aldehyde reactions.
Another possible reaction path for the formation of HNO_2 and
HNO_3 is a reaction of NO_x at aerosol surfaces most likely via
a surface catalyzed mechanism(KESSLER and PLATT,1984).
In contrast to the abovementioned reactions,the formation of
HNO_2 and HNO_3 in water droplets from NO_x absorbed is very small
under atmospheric conditions(LEE and SCHWARTZ,1981).
For a cloud liquid water content of 1 g/m^3 and typical atmospheric NO_x mixing ratios,NO_x removal rates of the order of
10^{-4} to 10^{-5}%/h were calculated(BEILKE,1983) using thermodynamic and kinetic data reported by LEE and SCHWARTZ.
Indications are strong that the rather high nitrate content in
rainwater has its origin in an absorption of gaseous HNO_3,
NO_3/N_2O_5 and/or in an incorporation of nitrate containing
aerosol and not in an absorption and chemical conversion of
NO_x in the droplet phase.

Except for wet deposition of nitrates,the other removal processes of NO_x from the atmosphere are not well understood.
As far as dry deposition of NO_2 is concerned deposition velocities between 0.01 - 0.8 cm/s are reported in the literature
(JUDEIKIS and WREN,1978;BÖTTGER et al.,1978;GRENNFELT,1983;
GARLAND,1983) over different surfaces with the higher values
measured over vegetation i.e. the dry deposition of nitrogen
oxides seem to occur as nitrogen dioxide since it is rapidly
taken up by vegetation(SJÖDIN and GRENNFELT,1984).
The possibility can not be ruled out that there may be a return flux of NO to the atmosphere(HILL,1971;BEILKE and
GRAVENHORST,1979;GRAVENHORST and BÖTTGER,1983).
There is growing evidence that dry deposition of HNO_3-gas may
considerably contribute to total removal of NO_x from the at-
mosphere.Measurements of HUEBERT(1983) over grassland under
different atmospheric stability conditions have resulted in

deposition velocities between 1.1 - 4.9 cm/s with a mean value
of 2.9 cm/s(standard deviation of 1.1 cm/s).
The limited number of dry deposition measurements does,however,
not permit a reliable quantification of these NO_x sink terms.
Another sink for NO_x is dry deposition of nitrate containing
aerosol.
As for NO_2 and HNO_3,deposition rates of particulate nitrate
are also problematic.Estimates of GARLAND(1983) as well as
field measurements of GEORGII et al.(1983) in the Federal
Republic of Germany and of FUHRER(1984) in Switzerland have
shown that dry deposition of NO_3^- - N was generally small as
compared to wet deposition(between 8-23% for 12 stations
in the FRG during September 1979-August 1981; 28% for one
rural station in Central Switzerland,1983).
However,there is not complete agreement that particle deposition of nitrate is generally small compared to wet deposition.
Extensive field studies of GRAVENHORST et al.(1983) have shown
that dry deposition of nitrate upon a spruce forest in the FRG
is not only of the same order of magnitude as wet deposition,
but exceeded the wet deposition by a factor of about 3.
If these results were representative for dry deposition of
particulate nitrate mainly(the possibility can not be ruled
out that a certain fraction of nitrate measured in the through-
fall has an additional source in an absorption and/or oxidation
of HNO_3 and NO_x on the branches),and if these results can be
applied to other European regions covered with forests,dry
deposition of particulate nitrate would be a significant sink
for atmospheric NO_x.

Examinations of the nitrate content in rainwater in rural areas
of the Northeast of USA and in Western Europe dating back to
the last century reveal a considerable increase(BRIMBLECOMBE
and STEDMAN,1982).
Furthermore,between 1956-1976 , 55 of 120 stations of the
European Atmospheric Chemistry Network(EACN) showed a statistically significant increase of nitrate levels with an average
increase rate of ca.6%/year(KALLEND,1983).
A similiar trend was also reported by GUICHERIT(1982) for the
western part of the Netherlands and by FUHRER(1984) for the
area of Bern in Switzerland.
Data from TNO show that the average concentration of NO_3^- in
deposition has more than doubled in the western part of the
Netherlands from the beginning of the sixties until 1982.
In the area of Bern(Switzerland) between 1957-1961 and 1983
the average NO_3^- - N deposition(bulk,dry plus wet)has increased
by more than a factor of 3(dry deposition:28% ;wet deposition:
72%) i.e.both the nitrates in aerosol and in rainwater should
have increased since that time.
An increase of nitrate aerosol has also been reported by COX
and PENKETT(1983) for a rural site in United Kingdom between
1957 and 1974.

In the context of increasing nitrate deposition it is interesting to note that a statistical analysis of the ozone data
of the German Democratic Republic has shown an increasing trend
in the annual means of the near-surface ozone at all stations
of the ozone network from the beginning of the measurements

(1952-1956) up to 1979(WARMBT,1979).
The increase of ozone occurred not only during the summer
months but with a lower rate also during the winter months.
The author interpreted the growing ozone trend in terms of an
increase of emissions of NO_x and reactive hydrocarbons by
mobile sources.
A similiar increase of near ground ozone concentrations was
also reported by ATTMANNSPACHER and HARTMANNSGRUBER(1977) for
Mt.Hohenpreissenberg(Bavaria,FRG) between 1971-1977.

Both the observed increase of nitrate deposition and ozone
near-surface concentrations over the last decades seem to be
large-scale phenomena rather than local ones.

3.Synthesis of elements in the regional NO_x cycle

Figure 1 shows a very simplified description of a relation
between emission,transformation,transport and deposition of
NO_x and its reaction products in the long-term average for a
central European area of ca. 3000 x 3000 km.
In this huge region NO_x is emitted to an overwhelming extent
as NO which is further oxidized to NO_2 mainly by ozone.
A small fraction of NO_x is directly emitted as NO_2.
A reliable quantification of the previously described NO_2
oxidation pathways is not possible yet.The same is true for
the different sinks.About 20% of the primarily emitted NO_x is
deposited as NO_3^- in wet form.
This figure is in fairly good agreement with the corresponding
value for the northeastern USA.On the basis of empirical data
GOLOMB(1983) concluded that ca. 20 % of the emitted NO_x is wet-
deposited in this large area in the long-term average.

The percentages for wet and dry deposition are long-term
averages for a large European area.Within this area the proportion of dry and wet deposition may vary quite significantly
and specific source-receptor relationships cannot be established with sufficient accuracy at the present time.

If we move from the very large European area to a smaller one,
the transport of NO_x including its reaction products out of
the reservoir becomes more important.
In this context it is interesting to compare measured long-
term average wet deposition of NO_3^--N with NO_x emissions in
some European countries.For example,for the Netherlands the
ratio was only ca. 14 $\pm$ 5 % in 1980(VAN AALST,1983).
For the Federal Republic of Germany the corresponding value
was about 17 $\pm$ 5 %(estimated from a publication of
GEORGII et al.,1983).

Even if the accuracy of the emission and wet deposition data
may leave something to be desired,science today recognizes
that NO_x emissions in the Northeast of USA and in Central
Europe are far from being balanced by wet deposition of
nitrate.
As the mean tropospheric residence time of NO_x and its conversion products is relatively short in Central Europe,only
a small fraction of NO_x is transported out of this huge

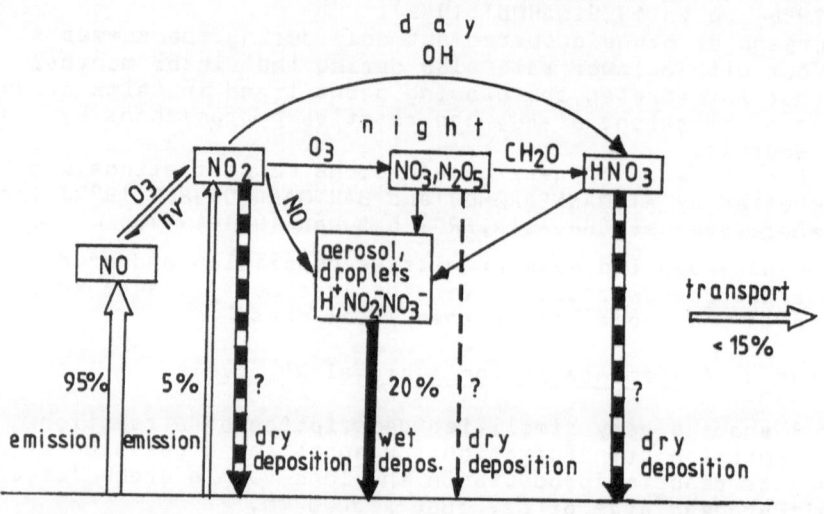

FIGURE 1 : Relation between emission,transformation and deposition
of NOₓ and its conversion products in the long-term
average for a central European region of ca.3000 x 3000 km.

reservoir($<$ 15%).In order to balance the regional NO_x budget
rather effective dry deposition removal processes are re-
quired.As specified in the previous section,a reliable
quantification of dry deposition of NO_2,HNO_3-gas,and of
nitrate aerosol is not yet possible.All dry deposition processes together should contribute to ca.60-70% to total
removal of NO_x.

Summarizing the present knowledge about the regional NO_x cycle,
some important knowledge gaps have to be filled in order to
come to a more realistic source-receptor relationship.
The most important research needs in the field of acid depo-
sition(emission,transformation,deposition,ect.) under special
consideration of the tropospheric N-chemistry were specified
by the CEC TASK FORCE"Acid Deposition"(Document AP/38/83 of
8 December 1983,Project COST 61a bis).
This list of research needs should provide a sound basis for
future work within the COST 61a bis follow-up programme.

4.References

ATTMANNSPACHER,W.and HARTMANNSGRUBER,R.(1977)
    On yearly ozone variations at different altitudes above
    Hohenpreissenberg(FRG) and near the ground.
    Proceedings Joint Symposium on Atmospheric Ozone,Dresden
    (GDR),1977,Vol.I,pp.277-288.

BEILKE,S.and GRAVENHORST,G.(1979)
    Cycles of pollutants in the troposphere.Proceedings:
    1.European Symposium"Physico-Chemical Behaviour of Atm.
    Pollutants",Ispra(Italy),16-18 October 1979.
    Editors:Versino,B.and Ott,H.;Project COST 61a bis.p.331.

BEILKE,S.(1983)
    Bildung von Säuren durch heterogene Reaktionen.
    VDI-Reports 500:Proceedings Symposium on "Acid Precipi-
    tation - Origin and Effects",Lindau(Germany),7-9 June
    1983(in German),pp.35-40.

BÖTTGER,A.;EHHALT,D.und GRAVENHORST,G.(1978)
    Atmosphärische Kreisläufe von Stickoxiden und Ammoniak.
    Bericht der Kernforschungsanlage Jülich,Jül.-1558,Nov.78.

BONIS,K.;MESZAROS,E.,and PUTSAY,M.(1980)
    On the atmospheric budget of nitrogen compounds over Europe.
    IDOJARAS,Vol.84,No.2,Mar-Apr.1980,Budapest,pp.57-68.

BRIMBLECOMBE,P.and STEDMAN,D.H.(1982)
    Historical evidence for a dramatic increase in the nitrate
    component of acid rain.Nature,Vol.298,29 July,1982.

BROLL,A.;HELAS,G.;RUMPEL,K.J.,and WARNECK,P.(1984)
    NOx background mixing ratios in surface air over Europe
    and the Atlantic Ocean.Paper presented at 3.European
    Symposium on "Physico-Chemical Behaviour of Atmospheric
    Pollutants",Varese(Italy),10-12 April 1984.
    Project COST 61a bis.

CALVERT,J.G.and STOCKWELL,W.R.(1983)
    Acid generation.Env.Sci.Techn.,Vol.17,No.9,pp.428A-443A.

CEC REPORT on ACID RAIN-A review of the phenomena in the CEC
    and Europe(1983)
    Published by Graham and Trotman Limited,London for the
    Commission of the European Communities.EUR.8684.

COX,R.A.and PENKETT,S.A.(1983)
    Formation of atmospheric acidity.Proceedings of the CEC
    workshop"Acid Deposition",September 9th,1982,Berlin.
    pp.56-81.Editors:S.Beilke and A.J.Elshout.
    Reidel Publishing Company,Dordrecht/Boston/Lancaster.

ELSHOUT,A.J.and BEILKE,S.(1984)
    Oxidation of NO to NO2 in flue gas plumes of power stations.
    Paper presented at 3.European Symposium on "Physico-Chemical
    Behaviour of Atmospheric Pollutants",Varese(Italy),10-12
    April 1984.Project COST 61a bis.

FUHRER,J.(1984)
    Study of the chemical characteristics of wet and dry depo-
    sition in Switzerland.Paper presented at 3.European Sym-
    posium on "Physico-Chemical Behaviour of Atmospheric
    Pollutants",Varese(Italy),10-12 April 1984.COST 61a bis.

GARLAND,J.A.(1983)
    Principles of dry deposition:Application to acidic species
    and ozone.VDI-Report 500:Proceedings of colloquium on
    Acid Precipitation,pp.83-95.Lindau(Germany),7-9June 1983.

GEORGII,H.W.and JOST,D.(1964)
    Untersuchung über die Verteilung von Spurengasen in der
    freien Atmosphäre.Pure and Appl.Geoph.59,p.217.

GEORGII,H.W.;PERSEKE,C.,and ROHBOCK,E.(1983)
    Trockene und nasse Deposition säurebildender Verbindungen.
    VDI-Report 500:Proceedings of colloquium on Acid Precipi-
    tation,Lindau(Germany),7-9 June 1983,pp.127-134.

GOLOMB,D.(1983)
    Acid Deposition-Precursor emission relationship in the
    Northeastern U.S.A.:The Effectiveness of regional emission
    reduction.Atm.Env.17,Nov.7,pp.1387-1390.

GRAVENHORST,G.and BÖTTGER,A.(1983)
    Field measurement of NO and NO_2 fluxes to and from the
    ground.Proceedings of the CEC workshop on "Acid Deposition",
    Berlin,9 Sept.1982.pp.172-184.
    Reidel Publishing Company;Dordrecht/Boston/Lancaster.
    Editors:S.Beilke and A.J.Elshout.

GRAVENHORST,G.;HÖFKEN,K.D.,and GEORGII,H.W.(1983)
    Acidic input to a beech and spruce forest.Proceedings of
    the CEC workshop on "Acid Deposition",Berlin,9 September
    1982;pp.155-171.Reidel Publishing Company;Dordrecht/Boston/
    Lancaster.Editors:S.Beilke and A.J.Elshout.

GRENNFELT,P.;BENGTSON,C.and SKÄRBY,L.(1983)
    Dry Deposition of Nitrogen Dioxide to Scots Pine Needles.
    Proceedings of 3.Symposium on Precipitation Scavenging,
    Dry Deposition,and Resuspension,Santa Monica(USA),29 Nov-
    3 Dec.1982.pp.753-761.Elsevier Publ.Eds:Pruppacher et al.

GUICHERIT,R.(1982)
    The global NO_x cycle.Proceedings:Air Pollution by Nitrogen
    Oxides.Elsevier Scientific Publishing Company,Amsterdam.
    Editors:Schneider,T.and Grand,L.

HILL,A.C.(1971)
    Vegetation: a sink for atmospheric pollutants.
    J.Air.Poll.Control Assoc.21,pp.341-346.

HUEBERT,B.J.(1983)
    Measurements of the dry deposition flux of nitric acid vapor
    to grasslands and forest.Proceedings of 3.Symposium on
    Precipitation Scavenging,Dry Deposition,and Resuspension.
    Santa Monica,USA,29 Nov-3 Dec.1982.pp.785-794.
    Elsevier Pubisher,Amsterdam.Editors:Pruppacher,Semonin and
    Slinn.

JUDEIKIS,H.S.and WREN,A.G.(1978)
    Laboratory measurements of NO and NO_2 deposition onto soil
    and cement surfaooo.Atm.Cnv.12,p.2315.

KALLEND,A.S.(1983)
    Trends in the acidity of rain in Europe:a re-examination of
    European Atmospheric Chemistry Network Data.Proceedings of
    the CEC workshop on "Acid Deposition",Berlin,9 Sept.1982.
    pp.108-113.Reidel Publishing Company,Dordrecht/Boston/
    Lancaster.Editors:S.Beilke and A.J.Elshout.

KESSLER,C.and PLATT,U.(1984)
     Nitrous Acid in Polluted Air Masses-Sources and Formation
     Pathways.Paper presented at 3.European Symposium on
     Physico-Chemical Behaviour of Atm.Poll.,Varese(Italy),
     10-12 April 1984.COST 61a bis.

LEE,Y.N.and SCHWARTZ,S.E.(1981)
     Evaluation of the Rate of Uptake of Nitrogen Dioxide by
     Atmospheric and Surface Liquid Water.J.Geoph.Res.,
     Vol.86,No.C 12(Dec.1981),pp.971-983.

OTTAR,B.(1983)
     Air pollution,emissions and ambient concentrations.
     Paper presented at CEC symposium on "Acid Deposition-A
     Challengefor Europe",Karlsruhe,19-21 Sept.1983.

PLATT,U.;PERNER,D.;SCHRÖDER,J.;KESSLER,C.,and TOENISSEN,A.(1981)
     The diurnal variation of NO_3.Paper presented at Aussois
     (France),April 1981. 2.meeting of working party 4 of project
     COST 61a bis.

REPORT of the National Academy of Sciences Panel on Atmospheric
     Transport and Chemical Transformation in Acid Deposition.
     Chairman:J.Calvert(June 29,1983).The full report is
     available from the National Academy of Sciences,USA.

SJÖDIN,Å.and GRENNFELT,P.(1984)
     Regional background concentrations of NO_2 in Sweden.
     Paper presented at 3.European Symposium on "Physico-Chem.
     Behaviour of Atm.Poll.",Varese(Italy),10-12 April 1984.
     COST 61a bis.

VAN AALST,R.M.(1983)
     Dry deposition of acid precursors in the Netherlands.
     VDI-Report 500:Proceedings of colloquium on Acid Preci-
     pitation,pp.97-102.Lindau(Germany),7-9 June 1983.

VAN DOP,H.(1983)
     The residence and transport of pollutants in the atmosphere-
     a meteorological problem.Paper presented at CEC symposium
     on "Acid Deposition-A Challenge for Europe",Karlsruhe,
     19-21 September 1983.

WARMBT,W.(1979)
     Ergebnisse langjähriger Messungen des bodennahen Ozons in
     der DDR(Results of long-term measurements of near-surface
     ozone in the GDR).Zeitschrift für Meteorologie,Band 29,
     Heft 1,1979.S.24-31(in German).

NO$_x$ BACKGROUND MIXING RATIOS IN SURFACE AIR OVER EUROPE AND

THE ATLANTIC OCEAN

A. Broll, G. Helas, K.-J. Rumpel[*], and P. Warneck
Max-Planck-Institut für Chemie (Otto-Hahn-Institut),
6500 Mainz, FRG

[*] affiliated with Umweltbundesamt, Meßstelle Deuselbach

Summary

An O$_3$-NO chemiluminescence detector combined with a
FeSO$_4$ converter was used to measure tropospheric NO$_2$ and
NO at the UBA station in Deuselbach, a rural community
about 90 km west of Mainz, and onboard research ships
during north-south cruises in the Atlantic ocean. The
mixing ratios of NO$_2$ found at Deuselbach had monthly mean
values of 4.5 ppbv in summer, and 16 ppbv in winter.
Mixing ratios of NO, by contrast, were by a factor of ten
lower. The winter maximum attests to the importance of
home heating as a source of NO$_x$. The data obtained so far
during ship cruises indicate minimum values of about 25
pptv in the tropical and polar regions with a pronounced
maximum of about 70 pptv in the northern hemisphere near
45 ^{0} latitude, and a weaker secondary maximum in the
middle latitudes of the southern hemisphere. Both maxima
are ascribed to anthropogenic NO$_x$ carried over the ocean
within the circumpolar belt of the westerlies.

1. Introduction

Despite the importance of the nitrogen oxides NO$_2$ and NO
(NO$_x$) as atmospheric pollutants, there are surprisingly few
measurements of their mixing ratios outside the cities. As a
consequence, our knowledge of NO$_x$ background levels on the
European continent is quite limited. The conversion of NO$_2$ to
HNO$_3$ by reaction with OH radicals and via the reaction mechanism involving O$_3$, NO$_3$, N$_2$O$_5$, and H$_2$O is fairly rapid leading
to lifetimes of NO$_2$ of the order of one day (1,2,3). Accordingly, one expects the NO$_x$ mixing ratio to exhibit large
variations in time and space, so that extensive measurement
series are required in order to obtain representative averages
of NO$_x$ mixing ratios. Another difficulty results from the
fact, that commercial NO$_x$ analyzers usually are not sufficiently sensitive for measurements outside the cities, so that
special instrumentation must be used. We have, for this purpose, constructed a NO$_x$ analyzer based on the well-known principle of the chemiluminescence released by the reaction of NO

with O_3. We have investigated NO_x mixing ratios for the period
of one year at a rural site in western Germany. Some of the
results obtained are presented here. We also have measured NO_x
onboard ships crossing the Atlantic ocean. These data provide
an indicator for the global background of NO_x with which the
continental mixing ratios may be compared.

2. Instrumentation and measurement site

The chemiluminescence NO_x analyzer has been described
previously (4). Compared with commercial designs, the sensitivity has been increased and the lower limit of detection
improved by enlarging the reaction chamber, by optimizing the
chamber pressure, by better cooling the photomultiplier needed
to monitor the $NO-O_3$ chemiluminescence, and by using advanced
photon counting techniques for signal processing. For a counting period of 100 s the lower limit of detection for NO in air
is about 0.01 ppbv, which is below that required for studies
of NO_x on the continent.

Whereas NO is monitored directly, NO_2 must first be
reduced to NO, before it can be detected with the chemiluminescence technique. Two types of converters were used for the
reduction: (i) $FeSO_4$ crystals at ambient temperature and (ii)
molybdenum chips heated to 350 ^oC. The latter type of converter is used in commercial instruments, but we found that it
frequently gives higher signals than the $FeSO_4$ converter,
presumably because it reduces to NO a number of atmospheric
nitrogen compounds other than NO_2 (5), such as PAN (6) and
HNO_3 (7). We believe therefore, that the $FeSO_4$ converter
yields the more representative values for NO_2 mixing ratios
and only these data will be discussed here.

Each converter was provided with a short air intake and
both were placed together with a magnetic switching valve in a
sampling head that was mounted about 5 m above ground. A
common line (stainless steel, 1 mm i. d.) connected the sampling unit with the instrument which was housed in the station
building.The instrument was automated to switch successively
within one hour through three sampling modes, one each for air
samples passing through the two converters giving sum signals,
and one entering the instrument directly for the determination
of NO. The first four minutes of data in each mode were dis-
carded to allow the gas flow to adjust to the new conditions.
The remaining 16 minutes of data were averaged and taken as
representative for the hourly mean. The instrument was calibrated every 25 hours with a standard of NO in nitrogen.

The measurement site chosen for the present study was the
monitoring station of the German Federal Environmental Agency
(Umweltbundesamt,UBA) in Deuselbach. The station is located in
a hilly country, 86 km west of Mainz, on top of a ridge next
to an isolated road. The building is heated electrically so
that local contamination is minimized. The effects of an
occasional vehicle on the road were clearly evident as spikes
in the data and were ommitted from the record before averaging. The surrounding area consists of 88% of farmland, 10% are
forests, the remaining area is occupied by roads and buildings

(8). There are no industries of significance in the neighbour-
hood. The village of Deuselbach (population about 300) is
located about 1 km to the south of the station, a highway with
moderate traffic lies about 2 km to the north-west. Larger
towns within 50 km distance are Metlach (population 13 000),
Witlich (16 000), Idar-Oberstein (38 000),St. Wendel (28 000),
and the city of Trier (95 000). Within 50 -100 km distance are
the cities of Luxembourg, Koblenz, Mainz, Ludwigshafen-
Mannheim, Kaiserslautern, and Saarbrücken, all with populat-
ions exceeding 100 000. The prevailing winds are from the
south-west, occuring with an average speed of 4 m/s.

Measurements in the boundary layer of the Atlantic ocean
were made onboard the German research vessel "Meteor". The
instrument used was of the same kind as in the continental
study, but was equipped with a more sensitive photomultiplier
yielding a lower limit of detection of ca. 5 pptv of NO. The
sampling head was mounted to the front mast 8.5 m above the
highest deck of the ship and approximately 20 m above sea
level.

3. Results and discussion

3.1 Deuselbach
Measurements in Deuselbach were made from September 1980
to August 1981 with a gap in January and February 1981 due to
instrument malfunction.
The hourly values were combined to derive daily averages
and these were further averaged to obtain monthly means. The
frequency distribution for the daily averages of the NO_2 and
NO mixing ratios show a lognormal behaviour with maxima near 4
and 0.4 ppbv respectively. The total range for NO_2 is 0.9 to
54 ppbv. The upper extremum occured once only during the
entire measurement period. 82% of the NO_2 values fall into the
range of 1 to 11 ppbv. The 82% range for NO is 0.1 to 1.1
ppbv, so that NO is only one tenth as abundant as NO_2, on
average. In this respect it is important to remember that in
the absence of direct sources NO derives from NO_2 by photodissociation and that a photostationary state is set up by
various back reactions, primarily that of NO with ozone. The
daily averages for NO include the night-time values which
theoretically should reduce to zero, if enough ozone is present. While the night-time values were indeed quite low, they
reached the zero level only in 40% of all observations, thus
indicating the existence of local sources of NO. In the
summer, when the mixing ratios of NO_x are the lowest, the
influence of the highway to the north-west of the station is
clearly evident in the data by an enhancement of NO mixing
ratios from that direction, as the wind diagram in figure 1
shows. The non-zero night-time NO values, however, are fairly
evenly distributed with wind direction, so that additional
nondirectional sources must exist. It is possible that part of
the night-time NO emanates from the soil due to microbial
activity (9). This process would be enhanced during the warm
period of the year. It is interesting, that the occurence of
non-zero NO values at night was most frequent during July and

August when the annual temperature reached its maximum. Alternatively, one must consider transport of NO toward the station from anthropogenic sources in the more distant surroundings. In this case, the night-time NO readings would represent remnants having escaped oxidation due to incomplete mixing with ozone-rich air.

Figure 2 shows the dependence of NO_2 mixing ratios on wind direction for the winter and the summer seasons. The hourly values were used to prepare these diagrams. Both indicate fairly even distributions. A slight enhancement occurs toward the south-east in winter and toward the north in summer. Hourly data are available for SO_2 also, which is measured routinely at the UBA station with an electrochemical method. The dependence of SO_2 mixing ratios on the wind direction is included in figure 2 and again a rather uniform distribution is evident. Both data sets demonstrate that the measurement site is not influenced by an isolated source of SO_2 and NO_x such as a coal- or oil-fired power plant. The locality appears to be immersed in a more or less uniform background of air pollutants originating from the urban centers in the distant surroundings as well as from multiple small sources scattered throughout the region. Further justification for classifying the measurement station as a regional continental background site with respect to NO_x is seen in the small ratio of NO to NO_2 when compared with those encountered in the cities.

Table I presents monthly means for the mixing ratios of NO and NO_2 as measured by us, and of SO_2 and O_3 as determined from the records of the UBA station. Monthly mean temperatures

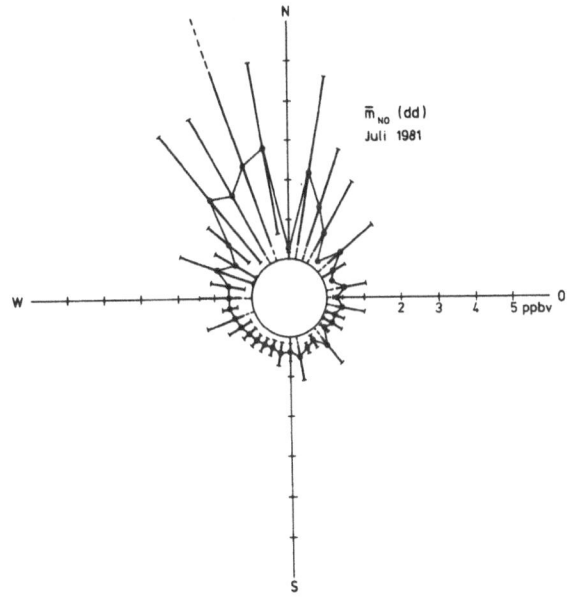

Figure 1   Dependence of NO mixing ratio on wind direction for the month of July,1981 at Deuselbach

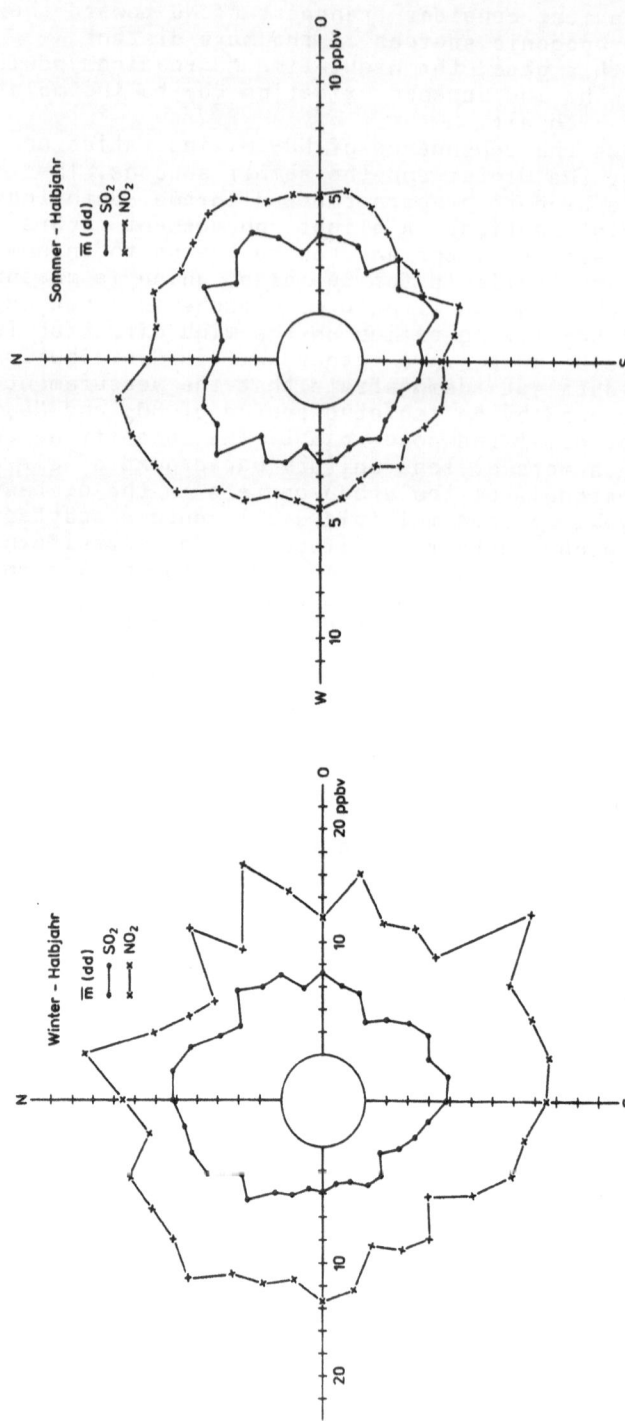

Figure 2   Dependence of NO_2 and SO_2 mixing ratios on wind direction for the measurement period of Sept. 1980 to Aug. 1981 in Deuselbach; left: winter half year, right: summer half year

Table I

Monthly averages of NO_2, NO, NO_x, O_3, SO_2, in ppbv and temperature in °C measured in Deuselbach for the period Sept. 1980 - Aug. 1981; * derived from Saltzman, see text

Month	NO_2	NO	NO_x	SO_2	O_3	TT
Sep	5.2	0.8	6.0	2.5	35	13.8
Oct	15.6	1.4	17.0	5.1	19	7.3
Nov	17.3	1.3	18.7	9.0	22	2.2
Dec	17.1	1.9	19.0	8.0	21	-0.2
Jan	24 *	-	-	9.7	20	-1.0
Feb	21 *	-	-	14.0	23	-1.0
Mar	7.7	1.0	8.7	4.5	33	6.9
Apr	11.2	2.1	13.3	6.5	35	7.3
May	4.6	0.5	5.1	3.1	42	11.7
Jun	4.0	0.3	4.3	2.2	40	13.9
Jul	3.9	0.6	4.5	2.2	36	15.2
Aug	4.9	0.8	5.7	3.2	36	15.6

Table II

Monthly averages of NO_2 and NO in the cities of Mainz (MZ), Ludwigshafen-Mannheim (LU), and Saarbrücken (SB) for the period Sept. 1980 - Aug. 1981, and NO_2, NO, NO_x, and SO_2 at Bottesford, UK for the period Sept. 1978 - Aug. 1979

Month	MZ		LU		SB		Bottesford, UK			
	NO_2	NO	NO_2	NO	NO_2	NO	NO_2	NO	NO_x	SO_2
Sep	21.3	24.5	37.2	57.1	20	50	7	8	15	13
Oct	16.0	32.6	16.0	48.9	20	60	12	12	24	11
Nov	16.0	48.9	31.9	57.1	10	50	8	13	21	11
Dec	26.6	32.6	31.9	57.1	10	40	17	22	39	13
Jan	37.2	32.6	37.2	65.2	20	40	19	15	34	26
Feb	26.6	24.5	31.9	32.6	20	40	11	5	16	15
Mar	21.3	24.5	26.6	24.5	20	30	8	7	15	12
Apr	26.6	16.3	31.9	16.3	10	40	6	4	10	12
May	26.6	16.3	26.6	16.3	10	20	7	3	10	8
Jun	21.3	8.2	21.3	16.3	10	20	7	4	11	9
Jul	21.3	8.2	21.3	16.3	10	30	6	4	10	7
Aug	26.6	16.3	26.6	16.3	10	40	4	4	8	8

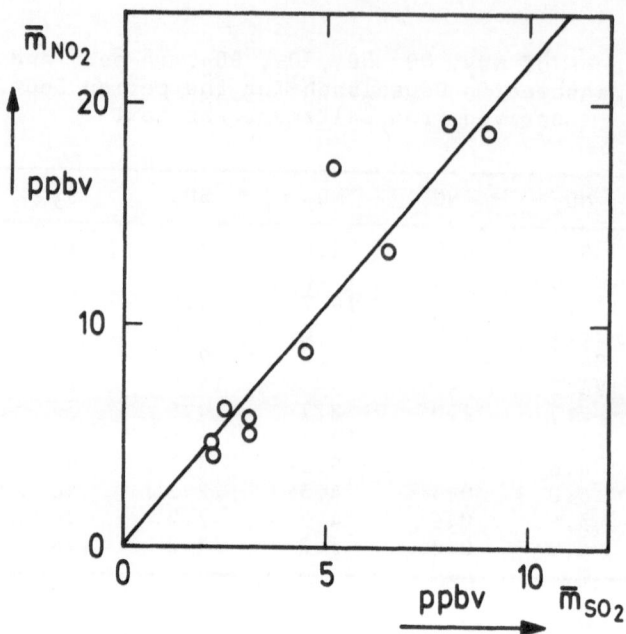

Figure 3   Monthly mean values of NO_2 mixing ratios
versus those of SO_2 in Deuselbach. The correlation
coefficient is 0.91 .

are shown in the table. The data reveal a seasonal variation
of the NO_x mixing ratios with the highest values occuring in
winter and the lowest in mid-summer. The monthly means are
clearly anticorrelated with temperature and, equally important, both NO_x and NO_2 correlate very well with SO_2. This
relation is shown in figure 3. There can be little doubt that
the winter maximum is due to the increase in fossil fuel
combustion associated with home heating during the cold season
of the year. At five of the stations within the UBA network,
including Deuselbach, daily averages for NO_2 mixing ratios are
measured by means of the (wet-chemical) Saltzman procedure
(8), the absorbing solution being exposed for 24 hours. The
record of these data extends back for a number of years and
the winter maximum shows up regularly at all five measurement
sites, although it is much weaker at the two mountain stations.
     Figure 4 compares monthly means of NO_2 derived from the
two measurement techniques, $NO-O_3$ chemiluminescence and Saltzman procedure. The data from both methods agree reasonably
well for the summer season but not in winter. It appears that
the 24 hour exposure time used in the wet-chemical analysis is
too long to accomodate NO_2 mixing ratios greater than 6 - 8
ppbv without saturation effects becoming important. A quantitative comparison of the data obtained with the two methods
nevertheless provides a correction curve which can be used to
estimated monthly mean NO_2 mixing ratios for the two month
(January and February) missing in our record. An alternative
way to derive an estimate for the missing data points is to

utilyze the correlation diagram in figure 3 between NO_2 and SO_2. Both methods yield fairly consistent values which are shown in figure 4 by dashed bars. They are also entered in table I and are there indicated by asterisks. The average NO_2 mixing ratio for the summer period, taken from May to September, is 4.5 ppbv. For the winter season, here taken from October to April, the average is 16.3 ppbv, i. e. almost four times higher than in summer.

Table II summarizes monthly mean values for NO_x mixing ratios in three cities of the region, Mainz, located 86 km to the east (10), Ludwigshafen-Mannheim, 91 km to the south-east (10), and Saarbrücken, 68 km to the south-west (11). While in comparison with Deuselbach the absolute values for NO_x in the cities are much higher, there is some indication also in these data for a seasonal trend with a maximum during the winter months. We present table II mainly to demonstrate the large contribution of NO to total NO_x. In the cities and presumably elsewhere on the European continent as well, most of the NO_x originates from the combustion of fossil fuels. Combustion sources emit NO_x primarily in the form of NO which is converted to NO_2 subsequently in the atmosphere by the reaction with O_3. Observation of large ratios of NO to NO_2 indicates, in our opinion, the vicinity of anthropogenic sources. The influence of local sources clearly exists in the cities, but apparently not in Deuselbach exept for the directional enhan-

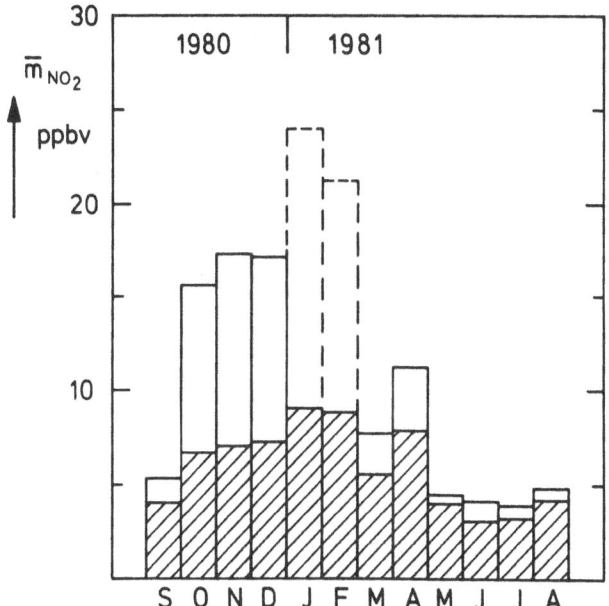

Figure 4   Annual variation of monthly mean NO_2 mixing ratios in Deuselbach; upper values are derived from measurements with the chemiluminescence technique with the dashed bars obtainded by extrapolation as described in the text. The lower values (shaded bars) were derived from measurements with the Saltzman method.

cement due to traffic as is discussed above. Also this observation leads us to characterize this station as a regional
background measurement site.

Martin and Barber (12) conducted a two year study of NO_x,
SO_2, and O_3 mixing ratios at Bottesford, a rural site in
central England 20 km east of Nottingham (population 300 000).
NO_x was measured with a commercial chemiluminescence analyzer.
To our knowledge, their study is the only one comparable to
ours with respect to choice of sampling site and duration of
observation period. Table II includes monthly mean values for
NO_x obtained by Martin and Barber for the period September
1978 to August 1979. Similar to our results, their NO_x data
are positively correlated with SO_2 mixing ratios, even though
the scatter is appreciable, and the data reveal again a seasonal variation with lower values occuring in summer and higher
ones in winter. The absolute values for NO_x are about twice of
those in Deuselbach. The major difference between both data
sets is the large fraction of NO_x present as NO in Bottesford
as compared to Deuselbach. In the light of the preceding
remarks, it appears that the Bottesford site may be influenced
to a considerable degree by local sources of NO_x. Judging from
the site description given by Martin and Barber, the region
surrounding Bottesford is quite densely populated, much more
than the region of Deuselbach, which suggests that the population density determines not only the average background level
of NO_x in a region, but also the ratio of NO to NO_2. Note that
this conclusion refers to daily averages of NO and NO_2 and
their ratios.

3.2. Measurements in marine air

Finally we discuss a limited number of measurements of NO_2
in the Atlantic ocean. Ultimately we wish to establish the
natural background level of NO_x in the marine atmosphere. The
task is beset with difficulties, however. The major problem
results from the fact, that the ship itself is a source of NO_x
and a criterion is needed by which one can differenciate air
samples that have suffered contamination from those that have
not. Additional difficulties arise from the contamination of
air by other ships encountered en route, and from the advection of continental air masses in regions not too distant from
the coasts.

In order to identify contaminated air samples and to
eliminate them, we have relied on the concept that under
natural conditions NO should be absent during the night. Since
combustion sources produce primarily NO, the observation of NO
during the dark hours of the day provides a convenient indicator for the presence of pollution. Accordingly, the data were
filtered by selecting only those night-time NO_2 values for
which NO was entirely absent (m(NO) smaller 5 pptv). A further
selection was then made to exclude data of measurements with
air of continental origin. The remaining data are taken to represent the marine background of NO_x. These results are shown
in figure 5. Also included in the figure is the night-time
level of NO_2 observed at Loop Head at the west coast of Ire-

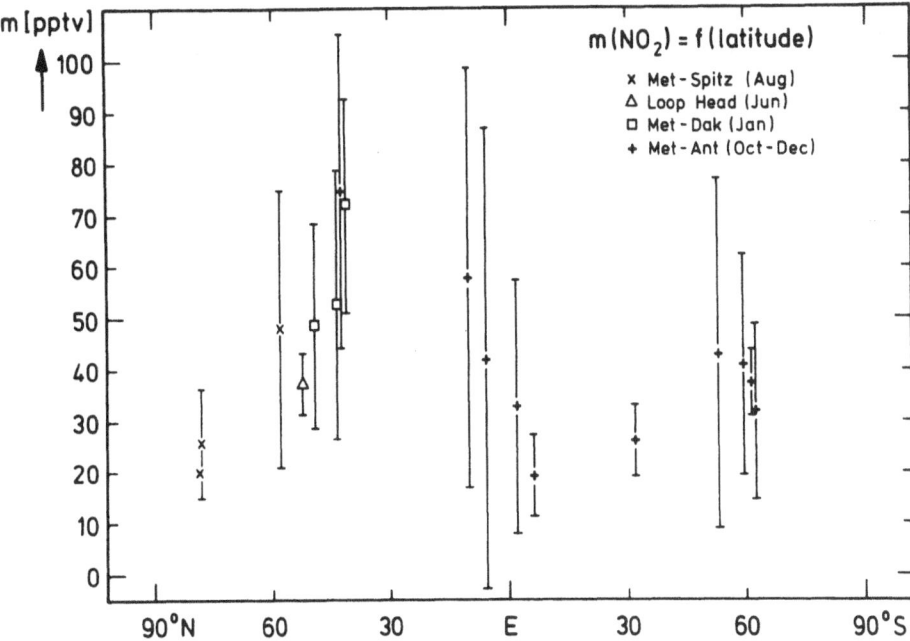

Figure 5   Background mixing ratios of NO_2 over the Atlantic ocean as a function of latitude

land in pure marine air masses (3). The variability of the NO_2 mixing ratios is indicated by the vertical bars (one standard deviation). Although the number of data points is as yet limited, they suggest the existence of a bimodal distribution with latitude, featuring a pronounced maximum in the northern hemisphere near 45 $^{\circ}$ latitude and a weaker secondary maximum in the middle latitudes of the southern hemisphere. These are the dominant regions of the westerlies. We thus attribute the two maxima to continental anthropogenic sources and assume spreading of the NO_x into the nearly source-free regions over the ocean. If this interpretation is correct, most regions of the Atlantic ocean must be considered polluted with respect to NO_x. The positions of the maxima coincide approximately with those given by Ehhalt and Drummond (13) for the latitudinal distribution of NO_x released from fossil fuel combustion. The relative strength is different, however. More than 90% of fossil fuels are estimated to be burned in the northern hemisphere. Due to the limited lifetime of NO_2, in the atmosphere the NO_2 mixing ratios in air masses of continental origin should decrease with increasing distance from the coast. Since most of our measurements in the northern hemisphere were made in the vicinity of the European continent in air masses originating from North America, it is probable that the NO_x mixing ratios in the western Atlantic of the northern hemisphere is still greater.

The lowest NO_2 mixing ratios were encountered near the equator and in the polar regions. They fall into the range of 20 to 30 pptv and these values may represent the natural background level. At present, we do not know which values would represent the true natural background of NO_x in the marine atmosphere. The decline of NO_x mixing ratios to values two orders of magnitude below continental ones is consistent with the short lifetime of NO_x in the troposphere.Another unknown is the magnitude of the natural NO_x background level in the boundary layer of the continents.

Ackowledgements
We thank the officers and the crew of the RV "Meteor" for assistance during the cruises. This study is part of the programme of the Sonderforschungsbereich 73 "Atmosphärische Spurenstoffe" and has received support from the Deutsche Forschungsgemeinschaft.

4. References

1) T. Y. Chang, J. M. Norbeck, and B. Weinstock
   Environ. Sci. Technol. 13,1534 - 1537 (1979)
2) C. W. Spicer
   Science 215,1095 - 1097 (1982)
3) G. Helas and P. Warneck
   J. Geophys. Res. 86,7283 - 7290 (1981)
4) G. Helas, M. Flanz, and P. Warneck
   Intern. J. Environ. Anal. Chem. 10,155 - 166 (1981)
5) A. Winer, J. W. Peters, J. P. Smith, and J. N. Pitts,Jr.
   Environ. Sci. Technol. 8,1118 - 1121 (1974)
6) R. A. Cox and M. J. Roffey
   Environ. Sci. Technol. 11,900 - 906 (1977)
7) D. W. Joseph and C. W. Spicer
   Anal. Chem. 50,1400 - 1403 (1978)
8) Umweltbundesamt,Texte: Großräumige Luftbelastung in der
   Bundesrepublik Deutschland, Berlin 1980
   Monatsberichte und Jahresberichte des Meßnetzes des
   Umweltbundesamtes, Berlin
9) J. E. Galbally and C. R. Roy
   Nature 275,734 - 735 (1978)
10) Monatsberichte über die Meßergebnisse des Zentralen
    Immissionsmeßnetzes für Rheinland-Pfalz, Landesgewerbe-
    aufsichtsamt für Rheinland-Pfalz, Mainz
11) Schriftenreihe, ed. Minister für Arbeit, Gesundheit und
    Sozialordnung, aus dem Staatlichen Institut für Hygiene
    und Infektionskrankheiten, Saarbrücken, Heft 14 und 15
12) A. Martin and F.R. Barber
    Atmos. Environ. 15,567 - 578 (1981)
13) D. H. Ehhalt and J. W. Drummond
    The tropospheric cycle of NO_x
    in Chemistry of the unpolluted and polluted troposphere,
    ed. H.- W. Georgii and W. Jaeschke, Reidel Publ. Comp.,
    Dordrecht, Holland 1982

REGIONAL BACKGROUND CONCENTRATIONS OF NO_2 IN SWEDEN

Åke Sjödin and Peringe Grennfelt
Swedish Environmental Research Institute

Summary

Regional background concentrations of NO_2 have been measured continously in Sweden since May 1981 at five locations far from main pollutant source areas. The monitoring stations are identical with those within the European Monitoring and Evaluation Programme (EMEP) Network, where also daily samples of SO_2, soot and particulate sulphate are collected. The NO_2 concentration is determined as a 24 hour mean value using a slightly modified version of the Saltzman method. The interference of ozone is eliminated by absorbing ozone on sodiumthiosulphate impregnated filters. The detection limit is better than 0.5 µg/m³ NO_2.
Data from two years of daily monitoring are presented. The results show a typical seasonal variation in the regional NO_2 background concentrations; the 6 months winter mean value being roughly two times the corresponding summer value. Moreover, the winter time is characterized by a large variation in the NO_2-concentration from one day to another, due to a high frequency of episodes with long range transport of air pollutants, resulting in "NO_x-peaks", primarily during the months December to February. Values during the summer months, especially June to August, reveal a more uniform concentration pattern, with concentrations close to the detection limit at several stations. The yearly average varies from 6.8 µg/m³ NO_2 at the station situated farthest to the south (Ekeröd) to 1.3 µg/m³ NO_2 at the station situated farthest to the north (Bredkälen). Studies of correlations to other air pollutants and between the different sampling sites show that the occurrence of NO_2 in rural air in Sweden to a large extent is caused by long range transport of polluted air masses.

1. INTRODUCTION

The occurence of nitrogen oxides in long range transported air masses over northwestern Europe has been observed in several studies (Grennfelt 1979, Hov 1983). However, during the transport the nitrogen oxides are involved in a complex chemistry which still is not very well understood. This chemistry results in several oxidized nitrogen species such as nitric acid, nitrous acid, peroxyacetyl nitrate and other

organic nitrates. Monitoring of these species in areas far
from the main source areas is necessary for a better understanding of the chemistry of oxidized nitrogen in the atmosphere.

Although nitrogen oxides are mainly emitted as nitrogen
monoxide, this compound will not occur in significant concentrations in remote areas due to the rapid oxidation of NO
by ozone. Instead transport of NO_x seems mainly to occur in
higher oxidized forms such as nitrogen dioxide, nitric acid
and PAN as observed in northwestern Europe (Ferm et al.,
1983). Of these species NO_2 seems to be predominant. Moreover,
nitrogen dioxide is of great importance as a driving agent
for ozone formation and other photochemical processes. A
substantial fraction of the dry deposition of nitrogen oxides
seems to occur as nitrogen dioxide, since it is rapidly
taken up by vegetation (Bengtson et al., 1982). Hence, with
special reference to the greatest regional air pollution
problems of today, the photochemical oxidant formation and
the acidification; it is important to know and to be able to
predict the regional distribution of NO_2 within a country. A
great part of this knowledge can be achieved by long term
continous measurements made at the same time at several
stations unaffected by local sources. In this paper we present
a two year NO_2 data set from a monitoring network of five
stations in non urban areas in Sweden. The first year of monitoring six stations were run.

2.   METHODS

Daily mean concentrations of nitrogen dioxide were determined by a slightly modified version of the Saltzman method
(Saltzman, 1954). The interference from ozone was eliminated
by absorbing ozone on a sodiumthiosulphate (0.2 M) impregnated
Whatman 40 filter (Ø 25 mm) before the midgetfritted bubbler
absorbing NO_2 (Figure 1). The absorption efficiency for ozone
was better than 98% at a flow rate of 200 ml min^{-1}, and at the
same conditions the NO_2-loss on the filter was less than 5%.
The sampling flow was 150-200 ml min^{-1} and was checked
every day by means of a rotameter. The sampling equipment is
shown by Figure 2. After the sampling, which was made from
6 to 6 am GMT, the samples were stored in refrigerators and
sent every week to the laboratory for immidiate analysis.
Since the formed azodye-complex is not stable, several tests
were made to study the degree of decompisition. These studies
indicate that the decomposition in almost all cases should
be less than 10%, since storage of samples at room temperature
for 10 days showed a mean decomposition of 5%. The detection
limit was about 0.4 µg/m^3 at 150-200 ml min^{-1} flowrate and
15 ml of Saltzman reagent.
Monitoring started in May 1981 and in this report we
present data for the period May 1981 - April 1983. During the
first year, measurements were made at six sites in Sweden
(Figure 3 and Table 1). Five of these are identical with the
monitoring stations within the European Monitoring and
Evaluation Programme (EMEP). From May 1981 the station Ryda

Kungsgård was no longer within the monitoring programme and measurements were only made at the five EMEP stations. At these stations daily samples of precipitation, SO_2 and particles are also collected.

3. RESULTS

The yearly mean of NO_2 varies between the different places from 6.8 $\mu g/m^3$ at Ekeröd, situated farthest to the south, to 1.3 $\mu g/m^3$ at Bredkälen, situated farthest to the north. There is a seasonal variation in the NO_2 concentration. The winter period mean value (November - April) normally exceeds the summer period mean value by a factor of 2 or 3 (Table 2). The monthly mean values from May 1981 to April 1983 for all stations are presented in Figure 4. The highest monthly mean, 23.7 $\mu g/m^3$, was monitored at Ekeröd in January, 1982. The highest daily value was monitored at Rörvik, January 22nd, 1982, and reached 59.2 $\mu g/m^3$. At most of the time, even at the sites with the lowest mean concentrations, the NO_2 concentration is above detection limit. This indicates that there is almost always a NO_2 background of at least 0.4 $\mu g/m^3$. During the summer months, June to August, the NO_2 concentrations pattern is very uniform and quite low, mostly 1-3 $\mu g/m^3$ (c.f. Figure 5). For instance, the standard deviation for the "cleanest" stations, Hoburg, Velen and Bredkälen, is less than 0.5 $\mu g/m^3$ in July.

During the winter months, especially December-March, episodes with high NO_2 concentrations occur. The variation in NO_2 concentration during these months is much higher compared to the summer months (c.f. Figure 6) and is very similar to what is observed for other pollutants such as SO_2, soot and particleborne sulphate (Figure 7).

Correlation coefficient calculations were made for the winter months December-February and the summer months June - - August for each station and are presented in Table 3. These calculations showed for the wintermonths in all cases a significant positive correlation between NO_2, SO_2 and soot, respectively. Although there was a positive correlation between these pollutants also during the summer, it was not significant in several cases.

The correlation between simultaneous NO_2 24-hour mean values at the different stations were also investigated. The station at Ekeröd was chosen as a reference. The results, presented in Table 4, show in most cases a significant positive correlation to the three closest stations during winters, but an insignificant correlation during summers.

The ratios of atmospheric nitrogen in the form of NO_2 to atmospheric sulphur as SO_2 and particleborne sulphate as a function of site location were also investigated and are presented in Figure 8. The highest ratios are found at Ekeröd and Rörvik, with a yearly mean ratio just below unity. At Hoburg, on the island of Gotland in the Baltic Sea, instead the sulphur exceeds the nitrogen by a factor of aobut five.

4. DISCUSSION

The Saltzman wet chemical method has been used earlier for the determination of NO_2 as well as NO_x in the background atmosphere (Galbally, 1975 and 1977). The wetchemical measurements fit well chemiluminescent measurements made at the same time, which has also been confirmed in this study (Figure 9). From this figure it can also be seen that the reproducibility within the method is acceptabel. One critical point is the decomposition of the formed azodyecomplex possibly resulting in an underestimation of the real NO_2-level. We believe that in our study this underestimation is less than 10%, e.g. within the measurement error, although, we cannot exclude that the absolute NO_2-levels are somewhat too low.

Despite of this, the results from the evaluation of the first two years background network monitoring of NO_2 in Sweden show that the regional concentration of NO_2 is markedly dependent on the transport of polluted air masses from the south or southwest to northern Europe. This is confirmed by the observed positive correlation at each station between NO_2 and other pollutants, especially during the winter episodes, and the positive correlation in simultaneous NO_2-values between different stations during winters.

Several studies so far have shown that the occurence of soot and particleborne sulphur in rural areas in Sweden is mainly caused by long range transport of these pollutants from the continent or British Isles (Brosset and Åkerström, 1972; Brosset and Nyberg, 1971). This clearly indicate that NO_2 is of almost the same origin as sulphur and soot. In this study wind trajectories have not been evaluated since these are only available for Rörvik and not for the other stations. A detailed evaluation with wind trajectories for Rörvik is given by Ferm et al. (1983).

However, Figure 6 alone supports the assumption of long range transported NO_x, clearest visualised during the winter episodes. In this figure all stations show almost the same NO_2-concentration pattern, if absolute values are not taken into account. The variation in the NO_2 concentration pattern between seasons (Figure 4) is thought to be the result of not only varying emissions of NO_x, but also a different chemistry, a different meteorology as well as different mechanisms of deposition during different parts of the year.

The comparatively high levels of NO_2 at Ekeröd during summers might to some extent be explained by local emissions of NO from aerable land. After the emission NO is rapidly converted to NO_2 by background ozone through the reaction:

$$NO + O_3 \longrightarrow NO_2 + O_2$$

Since the NO_2-levels are somewhat elevated during summers also at the other two sites in typical agricultural districts, Rörvik and Ryda Kungsgård, compared to the sites in non-agricultural districts, this indicates that emissions of NO from land might explain or contribute to the background level of NO_2 under certain conditions.

However, our measurements give no exact evidence for the assumption of an always present global tropospheric background of NO_2. Recently this theory has been questioned (Rittes et al., 1979) and earlier observed background tropospheric NO_x in very clean air (Junge, 1956; Lodge and Pale, 1966; Breeding et al.,1973; Moore, 1974; Cox, 1977) is referred to as NO_x concentrated in pollution clouds or close to the ground.

5. REFERENCES

Bengtson, C., Grennfelt, P., Boström, C-Å., Troeng, E., Skärby, L., Sjödin, Å. and Peterson, K. (1982) Deposition and uptake of nitrogen oxides in Scots pine neddles (Pinus Sylvestris L.). Swedish Environmental Research Institute, publication No. B 647.

Breeding, R.J., Lodge, J.P., Pale, J.B., Sheesley, D.C., Klonis, H.B., Fogle, B., Anderson, J.A., Englert, T.R., Haagenson, P.L., McBeth, R.B., Morris, A.L., Pouge, R. and Wartburg, A.F. (1973) Background trace gas concentrations in the central United States. J. Geophys.Res. 78, 7057-7064

Brosset, C. and Nyberg, A. (1971) Investigation of soot and particleborne sulphur in Sweden. Proc. Sec. Int. Clean Air Congress, Washington, December 6-11, 481-489

Brosset, C. and Åkerström, Å. (1972) Long distance transport of air pollutants - Measurements of black airborne sulphur in sweden during the period of September - December 1969. Atm. Env. 6, 661-673

Cox, R.A. (1977) Some measurements of ground level NO, NO_2 and O_3 concentrations at an unpolluted maritime site. Tellus 29, 356-362

Ferm, M., Samuelsson, U., Sjödin, Å. and Grennfelt, P. (1983) Long range transport of gaseous and particulate oxidized nitrogen compounds. 5th International Conference of the Commission on Atmospheric Chemistry and Globlad Pollution. August 28th - September 3rd 1983, Oxford, U.K.

Galbally, I.E. (1977) Measurement of nitrogen oxides in the background atmosphere. WMO (publ). 1977, 460 (Air Pollut. Meas.Tech.), Part II, 179-185

Galbally, I.E. (1975) Nitrogenoxides (NO_2 and NO_x) in the air of Aspendale and other places in Victoria. Clean Air, 9,12-15

Grennfelt, P. (1979) Oxidized nitrogen compounds in long range transported air masses. WMO symposium in the long range transport of pollutants and its relation to general circulation including stratospheri/tropospheri exchange processe. WMO-report 538, 199-206, Geneve

Hov, Ø (1983) Long-range transport of peroxyacetylnitrate
to Scandinavia. Norwegien Institute for Air Research.
Report No. 30/83.

Junge, C.E. (1956) Recent investigations in air chemistry.
Tellus 8, 127-139

Moore, H.E. (1974) Isotopoci measurement of atmospheric
nitrogen compounds. Tellus 26, 169-174

Ritter, J.A., Stedman, D.H. and Kelly, T.J. (1979) Ground-
level measurements of nitric oxide, nitrogen dioxide and
ozone in rural air. In "Nitrogenous air pollutants. Chemical
and biological implications". Ann. Arbor Science 1979.

Table 1.   Monitoring stations for NO_2

Place	Latitude	Longitude	Height over sea level m	Main characteristics of surrounding geography	Sampling inlet m above ground
Ekeröd	$55^\circ54'N$	$13^\circ43'E$	140	agricultural land	4
Rörvik	$57^\circ24'N$	$11^\circ56'E$	10	" , sea	5
Hoburg	$56^\circ55'N$	$18^\circ09'E$	58	sea	3
Velen	$58^\circ46'N$	$14^\circ18'E$	125	forest	3
Ryda Kungsgård	$59^\circ46'N$	$17^\circ18'E$	25	agricultural land	2
Bredkälen	$63^\circ51'N$	$15^\circ20'E$	404	forest	2

Table 2. 6 months mean values (summer and winter) of NO_2 from May 1981 to April 1983. Figures in parenthesis are max. daily mean values within each period.

Station	Period mean value of NO_2 ($\mu g/m^3$)			
	8105-8110	8111-8204	8205-8210	8211-8304
	s	w	s	w
Ekeröd	3.9 (12.6)	10.7 (55.2)	4.5 (12.6)	8.3 (29.2)
Rörvik	3.2 (10.5)	10.7 (59.1)	3.6 (12.3)	6.8 (28.9)
Hoburg	1.1 (4.1)	2.6 (23.2)	1.3 (9.8)	3.2 (17.2)
Velen	1.5 (8.4)	4.1 (25.8)	1.3 (6.5)	3.2 (20.6)
Ryda Kungsgård	2.3 (7.5)	3.9 (29.0)	-	-
Bredkälen	0.9 (2.6)	1.9 (7.1)	0.8 (2.1)	1.7 (5.3)

Table 3. Correlation between NO_2, SO_2 and soot for each station during summer months (S = June, July, August 1981 + 1982) and winter months (W = December, January, February 1981/82 + 1982/83).

Station	Correlation coefficients					
	$NO_2 \rightarrow SO_2$		$NO_2 \rightarrow$ soot		$SO_2 \rightarrow$ soot	
	S	W	S	W	S	W
Ekeröd	0.05	0.46	0.18	0.52	0.54	0.59
Rörvik	0.11	0.63	0.51	0.55	0.11	0.84
Hoburg	0.26	0.40	0.40	0.43	0.41	0.64
Velen	0.25	0.46	0.21	0.56	0.67	0.84
Bredkälen	0.01	0.32	0.25	0.37	0.26	0.58

Table 4. Correlations between NO$_2$-values (24-hours mean) monitored at Ekeröd and time corresponding values at the other stations.

Station	Correlation coefficient (period mean)	
	Summer 1981 and 1982 (June, July, August)	Winter 1982 and 1983 (December, January, February)
Rörvik	0.16	0.44
Hoburg	0.17	0.47
Velen	0.10	0.39
Bredkälen	-0.09	0.12

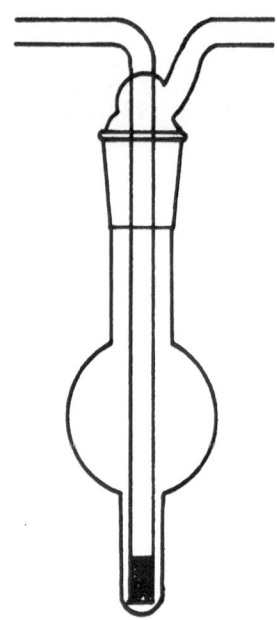

Figure 1. Sampler for NO$_2$. Fritted cylinder porosity is 20-40 μ .

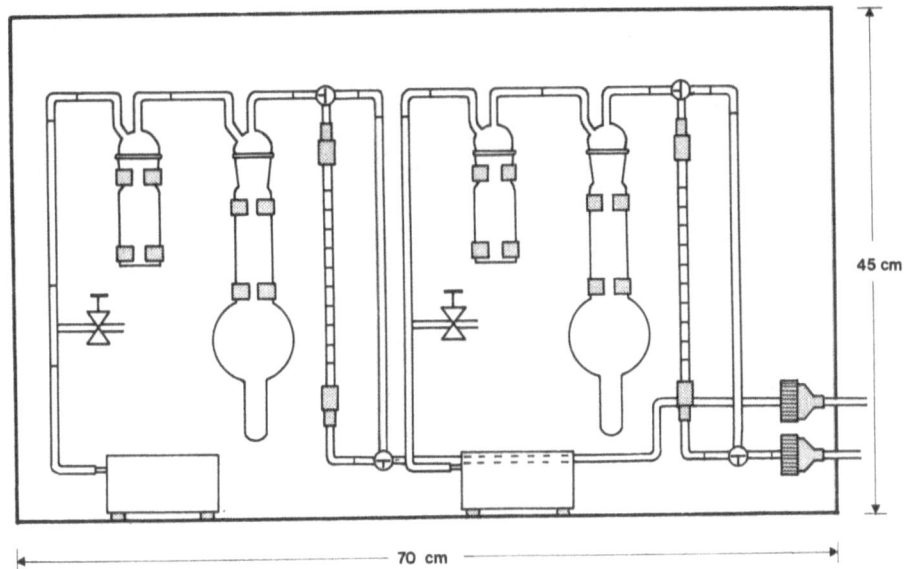

Figure 2.   Equipment for the 24 hours sampling of NO$_2$.
① Ambient air inlet, 1/4" teflon tubing ② Swinnex filter
holders with sodium-thiosulphate impregnated Whatman 40-filters
(ϕ = 25 mm) for removing ozone   ③ Three-way stopcock (sodium
glass) ④ Rotameter (meterate Bstst) ⑤ Rotameter by pass
(3/8" teflon tubing) ⑥ Midgetimpinger (see Figure 1)
⑦ Security vessel ⑧ Excess air inlet with critical needle
valve ⑨ Critical orifice ⑩ Membrane pump.

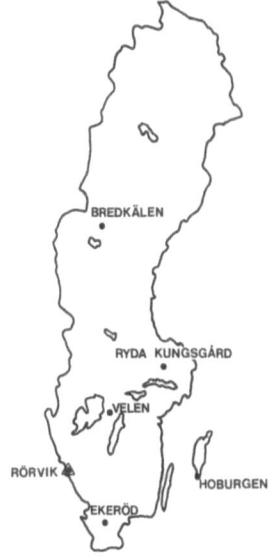

Figure 3.   Monitoring sites.

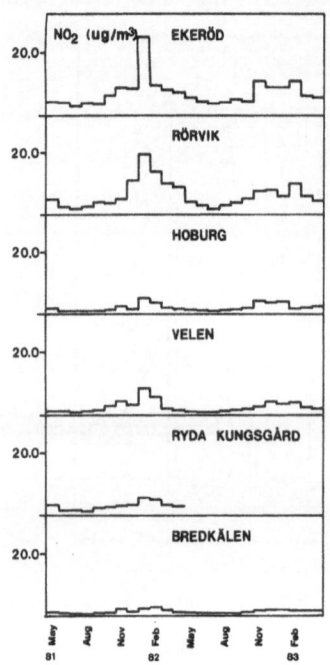

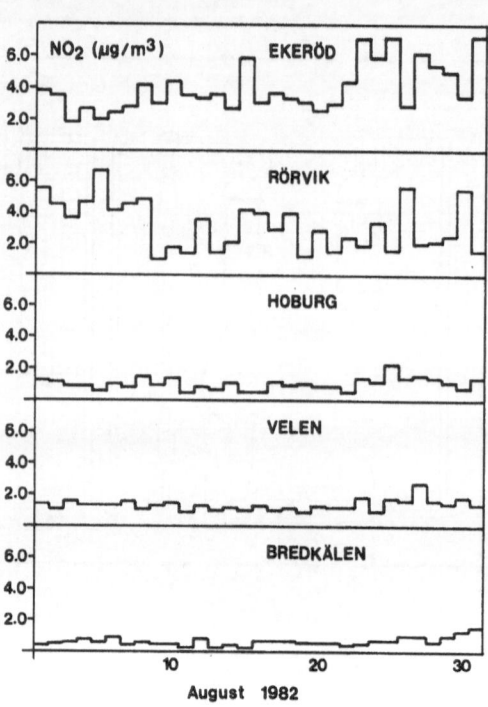

August 1982

Figure 4. Monthly mean values of the NO$_2$-concentration at the different sites from May 1981 to April 1983.

Figure 5. Daily mean concentrations of NO$_2$ (μg/m^3) at five different sites, August 1982.

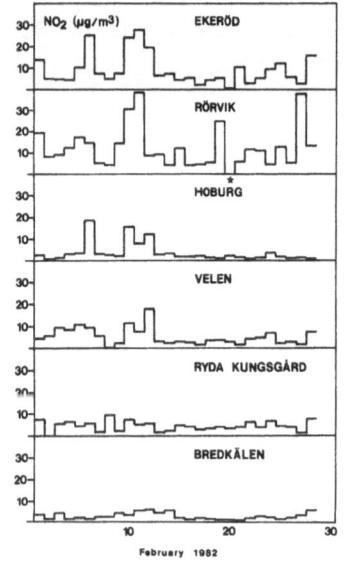

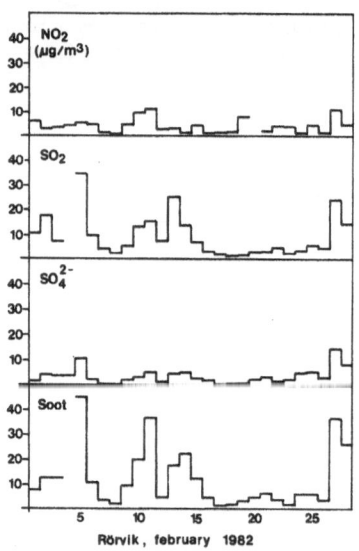

Figure 6. Daily mean concentrations of NO$_2$ (μg/m^3) at six different sites, February 1982. $*$ = missing value

Figure 7. Daily mean concentrations (μg/m^3) of NO$_2$, SO$_2$, particulate sulphate and soot, respectively at Rörvik, February 1982.

- 410 -

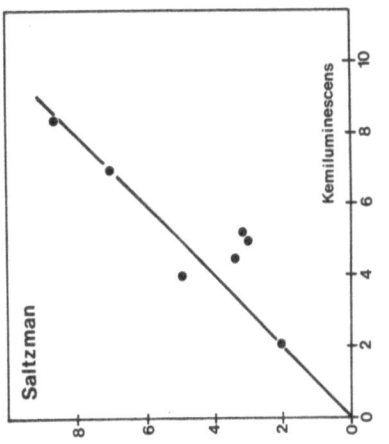

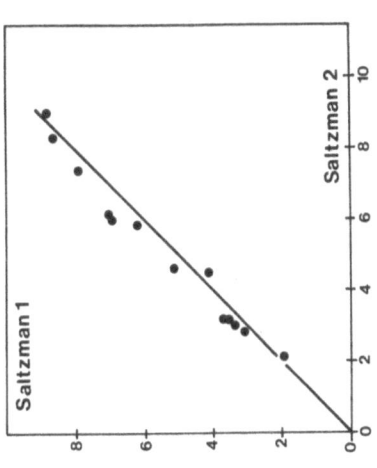

Figure 9. Reproducibility within the Saltzman method and comparison with chemiluminescent instrument for NO_2.

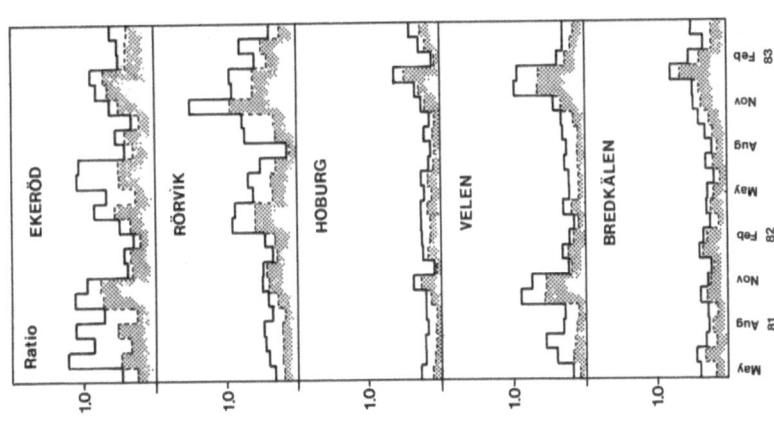

Figure 8. Ratios of nitrogen to sulphur. White fields represent the ratio $N-NO_2$ to $S-SO_2$, dotted fields the ratio $N-NO_2$ to the sum of $S-SO_2$ and $S-SO_4^=$.

NITROUS ACID IN POLLUTED AIR MASSES -

SOURCES AND FORMATION PATHWAYS

C. Kessler and U. Platt
Kernforschungsanlage Jülich GmbH
Institut für Chemie 3: Atmosphärische Chemie
Postfach 1913, D-5170 Jülich

ABSTRACT

Gaseous nitrous acid (HNO_2) has been observed in polluted air masses over Germany at concentrations up to 2.5 ppb comprising up to 5 % of total NO_x ($NO + NO_2$). Due to its role in smog chemistry it must be considered an important tropospheric nitrogen compound. The rates of HNO_2 formation observed after sunset cannot be explained on the basis of the published rate constants.

Simultaneous HNO_2, NO, and NO_2 measurements have been made in hot exhaust gas of gasoline and Diesel automobile engines and in the stack gas of a coal fired power station. They rule out these dominant NO_x emitters as significant HNO_2 sources. Field measurements along a frequently used highway gave further evidence that direct emission by automobiles is a less important HNO_2 source. However, HNO_2 is readily formed in automobile exhaust gas diluted by air. At NO_x concentrations of 10 to 160 ppm the HNO_2 concentration reached an average equilibrium value of 4.5 % (of NO_x) within 1 hour.

In the atmosphere as well as in laboratory investigations on the ppm scale the formation was found to follow first order overall kinetics with respect to NO_2.

INTRODUCTION

Since its first unambiguous detection in ambient air [Perner and Platt, 1979], considerable effort has been undertaken in the observation of gaseous nitrous acid. It was found at concentrations up to 2.5 ppb and production rates up to 0.4 ppb/h were determined [Kessler, 1984]. Thus it has to be considered as an important nitrogen species commonly present in NO_x - polluted air masses. The significance of nitrous acid is based upon three properties:

1. The photolysis of HNO_2 (1) may serve as a strong source

$$HNO_2 + h\nu \rightarrow NO + OH \qquad (1)$$

of hydroxyl radicals.

2. The formation of HNO_2 may remove NO_x from the atmosphere.

3. HNO_2 may react with amines to give cancerogeneous nitrosamines.

Its origin, however, is still not completely clear. The present paper describes a number of experiments performed to elucidate the formation mechanisms of nitrous acid in ambient air, and possible direct emission of the species by various sources.

EXPERIMENTAL

Techniques:

Nitrous acid, nitrogen oxides, sulfur dioxide, and formaldehyde were determined by differential absorption spectroscopy in the near ultraviolet. This technique has been described in more detail elsewhere [Platt et al., 1979], therefore only a short summary is given here.

The trace gases are identified by their characteristic absorption bands, and their concentration can be determined from the strength of these bands according to Beer's law. In the spectral region between 350 nm and 370 nm, where the strongest bands of HNO_2 occur, NO_2 also absorbs. This fact allows simultaneous measurement of both gases in this spectral region. SO_2 was measured around 300 nm.

Nitric oxide can be sensitively determined by the same technique. Suitable absorptions are the strong and narrow features of the $A\,^2\Sigma - X\,^2\Pi$ (0,0) transition at 226 nm (γ-band). Due to the strong atmospheric absorption at this wavelength, in the atmosphere only short path lengths of few hundred meters are possible. Therefore and because of easier handling atmospheric NO was usually measured by chemiluminescence (models 14A by Thermo Electron, or model PW 9762/02 by Philips), whereas the UV absorption technique was used for laboratory measurements with high NO and consequently short absorption paths.

The probing light beam for the spectroscopic measurements was usually produced by a Xe high pressure lamp (XBO 450 by Osram) equipped with a spherical mirror (f=250, 300 mm diameter). For laboratory investigations occasionally different types (XBO 75, XBO 150) and smaller mirrors were sufficient.

The $HNO_2 - NO_x$ - system was investigated in different types of experiments (Table 1) covering several orders of magnitude in NO_x [Kessler, 1984]. The path length was adapted to the respective range of concentrations.

Table 1

Location	NO_x ppm	path length m	HNO_2 detection limit, ppm
Atmosphere	0.02	700–3500	0.00003 – 0.00015
Highway	0.2	1000	0.0005
Smog chamber	2 – 50	6	0.04 – 0.06
Diluted exhaust gas	8 – 160	9	0.05
Undiluted exhaust gas	3000	0.15–1.5	0.3

Description of The Experiments:

- Ambient air observations at Deuselbach/Hunsrück, at Jülich and at Cologne. In order to obtain a widespread set of data, nitrous acid was measured for a period of 4 years under different meteorological conditions (temperature, relative humidity, precipitation, visibility range) at various locations (rural (< 10 ppb NO_x), moderately polluted (10 - 20 ppb), polluted (30 ppb)). The light source was placed at suitable heights above the ground (meteorological tower, prominent buildings) so that the probing beam ran through the air unaffected by obstacles. The average distance between the beam and the surface was 5 to 50 m, depending on the measuring site. In view of the low concentrations, path lengths of 700 m to 3.5 km were used. Thus sensitivities of 50 ppt HNO_2 and 300 ppt NO_2, respectively, could be achieved (3.5 km light path, 20 min temporal resolution).

The following atmospheric parameters were monitored: Temperature, relative humidity, and precipitation, total aerosol surface (nephelometer), and occasionally ozone (Dasibi).

- Measurements at a highway (Bundesautobahn A4, Cologne - Aachen). The light source for the spectroscopic measurement of HNO_2, NO_2, SO_2, and CH_2O was positioned on a highway bridge. The visible light of the probing beam was blocked above 380 nm by a UG 5 filter in order to avoid blinding of automobile drivers. The beam ran 6 m above the ground at a distance of 20 m downwind along the highway. The analytical arrangement was set up in a van on another bridge 1 km apart. At that place nitric oxide was monitored by chemiluminescence (model 14A by Thermo Electron), and additional measurements of temperature, relative humidity, wind direction, wind speed, and traffic frequency were made. The detection limits were

.5 ppb HNO_2, 1.5 ppb NO_2, 2 ppb NO, and 1 ppb SO_2, respectively.
Both lamp and analytical operations were fed by two Diesel generators standing 50 m
apart downwind. A possible contribution of NO_x from these sources proved to be
negligible.

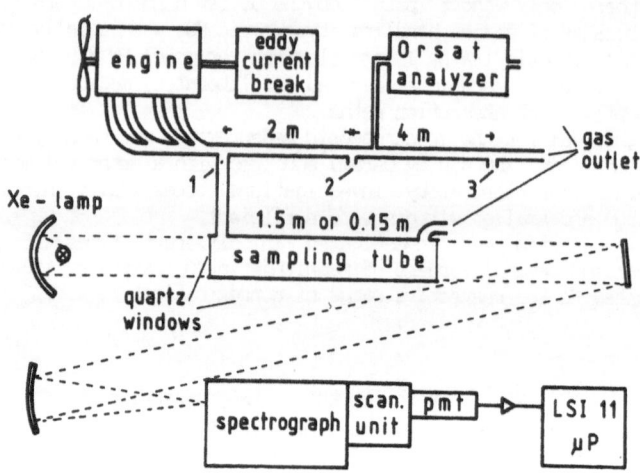

Figure 1. Experimental arrangement for the spectroscopic automobile engine
emission measurements. The sample gas was withdrawn at position 1. A short
sampling tube (15 cm) was only used for NO measurement.

- Emission measurements in the exhaust gas of automobile engines [Kessler et
al., 1984]. Measurements of NO, NO_2, and HNO_2 were performed at two stationary
engines, a Diesel engine, type OM 621.III manufactured by Daimler Benz AG
(1988 cm^3, 40 kw), and a spark-ignition (Otto) engine, type 80S YP by Audi NSU AG
(1588 cm^3, 63 kw). The effective power was absorbed and measured by an
eddy-current brake. The power as well as the number of revolutions per minute
(rpm) could be varied and thus certain defined states of operation could be
maintained during the time of measurement. Commercial fuel was used.
The hot undiluted exhaust gas was probed for several compounds by UV absorption
(Figure 1 and Table 1). For this purpose a small fraction (typically 50 l/min) of
the exhaust gas passed under its own pressure through aluminum tubes (100 mm inner
diameter) which carried quartz windows at both ends. One tube, 1500 mm long, was
used for the measurement of nitrous acid, nitrogen dioxide, sulfur dioxide,
formaldehyde, and methylnitrite in the region from 300 to 370 nm, while the main
fixed nitrogen compound, NO, was determined in a shorter tube (150 mm) at around
226 nm.
The sample gas was taken directly behind the engine (pos. 1 in Figure 1). To
prevent condensation the sampling tube, the connecting metal hose (1 m), and the
quartz windows were heated. Thus liquid phase reactions which might have altered
the results could be excluded. The gas was only admitted into the preheated cells
after the engine and the exhaust system had reached their operating temperature to
prevent condensed water to sputter against the cell windows during the warm up
phase of the engine.

In additon to the optical measurements, the NO mixing ratio in the Otto engine
exhaust gas was determined by chemiluminescence. For this purpose the sample gas

was taken at position 1 (see above) and diluted by a factor of 150 to 4,000 to fit the range of sensitivity of a commercial NO/NO$_x$ analyzer (model PW 9762/02 by Philips). Nitrogen was applied for dilution to prevent NO oxidation. Carbon monoxide (CO), carbon dioxide (CO$_2$), and oxygen were occasionally measured at the tailpipe by an Orsat gas analyzer. The fuel consumption was measured, or in some cases tabulated values were used. The air consumption was only measured for the Diesel engine. On the occasions when CO and O$_2$ were measured the air consumption was also determined via stoichiometry from the fuel consumption.

- Investigations of diluted automobile exhaust gas. A polyethylene bag of 35 m^3 (9 m long, 2.5 m diameter) was filled with laboratory air, and different volumes of exhaust gas from the gasoline engine (2,000 ppm NO) were added. The probing beam from a Xe high pressure lamp running axially through the bag was used for HNO$_2$ and NO$_2$ measurements. NO was measured with a commercial chemiluminescence device. For this purpose a negligible flux of air was withdrawn from the center of the bag and diluted with laboratory air. Additionally temperature and relative humidity in the bag were monitored.

- Emission from coal combustion. The stack gas of a coal fired heating plant was probed for HNO$_2$, NO and NO$_2$ [Kessler, 1984]. The same set-up was used as for the automobile engine measurements, and the sample gas was taken in a similar fashion.

- Smog chamber studies of NO$_x$ - synthetic air - mixtures. These experiments [Kessler et al., 1984] were performed in a 520 l glass chamber (surface to volume ratio 7.6 m^{-1}) designed for the study of the photodegradation of hydrocarbons. The experiments were performed in purified air and in oxygen, respectively. The gas mixture contained 2 to 80 ppm NO$_x$ with an initial NO/NO$_x$ ratio from 0.001 to 3, and non-methane hydrocarbons (NMHC) from 0.4 to 4 ppm.
The beam was folded to pass the chamber several times to increase sensitivity. The detection limit was 40 to 60 ppb HNO$_2$. The probing light was only applied for short periods of time to minimize HNO$_2$ photolysis. Nitric oxide and NO$_x$ were determined by chemiluminescence (PW 9762 by Philips). For this purpose the gas withdrawn from the chamber was diluted with air by a factor of typically 20. Measurements were only done occasionally to avoid diluton of the chamber content.
In 31 runs the relative humidity ranged from 30 % to 90 %, the chamber temperature was maintained at 300 ± 1 K.

RESULTS
Atmospheric Measurements.
    Whereas at rural continental and marine [Perner and Platt, 1979] sites nitrous acid was always found below the detection limit, it could regularly be observed in polluted air. Because of the rapid photolysis by sunlight, considerable concentrations, however, were only observed in the dark.
The time profile of HNO$_2$ was usually characterized by a steady increase after sunset, a period of stationary concentration for the second half of the nigth, and a fall-off due to photolysis after sunrise (Figure 2). From these plots the respective HNO$_2$ production rate was obtained, and with the assumption, that the stationary state was established by a chemical equilibrium, the destruction rate could be determined. In many nights, however, this ideal behaviour was modified by varying conditions (changing wind direction, changing total NO$_x$). In those cases transport could be neglected in the determination of the production rate which is a short time event, but could not be neglected for the stationary state and thus the destruction rate could not be calculated.
In Figure 3 the HNO$_2$ production rates as observed after sunset are plotted against the NO$_2$ concentration. It can be seen that the HNO$_2$ production rate depends approx. linear upon the NO$_2$ concentration. The slope corresponds to a NO$_2$ to HNO$_2$ conversion of 0.6 % per hour, with a correlation coefficient r = 0.73.

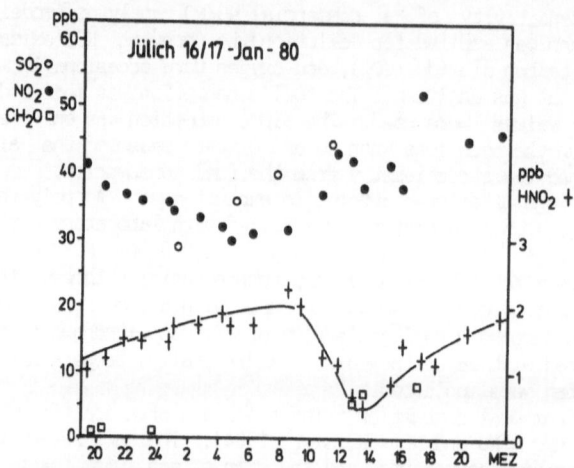

Figure 2. Temporal behaviour of the HNO₂ concentration in the atmosphere. Whereas in this case NO_2 and SO_2 remain rather constant, in the morning hours CH_2O increased, possibly indicating photochemical activity following HNO_2 photolysis.

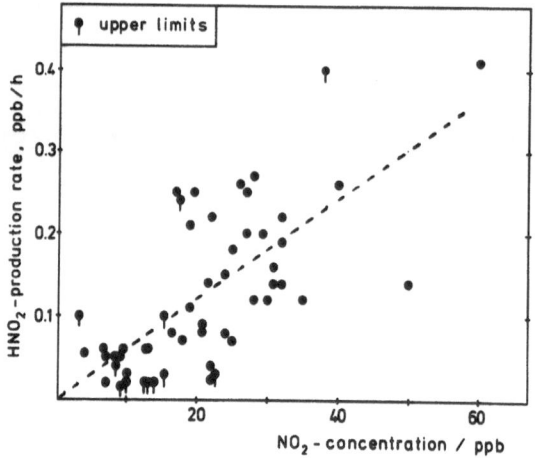

Figure 3. HNO_2 production rates observed at Jülich and at Cologne after sunset as a function of NO_2. The slope of the dashed line corresponds to a conversion rate of 0.6 % h⁻¹.

The dependence of the HNO_2 production rate upon the NO_2 concentration exceeded the correlation with other parameters by far. Especially the NO concentration did not strongly affect the formation of HNO_2, and even in the presence of ozone, which (at night) means complete absence of NO, considerable rates of HNO_2 increase were observed.
The steady state HNO_2 concentrations observed before sunrise are plotted in Figure 4 as a function of NO_2. Apparently low HNO_2 concentrations were found in

cases of precipitation (full circles in Figure 4). Disregarding these values the slope corresponds to a HNO_2/NO_2 of 3 %. From the production rate and the steady state concentration the average atmospheric life time of nitrous acid can be calculated as 5 hours. This life time appears to be independent of NO_2.

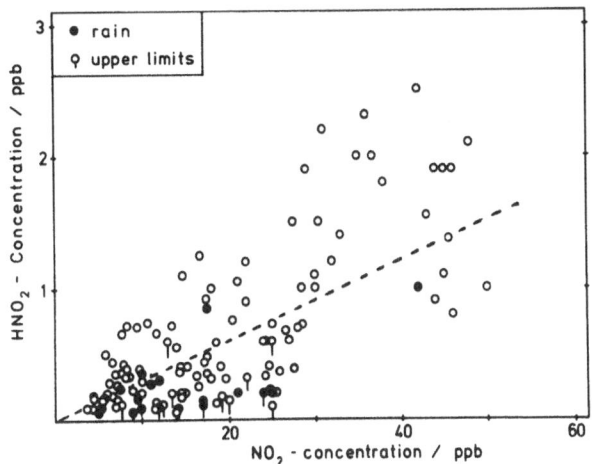

Figure 4. HNO_2 night time maximum concentrations observed before sunrise. Low concentrations were found in cases of precipitation (full circles). The slope of the dashed line corresponds to a HNO_2/NO_2 ratio of 3 %.

Measurements at various heights above the ground (5m, 25m, and 50m, respectively) gave no significant differences with respect to the relative production rate nor the steady state concentration.
Nitrous acid in ambient air could be due to two source mechanisms: Either HNO_2 is emitted directly into the atmosphere (presumably as by-product of NO_x sources) or it is formed in the atmosphere by chemical reactions. For two major NO_x sources – automobile traffic and coal combustion – the first possibility, i.e. the emission of HNO_2, was investigated.

Emission Measurements.
    At high load, about 3,000 ppm NO_x were found in the exhaust gas of the gasoline (Otto) engine, corresponding to 20 g fixed nitrogen per kg fuel combusted, but HNO_2 did not exceed 0.01 % (by volume) of the emitted NO_x. Under lean operation the percentage of HNO_2/NO_x increased and was then 0.15 %. The composition of the exhaust gas was found not to depend on the tail pipe position (Figure 1) where the sample was taken.
With the Diesel engine, only 6 g N were emitted per kg fuel, but in this case HNO_2/NO_x was about 1 %.
From the fractions of Diesel and gasoline fuel consumed (FRG, 1981; Diesel:gasoline = 2:3) and the measured HNO_2 emission rates an average traffic HNO_2 emission rate of 0.15 % (of NO_x) can be estimated. The results of those direct emission measurements were validated by field measurements of HNO_2 and NO_x downwind a frequently used highway. With NO concentrations up to 800 ppb and up to 50 ppb NO_2, the average HNO_2 concentration was 0.5 % of the total NO_x. Figure 5 gives the diurnal variation of the concentrations of NO, NO_2, and HNO_2 as measured beside the highway on Aug-2-83. From the observed fractions of trucks (Diesel engines) and passenger cars (90 % gasoline, 10 % Diesel engines) and the measured specific HNO_2

emission rates (see above) a HNO_2/NO_x ratio of 0.2 % was expected.

From the stack gas measurements a HNO_2/NO_x below 0.06 % was obtained. Thus coal combustion seems to be an even less important HNO_2 source.

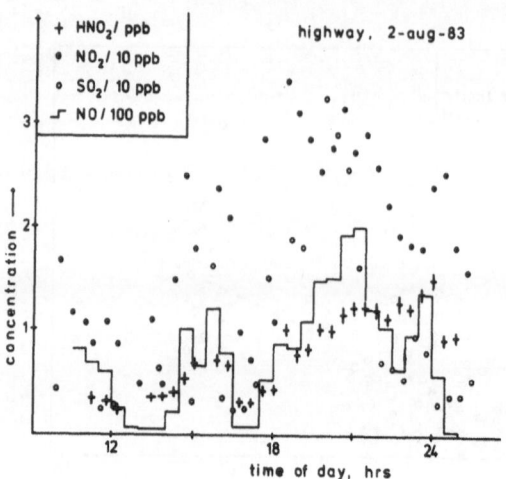

Figure 5. HNO_2, NO_x, and SO_2 diurnal profiles beside a highway. The large variance of the concentrations was caused by changing wind directions, occasionally background air was observed.

The HNO_2/NO_x ratios for the various sources are given in Table 2. Because in the atmosphere most of the emitted NO is oxidized to NO_2 these values may be understood as HNO_2/NO_2 and compared to the average HNO_2/NO_2 of 3 % observed in ambient air.

Table 2. HNO_2 percentage of some NO_x sources.

Source	HNO_2/NO_x %	remarks
Diesel engine	1.	
Gasoline engine	< 0.01	rich operation
	0.15	lean operation
Automobile traffic	0.15	estimated from emission rates and fuel consumption
Highway	0.5	observed
	0.2	expected from emission rates and traffic frequency
Coal combustion	< 0.06	
for comparison:		
Atmosphere	3.	average peak HNO_2/NO_2 before sunrise

Thus the emission measurements suggest, that the high nighttime HNO_2/NO_x ratios observed in ambient air (up to 5 %) can only be achieved by formation in the atmosphere.

Laboratory Studies

In the smog chamber the NO_x – air mixtures were investigated, and for time periods of about 10 hours the formation of nitrous acid was followed. The overall kinetics were first order with respect to NO_2 and the rate of HNO_2 formation was 0.2 to 2 % (of NO_2) per hour, slightly depending on the relative humidity. This rate could be proportionally increased when the chamber surface was enlarged. Thus it could be concluded that in this case the formation of HNO_2 proceeded at the chamber wall. However, beside its dependence on the NO_2 concentration, the humidity and the magnitude of surface, the mechanism of HNO_2 formation seemed to be subject to further parameters still unknown.

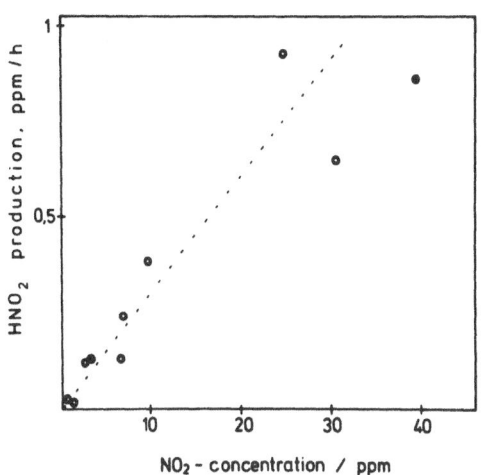

Figure 6. HNO_2 formation in diluted automobile exhaust gas as a function of NO_2. The slope of the dashed line corresponds to a conversion rate of 3 % h^{-1}. Two runs with increased relative humidity (full circles) gave similar results.

The engine exhaust gas which originally contained about 2000 ppm NO was diluted yielding mixtures of 8 to 160 ppm NO. Depending on the amount of exhaust gas, the resulting relative humidity was 42 to 62 %. These mixtures were observed for about 3 hours. The NO was rapidly oxidized at a rate expected from the three body reaction with molecular oxygen. Nitrous acid formation followed first order kinetics in NO_2. In Figure 6 the respective rates of HNO_2 formation are plotted vs the NO_2 concentration, the slope corresponds to a NO_2 to HNO_2 conversion of about 3 % per hour. In all cases HNO_2 reached a steady state and was then 3 % to 6 % of NO_2. At that time NO was below 10 % of total NO_x. In two runs (full circles in Figure 6) the humidity was increased by adding water vapour but this did not affect the relative HNO_2 formation rate nor the steady state concentration.

DISCUSSION

Several mechanisms have been proposed which could produce nitrous acid, namely the reactions of nitrogen oxides with gaseous or liquid water (2,3) and the reduction of nitric acid by nitric oxide (4).

$$NO + NO_2 + H_2O \rightarrow 2\ HNO_2 \qquad (2)$$
$$2\ NO_2 + H_2O \rightarrow HNO_2 + HNO_3 \qquad (3)$$
$$NO + HNO_3 \rightarrow NO_2 + HNO_2 \qquad (4)$$

However, with the published rate constants none of these reactions can explain the observed high concentrations of HNO_2. Moreover, in case of reactions (2) and (3) the HNO_2 production rate should be second order in NO_2, whereas first order was observed (see above).

Reactions including photochemical radicals such as OH (5) and HO_2 (6, followed by 5)

$$NO + OH \rightarrow HNO_2 \qquad (5)$$
$$NO + HO_2 \rightarrow NO_2 + OH \qquad (6)$$

might contribute to the formation of HNO_2 during the day. In the night these reactions (5,6) would rapidly deplete the radicals and thus would require strong radical sources. Recently it has been proposed [Stockwell and Calvert, 1983] that nitrate radicals mainly formed by (7) could react with formaldehyde (8) to produce such radicals (9) even in the dark.

$$NO_2 + O_3 \rightarrow NO_3 + O_2 \qquad (7)$$
$$NO_3 + CH_2O \rightarrow HNO_3 + HCO \qquad (8)$$
$$HCO + O_2 \rightarrow HO_2 + CO \qquad (9)$$

However, in the present case this mechanism could not really work because in many nights the presence of NO_3 or O_3, its precursor, could be excluded.

Another path of HNO_2 formation has been investigated by Lee and Schwartz [1983]: The reduction of NO_2 by bisulfite to give nitrite. They report the overall stoichiometry as (10)

$$2\ NO_2 + HSO_3^- \rightarrow 3\ H_3O^+ + 2\ NO_2^- + SO_4^=\qquad (10)$$

but the rate law has been observed to be first order with respect to NO_2, with a rate constant of $2 * 10^6 /[NO_2]\ M^{-1} sec^{-1}$. Despite the fairly good solubility and the high dissociation constant of SO_2, even under favourable conditions (10 ppb SO_2, 400 $\mu g\ m^{-3}$ liquid water content, pH = 5) in the atmosphere this mechanism (10) cannot effect NO_2 conversion rates of more than $5 * 10^{-3}$ % h^{-1}, whereas 0.6 % h^{-1} were observed. Thus for the measurements reported in this paper mechanism (10) proceeding in aerosol liquid water can be excluded as significant source of HNO_2. The correlation of the production rate d/dt HNO_2 with the quantity $[SO_2] * [NO_2]$ is poor (r = 0.15). But if the acidity were exclusively ruled by the SO_2 – HSO_3^- – system the HSO_3^- concentration would depend only on the square root of the gaseous SO_2 concentration and thus would the rate of reaction (10). Consequently the poor correlation does not disqualify that path (10).

On the other hand, in fog or in clouds considerable formation of HNO_2 could occur. For example, with a liquid water content of 1 g m^{-3} and the above conditions, mechanism (10) would give a NO_2 conversion of about 10 % h^{-1}. However, the spectroscopic technique used in the present study did not allow measurements under such conditions.

If that mechanism (10) should proceed at the earth's surface under the above conditions (10 ppb both NO_2 and SO_2, pH=5), a water layer of 2.5 μm were required to fill up a boundary layer of 50 m at the observed rate. In certain cases this condition might be fulfilled such that mechanism (10) proceeding at the earth surface were sufficient. However, observations of considerable HNO_2 production rates preferably under dry conditions seem to exclude mechanism (10).

Consequently either a different, still unknown mechanism must be active or one of the above mechanisms proceeds with an effective rate constant actually much larger under atmospheric conditions.

As in the atmosphere first order kinetics with respect to the NO_2 concentration were also found in laboratory experiments with much higher NO_x concentrations.

Thus despite the fact that the NO_x concentrations were different by 4 orders of magnitude the formation of HNO_2 was found to be quite similar in the atmosphere and in the two artificial systems. In all cases relative rates of NO_2 to HNO_2 conversion were comparable.

These observations suggest that the HNO_2 formation mechanism was possibly the same in the three investigated systems. Thus this mechanism might proceed at the respective surface in all cases as was observed in the case of the glass chamber experiments.

In the atmosphere two types of surface are available: the earth and airborne surface, i.e. aerosol particles. The latter was monitored and was found not to severely influence the HNO_2 formation, the correlation coefficient of (d/dt HNO_2)/NO_2 and the aerosol surface was only 0.35.

The influence of the earth's surface has not yet been investigated. In view of the comparably long HNO_2 life time (5 h), the observation that HNO_2 measurements at several heights above the ground gave similar results (see above) does not disqualify HNO_2 formation at the ground.

For comparison, the HNO_2 production rates and the surface / volume ratios of the respective system are given in Table 3.

Table 3. Nitrous acid production in the atmosphere and in two artificial systems.

	NO_x ppm	(d/dt HNO_2)/NO_2 % h^{-1}	surface/volume m^{-1}
Boundary layer (50 m)	0.004-0.06	0.2 - 1.5	0.02 (ground surface) 0.002 (aerosol surface)
Polyethylene bag	8. - 160.	2. - 3.5	2.
Glass chamber	2. - 50.	0.2 - 2.	7.6

CONCLUSION

The formation of nitrous acid was found to follow first order kinetics with respect to NO_2 in all cases investigated. Additionally the observed rate of NO_2 to HNO_2 conversion was quite similar in magnitude, the average conversion rates ranging from 0.6 to 3 % h^{-1}. Several suggested formation mechanisms including gas / liquid interactions were considered but with the published rate constants none of those could explain the observations. Only with sufficient liquid water available, e.g. in fog or in clouds, where unfortunately no measurements could be made, considerable rates of HNO_2 formation would be expected even on the basis of the published rate constants.

Laboratory measurements suggest that nitrous acid is formed by a yet unknown mechanism at moist surfaces. If in the atmosphere the same mechanism of HNO_2 formation is active it could preferably proceed at the moist ground surface rather than in the aerosol liquid water.

However, the present understanding of the observations appears unsatisfactory. In view of the prominent importance for the chemistry of the polluled atmosphere, the origin of nitrous acid requires further investigation.

REFERENCES

Kessler, C., D. Perner and U. Platt,
Spectroscopic Measurements of Nitrous Acid and Formaldehyde -
Implications for Urban Photochemistry, in: Physico-Chemical Behaviour
of Atmospheric Pollutants, B. Versino and H. Ott (ed.), 393-400, 1981

Kessler, C.,
Gasförmige salpetrige Säure (HNO_2) in der belasteten
Atmosphäre, Dissertation, Univ. Köln, 1984

Kessler, C., D. Perner, U. Platt, P. Bruckmann and P. Eynck,
Wall Reactions in Smog Chamber Studies, In preparation, 1984

Kessler, C., D. Perner, U. Platt and K.H. Ertl,
The Emission of Nitrous Acid and Nitrogen Oxides by Automobile
Engines - A Spectroscopic Study, In preparation, 1984

Lee, Y.N. and S.E. Schwartz,
Kinetics of Oxidation of Aqueous Sulfur(IV) by Nitrogen Dioxide,
in: Precipitation Scavenging, Dry Deposition, and Resuspension,
Vol. 1, H.R. Pruppacher, R.G. Semonin, W.G.N. Slinn (Ed.),
Elsevier, 453-470, 1983

Perner, D. and U. Platt,
Detection of Nitrous Acid in the Atmosphere by Differential
Optical Absorption, Geophys. Res. Lett., 6, 917-920, 1979

Platt, U., D. Perner and H. Paetz,
Simultaneous Measurement of Atmospheric CH_2O, O_3, and NO_2 by
Differential Optical Absorption, J. Geophys. Res., 84, 6329-6335, 1979

Stockwell, W.R. and J. Calvert,
The Mechanism of NO_3 and HNO_2 Formation in the Nighttime
Chemistry of the Urban Atmosphere,
J. Geophys. Res. C, 88, 6673-6682, 1983

STUDY OF THE CHEMICAL CHARACTERISTICS OF WET AND DRY DEPOSITION IN SWITZERLAND

J. FUHRER

Institute of Plant Physiology, University of Bern, Switzerland

Summary

Because of the growing concern about the deleterious effect acid deposition may have on forest ecosystems, a sampling program was started in central Switzerland to determine the inorganic composition of wet and dry deposition. The methods used, including a rain-activated wet/dry precipitation collector, are presented and discussed. Preliminary data from a rural sampling site near the city of Bern are reported and compared with results from sampling sites in W.Germany. Deposition rates for sulfate and nitrate contained in wet and dry fallout are compared with those for SO_2-S and NO_2-N as calculated from ambient concentrations.

1. INTRODUCTION

Large-scale die-back of forests in Europe has attracted great research and public interest in the last years. A clear cause-effect relationship, however, could not yet be established. There is evidence that the deposition of acidic components in wet and dry form can interact with forest ecosystem processes (1). This may lead to destabilization and reduced flexibility of forests (2). The evaluation of acidic deposition as a driving force in this process must be based upon the determination of the chemical species deposited on the forest and the rates at which such deposition occurs at a specific location.

Deposition processes to be considered are (i) wet deposition by rain and snow, and dry deposition by sedimentation of particles (= precipitation deposition), as well as impactation of aerosols and mist, fog or cloud droplets together with the absorption of gases on wet surfaces or inside the stomata of leaves and needles (= interception deposition). Because of the growing concern about the increasing loss of forest trees in Switzerland since 1982, a sampling program was started in 1983 to determine the chemical characteristics of wet and dry deposition in the central part of the country. In this paper, preliminary data obtained in 1983 from a rural station are presented and compared with information on gaseous pollutants (SO_2, NO_2) from the National Air Pollution Monitoring Network (NABEL), and on deposition data from sites in W.Germany.

2. MEASUREMENTS

2.1. Collection site:

The collection of wet and dry deposition (precipitation deposition) was carried out during 1983 at a rural sampling station south of the city of Bern at 540 m a.s.l. (46°70'N, 7°31'E) (Fig. 1). The site is surrounded by agricultural land. The collector was placed approximately 4 m above the ground in order to minimize contamination due to nearby agricultural activity.

2.2. Collection procedure:

A wet/dry precipitation collector (Aerochem Metrics Florida, Model 301) was used which exposes separate containers during periods of wet (rain, snow) and dry (sedimentation of particulates) deposition. To exclude dry from wet deposition, a rain-activated motor automatically moves a cover from the container receiving wet deposition to that receiving dry deposition at the beginning of each rainfall. The base-plate of the rain sensor (the latter consists of a 34 cm^2 stainless steel grid) is heated during the rainfall and dries off rapidly as the rainfall stops. The cover moves in the other direction with a lagtime of about 3 to 5 minutes. Both collection containers are formed from 0.2 cm thick, high density polyethylene. They have an open surface of 6.29 dm^2 and a capacity of 13 liters. The total annual amount of collected water was within 1% of the amount collected by a standard rain gauge. Wet and dry samples were collected on a weekly basis. Because of the location and the frequent rainfall during certain periods (up to 40-50 events per week), sampling on an event basis was not possible and would certainly not be feasible at more remote locations, for example in mountain areas. A study on the stability of the samples collected with this specific type of collector showed that there is no significant change in their inorganic composition during this sampling interval (3).

2.3. Sample analysis:

Conductivity and pH were measured in wet samples within 1 hour after they arrived at the laboratory. Dry samples were dissolved in 250 ml dist. water and analyzed as wet samples after an equilibration period of 24 hours (4). H^+-concentrations were calculated from the pH-measurements. Wet and dissolved dry samples were filtered through 0.45 μm filters (Sartorius Type SM) prior to further analysis or storage. Anions (sulfate, nitrate and chloride) were quantified by ion chromatography (Dionex Anion Separator). Cations (Ca^{2+}, Mg^{2+}, K^+ and Na^+) were quantified in a 20 ml-aliquot acidified with HNO_3 (1%) by atomic absorption spectrophotometry. Ammonium was determined with a specific electrode (WINTION 2060 D) fitted to a pNH_4^+-meter (WINTION 3D 911W). The determination of pNH_4^+-values was carried out simultaneously with the pH-measurement at 20°C. The residue of the samples was stored in the cold (2-4°C) or frozen.

2.4. Quality control:

The quality of the analytical data from each sample was tested by examining the balance between positive and negative ions. Samples with an ion imbalance of more than 10% of the total

number of ions present were not considered (5). The quality was
further tested by comparing the measured conductivity with the
conductivity calculated from the ion equivalents present in the
sample.

2.5. Additional data:
For the purpose of comparison, the 1978 deposition data from
the BAPMoN station "Jungfraujoch" (3500 m a.s.l., Fig. 1) are
included in this paper (6). A bulk-collector is used at this
site. Dry deposition should be very small at this altitude. The
data can thus be compared with wet-only data sets from sites
at lower altitude. Also included is a 1978 data set from the
station located at Payerne (460 m a.s.l., Fig. 1) (6). In order
to place the rates of wet and dry deposition of sulfate and
nitrate measured at Bern in relation to the occurrence of SO_2
and NO_2, respectively, in ambient air and their respective
dry deposition rates, annual average concentrations of these
gases were taken from rural stations of the National Air Pollution Monitoring Network (NABEL) (Fig. 1). Because the data
from the 1983 measurement period were not yet available, average values were calculated from the 1981 and 1982 data (7,8).

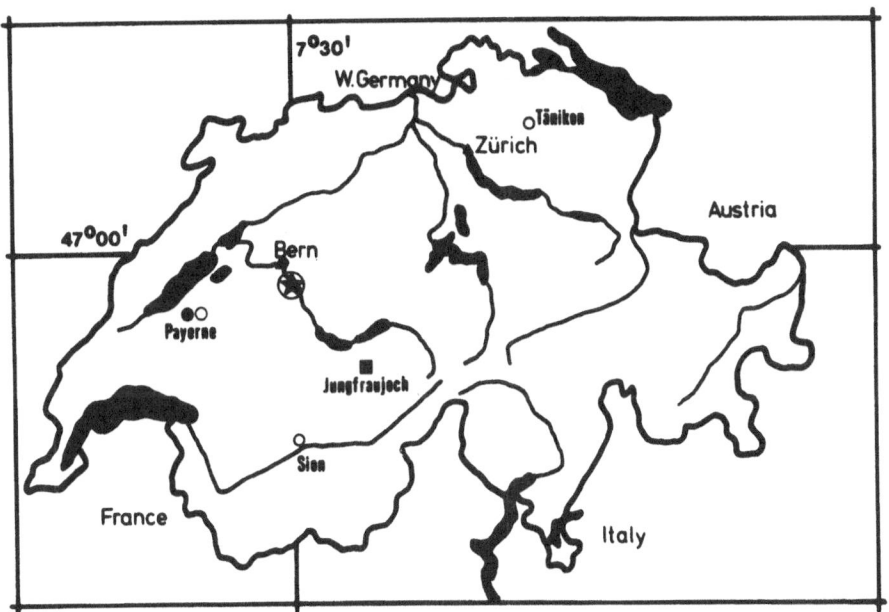

Fig. 1: Location of measuring sites in Switzerland.

⭐Bern, wet/dry deposition; ■ Jungfraujoch, bulk-
deposition; ● Payerne, wet-only deposition.

O Rural stations of the National Air Pollution
Monitoring Network (NABEL).

3. RESULTS AND DISCUSSION

3.1. The composition of wet-only samples:
Volume-weighted average concentrations of the major inorganic ions, their concentration ranges, the specific conductivity (measured and calculated) and precipitation amounts are given in Table I on a seasonal basis. Highest SO_4^{2-}-concentrations were observed during the summer months (June-August) when the precipitation amount was low. In spring (March-May), higher amounts of precipitation apparently reduced the average SO_4^{2-}-concentration. On the other hand, NO_3^--concentrations did not increase considerably from spring to summer. This different behaviour indicates the differences in source and sinks for the respective acid precursors (SO_2, NO_2) (9). Concentrations of H^+ were highest in the spring and decreased during summer and fall (September-November). High SO_4^{2-}-levels in the summer were thus most likely associated with NH_4^+.

Table I: Volume-wighted average concentrations of major inorganic ions, concentration ranges, specific conductivity and the amount of precipitation during the four seasons in 1983 at Bern. Ion concentrations are given in $\mu eq \ l^{-1}$, conductivity in $\mu S \ cm^{-1}$ and the amount of precipitation in $1 \ m^{-2}$.

	WINTER	SPRING	SUMMER	FALL
H^+	22.6	32.2	11.6	7.4
	8.3-177.8	1.0-63.1	0.2-35.5	2.0-39.8
Na^+	4.3	4.3	4.7	0.8
	0.4-92.3	0.1-49.8	0.1-15.2	0.1-14.8
K^+	1.2	1.4	2.2	0.8
	0.6-8.4	0.1-15.4	0.1-22.4	0.8-7.4
Mg^{2+}	2.5	1.9	2.3	0.3
	0.0-2.8	0.0-10.7	0.0-10.7	0.0-7.5
Ca^{2+}	10.9	17.4	23.5	6.0
	1.5-76.8	0.1-86.6	4.3-117.0	1.5-61.0
NH_4^+	20.4	41.6	59.6	23.8
	5.0-92.3	16.7-138.9	12.0-100.6	3.8-151.1
SO_4^{2-}	23.8	43.7	62.5	24.3
	15.5-161.6	12.3-109.4	28.8-131.9	14-262.9
NO_3^-	15.0	32.9	30.2	9.3
	7.2-166.4	16.1-100.2	17.1-52.3	3.2-175.7
Cl^-	14.2	9.0	8.0	5.6
	5.5-168.1	2.5-73.5	2.3-29.0	1.4-50.4
Sum of anions	53.0	85.6	100.7	39.2
Sum of cations	61.9	98.8	103.9	39.1
Cond. (measured)	14.6	22.2	22.9	10.8
Cond. (calculated)	12.8	19.7	15.8	6.9
Precipitation amount	145	297	152	303

To show the change in the relative importance of H^+ and NH_4^+, their contribution to the total cation composition of wet samples is plotted in Fig. 2 for the different seasons. The coefficient for the correlation between $(H^+ + NH_4^+)$ and $(SO_4^{2-} + NO_3^-)$ was 0.83 (significant at p=0.1%).

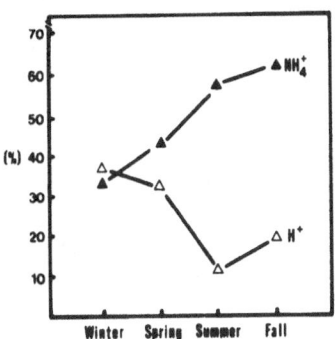

Fig. 2: Change in the relative contribution of H^+ and NH_4^+ to the total cation composition of wet samples at Bern.

3.2. Comparison of the ranges of pH values and SO_4^{2-}-S concentration determined at Bern and Deuselbach (W.Germany):
The range of pH values and of ion concentrations in wet samples and the difference between different stations can best be shown with a frequency distribution. In Fig. 3A the frequency distribution of the pH value in wet samples collected at Bern is presented together with that obtained from 1979 to 1981 in a relatively unpolluted area in W.Germany (Deuselbach) (10). In Fig. 3B a similar comparison is made for SO_4^{2-}-S concentrations observed at these two stations.

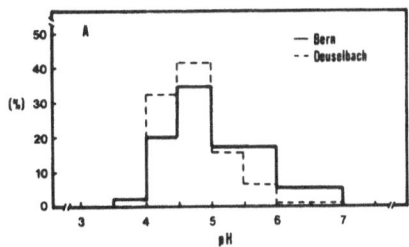 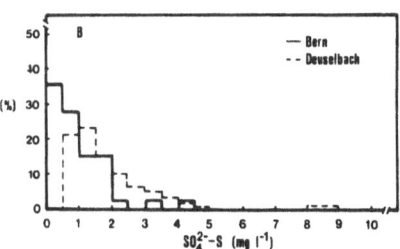

Fig. 3: Histograms for the relative frequency of pH values (A) and of SO_4^{2-}-S concentration (B) in wet-only samples collected at Bern (Switzerland) and Deuselbach (W.Germany). Data for Deuselbach are from Ref. 10.

The histogram for the relative frequency of pH values for Bern is clearly shifted to higher values, as compared to Deuselbach. At Bern, 50% of all values were below 4.8 and below 4.4 at Deuselbach. The SO_4^{2-}-concentration was significantly higher in rain collected at Deuselbach. 50% of all values were below 15 µeq l^{-1} and 41.7 µeq l^{-1} at Bern and Deuselbach, respectively. The relatively low sulfate content in rain as observed at Bern can be explained by the fact that the central part of Switzerland is distant from the centers of sulfur emission in Europe.

Nitrate on the other hand is formed from NO_2 faster than is sulfate from SO_2 (9). The NO_3^--concentration in rain is therefore influenced by sources located near the receptor site. In areas with heavy automobile traffic, such as the central part of Switzerland, NO_3^--concentrations are thus relatively high, as compared to SO_4^-. This is underlined by the comparison between the concentrations of sulfate and nitrate, as well as their ratio, in wet samples collected at three different stations (Table II).

Table II: Annual volume-weighted average concentrations of sulfate and nitrate (both in µeq l^{-1}), and their ratio in wet-only samples from Bern (Switzerland), Deuselbach (W.Germany, Ref. 10) and De Bilt (Netherlands, Ref. 11).

	BERN	DEUSELBACH	DE BILT
SO_4^{2-}	39	69	100
NO_3^-	22	29	48
NO_3^-/SO_4^{2-}	0.56	0.42	0.48

The $NO_3^-:SO_4^{2-}$ ratio in samples at Bern changed during the year with highest values (0.95) in April and lowest values (0.35) in September.

3.3. The influence of the altitude on ion concentrations: An influence of the altitude on the composition of wet samples is shown by the comparison between data from the BAPMoN stations "Jungfraujoch" and "Payerne", and those obtained at Bern. (Table III).

Table III: Annual volume-weighted average concentrations of ions in samples collected at 3500 m (Jungfraujoch, 1978) (6), 540 m (Bern, 1983) and 460 m (Payerne, 1978) (6). Concentrations are given in µeq l^{-1}.

	pH	H^+	Na^+	K^+	Ca^{2+}	Mg^{2+}	NH_4^+	NO_3^-	SO_4^{2-}	Cl^-
Jungfraujoch[1]	5.4	4	10	5	28	5	11	9	16	17

Table III, Continuation:

	pH	H^+	Na^+	K^+	Ca^{2+}	Mg^{2+}	NH_4^+	NO_3^-	SO_4^{2-}	Cl^-
Bern[2]	4.7	18	4	2	15	2	36	22	39	9
Payerne[2]	4.8	15	12	5	31	15	46	27	45	48

1) Bulk-samples, 2) Wet-only samples

The data in Table III show decreased levels of SO_4^{2-}, NO_3^-, NH_4^+ and H^+ at the highest location (3500 m). The concentration of these ions are quite similar in samples collected in 1978 at Payerne and 1983 at Bern, both being rural stations. No significant influence of altitude was found on concentrations of Na^+, K^+, Ca^{2+} and Mg^{2+}. Further studies on the chemical composition of wet samples at intermediate altitudes (around 1000 m) are in progress.

3.4. Rates of wet and dry deposition:
The rates of wet and dry deposition of the various ions analyzed are presented in Fig. 4 on a monthly basis. Only K^+, Ca^{2+} and Mg^{2+} showed a higher rate of dry deposition during the summer than during the winter. The rate of dry deposition of all other ions were low during the summer, relative to the rate of wet deposition. Highest total deposition rate (wet + dry) of SO_4^{2-}-S occurred during periods with highest rainfall amounts. This is consistent with the observation that the regional distribution of sulfate wet deposition is determined by the distribution of the annual rainfall amount (10). Maximum total deposition (wet + dry) for SO_4^{2-}-S was 3.20 mg m^{-2} day^{-1} (May), and the annual average rate was 2.27 mg m^{-2} day^{-1}. This rate was significantly lower than the rate determined from 1979 to 1981 at Deuselbach (W.Germany) (8.73 mg m^{-2} day^{-1}) and at the mountain station Schauinsland (W.Germany) (6.02 mg m^{-2} day^{-1}) (10). Clearly, the ratio between wet and dry deposition rates of SO_4^{2-}-S dependet on the time of the year. Dry deposition of NO_3^--N was generally low, as compared to wet deposition. A similar observation was also made in W.Germany (10). Wet deposition as the predominant sink for nitrate did not depend only on the rainfall amount. The processes involved in nitrate formation and its incorporation into the precipitation elements seemed to be of importance as well. Highest total nitrate deposition (wet + dry) occurred in April (1.78 mg NO_3^--N m^{-2} day^{-1}), and the annual average rate was 0.93 mg m^{-2} day^{-1} (dry deposition: 0.26 mg m^{-2} day^{-1}; wet deposition: 0.67 mg m^{-2} day^{-1}). This rate was lower than the rate determined at Deuselbach (1.57 mg m^{-2} day^{-1}) (10). Between 1957 and 1961, an average NO_3^--N deposition (bulk) of only 0.27 mg m^{-2} day^{-1} was measured in the area of Bern (12). This difference underlines the dramatic increase in NO_3^--N deposition which was also reported earlier for other European regions and the USA (13).

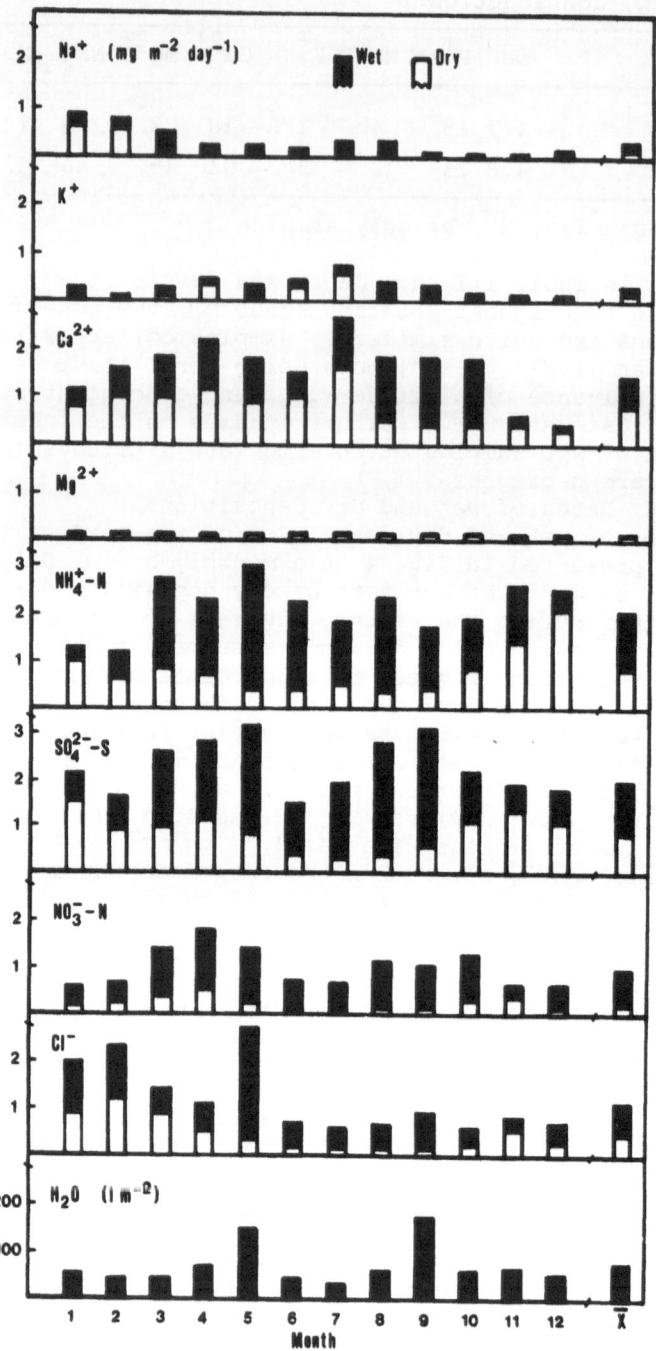

<u>Fig. 4</u>: Combined wet and dry deposition rates for each month in 1983 at Bern. Deposition rates are given in mg m^{-2} day^{-1}, the precipitation amount in l m^{-2}.

3.5. Comparison between the rates of different S- and N-fluxes to·the ground:

The rates of three different element fluxes to the ground for sulfur and nitrogen, as calculated for three different locations, are presented in Fig. 5. It was assumed that the average deposition velocity over ground covered with vegetation for SO_2-S and NO_2-N are 0.7 and 0.5 cm sec^{-1}, respectively.

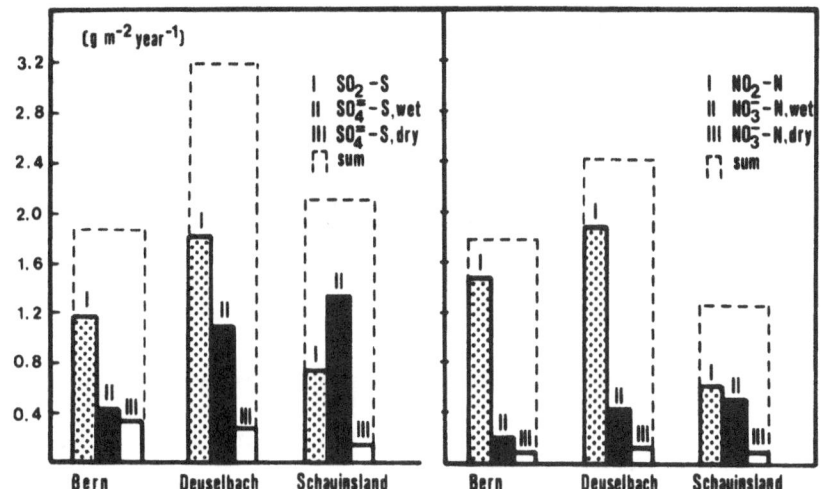

<u>Fig. 5:</u> Comparison between estimated rates of SO_2-S and NO_2-N deposition and measured rates of wet and dry SO_4^{2-}-S and NO_3-N deposition, respectively, at three locations. Annual average concentrations for SO_2 and NO_2 were taken from Refs. 7 and 8 for rural Switzerland, and from Ref. 14 for the two German locations. All rates are given in g m^{-2} $year^{-1}$.

At Bern and Deuselbach, dry deposition of SO_2-S is the major sink for sulfur. Wet deposition becomes more important at elevated locations, such as "Schauinsland" in the Black Forest (1284 m a.s.l.) where precipitation amounts are high and the ambient concentration of SO_2 is low. The total of the three S-fluxes was about 40% less in central Switzerland than at Deuselbach.

Dry deposition of NO_2-N exceeds both wet and dry deposition of NO_3-N by far, at Bern and Deuselbach, and still at the remote mountain station "Schauinsland". With respect to Bern, the data in Fig. 5 show that the rates of total sulfur and nitrogen deposition are about equal. This is not the case in W.Germany, where S-deposition exceeds N-deposition at Deuselbach, which is located in an area with relatively low level air pollution, and even at "Schauinsland".

4. CONCLUSIONS

The major cations in rain collected at Bern in 1983 were NH_4^+ and H^+, and the major anions were SO_4^{2-} and NO_3^-. Wet deposition was the major sink of most ions analyzed. The average concentrations of sulfate and nitrate were lower by 44% and 24%, respectively, at Bern as compared to Deuselbach (non-industrialized area in W.Germany), and so was the total deposition of sulfur and nitrogen. Highest rates of deposition in wet and dry form was measured in spring and early fall for SO_4^{2-}-S, and for NO_3^--N in spring. During summer and fall, nitrogen seemed to be deposited predominantly as NO_2-N. Averaged over the year, wet plus dry deposition of NO_3^--N was low, as compared to calculated rates of NO_2-N flux to the ground. At the German stations Deuselbach and Schauinsland, the total S-deposition clearly exceeds the total N-deposition, while at Bern, the respective rates are almost equal. This difference between the situation in Switzerland and in Germany reflects the spatial variability of the importance of the different pollutants with respect to their ability to acidify the environment.

5. ACKNOWLEDGMENT

This work was supported by the Bundesamt für Bildung und Wissenschaft. I wish to thank U. Nyffeler and U. Feller for their valuable help with the chemical analysis, and Prof. K.H. Erismann for his constant support.

6. REFERENCES

(1) Fuhrer, J. and Fuhrer-Fries, C., 1982, Europ. J. Forest Pathol. 12, 377.
(2) Ulrich, B., 1983, Allg. Forst Zeitschr. 26/27, 670.
(3) Madsen, B.C., 1982, Atmosph. Environ. 10, 2515.
(4) Stensland, G. et al, 1983, NADP Quality Assurance Report. Central Analyt. Res. Lab., Illinois State Water Survey.
(5) Kallend, A.S. et al., 1983, Atmosph. Environ. 17, 127
(6) WMO Report, 1981, BAPMoN Data for 1978.
(7) Bundesamt für Umweltschutz, Bern, 1982, Luftbelastung 1981.
(8) Bundesamt für Umweltschutz, Bern, 1983, Luftbelastung 1982.
(9) Fuhrer, J., 1984, Experientia, in press.
(10) Georgii, H.W., Perseke, C. and Rohbock, E., 1982, Bericht Univ. Inst. f. Meteorologie und Geophysik, Frankfurt a.M.
(11) Ridder, T.B. and Frantzen, A.J., 1983, In: Acid Deposition, Proc. CEC Workshop, Berlin, Sept. 9, 1982 (S.Beilke and A.J. Elshout, Eds.), 123.
(12) Zuber, R., 1962, Eidg. Gesundheitsamt Bern.
(13) Brimblecomb, L. and Stedman, D.H, 1982, Nature 298, 460.
(14) Umweltbundesamt, 1983, Monatsberichte aus dem Messnetz, Nr. 2.

VERTICAL AND HORIZONTAL PROFILES OF HYDROGEN CHLORIDE IN THE MEDITERRANEAN REGION

B. VIERKORN-RUDOLPH, J. RUDOLPH, F.X. MEIXNER
Institut für Chemie 3: Atmosphärische Chemie der
Kernforschungsanlage Jülich GmbH, P.O.Box 1913, D-5170 Jülich, F.R.G.

K. BÄCHMANN, B. SCHWARZ
FB 8 (Anorganische Chemie und Kernchemie)
Technische Hochschule Darmstadt, Hochschulstr. 4, 6100 Darmstadt, F.R.G.

Summary

Vertical and horizontal profiles of gaseous hydrogen chloride have
been measured in the Mediterranean Region in early September 1981.
The samples were collected over the Mediterranean Sea and over northeastern Spain between 0.1 and 7 km altitude on December, the 6th and
8th 1981 and along the eastern coast of the Iberian Peninsula at an
altitude of 370 m on December, the 5th 1981. During the flight on
December, the 5th parallel to some of the HCl samples whole air samples were also collected and analyzed for CO, CO_2 and several light
hydrocarbons.

1. INTRODUCTION

One of the acidic components of air which is present in polluted as
well as unpolluted areas is gaseous hydrogen chloride. Despite the interest in measurements of hydrogen chloride the data base on tropospheric
HCl mixing ratios especially in unpolluted areas is rather limited. Sources
of gaseous hydrogen chloride are the formation of HCl by the reaction of
sea-salt aerosols with acidic components of the air like H_2SO_4 or HNO_3
(1) (2) or emissions from volcanoes (3). Further possible sources of HCl
are the combustion of coal (4) (5) (6) and emissions from incinerators (7).
Effective sinks for gaseous hydrogen chloride are in-cloud and below-cloud
scavenging and dry deposition (8). However the strength of the different
sources and sinks of gaseous hydrogen chloride in the troposphere are not
yet well known. The tropospheric life-time of HCl is estimated to be between a day and a week (9) (10) (11) (12).
    In 1981 Cicerone (8) published a review of measurements of gaseous
hydrogen chloride. For the marine atmosphere values between 0.5 to 1.5 ppb
are reported for total inorganic gaseous chlorine. Sometimes lower values
are found for continental areas. Ground-base spectroscopic measurements of
HCl from Farmer et al (13) are in the range of 1 to 2 ppb for continental
as well as coastal regions in contrast to values of 1 to 100 ppt from
Marché et al (14).
    It is not clear whether these large fluctuations in the HCl mixing
ratios are systematic and due to variations in location, seasons etc or
due to some other factors which influence the mixing ratios of hydrogen
chloride. Also the possibility of systematic experimental errors cannot be
excluded especially since different methods for sampling and analysis have
been used by the various authors. Since the altitudinal resolution of all
the existing HCl measurements is only limited the vertical distribution of
gaseous HCl is not really known.

2. EXPERIMENTAL

For sampling of HCl on board of airplanes a twelve-fold adsorptive air
sampler with a teflon inlet line was built. Hydrogen chloride was collected
on denuder-tubes wall-coated with porous silica. Gaseous hydrogen chloride
is separated from chloride on aerosols due to the different diffusion velocities of gaseous hydrogen chloride and aerosols. Thus only gaseous HCl is
collected. Details of the design of the adsorptive air sampler and of the
sampling procedure are described in (15) (16).

For the determination of HCl, derivatization with 7-oxabicyclo (4.1.0)-
heptane and gas chromatographic separation and detection of the reaction
product - 2-chlorocyclohexanol - was employed. For the gas chromatographic
determination both packed and capillary glass columns with highly polar stationary phases (Carbowax 20 M, WG 11) can be used in combination with flame
ionization or electrolytic conductivity detection. Details of the derivatization procedure in small volumes and the gas chromatographic separation are
described elsewhere (15) (16) (17). The detection limit (3 δ) of this method
corresponds to atmospheric HCl mixing ratio of 25 ppt. The error near the
detection limit is 10 ppt. For higher HCl mixing ratios (> 100 ppt) the precision of the measurements is better than 15 %.

The samples were collected on board of a Piper Navajo, a propeller-driv-
en airplane. For the vertical profiles 15 samples were collected between 0.1
and 7 km altitude. The sampling time was 30 minutes per sample. Usually two
samples were collected simultaneously in order to check the precision of the
measurements. For the flights an equilateral triangle (side length ∿ 150 km)
was used as flight route to prevent the airplane from crossing its own exhaust trail. As the end of one side of the triangle is reached the airplane
climbs to the next altitude level. The flight locations are shown in Fig. 1.
One profile ("Ibiza") was measured over the Mediterranean Sea south of

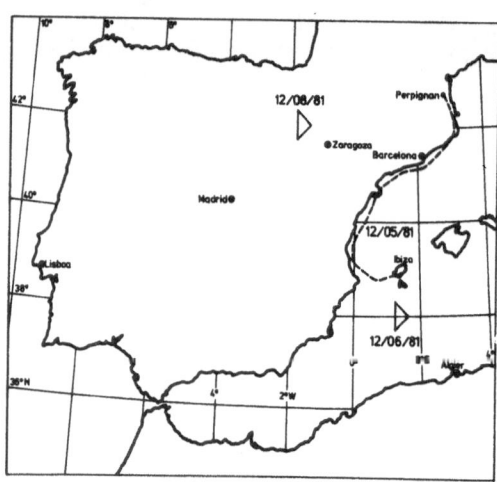

Fig. 1: Flight locations for the measurements of vertical and horizontal
profiles of gaseous hydrogen chloride.

Balearian Islands on December, the 6th 1981, the other profile ("Zaragoza") over northern Spain near Zaragoza on December, the 8th 1981. For the horizontal profile the samples were collected at an altitude of 370 m along the eastern coast of the Iberian Peninsula on December, the 5th 1981. In this case, the sampling time was only 15 minutes per sample in order to obtain a better spatial resolution. On December, the 5th parallel to some of the HCl samples whole air samples were also collected in evacuated 2 dm³ stainless steel containers and were analyzed for CO, CO_2 and several light hydrocarbons by gas chromatography. Details of the gas chromatographic analysis and the sample collection procedure are described in (18).

3. RESULTS AND DISCUSSION

The measured vertical profiles of gaseous hydrogen chloride are shown in Fig. 2. The profile on December, the 6th 1981 (Ibiza) shows a pronounced maximum of the HCl mixing ratios at approximately 1 km altitude in contrast

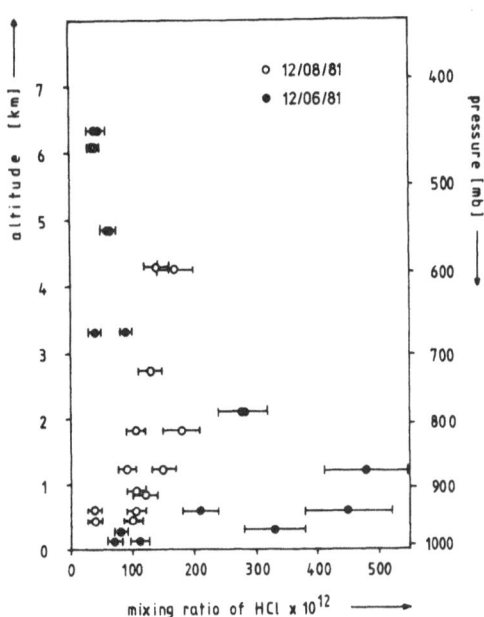

Fig. 2: Vertical profiles for the mixing ratios of gaseous HCl
● December, the 6th 1981 - profile Ibiza
flight location: Mediterranean Sea south of the Balearian Islands
O December, the 8th 1981 - profile Zaragoza
flight location: northern Spain near Zaragoza.

to the profile on December, the 8th 1981 (Zaragoza) with more uniformly distributed HCl mixing ratios over almost the whole altitude range. The mixing ratios of HCl at about 6 km altitude are approximately 40 ppt for each of the profiles.

The HCl-mixing ratios which have been measured along the spanish coast are shown in Fig. 3. The mixing ratios of two of the simultaneously measured light hydrocarbons are shown in Fig. 4. For a more detailed discussion of the measured data the meteorological conditions and horizontal and vertical transport have to be considered also. Therefore we calculated backward iso-

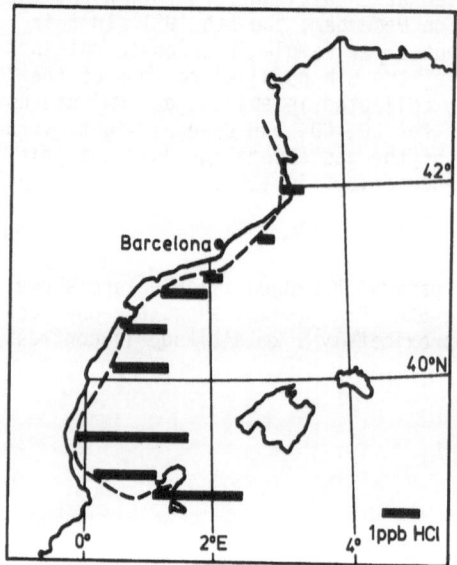

Fig. 3: Horizontal distribution of the mixing ratios of gaseous hydrogen chloride on December, the 5th 1981.

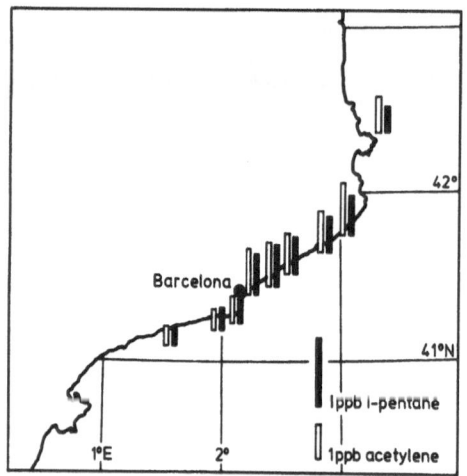

Fig. 4: Horizontal distribution of the mixing ratios of acetylene and i-pentane on December, the 5th 1981.

baric trajectories for the surface, 850 mb, 700 mb and 500 mb levels (Fig. 5) on the basis of the air sounding data and - over the Atlantic ocean where only few radio-sounding stations exist - mainly from the geostrophic wind approximation. The details of this calculation procedure aré given in (19).

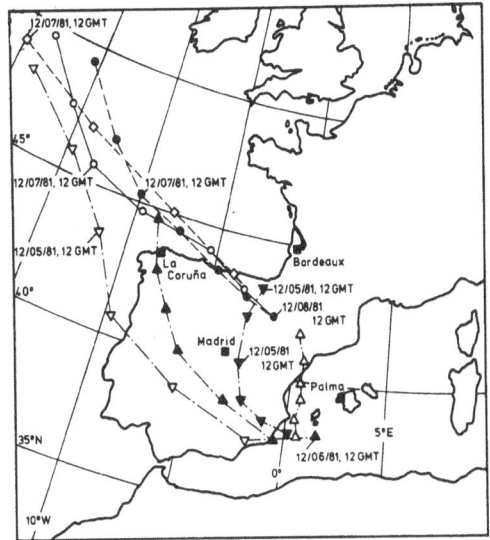

Fig. 5: Isobaric backward trajectories
for the profile Ibiza
△ surface   ▼ 850 mb
▲ 700 mb   ▽ 500 mb
for the profile Zaragoza
● 850 mb   ○ 700 mb
◇ 500 mb
time interval between two points: 6 hours

For the profile "Zaragoza" the air masses traveled from the north-western Atlantic ocean to the flight location and traverses continental areas for 6 to 12 hours (Fig. 5). Also for the profile "Ibiza" the air mass trajectories on the 500 mb- and 700 mb-level passed from the Atlantic ocean via the Iberian Peninsula to the flight location, whereas the air masses on the 850 mb-level originate from over continental areas (see Fig. 5). The corresponding surface level trajectory shows a rather slow transport from the spanish coast over the Mediterranean Sea to the flight location. For the profile "Ibiza" a temperature inversion at approximately 2 km altitude prevailed for the time of the passage of the air masses over the continent. Thus the continental influence is reduced above 3 km altitude due to this inversion layer and the HCl mixing ratios are rather low which can be expected for clean air masses from the Atlantic ocean. Below 3 km the mixing ratios are 5-10 times higher than above this altitude. Considering the course of the air mass trajectory on the 850 mb-level these higher HCl-mixing ratios indicate the existence of significant sources of either HCl itself or its precursors over continental spain.
    The HCl mixing ratios which have been measured along the spanish coast show an increase from 50 ppt to 130 ppt in the neighbourhood of Barcelona

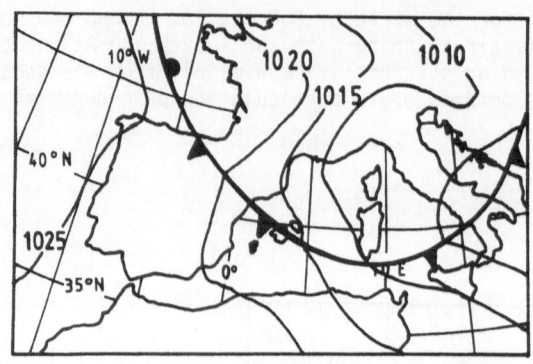

Fig. 6: Section of the synoptic surface analysis on December the 5th 1981, 12 GMT

(∿ 41 °N). At this area, the flight track crosses a cold front (see Fig. 6). Thus the observed stepwise change in the HCl mixing ratios seems reasonable. A similar change can also be expected for other trace gases. Indeed also the light hydrocarbons and CO exhibit a significant horizontal gradient over this area (see Fig. 3 and Tab. I).

Table I: Horizontal distribution of the mixing ratios of CO and some light hydrocarbons measured at a flight along the spanish coast on Dec. the 5th 1981

Coordinates	CO	C_2H	C_2H_4	mixing ratios [ppb] C_3H_8	$i-C_4H_{10}$	$n-C_5H_{12}$
42°15'N/3°20E	335	5.10	-	1.58	0.59	0.38
41°45'N/3°E	397	3.24	1.41	1.40	0.85	0.65
41°40'N/2°50'E	343	6.53	1.50	2.01	0.70	0.39
41°32'N/2°35'E	344	6.47	1.50	1.95	0.76	0.45
41°27'N/2°25'E	358	7.10	1.85	2.28	0.86	0.49
41°22'N/2°13'E	339	7.60	1.32	1.98	0.76	0.46
41°15'N/2°10'F	297	4.37	0.76	1.16	0.39	0.24
41°13'N/1°55'E	244	3.73	0.52	0.98	0.32	0.19
41°08'N/1°30'E	228	2.35	0.48	0.85	0.29	0.21

However the hydrocarbon mixing ratios decrease from north to south whereas the HCl mixing ratios increase. From the hydrocarbon and CO mixing ratios it seems evident that the level of the anthropogenic pollution is signifi-cantly higher in the air masses north of the cold front than south of it. This rather surprising observation of an anticorrelated change of the mixing ratios of HCl and the hydrocarbons may be an indication that the main sources of gaseous HCl, at least in this case, are not direct anthropogenic emissions of HCl. These measurements seem to support the suggestion that the release of HCl from sea salt-aerosols by means of acidic components like H_2SO_4 or HNO_3 is the major source of gaseous hydrogen chloride. For a more detailed discussion further informations about the mixing ratios of precursors of HCl such as gaseous HNO_3 and sodium, chloride and sulfate in aerosols are necessary.

4. CONCLUSION

The presented data give some idea of the HCl mixing ratios in the Mediterranean area on December 1981. Our measurements are far from being representative, however it is interesting to notice that air masses from the Atlantic ocean which are uninfluenced by continents show rather low mixing ratios of 50-100 ppt for gaseous hydrogen chloride. On the other side there seems to be significant sources of gaseous HCl or its precursors over continental spain. There is some evidence that - at least for one of the situations investigated in this work - the release of HCl from sea-salt aerosols by means of acidic components is the more important source of gaseous hydrogen chloride compared with direct anthropogenic emissions.

Acknowledgement

The HCl measurements have been made at the "Technische Hochschule Darmstadt" as part of a research programme which has been support by the "Bundesministerium für Forschung und Technologie" under grant No. FKW 26.

References

(1) Horvath, L., E. Meszaros, E. Antal, A. Simon, On the sulfate, chloride and sodium concentration in maritime air around the Asian continent, Tellus 33, 382-386 (1981)

(2) Martens, C.S., J.J. Wesolowski, R.C. Harris, R. Kaifer, Chlorine Loss from Puerto Rican and San Francisco Bay area marine aerosols, J. Geophys. Res. 78, 8778-8792 (1973)

(3) Cadle, R.D., A Comparison of Volcanic With Other Fluxes of Atmospheric Trace Gas Constituents, Rev. Geophys. and Space Phys. 18(4), 746-752 (1980)

(4) Smith, W.S. and C.W. Gruber, Atmospheric emissions from coal combustion: An inventory guide, Environ. Health Ser. 999-AP-24, 112 p., U.S. Public Health Serv., Cincinnati, Ohio (1966)

(5) Robinson, E. and R. Robbins, Emissions, Concentrations, and Fate of Gaseous Atmospheric Pollutants, 110 pp., SRI International, Menlo Park, Calif. (1968)

(6) Block, C., R. Dams, Inorganic composition of Belgium coals and coal ashes, Environ. Sci. Technol. 9, 146-150 (1975)

(7) E.P.A. No. AP-100, 1971, Hydrochloric Acid and Air Pollution: An Annotated Bibliography, Research Triangle Park, North Carolina

(8) Cicerone, R.J., Halogens in the Atmosphere
Rev. Geophys. and Space Phys. 19(1), 123-129 (1981)

(9) Kritz, M.A., J. Rancher, Circulation of Na, Cl and Br in the tropical
marine atmosphere,
J. Geophys. Res. 85, 1557-1560 (1980)

(10) Augustsson, T.R., J.S. Levine and D.A. Brewer, The chlorine budget of
the Troposphere, 2nd Symposium: "Composition of the Non-urban Troposphere" May 25-28, 1982, Williamsburg, Va. S. 79-84 (1982)

(11) Wofsy, S.C., M.B. McElroy, HO_x, NO_x and ClO_x: Their role in atmospheric
photochemistry,
Can. J. Chem. 52, 1582-91 (1974)

(12) Stedman, D.H., W.L. Chameides, R.J. Cicerone, The vertical distribution of soluble gases in the troposphere,
Geophys. Res. Lett. 2, 333-336 (1975)

(13) Farmer, C.B., O.F. Raper, R.H. Norton, Spectroscopic Detection and
Vertical Distribution of HCl in the Troposphere and Stratosphere,
Geophys. Res. Lett. 3(1), 13-16 (1976)

(14) Marché, P., A. Barbe, C. Secroun, J. Corr and P. Jouve, Ground based
spectroscopic measurements of HCl,
Geophys. Res. Lett. 7(11), 869-872 (1980)

(15) Vierkorn-Rudolph, B. and K. Bächmann, Determination of Trace Amounts
of Hydrogen Chloride in the Atmosphere by Collection in Sampling Tubes
and Subsequent Derivatization and Gas Chromatography,
J. Chromatogr. 217, 311-316 (1981)

(16) Bächmann, K., P. Matusca, B. Vierkorn-Rudolph and M. Kontos, Bestimmung von HCl im Sub-Nanogramm-Bereich in Luft,
Fresenius Z. Anal. Chem. 310, 89-97 (1982)

(17) Vierkorn-Rudolph, B., M. Savelsberg, and K. Bächmann, Determination of
Trace Amounts of Hydrogen Chloride by Derivatization with Epoxides and
Gas Chromatographic Separation,
J. Chromatogr. 186, 219-226 (1979)

(18) Rudolph, J., D.H. Ehhalt, A. Khedim, C. Jebsen, Determination of C_2-C_5-
hydrocarbons in the atmosphere at low parts per 10^9 to high parts per
10^{12} levels,
J. Chromatogr. 217, 301-310 (1981)

(19) Meixner, F.X., Die vertikale Verteilung des atmosphärischen Schwefel-
dioxids im Tropopausenbereich, Dissertation, J.W. Goethe-Universität,
Frankfurt/M., FRG (1981)

BILAN IONIQUE ET ACIDITE DE LA PRECIPITATION ANTARCTIQUE

M. LEGRAND, R.J. DELMAS and F. ZANOLINI
Laboratoire de Glaciologie et Géophysique de l'Environnement
B.P. 68   38402 St Martin d'Hères Cedex (France)

Summary

More than a thousand snow samples collected at several Antarctic locations have been analysed after melting by ion chromatography and acid base titrimetry. The ionic balance of major impurities present in Antarctic precipitation has been obtained for the first time. The concept of ionic budget is presented along with the way to calculate it and the various parameters which are involved are illustrated by results obtained at three selected Antarctic sites : South Pole, Dome C and D 57. The predominant role played by gas derived atmospheric acids (H_2SO_4, HNO_3 and HCl) is emphasized.

1. INTRODUCTION

L'Antarctique présente certaines spécificités intéressantes dans le domaine de l'Environnement. Situé aux hautes latitudes de l'hémisphère Sud et très isolé par rapport à l'Australie, l'Amérique du Sud et l'Afrique, ce vaste continent (14 millions de km^2) peut être considéré comme une île dont la quasi totalité de la surface serait recouverte d'une épaisse couche de glace (jusqu'à 4000 m dans les régions centrales). Sur les rares terres libres de glace, aucune végétation importante n'a pu se développer. Enfin la présence humaine y est récente et très circonscrite. De toute façon plus de 90 % des industries du globe sont localisées dans l'hémisphère Nord. On peut donc considérer le continent Antarctique comme encore vierge vis à vis des interactions toujours grandissantes entre l'Homme et son Environnement.

La situation géographique du continent Antarctique et les caractéristiques actuelles de la circulation atmosphérique sont telles que les seules contributions notables à l'aérosol atmosphérique de ces régions proviennent de la conversion gaz-particules et des embruns marins. De plus, lorsqu'on quitte les zones côtières en direction de l'intérieur, la composante marine s'affaiblit considérablement. Le plateau central Antarctique est donc un lieu privilégié pour étudier l'aérosol dit de "ligne de base". On atteint en quelque sorte dans ces régions le "bruit de fond des bruits de fond" de l'aérosol atmosphérique (Cunninghamn and Zoller, (1981) ; Delmas and Gravenhorst (1982)).

La glaciochimie est l'étude des informations chimiques dans les neige et glace polaires en relation avec la composition de l'atmosphère. Les couches de neige s'étant déposées au cours des millénaires ont archivé une histoire de la composition chimique de l'atmosphère de ces régions. Si l'on considère que cette dernière est en relation permanente avec l'atmosphère globale, les analyses chimiques réalisées sur les échantillons de glace fossile antarctique fournissent en principe des informations sur tout évènement ayant affecté l'atmosphère terrestre dans le passé en particulier les changements climatiques avec leurs conséquences sur la circulation atmosphérique et l'activité biologique terrestre ou marine ainsi que les grands cataclysmes naturels tels que les éruptions volcaniques ou les

chutes de météorites géantes. Dans notre laboratoire, plusieurs milliers
d'échantillons ont déjà été analysés en détail et des résultats nouveaux et
originaux ont été obtenus en ce qui concerne la composition chimique
présente et passée de l'aérosol secondaire Antarctique. A l'aide de
quelques exemples, nous exposerons la méthodologie suivie et présenterons
les informations que l'on peut en tirer pour mieux appréhender la
composition chimique de l'aérosol de bruit de fond global présent et passé.

2. METHODES EXPERIMENTALES

D'une façon quelque peu arbitraire mais qui correspond assez bien à la
réalité nous distinguerons des impuretés majeures et des impuretés
mineures, les majeures étant celles (SO_4^{--}, NO_3^-, Cl^-, NH_4^+, Mg^{++}, H^+) dont
les concentrations moyennes dans la neige polaire sont supérieures à 0,1
µEq. 1^{-1}, les mineures restant inférieures à ce chiffre et participant
donc peu au bilan ionique de la glace (budget ionique total en général
dans le domaine 5-10 µEq. 1^{-1}). Le cas des particules insolubles ne sera
pas abordé ici.

L'analyse des ions ou composés mineurs nécessite des précautions et des
méthodes particulièrement raffinées. La détermination des majeurs présente
elle aussi certaines difficultés et exige l'emploi de tout un protocole
spécial tant au niveau des prélèvements que de celui du travail de
laboratoire.

Les échantillons de neige récente (jusqu'à quelques dizaines d'années)
sont prélevés dans des puits à l'aide de godets en polycarbonate de 30 ml
(type "Accuvettes" vendues par Coulter) que l'on enfonce directement dans
la neige quand elle est suffisamment molle ou avec lesquels on racle la
paroi lorsque le névé devient trop dur. Les récipients doivent etre prénet-
toyés avant l'emploi. Les transports se font à l'état solide (donc à température constamment négative) et sous double gaine plastique soudée.

Les échantillons de névé profond ou de glace sont obtenus par
carottage. Les carottes sont transportées et conservées à Grenoble où
ensuite elles doivent être décontaminées par recarottage. Cette opération
s'effectue à - 15° C et en pièce sans poussières à l'aide d'un petit
carottier électrique ou d'une sonde thermique suivant qu'il s'agit de névé
ou de glace (Legrand et al., 1984).

Les concentrations des ions majeurs (à l'exception de celle du proton
titré selon une méthode développée au Laboratoire (Legrand et al., 1982) et
de Mg^{++} calculée à partir de Na^+)) sont mesurées par chromatographie
ionique (appareillage Dionex, modèle 10) qui est la technique de choix pour
la détermination des ions au niveau du nanogramme par gramme
(ng g^{-1} = 10^{-9} g g^{-1} = PPB). La quantité de liquide nécessaire pour
chaque détermination (anion, cation et acidité) est de 5 ml.

Les précisions sont de l'ordre de ± 10 % en moyenne sauf pour les
concentrations proches des limites de sensibilité de la méthode (cas fréquent
pour NH_4^+). Les prélèvements analysés proviennent des différens sites
suivants : Pôle Sud (90° S, altitude 2800 m), Dôme C (74° S, 124° E,
altitude 3240 m) et D 57 (75° S, 152° E, altitude 2053 m).

3.RESULTATS ET DISCUSSION

3.1. Généralités

Le budget ionique B_i d'une solution s'établit en faisant la somme des
concentrations de tous les ions dissous :

$$B_i = C_{anions} + C_{cations} \quad \text{(en équivalents par litre)}$$

Dans la précipitation antarctique fondue les valeurs de ce paramètre sont extrêmement basses par rapport à ce que l'on trouve dans les pluies des régions tempérées. La différence atteint un à deux ordres de grandeur en fonction de la charge en aérosols de l'atmosphère. Dans une étude précédente concernant le Pôle Sud (Delmas et al., 1982) nous avions montré que les ions présents dans la neige avaient pour origine d'une part des embruns marins, d'autre part l'aérosol secondaire provenant de la conversion de certains gaz traces atmosphériques. Si la contribution de la source marine est relativement simple à interpréter, celle concernant l'aérosol secondaire est beaucoup plus complexe et son étude mérite donc plus d'attention.

Le bilan ionique de la précipitation Antarctique n'avait jamais été établi en totalité. Nous l'avons réalisé pour la première fois au Pôle Sud en complétant les résultats publiés auparavant (Legrand and Delmas, sous presse). Les analyses portent sur une série de 100 échantillons prélevés en séquence continue dans un puits creusé en 1977 dans le secteur propre de la base Amundsen-Scott. Le taux des précipitations à cet endroit (env. 8 g. cm^{-2}. an^{-1}) est suffisamment important pour qu'une excellente datation stratigraphique (visuelle et isotopique, Jouzel et al.,1983) des couches de neige soit possible. Les échantillons analysés (entre 2 et 4 m de profondeur) représentent les années de 1959 à 1969. Cette période est intéressante car elle comprend les années 1963-66 au cours desquelles la chimie atmosphérique a été profondément marquée par l'éruption du volcan Agung (Mars 1963 à Bali).

Pour réaliser les calculs qui vont suivre, on fait une série d'hypothèses quant à la nature et à l'origine de chacun des composés ou ions présents dans la neige. Ces hypothèses sont basées sur un certain nombre de résultats antérieurs et on verra comment les résultats que nous exposons ici confirment la plupart des hypothèses de départ. Ces hypothèses étaient les suivantes : "Les ions majeurs proviennent de l'inclusion ou du dépôt dans la neige de sel de mer et d'un mélange des 3 acides atmosphériques H_2SO_4, HNO_3 et HCl partiellement neutralisés par NH_3. le sel de mer est responsable de la présence de la totalité des ions Na^+, K^+, Mg^{++} et d'une partie des ions $SO_4^=$ et Cl^-. On supposera que la neutralisation par NH_3 (ion NH_4^+) est très faible et que la seule source de NO_3^- est HNO_3. On appellera sulfate en excès ($SO_4^{=}{}_{exc.}$) le sulfate sous forme de H_2SO_4 et chlore en excès ($Cl^-{}_{exc.}$) le chlore sous forme de HCl. (La terminologie "en excès" est celle employée couramment dans la littérature pour qualifier le sulfate et le chlore ne provenant pas du sel de mer).

On peut écrire (concentrations exprimées en µEq. l^{-1}) :

$$[Na^+] + [K^+] + [Mg^{++}] + [NH_4^+] + [H^+] = [SO_4^=] + [NO_3^-] + [Cl^-] \quad (1)$$
$$\text{balance ionique}$$

$$[H^+] = [SO_4^=]_{exc.} + [NO_3^-] + [Cl^-] \quad (2)$$

$$[SO_4^=]_{exc} = [SO_4^=] - 0,12 \, [Na^+] \quad (3)$$

$$[Cl^-]_{exc.} = [Cl^-] - 1,17 \, [Na^+] \tag{4a}$$

$$[Na^+]_{exc.} = [Na^+] - 0,85 \, [Cl^-] \tag{4b}$$

Les valeurs de 0,12 et 1,17 apparaissant dans les équations (3) et (4a) sont celles des rapports $[SO_4^-] / [Na^+]$ et $[Cl^-] / [Na^+]$, respectivement, dans l'eau de mer (Stumm and Morgan, 1970) (0,85 est l'inverse de 1,17). $Na^+_{exc.}$, le sodium en excès, est le sodium non lié au chlore.

Les équations 4a et 4b correspondent à deux hypothèses relatives à la réaction (5) :

$$H_2SO_4 + 2NaCl_{(sel\ de\ mer)} \longrightarrow 2\ HCl + Na_2SO_4 \tag{5}$$

Il est en effet admis que cette réaction peut se passer dans l'atmosphère, l'acide sulfurique attaquant les particules de sel de mer avec libération de HCl gazeux. Nous avons discuté récemment les arguments montrant que cette réaction avait probablement lieu en Antarctique (Legrand et Delmas, 1983). On aurait alors dépôt dans certaines zones de l'Antarctique d'un excès de chlore sous forme de HCl (équation 4a), dans d'autres un excès de sodium sous forme de Na_2SO_4 (équation 4b). Ces deux éventualités semblent effectivement se produire en Antarctique, la première à Dôme C et Pôle Sud par exemple, la seconde à Vostok durant l'Holocène (De Angelis et al., sous presse).

3.2. Au Pôle Sud

Examinons maintenant les résultats des cent échantillons du Pôle Sud à l'aide des équations (1), (2), (3) et (4). La figure 1 représente les pourcentages de chacun des ions majeurs dans le bilan ionique moyen (8,54 µEq. 1^{-1}). Cette répartition moyenne, valable pour cette série de prélèvements, c'est-à-dire au Pôle Sud et pour la période de temps 1959-1969, est un peu différente de celle que nous avions calculée précédemment (Delmas et al., 1982) en raison surtout des erreurs qui existaient dans les données des composés azotés NO_3^- et NH_4^+ (Legrand et al., 1984).

La figure 1 illustre bien l'importance prépondérante des acides dans le bilan ionique de la neige du Pôle Sud. En effet les acides apportent près de 80 % des ions (39,4 x 2).

Le fait que les sommes des anions et des cations soient sensiblement égales est une bonne vérification que tous les ions majeurs ont bien été mesurés. La méthode chromatographique utilisée ici permet d'ailleurs d'identifier la plupart des anions et des cations entrant dans le bilan ionique, hormis le proton qui doit être déterminé à part. On peut simplifier la vérification de la balance ionique en ne considérant que l'équation (2). C'est ce que nous avons effectué sur chacun des cent échantillons du Pôle Sud Le graphique de la figure 2 visualise en détail le résultat de ce calcul. En effet la somme $[SO_4^-]_{exc} + [NO_3^-] + [Cl^-]_{exc.}$ c'est-à-dire l'acidité calculée $[H^+]_{calc}$ (moyenne glissante sur 3 échantillons) a été comparée à l'acidité mesurée $[H^+]$ représentée sur la figure 2 par les points expérimentaux des titrations.

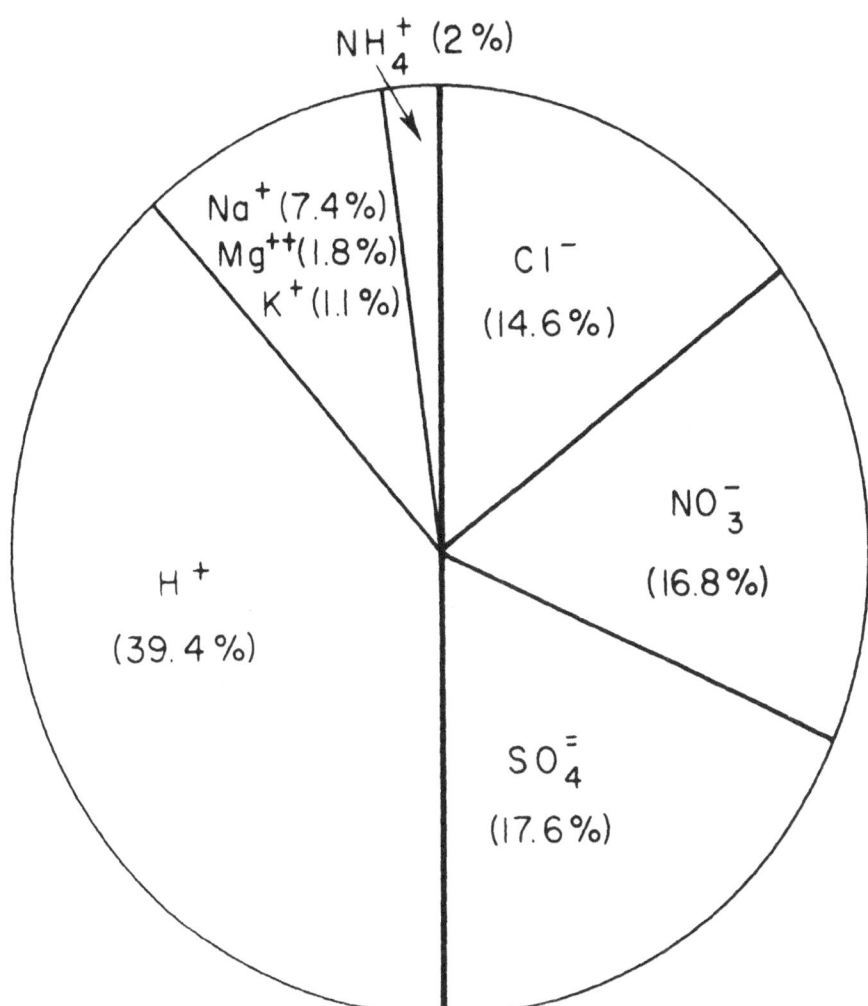

Figure 1 - Station Pôle Sud. Bilan ionique moyen de la neige pour la
période 1959-1969. Les pourcentages sont calculés surt les concentrations
en µEq. l⁻¹ .

Le bon accord entre $\left[H^{+}\right]_{calc}$ et $\left[H^{+}\right]$ confirme l'hypothèse de départ
des trois acides H_2SO_4, HNO_3 et HCl.
Au cours des deux années (1964 et 1965) pendant lesquelles l'influence
des retombées acides d'Agung s'est fait sentir, la part de H_2SO_4 a
considérablement augmenté, la concentration du sulfate total augmentant
d'un facteur 3 environ dans la neige (Legrand and Delmas, sous presse). Cet
effet est clairement visible sur la figure 2. C'est ainsi que des éruptions
volcaniques se produisant même aux basses latitudes de l'hémisphère Sud
peuvent avoir un impact très sensible sur la chimie atmosphérique
Antarctique (pourvu qu'elles soient suffisamment puissantes, bien sûr).
Mais des évolutions à plus long terme ont aussi été mises en évidence, sans
stant une explication satisfaisante à de telles variations
puisse être avancée.

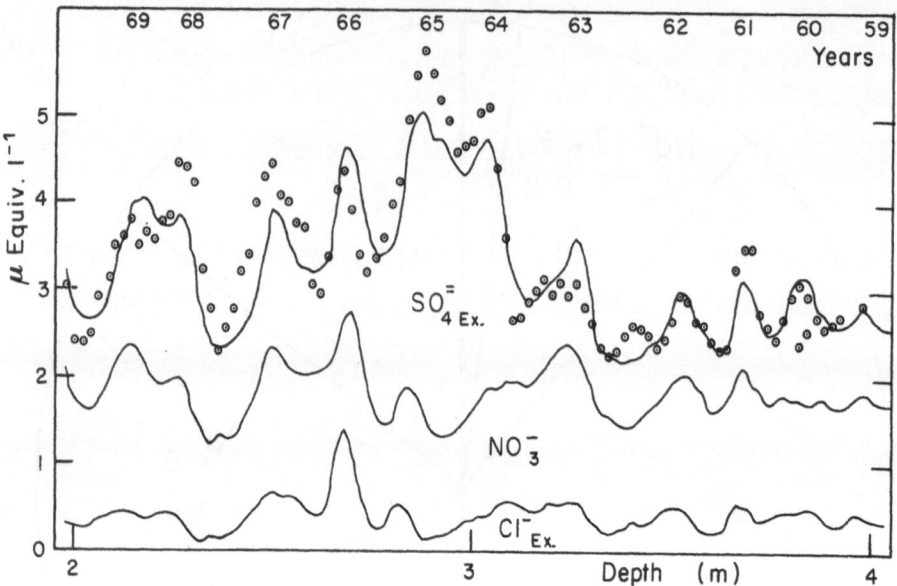

Figure 2 - Reconstitution de l'acidité de la neige du Pôle Sud pour la période 1959-1969. Les courbes continues ont été obtenues par lissage sur 3 points expérimentaux. La courbe du bas représente Cl^-_{ex}, celle du milieu $Cl^- + NO_3^-$ et celle du haut la somme $Cl^-_{exc.} + NO_3^- + SO_4^=_{exc.}$ Les points ronds sont les acidités expérimentales.

3.3. Au Dôme C

Nous avons établi le bilan ionique moyen au Dôme C, situation située en Antarctique de l'Est. La comparaison représentée figure 3 porte sur deux périodes de temps de 25 ans situées à environ un siècle de distance : 1955-1980 et 1870-1895. Ces deux séquences encadrent les éruptions volcaniques d'Agung (1963) et du Krakatoa (1883). Ce diagramme appelle deux commentaires :

    - Si on compare ces bilans ioniques avec celui du Pôle Sud (Figure 1) il apparaît que l'importance relative de l'acide nitrique est beaucoup plus faible au Dôme C qu'au Pôle Sud et cela pour les deux périodes de temps considérées,

    - L'acidité était deux fois plus faible au siècle dernier qu'actuellement, en relation avec l'absence complète de HCl il y a un siècle.

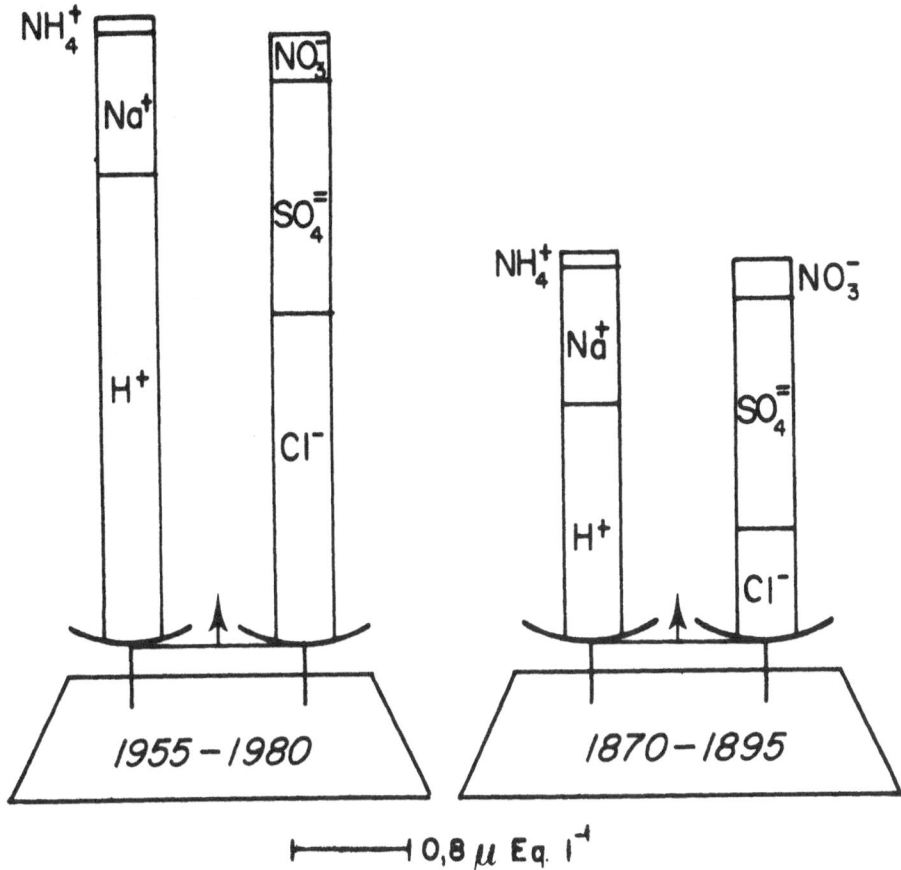

Figure 3 - Evolution du bilan ionique moyen de la neige du Dôme C au cours des 100 dernières années pour deux périodes de temps de 25 ans (1955-1980 et 1870-1895). On s'aperçoit que l'essentiel de la différence observée provient des variations des teneurs en HCl.

Ces deux observations sont importantes mais ne seront pas commentées dans le cadre de cet article car les phénomènes atmosphériques impliqués sont vraisemblablement complexes et encore incomplètement expliqués. Des résultats obtenus en d'autres lieux antarctiques confirment que les retombées de HNO_3 sont très inégales sur le continent Antarctique et que d'autre part la présence de HCl n'est pas une caractéristique générale de la précipitation Antarctique. Il faut se garder en particulier d'attribuer les retombées récentes de HCl au Dôme C et au Pôle Sud à une quelconque influence de la pollution globale anthropogène (Legrand and Delmas, 1983).

3.4. A D 57

Plusieurs séquences d'un forage profond réalisé en Terre Adélie à D 57 en un site situé à 200 km de la côte ont été analysés dans le but d'y détecter des niveaux acides exceptionnels. Les mesures chimiques ont porté essentiellement sur les ions H^+, $SO_4^=$ et NO_3^-. En ce site l'ion Cl⁻

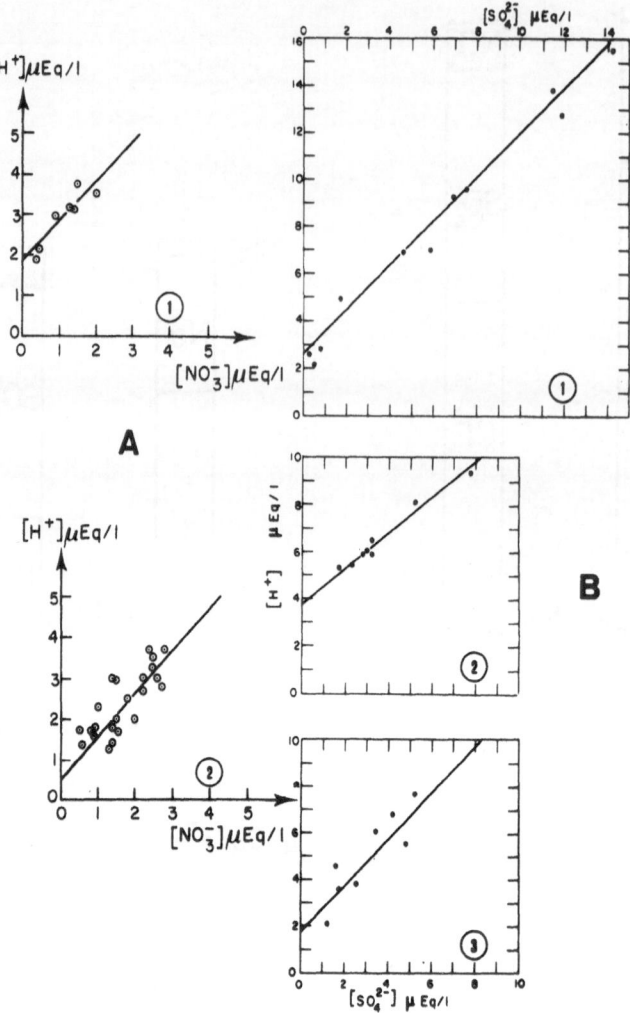

Figure 4 - Corrélations obtenues à D 57 (Terre Adélie) entre l'acidité et les teneurs en nitrate (A) ou sulfate (B) pour différents "pics" d'acidité détectés à plusieurs profondeurs le long d'une carotte de glace de 200 m couvrant environ 500 ans. Chaque évènement, couvrant environ une année, a été découpé en "tranches" représentant environ un mois de précipitation. Les valeurs de H^+, NO_3^- et $SO_4^=$ ont été mesurées sur chacune de ces tranches après fusion.
Les équations des droites obtenues sont les suivantes :

A1 : $[H^+] = 0,94 [NO_3^-] + 1,9$

A2 : $[H^+] = 1,02 [NO_3^-] + 0,5$

B1 : $[H^+] = 0,92 [SO_4^=] + 2,67$

B2 : $[H^+] = 0,75 [SO_4^=] + 3,87$

B3 : $[H^+] = 1,02 [SO_4^=] + 1,67$

provient uniquement de l'aérosol marin. Chaque échantillon correspondait à une fraction d'année; ce qui a permis de mettre en évidence des phénomènes naturels d'une durée de l'ordre de l'année. Par une méthode électroconductimétrique mise en oeuvre directement sur les carottes de glace (Zanolini, 1982), plusieurs évènements ont d'abord été repérés tout au long des 200 m de la carotte. Nous avons pu montrer ensuite que la majorité de ces "accidents" ou "pics" sont liés à des retombées inhabituelles, soit de H_2SO_4, soit de HNO_3. L'identification de l'acide s'effectue en corrélant pour chacun des "pics" les teneurs en $[H^+]$ aux teneurs soit de $[SO_4^=]$ soit de $[NO_3^-]$ (exprimées en µEq. 1^{-1}). Une pente de 1 ou proche de 1 est une indication qu'il s'agit d'un acide pur et non pas d'un sel (sulfate ou nitrate). OPn donne sur les figures 4 A et 4 B des exemples de ces corrélations respectivement pour des pics de HNO_3 et H_2SO_4. Les pics de sulfate (le plus fort atteint 13,7 µEq. 1^{-1}) sont caractéristiques des éruptions volcaniques (Delmas and Boutron, 1980) et sont relativement bien expliqués. Par contre l'origine des brusques augmentations des teneurs des teneurs en nitrate (les plus fortes augmentations n'atteignent cependant que 2 µEq.1^{-1}) est encore difficile à déterminer, comme est encore inexplicable la différence de concentration existant pour ce composé entre les stations de Dôme C et Pôle Sud (voir plus haut).

4. CONCLUSION

Ces trois exemples ont montré qu'il est possible d'obtenir un bilan équilibré des ions présents dans la précipitation Antarctique et que les trois acides minéraux HSO_4, HNO_3 et HCl y jouent un rôle fondamental. La contribution respective de ces 3 acides à l'acidité totale varie d'un point à un autre et dépend aussi de la priode de temp étudiée. Si on commence à bien connaître la composition chimique des impuretés de la précipitation neigeuse en différents sites du continent Antarctique, il reste à expliquer l'origine des variations spatiotemporelles observées.

5. REMERCIEMENTS

Ces travaux ont été effectués avec l'appui des Expéditions Polaires Françaises, des Terres Australes et Antarctiques Françaises, du Ministère de l'Environnement, du CNRS et de la NSF (Etats-Unis).

REFERENCES

De Angelis M., Legrand M., Petit J.R., Barkov N.I., Korotkevich Y.S., and Kotlyakov V.M. (sous presse). Soluble and insoluble impurities along the 950 m deep Vostok ice core (Antarctica) - Climatic implications. Journal of Atmospheric Chemistry.

Delmas R.. and Boutron C., (1980), Are the past variations of the stratospheric sulfate burden recorded in central Antarctic snow and ice layers ?, J. Geophys. Res., 85, 5645-5649.

Delmas R.J., Briat M., and Legrand M., (1982), Chemistry of South Polar snow, J. Geophys. Res., 87, 4314-4318.

Delmas R.J., and Gravenhorst G., (1982), Background precipitation acidity, Proc. C.E.C. Workshop "Acid deposition", Berlin, 9-10 Sept. 1982, 82-107.

Jouzel J., Merlivat L., Petit J.R., and Lorius C. (1982), Climatic information over the last century deduced from a detailed isotopic record in the South Pole snow. J. Geophys. Res., 88, 2693-2703.

Legrand M., Aristarain A.J., and Delmas R.J., (1982), Acid titration of polar snow. Analyt. Chem., 54, 1336-1339.

Legrand M., De Angelis M., and Delmas R.J., (1984), Ion chromatographic determination of major ions and ultratrace levels in Antarctic snow and ice, Analyt. Chim. Acta, 156, 181-192.

Legrand M and Delmas R.J., (sous presse), The ionic balance of Antarctic snow : a 10 yr detailed record, Atmospheric Environment.

Legrand M., and Delmas R., (1983). Spatiotemporal variations of the Cl/Na ratio in Antarctic snow, papier présenté à la 5e conférence internationale de la Commission sur la Chimie Atmosphérique et la Pollution Globale (CACGP), Oxford 28/08-3/09-1983.

Stumm W., and Morgan J.J., (1970), Aquatic Chemistry, Wiley J. and Sons Ed., 583 p.

Zanolini F., (1983), Conductimétrie et Chimie de la glace à D 57 (Terre Adélie), Application à la recherche du paléovolcanisme. Bulletin PIRPSEV N° 76 - CNRS Paris - pp. 84.

ETUDE DE L'INFLUENCE D'UNE SOURCE LOCALE NATURELLE

INTENSE DE COMPOSES ORGANOSOUFRES SUR LA CHIMIE DE

LA TROPOSPHERE EN MILIEU NON POLLUE.

P. CARLIER, C. LUCE, R. GIRARD, G. MOUVIER, J. MORELLI
L. GIRARD-REYDET, T. MARCHAL et S. CADENE
Université PARIS VII - Laboratoire de Physico-Chimie
Instrumentale et Laboratoire de Chimie Minérale des
Milieux Naturels, associés au CNRS - 2, place Jussieu
75251 PARIS CEDEX 05.

Summary

In preparation to the study of mechanisms of photooxidation of biological-
ly generated organosulfur compounds in natural conditions, we have measured the concentration of different minor reactive compounds in the atmosphere in the case of a simple model, that of an intense local source in
an "unpolluted" environment. The site studied, Penmarc'h in Britany, has
a large field of algae which generates a very large quantity of dimethyl
sulfide. In order to path of chemical transformation of this compound
in the atmosphere, analyses for different intermediate and related compounds were carried out: organosulfur compounds, sulfur dioxide, ozone,
nitrogen oxides and carbonyl compounds. Ozone and nitrogen oxides were
monitered continuously by means of commercial analyzers whereas the organosulfur compounds, the sulfur dioxide and the carbonyl compounds were
monitered by means of chemical analyses carried out every 6 hours. It
was necessary to develop new sampling and analysis techniques with high
sensitivity in order to study the carbonyl and organosulfur compounds.

1. INTRODUCTION

Depuis la mise en évidence par Lovelock et al. (1) du rôle du sulfure
de diméthyle (DMS), dans le cycle biogéochimique du soufre, ce composé a
fait l'objet de recherches dont le rythme s'est accéléré ces dernières
années. On peut, en première analyse, classer ces travaux en deux types :
des recherches à dominante géochimique qui visent essentiellement à préciser le flux d'émission et la répartition géographique du DMS (2-19) et des
recherches à dominante physico-chimique en vue d'élucider le mécanisme de
la photooxydation atmosphérique de ce composé, soit par des études cinétiques de processus élémentaires (réaction DMS + O·, (20-24) ; DMS + OH·,
(25-31) ou DMS + O_3 (32)), soit par des études en chambre de simulation
de dimensions variables (33-39).

Pour notre part, nous avons tenté d'aborder l'étude directe des mécanismes de photooxydation des composés organosoufrés d'origine biogénique
en milieu naturel, par la caractérisation de l'influence d'une source
locale intense sur la chimie de la troposphère. Le site retenu est celui

de la Pointe de Penmarc'h (53,1 grN - 7,4 grO) à l'extrémité sud-ouest de la Bretagne. Un vaste champ d'algues y émet des quantités très importantes de sulfure de diméthyle dans une atmosphère de type océanique que l'on peut considérer comme non polluée par régime de vent d'ouest. Une campagne de mesures a eu lieu fin Septembre 1983.

2. CONCEPTION GENERALE DE LA CAMPAGNE

2.1. Points de mesures

La Pointe de Penmarc'h est une côte rocheuse, basse, très favorable au développement des algues.

Les prélèvements et mesures ont eu lieu en trois points (Figure 1):
1) au sémaphore de la Marine Nationale, à l'altitude de 5 m directement à la limite du champ d'algues.
2) à l'ancien phare de Penmarc'h, à l'altitude de 40 m, distant de 20 m du champ d'algues.
3) au lieu-dit "Kergadien", situé à environ 4 km à l'E.N.E. de la Pointe de Penmarc'h.

Quelques mesures de référence ont également été faites durant la même période au sémaphore de la Pointe du Raz (altitude 80 m) dans une zone de côte rocheuse, très escarpée, exempte d'algues.

2.2. Techniques de mesures

Certaines mesures ont pu être faites à l'aide d'appareils commerciaux automatiques, d'autres par des techniques éprouvées, d'autres enfin ont nécessité la mise au point de chaînes de mesure originales.

a) Oxydes d'azote : mesure en continue par analyseur AC 30M (Environnement SA).Un point de mesure : Kergadien.
b) Ozone : mesure en continu par analyseur 1003 AH (Environnement SA). Deux points de mesure : Ancien phare de Penmarc'h et Kergardien).
c) Dioxyde de soufre : mesuré par la méthode de West et Gaeke (40). Durée de prélèvement 6 h. Trois points à Penmarc'h et point de référence à la pointe du Raz.
d) Composés organo-soufrés : mesurés par une chaîne de mesure originale décrite précédemment ayant un seuil de détection de l'ordre de 1 ng/m³ (41). Trois points de prélèvements à Penmarc'h et un point de référence à la pointe du Raz. Durée d'échantillonnage : 6 h.
e) Composés carbonylés : Mesurés par une chaîne de mesure originale basée sur la capture des composés carbonylés par la Dinitro-2,4 phenylhydrazine et analyse des Dinitro-2,4 phényldrazones formées par chromatographie en phase gazeuse avec détecteur à capture d'électrons. Un point de mesure : sémaphore de Penmarc'h. Durée d'échantillonnage : 24, 12 ou 6 h. Les résultats aujourd'hui encore trop fractionnaires seront discutés ultérieurement (42).

Durant la même campagne des prélèvements d'aérosols (par filtration et par impaction) ont été effectués par une équipe du Laboratoire de Chimie Minérale du milieu naturel. Certains résultats font l'objet de l'article suivant.

2.3. Données météorologiques

Les données météorologiques classiques enregistrées aux stations de Penmarc'h et de la pointe du Raz nous ont été communiquées par la Marine Nationale. En plus, nous avons enregistré la puissance solaire totale au

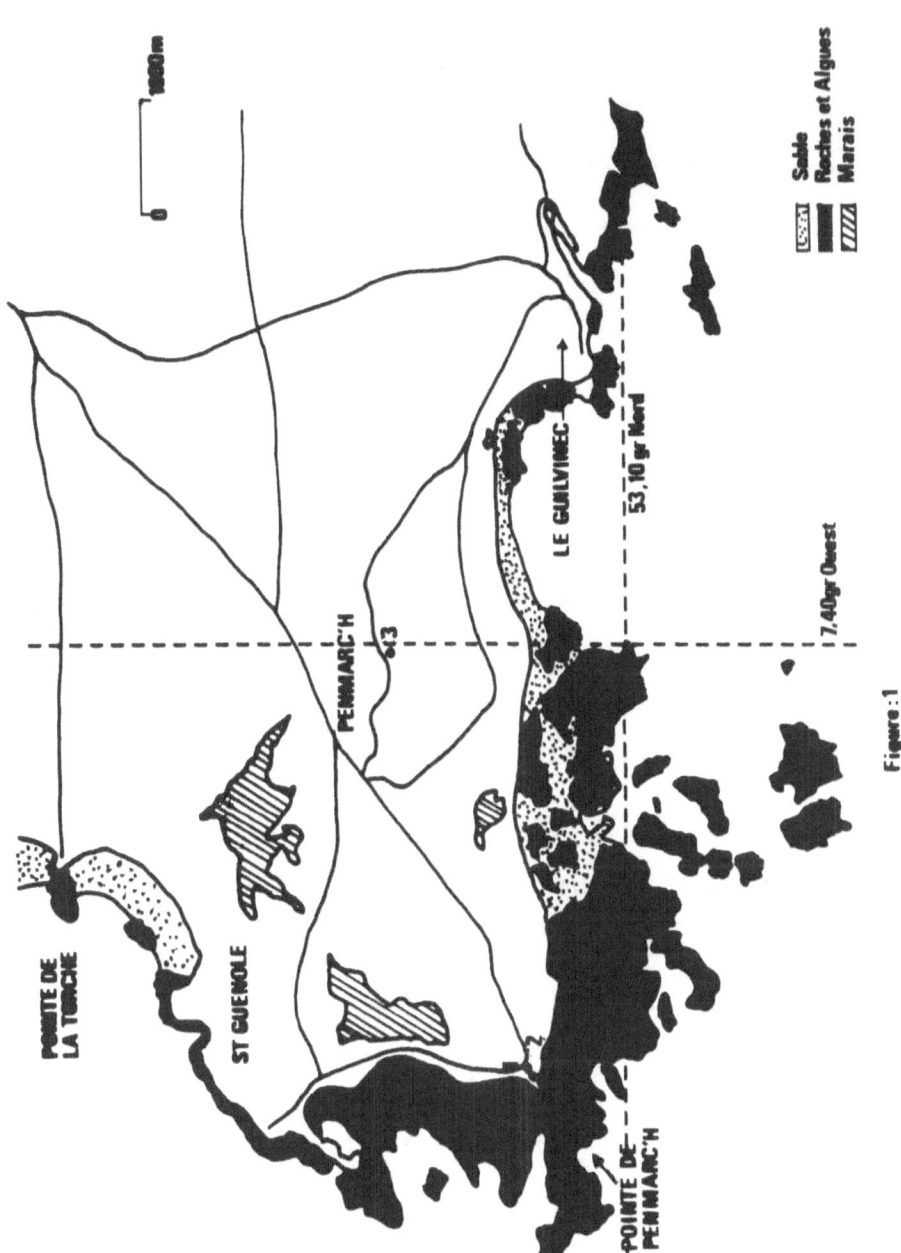

POINTE DE
LA TORCHE

ST GUENOLE

PENMARC'H

43

LE GUILVINEC

POINTE DE
PENMARC'H

53,10 gr Nord

7,40gr Ouest

0   1000 m

Sable
Roches et Algues
Marais

Figure : 1

semaphore de Penmarc'h à l'aide d'un pyronomètre à thermopiles.

3. DONNEES EXPERIMENTALES

Nous présentons ici les résultats relatifs aux journées des 17, 18 et 19 Septembre 1983, caractérisées, par un régime de vent local de S.S.O à N.O. une atmosphère assez instable (pression atmosphérique de 1 017 à 1 033 m.bar), des passages nuageux, et un long épisode pluvieux continu le 18 Septembre de 8 h. à 17 h TU. La direction moyenne des vents correspondait à l'alignement des points 1 et 2 avec le point 3.

3.1. Données météorologiques

Les données météorologiques relevées à la station de Penmarc'h (direction et vitesse du vent, pression, hygrométrie) et la puissance de l'irradiation solaire sont reportées sur la figure 2. Les trajectographies des masses d'air (à 925 m bar) sont données sur la figure 3. La période étudiée était particulièrement favorable car la zone de prélèvement était sous l'influence d'un régime de vents purement océaniques.

3.2. Enregistrement des teneurs en NO, NO_x et ozone.

Les teneurs en NO, NO_x et ozone enregistrées au point 3 sont également portées sur la figure 2. Les teneurs en NO et NO_x sont très stables (environ 20 et 40 ppb). Les teneurs en ozone oscillent autour d'une valeur moyenne de 35 ppb avec quelques pointes jusque vers 50-60 ppb

3.3. Mesure des teneurs en espèces soufrées gazeuses.

Les résultats des mesures en espèces soufrées gazeuses sont rassemblés dans le tableau 1. Les mesures de références effectuées à la pointe du Raz du 17 Septembre 15 h. au 19 Septembre 12 h. en sommant les tranches horaires sur les deux jours donnent :

	$SO_2 \mu g/m^3$	DMS ng/m^3
6 - 18 h TU	0,14	< 1
18 - 6 h TU	0,14	37

Aucun autre composé organosoufré n'a été détecté à la pointe du Raz, alors qu'à Penmarc'h outre le DMS ont été détectés durant la période du 17 au 19 Septembre 1983 du Disulfure de diméthyle et un autre composé soufré qui n'a pu être identifié.

On observe par ailleurs sur les chromatographes un pic souvent très important correspondant au CS_2. Le pic n'est par contre pas reproductible. Nous n'avons trouvé dans la littérature aucune référence sur l'existence d'un complexe stable mercure-CS_2. On ne peut donc espérer utiliser nos expériences pour estimer même très approximativement la concentration de CS_2 dans l'air.

4. DISCUSSION

Les teneurs en DMS enregistrées près du champ d'algues sont très importantes comparées au fond océanique tel que l'on peut le mesurer à la pointe du Raz ou tel que l'a caractérisé d'autres auteurs. Pour l'ensemble des mesures aux trois points de Penmarc'h, on peut donc considérer que l'on

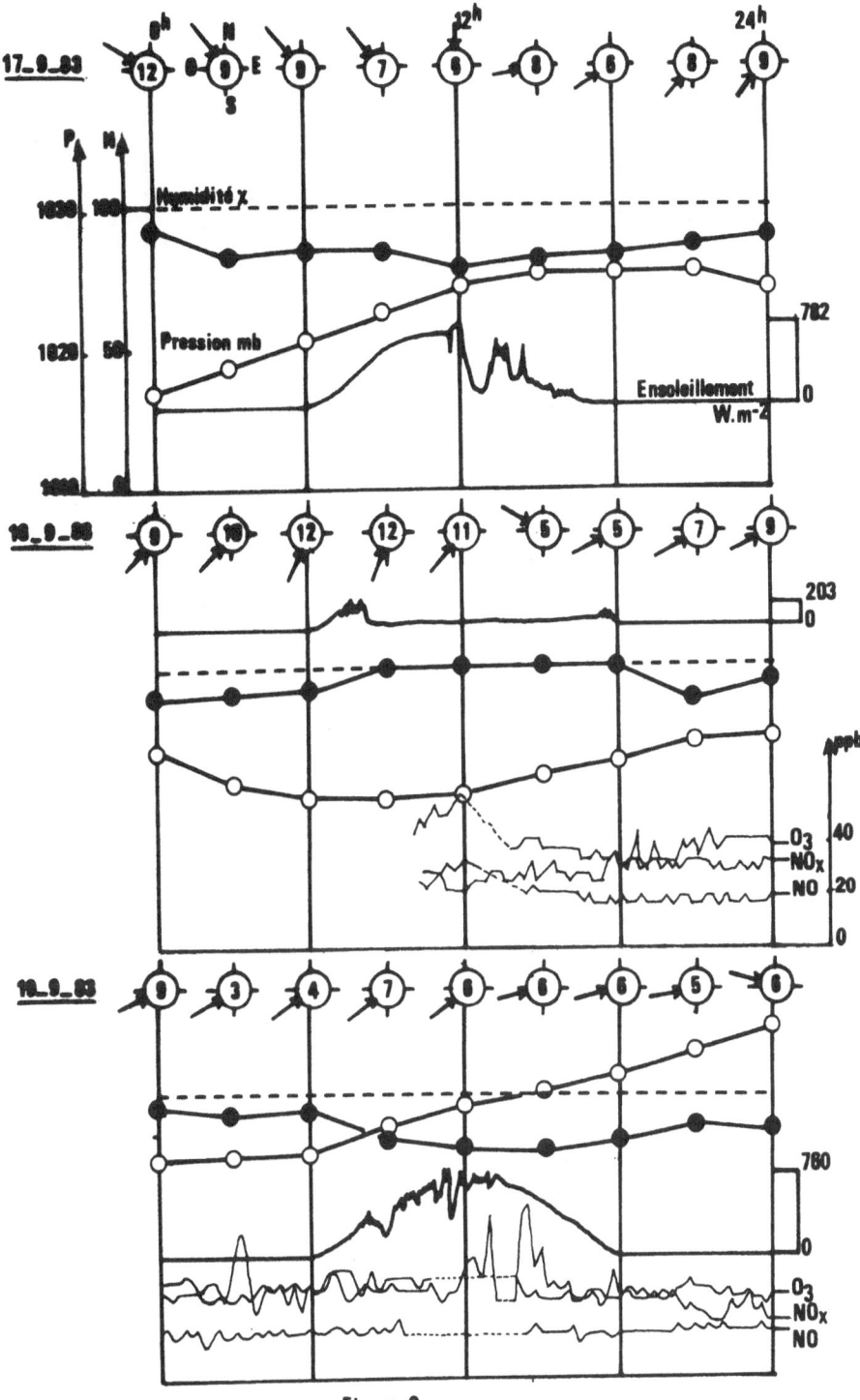

Figure: 2

DMS – DMDS – SO$_2$

TABLEAU 1

		16-9	17-9				18-9				19-9		
		18-24	0-6	6-12	12-18	18-24	0-6	6-12	12-18	18-24	0-6	6-12	12-18
SO$_2$ µg/m³	Semaphore 1	0,49	0,46	0,96	0,49	0,17	< 0,1	0,85	< 0,1	1,11	0,81	1,36	0,80
	Phare 2	< 0,1	< 0,1	< 0,1	-	-	< 0,1	< 0,1	< 0,1	< 0,1	3,56	1,14	< 0,1
	Kergadien 3	< 0,1	< 0,1	0,30	< 0,1	< 0,1	0,3	< 0,1	< 0,1	0,2	< 0,1	0,25	0,39
DMS ng/m³	Semaphore 1	<	60	800	665	76	640	6200	318	567	58	10900	2450
	Phare 2	-	27	195	2315	-	<	<	6	<	<	715	1470
	Kergadien 3	-	88	240	2290	-	13000	-	135	<	<	-	<
DMDS ng/m³	Semaphore 1	<	<	42	5	<	26	121	30	<	<	1090	45
	Phare 2	<	<	10	360	-	<	<	5	<	<	<	49
	Kergadien 3	-	35	23	115	-	190	-	5	<	<	-	<
X ng(S)/m³	Semaphore 1	-	<	<	<	<	<	<	<	1780	76	153	930
	Phare 2	-	<	<	<	<	<	170	<	220	<	750	960
	Kergadien 3	-	<	<	<	<	190	-	<	120	220	-	<
	Vent	-	NO	NO	0	SO	SO	SO	SO	SO	SO	SO	0
	Pluie	+	++	0	0	0	0	++	++	0	0	0	0

0   pas de pluie
+   pluie intermitante
++   pluie continuelle

<   teneur inférieure au seuil de détection

teneurs en composés soufrés à Penmarc'h du 17 au 19 septembre 1983

Figure: 3

Trajectographies des masses d'air à 925 mbar , avec un pas de 6h , arrivant à la pointe de Penmarc'h entre le 17 septembre 0h T.U. et le 19 septembre 24h T.U.. Les trajectographies successives (n°1 à 7) sont séparées par des intervalles de temps de 12 heures .

est sous l'influence prépondérante du champ d'algues pour les composés organo-soufrés. On observe en général un fort maximum pendant la matinée et ce, bien que l'atmosphère soit normalement plus instable pendant cette période que la nuit. Ceci indique alors une production de DMS très nettement supérieure. Ces variations ne sont pas conformes aux prévisions de Graedel (6) car celui-ci postulait implicitement que les emmissions étaient constantes pendant la journée. Les deux autres composés sont surement des composés secondaires car leur abondance relative par rapport au DMS croît quand on s'éloigne de la source, soit vers le haut (point 2), soit vers l'intérieur des terres (point 3).

L'évolution très rapide des rapports d'abondance entre le DMS et les produits secondaires indique un temps de séjour très faible du DMS : quelques minutes tout au plus. En effet le temps du transport entre le point 1 et le point 3 éloigné de 4 km est compris entre 22 et 5,5 minutes (vents de 3 à 12 m/s pendant cette période) et la conversion du DMS est quasi totale au point 3. L'estimation de la vitesse de migration verticale en l'absence de profil vertical de température est très problématique. Dans l'atmosphère assez instable de la période considérée, il est cependant raisonnable de penser que ce temps n'excède pas la minute. Ce qui confirme très qualitativement le résultat énoncé. Il faut noter également que la disparition du DMS est très rapide pendant la nuit.

Les très faibles valeurs de SO_2 enregistrées sont logiques. Dans les périodes pluvieuses ou brumeuses, le SO_2 n'est détectable qu'au bord du champ d'algues où sa production est maximum à partir du DMS et le nettoyage de l'air par la pluie ou la brume élimine le SO_2 qui ne peut plus etre détecté ailleurs.

5. CONCLUSION

Dans les conditions océaniques pures de la période étudiée ici, il apparaît que le temps de résidence du DMS dans l'atmosphère est très court, ce qui confirme les résultats obtenus par d'autres équipes en pleine mer. Ce résultat est encore difficile à interpréter dans la mesure où l'estimation de ce temps de résidence à partir de données cinétiques est de l'ordre de 20 h en estimant une concentration de OH' égale à 2.10^6 radicaux par cm^3. Les résultats des mesures d'aldéhydes , qui par photolyse génèrent des radicaux OH' permettront de préciser si cette hypothèse est réaliste. Il faudra alors chercher une hypothèse mécanistique permettant d'expliquer le désaccord entre les mesures sur le terrain et les données cinétiques.

BIBLIOGRAPHIE

(1) J.E. LOVELOCK, R.J. MAGGS et R.A. RASMUSSEN, Nature 237 (1972) 452.
(2) B. BONSANG, N'GUYEN B.C., J.Y. PANGAM, C.R.Acad.Sci. Paris Ser.D, 283, 1285.
(3) P.J. MAROULIS et A.R. BANDY, Science 196 (1977) 647.
(4) N'GUYEN B.C., A. GAUDRY, B. BONSANG et G. LAMBERT, Nature 275 (1978) 637.
(5) M.R. MOSS, Sources of Sulfur in the Environment : The global sulfur cycle Dans : Sulfur in the Environment - Part 1 , p. 23 - J.O. NRIAGU Ed. Willey Intersc. (1978).
(6) T.E. GRAEDEL, Geophys. Res.Lett., 6 (1979) 329.

(7) D.F. ADAMS, SO. FARWELL, M.R. PACK et W.L. BAMESBERGER, J. Air Poll Control Assoc., 29 (1979) 381.

(8) V.P. ANEJA, J.M. OVERTON Jr, L.T. CUPITT, J.L. DURAAM et W.E. WILSON Tellus 31 (1979) 174.

(9) B. BONSANG, These de Doctorat ès Sciences - Université de Picardie 24 Mars 1980.

(10) C.F. CULLIS et M.M. MIRSCHLER, Atmospheric Environment 14 (1980) 1263.

(11) B. BONSANG, N'GUYEN B.C., A. GAUDRY et G. LAMBERT, J.Geophys. Res., 85 (1980) 7410.

(12) D.F. ADAMS, S.O. FARWELL, E. ROBINSON, M.R. PACK et W.L. BAMESBERGER, Environm. Sci. Technical 15 (1981) 1493.

(13) D.F. ADAMS, S.O. FARWELL, M.R. PACK et E. ROBINSON, J.Air Poll. Contr Assoc., 31 (1981) 1083.

(14) V.P. ANEJA, J.M. OVERTON et A.P. ANEJA, J. Air Poll Contr Ass., 31 (1981) 256.

(15) P. NADAUD, These de 3ème Cycle - Université de Paris VII - 21 Décembre 1981.

(16) W.R. BARNARD, M.O. ANDREAE, W.E. WATKINS, M. BINGEMER et H.W. GEORGII J. Geophys. Res., 87 (1982) 8787.

(17) B. BONSANG , La Recherche 13 (1982) 1132.

(18) M.O. ANDREAE, W.R. BARNARD et R.J. FEREK CACGQ Symposium on Tropospheric Chemistry - Oxford 28 Aout - 3 Septembre 1983 - Communication II-8.

(19) H.G. BINGEMER et H.W. GEORGII CACGP Symposium on Tropospheric Chemistry - Oxford 28 Aout - 3 Septembre 1983 - Communication III-2.

(20) I.R. SLAGLE, R.E. GRAHAM, D. GUTMAN, Int.J.Chem.Kinet., 8 (1976) 451.

(21) J.H. LEE, R.B. TIMMONS et L.J. STIEF, J.Chem.Phys., 64 (1976) 300.

(22) J.H. LEE, I.N. TANG, R.B. KLEMM, J.Chem.Phys., 72 (1980) 1793.

(23) W.S. NIP, D.L. SINGLETON, R.J. CVETANOVIC, J.Amer.Chem.Soc., 103 (1981) 3526.

(24) R.J. CVETANOVIC, D.L. SINGLETON, R.S. IRWIN, J.Amer.Chem.Soc., (1981) 3530.

(25) M.J. KURYLO, Chem.Phys.Letters, 58 (1978) 233.

(26) R.A. COX et D. SHEPPARD, Nature 284 (1980) 330.

(27) P.M. WINE, N.M. KREUTTER, G.A. GUMP et A.R. RAVISCHANKARA, J.Phys. Chem., 85 (1981) 2660.

(28) J.M. LEE et N. TANG, J.Chem.Phys., 78 (1983) 6646.

(29) H. McLEOD, G. POULET et G. LEBRAS, J.Chim.Phys., 80 (1983) 287.

(30) H. McLEOD, J.L. JOURDAIN, G. POULET et G. LEBRAS, CACGP Symposium on Tropospheric Chemistry - Oxford 28 Aout - 3 Septembre 1983 - Communication VI-2.

(31) H. McLEOD - Thèse de 3ème Cycle - Université de PARIS VII - 27 Octobre 1983.

(32) R.I. MARTINEZ et J.T. HERRON, Inter J.Chem.Kinet., 10 (1978) 433.

(33) M.B. BENTLEY, I.B. DOUGLASS, J.A. LACADIE et D.R. WHITTIER, J. Air Poll. Control Assoc., 22 (1972) 359.

(34) R.A. COX et F.J. SANDALLS, Atmos. Env 8 (1974) 1269.

(35) R. ATKINSON, R.A. PERRY et J.N. PITTS Jr., J.Chem.Phys., 66 (1977) 1578.

(36) D. GROSJEAN et R. LEWIS, Geophys. Res. Letters 9 (1982) 1203.

(37) S.H. ATAKEYMA, M. OKUDA et H. AKIMOTO, Geophys. Res. Letters 9 (1982) 583.

(38) H. NIKI, P.D. MAKER, C.M. SAVAGE et L.P. BREITENBACH, Int.J.Chem. Kinet., 15 (1983) 647.

(39) I. BARNES, K.N. BECKER, E.H. FINK CACGP Symposium on Tropospheric Chemistry - Oxford 28 Aout - 3 Septembre 1983 - Communication VI-1.

(40) P.W. WEST et G.C. GAEKE, Anal.Chem., 28 (1956) 1816.

(41) C. LUCE, P. CARLIER, R. GIRARD, H. HANNACHI, P. FRESNET et G.MOUVIER Analusis (1984) à paraître.

(42) P. FRESNET, C. LUCE, R. GIRARD, P. CARLIER et G. MOUVIER, en préparation.

FLUCTUATIONS TEMPORELLES FINES DE LA COMPOSITION
CHIMIQUE DE L'AEROSOL COTIER

par

Jacques MORELLI[1], Laurence GIRARD-REYDET[1], Thierry MARCHAL[1], Pierre
MASNIERE [2], Michel. FEDOROFF[3], Jean-Claude ROUCHAUD[3], Lucienne
DEBOVE[3] et Patrick CARLIER[4].

(1)Laboratoire de Chimie Minérale des Milieux Naturels - ERA CNRS 889 -
   Université Paris 7, 2 Place Jussieu, 75251 Paris Cedex 05, France.

(2)Direction des Etudes et Recherches, Electricité de France, 78400 Chatou.

(3)Centre d'Etudes de Chimie Métallurgique du CNRS, 94400 Vitry sur Seine.

(4)Laboratoire de Physico-Chimie Instrumentale - ERA CNRS 889 -
   Université Paris 7, 2 Place Jussieu, 75251 Paris Cedex 05.

RESUME

     Une étude de la géochimie de l'aérosol côtier a été développée dans
une zone peu industrialisée du littoral Atlantique, à la pointe de la
Bretagne. On s'intéresse ici aux fluctuations temporelles fines des concentrations atmosphériques d'une dizaine de constituants particulaires observées en avril 1982 et septembre 1983. Les échantillons, prélevés par filtration d'air ont été analysés en utilisant la spectrométrie de fluorescence X (mesures en Cl, S, Al, Si, Fe, K, Ca, P) et l'activation neutronique
(mesures en Na et Mn). Les résultats font apparaitre de nettes différences
de composition chimique de l'aérosol collecté, selon la provenance des
masses d'air. Ces différences peuvent être reliées aux influences antago-
nistes des sources continentales et océaniques. L'impact du transport à
grande échelle de la pollution soufrée se fait notamment sentir lorsque
les masses d'air ont survolé diverses régions industrielles européennes.
On observe alors souvent des déficits marqués en chlore dans les particules
de sels marins.

ABSTRACT
     A geochemical study of coastal aerosol was performed in a low indus-
trialized area of the Atlantic Seaside (Brittany). Attention is focused
here on short time fluctuations of atmospheric concentrations of ten particulate components observed in april 1982 and september 1983. The samples,
collected by air filtration, were analyzed using X ray fluorescence spectrometry for Cl, S, Al, Si, Fe, K, Ca, P and neutron activation for Na and
Mn. Results show strong differences in chemical composition of coastal
aerosol according to the origin of air masses. These differences can be
related to the antagonistic influences of continental and oceanic sources.
A marked effect of a large scale transport of sulfur pollution appears when
air masses passed over various european industrial regions. In this situation, loss of chlorine from sea-salt particles is generally observed.

I. CONTEXTE DES RECHERCHES EFFECTUEES ET OBJECTIFS

Depuis la fin 1980, le groupe de Physico-Chimie de l'Atmosphère de l'Université de Paris 7 a développé un programme de recherche sur la géochimie des particules en suspension dans l'air, à l'extrémité de la Bretagne. Il s'est avéré que cette région, encore peu industrialisée et relativement éloignée des grandes sources de pollution atmosphérique de l'Europe Occidentale, consitue un bon observatoire de la qualité de l'air en milieu côtier sur la façade Atlantique de notre continent. Les travaux entrepris ont d'abord été axés sur l'étude de la composition chimique de l'aérosol cotier, avec et sans distinction de sa distribution granulométrique, en relation avec les fluctuations des paramètres météorologiques (1,2,3, 4). D'autres types d'observation sont progressivement venus se greffer sur le programme initial. Ils ont donné lieu à des coopérations interéquipes et ont entraîné l'organisation de campagnes conjointes. Ainsi, les mesures ont été étendues aux précurseurs gazeux du soufre particulaire, en vue notamment de la caractérisation d'émissions naturelles de composés organosoufrés (5) et de l'étude de leur évolution compte tenu de leur interaction avec d'autres espèces gazeuses telles que l'ozone et les NO_x (6). Un échantillonnage séquentiel des pluies récemment effectué et couplé à des prélèvements particulaires et gazeux devrait par ailleurs apporter d'utiles renseignements sur les mécanismes responsables de leur composition chimique.

Les résultats présentés ici reposent sur un suivi des variations fines des concentrations atmosphériques d'une dizaine de constituants de l'aérosol côtier collecté par filtration d'air en avril 1982 et septembre 1983. Dans ce qui suit, on se propose d'en dégager quelques grandes tendances et de répondre aux objectifs suivants :
- fournir des indications sur les ordres de grandeur des concentrations dans l'air des éléments à l'état particulaire analysés (Cl, Na, S, Al, Si, Fe, Mn, K, Ca, P) et illustrer par quelques exemples l'évolution temporelle fine de certains d'entre eux;
- relier ces données à l'influence des conditions météorologiques rencontrées (provenance des masses d'air);
- évaluer les contributions de différentes sources aux concentrations élémentaires observées en examinant tout spécialement le cas du soufre, polluant majeur de l'atmosphère.

II. CONDITIONS DE PRELEVEMENT ET D'ANALYSE

L'échantillonnage a été réalisé en continu à la pointe de Penmarc'h pendant les périodes du 11 au 27 avril 1982 et du 15 septembre au 1er octobre 1983. Il a comporté des filtrations d'air successives d'une durée de 6 heures pour la première campagne (63 prélèvements à 5 mètres au dessus de la mer) et de 12 heures pour l'autre campagne (32 prélèvements au sommet d'un phare, à 42 mètres d'altitude). Dans les deux cas, les débits d'air moyens étaient voisins de 15 l.min^{-1} et les dépôts particulaires ont été collectés sur des filtres Nuclépore de porosité 0,4 μm.

La figure 1 permet de situer les zones d'où provenaient les masses d'air lors de l'échantillonnage. Elle s'appuie sur une étude de leur trajectographie à 850 ou 925 mbar effectuée grâce au concours de la Météorologie Nationale Française, l'origine des trajectoires étant prise 3 jours avant leur arrivée à l'extrémité de la Bretagne. En avril 1982, on a bénéficié d'un temps sec, calme et ensoleillé, avec en quasi permanence un vent local soufflant de la terre. Les masses d'air s'étaient alors préalablement déplacées au dessus de la Grande Bretagne et de diverses régions du continent européen (secteur A sur la figure 1). La première partie de l'autre

campagne (15 au 23 septembre 1983)a par contre été dominée par une forte
influence océanique avec une circulation d'ouest très active et parfois des
averses. Elle correspond au secteur B de la figure 1. Par la suite, les
trajectoires des masses d'air se sont infléchies vers le Sud avec des tran-
sits d'abord au dessus de la Péninsule Ibérique et du Golfe de Gascogne
puis ont balayé une zone comprise entre l'Ouest de la France et la mer du
Nord (secteur C sur la figure 1).

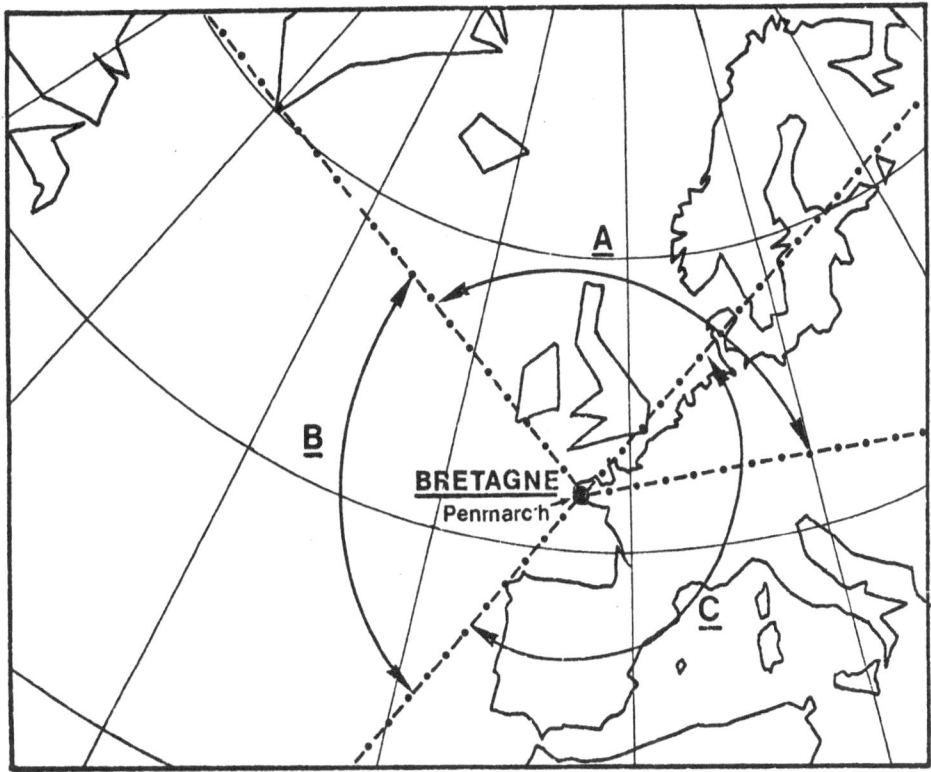

Figure 1 : Secteurs géographiques balayés par les masses d'air
arrivant sur la zone d'échantillonnage.

Les prélèvements sur filtres ont été directement analysés sur l'installation de spectrométrie de fluorescence X du Laboratoire de Chimie Minérale
des Milieux Naturels en vue de la détection des éléments suivants : Cl, S,
Al, Si, Fe, K, Ca, P. Des mesures complémentaires en Na et Mn ont été effectuées ultérieurement sur les mêmes filtres au moyen de l'analyse par activation neutronique. Les échantillons avaient à cet effet été irradiés 2 minutes dans un flux de 10^{14} neutrons $cm^{-2}.s^{-1}$ puis fait l'objet de comptages γ
sur un détecteur Ge-Li au Centre d'Etude de Chimie Métallurgique du CNRS à
Vitry sur Seine, les spectres obtenus étant traités selon une méthode mise
au point par ce laboratoire (7).

III. RESULTATS ET DISCUSSION

. Etude de l'évolution des concentrations atmosphériques mesurées

On a reporté dans le tableau 1 les moyennes géométriques des concentrations atmosphériques élémentaires observées lors des campagnes d'Avril 1982 et de Septembre 1983.

Eléments	Na	Cl	S	K	Ca	Al	Si	Fe	Mn	P
Concentrations atmosphériques	A – Campagne d'Avril 1982									
	1200	1400	2900	550	900	850	3000	650	15	40
élémentaires	B – Campagne de Septembre 1983 (1ère partie)									
moyennes en ng.m^{-3}	4400	8600	750	200	250	30	50	40	1	6
	C – Campagne de Septembre 1983 (2ème partie)									
	1200	400	3900	450	450	400	800	250	20	30

Tableau 1.

On note de nettes variations des concentrations mesurées selon la provenance des particules collectées (secteurs A, B et C de la figure 1). Ainsi, les plus fortes concentrations en Na et Cl sont celles correspondant à la première partie de la campagne de septembre 1983, constamment soumise à l'influence de vents soufflant du large (secteur B). Des concentrations extrêmement faibles en Al, Si, Fe, Mn et P leurs sont associées. La situation est inversée en avril 1982 et pendant la deuxième partie de la campagne de septembre 1983, lorsque l'influence continentale se fait sentir (secteur A et B). Les concentrations en S, K et Ca sont alors maximales.

En choisissant les exemples bien caractéristiques de Na, Al et S on a reconstitué sur les figures 2 et 3 les fluctuations temporelles fines des concentrations atmosphériques de ces éléments lors des campagnes d'avril 1982 et de septembre 1983. Ces concentrations sont données en ordonnées logarithmiques et dans un souci d'homogénéisation de la présentation des figures, on a pris dans les deux cas des pas de temps de 12 heures.

L'examen de la figure 2 fait apparaître des fluctuations journalières de forte amplitude des concentrations en Al et Na, atteignant parfois un ordre de grandeur. Les concentrations de ces deux éléments, qui sont ainsi très dispersées de part et d'autre de leur valeur moyenne donnée dans le tableau 1, sont fréquemment anticorrélées. Dans le cas de Al – comme d'ailleurs dans celui de Si, Fe, Ca et K –, les concentrations augmentent en fonction de la vitesse du vent local (2,4) et sont dans une large mesure associées à de grosses particules – d>2μm – (3). Ce résultat peut être relié à un effet de l'érosion éolienne du sol au voisinage de la zone d'échantillonnage. Quant aux concentrations en S, elles présentent des variations de moindre amplitude que celles relatives à Al et Na et se maintiennent à des niveaux élevés (en général 2000 – 4000 ng.m^{-3}).

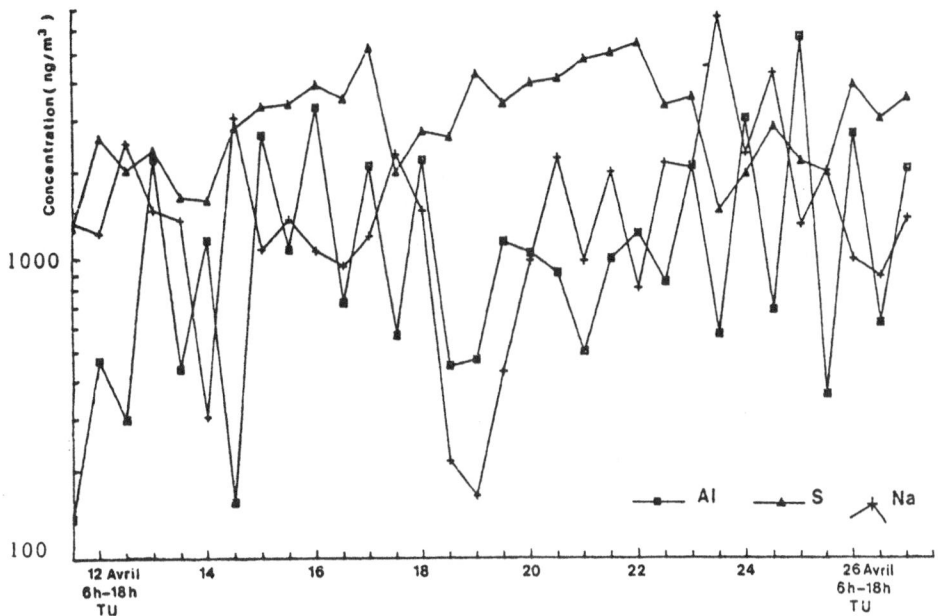

Figure 2 : Variations temporelles fines des concentrations atmosphériques en aluminium, soufre et sodium lors de la campagne d'avril 1982.

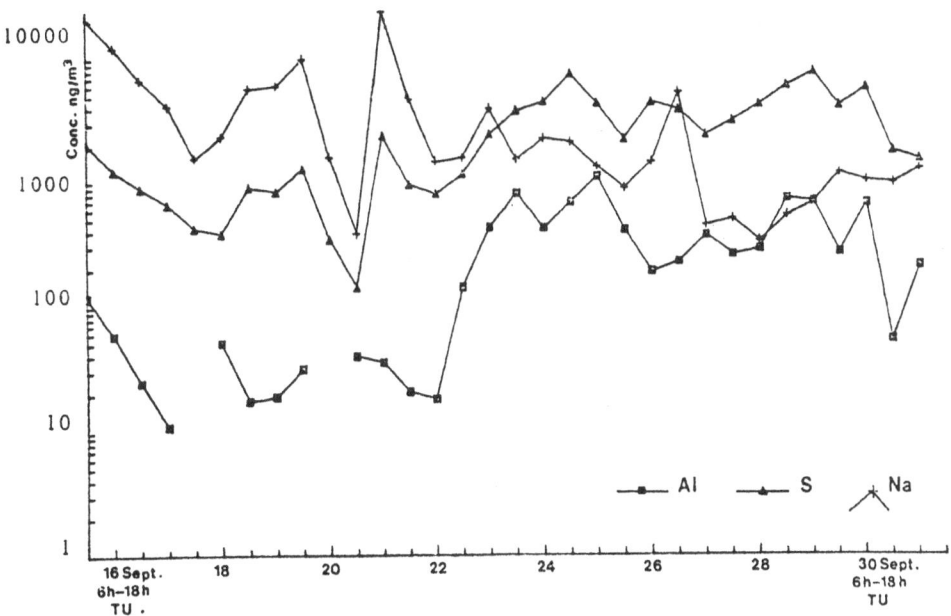

Figure 3 : Variations temporelles fines des concentrations atmosphériques en aluminium, soufre et sodium lors de la campagne de septembre 1983.

On voit d'autre part sur la figure 3 que les concentrations en Na et S évoluent de manière similaire pendant la période à influence océanique (15-23 septembre 1983). Les valeurs minimales observées le 18 septembre et dans la nuit du 20 au 21 septembre peuvent être expliquées par un "nettoyage" des particules sous l'effet de fortes précipitations. A partir du 23 septembre, les concentrations en S deviennent supérieures à celles en Na, dépassant parfois 7000 ng.m^{-3}. Au cours de cette période largement soumise à l'influence continentale, on note certaines analogies entre l'évolution des concentrations en S et Al. En ce qui concerne ce dernier élément, on remarque enfin que l'amplitude des variations journalières de ces concentrations est beaucoup moins grande que lors de la campagne d'avril 1982 (voir figure 3). Ceci est probablement lié aux conditions d'échantillonnage, les prélèvements en haut d'un phare (septembre 1983) étant moins sensibles à l'effet des plus grosses particules d'origine proche éliminées dans les premiers mètres de l'atmosphère.

. Etude des origines des constituants particulaires analysés

Si C_{Al} et C_{Na} sont les concentrations en aluminium et sodium mesurées, on peut calculer pour un élément X des concentrations théoriques C_{XT} et C_{XM} dues aux apports terrigène et marin, au moyen des relations suivantes :

(a) $C_{XT} = (X/_{Al})_{\text{croûte}} \times C_{Al}$ (b) $C_{XM} = (X/_{Na})_{\text{eau}} \times C_{Na}$
terrestre de mer

On suppose dans ce type de calcul que tout l'aluminium collecté est d'origine crustale et que l'essentiel du sodium provient de la source marine. Pour évaluer C_{XM}, il convient en toute rigueur de considérer dans la relation (b) un terme C_{Na} diminué d'une concentration théorique en sodium terrigène évaluée par la relation (a). Après avoir effectué cette correction, on a calculé les valeurs de C_{XT} et C_{XM} des divers éléments analysés en prenant respectivement dans les relations (a) et (b) le rapport crustal et marin de référence des modèles de MASON (8) et de BREWER (9). On a ensuite déduit les contributions respectives terrigène T et marine M à la concentration de l'élément X effectivement mesurée. Le tableau 2 donne la moyenne arithmétique de ces contributions théoriques exprimées en % des quantités élémentaires collectées.

Eléments		Na	Cl	S	K	Ca	Al	Si	Fe	Mn	P
Contribution des sources	A - Campagne d'Avril 1982										
	T %	25	≈0	≈0	62	56	100	≈100	≈100	≈60	37
	M %	75	>100	5	12	10	≈0	≈0	≈0	≈0	≈0
	B - Campagne de Septembre 1983 (1ère partie)										
	T %	0,5	≈0	0	7	5	100	>100	64	39	7
	M %	99,5	85	50	59	68	0	≈0	≈0	≈0	0,6
	C - Campagne de Septembre 1983 (2ème partie)										
	T %	19	0,6	≈0	37	51	100	>100	100	28	19
	M %	81	>100	3,5	11	11	0	≈0	≈0	≈0	≈0

Tableau 2.

Comme on pouvait s'y attendre, les valeurs du tableau 2 permettent de caractériser un groupe d'éléments typiquement terrigènes (Al, Si, Fe), pour lesquels la contribution de la source océanique est pratiquement nulle et une famille essentiellement marine (Na, Cl), associée aux particules salines produites par le pétillement de la mer (10).

Pour cette dernière famille, on note que dans les situations à influence continentale, le rapport Cl/Na peut devenir nettement inférieur à la valeur 1,8 de référence de l'eau de mer (4). On comprend dès lors que l'on ait dans le tableau 2 des contributions marines en chlore supérieures à 100%, le calcul conduisant pour cet élément à des concentrations théoriques souvent 2 à 5 fois plus grandes que celles effectivement mesurées. Un tel déficit en chlore, également observé par d'autres auteurs, a été attribué au passage sous forme gazeuse d'une partie de ce constituant des particules de sel marin, à la suite de réactions avec des composés acides associés à des masses d'air polluées (11, 12, 13, 14, 15).

Le potassium et le calcium présentent d'autre part un caractère mixte, étant tantôt à dominante terrigène (avril 1982) et fin de la campagne de septembre 1983), tantôt à dominante marine (début de la campagne de septembre 1983). La contribution terrigène théorique ne suffit pas par ailleurs à rendre compte des concentrations en Mn et P observées, la contribution marine étant alors quasi nulle. Pour ces deux constituants, il n'est pas exclu que les quantités collectées s'expliquent en partie par des apports anthropogéniques -effet de l'utilisation d'engrais phosphatés dans les zones cultivées (3), éventuelle influence du Mn résultant d'émissions industrielles-.

En ce qui concerne le soufre, la composante terrigène de cet élément est toujours infime. Dans le cas des situations à influence continentale, sa composante marine due à l'aérosol formé par éclatement de bulles à la surface de l'océan est faible (contribution moyenne de 5% en avril 1982). Cette composante devient importante dans les conditions océaniques du début de la campagne de septembre 1983 (contribution moyenne de 50%). Le tableau 3 apporte des compléments d'information sur les caractéristiques du soufre particulaire lors de nos campagnes.

Période d'échantillonnage	Situation météorologique	Rapport S/Na (ensemble des particules collectées par filtration d'air)	Contribution en masse au soufre particulaire collecté -d'après(3,16)-	
			d > 2 μm	d < 2 μm
Avril 1982	A. influence continentale	2,5	≃10 %	≃90 %
15-23 Sept. 1983	B. influence océanique	0,10 - 0,20	45 %	55 %
23-30 Sept. 1983	C. influence continentale	3,2	4 %	96 %

Tableau 3.

On remarque que la contribution des grosses particules dans la gamme de taille des "jets drops" issues du pétillement de la mer (17) est de loin la plus élevée lors de la période océanique de septembre 1983 (contribution à la masse du soufre collecté de 45% pour d > 2 μm). Pour cette même période, le rapport S/Na de nos échantillons est en moyenne

égal au double du rapport de référence de l'eau de mer (S/Na = 0.084) et
correspond aux valeurs généralement observées au-dessus de zones océaniques
non polluées de l'hémisphère Sud [18]. Bien que l'on ne puisse complètement
écarter un éventuel effet d'une pollution soufrée, de telles valeurs semblent surtout traduire la présence d'un aérosol fin résultant de la conversion du SO_2 produit après oxydation de gaz organosoufrés biogéniques émis
par une source marine [18,19,5]. Si à cet égard, l'existence d'un champ
d'algues à la pointe de Penmarc'h est à l'origine d'émissions intenses de
composés de ce type [5], il parait difficile de pouvoir recueillir locale-
ment en quantité notable l'aérosol soufré provenant de leur transformation.
En effet, le temps de résidence du SO_2 mesuré dans l'atmosphère marine [20]
ou déterminé à partir de modèles cinétiques [21], est au moins de l'ordre
de la demi-journée. Par ailleurs, les concentrations en soufre particulaire
observées ne sont guère plus élevées que celles trouvées simultanément en
l'absence de champ d'algues à environ 50 km de Penmarc'h sur le site de la
pointe du Raz [4]. L'écart au rapport marin S/Na semble alors principalement dû à l'influence de la source océanique biogénique diffuse.

Dans les situations à influence continentale, le rapport S/Na devient en moyenne 30 à 40 fois plus grand que le rapport marin de référence.
L'excédent de soufre, qui est essentiellement porté par de fines particules, est responsable des fortes concentrations observées pour cet élément
-en moyenne près de 3000 ng.m^{-3} en avril 1982 et de 5000 ng.m^{-3} pendant
la seconde partie de la campagne de septembre 1983, alors que la valeur
moyenne n'était que de 750 ng.m^{-3} durant la période océanique de cette
campagne-. Un tel résultat peut être attribué à un effet de la contamination anthropogénique, compte-tenu de l'arrivée en Bretagne de masses d'air
ayant survolé diverses régions industrialisées (Grande-Bretagne, Nord du
continent Européen, Nord de l'Espagne). L'étude fine de l'évolution de la
trajectographie à grande échelle des masses d'air échantillonnées vient
renforcer cette explication [1, 3, 4, 16].

CONCLUSION

La présente étude a apporté des données d'intérêt régional sur les
teneurs dans l'air d'une dizaine de constituants de l'aérosol côtier collecté sur le littoral Atlantique à la Pointe de la Bretagne. Il est apparu
que leur concentration dépend dans une large mesure des conditions météorologiques rencontrées. Elle peut présenter des variations temporelles fines
de grande amplitude, atteignant parfois un ordre de grandeur (cas de la
campagne d'avril 1982), qui auraient été masquées si l'on avait utilisé
des durées de prélèvement égales ou supérieures à 24 heures. Selon la provenance des masses d'air arrivant sur la zone d'échantillonnage, on assiste
à une compétition entre les influences des sources océaniques et continentales. Une influence océanique prépondérante a été observée au début de la
campagne de septembre 1983. Les concentrations en soufre mesurées à cette
époque s'expliquent en partie par un apport de grosses particules salines
résultant du pétillement de la mer et vraisemblablement aussi par une
contribution marine d'origine biogénique. Dans des situations dominées par
l'influence de l'environnement continental (avril 1982 et fin de la campagne de septembre 1983), des concentrations en soufre particulaire très élevées ont été mises en évidence. Elles reflètent dans une très large mesure
l'impact d'une importante pollution soufrée en provenance de diverses zones
industrielles de l'Europe Occidentale (Grande-Bretagne, Nord de l'Europe
Continentale, Nord de l'Espagne). Dans ce type de situation, on observe du-
souvent des déficits marqués en chlore, qui sont généralement attribués au
passage sous forme gazeuse d'une partie de ce constituant des particules de
sel marin. [11,12,13,14,15].

Nos recherches ont bénéficié d'un soutien de la Direction des Etudes et Recherches d'Electricité de France (Convention EDF/AEE287). Elles se poursuivent avec une aide du Secrétariat d'Etat chargé de l'Environnement et de la Qualité de la Vie.

REFERENCES

(1) MORELLI J., MARCHAL T., GIRARD-REYDET L., CARLIER P., PERROS P., LUCE C. et GIRARD R., 1983, "Variations des concentrations en soufre particulaire dans un environnement atmosphérique côtier en relation avec le changement d'origine des masses d'air", J. Rech. Atmos. 17, n° 3, 259-271.

(2) GIRARD-REYDET L., MARCHAL T., MORELLI J. et MASNIERE P., 1983, "Etude de la composition chimique de l'aérosol atmosphérique en bordure du littoral breton", Convention EDF/AEE287, Rapport final, Département Environnement Atmosphérique et Aquatique, Direction des Etudes et Recherches, Electricité de France.

(3) MARCHAL T., 1983, "Contribution à l'étude physico-chimique de l'aérosol atmosphérique en milieu côtier", Thèse de Doctorat de 3ème cycle, Université de Paris 7.

(4) GIRARD-REYDET L., Thèse de Doctorat de 3ème cycle, Université de Paris 7, soutenance prévue en 1984.

(5) CARLIER P., LUCE C., GIRARD R., MOUVIER G., MORELLI J., GIRARD-REYDET L., MARCHAL T. et CADENE S., 1984, "Etude de l'influence d'une source locale naturelle intense de composés organo-soufrés sur la chimie de la troposphère en milieu non pollué", Proceedings of the Third European Symposium on Physico-Chemical Behaviour of Atmospheric Pollutants (Varese, Italy, 10-12 april 1984), Ed. B. Versino & H. Ott, This issue.

(6) CARLIER P., 1984 "L'ozone serait-il l'oxydant principal des composés organo-soufrés en milieu océanique ?", Communication au "Quadriennal Ozone Symposium" organisé par IAMAP, (Kassandra-Halkidili, Grèce, 3-7 septembre 1984).

(7) FEDOROFF M., 1979, "Simple data processing for gamma spectroscopy in activation analysis", In computers in activation analysis and gamma ray spectroscopy, Proceedings of the American Nuclear Society, Puerto Rico (april 30 - May 4, 1984), Published by US Department of Energy, 653-663.

(8) MASON B, 1966, Principles of geochemistry, 3rd Edition, Wiley & Sons, New York.

(9) BREWER P.G., 1975, Chemical Oceanography, 2nd Edition, Ed. Riley and Skirrow, Vol. I, Academic Press.

(10)BLANCHARD D.C,, 1963, "The electrification of the atmosphere by particles from bubbles in the sea", In Progress in Oceanography, 1, 71-202, Ed. Mary Sears, Pergamon Press, Oxford.

(11)CHESSELET R., MORELLI J. and BUAT-MENARD P., 1972, "Some aspects of the geochemistry of marine aerosols", In The changing chemistry of the Oceans, Nobel Symposium 20, Ed. D. Dyrssen & D. Jagner, Stockholm, 93-114.

(12)BERG W.W. Jr. and WINCHESTER J.W., 1978, "Aerosol Chemistry of the Marine Atmosphere", In Chemical Oceanography, Ed. Riley and Chester, 2nd Ed., Academic Press, Vol. 7, 173-231.

(13) TEN BRINK H.M., MALLANT R.K.A.M., GOUMAN J.M., KOS G.A. and VAN DE VATE J.F., 1980, "SO_2 conversion in a marine atmosphere", Proceedings of the First European Symposium on Physico-Chemical Behaviour of Atmospheric Pollutants (Ispra, Italy, 16-18 october 1979), Ed. B. Versino & H. Ott, 298-306.

(14)METTERNICH P., GEORGII H.W. and GROENEVELD K.O., 1984 "Elemental composition and size distribution of atmospheric aerosols during long range
transport", Proceedings of the Third European Symposium on Physico-Chemical Behaviour of Atmospheric Pollutants (Varese, Italy, 10-12 april
1984), Ed. B. Versino & H. Ott, This issue.

(15)VIERKORN-RUDOLPH B., RUDOLPH J., MEIXNER F.X., BACHMANN K. and SCHWARZ
B., 1984, "Vertical and horizontal profiles of hydrogen chloride in the
Mediterranean region", Proceedings of the Third European Symposium on
Physico-Chemical Behaviour of Atmospheric Pollutants (Varese, Italy,
10-12 april 1984), Ed. B. Versino & H. Ott, This issue.

(16)ROMERO J., Rapport d'un stage effectué dans le cadre du Diplôme d'Etudes
Approfondies "Chimie de la Pollution" de l'Université de Paris 7 (en
préparation).

(17)MACINTYRE F., 1974, "Chemical fractionation and sea-surface microlayer
processes", In The Sea, Vol. 5, Chapter 8, Ed. E.D. Goldberg, John
Wiley & Sons, New-York, 245-299.

(18)BONSANG B., NGUYEN B.C., GAUDRY A. and LAMBERT G., 1980, "Sulfate enrichment in marine aerosols owing to biogenic gaseous sulfur compounds",
J. Geophys. Res., 85, 7410-7416.

(19)NGUYEN B.C., GAUDRY A., BONSANG B. et LAMBERT G., 1978, "Reevaluation
of the role of dimethylsulfide in the sulfur budget", Nature, 275, 637.

(20)NGUYEN B.C., BONSANG B., LAMBERT G. and PASQUIER J.L., 1975, "Residence
time of sulfur dioxide in the marine atmosphere, Pure and Appl. Geophys., 123, 489-500.

(21)CALVERT J.G. and STOCKWELL W.R., 1983, "The mechanism and rates of the
gas phase oxidations of sulfur dioxide and nitrogen oxides in the atmosphere", In Acid Precipitation : SO_2, NO and NO_2 oxidation mechanisms,
atmospheric considerations, Chapter 1, Ann Arbor Science, An Arbor,
Michigan.

TOXIC METALS AND METALLOIDS IN HIGH ALTITUDE ALPINE GLACIERS SNOW AND ICE

F.M. BATIFOL and C.F. BOUTRON
Laboratoire de Glaciologie et Géophysique de l'Environnement
du CNRS, 2, rue Très-Cloîtres 38031 GRENOBLE, France

Abstract

Assessing the occurence of toxic metals and metalloids such as Pb, Hg, Sb, Cd, Ag, Se, As, Zn and Cu in the successive snow and ice layers in high altitude alpine glaciers is very interesting in order to investigate historical variations of the atmospheric concentrations of these elements from pre-industrial times to present in populated temperate areas on a regional scale. The available data are unfortunately very sparse, and most of them appear to be highly questionable because of improper sampling site selection and because of not fully solved contamination problems during field sampling, laboratory analysis or both.

1. INTRODUCTION

During recent years, numerous though often questionable data have been published on the occurence of toxic metals and metalloids (1) (2) such as Pb, Hg, Sb, Cd, Ag, Se, As, Cu and Zn in the successive dated snow and ice layers deposited in the large remote Antarctic and Greenland polar ice caps. The major goal is to get historical records of background atmospheric concentrations of these metals and metalloids in the most remote areas of both hemispheres from prehistoric times to present. These data have been recently critically reviewed and assessed (3).

Surprisingly, there are on the other hand presently very few data on the occurence of toxic metals and metalloids in the successive snow and ice layers deposited in high altitude alpine glaciers both in Europe and in other temperate areas of the world. These glaciers are less remote from populated and industrialized areas of the world than are polar ice caps, and therefore have considerable potential to provide investigators with historical records of atmospheric concentrations of these metals and metalloids in temperate areas on a regional scale. Such glacier data are greatly needed, especially since data from precipitation chemistry sampling networks are often very questionable and do not allow to determine clear regional patterns and time trends mainly because of variability in the methods of collection and in the analytical techniques (4).

In this paper, we shall first succintly review the various criteria which should be taken into account when selecting alpine glacier sampling sites for the collection of snow and ice samples suitable for obtaining reliable historical records of atmospheric concentrations of toxic metals and metalloids. The field sampling techniques and laboratory analytical procedures will then be discussed. The toxic metals and metalloids data presently available for alpine glaciers will then be rapidly reviewed.

2. FIELD SAMPLE COLLECTION AND LABORATORY ANALYSIS

Sampling sites

Proper selection of sampling sites is of paramount importance. An ideal site should especially meet the following conditions : (a) High enough elevation in order to have zero to negligible surface melting even during summer months, in order to get snow or ice layers unaffected by post-depositional percolation (b) Location not too wind exposed in order to avoid any strong mechanical mixing of the successive snow layers after deposition (c) Mean annual snow accumulation rate as low as possible and glacier thickness as large as possible so that long time series can be obtained (d) Rather flat location with simple firn and ice flow patterns in order to get reliable dating of the successive snow and ice layers (e) Location remote from any local contamination sources (mountaineers, helicopters...).

It is unfortunately very difficult to find sampling sites which meet all these conditions. This explains why all presently available data on toxic metals and metalloids in high altitude temperate glaciers have been obtained from the analysis of samples collected at rather imperfect sites. Post depositional effects by percolation and wind are especially often very significant so that the datation of the samples and interpretation of the data are difficult. Annual accumulation rates are moreover very high : as an illustration about 35 g H_2O cm^{-2} yr^{-1} at Colle Gnifetti (4450 m a.s.1) in Switzerland where Wagenbach and coworkers collected their samples (5)(6)(7)(8) and at Col du Gouter (4300 m a.s.1) in France where Briat, Boutron and coworkers collected most of their samples (9)(10)(11) ; and more than 250 g H_2O cm^{-2} yr^{-1} at the summit of Mont Blanc where Batifol and Boutron obtained part of their samples (9). The time series which can be obtained are therefore rather short, 50 to 100 years at best.

Field sampling procedures

Contrary to what was anticipated by many investigators, concentrations of toxic metals and metalloids in high altitude temperate snow and ice are extremely low, in the 10^{-8} g/g to 10^{-11} g/g range, i.e. concentrations close to the ones measured in present day Greenland snows (3). Extreme care must then be taken to avoid to get data which would be unreliable because of not fully solved contamination problems during field sampling, laboratory analysis or both.

Satisfying ultraclean sampling techniques similar to the ones developped for Antarctic and Greenland studies (3) can be used to collect surface or near surface snow samples. The samples are collected from the walls of shallow hand dug pits. After cutting back the upwind vertical wall of the pit to a distance of a few decimetres with acid cleaned plastic shovels, the samples are collected by operators in clean room garb by forcing acid cleaned high purity containers into the snow (7)(9)(10)(12). Due to the high annual snow accumulation rates, the time series provided by such ultraclean pit samples are unfortunately very short, a few years at best.

The situation is much more uneasy if one wants to get longer time series by collecting deeper samples. An interesting possibility consists in collecting the samples from the vertical walls of deep crevasses or from ice cliffs. This has been done by Jaworowski and coworkers in the accumulation areas of various glaciers in Europe, Alaska, Nepal, Peru and Uganda (13)(14)(15). Jaworowski and coworkers have unfortunately used very dirty sampling techniques, so that the samples they obtained are probably highly contaminated (16). In the future, it should be possible to obtain ultraclean samples from the wall of crevasses or from ice cliffs by having operators in clean room garb excavate blocks of snow or ice from various levels using acid cleaned tools and containers, as was done in Greenland by Patterson and coworkers in deep firn trenches and in ice tunnels (17).

Another possibility to get deep old samples is to use hand operated mechanical metallic augers (10) or electromechanical drills (5)(6)(7). The outside of the small diameter (about 7 cm) medium depth cores so obtained is unfortunately always heavily contaminated, and the laboratory decontamination of such small diameter cores is extremely difficult, especially for poreous snow cores, since any outside contamination can easily diffuse into the interior. Anyway, one can try to decontaminate such core samples by mechanically removing the outside contaminated veneer layers using techniques similar to the ones recently developped for Antarctic and Greenland studies (18)(19). In any case, it is mandatory to evaluate the cleanliness of the inner part of the cores by drawing curves representing the variations of the measured concentrations as a function of radius : a clear plateau must be obtained in the inner part of the cores (3) (18) (19). To our knowledge, such carefully controlled laboratory decontamination of snow or ice cores collected in high altitude temperate glaciers has never been performed : the investigators who tried to decontaminate such cores (7)(8)(10) have indeed removed the outside of their samples without using ultraclean procedures, and have failed to control the cleanliness of the inner part they kept for analysis.

Several investigators have also tried to get old ice samples by excavating blocks of ice in terminus ablation areas of glaciers either directly from the surface or from shallow crevasses (13)(14)(15)(20). The ice in such ablation areas is indeed very old, up to more than 1000 years, as confirmed by several direct C^{14} or Si^{32} dating measurements (21)(22) and by ice velocity data. It should be possible to get very clean blocks using procedures cleaner than these used in (13) (14) (15) (20). A major problem is however that concentrations of toxic metals and metalloids in such ice are possibly strongly disturbed by various processes, mainly summer percolation, which results in questionable interpretations of the data so obtained.

Laboratory analytical procedures

Due to the very low concentrations, in the 10^{-8} to 10^{-11} g/g range, to be measured, the key problem associated with these procedures is contamination. The reliability of the data obtained will depend primarily on the control of contamination during the successive steps of the analytical procedures.

Contrary to what is thought by many investigators, the use of clean laboratories flushed with filtered air, or at least of high quality laminar flow clean benches, is mandatory to get reliable data on toxic metals and metalloids in snow and ice from high altitude temperate glaciers. All data obtained from analyses performed in conventional laboratories ((13)(14)(15) and several others) are then more or less questionable. Of paramount importance is also the choice of materials used to contact melted samples and reagents (quartz, various teflons and conventional polyethylene are generally the less contaminating materials), and the use of extensive cleaning procedures for all laboratory equipment (23). It is finally essential to be able to estimate contamination introduced during the various steps of the analytical procedure, including laboratory decontamination of the samples if any, by performing carefully controlled blank determinations.

Most toxic metals and metalloids to be measured are easily soluble using dilute acid methods. This is not the case of the various crustal reference elements such as Al, Fe and Mn whose measurement is necessary if one wants to calculate crustal enrichment factors (24) for the investigated toxic metals and metalloids. Reliable concentrations data for these crustal elements in temperate snow and ice can be obtained only if careful solubilisation of all solid particles is performed using concentrated hot acids, or if measurements are performed by Instrumental Neutron Activation of the frozen samples without any melting.

3. AVAILABLE DATA

The various metals and metalloids will be discussed successively, according to decreasing atomic weights. The discussion will not include data obtained by the analysis of winter snow pack in unglaciated areas.

Lead (207)

Available data on the occurence of this heavy metal in snow and ice at various glacier sites in the world are summarized in Table I. It can be seen that the data published by Jaworowski and coworkers (13)(14)(15) represent a large part of the data shown in Table I. These data have however been recently questionned by Patterson and coworkers (16), who convincingly showed that the concentration values obtained by Jaworowski and coworkers for Kahiltna Glacier in Alaska, Table I, are erroneous by about one order of magnitude because of improper highly contaminating field sampling techniques and laboratory procedures. All data by Jaworowski and coworkers (13)(14)(15) given in Table I must hence unfortunately probably be considered suspect.

The remaining data are very sparse and concern only a few sampling locations. They are moreover very scattered, from about 0.1 to 10 x 10^{-9} g Pb/g, and their quality is often probably questionable, with the exception of data published by Hinkley (12) for Kahiltna and Ruth glaciers in Alaska which are thought to be of very high quality. It is our feeling that it might turn out in the future that only the lowest concentration values in the 0.1 to 1 x 10^{-9} g Pb/g range or so are reliable and that higher concentration values are artificially too high because of not fully solved contamination problems.

From these available data, it is difficult to infer whether there has been any clear increase trend of the concentrations of Pb in snow or ice in temperate areas from pre-industrial times to present, especially during the last decades, or not. All available data for ancient ice several centuries old, whose analysis should have providen with reference baseline values for pre-industrial times, were indeed obtained by analysing ice collected in terminus ablation areas where strong summer melting occurs so that the ice samples might have been contaminated with modern snow or rain rich in anthropogenic Pb. Regarding now the last few decades, the only available time serie besides the questionable ones published by Jaworowski and coworkers is the one published by Briat (10) for the years 1948-1974 at Col du Gouter (4250 m a.s.l), France. Briat argues that there has been a 2 fold increase in Pb during that period, but much temporal variation occurs in the data possibly because the decontamination procedure used by Briat led her with analysed parts of her contaminated poreous snow cores still

Table I.

Location	Elevation (m a.s.l.)	Description of the samples	Pb measured concentrations (10^{-9} g Pb/g)	Reference	Comment
Alaska, Kahiltna Glacier	2450	7 samples 1950-1977	0.11-0.53 (mean 0.24) no clear time trend	(12)	Ultraclean sampling and analytical procedures
		23 samples 1950-1977	0.55-3.0 (mean 1.0) no clear time trend	(15)	Values questionned in (16)
Alaska, Ruth Glacier	550	1 sample 400 years old	0.12	(12)	Ablation area
Norway, Storbreen Glacier	1750	17 samples 1954-1972	1.6-20.0 (mean 8.6)	(13) (14) (15)	Values questionned in (16)
	1400	10 samples 100-700 years old	1.0-9.0 (mean 3.7) about 5 fold increase during the last 700 years	(13) (14) (15)	Ablation area, values question-ned in (16)
France, Col du Goûter	4250	Numerous samples 1948-1974	0.4-10.0 (mean 2) possibly 2 fold increase during the last 30 years	(10)	Decontaminated snow cores
France, Col du Midi	3600	4 samples 1973	1.7-9.3 (mean 4.0)	(9)	
France, Mont-Blanc	4800	7 samples 1973	1.8-3.6 (mean 2.5)	(9)	
Switzerland, Tsanfleuron Glacier	2500	3 samples recent snow 1974	1.1-1.5 (mean 1.4)	(20)	
		2 samples ice ⩾ 80 years old	0.3	(20)	Ablation area
Austria, Gurgler Ferner	3180	25 samples 1950-1974	1.6-17.0 (mean 8.8)	(15)	Values questionned in (16)
	2970	3 samples 200 years old	11.8-15.5 (mean 13) no clear time trend	(15)	Ablation area ; values question-ned in (16)
Nepal, Cherku Langtang Glacier	5350	15 samples 1957-1972	1.8-8.0 (mean 1.8)	(14) (15)	Values questionned in (16)
	4500	5 samples 700 years old	2.6-6.4 (mean 4.4) no clear time trend	(14) (15)	Ablation area ; values question-ned in (16)
Uganda, East Stanley Glacier	4755	12 samples 1960-1974	8.5-32.8 (mean 19.6)	(15)	Values questionned in (16)
Uganda, Elena Glacier	4511	4 samples 100 years old	12.8-54.6 (mean 21.6) no clear time trend	(15)	Ablation area, values question-ned in (16)
Peru, Jatunjampa Glacier	5350	21 samples 1955-1975	6.9-29.7 (mean 13.6)	(15)	Values questionned in (16)
	5050	4 samples 250 years old	12.3-37.3 (mean 22.8) no clear time trend	(15)	Ablation area ; values question-ned in (16)

Table I. Published data on the occurence of Pb in alpine glaciers in temperate areas.

contaminated at variable amounts.

A promising way to investigate the origin of Pb in temperate glaciers snow or ice is to look at its isotopic composition (20). This was done by Petit, Picciotto and coworkers for several samples collected at Tsanfleuron glacier in Switzerland (20)(25). For recent surface snow, they found isotopic ratios Pb^{206}/Pb^{204} Pb^{207}/Pb^{204} and Pb^{208}/Pb^{204} close to the mean ones observed in gasoline Pb emitted in Switzerland. For rather old ice more than 80 years old, these isotopic ratios were significantly higher and therefore closer to the ones in pre-industrial natural Pb. This suggests that at least part of Pb in recent snow in this glacier in Switzerland is derived from anti-knock Pb alkyl additives in automotive gasoline.

Hg (201)

To our knowledge, the only available data for this very interesting heavy metal are the ones published by Jaworowski and coworkers (15) for the various glaciers, listed for Pb in Table I, studied by these authors. They found Hg concentrations in the 40 to 2000 x 10^{-12}g Hg/g range, without any increase trend from pre-industrial times to present. All these Hg concentration values by Jaworowski and coworkers have however been strongly questionned by Patterson and coworkers (16), as was already the case for Pb.

Sb (122)

The only available data for this metalloid have been published by Wyttenbach and coworkers (26), who found about 15 to 100 x 10^{-12} g Sb/g in recent ice collected in a percolation area at Jungfraujoch (3470 m a.s.1) in Switzerland, and by Wagenbach (8) who measured about 2 to 200 x 10^{-12} g Sb/g in recent snow collected at Colle Gnifetti (4450 m a.s.1) in Switzerland. The reliability of these Sb data is difficult to assess.

Cd (112)

Jaworowski and coworkers have measured the concentrations of this metal both in old ice and in recent snow in the various glaciers listed for Pb in Table I (13)(14)(15). The measured concentrations range from 50 to 35000 x 10^{-12} g Cd/g, without any clear time trend from pre-industrial times to present. As was already the case for Pb and Hg, these Cd data by Jaworowski and coworkers are unfortunately probably suspect (16).

Other sparse and possibly questionable Cd values have been published for recent ice collected in a percolation area at Jungfraujoch (3470 m a.s.1) in Switzerland, about 30 to 140 x 10^{-12} g Cd/g (26), for recent snow collected at Col du Midi (3600 m a.s.1), Col du Gouter (4250 m a.s.1) and summit of Mont Blanc (4800 m a.s.1) in the french Alps, about 20 to 200 x 10^{-12} g Cd/g (9), and for snow cores covering the 1948-1974 time period collected at Col du Gouter, about 50 to 200 x 10^{-12} g Cd/g with possibly a 2 fold increase from 1948 to 1974 (10).

Ag (108)

To our knowledge, the only available data for this metal are the sparse tentative ones published by Batifol and Boutron (9) for recent snow collected at Col du Midi (3600 m a.s.1), Col du Gouter (4250 m a.s.1) and summit of Mont Blanc (4800 m a.s.1) in the french Alps. They obtained very low concentration values in the 10 to 33 x 10^{-12} g Ag/g range.

Se (79)

The only tentative concentration values for this interesting metalloid are the few ones published by Wagenbach (8) for recent snow collected at Colle Gnifetti (4450 m a.s.l) in Switzerland : about 10 to 25 x 10^{-12} g Se/g.

As (75)

The only tentative concentration values for this metalloid are the ones by Wyttenbach and coworkers (26) : about 10 to 100 x 10^{-12} g As/g for recent ice collected in a percolation area at Jungfraujoch (3470 m a.s.l) in Switzerland.

Zn (65)

There are some data on the occurence of this toxic metal at several glacier sites in the Alps. Wagenbach found about 1.4 to 2.8 x 10^{-9} g Zn/g in recent snow collected at Colle Gnifetti (4450 m a.s.l) in Switzerland (8). Briat obtained concentration values in the 1 to 9 x 10^{-9} g Zn/g range in snow cores covering the years 1948-1974 collected at Col du Gouter (10) (4250 m a.s.l) in the french Alps, without any significant increase trend during the last three decades. Finally, Batifol and Boutron found 0.9 to 6 g Zn/g in recent snow at Col du Midi (3600 m a.s.l) and 0.7 to 2.4 x 10^{-9} g Zn/g at summit of Mont Blanc (4800 m a.s.l) in the french Alps (9). The reliability of all these Zn values is difficult to assess and will need to be confirmed in the future.

Cu (64)

There are only few tentative data for this metal. Wyttenbach and coworkers obtained rather high concentration values in the 3 to 5 x 10^{-9} g Cu/g range for recent ice collected in a percolation area at Jungfraujoch (3470 a.s.l) in Switzerland (26). Briat found 0.2 to 2.5 x 10^{-9} g Cu/g for the snow layers deposited from 1948 to 1974 at Col du Gouter (4250 m a.s.l) in the french Alps, without any clear time trend during the investigated time period (10). Recently, Batifol and Boutron obtained low concentration values (0.1 to 0.3 x 10^{-9} Cu/g) for recent snow collected at Col du Midi (3600 m a.s.l) and at summit of Mont Blanc (4800 m a.s.l) in the french Alps (9).

4. CONCLUSIONS

From this rapid review, it appears that our understanding of the occurence of toxic metals and metalloids in alpine glaciers snow and ice is presently still extremely limited, and provide neither a clear insight on time variations of the concentrations of these elements from pre-industrial times to present nor an adequate picture as to the geographical variations of these concentrations on a regional scale.

Much work then still remains to be done in the future in this very promising field of interest. This work will however be useless if the investigators fail to focus their attention on the key problem of contamination during field sampling and laboratory analysis. They must use

ultraclean field sampling techniques and analytical procedures similar
to the ones recently developped for Greenland and Antarctic studies, in
order to get trustworthy fully reliable data.

Another critical point will be in proper selection of high altitude
sampling sites, expecially in order to get pristine firn and ice samples
unaffected by post-depositional effects such as melting. Such sites will
however provide with short-term time series only, 50 to 100 years or so
at best. It will therefore be necessary to clearly assess to what extent
very old ice collected in terminus ablation areas of glaciers is plagued
by summer percolation, and then to evaluate whether the analysis of such
ice can be used to get pre-industrial baseline concentrations of the
investigated metals and metalloids in temperate areas or not.

It will be also of utmost importance to initiate simultaneous air-snow
sampling programmes at various high altitude temperate alpine glacier
sites in order to investigate to what extent chronological changes observed
in snow and ice reflect parallel chronological changes in the local
atmosphere.

REFERENCES

(1) Wood J.M. (1974). Biological cycles of toxic elements in the
    environment. Science 183, 1049-1052.
(2) EPA (1976). National interim primary drinking water regulations.
    Rept. 570/9-76-003, U.S. Environmental Protection Agency,
    Washington D.C., USA.
(3) Boutron C. (1984). Atmospheric toxic metals and metalloids in the
    snow and ice layers deposited in Greenland and Antarctica from
    prehistoric times to present. In "Toxic Metals in the Air",
    Davidson C.I. and Nriagu J.O. Eds., Advances in Environmental
    Science and Technology Series, Wiley, New York.
(4) Lyons W.B. and Mayewski P.A. (1983). Glaciochemical investigations
    as a tool in the historical delineation of the acidic precipitation
    problem. In "The acidic deposition phenomenon and its effects :
    critical assessment review papers", Altshuller A.P. and Linthurst
    R.A. Eds., EPA Rept. 600/8-83-016 A, pp. 8-70 to 8-98.
(5) Oeschger H., Schotterer U., Stauffer B., Haeberli W. and Röthlis-
    berger H. (1977). First results from alpine core drilling projects.
    Zeit.Gletscherkunde und Glazialgeologie 13, 193-208.
(6) Gäggeler H., Von Gunten H.R., Rössler E., Oeschger H. and
    Schötterer U. (1983). Pb^{210} dating of cold alpine firn/ice cores
    from Colle Gnifetti, Switzerland. J. Glaciology, 29, 165-177.
(7) Wagenbach D. and Schötterer U. (1983). Charakterisierung der
    Aerosol deposition auf dem Colle Gnifetti aus Spurenelement-unter-
    suchungen von Firnproben. Zeit. Gletscherkunde und Glazialgeologie,
    in press.
(8) Wagenbach D. (1981). Pilotstudie zur aerosoldeposition auf einer-
    hochalpinen "kalten" firndecke. PhD thesis, Universität Heidelberg,
    FRG.
(9) Batifol F. and Boutron C. (1984). Atmospheric trace metals in high
    altitude surface snows from Mont Blanc, french Alps. Atmospheric
    Environment, Sub.
(10) Briat M. (1978) Evaluation of levels of Pb, V, Cd, Zn and Cu in
    the snow of Mt Blanc during the last 25 years. In "Studies in
    Environmental Science Vol. 1", Benarie M.M. Ed., Elsevier,
    Amsterdam.

(11) Jouzel J., Merlivat L. and Pourchet M. (1977). Deuterium, tritium and β activity in a snow core taken on the summit of Mont Blanc, French Alps. Determination of the accumulation rate. J. Glaciology, 18, 465-470.

(12) Hinkley T.K. (1980). Identity of natural and polluted dusts in snow packs at widely distributed sites. U.S. Geol. Surv. Open File Report 78-701-179-80.

(13) Jaworowski Z., Bilkiewicz J. and Dobosz E. (1975). Stable and radioactive pollutants in a Scandinavian glacier. Environ. Pollut. 9, 305-315.

(14) Jaworowski Z., Kownacka L., Bilkiewicz J. and Oakley D.T. (1977). Stable and radioactive pollutants in some Northern Hemisphere glaciers. In "Isotopes and Impurities in Snow and Ice", Proceedings of the UGGI Symposium, Grenoble August-September 1975, IASH Publ. 118, 112-115.

(15) Jaworowski Z., Bysiek M. and Kownacka L. (1981). Flow of metals to the global atmosphere. Geochim. Cosmochim. Acta 45, 2185-2199.

(16) Patterson C.C. (1983). Criticism of "Flow of metals into the global atmosphere" by Jaworowski et al., Geochim. Cosmochim. Acta 47, 1163-1168.

(17) Murozumi M., Chow T.J. and Patterson C.C. (1969). Chemical concentrations of pollutant lead aerosols, terrestrial dusts and sea salts in Greenland and Antarctic snow strata. Geochim. Cosmochim. Acta 33, 1247-1294.

(18) Ng A. and Patterson C.C. (1981). Natural concentrations of lead in ancient Arctic and Antarctic ice. Geochim. Cosmochim. Acta 45, 2109-2121.

(19) Boutron C. and Patterson C.C. (1983). The occurence of lead in Antarctic recent snow, firn deposited over the last two centuries and prehistoric ice. Geochim. Cosmochim. Acta 47, 1355-1368.

(20) Petit D. (1977). Etudes sur la pollution de l'environnement par le plomb en Belgique. Les isotopes stables du plomb en tant qu'indicateurs de son origine. Thèse, Université Libre de Bruxelles, Belgique.

(21) Clausen, H.B., Buchmann B. and Ambach W. (1967). Si^{32} dating of an alpine glacier. In "Commission of Snow and Ice", Proc. IUGG Symposium, Bern, September-October 1967, IASH Publ. 79, 135-140.

(22) Dansgaard W. and Clausen H.B. (1966). The Si^{32} fallout in Scandinavia. Tellus, 18, 187-191.

(23) Patterson C.C. and Settle D.M. (1976). The reduction of orders of magnitude errors in lead analyses of biological materials and natural waters by evaluating and controlling the extent and sources of industrial lead contamination introduced during sample collection and analysis. In "Accuracy in Trace Analysis", La Fleur P. Ed., National Bureau of Standards Special Publication 422, 321-351.

(24) Rahn K.A. (1976). The chemical composition of the atmospheric aerosol. Technical Report, University of Rhode Island, USA, 273 pp.

(25) Picciotto E.E. (1977). Sur l'origine des impuretés minérales de la glace. In "Isotopes and Impurities in Snow and Ice", Proceedings of the UGGI Symposium, Grenoble, August-September 1975, IASH Publ. 118, 78-87.

(26) Wyttenbach A., Rauter R., Stauffer B. and Schötterer U. (1977). Determination of impurities in ice cores from the Jungfraujoch by neutron activation analysis. J. Radioanal. Chem. 38, 405-413.

RESULTS OF MANY YEARS' ANALYSES OF PRECIPITATION CHEMISTRY ON SAMPLES
OBTAINED SIMULTANEOUSLY AT 3.0 KM, 1.8 KM, AND 0.7 KM ASL

R. REITER, K. PÖTZL, and K. MUNZERT
Fraunhofer Institute for Atmospheric Environmental Research
Garmisch-Partenkirchen, FRG

Summary

Since 1964 up to now precipitation samples have been obtained synchronously at 3 neighboring mountain-observatories: Zugspitze (2964 m ASL),
Wank peak (1780 m ASL), and valley floor (740 m ASL). These samples
have been collected at high frequency (sometimes less than 1 day , depending on precipitation intensity). They have been measured by chemical methods for NO_3^-, NH_4^+, $SO_4^=$, Cl^- (and further ions) and recently, in
addition, for K^+, Na^+, Ca^{++}, as well as for pH and electrical conductivity. The data obtained until today are analyzed as time series for
recognition of trends or irregularities and with regard to the frequency distribution of specific concentrations. Furthermore is represented per m^2 the amount of trace substances added to the soil. A more
intensive statistical fine analysis and especially parameterization
with meteorological parameters recorded simultaneously at the observa-
tories are under way and will be published in the near future.

1. BASIC COMMENTS

Dealing with the nowadays so important problem of environmental pollution through acidity and chemical composition of precipitation requires the
following fundamental actions: Uninterrupted periods of sampling over as
many years as possible; sampling at very short intervals (possibly less
than 1 day) as a function of the type of precipitation; if possible, simultaneous collection of precipitation at different altitudes at mountain stations of small horizontal distance; if possible, detection of all acid-
forming ions and the most important cations besides the anions and, additionally, of pH and conductivity Λ; precise and consistent supply of meteorological data from each station, irrespective of precipitation periodic-
ity; if possible, comparison with the chemistry of aerosol particles.

2. SUDIES PERFORMED

2.1. Stations

Samples have been taken each synchronously at the following own obser-
vatories: Zugspitze peak (Z, 2964 m ASL), Wank peak (W, 1780 m ASL), and
valley institute Garmisch (G, 740 m ASL). All chemical analyses were made
immediately at the valley institute to which the precipitation samples have
been brought as quickly as possible. The pH and Λ-measurements were made
right after collection directly at the stations. Duration of sampling: up
to 1 day maximum during showers and thunderstorms and, depending on the respective circumstances, 1 to 3 days during persistent, uniform precipitation from an upsliding air mass. The type of precipitation (snow, graupel,
rain, etc.) has been observed and noted at each station.

2.2. Time Series

NO_3^- and NH_4^+ from 1964 until today; $SO_4^=$ and Cl^- from 1977 up to now; determination of Na^+, K^+, Ca^{++}, pH and Λ was started at the beginning of 1982. The very early determination of NO_3^- and NH_4^+ in precipitation was motivated by atmospheric-electrical problems. For, it was very soon and definitely shown that production of nitrous gases takes place in the atmospheric range of precipitation formation and cloud turbulence due to electrical discharges on precipitation particles so that we are facing here a natural, additional component (1).

2.3. Detection Methods

Methods of analyses on the basis of spectral-photometry. Determination of metal ions by flame-photometry.

3. BRIEF SURVEY OF LITERATURE

A complete survey of literature cannot nearly be expected here. Mentioned be only the following publications: Papers (2-4) deal with measurement methods and stations installed by the WMO. Comprehensive references to the problem of acid deposition are given in (5). Results of studies on dry and wet deposition, especially of acid-forming precipitation are presented in (6-8). The global distribution of acid rain is discussed by (9). For instructive studies with comprehensive summary of results of different authors see (10-12). Causes and effects of acid rain are dealt with in (13). Extensive chemo-theoretical reflections in connection with disturbances in hydro-geochemical circles are found in (14). Current views regarding effects of acid rain in the soil see in (15). In the recent past L. Machta dealt with the question of whether acid rain might some day be controllable.

In view of the relevant literature we can summarize that in spite of an ever increasing spate of worldwide discussions in recent years there are rarely any fundamental long-term measuring series found in the publications. However: the presumably longest pH-measurements in precipitation were made in the USA but lacking are sufficiently long analyses of the acid-forming components. Although in Sweden the known network for precipitation chemistry (Bolin) was installed very early, only total monthly samples have been taken and it is questionable to what extent such are to be regarded as representative and clarifying. We must consider that the pH by itself cannot be an informative quantity for an extremely diluted, unbuffered solution of different anions and cations. Decisive is always the real chemical composition of precipitation.

4. RESULTS

4.1. Time Series for Chemical Precipitation Components

Results are presented in Figs. 1-17 and speak for themselves so that a detailed description is hardly necessary. Figs. 1-8 show the time trends through lining-up of seasonal mean values of components NO_3^-, NH_4^+, $SO_4^=$ and Cl^-. The concentrations in precipitation are presented each in pairs in mg/l and mg/m², respectively. Figs. 1-8 are plotted all in analogous manner: abscissa = time given in years, ordinate = concentration or soil burden (arithmetical seasonal means; Z = Zugspitze; W = Wank; G = Garmisch (see 2.1.).

If we look at the long rows in Figs. 1-4 we note that at least from 1974 up to now a significant trend is recognized neither for NO_3^- nor for NH_4^+. This holds for all 3 stations. Remarkable is only that the concentra-

tion of NO_3^- increased from 1967 to 1975 by about factor 2 at all 3 stations
(Fig.1). This increase is weaker for soil burden (Fig.2). In the same interval (1966-1975) we find also a doubling of concentration and soil burden
for NH_4^+. But after this increase there is with certainty no further trend
observable. We, of course, have always seasons with extreme values towards
maximum or minimum. The high NO_3^- deposition on the soil in G in summer
1983 (cf. Fig.2) is due to a high precipitation rate. Since NO_3^- and NH_4^+
compensate in precipitation regarding acid formation, the mentioned increase in the acid-forming NO_3^- in G (Fig.2) cannot be regarded as critical.

Let us look at the SO_4 data shown in Figs. 5 and 6. It is remarkable
that concentration as well as soil burden increased continuously from 1977
to 1982 and by about a factor 4. An analogous increase or a general significant variation of the SO_4 content in the aerosol (18,19) could not be observed from 1972 to 1982. This is in contrast to (17) where a creeping increase of SO_4 and NH_4 is postulated on page 303. Of particular interest,
too, is the behavior of SO_4 on Figs. 5 and 6 in the years 82/83. In winter,
we find a sharp decrease of concentration and soil burden to values from
77/78. But immediately thereafter the concentration and also soil burden
rose extremely in summer 83 followed by a decline towards fall 1983. Thus,
we have to deal here with extreme irregularities that interrupt the rising
trend. Regarding frequency, the SO_4 concentration (mg/l) can easily be sub-
divided into 4 source areas of the air mass from which precipitation falls,
as shown by Table 1. It is clearly evident that the lowest $SO_4^=$ concentrations occur, in all, most frequently in hemispheric-maritime and continental air masses (more often at Wank than in Garmisch) whereas higher $SO_4^=$
concentrations of regional origin are generally rarer. Practically they are
no more observable at station Wank but only yet in the valley.

Table 1: Rel. frequency given in % per .5 mg/l

W:	11 - 20	3 - 10	0.7 - 2	0
G:	3 - 12	2 - 7	1 - 6	0.5 - 3
Source:	hemispheric maritime	continental	regional	only local

It is a fact that we have to consider not only mean values over longer
intervals but also the fine structure as a function of time and meteorology.
So the seasonal means presented here are still too rough. The values of the
single expositions are now being analyzed - as a function of meteorological
parameters - for the spectral frequency distribution of the individual
precipitation components. In this way it is possible to draw significant
conclusions as to biological effects.

When we now turn to Cl^- in Figs. 7 and 8, we cannot observe any trend
but there are some peak values in 81 or 82, which do not occur in like
manner at all stations. The values from 1983 are again extremely low.
(There is evidence showing that Cl^- and Na^+ come mostly from the maritime
area).

4.2. Cumulative Frequencies

In Figs. 9-12 are plotted, separately for the 3 stations (arranged horizontal to each other), the cumulative frequencies in % for different

stages of the chemical component in mg/l (= numbers at the curves). Fig.9 clearly shows how the frequency of very low concentrations of NO_3 decreases continuously from Z through W (and less pronounced to G). Thus, it seems to be obvious that there exists for nitrous gases a pronounced washout and an active atmospheric-electric source of formation in the cloud region between Z and W (1). Also Fig.10 reveals a distinct washout for NH_4^+ from Z through W to G (frequency of weak concentration stages increasing with height from 1-3). A completely different behavior shows SO_4 in precipitation: from 1978 to 1982 the frequency of low concentrations decreased constantly which means, in turn, a continuous increase in concentration. An irregular trend with time can be seen in Fig.12 for Cl^- (dependence on the occurence of maritime air masses).

Figs.13-16 represent the cumulative frequencies for NO_3^-, NH_4^+, $SO_4^=$, Cl^- in precipitation (numbers mg/l at the curves) as a function of season averaged over all years and for all 3 stations. For NO_3^- we find in spring a minimum of the frequency of low concentrations, the same applies to NH_4^+ (Fig. 14) but for summer and fall. Agreement from station to station is very good. In contrast, we observe in Fig.15 an inverse variation for SO_4 between G and W with regard to the dependence on season of weak concentrations: highest frequency in the valley in spring and at the Wank in summer. Furthermore it can be seen in Fig.15 that the cumulative frequency for low concentrations is also in winter at W essentially less than in G. This is evidence for accumulation of SO_2-gas under inversions during the heating periods. In summer, however, the frequency for mg/l is the same at both stations due to the intense vertical exchange which carries SO_2 and its secondary compounds far beyond the level of W. The cumulative frequency distributions of the Cl^- concentration in precipitation of G and W resemble one another in the annual variation (Fig.16). Both differ drastically from the behavior at station Z: there the frequency is almost independent of the season. This is due to the fact - also with regard to our aerosol-chemical studies - that Cl^- is to a considerable extent of maritime origin (air masses causing persistent precipitation). Therefore, the Cl^- content in precipitation at Z is determined by that in maritime air masses and is less dependent on seasons and vertical mixing conditions.

4.3 The pH of Precipitation

Finally a word to the frequency distribution of pH-values in precipitation at G and W (Fig.17). We do not find any significant differences in the pH-value of precipitation at the two stations. The highest frequency lies closely below 5 pH. That means: the most frequent pH-value is not much lower than that to be expected for precipitation from a completely unpolluted atmosphere (5.6), however with a CO_2 concentration of 350 ppm.

5. CONCLUSIONS

The only remarkable long-term variation of that component of precipitation which determines - along with others - the degree of acidity is the $SO_4^=$. From 1977 to 1982 we find at 740 m as well as at 1780 m ASL a significant increase in concentration and supply to the soil. In the recent past extreme fluctuations are observed. The frequency of low SO_4 concentrations decreases also continuously and in a layer of at least 1 km thickness. Nevertheless the pH in precipitation does not yet assume a critical value of acidity. NO_3^- and NH_4^+ show only slight variations over 20 years. Especially in the last 10 years there is no significant increase. Continuation of studies and detailed analyses of results are required. A complete collection of all results until spring 1983 can be found in (18).

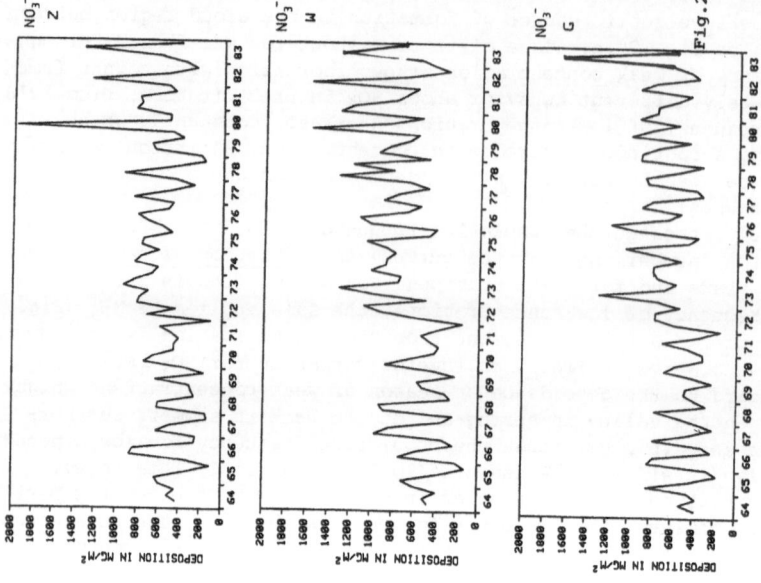

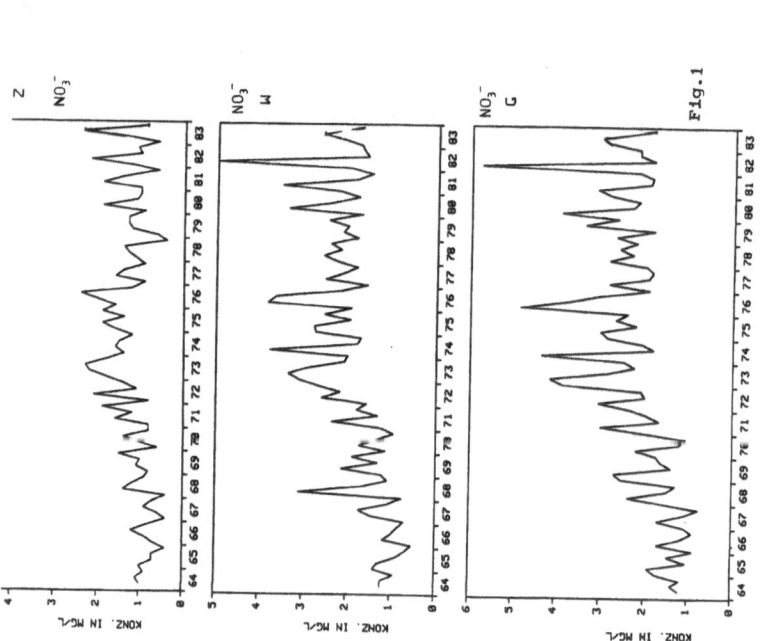

Figs. 1 + 2: Concentration of NO_3^- in precipitation (mg/l) and supply to the soil (mg/m²) since 1964. The increase from 1964 to 1976 is in good agreement with the increase of the NO_x emission in the FRG from 1966 to 1978 of 2,0 to 3,1 Mt/a (Umweltbundesamt (Federal Environmental Agency),1984). Thereafter no trend observable any more (NO_x emission 1978/1982 constant!). (Seasonal Means).

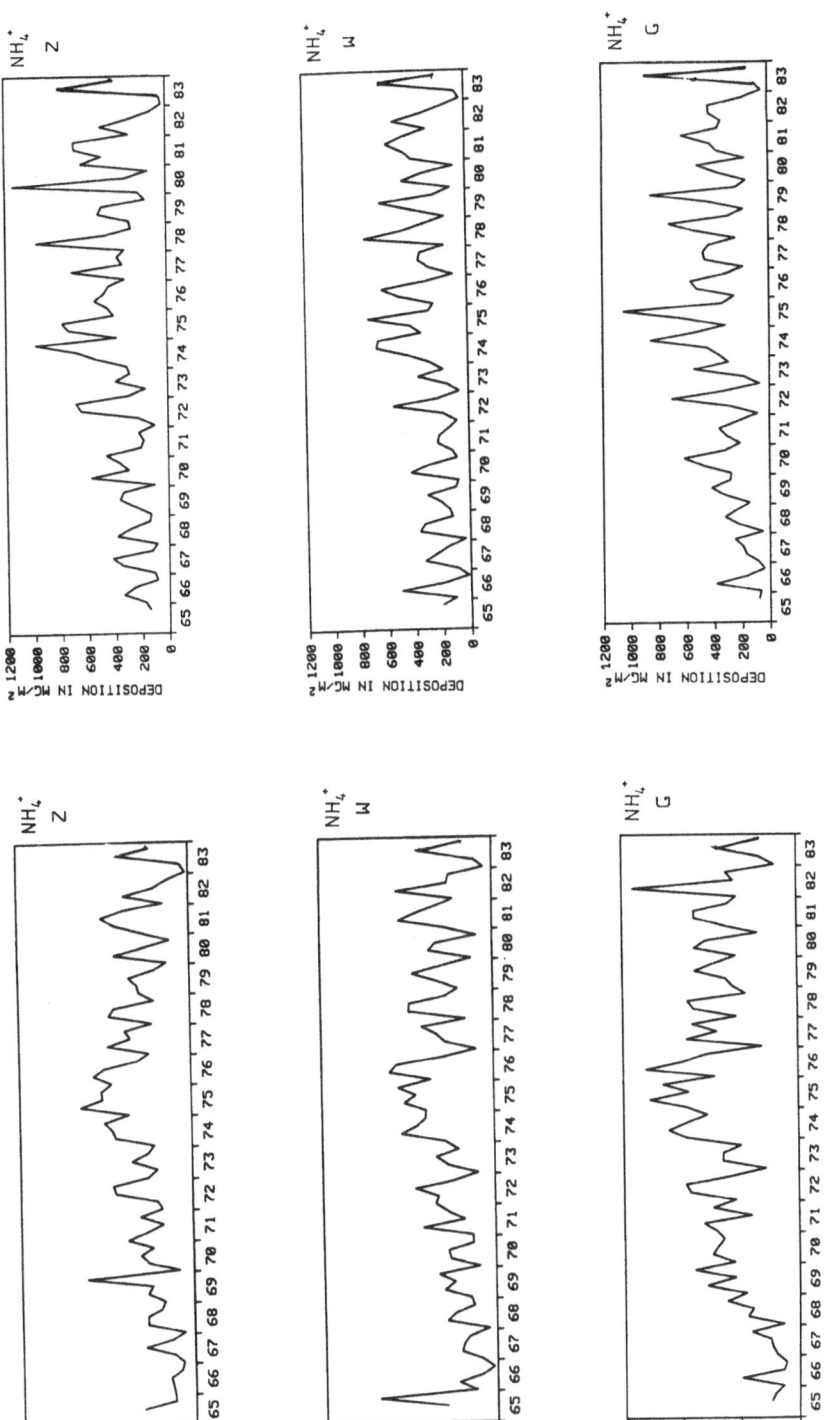

Fig. 3

Fig. 4

Figs. 3 + 4 : The behavior of NH_4^+ in precipitation, plotted analogous to figures 1 and 2. Behavior almost identical with that of NO_3^- (seasonal means)

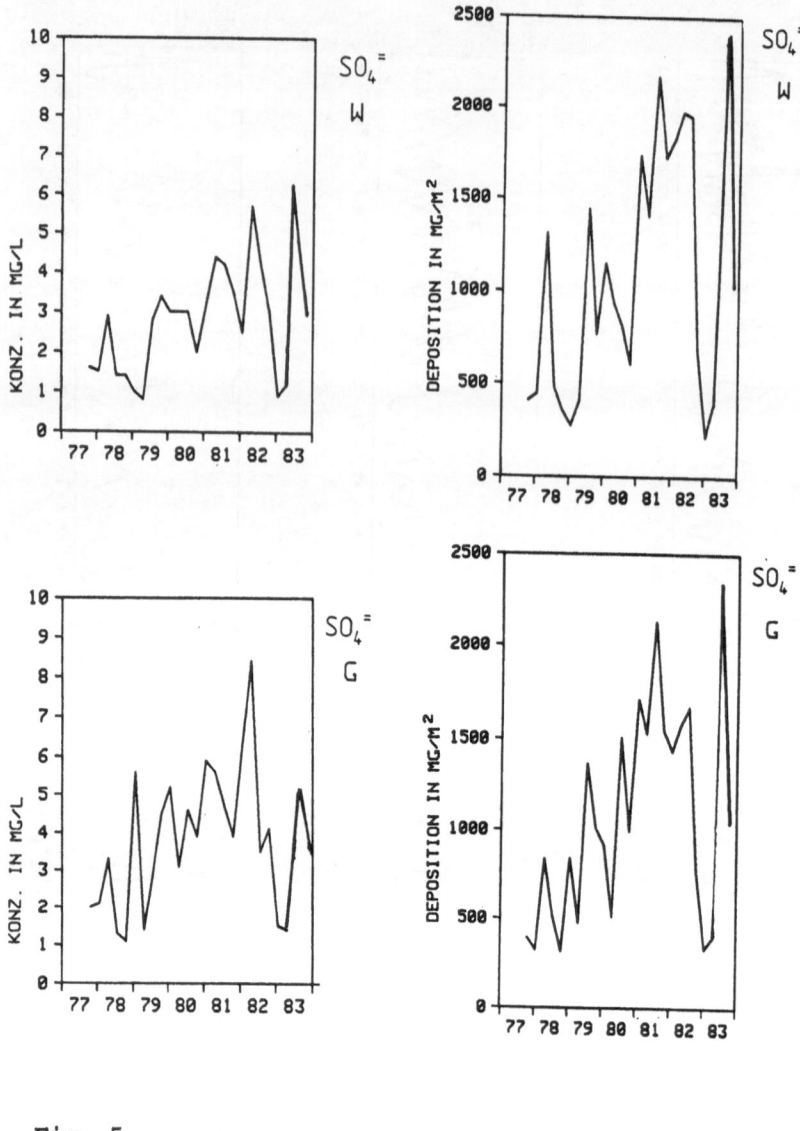

Fig. 5   Fig. 6

Figs. 5 + 6 : Concentration of $SO_4^=$ in precipitation
(mg/1) as well as supply to the soil both increasing
since 1977. Data superposed by strong irregularities
(factor 4) the minimum being however remarkable in
winter/spring 1983.

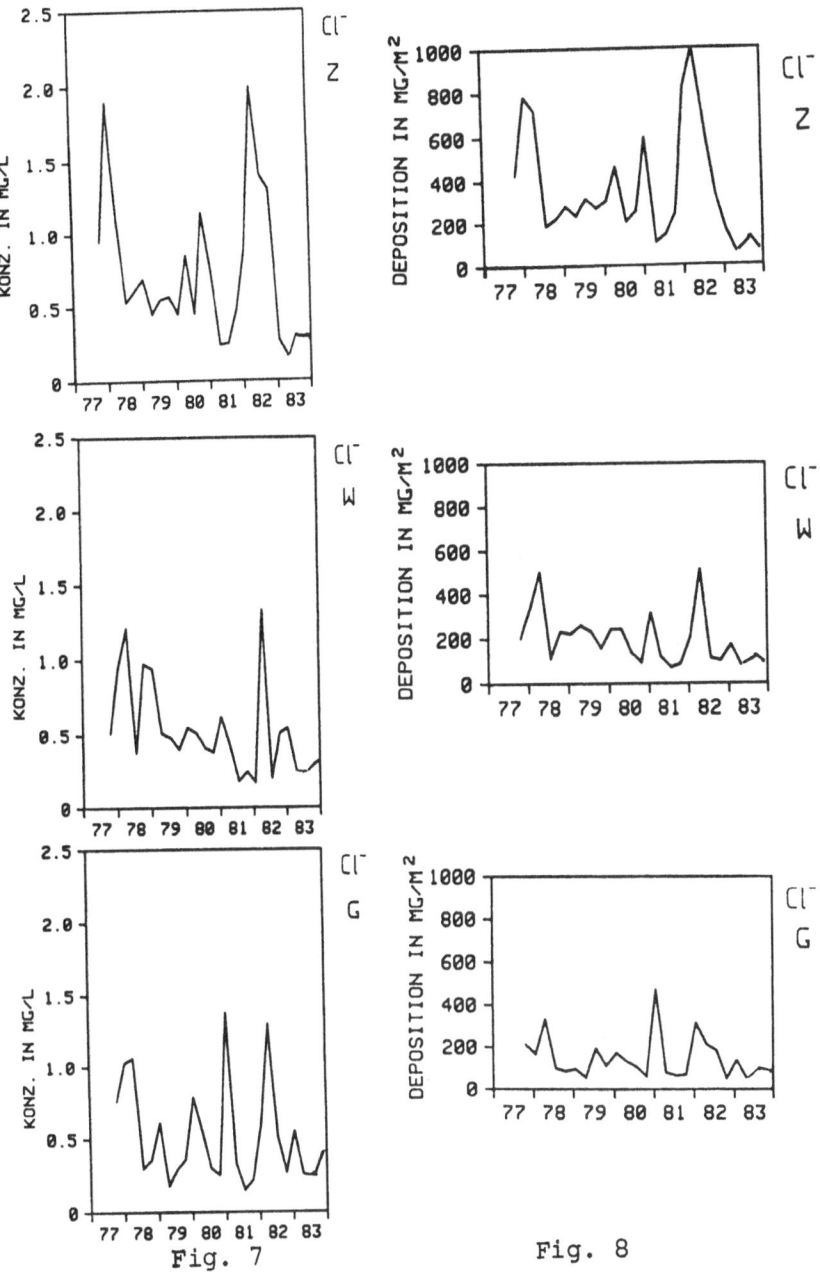

Fig. 7

Fig. 8

Figs. 7 + 8 : Neither concentration in precipitation
(mg/l) nor supply to the soil (mg/m^2) reveal a trend for
Cl$^-$ but strong irregularities are superposed.

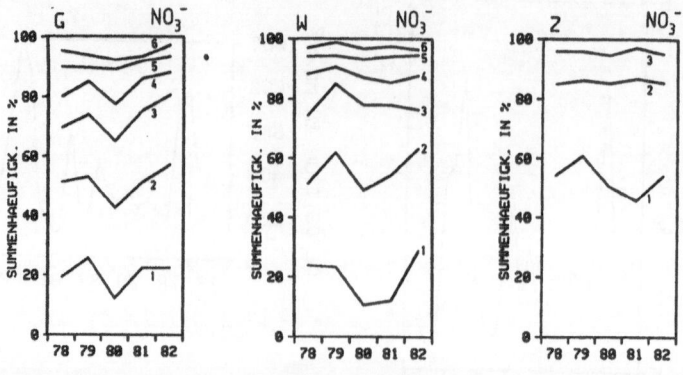

Fig. 9   Summenhäufigkeit ≑ cummulative frequency

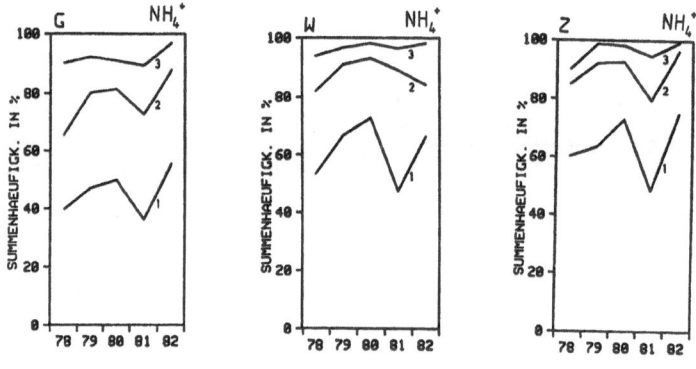

Fig. 10

Figs. 9 + 10 : Cumulative frequency in % for specific
concentration values (mg/1 = numbers indicated at the
curves) of NO_3^- and NH_4^+ at the three stations from
1978 to 1982.

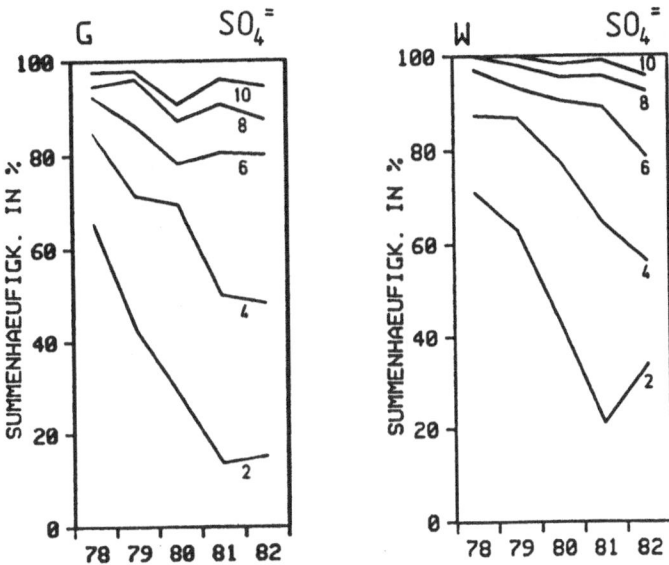

Fig. 11 : Cumulative frequency in % for specific
concentration values (numbers at curves), given for
concentrations of SO₄⁼ at the two stations.

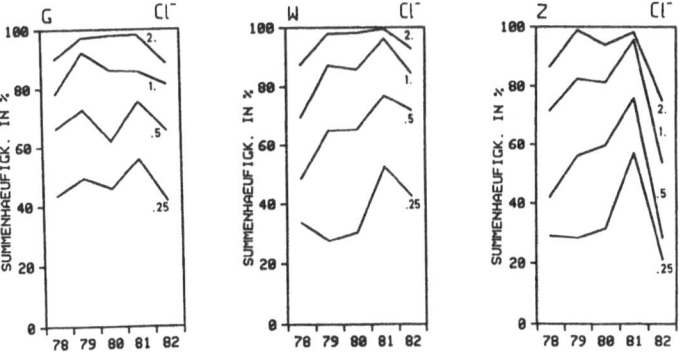

Fig. 12 : Cumulative frequency in % for specific
concentration values of Cl⁻ at the three stations.

- 489 -

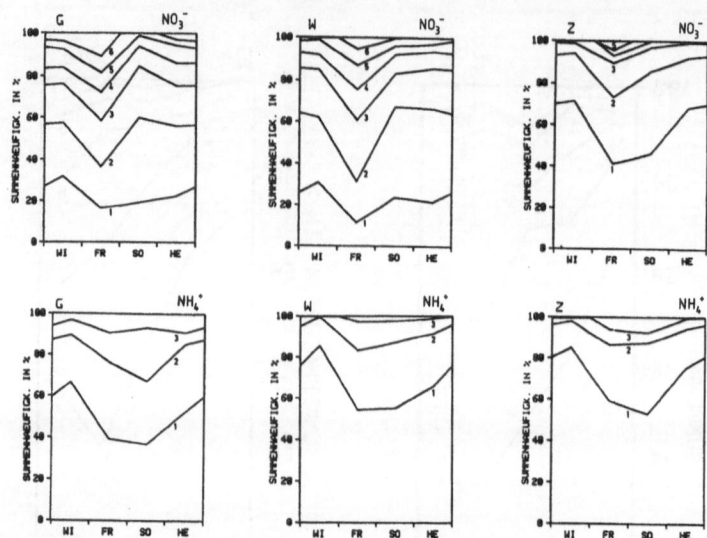

Figs. 13 + 14 : Cumulative frequency of NO_3^- and NH_4^+ over all years, grouped by seasons (WI = winter, FR = spring, SO = summer, HE = fall)

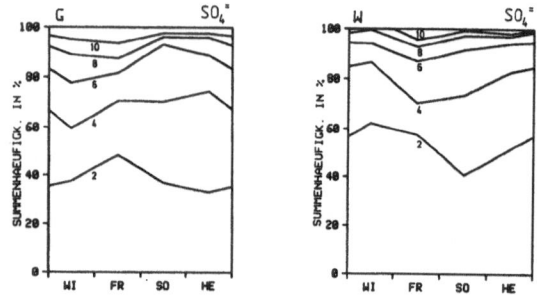

Fig. 15 : The cumulative frequency of SO_4^+ for low concentrations (2mg/l) at the two stations 1.8 and 0.7km ASL, given for the four seasons (WI = winter, FR = spring SO = summer, HE = fall).

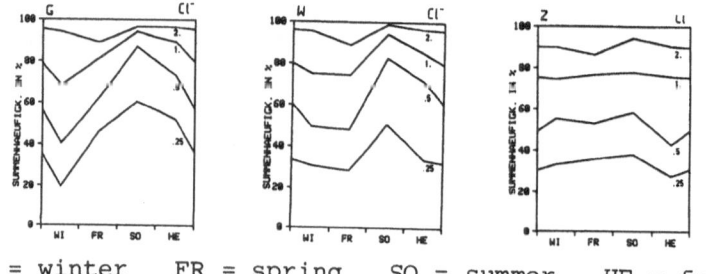

WI = winter   FR = spring   SO = summer   HE = fall

Fig. 16 : Dependence on the season (abbreviations   see figure 15) of Cl^- (concentration in mg/l ÷ values at the curves)

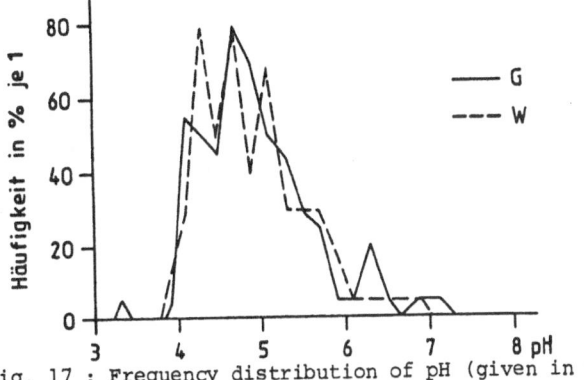

Fig. 17 : Frequency distribution of pH (given in %
per 1 pH unit). No significant difference in the
precipitation samples from the two stations with
1 km height difference can be observed. The acidity
is practically normal.

References

1) Reiter, R.: On the
   Causal Relation Bet-
   ween Nitrogen-Oxygen
   Compounds in the Tro-
   posphere and Atmos-
   pheric Electricity.
   Tellus, XXII, 122,
   (1970)

2) World Meteorological
   Organization: Air
   Pollution Measure-
   ment Techniques.
   Special Environmental
   Report No. 10, WMO-
   No. 460, (1977)

3) Georgii, H.W.: World Meteorological Organization, Environmental Pollution Monitoring Program (1982)

4) Pack, D.H. and Shepherd, A.A.: Proceedings; Advisory Workshop on methods for comparing precipitation chemistry data. UAPSP-100 Workshop
   Proceedings, New York (1982)

5) Department of Energy: Acid Precipitation, a Bibliography. DOE/TIC 3399
   (DE83008750), (1983)

6) Georgii, H.W. and Jaeschke W.: Chemistry of the unpolluted and polluted troposphere. D. Reidel Publishing Company (1982)

7) Georgii, H.W., Perseke, C. et al: Umweltforschungsplan des Bundesmini-
   sters des Innern, Luftreinhaltung: Forschungsprojekt 104 02 600, I Ab-
   schlußbericht / II Datenband (1982)

8) Perseke, C.: Die trockene und feuchte Deposition säurebildender Spu-
   renelemente. Ber. d. Inst. f. Met. u. Geophys. d. U. Frankfurt (1982)

9) Georgii, H.W. and Pankrath J.: Global Distribution of the Acidity in
   Precipitation; D. Reidel Publishing Company (1982)

10) Hales J.M. and Lodge, J.P. et al: Precipitation Chemistry. Atmosph.
    Env. 16, Nr. 7, (1982)

11) U.S. Department of Energy: Acid Rain Information Book. Dok. No. DOE/
    EP-0018/1 (1983)

12) Beilke, S. and Elshout, A.J.: Acid Deposition. D. Reidel Publishing
    Company (1983)

13) VDI Kommission Reinhltg. d. Luft: Saure Niederschläge – Ursachen und
    Wirkungen. Int. Koll., Lindau/Bodensee (1983)

14) Stumm, W. et al: Saurer Regen, eine Folge der Störung hydrogeoche-
    mischer Kreisläufe. Naturwiss. 70, (1983)

15) Krug, E.C. and Frink, C.R.: Acid Rain on Acid Soil. Science 221, 520
    (1983)

16) Machta, L.: Acid Rain: Controllable? EOS, 64 No. 48 (1983)

17) Georgii, H.W. and Jaeschke W.: The Atmospheric Sulfur-Budget. Chemis-
    try of the Unpolluted and Polluted Troposphere, D. Reidel Publishing
    Company (1982)

18) Reiter, R. et al: Bericht zum Problem "Waldschäden im Bayerischen Vor-
    alpenraum", (1983); Bayerisches Staatsministerium f. Landesentwicklung
    und Umweltfragen; Vertrag 8272 - 623 - 27768

19) Reiter, R. et al. WMO: Environmental Poll. Monitoring and Res. Progr.
    No. 20. TECOMAC Conf. Vienna (1983)

MEASUREMENTS OF THE LATITUDINAL DISTRIBUTION OF LIGHT HYDROCARBONS AND HALOCARBONS OVER THE ATLANTIC

J. Rudolph, C. Jebsen, A. Khedim and F.J. Johnen
Institut für Chemie 3: Atmosphärische Chemie der
Kernforschungsanlage Jülich GmbH, P.O. Box 1913, D-5170 Jülich, FRG

Summary

During the research and supply cruise to Antarctica in 1982/83 of
the "DPFVS Polarstern" more than 50 whole air samples were collected.
In addition about 15 air samples were taken at the German antarctic
research station. These samples were analysed in our institute by a
number of gaschromatographic techniques for a variety of trace gases:
carbonmonoxide, light aliphatic hydrocarbons, benzene, toluene, chlo-
romethane, dichloromethane, chloroform, carbontetrachloride, 1.1.1.-
trichloroethane, trichloroethene, tetrachloroethene, F-11, F-12, F-113,
carbonylsulfide and carbondisulfide. Although the limited air volume
in the individual samples did not allow for the analysis of all of
these species in each sample, our measurements enable us to obtain
latitudinal profiles for these species between 77° S and 45° N. Several examples of these latitudinal trace gas distributions will be
presented and discussed with respect to the source and sink distribution for those trace gases and their atmospheric lifetimes.

1. INTRODUCTION

It is by now well known that the atmosphere contains a large number
of different kinds of organic gaseous species at trace levels. In the remote, unpolluted troposphere the mixing ratios of most of these trace gases
are very low - in the ppb and ppt range. Nevertheless, many of these organic trace gases are of considerable interest for the chemistry of the atmosphere for a number of different reasons.

Several of the organic trace gases act as precursors for important atmospheric components. For example longlived chlorinated species (e.g. chlorofluorocarbons, CCl_4, CH_3Cl) can be transported into the stratosphere and
thus are a source for stratospheric chlorine, an important species for the
stratospheric ozone chemistry (1). Organic sulfur compounds - e.g. dimethyl
sulfide, CS_2 etc. - are decomposed in the atmosphere and yield SO_2 which
in turn can be further oxidized to sulfuric acid or sulfate (2).

Many organic molecules - especially hydrocarbons and similar species -
react with OH radicals and thus participate directly in the atmospheric radical reaction chains (3). Since some of the organic molecules react quite
fast with OH radicals, they may be of considerable importance even at very
low (ppb) levels (4). The photochemical oxidation of hydrocarbons produces
other organic compounds such as aldehydes and ketones. These species in
turn react further - with OH radicals or photolytically - to peroxyacyl radicals which - in the presence of NO_2 - subsequently form peroxyacyl nitrates. The most important of the peroxyacyl nitrates, peroxyacetyl nitrate
(PAN), may be an important reservoir for nitrogen oxides (5).

Measurements of trace gas distributions have also been used to derive
information on the distribution of sources and sinks and atmospheric removal

processes. Thus average tropospheric OH concentrations have been estimated from budgets of CH_3CCl_3 or 14-CO (6, 7, 8).

However the analysis of such complex mixtures, with many of the trace constituents at levels of only a few ppb or fractions of a ppb, presents considerable analytical and technical problems: Sample collection without contamination, stability of the collected sample, difficult logistics for measurements at remote sites, reliable identification and quantification of the various trace gases.

It is obvious that the variety of organic compounds in the atmosphere is far too complex to be completely covered within a few series of investigations. In our work we concentrate on the measurement of the latitudinal dependence of light and medium molecular weight hydrocarbons and halocarbons. We also made some measurements of carbonylsulfide and carbondisulfide at different latitudes. In this paper we present a brief description of the sampling and measurement techniques (gaschromatography) and some examples of latitudinal trace gas profiles.

2. EXPERIMENTAL

Approximately 50 whole air samples were collected during a research and supply cruise of DPFVS "Polarstern" to Antarctica in February, March and April 1983. In addition about 15 samples were taken at the antarctic research station of the Federal Republic of Germany "Georg von Neumaier" between April 1982 and January 1983. The antarctic station and the cruise track of the ship together with the sampling locations are shown in this figure.

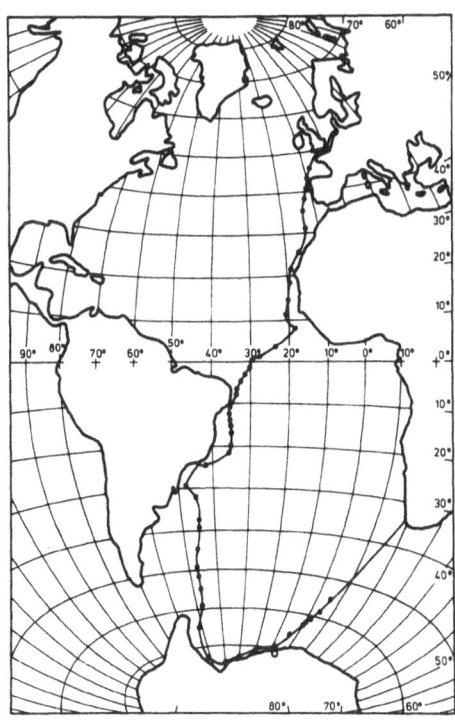

Fig. 1: Cruise track of DPFVS Polarstern in February - April 1983 and sample collection sites, open circle : antarctic station.

Most of the ship samples were collected on a boom which extended approximately 10 m ahead of the bow of the ship at 18 m above the ocean surface. Samples were only collected if relative wind direction and speed eliminated any possibility of sample contamination from trace gas sources on board the ship.

The samples were collected in evacuated (p $\leq$ 5x10^{-5} mbar) internally electropolished stainless steel containers (2 dm³ or 10 dm³) with metal bellow sealed valves. Simple grab samples were taken as well as pressurized samples. For the compression of the sample air the following equipment was used in order to minimize all possible effects of sample degradation. The following figure shows a schematic drawing of the apparatus.

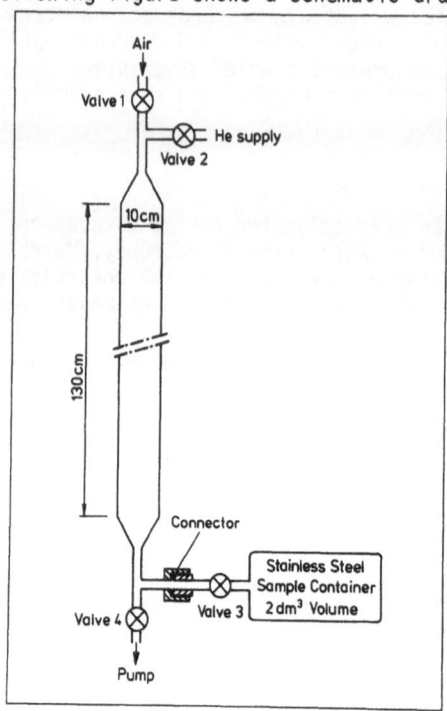

Fig. 2: Schematic drawing of equipment for the collection of pressurized whole air samples.

It consists of a stainless steel tube of 1.3 m length and 10 cm diameter with two metal bellow valves at each end of the tube. The ends of the tube are conically shaped in order to reduce turbulences and avoid dead volumes. At one end one of the valves is connected to a supply of high purity helium, at the other end to one of the sample containers. At first the tube is flushed with ambient air for about 15 min at a flow rate of more than 10 l min^{-1} by means of a pump (valves 1 and 4 open, valves 2 and 3 closed). Then valve 4 is closed and valve 3 opened. In order to pressurize the sample, valve 1 is closed and helium added with a flow rate of $\approx$ 1.5 l min^{-1} by opening valve 2. The air sample is compressed to a final pressure of $\approx$ 3.5 bar within 20 minutes. Then valves 2 and 3 are closed and the sample container replaced. During the pressurization of the sample, the tube is kept in an upright position in order to avoid convective mixing between helium and sample gas.

By this procedure only 5 l of the total of 10 l in the tube are transferred into the sample container and thus the mixed zone at the interface between helium and sample air remains outside the sample container. Laboraratory test showed that the dilution of the air sample by helium is less than 1 %. Comparisons between pressurized and unpressurized samples proved that - within the precision of the measurements - no further changes in the sample composition is caused by this procedure as far as the trace gases investigated in this work are concerned.

The trace gas measurements were made at our laboratory for gas chromatography at Jülich with a number of different instruments. A short description of these techniques is presented in table 1.

The measurements are quantitated by comparisons with secondary air standards which in turn are calibrated with mixtures of the pure substances in synthetic air. These mixtures are prepared by a two or three step static dilution procedure. The reproducibility and accuracy of the absolute calibration depends on the mixing ratios of the individual trace gases in background air and the substance itself. For low molecular weight compounds with quite high atmospheric mixing ratios (e.g. CH_4 or CO) the reproducibility of the calibration is better than 1 % and the estimated accuracy 1 % or 2 %. The reproducibility for the calibration of heavier species such as toluene or perchloroethylene is only 20 % and the estimated accuracy 30 %.

Due to several technical reasons (available sample volume and occupation of the instruments by other measurements etc.) it was not possible to analyse all of the samples for all the species listed in table 1. Still we were able to obtain latitudinal profiles for a considerable number of trace gases.

3. RESULTS AND DISCUSSION

In this paper we present the latitudinal profiles of six trace gases with significant anthropogenic sources: CO, C_2H_2, C_3H_8, C_6H_6, C_7H_8, and C_2Cl_4. They are shown in fig. 3.

The data points with errorbars at 70° S represent the average and its error of several measurements which were made between April 1982 and February 1983. The other samples were collected in early 1983. Thus there may be a systematic difference due to possible seasonal variations of the trace gas mixing ratios. However - in view of the restricted number of measurements and the general variability of the trace gas mixing ratios - it seems premature to draw any inferences on systematic seasonal cycles from the data in fig. 3.

It should be considered that several of the air samples were collected quite near the south American or African continent (see fig. 1). Therefore it cannot be ruled out that some of the air samples were influenced by continental air masses. All the six latitudinal profiles shown in fig. 3 have one prominent feature in common: rather low mixing ratios in the southern hemisphere and a strong increase between the equator and mid northern latitudes. Such hemispheric assymetries are well known for several trace gases which are predominantly man made (13, 14). This assymetry simply reflects the impact of the industrialized belt of the northern hemisphere on the composition of the atmosphere.

For most of the trace gas profiles in fig. 3, we also observe several points with significantly enhanced mixing ratios north of $\approx$ 35° S compared to the latitude range from 75° S to 35° S. This is most probably due to the vicinity of the south American continent to the sampling locations for the latitudes between 35° S and the equator.

Apart from some similarities in the general shape of the individual profiles, there are also several significant differences. Not only the absolute values of the mixing ratios differ by orders of magnitude between the various trace gases, but also the relative gradients between the nor-

Table 1: Parameters for Gaschromatographic Trace Gas Measurements

Compound	Sample Injection Procedure	Separation Conditions	Detector	Precision	Detection Limit	Estimated Accuracy	Remarks
CH_4 CO	$2cm^3$ (STP), sample loop direct injection	Molecular sieve 5A 1/4" x 2m, 100 °C	FID	0.5 % 1.5 %	5-10ppb	1 % 2 %	Catalytic conversion of CO to CH_4 (see (9))
CO_2	$1cm^3$ (STP), as above	Silicagel 1/4" x 1m 100 °C	FID	0.2 %	10 ppb	1 %	Catalytic conversion of CO_2 to CH_4
N_2O CF_2Cl_2 $CFCl_3$	$2cm^3$ (STP), sample injection as above	Porasil C, 1/8" x 3m -30 °C to 150 °C temperature program	ECD	3 % 2 % 3 %	5 ppb 1 ppb 1 ppb	5 % 5 % 10 %	Method similar to that described in (10)
$C_2 - C_5$ hydro-carbons	$0.5dm^3 - 1dm^3$ (STP), adsorptive preconcentration on porous glass-beads at liquid N_2 temperature and 300 mb pressure	Porasil-n-Octane or Spherosil XOB 1/16" x 10 m -90 °C to 65 °C temperature program	FID	≈ 5 %	10-25 ppt	≈ 10 %	Method similar to that described in (11)
$C_2 - C_5$ hydro-carbons, $C_1 - C_2$ halo-carbons, CS_2 COS $C_2F_3Cl_3$	0.2 - 0.5 dm^3 (STP) preconcentration as above	Porasil C-n Octane 1/16" x 10 m	PID ECD and FID in series	between 3 % and 15 % depend-ing on type of com-pound and concen-tration	between ≈ 20-30ppt for FID, 0.5-3 ppt for alkenes and ben-zene with PID, 0.1-1 ppt for most halocar-bons with ECD	between 10 % and 30 %	Instrument des-cribed in (12)

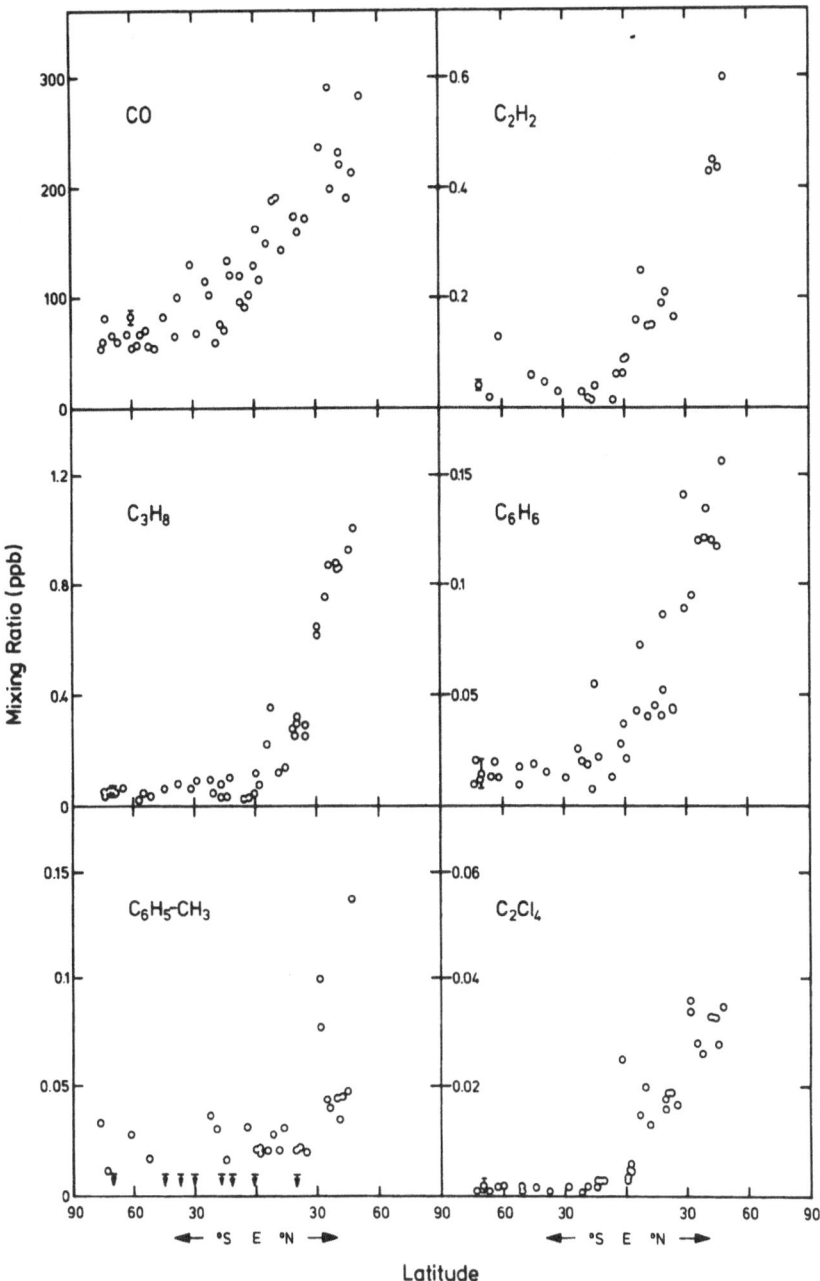

Fig. 3: Latitudinal profiles of several trace gases. The points with error bars at 70° S represent the average and its error of several measurements at the "Georg von Neumaier" station between April 1982 and February 1983. The downward arrows denote the lower limit of detection for those measurements where the trace gas could not be detected.

thern and southern hemisphere vary considerably. This is hardly surprising if we consider the many different sources for the individual trace gases (15), their geographical distributions as well as the different atmospheric removal rates, especially for the reaction with OH radicals (see e.g. 14).

For CO we observe mixing ratios around 60 ppb at far southern latitudes, somewhat higher values between 35° S and the equator and an increase to 200 ppb and more at mid northern latitudes. There are several points of elevated mixing ratios which are most probably - as mentioned before - due to continental influences. Apart from these straggling points, our latitudinal CO profile compares well with other published measurements over the Atlantic (16, 17). The difference between mid northern and far southern CO mixing ratios is roughly a factor of 3 - 4. The profiles of C_2H_2, C_3H_8, C_6H_6, and C_2Cl_4 exhibit much larger relative differences between the northern and the southern hemisphere.

Southern hemispheric C_2H_2 values are around 0.04 ppb, ranging from 0.02 to 0.06 ppb with the exception of one straggling point at 61° S. From the equator towards mid northern latitudes the C_2H_2 mixing ratios increase to ≈ 0.45 ppb around 45° N with one point of 0.6 ppb at 49° N.

For C_3H_8 the southern hemispheric mixing ratios range around 0.06 ppb with several very low values at high southern latitudes and between 20° S and the equator (0.03 - 0.05 ppb). North of the equator the C_3H_8 mixing ratios increase up to ≈ 1 ppb around 45° N.

Rudolph and Ehhalt (18), Rudolph et al. (19) and Singh and Salas (20) report rather similar hydrocarbon mixing ratios from over the Atlantic and Pacific with the exception of the southern hemispheric data of Singh and Salas. They report values of 0.16 ppb between 0° and 10° S, 0.33 ppb between 10° S and 20° S and 0.13 ppb between 20° S and 30° S for C_2H_2 over the eastern Pacific. Their corresponding C_3H_8 mixing ratios are 0.31 ppb, 0.20 ppb, and 0.11 ppb. These values are substantially higher than our results.

There are also a few published data on the latitudial distribution of benzene which can be compared to our measurements. In 1983 Rasmussen and Khalil (21) published the results of measurements at several remote mainly continental sites at different latitudes. Their values range around 0.03 ppb to 0.05 ppb in the southern hemisphere (with the exception of the south pole where C_6H_6 was below their lower limit of detection of 0.005 ppb) and 0.1 to 0.26 ppb at northern latitudes. Our values are lower, 0.015 ppb at southern latitudes, 0.02 - 0.04 ppb near the equator increasing up to 0.15 ppb at 49° N. Much better agreement with our results are shown by the data published by Penkett (14) of 0.066 ± 0.032 ppb (average of five measurements) for ≈ 35° N.

Our toluene data exhibit a considerable scatter with several measurements - mainly in the southern hemisphere - below the detection limit of ≈ 0.01 ppb. The difference between northern and southern hemisphere is not very well defined but still the data suggest higher toluene mixing ratios in the northern hemisphere than south of the equator. Compared to the other trace gases shown in fig. 3, toluene reacts quite fast with OH radicals and thus has a much shorter atmospheric residence time than the other trace gases shown in this figure. Consequently we can expect a considerably larger scatter since the influence of regional sources and transport will be much more pronounced in this case than for trace gases which are longlived enough to be well mixed in the atmosphere.

C_2Cl_2 shows extremely low mixing ratios, ≈ 1 ppt in the southern hemisphere - with a few values up to 4 ppt at lower latitudes - and an increase north of the equator to 30 - 35 ppt in the latitude range from 30° N to 50° N. Comparable values were reported by Penkett (14): 34 + 8 ppt at roughly 35° N, whereas Rasmussen and Khalil found significantly larger C_2Cl_4 mixing ratios: 10 - 20 ppt in the southern hermisphere and up to 70 ppt around 40° N.

Finally we will compare the profiles of C_2Cl_4 and C_2H_2 with each other. Both species are predominantly man made and thus their sources are mainly located in the industrialized zone of the nothern hemisphere. Since C_2Cl_4 reacts slower with OH radicals than C_2H_2 (14) its atmospheric lifetime should be longer and consequently its relative interhemispheric gradient weaker if all C_2H_2 and C_2Cl_4 sources are located in the northern hemisphere. However it can be seen from figure 3 that just the opposite is the case. If we exclude the possibility that there exists some unknown sink mechanism for C_2Cl_4, this is a strong evidence for the existence of small but significant sources for C_2H_2 in the southern hemisphere. However it should be noticed that the presented latitudinal profiles do not present representative latitudinal averages and thus all inferences drawn from these few measurements are still rather uncertain.

4. CONCLUSIONS

The analytical techniques and the sampling procedures described here have been shown to be very useful and adequate for the measurement of a number of different light and medium weight organic molecules in the atmosphere, even in remote areas. The relatively simple procedures for sampling air at ambient or increased pressure in stainless steel containers enable measurements in very remote areas. The contamination free collection of pressurized air samples provides sufficiently large air volumes to allow the analysis of a considerable number of different trace gases at very low concentrations.

Data on the concentration distributions of organic trace gases allow inferences on their source and sink distribution and their chemical reactions in the atmosphere. At present only for few trace gases the latitudinal distributions are well established. Even less is known on the altitudinal and seasonal variability of organic trace species. It can be expected that systematic measurements of the space-time distribution of organic trace gases in the atmosphere will provide very valuable information on the source and sink distribution of these compounds as well as on the key removal mechanisms.

5. ACKNOWLEDGEMENT

We would like to thank the "Alfred-Wegener-Institut für Polarforschung" in Bremerhaven for the possiblity to participate on the Polarstern cruise in 1983 from Rio de Janeiro to Bremerhaven and the crew of the Polarstern and Captain Suhrmeier for their support and assistance on board of the ship. We are also grateful to G. Helas from the "Max-Planck-Institut" in Mainz and D. Wagenbach from the "Institut für Umweltphysik" in Heidelberg who collected some of the air samples.

6. REFERENCES

(1) M.J. Molina, F.S. Rowland, Stratospheric sink for chlorofluoromethanes: chlorine atom catalysed destruction of ozone, Nature 249, 810 (1974)

(2) T.E. Graedel, The homogeneous chemistry of atmospheric sulfur, Rev. Geophys. Space Phys. 15, 421 (1977)

(3) D.A. Brewer, T.R. Augustsson, J.S. Levine, The photochemistry of anthropogenic nonmethane hydrocarbons in the troposphere, J. Geophys. Res. 88, 6683 (1983)

(4) J. Rudolph, D.H. Ehhalt, Measurements of $C_2 - C_5$ hydrocarbons over the north Atlantic, J. Geopyhs. Res. 86, 11959 (1981)

(5) H.B. Singh, P.L. Hanst, Peroxyacetyl nitrate (PAN) in the unpolluted atmosphere: an important reservoir for nitrogen oxides, Geophys. Res. Lett. 8, 941 (1981)

(6) J.E. Lovelock, Methyl chloroform in the troposphere as an indicator of the OH radical abundance, Nature 267, 452 (1977)

(7) H.B. Singh, Atmospheric halocarbons: evidence in favor of a reduced average hydroxyl radical concentration in the troposphere, Geophys. Res. Lett. 4, 101 (1977)

(8) A. Volz, D.H. Ehhalt, R.G. Derwent, Seasonal and latitudinal variation of 14-CO and the tropospheric concentration of OH radicals, J. Geophys. Res. 86, 5163 (1981)

(9) K. Porter, D.H. Volman, Flame ionization detection of carbonmonoxide for gaschromatographic analysis, Anal. Chem. 34, 748 (1962)

(10) A.L. Schmeltekopf, D.L. Albritton, P.J. Crutzen, P.D. Goldan, W.J. Harrop, W.R. Henderson, J.R. McAfee, M. McFarland, H.I. Schiff, T.L. Thompson, D.J. Hoffmann, N.T. Kjome, Stratospheric nitrous oxide altitude profiles at various latitudes, J. Atm. Sciences 34, 729 (1977)

(11) J. Rudolph, D.H. Ehhalt, A. Khedim, C. Jebsen, Determination of $C_2 - C_5$ hydrocarbons in the atmosphere at low parts per 10^{-9} to high parts per 10^{-12} levels, J. Chromatography 217, 301 (1981)

(12) J. Rudolph, C. Jebsen, The use of photoionization, flameionization and electron capture detectors in series for the determination of low molecular weight trace components in the non urban atmosphere, Intern. J. Environ. Anal. Chem. 13, 129 (1983)

(13) R.A. Rasmussen, M.A.K. Khalil, Latitudinal distributions of trace gases in and above the boundary layer, Chemosphere 11, 227 (1982)

(14) S.A. Penkett, Non-methane organics in the remote troposphere, in: Atmospheric Chemistry, E.D. Goldberg (Editor) Physical and Chemical Sciences Research Report 4, Report of the Dahlem workshop on atmospheric chemistry, Berlin 1982, May 2-7. Springer, Berlin-Heidelberg-New York, 1982

(15) T.E. Graedel, Chemical compounds in the atmosphere, Academic Press, New York, 1978

(16) W. Seiler, The cycle of atmospheric CO, Tellus XXVI, 116 (1974)

(17) U.Schmidt, A. Khedim, F.J. Johnen, J. Rudolph, D.H. Ehhalt, Two dimensional meridional distribution of CO, CH_4, N_2O, $CFCl_3$ and CF_2Cl_2 in the remote troposphere over the Atlantic Ocean. Preprint Volume: Second Symposium on the composition of the nonurban troposphere, May 25-28, 1982, Williamsburg, Va, American Meteorological Society, Boston, Mass. p. 52-55

(18) J. Rudolph, D.H. Ehhalt, Measurements of C_2 - C_5 hydrocarbons over the North Atlantic, J. Geophys. Res. <u>86</u>, 11959 (1981)

(19) J. Rudolph, D.H. Ehhalt, A. Khedim, C. Jebsen, Latitudinal profiles of some C_2 - C_5 hydrocarbons in the clean troposphere over the Atlantic, Preprint Volume: Second Symposium on the composition of the nonurban troposphere, May 25-28, 1982, Williamsburg, Va, American Meteorological Society, Boston, Mass., p. 284-286

(20) H.B. Singh, L.J. Salas, Measurement of selected light hydrocarbons over the Pacific Ocean: latitudinal and seasonal variations, Geophys. Res. Lett. <u>9</u>, 842 (1982)

(21) R.A. Rasmussen, M.A.K. Khalil, Atmospheric benzene and toluene, Geophys. Res. Lett. <u>10</u>, 1096 (1983)

SESSION 5

TRANSPORT AND MODELLING FIELD EXPERIMENTS

Summary by the Chairman
   A.J. ELSHOUT

Composition and origin of cloudwater solutes

Air sampling flights round the British Isles at law
altitudes : SO_2 oxidation and removal rates

Oxidation of NO to NO_2 in flue gas plumes of power
stations

Campagne experimentale en vue de la modélisation de
panache réactif

The atmospheric significance of liquid phase oxidation
of SO_2 to sulphate by O_3 and H_2O_2

Some possibilities of modelling data from "FOS" European
field experiment (June 6/15, 1983)

Atmospheric tracer dispersion experiments (at the Mol
Nuclear Energy Research Center)

A sampling network for the assessment of the heavy metal
pollution originating from a municipal incinerator plant
in Belgium

Evaluation of the information from a continuously
working precipitation monitor

Production et transfert d'ozone sur le bassin de
Fos-Berre

Comparison between the scavenging ratios for nitrate and
sulphate at a rural site

Identification des sources d'hydrocarbures aromatiques
polycycliques particulaires dans l'atmosphère urbaine

Photochemical air pollution in Denmark. Weekday effects
and evidence of large-scale formation

Classement automatique des trajectoires du panache de
l'Etna - Etude climatologique

DISPERSION AND TRANSPORT - MODELLING AND FIELD EXPERIMENTS

Summary by the chairman

A.J. ELSHOUT

N.V. KEMA, Arnhem

In the scope of the Concerted Action the European Symposia should permit an overall review of the progress of the research projects ongoing in the participating countries and give the possibility to collect and disseminate the results. It is my opinion that we succeeded very well in the preceding two symposia and in this third symposium on the physico--chemical behaviour of atmospheric pollutants, in giving an overview of the important achievements within the framework of this action. The results shall be summarized in the final activity report.

In the follow-up of this action one of the main interests is directed to a better understanding of the impact of emitted air pollutants on the environment, especially in view of the "acid deposition" problem.

The chemical content of wet and dry deposition is determined by the emission, transport and transformation of anthropogenic and natural substances. In recent years our understanding of these processes has increased considerably, also as a consequence of the results of research projects going on in our COST action 61a-bis.

A part of our knowledge gathered in this action has already been reported conveniently in a number of lectures, as for example at the CEC symposium "Acid deposition; a challenge for Europe", which was held in Karlsruhe in September 1983 by Cox and Penkett with the lecture "Formation of atmospheric acidity" and by Beilke in his report on the session "Origin, transport, conversion and deposition of air pollutants". They need no repetition here.

Though we understand more, we still do not know at all what determines chemical deposition at a particular location, or how that deposition would be influenced by a change in emissions. Under these circumstances research needs have to be defined in view of our missing knowledge. Within the framework of COST 61a-bis some needs have been established already and for a large part summarized in the contributions of COST 61a-bis participants to the Karlsruhe symposium already mentioned and later on confirmed in the COST document AP/38/83 dated 8.12.1983.

The activities in the folluw-up of this COST action should for an important part be directed to the improvement and development of the communication between scientists engaged in field, laboratory and modelling research, so that each group could be aware of and contribute to the other's activities, capabilities and information needs.

To do this well, in the structure of the follow-up programme of this COST action a group is needed which has the capability to determine which type of new information is required, how existing data can be utilized and how these needs fit into an overall description of the problem. In doing so it should be remembered that whereever possible these activities should also be co-ordinated with other co-ordinated projects or actions, e.g. in the United States. One of the main points of our follow-up programme is the topic of "deposition chemistry", where the practical goal of research has to be the delivery of knowledge for the formulation of numerical models. With this knowledge we can quantitively answer questions about the dependence of acid deposition on emissions. The performance of such models depends on the profound understanding of the individual chemical and physical processes obtained from field and laboratory research.

In several countries there are discussions on research priorities and criteria to establish factors which govern deposition chemistry. They all come in different detailed descriptions to the same type of recommendations. Looking to the aspects of working group 5 of our action "Dispersion and Transport - Modelling and Field Experiments" it can be concluded that there is still a clear need to enlarge the performance of models to describe the long-range transport of pollutants. In this field for example there is a need for validating the transport and diffusion portion of the models. Typical cases have to be studied more in detail such as the vertical transport of substances out of the polluted layer, e.g. by convective clouds, which strongly influences the horizontal transport and the ultimate fate of pollutants. Such studies should aim at the proper treatment of the effect in models.

There is a strong recommendation for research on aqueous phase reactions. To be applicable to cloud chemical processes, reaction studies should be done also in actual precipitation water including in the ice phase. More laboratory and field work is also needed about the transfer rates of various substances between the gas phase and cloud drops.

Specific field investigations shall help to improve the current understanding of sulphate and nitrate formation in stratus clouds and the more complex situations of summer convective clouds. By comparing daytime and nighttime measurements in clouds and precipitation an impression may be obtained of the importance of photochemical mechanisms by investigation of diurnal variation in transformation rates. The ratios of sulphate to nitrate ions in cloud water and precipitation can give information on transport and transformation rates when combined with emission data.

Aircraft field studies are of special interest for providing information on the aspects mentioned and are necessary for the proper treatment of their effects in models. For the moment these actions require an improvement of instruments and techniques which can be used in this type of operation, especially for cloud water sampling and for more adequate measurement of e.g. OH-radicals, hydrogen peroxide, nitric acid, ammonia gas, hydrogen chloride and some specific hydrocarbons and aerosols.

This all gives strong needs for more financial investments and above all for a good co-ordination to optimize the programmes and the use of the results. I express the wish that the follow-up of this concerted action, physico-chemical behaviour of atmospheric pollutants, also as a part of the CEC work on the "acid rain" problem, can meet these demands.

COMPOSITION AND ORIGIN OF CLOUDWATER SOLUTES

G.P. GERVAT, A.S. KALLEND and A.R.W. MARSH
Central Electricity Research Laboratories, Leatherhead, UK

Summary

Measurements of the chemical composition of cloud water are
reported for samples collected from both aircraft and from a
ground-based sampler in both clean and polluted air masses.
Using the measured cloud liquid water content, the equivalent dry
air composition is evaluated and the relative contributions of
the strong acids HCl, HNO_3 and H_2SO_4 to the free acidity of the
samples is deduced. The results indicate that anions in excess
of those derived from sea salt can confer free acidity on cloud
water even in maritime air masses not recently associated with
pollutant sources. High concentrations observed on hilltop sites
confirm the view that direct transfer from cloud can be
significant compared with normal wet and dry deposition for such
sites.

1. INTRODUCTION

Physical and chemical processes during the nucleation of natural
clouds and subsequently at the surface of and within cloud droplets, form
an essential part of the pathway linking atmospheric emissions to the wet
deposition of pollutants in precipitation. Recently, particular attention
has been focussed on the relative importance of nucleation as opposed to
gas dissolution in determining the composition of cloud water. In this
paper, we present measurements of the composition of cloud water collected
using both aircraft and ground-based collectors and attempt to interpret
the measurements through the application of simple concepts of mass balance
and chemical and phase equilibria. Specifically we attempt to deduce the
chemical origin (e.g. aerosol, gas phase species etc.) of the major ions
NH_4^+, SO_4^{2-}, Cl^-, and NO_3^-. This leads to an estimate of the relative
contributions of sulphate, nitrate and chloride to the observed acidity of
cloud water.

2. EXPERIMENTAL

Cloud water samples were collected sequentially during flights of the
Meterological Office C130 (Hercules) aircraft using a system based on the
C.E.R.L. centrifugal cloud water collector which has been described
elsewhere (1). Extensive wind-tunnel tests, as well as direct simultaneous
in-flight comparison with a completely different design of separator (2),
confirmed the integrity of the collecting method and, in particular, the
absence of significant contamination or evaporation. Comparison of the
volume of collected water with the time-integrated cloud liquid water
content (L.W.C.) indicated a collection efficiency of at least 60% for a
cloud L.W.C. of ∼0.5 $g.m^{-3}$. Liquid water content was continuously

monitored on board the aircraft using a Johnson-Williams hot wire detector. Cloud water pH was also continuously monitored on the aircraft during periods of collection using a glass electrode.

Low cloud and precipitation were collected on the Pennine Hills in Northern England during the period 13th to 28th April 1983. The exact location of the sampling site was just to the south of the front wall of Chew Reservoir, altitude 490 m, to the east of Manchester. The cloud water collector is based on the design described by Falconer and Falconer (3). Cloud liquid water content was estimated from the measured wind speed and the calculated efficiency of the collector for different wind speeds. Sequential samples were stored for subsequent chemical analysis but pH was always measured within 48 hours of collection.

Samples from both the aircraft and from the ground-based collector were analysed in the laboratory for the major inorganic ions by ion-chromatography. In some instances particle loading was sufficiently high to necessitate filtration prior to analysis.

3. RESULTS AND DISCUSSION

In considering the origin of the solutes observed in cloud water we note, fist of all, that nucleation by aerosol particles and then gas dissolution are probably the major processes involved and, in particular, that particle attachment after nucleation is probably unimportant (4,5). It follows that, unless significant chemical transformation takes place in cloud droplets on a time-scale comparable with the droplet growth and evaporation cycle (e.g. SO_2 oxidation to sulphate), then the measured cloud water composition should be simply related to the gas/aerosol composition of the associated neighbouring air mass. The latter may be considered to be the equivalent dry air composition or the "below-cloud" air composition for a well mixed system. Alternatively the cloud may be considered as having a potential partial pressure of various components (which may be gases or particles). It is on this basis that the following discussion is developed. This approach has been described in some detail by Marsh (6).

The species of interest in dry air are aerosol particles containing H_2SO_4, $(NH_4)_2SO_4$, NH_4Cl and NH_4NO_3 and the gases NH_3, HCl and HNO_3. An estimate of the vapour pressures of NH_4Cl and NH_4NO_3 in the presence of partially neutralised sulphate cannot be accurately calculated because of the lack of fundamental data including activity coefficients. In the limit, at low relative humidities <65% at 25°C and <78% at 0°C, the maximum volatility of these compounds is that of the pure component. At higher humidities the vapour pressure of the pure component drops as a delequescent solution forms. In a real system at these higher humidities the vapour pressure will be higher than that of the pure component because of the presence of sulphate, the exact value depending on the relative ionic strengths, Stelson and Seinfeld (7) and Tang (8). For the purposes of estimating the potential 'acidity' loading of the dry air it is assumed that the relevant vapour pressures are those of the pure components. If the concentrations of the ionic species in cloud water are expressed in equivalents per unit volume of air then the following mass balance equations may be written to relate the cloud concentrations to the equivalent concentrations in dry air through the measured cloud liquid water content.

(1) $SO_4^{2-} = H_2SO_4 + (NH_4)_2SO_4$

(2) $NO_3^- = HNO_3 + NH_4NO_3$

(3) $Cl^- = HCl + NH_4Cl$

(4) $NH_4^+ = NH_3 + (NH_4)_2SO_4 + NH_4Cl + NH_4NO_3$

(5) $H^+ = H_2SO_4 + HCl + HNO_3$

These equations together with the vapour pressure relationships can be solved, given the concentration of $NH_3(g)$. (It is necessary to give an ammonia concentration because the full set of vapour pressure relationships cannot be solved accurately.) Solutions to the equations have been made for different $NH_3(g)$ concentrations, typically in the range 0.01 to 1 ppb. Provided the temperature is greater than ~10°C it is usually predicted that the chloride and nitrate ions in the cloud water will yeld $HCl(g)$ and $HNO_3(g)$ in dry air, whatever the ammonia concentration chosen. At lower temperatures and for some combinations of concentrations, the ammonium chloride and nitrate aerosols can be stabilised.

We consider first the results of aircraft measurements made on 28th January 1981 in stratiform cloud within the atmospheric boundary layer over the North Sea. Trajectory analysis indicated that the air had travelled around the perimeter of an anticyclone centred over Northern France before turning north-east across northern England and the North Sea towards Denmark. A layer of stratus cloud persisted throughout passage over the North Sea where the cloud base was within about 50 m of the sea surface. The vertical profile of liquid water content was close to adiabatic, rising to about 0.6 $g.m^{-3}$ at 350 m. The boundary layer was capped by a strong subsidence inversion which confined all emissions. Concentration profiles indicated that the boundary layer was very well mixed vertically. This, coupled with an accelerating air flow over the sea led to low dispersion of pollutant plumes so that high concentrations persisted for several hundred km (9,10). Oxidant concentrations were generally low and very little oxidation of SO_2 within the identified plumes occurred over a period of 24 h (9,10).

Fig. 1 records measurements made during a cross-wind traverse encompassing the plumes from major sources in Mid- and Northern England. The traverse was made from south to north at an altitude of 300 m roughly parallel to the coast, passing ca 100 km downwind of the northern source area. The pH, which was continuously monitored during the traverse, is also shown in Fig. 1 along with a table showing ionic concentrations of the various major ions corrected for the sea salt contributions by reference to the sodium concentrations. The vertical divisions in the table indicate the periods over which the individual samples were collected. The final row is the pH calculated from the ion balance, which compares satisfactorily with the in-flight measurements.

Taking average values from the table in Fig. 1 and the measured cloud liquid water content of 0.60 $g.m^{-3}$, solution of equations (1) to (5) yields the composition matrix in Table 1 for a relative humidity of <~70%. It predicts that most of the sulphate would be present as ammonium sulphate aerosol and that the contributions to the acidity from HCl, HNO_3 and H_2SO_4 are, respectively, 71%, 24% and 5%. Thus, on this particular occasion the acidity appears to have been dominated by HCl and this conclusion is not substantially altered for higher values of relative humidity up to ~95%. The predicted concentration of HCl is consistent with the observed concentration of SO_2 (Fig. 1) and the known ratio of HCl to SO_2 in emissions. While ammonia was not measured, its ambient concentration over the sea would be expected to be low, <<1 ppb.

Table 1

28.1.81 Mean of Samples in Yorkshire Plume
Liquid water content is 0.60 g.m^{-3}

Composition in and below cloud in µeqs/l for (1), liquid,
ppb for (g), gases, and µg/m^3 for (s), aerosols.

Cloud	Below Cloud		
SO$_4$$^-$(1) 2497	H$_2$SO$_4$(s) 3	(NH$_4$)$_2$SO$_4$(s) 69	
Cl$^-$(1) 1905	HCl(g) 22	NH$_4$Cl(s) 6	
NO$_3$$^-$(1) 842	HNO$_3$(g) 7	NH$_4$NO$_3$(s) 11	
NH$_4$$^+$(1) 2961	NH$_4$Cl(s) 6	NH$_4$NO$_3$(s) 11	(NH$_4$)$_2$SO$_4$(s) 69
H$^+$(1) 2282	HCl(g) 22	HNO$_3$(g) 7	H$_2$SO$_4$(s) 3

pH of Cloud Sample 2.78
% Sea Salt   13
% (NH$_4$)$_2$SO$_4$ of SO$_4$$^-$ 95
% NH$_4$Cl   of Cl$^-$ 15
% NH$_4$NO$_3$ of NO$_3$$^-$ 34

% HCl   of H$^+$ 71
% HNO$_3$ of H$^+$ 24
% H2SO4   of H$^+$ 5

By contrast, the measurements presented in Table 2 refer to samples
taken by aircraft in stratiform cloud under quite different conditions.
These were collected in an unpolluted westerly air stream off the west
coast of the United Kingdom. The observed pH averaged 4.26 and the mean
cloud L.W.C. was 0.4 g.m^{-3}. The composition was dominated by sea salt
which contributed about 78% of the inorganic ions. Because of the low

levels of SO2 present (<2 ppb), it is again argued that sulphate production
was negligible within the condensation cycle of a cloud droplet. Tables 2a
and 2b present the estimated dry air composition matrix. Making the sea
salt correction by reference to sodium, the analysis suggests that

Table 2

2.12.80 West Coast Samples. Liquid Water Content is 0.50 g/m³.

(a) Excess Chloride Present

Composition in and below cloud in μeqs/l for (1), liquid, ppb for (g), gases, and μg/m³ for (s), aerosols.

Cloud	Below Cloud			
$SO_4^=$(1) 152.1	H_2SO_4(s) .6	$(NH_4)_2SO_4$(s) 3.0		
Cl^-(1) 100.2	HCl(g) 1.1	NH_4Cl(s) 0.0		
NO_3^-(1) 73.0	HNO_3(g) .8	NH_4NO_3(s) 0.0		
NH_4^+(1) 126.0	NH_4Cl(s) 0.0	NH_4NO_3(s) 0.0	$(NH_4)_2SO_4$(s) 3.0	
H^+(1) 199.3	HCl(g) 1.1	HNO_3(g) .8	H_2SO_4(s) .6	

pH of Cloud Sample 4.26
% Sea Salt 78
% $(NH_4)_2SO_4$ of $SO_4^=$ 83
% NH_4Cl of Cl^- 0
% NH_4NO_3 of NO_3^- 0

% HCl of H^+ 50
% HNO_3 of H^+ 37
% H_2SO_4 of H^+ 13

(b) No Excess Chloride Present

Composition in and below cloud in μeqs/l for (1), liquid, ppb for (g), gases, and μg/m³ for (s), aerosols.

Cloud	Below Cloud			
$SO_4^=$(1) 141.8	H_2SO_4(s) .4	$(NH_4)_2SO_4$(s) 3.0		
Cl^-(1) 0.0	HCl(g) 0.0	NH_4Cl(s) 0.0		
NO_3^-(1) 73.0	HNO_3(g) .8	NH_4NO_3(s) 0.0		
NH_4^+(1) 126.0	NH_4Cl(s) 0.0	NH_4NO_3(s) 0.0	$(NH_4)_2SO_4$(s) 3.0	
H^+(1) 88.8	HCl(g) 0.0	HNO_3(g) .8	H_2SO_4(s) .4	

pH of Cloud Sample 4.26
% Sea Salt 86
% $(NH_4)_2SO_4$ of $SO_4^=$ 89
% NH_4Cl of Cl^- 0
% NH_4NO_3 of NO_3^- 0

% HCl of H^+ 0
% HNO_3 of H^+ 82
% H_2SO_4 of H^+ 18

"excess" sulphate, nitrate and chloride are all present. The presence of strong acid is confirmed by the pH measured both on board the aircraft and subsequently in the stored samples. Of the "excess" sulphate about 70% would be derived from $(NH_4)_2SO_4$ aerosol. The analysis also suggests that all three strong acids contributed to the free acidity although the errors in this instance could be large because of the high proportion of sea salt in the original samples. Even if the "excess" chloride were an artifact of the sea salt correction, the contribution of nitric acid is certainly significant (Table 2). For either acid, an average of approximately 1 ppb in clean air reaching Europe from the Atlantic represents a substantial flux which could have a significant influence on atmospheric chemistry during long range transport, particularly through its influence on oxidation kinetics in clouds.

Fig. 2 shows the results of ground-based measurements of cloud water samples collected over a five hour period at the Chew Reservoir site on 27th April 1983. Geostrophic air trajectories indicated that the air reaching the site in the morning had probably passed over Holland before crossing the North Sea and turning to arrive at the site from the east to the north east. The trajectory was more northerly later in the day. Tables 3a and 3b give the dry air concentration matrix for the sample collected between 0840 and 0847 hours, again corrected for sea salt. The concentrations are of the same order as those observed from the aircraft in polluted air but, in this instance the sea salt correction is negligible contributing only about 1 per cent of the total inorganic ions on the basis of the sodium concentration. On this occasion the calculated composition of dry air is sensitive to the ammonia gas concentration in the range 0.01 to 1 ppb. The volatility of NH_4Cl implies a 50/50% mixture of gaseous and aerosol chloride at 0.125 ppb $NH_3(g)$. However, an upper limit to the ammonia gas concentration of 0.095 ppb is imposed by the mass balance equations. Consequently, the contribution of $HCl(g)$ to the acidity of the dry air is in the range 40-59%.

The sea salt contribution based on Na^+ remained low throughout the day despite the long air passage across the North Sea. Later in the day the concentration of chloride gradually fell and by 2100 hours was less than one fifth of the value shown in Table 3. Sulphuric and nitric acids then became dominant strong acids and substantial particulate content became evident although the site was about 20 km from Huddersfield, the nearest major source area upwind.

Cloud samples were also collected at the Chew Reservoir site on 13th April 1983 in a westerly airstream that veered to north westerly as a cold front passed through. Precipitating stratocumulus cloud was sampled during the passage of the front. Rain started at about 1330 hours and precipitating cloud descended until the site was enveloped by 1615 hours. During the period of precipitation, the cloud water collector received both rain and cloud water droplets. Observed concentrations were generally low during the period of rain. For example, the chloride concentration fell to below 100 $\mu equ.l^{-1}$ and the nitrate to about 30 $\mu equ.l^{-1}$. After precipitation ceased the concentrations generally increased by an order of magnitude. This phenomenon has been observed elsewhere (11). The concentration matrix for a sample collected after precipitation had ceased is shown in Table 4.

Table 3

27.4.83 Stratus Cloud Samples from Chew Reservoir. Liquid Water Content is 0.19 g/m³.

(a) $[NH_3] = 0.05$ ppb

Composition in and below cloud in μeqs/l for (1), liquid, ppb for (g), gases, and μg/m³ for (s), aerosols.

Cloud	Below Cloud	
$SO_4^=(1)$ 1288	$H_2SO_4(s)$ 4	$(NH_4)_2SO_4(s)$ 8
$Cl^-(1)$ 1598	$HCl(g)$ 7	$NH_4Cl(s)$ 0
$NO_3^-(1)$ 998	$HNO_3(g)$ 3	$NH_4NO_3(s)$ 4
$NH_4^+(1)$ 1175	$NH_4Cl(s)$ 0	$(NH_4)_2SO_4(s)$ 8
$H^+(1)$ 2709	$HCl(g)$ 7	$H_2SO_4(s)$ 4

pH of Cloud Sample 2.58
% Sea Salt 1
% $(NH_4)_2SO_4$ of $SO_4^=$ 65
% NH_4Cl of Cl^- 0
% NH_4NO_3 of NO_3^- 34
% HCl 59 of H^+
% HNO_3 24 of H^+
% H_2SO_4 17 of H^+

(b) $[NH_3] = 0.095$ ppb

Composition in and below cloud in μeqs/l for (1), liquid, ppb for (g), gases, and μg/m³ for (s), aerosols.

Cloud	Below Cloud	
$SO_4^=(1)$ 1288	$H_2SO_4(s)$ 12	$(NH_4)_2SO_4(s)$ 0
$Cl^-(1)$ 1598	$HCl(g)$ 5	$NH_4Cl(s)$ 3
$NO_3^-(1)$ 998	$HNO_3(g)$ 1	$NH_4NO_3(s)$ 8
$NH_4^+(1)$ 1175	$NH_4Cl(s)$ 3	$NH_4NO_3(s)$ 8
$H^+(1)$ 2709	$HCl(g)$ 5	$HNO_3(g)$ 1	$H_2SO_4(s)$ 12

pH of Cloud Sample 2.58
% Sea Salt 1
% $(NH_4)_2SO_4$ of $SO_4^=$ 0
% NH_4Cl of Cl^- 32
% NH_4NO_3 of NO_3^- 65
% HCl 40 of H^+
% HNO_3 13 of H^+
% H_2SO_4 47 of H^+

Table 4

13.4.83 Stratocumulus cloud sample from Chew Reservoir.
Liquid water content is 0.10 g.m^{-3}

Composition in and below cloud in µeqs/l for (l), liquid,
pph for (g), gases, and µg/m^3 for (s), aerosols.

Cloud	Below Cloud		
$SO_4^=$(l)	H_2SO_4(s)	$(NH_4)_2SO_4$(s)	
252.4	.4	.8	
Cl^-(l)	HCl(g)	NH_4Cl(s)	
186.4	.4	0.0	
NO_3^-(l)	HNO_3(g)	NH_4NO_3(s)	
118.0	.1	.3	
NH_4^+(l)	NH_4Cl(s)	NH_4NO_3(s)	$(NH_4)_2SO_4$(s)
215.0	0.0	.3	.8
H^+(l)	HCl(g)	HNO_3(g)	H_2SO_4(s)
341.8	.4	.1	.4

pH of Cloud Sample		3.54
% Sea Salt		38
% $(NH_4)_2SO_4$	of $SO_4^=$	63
% NH_4Cl	of Cl^-	0
% NH_4NO_3	of NO_3^-	47
% HCl	of H^+	55
% HNO_3	of H^+	18
% H2SO4	of H^+	27

The measurements described here on samples collected by two totally
different methods confirm that non-precipitating cloud can accumulate large
concentrations of inorganic ions even at sites reasonably far distant from
major sources. Such high values are more usually associated with urban
fogs following episodes of high pollution. On the two occasions when the
air trajectories had passed near major sources of pollution, HCl made a
significant contribution to the free acidity of the cloud and indeed was
the major acid component in most samples.

It is of interest that samples with a predominantly maritime trajectory also contained free strong acids and that again HCl made a significant contribution. The air sampled from the aircraft on 2 December 1980 had certainly not passed over any land source for several days and yet the cloud samples contained excess chloride equivalent to about 1 ppb in the gas phase. This, however, is consistent with limited observations of gas phase HCl made in clean Atlantic air (12). HCl is not normally observed in rain at remote sites but has been observed to contribute to precipitation composition even on an annual average basis in some places (13).

A preliminary estimate of the importance of the transfer of pollutant material through direct deposition from clouds at an upland site in the same general area has been made by Dollard et al. (14). They assumed that only air from the south to east sector was significant. The present measurements show that, even in air from a predominantly maritime north westerly trajectory, concentrations can be an order of magnitude or more greater in cloud water than in average precipitation. Dollard et al. estimated that direct deposition from cloud could lead to deposition ~20% higher than might be indicated by rain gauge data. The present measurements, which show that much higher concentrations are possible, suggest that this may be an underestimate for this and similar sites.

5. REFERENCES

1. Walters, P.T., Moore, M.J. and Webb, A.H., (1983), A separator for obtaining samples of cloud water in aircraft. Atmos. Environ., 17, 1083.
2. Winter, W., Hogan, A., Mohnen, V. and Barnard, S., (1979), A.S.R.C. Airborne Cloud Water Collection System. A.S.R.C. Publication No. 728, State University of New York, N.Y.
3. Falconer, R.E. and Falconer, P.D., (1979), Determination of Cloud Water Acidity at a Mountain Observatory in the Adirondack Mountains of New York State. A.S.R.C. Publication No. 741, State University of New York, N.Y.
4. Mason, B.J., (1971), The Physics of Clouds, Oxford University Press, 2nd Edition.
5. Goldsmith, P., Delafield, H.J. and Cox, K.C., (1963), The Role of Diffusophoresis in Scavenging Radioactive Particles in the Atmosphere. Q.J. Roy. Met. Soc., 89, 43.
6. Marsh, A.R.W., (1982), Proceedings of C.E.C. Workshop on Acid Deposition, D. Reidel 1983.
7. Stelson, A.W. and Seinfeld, J.H., (1982), 'Thermodynamic prediction of the water activity, NH_4NO_3 Dissociation Constant, density and refractive index for the $NH_4NO_3-(NH_4)_2SO_4-H_2O$ system at 25°C. Atmos. Envir. 16, p. 2507-2514.
8. Tang, I.N., (1980), 'On the equilibrium partial pressures of nitric acid and ammonia in the atmosphere". Atmos. Envir. 14, p. 819-828.
9. Cocks, A.T., Kallend, A.S. and Marsh, A.R.W., (1983), Dispersion Limitations of Oxidation in Power Plant Plumes During Long-Range Transport. Nature, 305, 122.
10. Clark, P.A., Fletcher, I.S., Kallend, A.S., McElroy, W.J., Marsh, A.R.W. and Webb, A.H., (1984), Observations of Cloud Chemistry During Long Range Transport of Power Plant Plumes, Atmos. Environ. To be published.

11. Falconer, P.D. and Kadlecek, J.A., (1980), Cloud Chemistry and Meteorological Research at Whiteface Mountain, Summer, 1979, A.S.R.C. Publication No. 748, State University of New York, N.Y.

12. Buat-Menard, P. and Chesselet, R., (1971), Sur la Présence de Chlore (gaseux) a'irigube Narube dans l'atmosphere, C.R. Acad. Sc. Paris, 272, 1330.

13. Martin, A. and Barber, F.R., (1978), Some Observations of Acidity and Sulphur in Rainwater from Rural Sites in Central England and Wales, Atmos. Env., 12, 1481.

14. Dollard, G.J., Unsworth, M.H. and Harve, M.J. (1983), Nature, 302, 241.

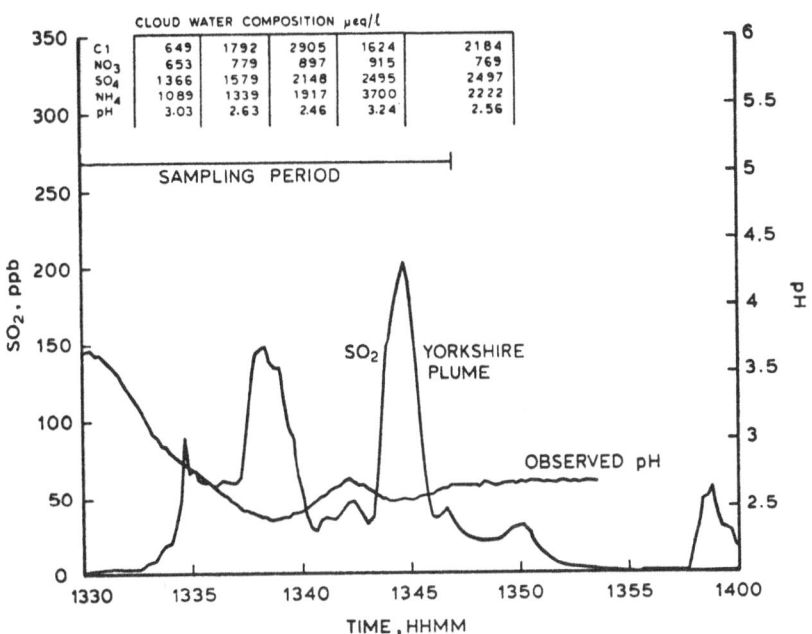

CLOUD WATER COMPOSITION µeq/l

Cl	649	1792	2905	1624	2184
NO₃	653	779	897	915	769
SO₄	1366	1579	2148	2495	2497
NH₄	1089	1339	1917	3700	2222
pH	3.03	2.63	2.46	3.24	2.56

Fig. 1. A COMPARISON OF OBSERVED CLOUD WATER COMPOSITION WITH CALCULATIONS ON 28.1.81

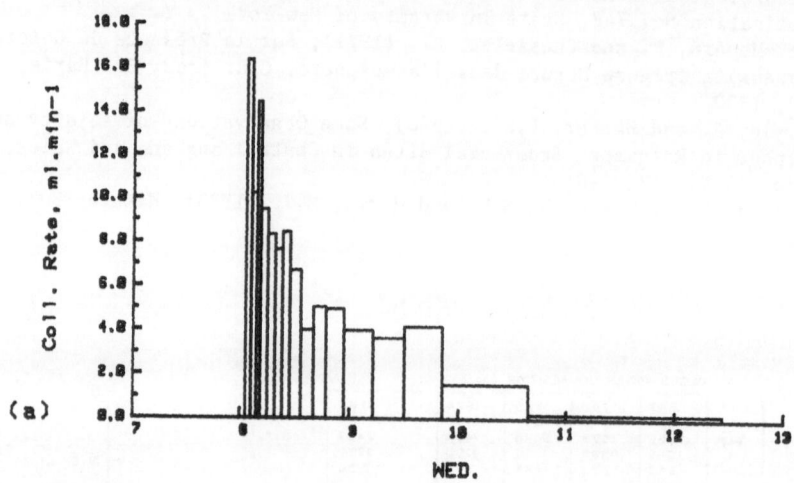

(a)

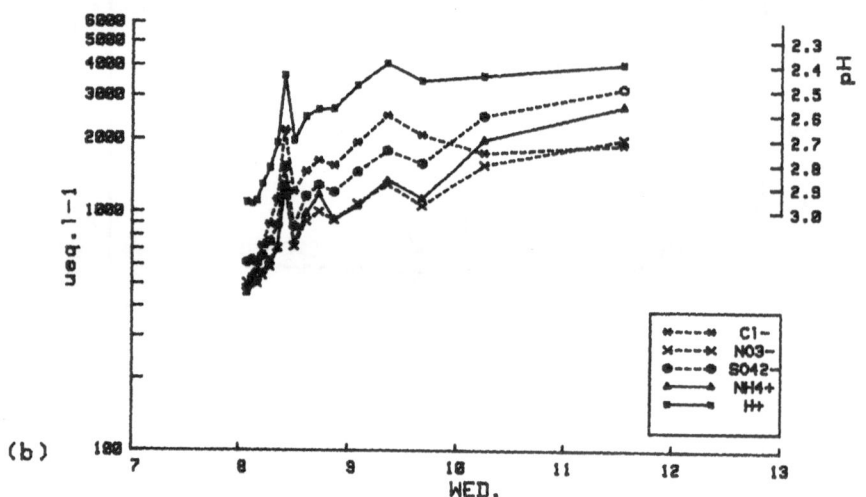

(b)

Fig. 2   (a) Sample collection rate from the
              cloudwater collector, 27.4.83

         (b) Concentrations of major inorganic ions
              in the samples.

AIR SAMPLING FLIGHTS ROUND THE BRITISH ISLES AT LOW ALTITUDES:

SO_2 OXIDATION AND REMOVAL RATES

D.J. Bamber*, P.A. Clark**, G.M. Glover**, P.G. Healey*,
A.S. Kallend**, A.R.W. Marsh**, A.F. Tuck* and G. Vaughan*

* Meteorological Office, London Road, Bracknell, Berkshire, RG12 2SZ

** Central Electricity Generating Board, Central Electricity Research
Laboratories, Kelvin Avenue, Leatherhead, Surrey, KT22 7SE

ABSTRACT

The results are reported of measurements made in the
atmospheric boundary layer during aircraft flights in which plumes from
major point sources of SO_2 in N. England were intercepted on their way
across England and N. Wales to the Irish Sea in July 1982. Over this
part of Britain the boundary layer was well mixed, and capped by a strong
subsidence inversion at about 1200 m. The mean wind speed in the
boundary layer was 9 ms^{-1} from 070°. A layer of quasistationary
stratiform cloud was present over Wales during the whole period of
daylight.

SO_2 and $SO_4{}^{2-}$ flux measurements made upwind of the cloud system
are consistent with an average oxidation rate for SO_2 of ~2% h^{-1} and a
deposition velocity of ~2 cms^{-1} over ~4 h travel. A much higher SO_2 loss
rate of 27% h^{-1} was observed between two sampling points encompassing the
cloud system of which measured sulphate aerosol downwind of the cloud
accounted for 5.3% h^{-1}. It is concluded that a major loss mechanism
within the cloud system was through impaction of cloud droplets on the
Welsh Mountains.

The cloud oxidation rate was estimated to be in the range 3.8
to 29% h^{-1} and it is concluded that ammonia emissions from Wales play a
vital role in sustaining the high oxidation rate.

These results may be contrasted with those of other flights
over the N. Sea in which no substantial oxidation was observed during
plume interaction with cloud over trajectories extending several hundred
km. Unlike these previous flights, oxidation on the present occasion may
have been facilitated by the free availability of oxidant precursors and
by the high levels of atmospheric ammonia that are characteristic of
airstreams over the UK from easterly or north easterly directions.

1. INTRODUCTION

During long range transport of pollutants within the atmospheric boundary layer primary emissions can be removed by dry deposition on land or sea surfaces or by precipitation scavenging. They can also undergo chemical transformations into secondary products which can then be removed by the same processes. In the case of SO_2 transported in clear air, a dry deposition velocity ~1 cm s^{-1} is equivalent to a loss rate of order 3% h^{-1} whilst typical measurements of the rate of transformation into sulphate aerosol indicate an average rate of ~1% h^{-1} at mid latitudes in summer which can be explained by the known rates of individual gas phase reactions (Rodhe, 1981; Cocks, Fletcher and Kallend, 1983).

The interaction of boundary layer air with cloud offers the possibility of substantially higher removal rates through aqueous phase oxidation within cloud droplets. However, this prediction rests more on the results of laboratory studies and regional mass balance calculations than on the relatively few attempts to directly demonstrate the occurrence of in-cloud reaction.

This paper reports measurements made on aircraft flights during July 20-21 1982. On the 21st, the Meteorological Office Hercules C130 W MK 2 aircraft circumnavigated mainland Britain at an altitude of 150 m and an average offshore distance of about 20 km, while the Cranfield Institute of Technology Jetstream aircraft followed flight tracks at different altitudes, mainly in the boundary layer, between the Isle of Man, Liverpool and Cranfield (Fig. 1). The Jetstream had also flown in the boundary layer off the E. coast of England on the 20th. Each aircraft is equipped with instruments to measure chemical and meteorological parameters of the atmosphere; the details of which have been described elsewhere (Nicholls, 1978; James and Nicholls, 1976; Axford, 1972; Marsh, Crabtree and Hatton, 1982 and Kallend et al., 1983).

Gas Chemistry Instrumentation - Both Aircraft		
Freons and chlorocarbons		3 Channel Electron Capture gas chromatograph (Hercules only)
Sulphur dioxide		Meloy labs. SA 285 flame photometric analyser. K_2CO_3 impregnated filter behind particulate filter.
Oxides of nitrogen	(Hercules)	Thermoelectron model 14T chemiluminescent detector
	(Jetstream)	Monitor labs 8440 chemiluminescent detector.
Ozone		Bendix chemiluminescent detector.
Cloud water		CERL cloud water separator.

Table 1: Aircraft Instrumentation

Both aircraft measured background air entering the UK and then intercepted westward moving plumes of SO_2 and NO_x between 51°N and 54°N which were of UK origin, part of whose trajectories coincided with a stationary stratiform cloud system over Wales. This paper presents an interpretation of the measurements in terms of the known source distribution of these gases and the meteorological and chemical processes acting on them.

2. METEOROLOGY

Synoptic Situation

The surface synoptic chart for 1200 on 21 July 1982 is shown in Fig. 2. An anticyclone centred off western Scotland had been slow moving in the vicinity of the British Isles for 4 days. On the day in question there was a steady easterly flow across central and S. England, and across Wales; trajectory analysis showed that this air had previously moved south over the North Sea and Scandinavia, and did not have a long easterly fetch from continental Europe. The mean wind speed in the planetary boundary layer between 51° 30'N and 53°20'N over mainland Britain was 9 m s^{-1} from 070°.

Boundary Layer

The midday radiosonde ascents at Aughton, (53°31'N, 2°55'W), Camborne (50°12'N, 5°19'W) and Hemsby (52°42'N, 1°41'E) all showed a boundary layer capped by a strong subsidence inversion. The profiles of temperature and humidity from the surface to 750 mb are shown in Fig. 3. These profiles also show that the boundary layer was convectively unstable with respect the inversion, favouring good vertical mixing from the surface to its top. The Camborne and Hemsby profiles suggest the presence of cloud in the upper levels of the boundary layer, consistent with both observations and satellite images. The profiles do not suggest sufficient surface heating to supply enough energy for convective penetration of the capping inversion, a conclusion again supported by satellite evidence. No convective cloud was visible in the area of interest. The mean height of the top of the boundary layer was 1200 m at the 3 stations, a value which was taken to be representative of mainland Britain between 51°30'N and 53°20'N.

Clouds and Precipitation

Cloud images in the visible and infrared were available from three satellites: the polar orbiters NOAA-6 and NOAA-7 and the geostationary Meteosat. The half-hourly sequence of visible images from Meteosat shows England and Wales covered by a layer of stratiform cloud at 0755 GMT; by midday this layer was restricted to England south of 52°40'N and Wales, and by 1655 GMT had thinned noticeably and was restricted to Wales, and England south of about 51°30'N. The infrared images confirm that the stratiform cloud layer was at temperatures corresponding to low altitudes, i.e. at and below the capping inversion. The visible image from the AVHRR on NOAA-7, taken at 1312 GMT, is shown in Fig. 4. The important point is that many of the power station sources of SO_2 are situated in areas which were free of cloud by midday. The evidence of wave cloud over N. Wales in Fig. 4 reinforces the view from the Meteosat sequence that the layer of stratiform cloud was quasi-stationary over Wales during the day.

Precipitation records for the day were examined. Out of 86 recording raingauges reporting on mainland Britain, 18 reported some precipitation. Of these 18, 16 reported a trace (defined as less than 0.1 mm total) during the whole 24 h from midnight to midnight, occurring almost wholly in short periods between 0600Z and 1100Z. Fig. 5 shows that these stations were located towards the east coast of England. The 2 remaining stations reporting more than 0.1 mm were in North Lincolnshire and South Yorkshire, which accumulated respectively 0.2 mm during 0.9 h between 0500 and 0800 GMT and 0.2 mm during 0.2 h between 0200 and 0300 GMT. No precipitation was reported from recording raingauges anywhere west of 1°W during the period. The daily gauges reporting for the 24 h period from 0900, 21.7.82 and 0900, 22.7.82 showed a similar picture, except that 3 gauges in South Wales reported total amounts of 0.2 mm, 0.4 mm and 1.1 mm at altitudes of 434 m, 198 m and 311 m respectively. The remainder of the 550 or so daily gauges in Wales, some of which are sited at higher altitudes than these 3 stations, reported zero precipitation.

3. RESULTS

Hercules Flight Measurements

The records of SO_2 and O_3 along the Hercules flight track are shown in Fig. 6. The easterly incoming flow to mainland UK had an SO_2 content at or below the limit of detectability for the instrument operating at 120 s intervals, namely ~2 ppbv. The O_3 amounts were remarkably steady, at about 16–20 ppbv. Elevated SO_2 mixing ratios, up to 16 ppbv, were encountered from 1336–1402 GMT, corresponding to the region between Glasgow and Belfast. In this area the winds were very light and variable, and no plume interpretation or trajectory analysis was attempted. From 1404 to 1432 GMT, the SO_2 was again at or below 2 ppbv, while the O_3 ranged from 16 to 26 ppbv, averaging 24 ppbv. From 1432 to 1457, four SO_2 plumes were resolved, with mixing ratios ranging up to 13 ppbv. The O_3 was generally in the range 9–18 ppbv during this period, but was in the range 26 decreasing to 20 ppbv during the first 150 s. During the plume traverse period measurements were logged at 12.5 s intervals.

In addition to gaseous species, aerosols were collected on 8 filters (4 pairs) exposed from the aircraft along its flight track. After correction by blank assays, amounts of Cl^-, NO_3^-, SO_4^{2-}, NH_4^+, Na^+, K^+, Ca^{2+} and Mg^{2+} were determined. The data are listed in Table 2. Filter exposure times ranged from 45 min in background air to about 11 min in polluted air. The amount of non sea salt sulphate measured off the East coast in the inflow to mainland Britain was 9 nmol m^{-3} (1.0 µg m^{-3}), fully neutralized by 19 nmol m^{-3} (0.5 µg m^{-3}) of NH_4^+. By contrast the plume downwind of Wales (filters H_5 and H_6) contained ~110 nmol m^{-3} (10–12 µg m^{-3}) of SO_4^{2-} again almost neutralized by ~205 nmol m^{-3} (3.5 – 3.9 µg m^{-3}) of NH_4^+.

An onboard 3-channel electron capture gas chromatograph was also operative, and recorded chromatograms at 5 min intervals. The species detected were $CFCl_3$, CF_2Cl_2 CCl_4, CH_3CCl_3 on all chromatograms, and $CHCl_3$ and C_2HCl_3 in polluted air. The data for $CFCl_3$ and CCl_4 are

shown in Fig. 7; the absolute values were obtained by comparison with
35 1 standard air sample #063 supplied by R.A. Rasmussen of Oregon
Graduate Center. Previous data (unpublished data) from low level flights
round the British Isles in westerly flows, when combined with back-
trajectory analysis, have shown the above chlorocarbons to be excellent
tracers for low emission height sources well correlated with population
density.

Jetstream Flight Measurements

On the 20 July between 1100 and 1230 GMT, the Jetstream
aircraft made background measurements off the east coast of England from
53°30'N, 0°25'E to 54°42'N, 0°27'W. The observations, made at altitudes
of 150 and 1500 m, indicated SO_2 and NO_x concentrations below the
detection limit of about 2 ppbv. The aerosol concentrations, listed in
Table 1, were fairly low although rather higher than the samples taken by
the Hercules in the same area on the following day. It is also notable
that the Jetstream samples were nearly neutral with sulphate averaging
25 nmol m^{-3} (2.35 µg m^{-3}) and ammonium 47 nmol m^{-3} (0.85 µg m^{-3}) - i.e.
96% neutralization of sulphate by ammonium.

On the following day between 1030 and 1330 GMT four further
runs were made over the W. Midlands and N. Wales at altitudes of 400 m,
1100-1300 m, 1200 m and 1100 m. The SO_2 and O_3 data shown in Fig. 6 are
from run 1 at 400 m, which extended over the W. Midlands from 1045 to
1115 GMT, for 150 km approximately normal to the air flow. Well defined
plumes of SO_2 were seen, at the higher levels as well as 400 m, providing
evidence that the boundary layer was well mixed, as expected from the
radiosonde temperature profiles in Fig. 3. The O_3 showed signs of
depletion at several points along the Jetstream track in Fig. 6; this is
consistent with the NO_x plumes (not shown) and well correlated with the
SO_2 plumes detected by instruments aboard the aircraft. The aerosol
concentrations are again shown in Table 2. At 400 m, the sulphate
concentration was 52 nmol m^{-3} (5.0 µg m^{-3}) and the ammonium concentration
66 nmol m^{-3} (1.19 µg m^{-3}). Comparing this with the background data
measured on the previous day indicates that about 21 nmol m^{-3} of sulphate
had been added to the air during its passage across the country. Since
the ammonium level had not increased correspondingly, the aerosol was at
this stage more acidic with a 63% neutralisation of sulphate by
ammonium.

About 7 ml of cloud water were obtained by the Jetstream
aircraft intermittently between 11:16 and 11:57 at the upwind edge of the
stratiform cloud deck. This sample contained ~370 µmol l^{-1} SO_4^{2-} and
260 µmol l^{-1} of NH_4^+ and had a pH of 3.6. The nominal 35% neutralisation
is less than the aerosol immediately upwind of the same area. Using a
nominal cloud water loading of 0.5 g m^{-3} these concentrations can be
expressed as equivalent aerosol concentrations, and are shown in Table 1.
It is interesting to note that the apparent sulphate loading would be of
the same order as, but generally higher than, the last 4 Hercules filter
measurements.

4. MODEL CALCULATIONS

The Model

A Gaussian plume model (Fisher, 1973) was used, employing the simplest formulation of cross-wind spread σ_y, namely

$$\sigma_y = \sigma_\theta x$$

where σ_θ is a constant angular spread and x the distance from source. Vertical diffusion is represented by a vertical diffusivity, set arbitrarily to a value of 30 m^2 s^{-1} in this case. This value is on the high side of the range of values possible, and is justified by the evidence for the efficiency of vertical mixing. An altitude of 400 m (post plume rise) was chosen for the high level source, but the model is not sensitive to this choice because of the rapid vertical mixing. Some of the other parameters in the model were specified on the basis of observations for 21.7.82: the depth of the boundary layer was set to 1200 m, and the mean wind speed to 9 m s^{-1}.

Sources of SO_2

The distribution of low altitude sources of SO_2 for 1976 (SARU, 1980) was scaled using an annual emission figure of 1.39 Mt a^{-1} of SO_2 for 1982 (Weatherley, private communication), with further seasonal and diurnal corrections by the method of Fisher (1973). These low level emissions are somewhat uncertain, but are a minor component compared to the high level sources (i.e. power station stacks) in the particular area of interest. The power station emissions were estimated using the actual load figures for 21.7.82; the major uncertainty is the sulphur content of the fuel. The low level sources were aggregated into lines, to facilitate computation. The total SO_2 source distribution is shown in Fig. 8.

Calculation

In order to minimize the number of free parameters, dry deposition of particulate SO_4^{2-} was assumed to be negligible. Two SO_2 loss mechanisms were represented: dry deposition at a uniform velocity V_g and oxidation by a pseudo-first order rate coefficient k_{ox}.

The value of σ_θ was selected by reference to the observed, well separated plume structure on the Jetstream track, and by noting that a distinct plume was observed on the Hercules traverse at about $52°52'N$, for which the only possible large enough upwind source was Fiddler's Ferry power station and the Liverpool conurbation. From these observations and assumptions, a value of $\sigma_\theta = 0.03$ was derived. Given this, the boundary layer depth of 1200 m and the mean wind velocity of 9 ms^{-1}, it remained to fit V_g and k_{ox} to give the best agreement with the available observations.

The model was first applied to the air passing from the east coast Jetstream track to the inland Jetstream track. Using $V_g = 2$ cm s^{-1} and $k_{ox} = 2\%$ h^{-1}, calculations for the inland track predicted a flux of 166 t h^{-1} of SO_2 (corresponding to 83 t h^{-1} of S) and an increase above the background level of 3.3 μg m^{-3} in the SO_4^{2-} concentration (corresponding to 1.1 μg m^{-3} of S). These predictions are in good agreement with the Jetstream observations (Table 3). However, when the same values of V_g and k_{ox} were applied to the air passing between the Jetstream track and the Hercules track off the west coast, considerably more SO_2 and less SO_4^{2-} were predicted than observed. In fact, a mean V_g of 5 cms^{-1} and k_{ox} of 4% h^{-1} would need to be invoked, from the source, to obtain agreement with the Hercules observations. A comparison of the calculated fluxes and concentrations with those estimated from the observations is shown in Table 2.

Clearly the SO_2 removal and oxidation rates were much higher downwind than upwind of the Jetstream track.

Sensitivity to Errors and Uncertainties

The sensitivity of the SO_2 measuring instruments on the Hercules and Jetstream aircraft is such that 2 ppbv of the gas is the lowest mixing ratio which can be detected. This leads to an uncertainty in the flux off the E. coast of mainland Britain of about 16 t h^{-1}. The error on the background and plume SO_4^{2-} concentrations are respectively ± 10% and ± 5%.

The error in the magnitude of the SO_2 source strengths originates primarily in the figure for the S content of the fuel, and is estimated at ± 10%. The error in the depth of the boundary layer is at most ± 15%, and in the transit times of the air between the aircraft tracks is ± 20%. A key assumption, affecting the whole model analysis and interpretation, is that vertical mixing was efficient. The sensitivity of the flux estimates to the plume spread parameter σ_θ is not great, provided a reasonably low value is used, since only the degree of admixture of other sources changes.

Given the observations and uncertainties, the values of V_g and k_{ox} for the airflow between the source and the Jetstream track could take values in the ranges 2 ± 1.6 cm s^{-1} and 2 ± 0.8% h^{-1} respectively. For the airflow between the source and the W. coast Hercules track these values would be 5 ± 1.6 cm s^{-1} and 4 ± 1% h^{-1} respectively.

One final uncertainty is the constancy of the composition of the airflow as sampled by the Jetstream and the Hercules. The transit time of the air between the Jetstream track and the W. coast leg of the Hercules track, at the location of the plumes, was calculated from trajectory analysis to be 4 2/3 h; the actual air in which the Jetstream flew was still about 40 min upwind of the Hercules when it intercepted the plumes off the W. coast.

5. DISCUSSION AND CONCLUSIONS

The conceptual view being presented here is the simplest possible view consistent with the observations available. In cloud free air over central England, the dry deposition velocity and pseudo first order rate constant for oxidation of SO_2 were found to be 2 cms^{-1} and 2% h^{-1} respectively, i.e. on the high side of the values generally found in the literature.

Comparison of the estimated fluxes based on the observations of both aircraft shows that 49 ± 16 tonnes of sulphur, (27 ± 8% h^{-1} SO_2 removal rate), were lost from an area of about 150 × 150 km in 4 2/3 h from a boundary layer 1200 m deep. This layer contained a stratiform deck of cloud about 150 × 100 km which could have been 300–400 m thick based on the Camborne radiosonde ascent.

The simple model described above only applies to dry conditions. Between the two aircraft tracks over Wales, the rate of deposition and oxidation increased almost certainly because of the presence of cloud. In order to describe the changes over Wales at least two further terms are required to account for the wet deposition loss of SO_2 and SO_4^{2-}.

The minimum oxidation rate of SO_2 over Wales required to account for the SO_4^{2-} observed by the Hercules is 5.3 ± 1.5% h^{-1}. However, this value requires the total loss ofsulphur to be in the form of SO_2. If the same dry deposition velocity applied as the model suggests for the clear dry air between the source and the Jetstream track, 2 cm s^{-1}, then 20.2 ± 3.4 tonnes of sulphur would be lost across Wales, leaving 28 ± 15 tonnes of sulphur to be accounted for by the wet deposition of SO_2. Alternatively, the 28 tonnes of sulphur could be lost by the wet deposition of SO_4^{2-} assuming, as in the model, that dry deposition of SO_4^{2-} is negligible. In this case a maximum oxidation rate of SO_2 of 21 ± 8% h^{-1} is required.

If the mean concentration of sulphate measured in the cloud water sampled by the Jetstream was appropriate, the wet deposition required to remove 28 t of S would be of the order of 0.025 mm h^{-1} over the entire area of the cloud deck. The actual area is much more likely to be confined to the highest ground in Wales but even so the required loss of water from the cloud to the surface is quite credible. On the highest hills, there are few rain gauges and the absence of recorded rainfall does not necessarily preclude the removal of sulphur by wet deposition by either direct droplet impaction or in drizzle.

An oxidation rate between 5.3 ± 1.5 and 21 ± 8% h^{-1} for SO_2 is a high value, and there are few available literature measurements of a similar nature with which to compare the results for the stratiform cloud layer; the results of Hegg and Hobbs (1981, 1982) refer to individual wave clouds on smaller time and space scales than in our case. Their analysis also concentrated on measurements of the ionic composition of cloud droplets. They observed oxidation of SO_2 to SO_4^{2-} in 16 of 28 cases, and obtained an empirical rate law from the equation:-

$$\frac{d\left[SO_4^{2-}\right]}{dt} = k\left[H^+\right]^\alpha \left[SO_3^{2-}\right] \exp\left(-\beta/T\right)$$

We can make no direct comparison, because on this occasion, for reasons of available aircraft time, the Jetstream cloud measurements could not be directly compared to aerosol composition except in a qualitative way as shown in Table 2.

The high oxidation rate for SO_2 can only be explained on the basis of aqueous oxidation of SO_2. The most probable routes involve H_2O_2 and O_3. The H_2O_2 mechanism is fast but it is difficult to see with present understanding how it can be sustained to consume more SO_2 than the original H_2O_2 mixing ratio estimated to be <1 ppbv (e.g. Cocks, Kallend and Marsh, 1983). The oxidation by O_3 is not restricted by its availability but by the acidity of the aqueous phase, and it is relevant to note the high degree of neutralisation, 96%, by ammonium ions in the Hercules filter samples as compared with ~63% for the Jetstream filter samples. The presence of NH_3 would enhance both the solubility of SO_2 and the oxidation in solution by sustaining a higher pH.

The observed SO_2 concentration and cloud water pH would yield an equilibrium dissolved SO_2 concentration of only ~ 2 μmol 1^{-1} whereas in the presence of a nominal 1 ppbv of NH_3, the equilibrium dissolved SO_2 concentration would be ~400 μmol 1^{-1}. Such a value of dissolved SO_2 would imply both a high oxidation rate and a high wet removal rate for SO_2. At the cloud water pH value observed by the Jetstream at the upwind edge of the cloud deck the rate of oxidation of dissolved SO_2 by ozone would be <0.1% h^{-1}, but if a pH of 5 were appropriate because of the presence of ammonia, the measured O_3 and SO_2 concentrations could sustain an oxidation rate of ~6% h^{-1}, while if equilibrium with 1 ppbv of NH_3 were maintained the rate could be as high as 63% h^{-1} (see for example Penkett, Jones and Brice, 1977). The role of the high ground in Wales in reducing the vertical travel distance between the ground level sources of ammonia and the clouds may have been significant on this occasion.

The increase in the flux of ammonium in aerosol between the aircraft tracks was roughly 15 t h^{-1}. This figure is a minimum estimate, compatible with the 5% h^{-1} oxidation rate of SO_2, because it assumes no deposition loss of ammonium ions in association with sulphate. On the other hand, if the higher oxidation rate for SO_2 is correct and all the sulphate lost by wet deposition was ammonium sulphate, (as inferred from the aerosol samples collected by the Hercules), then an extra flux of about 31 t h^{-1} of NH_3 is required from this area of Wales. Estimates of the ammonia emission from Northern and Central Wales using the annual emission factors recommended by Boris, Meszaros and Putsay (1980) indicate a flux of ~29 kt a^{-1}. This is equivalent to ~3.3 t h^{-1}, 1/5 of the required minimum ammonia flux. Given the seasonal and diurnal variability of ammonia emissions, it would not be unreasonable to increase this hourly estimate by a factor of 4, especially as the measurements were made in the early afternoon of midsummer. However, the resulting flux of 13.2 t h^{-1} is less even than the minimum estimate derived from the aircraft data. This accords with experience from previous flights, and suggests that the ammonia emission factors quoted above may be seriously underestimated.

The high rates of oxidation of SO_2 observed on this flight contrast markedly with those reported by Clark et al. (1984) and Cocks, Kallend and Marsh (1983). In those flights an extremely low oxidation rate of <<0.1% h^{-1} was found but the very low dispersion of the plume significantly reduced the available O_3 and other oxidant levels.

Our conclusions are critically dependent on two assumptions: that the Hercules did not miss the plumes on its W. coast leg, and that the boundary layer was well mixed with a strong capping inversion. The evidence that these assumptions hold is that the plume structure was observed by the Hercules, and was seen in both chlorocarbons (which have surface sources) and SO_2 (which comes mainly from high power station stacks), an observation which also tends to support the idea that the boundary layer was well mixed. The Jetstream also saw similar plume structure at 400 m and 1100 m. Radiosonde temperature profiles and satellite images both confirm that the SO_2 and sulphate could not have escaped into the free troposphere in clouds penetrating the inversion at the top of the boundary layer.

Acknowledgements

The authors are indebted to a large number of their colleagues at the Meteorological Office, the Meteorological Research Flight and Central Electricity Research Laboratories who participated in this experiment.

This work was carried out at the Central Electricity Research Laboratories and the Meteorological Office and is published by permission of the Central Electricity Generating Board and the Meteorological Office.

6. REFERENCES

Axford, D.N., 1972, Brief summary of the Meteorological Research Flight digital magnetic-tape data-recording and data-processing system. Meteorological Magazine 101, 329-339.

Boris, K., Meszaros, E. and Putsay, M., 1980, On the atmospheric budget of nitrogen compounds over Europe. Idojarus 84, 57-68.

Clark, P.A., Fletcher, I.S., Kallend, A.S., McElroy, W.J., Marsh, A.R.W. and Webb, A.H., 1984, Observations of cloud chemistry during long range transport of power plant plumes, this issue.

Cocks, A.T., Fletcher, I.S. and Kallend, A.S., 1983, Chemical modelling studies of the long range dispersion of power plant plumes, in Air pollution modelling and its application, Plenum, London.

Cocks, A.T., Kallend, A.S. and Marsh, 1983, Dispersion limitations of oxidation in power plant plumes during long range transport. Nature 305, 122-123.

Fisher, B.E.A., 1973, The deposition velocity and transport of sulphur dioxide over the United Kingdom, Central Electricity Generating Board report RD/L/N 153/73.

Hegg, D.A. and Hobbs, P.V., 1981, Cloud water chemistry and the production of sulphates in clouds, Atmospheric Environment 15, 1597-1604.

Hegg, D.A. and Hobbs, P.V., 1982, Measurements of sulphate production in natural clouds. Atmospheric Environment 16, 2663-2668.

James, D.G. and Nicholls, S., 1976, Current work of the Meteorological Research Flight. Meteorological Magazine 105, 86-99.

Kallend, A.S., Clark, P.A., Cocks, A.T., Fisher, B.E.A., Glover, G.M., Marsh, A.R.W., Moore, D.J., Sloan, S.A. and Webb, A.H., 1983, The fate of atmospheric emissions along plume trajectories over the North Sea: Final Report to EPRI, Contract RP1131-1. Volume 4, Central Electricity Research Laboratories, Leatherhead.

Marsh, A.R.W., Crabtree, J. and Hatton, D.B., 1982, Instrumentation of the Hercules aircraft of the Meteorological Research Flight, MRF Internal Note No. 14. Available from the Assistant Director, MRF, RAE Farnborough.

Nicholls, S., 1978, Measurements of turbulence by an instrumented aircraft in a convective atmospheric boundary layer over the sea. QJR Met. Soc. 104, 653-676.

Penkett, S.A., Jones, B.M.R. and Brice, K.A., 1977, Rate of oxidation of sodium sulphite solution by oxygen, ozone and hydrogen peroxide, and its relevance to the formation of sulphate in cloud and rainwater, AERE R-8684, Harwell.

Rodhe, H., 1981, Formation of sulphuric and nitric acid in the atmosphere during long-range transport. Tellus, 33, p 132-141.

SARU, 1980, Current geographical distribution of sulphur emissions, Department of Environment Report CENE(80) 16.

TABLE 1

Filter analyses, 21 July 1982

Filter	time	Cl⁻ µg m⁻³	NO₃⁻ µg m⁻³	SO₄²⁻ µg m⁻³	Na⁺ µg m⁻³	NH₄⁺ µg m⁻³	K⁺ µg m⁻³	Ca²⁺ µg m⁻³	Mg²⁺ µg m⁻³	SO₂(impr)** ppbv
H1	0932-1017	5.6	0.4	1.7	3.4	0.3	0.1	0.09	0.43	0.08
H2	0932-1017	3.8	0.3	1.7	2.3	0.4	0.2	0.11	0.27	0.08
H3	1248-1336	0.15	not det	0.49	0.1	0.18	0.0	0.0	0.0	0.0
H4	1248-1336	0.08	not det	0.54	0.08	0.24	0.07	0.15	0.0	0.13
H5	1416-1503	0.20	0.68	10.0	0.33	3.5	0.0	0.94	0.07	2.2
H6	1416-1503	0.12	0.30	11.5	0.35	3.9	0.13	0.87	0.04	2.9
H7	1513-1524	3.8	3.3	11.0	3.1	3.5	0.61	1.5	0.43	0.44
H8	1513-1524	2.6	not det	14.8	2.0	4.0	0.44	2.4	0.25	not det
J1	1102-1132 20/7/82	0.05	0.17	2.9	0.08	1.08	0	0.17	0	0.12
J2	1049-1118 21/7/82	0	0.1	5.0	0.02	1.19	0	0	0	0.32
J3*	1216-1250 21/7/82	4	5	18	1	2	1	4	0	-

Cloud water sample

* equivalent aerosol concentrations assuming 0.5 g m⁻³ of cloud water.
** K₂CO₃ impregnated filter measurement "not det" means below detection level.
H in column 1 indicates Hercules
J indicates Jetstream

Table 2: Model calculations of fluxes and concentrations of SO_2, SO_4^{2-} and NH_4^+ compared with aircraft observations

	SO_2 flux tonnes h⁻¹ of S		SO_4^{2-} flux tonnes h⁻¹ of S		NH_4^+ flux tonnes h⁻¹		SO_4^{2-} conc μg m⁻³		NH_4^+ conc μg m⁻³	
	OBS	CAL	OBS	CAL	OBS	CAL	OBS	CAL	OBS	CAL
Jetstream	84 ± 14	83	5.9 ± 1.6	6.1	3.5 ± 1.4	9.7	1 ± 0.2	1.1	0.6 ± 0.2	1.7
Hercules	24 ± 8	27	17.5 ± 3.3	18.7	18 ± 3.6	13	3 ± 0.4	3.0	3.1 ± 0.2	2.2
Change	60 ± 16		11.7 ± 2.1		14.7 ± 2.4		2 ± 0.2		2.5 ± 0.3	

Note: The observed values are presented as increases over assumed background levels of zero for SO_2, 0.6 ± 0.3 μg m⁻³ of S for SO_4^{2-} and 0.6 ± 0.3 μg m⁻³ of NH_4^+. Calculations for the Jetstream flight used V_g = 2 cm s⁻¹ and k_{ox} = 2% h⁻¹; those for the Hercules flight used V_g = 5 cm s⁻¹ and k_{ox} = 4% h⁻¹. Zero deposition was assumed for the NH_4^+ calculations.

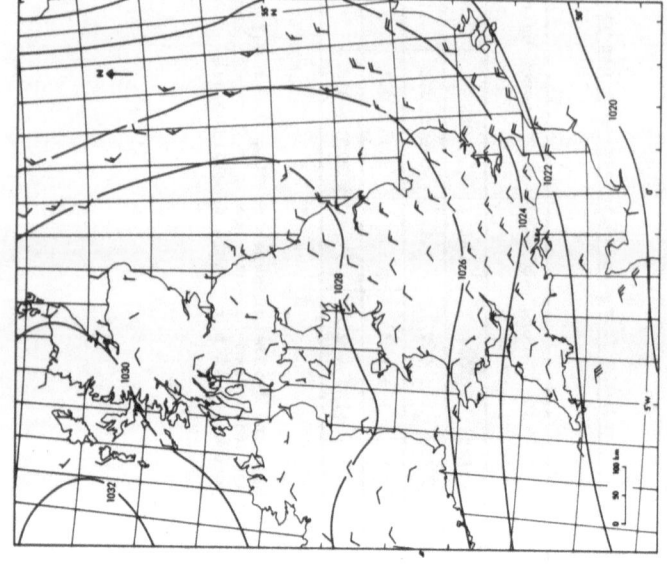

Fig. 2 – Surface synoptic chart 1200Z, 21 July 1982. Surface winds and isobars only are shown. Pressures are given in units of mb, wind vectors in knots (one half 'feather' corresponds to 5 kt).

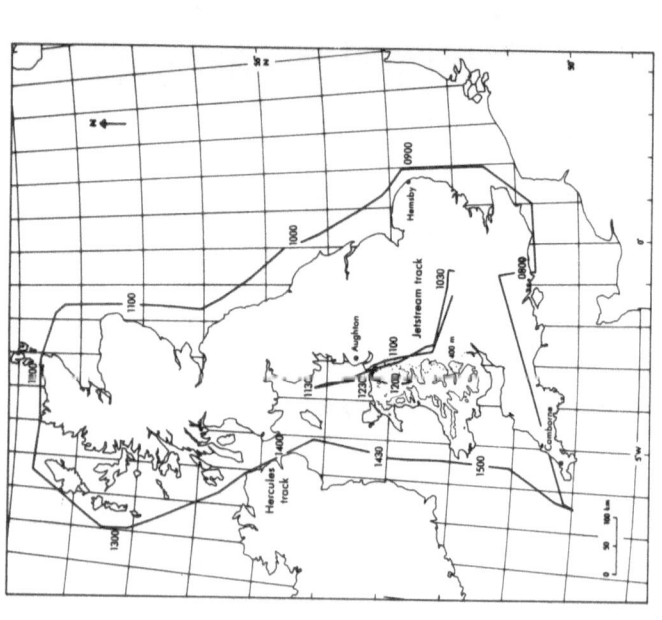

Fig. 1 – Aircraft tracks, with times (GMT) on 21 July 1982. Radiosonde stations whose 1 200 Z ascents on 21 July 1982 were used are marked, as is the 400 m land contour over Wales.

Aughton 03322

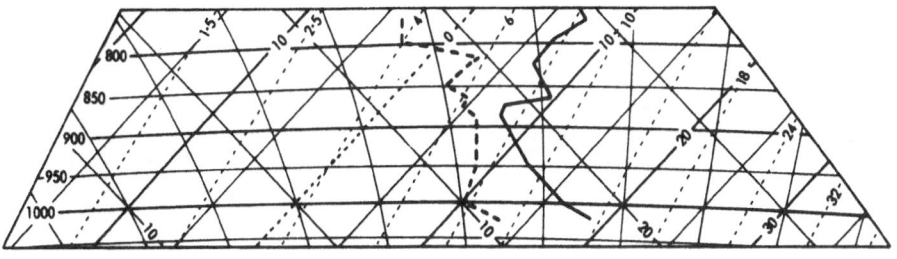

Pressure (mb)	700	850
Wind direction	075°	050°
Wind speed (ms^{-1})	7	9

Hemsby 03496

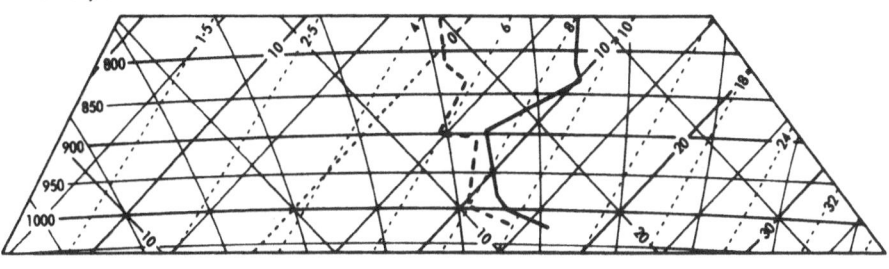

Pressure (mb)	700	850
Wind direction	065°	040°
Wind speed (ms^{-1})	5	7

Camborne 03808

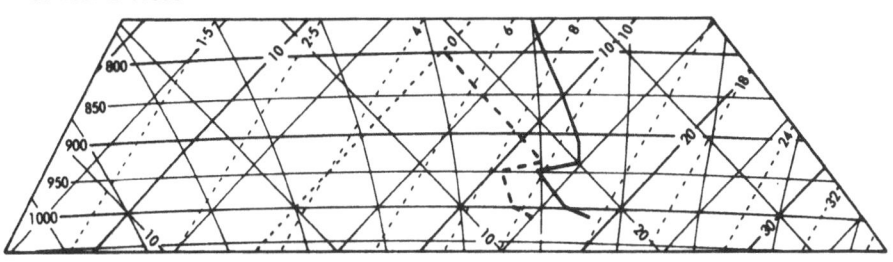

Pressure (mb)	700	850
Wind direction	115°	100°
Wind speed (ms^{-1})	6	7

Fig. 3 - Tephigrams, 1050-750 mb, 1200 A, 21 July 1982. Solid line is temperature, dashed line is dew point in deg C, pressure in mb.
a. Aughton   b. Hemsby   c. Camborne

Fig. 4 - Satellite image, visible wavelength channel, AVHRR on
NOAA-7, 1312 GMT, 21 July 1982

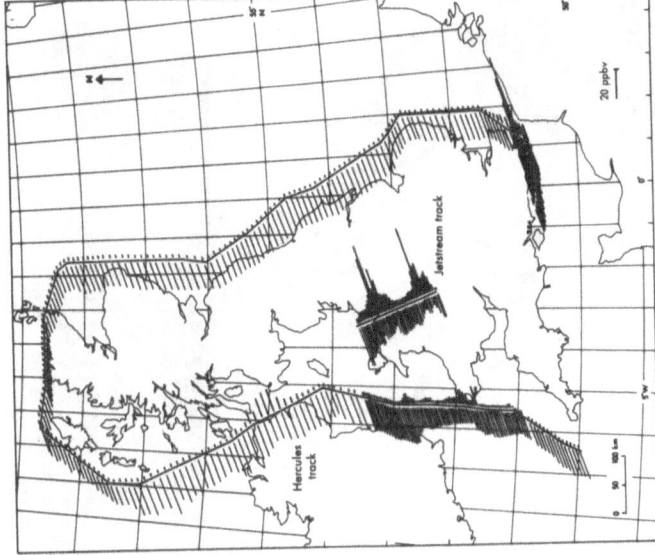

Fig. 6 – Sulphur dioxide (to the east of the aircraft tracks) and ozone (to the west of the aircraft tracks). The common scale is given on the figure. Where the lines whose heights represent the SO2 and O3 mixing ratios are more closely spaced a shorter data sampling interval was used on the instruments. The directions of these lines correspond roughly to the direction of the mean wind in the boundary layer between 51°30'N and 53°20'N.

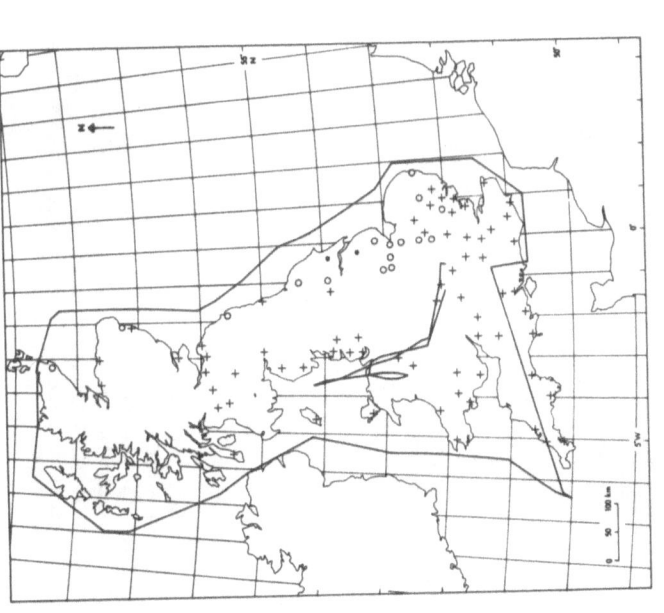

Fig. 5 – Map showing recording raingauge stations reporting during the 24 hours from 000 Z to 2400 Z, 21 July 1982. Open circles are stations reporting precipitation up to 0.1 mm total during the period, crosses are stations reporting zero precipitation. The two stations marked by filled circles are discussed in the text.

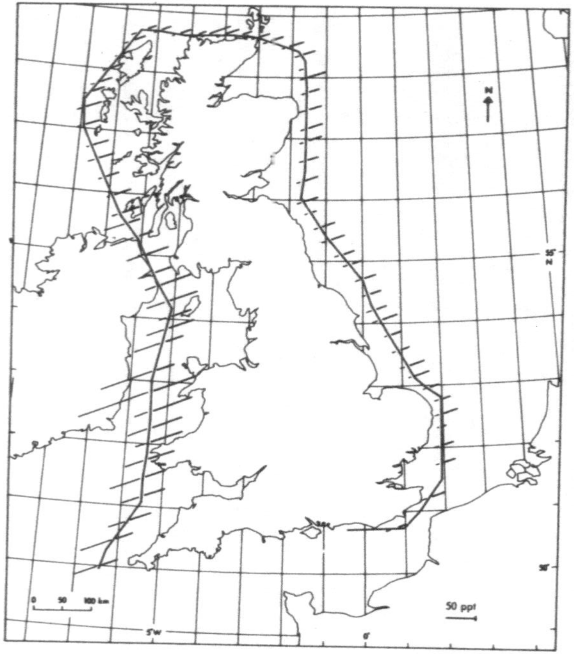

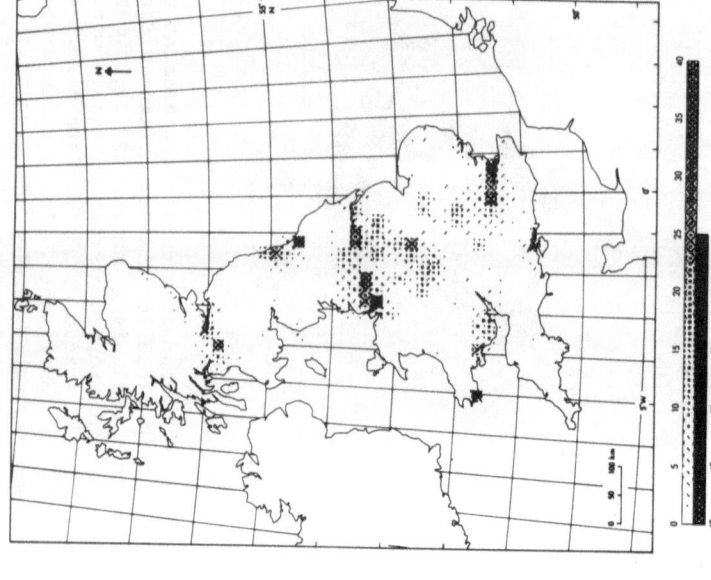

Fig. 7 – Measurements of CFCl₃ (to the west) and CCl₄ (to the east of the aircraft tracks) made by a 3-channel electron capture gas chromatograph on the Hercules. Note that constant background levels of 209 and 141 pptv have been subtracted respectively; otherwise the data presentation is as in fig. 6

Fig. 8 – A map of SO₂ sources (tonnes h⁻¹) for 21 July 1982, based on power stations loads for the day and annual data corrected for daily and seasonal variations for non-power station sources.

OXIDATION OF NO TO NO$_2$ IN FLUE GAS PLUMES OF POWER STATIONS

A.J. Elshout
N.V. KEMA, Arnhem, the Netherlands
and
Dr. S. Beilke
UBA, Frankfurt, West Germany

Summary

The oxidation of NO to NO$_2$ in flue gas plumes takes place after release in the atmosphere as a result of the reaction with atmospheric oxygen and atmospheric ozone. The conversion rate is in a large measure a function of the turbulent mixing rate of flue gas plume and atmospheric air. The effects of several parameters (such as given by the dispersion conditions, ozone concentration in the atmospheric air and photo-stationary equilibrium) are illustrated with the measuring results in flue gas plumes emitted from power stations in the Netherlands and Germany. Besides, on the basis of the results of model calculations, the dependence of the conversion rate on the various parameters is discussed. The inhomogeneity of mixing of the flue gas plume with the surrounding air, just as the oxidation with oxygen, appears to play a role in the near field. For a more realistic calculation of the NO-oxidation in plumes, a model has to be developed, which incorporates the inhomogeneous mixing and both oxidation routes of NO.

INTRODUCTION

The oxidation of NO to NO$_2$ in the atmosphere and especially in plumes from large point sources such as power stations is a problem of current interest. Nitrogen dioxide (NO$_2$) plays an essential part in many chemical processes in the atmosphere. This is illustrated by the formation of ozone (O$_3$) and other secondary air pollutants such as nitric acid (HNO$_3$) and peroxyacetylnitrate (PAN). In higher concentrations NO$_2$ is yellowish brown and gives cause for the occurrence of brown plumes. It is considered more toxic than nitrogen monoxide (NO) as it is emitted primarily from the combustion process. Therefore, it is of importance to know the ground level concentrations of NO$_2$ downwind from large NO emission sources under a wide range of conditions for dispersion and oxidation.

The oxidation of NO to NO$_2$ in plumes is a complicated process depending on a series of chemical and physical parameters. Nitrogen oxide (NO) is oxidized in three reactions:

$$2\ NO + O_2 \rightarrow 2\ NO_2 \quad\quad (1)$$
$$NO + O_3 \rightarrow NO_2 + O_2 \quad\quad (2)$$
$$NO + RO_2 \rightarrow NO_2 + RO \quad\quad (3)$$

In the "near field" of plume dispersion the oxidation of NO by peroxiradicals (reaction 3) is of minor importance. It only plays a role in a polluted atmosphere containing high concentrations of reactive hydrocarbons in the presence of high uv radiation.

The reaction 1 is second order in NO concentration and the rate constant is such that this reaction is only of importance if the NO concentrations are sufficiently high. Such conditions can exist in the plume close to the stack and further on in plume parcels that are not well mixed.

For most situations occurring in the lower atmosphere, the oxidation of NO by O_3 (reaction 2) is the most effective oxidation mechanism. During the day, in the presence of sunlight, absorption of ultraviolet radiation by nitrogen dioxide (NO_2) leads to the formation of ozone:

$$NO_2 \xrightarrow{UV} NO + O \qquad\qquad (4)$$
$$O + O_2 + M \rightarrow O_3 + M \qquad\qquad (5).$$

Reaction 5 is so fast that it can be combined with reaction 4 to:

$$O_2 + NO_2 \xrightarrow{UV} O_3 + NO \qquad\qquad (6).$$

The lumped reaction 6 is the reverse of reaction 2. Reactions 2 and 6 are much the fastest known reactions in which NO, NO_2 and O_3 participate. Both reactions should therefore control the short-term dynamics of the $NO/NO_2/O_3$ system, and concentrations should, in a well mixed situation, quickly approach a photostationary state in which, with $k_6 = k_4$:

$$[O_3][NO] / [NO_2] = k_4/k_2 \qquad\qquad (7),$$

in which k_4 and k_2 are the reaction constants of the reactions 4 and 2 respectively. In the case of NO oxidation by O_3 there is turbulent mixing of the plume (NO) and the surrounding air (O_3). The rate at which the oxidation of NO in the plume occurs consequentely depends on complex interactions between turbulent mixing and chemical kinetics.

In an instantaneous "plume" a heterogeneous mix of plume-rich parcels and air-rich parcels will be formed, with more air-rich parcels on the outside of the "plume" and more plume-rich parcels in the inside of the "plume". It is ultimately the molecular diffusion, which at the boundaries of the eddies leads to local mixing, that makes reaction between NO from the plume and O_3 from the ambient air possible. In situations where the time scale for reaction is comparable with or smaller than the time scale for mixing the chemistry in the plume is controlled by the rate of mixing.

This paper deals with results of our measurements in power plant plumes and the state of model calculations of NO oxidation in plumes.

MEASUREMENTS

Field studies on the oxidation rates of NO in power plant plumes were carried out by UBA with Barringer remote sensing spectrometers (COSPEC) installed in a car [1], and by KEMA with direct measurement of the concentrations of the gaseous components in the plume, using a Piper Navajo Chieftain aircraft, equipped with instruments [2] or a van with the same instruments.

In the period 1980-1982 measurements were carried out by UBA in co operation with Staudinger power station in the plume of this coal-fired power station. In the Netherlands measurements were done in the plumes of several power stations fired with gas, oil and coal from 1975 on. The results given here refer to the oil and gas-fired power stations Maasvlakte and Flevo. In May 1982 combined measurements were carried out by UBA and KEMA in the plume of Wilhelmshaven power station.

In figure 1 the results are given of 29 measurement campaigns in the Staudinger plume. These 29 days enclose a broad range of physical and chemical parameters. The figure shows highest oxidation rates in situations of unstable atmospheric conditions (Pasquill stability classes B and C) and lower rates in situations with more stable conditions (Pas-

quill stability classes D for neutral and E for light stable). For the
unstable situations (Pasquill classes B and C), the 50% NO oxidation (integrated over the plume cross-section) was found between 12 min and 70
min of plume travel time. For the neutral situations (Pasquill class D),
here with mean transport velocities of about 5 to 6 m/s, the 50% NO oxidation was found (by extrapolation) at plume travel times larger than 45
min and for most of the days with these situations likely in the order of
some hours.

If we use the plume width as a measure of plume dilution, the results
of all the measurements can be arranged in the form of percentage oxidation $NO_2/NO+NO_2$ as a function of plume width. For the situations with
a measurable concentration of ozone (ozone concentrations in Frankfurt
higher than 15 ppb) the regression line can be determined. The resulting linear regression line is $y = 2.84+0.0096x$ ($y = NO_2/NO+NO_2$ in %,
x is plume width in m), with a correlation coefficient of 0.92. This significant correlation gives reason for the conclusion of the diffusion-
controlled NO oxidation.

It shall be emphasized, however, that the plume width needs not be an
adequate measure for the degree of mixing in the plume. In the plume from
Wilhelmshaven power station only a relatively slight NO conversion could
be observed on 13th May 1982 despite considerable plume width, because in
this case there was very little vertical mixing. This effect has also
been found with other power stations on the North Sea and is caused by
the perculiarities of the land-seawind system.

The results of the airborne measurements and the measurements with a
van carried out by KEMA in the plumes of power stations fairly agree with
the results as represented here for the Staudinger power plant. For the
unstable situations (Pasquill classes B and C) and ozone concentrations
up to 60 ppb ozone in the surrounding air, the 50% NO oxidation (integrated over the plume cross-section) was found between 5 min and about 50
min of plume travel time. In a situation of 125 ppb ozone, the 50% oxidation was already found after 2.4 minutes.

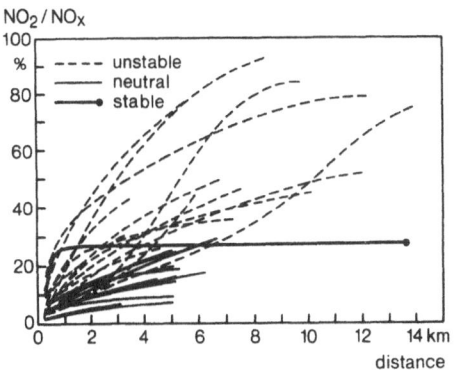 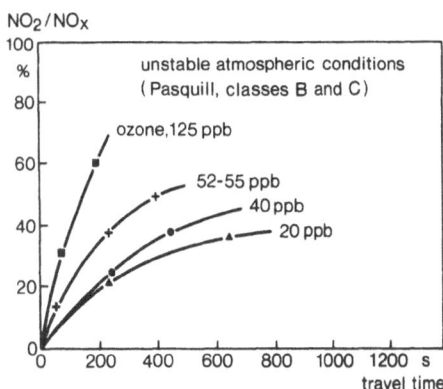

Figure 1 The NO oxidation percentages integrated over the whole plume
cross-section as measured in 29
campaigns in the plume of Staudinger power station as a function of
distance.

Figure 2 Percentage oxidation ratio
NO_2/NO_x in plumes of power stations Flevo and Maasvlakte integrated over the whole plume cross-section as a function of travel time
for different ozone concentrations
in the surrounding air and unstable
atmospheric conditions.

In figure 2 results from airborne measurements are given for the unstable situation at different ozone concentrations. This figure shows the
influence of the ozone concentrations in unstable situations on the NO
oxidation in the plumes. For neutral situations the airborne measurements
give values from 20 min to more than 60 min for transport velocities of
the plume lower than 10 m/s, for the 50% NO oxidation times. At higher
transport velocities of the plume this 50% NO oxidation time decreased
for example to a value of about 4 min at a transport velocity of 15 m/s
as shown in figure 3. In this figure also the results are given of ground
level measurements (with the van) in the plume in neutral situations with
different transport velocities. This illustrates the effect of the wind
velocity on the dilution of the plume after release from the stack.

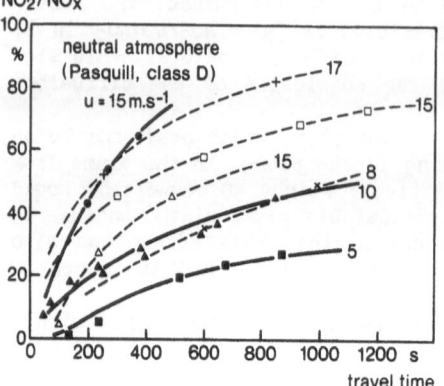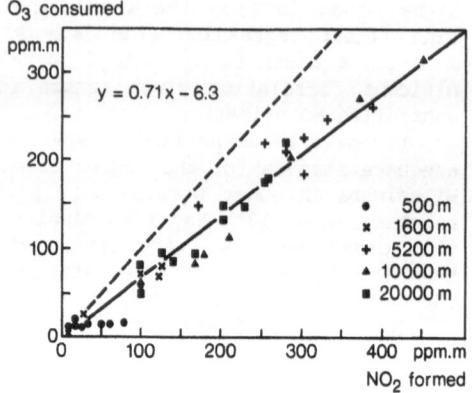

Figure 3 Percentage oxidation ratio
NO_2/NO_x in plumes of power stations Flevo and Maasvlakte integrated over the whole plume cross-section (airborne measurements) and
over the plume cross-section on
ground level (van measurements) at
different transport velocities on
plume height in a neutral atmosphere.

———— integrated over the plume
      cross-section   (airborne
      measurements)
- - - - integrated over the plume
      cross-section   at   ground
      level (van measurements)

Figure 4 The relationship between
O_3 consumed and NO_2 formed in
the Wilhemshaven plume on 13 May
1982.

The relative importance of the two NO oxidation routes (reactions 1
and 2 in the introduction), is apparant from figure 4. This figure is on
airborne measurements carried out in the plume of Wilhelmshaven power
station and in it the amount of NO_2 observed for each pass through the
plume has been plotted against the corresponding O_3 deficit. The excess
of NO_2 with respect to the corresponding O_3 deficit can be assigned
to the termolecular path, the NO oxidation with O_2. The other part of
NO_2 is formed from the reaction of NO with O_3.
    In figure 5 the oxidation percentage $NO_2/NO+NO_2$ in the plume is
given as a function of distance, together with the percentage excess
NO_2 defined by $NO_2-(-O_3)/NO+NO_2$. From this figure it can be con-

cluded that the NO oxidation by O_2 cannot be neglected. At a distance of about 10 km the oxidation percentage by O_2 oxidation is in this situation 18% at a total of 37%.

Another effect can be remarked, viz a decrease in the $NO_2-(^-O_3)/NO_x$ ratio in the afternoon at distances larger than 10 km. This can be attributed to the disappearance of NO_2 as a result of the reaction $NO_2+OH\cdot\rightarrow HNO_3$. Similar results are given by Melo and Stevens [3] and Meagher et al. [4].

Table 1   Results of measurements of UBA and KEMA in the plume of Wilhelmshaven power station

	date	$NO_2/NO+NO_2$ (%)	
		distance 0.5 km	20 km
UBA	82-05-11	8	32
	82-05-12	8.8	40.8
	82-05-13	6.5	NM
KEMA	82-05-13	7.0	41

(NM = not measured)

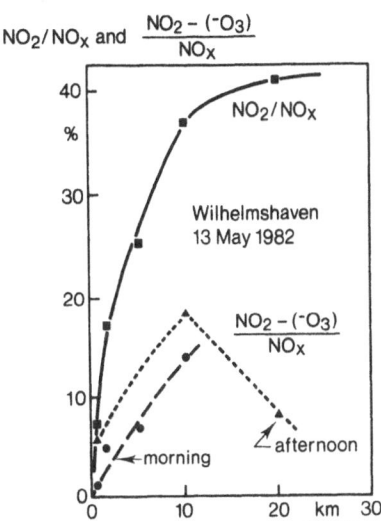

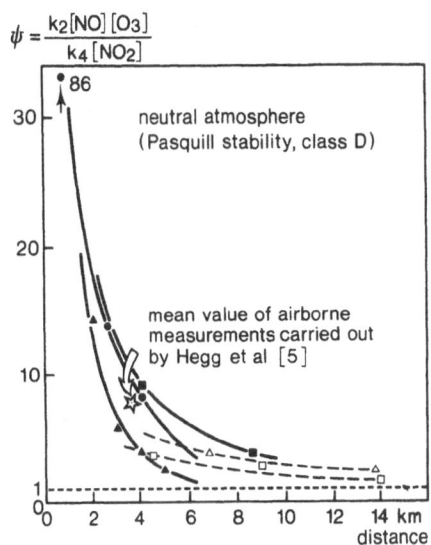

Figure 5 The NO oxidation percentage integrated over the whole plume cross-section as a function of distance from the stack of Wilhelmshaven power station, together with the difference between NO_2 formed and O_3 consumed as a percentage of the total nitrogen oxides also integrated over the whole plume cross-section.

Figure 6 The photostationary state relationship in the plume of Flevo power station as a function of distance from the stack.
_____ integrated over the whole plume cross-section (airborne measurements)
------- integrated over the plume cross-section on ground level (van measurements)

Of the measurements of UBA and the combined measurement with KEMA in the plume of Wilhelmshaven power station, the results are given in table 1.

On 83-05-13 the atmospheric stability was neutral and the transport velocity on plume height was about 7 m/s. The ozone concentrations on plume height in the surrounding air were 50-60 ppb on that day.

A direct comparision was possible on 82-05-13 at a distance of 0.5 km from the stack with the result given in the table. The weather situation and the emissions on the concerning three days were comparable and there-fore, it can be concluded from the rest of the table that there is a sa-tisfactory similarity between the results obtained with two totally dif-ferent measurement methods.

The pseudo steady-state approximation for the NO_x-O_3 system as given by equation 7 requires that $\psi = k_2(O_3)(NO)/k_4(NO_2) = 1$. Observed values of ψ are generally > 1 (Hegg et al. [5], Kewley and Post [6]). These results are generally assumed to be indicative of the mixing of segregated packets of O_3-rich and NO-rich air. In figure 6 the calculated values are given of ψ for three of the flights and two runs with the van as a function of distance from the stack of the power station, in neutral atmospheric situations (Pasquill class D). The re-sults clearly show the deviation of ψ from the value 1 and the decrease of the deviation with larger distances. The deviation can be used as a measure of the inhomogeneous mixing. From the results of the plume measu-rements on ground level (made by the van) the figure shows that there still exists a situation of inhomogeneous mixing of the plume and the surrounding air.

MODEL CALCULATION

In recent years a number of studies have been published concerning the modelling of chemical reacting plumes. In general the diffusion equa-tion for the concentration of NO in a plume dispersing in an O_3-con-taining atmosphere, can be written as:

$$u\delta[NO]/\delta_x = K_y \delta^2[NO]/\delta_y^2 + K_z \delta^2[NO]/\delta_z^2 - k[NO][O_3] - k[NO]^*[O_3]^* \quad (8),$$

in which u is the mean wind velocity in the x direction; [NO] is the mean concentration of NO; K_y and K_z are the turbulent diffusivities in the y and z directions; k is the rate coefficient given by the reaction $NO+O_3 \rightarrow NO_2+O_2$, and $[NO]^*$ and $[O_3]^*$ are the turbulent fluc-tuations of the NO concentration and O_3 concentration respectively.

A plume of NO interacting with ambient air containing O_3 may be viewed as a bimolecular reaction with initially unmixed reactants in a turbulent fluid. For modelling an important point is the ratio of the time scale for chemical kinitics τ_k, to the time scale characteris-tic of turbulent mixing τ_m [5]. In case of slow reaction $(\tau_k/\tau_m>1)$, the turbulence induces chemical homogeneity before any significant reaction occurs and the reaction rate can be described in terms of the mean concentration. In case of rapid ractions $(\tau_k/\tau_m<1)$, the reaction is controlled by the rate of mixing and also depends on how the reactants are introduced, in this case segregated.

In the situation of $\tau_k/\tau_m>1$, the solution of the first four terms can be given by the gaussian plume model. Describing the second order reaction $NO+O_3 \rightarrow NO_2+O_2$ as a pseudo first order reaction, starting from the assumption that $[O_3]$ = constant, $\tau_{1/2} = \ln 2/k_1$ is the half life time of NO. For the dispersion of a point source with emission strength Q, the dispersion formula of the gaussian plume model can, in that situation be modified by replacing Q by Q $\exp(-k_1 t)$ or for the distance x = ut by Q exp(-xln2/uτ$_{1/2}$). This concept has been used by Binaris [7] in his calculations to mention an example.

The dispersion of NO from a stack in an O$_3$ atmosphere, however, cannot be described by the modified gaussian plume model. This is caused by the following three interrelated facts [8]:

- the ratio τ_k/τ_m is < 1. For the measurements in plumes of four power stations Hegg et al. [6] deduced values for the ratio τ_k/τ_m of 0.005 to 1.17, with a mean value of 0.33 for travel times between 1 and 100 min;
- the concentration of O$_3$ is not constant through the plume;
- the term k[NO]* [O$_3$]* in equation 8 is not zero.

Looking to the existing reactive plume models, the models using the concept of one single expanding well mixed box can be mentioned. This type of models can be used for the far field chemistry and can incorporate a relatively large reaction scheme. Main emphasis in these models is on chemistry not on dispersion. Examples are the models developed by Liu [9], Varey et al. [10], Isaksen et al. [11] and Forney and Giz [12]. A simple model based on the mixing equation $\delta V/\delta t = f(t)$ has also been given by Schurath and Ruffing [13]. These models do not take into account inhomogeneous mixing, nor the correlation term in equation 8, and will consequently not be able to predict the retardation of both phenomena.

Of the models mentioned the CERL model of Varey et al. and that of Schurath et al. are the only models which include the contribution of the third order reaction with ambient oxygen.

The CERL model shows that for example for an emission concentration of (NO)$_o$ = 450 ppm and an O$_3$ concentration of 30 ppb in the sourrounding air, after a travel time of 500 s for a Pasquill class D, the percentage oxidation of NO is mainly a consequence of the reaction with O$_2$ and for Pasquill class B the reaction with O$_3$ is the most important oxidation path. In real plumes with their inhomogeneous mixing, the result with respect to these model results would be to slow down the reaction with ozone, but it could enhance the oxidation through the third order reaction by maintaining the local (instantaneous) concentration of NO on levels higher than predicted.

Modifications of the well-mixed models are given by Peters and Richards [14] and White [15], but while also in these models there is a complete separation of chemistry and dispersion, these models are likewise inadequate to describe the conversion of NO in a plume.

More promising is the concept of the "divided" gaussian plume. The basis is the dispersion of an inert species according to the gaussian plume model. To calculate chemical transformations, the gaussian plume profiles are divided, either in boxes (Liu [16]), or in elliptical rings (Lusis [17]). A few comparisons made between the model and field studies (light stable atmospheric situation, Pasquill class D/E and at distances of 40-100 km and travel times of 1-2 h) show a reasonable agreement [17].

It should be noticed that these models do not take into account the correlation term of the turbulent concentration fluctuations. This correlation term is incorporated in reactive plume models described by Lamb [18] and Shu [19]. Calculations made with these models clearly show the retardation.

A model (incorporating, however, only the oxidation reaction with O$_3$), based on a mixing reaction in series is given by Carmichael and Peters [20]. This model relates the mixed and unmixed fractions to a mixing intensity factor k_m, which is valuated by mixing parameters cal-culated by Shu [19] from data generated by numerical turbulence models.

DISCUSSION AND CONCLUSIONS

The results of the measurements presented here give rise to the fol-lowing conclusions:
- The oxidation percentage in the plume is strongly dependent on the mixing of the plume with ambient air. In more unstable atmospheric condi-tions the oxidation rate is significantly larger than in neutral and sta-ble situations. In most situations the oxidation of NO and NO_2 in plumes is a diffusion-controlled process.
- In the near field and further on in situations of a low degree of mixing with the surrounding air, as for example in stable situations, the oxidation of NO through the termolecular oxidation reaction with oxygen can contribute to an important part of the total NO_2 in the plume. This conclusion is supported by the results of model calculations.
- From the results gathered in unstable atmospheric situations a clear dependence of the NO oxidation percentage on the O_3 concentrations can be proved.
- The inhomogeneous mixing of the plume can be illustrated by the de-viation from the photostationary state.
- The 50% NO oxidation times of the measurements presented here are in the same order as already given by Beilke et al. [1] in a review of the literature. The variance in the results can be explained by the mixing intensity in the atmosphere (to some degree indicated by the stability classes), the wind velocity and the ozone concentrations in the surroun-ding air. For the most frequent atmospheric situation (neutral, Pasquill stability class D), the 50% NO oxidation times are with mean transport velocities of about 5 to 6 m/s, mostly larger than 1 h and the corres-ponding distances larger than 20 km.
- Although the NO oxidation in plumes cannot be described by a first order reaction, the approximation as given by Binaris [7] by extending the gaussian formula by an NO depletion term, can be used as a first rough approximation for the calculation of the long-term mean concentra-tion of NO_2 on ground level. It should be borne in mind that the half life times of NO used for this calculation as such incorporate the ef-fects of mixing intensity, wind velocity, O_3 concentration in the sur-rounding air, both oxidation routes by reaction with O_2 and O_3 res-pectively and also the NO emission concentration. For this purpose there is a need for a better statistic of half life times of NO in plumes for the different stability classes, also as a function of NO emission con-centration.

The model as described by Schurath and Ruffing [13], based on homoge-neous mixing gives incorrect results as already shown by the measurements and modelling results, if in that case the gaussian concentration distri-bution, which is a statistic mean one, will be used for the determination of concentrations in the plume axis and on ground level. The concentra-tions based on gaussian dispersion are time mean values of the concentra-tions of successive parcels of NO rich air (plume parcels) and O_3 rich air (air parcels). The concentrations on the boundaries of these parcels determine the oxidation rate.
- From the measurements in the near field it can be deduced that the degree of oxidation as found by integrating over the whole plume cross-section is comparable with the degree of oxidation as found by in-tegrating over the plume cross-section on ground level.

- For a more realistic calculation of the NO oxidation in plumes, a model has to be developed, which incorporates the inhomogeneous mixing and both oxidation routes of NO.

REFERENCES

1   Beilke,, S., Markusch, H. and Jost, D. Proceedings of the Second European Symposium, Varese, Italy (1981). D. Reidel Publ. Cie, pp. 448-459
2   Elshout, A.J. VGB Kraftwerkstechnik 62 (1982), pp. 295-303
3   Melo, O.T. and Stevens, R.D.S. Atm. Environ. 15 (1981), pp. 2521-2529
4   Meagher, J.F., Stockburger. L., Bonano, R.J., Bailey., E.M. and Luria, M. Atm.. Environ. 15 (1981). pp. 749-762
5   Hegg, D., Hobbs, P.V., Radke. L.F. and Harrison, H. Atm. Environ. 11 (1977), pp. 521-526
6   Kewley, D.J. and Post, K. Atm. Environ. 12 (1978), pp. 2179-2184
7   Binaris, S. VGB Kraftwerkstechnik 62 (1982), pp. 500-505
8   Builtjes,, P. A comparison between chemically reacting plumes and wind tunnel experiments. Presented at the 12th International Technical Meeting on Air Pollution and its Applications, Palo Alto, USA, 1981
9   Liu. M.K. Proceedings of the fourth international Clean Air Congress. Tokyo, Japan (1977), pp. 31--314
10 Varey, R.H., Sutton, S. and Marsh, A.R.W. The oxidation of nitric oxide in power station plumes. CERL Laboratory Note RD/L/N 184/78 (1978)
11 Isaksen, I.S.A., Hesstvedt,, E. and Hov, W. Atm. Environ. 12 (1978, pp. 599-604
12 Forney, L.T. and Giz, Z.G. Atm. Environ. 15 (1981), pp. 345-352
13 Schurath, U. and Ruffing, K. Staub - Reinh. d. Luft 41 (1981), pp. 277-281
14 Peters, L.K. and Richards, L.W. Atm. Environ. 11 (1977), pp. 101-108
15 White, W.H. Env. Sc. and Techn. 11 (1977), 995-1000
16 Liu. M.K. c.s. An improved version of the reactive plume model. Presented for the 9th International Technical Meeting on Air Pollution and its Applications, Toronto. Canada (1978)
17 Lusis, M. Atm. Environ. 12 (1978). pp. 2429-2437
18 Lamb, R.G. and Shu, W.R. Atm. Environ. 12 (1978), pp. 1685-1694
19 Shu, W.R. Atm. Environ. 12 (1978), pp. 1695-1704
20 Carmichael, G.R. and Peters. L.K. Atm. Environ.15 (1981), pp. 1069-1074
21 Davis, E.E., Smith, G. and Klauber, G. Science 186 (1974), pp. 733-736

CAMPAGNE EXPERIMENTALE EN VUE DE LA MODELISATION DE PANACHE REACTIF

G. MAFFIOLO, E. JOOS and A.E. SAAB
Electricité de France
Direction des Etudes et Recherches
6, quai Watier - 78400 CHATOU

Summary

ELECTRICITE DE FRANCE is engaged in a significant experimental program designed to develop an improved understanding of physico-chemical behaviour of power plant plumes and to validate advanced reactive plume models. In October 1983, a first campaign including source emissions and intensive air quality measurements was carried out on a 1 000 MW coal-burning power plant site. During this program, an in-plume chemistry aircraft, a lidar system, a mobile correlation spectrometer (COSPEC) and multiparameter ground monitoring laboratories were fielded to collect detailed information regarding the behaviour of the plume. The physicochemical measurements included SO_2, NO/NO_x, O_3 as well as aerosol compounds. This paper describes the EDF equipment, the methodology used during these field experiments and provides the first experimental results which are necessary for the development of improved modeling capabilities.

1. INTRODUCTION

L'étude des processus d'évolution et de transformations physico-chimiques des polluants primaires des panaches de centrales thermiques classiques, a conduit ELECTRICITE DE FRANCE à initier un important programme de recherches comportant deux approches complémentaires :
- une approche expérimentale conduite sur des sites de centrales thermiques à charbon permettant de recueillir des informations sur les processus mis en jeu aussi bien dans l'évolution du panache à l'échelle locale que dans la phase de transformations physicochimiques des espèces dans l'atmosphère, à l'échelle locale et régionale,
- une approche numérique portant sur la mise au point de modèles réactifs de panaches prenant en compte les processus de transformations physicochimiques et de dépôts secs et humides, ainsi que sur leur validation à partir des résultats des campagnes expérimentales.
Cette communication présente essentiellement les différents moyens mis en oeuvre lors d'une première campagne expérimentale réalisée en octobre 1983 sur un site d'une puissance thermique de 1000 MW localisé dans une région faiblement industrialisée, ainsi que la méthodologie suivie lors du déroulement des essais. Les premières informations ainsi recueillies nous permettent d'ores et déjà d'engager une réflexion critique sur l'amélioration des moyens de mesure et sur la définition de critères quantitatifs de validation des modèles réactifs de panache.

2. MOYENS MIS EN OEUVRE

Depuis plusieurs années, ELECTRICITE DE FRANCE réalise des campagnes de mesure de courte durée ayant pour but la mise en évidence de l'influence d'une source ponctuelle (centrale le plus souvent) sur la pollution d'un site.

La méthodologie suivie (1) met en oeuvre très généralement des analyseurs de gaz (dioxyde de soufre et oxydes d'azote notamment) et des préleveurs de particules (en vue d'analyses pondérale, granulométrique et chimique) équipant :

- un ou plusieurs postes fixes implanté(s) autour de la source de pollution,

- un poste semi-mobile placé en fonction des conditions météorologiques du moment,

- un laboratoire mobile opérant en roulant et comprenant également un spectromètre de corrélation qui permet la localisation dans l'espace de panaches contenant du dioxyde de soufre.

L'interprétation des résultats obtenus par ces trois types de postes de mesure est réalisée à l'aide de données météorologiques relevées simultanément.

Ces mesures dans l'atmosphère sont précédées d'analyses de gaz et particules avant le rejet des effluents de la source.

Toutefois, ces équipements bien adaptés pour établir une cartographie au sol des niveaux de pollution, sont incomplets pour étudier l'évolution des concentrations en polluants primaires et leur transformation par oxydation notamment.

Des mesures en altitude s'avéraient donc nécessaires.

A cette fin, l'avion HUREL-DUBOIS HD 34, de l'INSTITUT GEOGRAPHIQUE NATIONAL (IGN) évoluant à 60 m/s et utilisé ces dernières années pour l'exploration de panaches d'aéroréfrigérants a été équipé, avec le concours de l'INSTITUT NATIONAL D'ASTRONOMIE ET DE GEOPHYSIQUE (INAG), d'analyseurs physicochimiques qui ont été associés aux capteurs physiques (température, humidité, vitesses de l'air en particulier) existants.

Il est apparu également nécessaire de localiser le plus précisément possible le panache en complétant les informations recueillies à l'aide du spectromètre de corrélation par des mesures LIDAR permettant de connaître l'altitude du panache.

Enfin, l'interprétation de tous ces résultats requérant la disponibilité de données météorologiques en altitude, des radiosondages avec poursuite radar ont été réalisés. En résumé, l'ensemble des équipements utilisés est présenté sur le schéma I.

Les différentes méthodes d'analyses mises en oeuvre dans ces équipements sont rassemblées dans le tableau I qui fait notamment apparaître le choix de 3 types de polluants primaires (SO_2, NO_x et granulométrie des particules) et de deux espèces non émises (O_3 et soufre élémentaire particulaire).

Pour cette dernière espèce, des analyses comparatives réalisées par fluorescence X et par chromatographie ionique (sur la solution de lavage des filtres) ont montré que le soufre élémentaire provenait essentiellement des sulfates ; il apparaît donc raisonnable d'assimiler la teneur en soufre élémentaire telle que déterminée par fluorescence X à la teneur en sulfate.

En ce qui concerne les analyseurs de gaz calibrés régulièrement à l'aide de mélanges de gaz SO_2 et NO en bouteilles (2), ils sont de type automatique, continu. Les caractéristiques de temps de réponse et gamme de mesure sont classiques pour les analyseurs opérant au sol ; du fait des

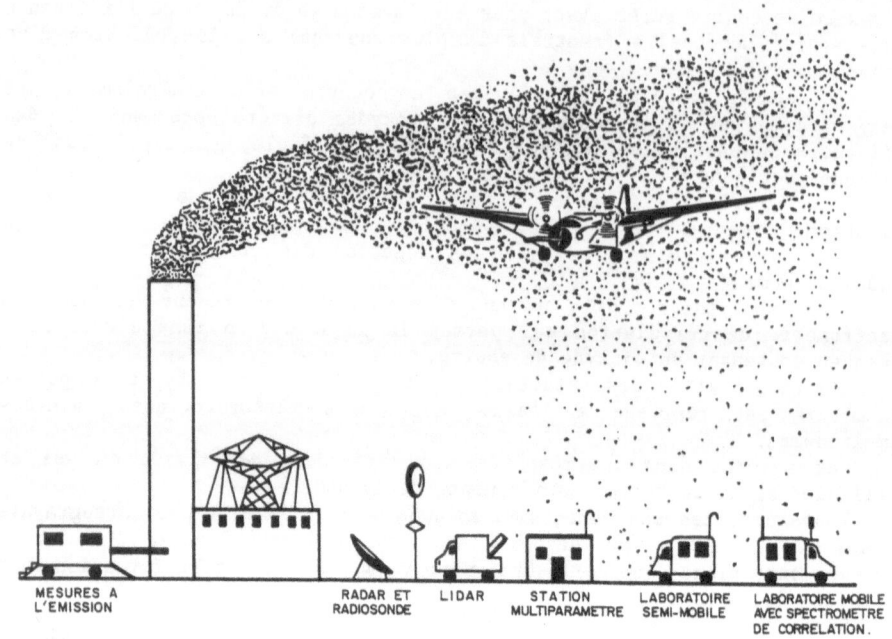

RADAR ET
RADIOSONDE

LIDAR

STATION
MULTIPARAMETRE

LABORATOIRE
SEMI-MOBILE

LABORATOIRE MOBILE
AVEC SPECTROMETRE
DE CORRELATION.

SCHEMA I
MOYENS MIS EN OEUVRE

conditions de mesure particulières, ces caractéristiques ont été modifiées pour les analyseurs équipant l'avion.

Les comptages de particules par diffusion de lumière sont également réalisés à l'aide d'appareils automatiques.

3. UTILISATION DES EQUIPEMENTS DE MESURE

Afin de caractériser la source de pollution, les mesures à l'émission ont été réalisées préalablement à toutes analyses dans l'air ambiant. Une seule des quatre tranches thermiques a fait l'objet d'analyses complètes ; les différences de fonctionnement entre les tranches ont été déterminées par mesures de la teneur résiduelle en oxygène.

Outre la connaissance des diverses concentrations, ces analyses ont permis d'accéder au rapport des teneurs SO_2/NO_x utile à l'identification de la source responsable des résultats relevés dans l'atmosphère.

Dans le cadre de la campagne décrite, une seule station fixe a été implantée sous les vents dominants, à 5 km de la centrale. Elle a fonctionné d'une manière continue pendant les 15 jours de l'expérimentation dans l'air ambiant.

Les emplacements de la station semi-mobile ont été choisis quotidiennement afin de placer le poste dans les retombées des effluents de la centrale ; une dizaine d'emplacements ont ainsi été étudiés à une distance maximale de 10 km de la source.

TABLEAU I

DESCRIPTIF DES MÉTHODES D'ANALYSES UTILISÉES

Paramètres mesurés / Poste de mesure	DIOXYDE DE SOUFRE (SO_2)	OXYDES D'AZOTE (NO et NO_x)	OZONE (O_3)	SOUFRE ÉLÉMENTAIRE PARTICULAIRE	GRANULOMÉTRIE DES PARTICULES
EMISSION	Iodométrie manuelle	Chimiluminescence (gamme de mesure : 1 000 ppm V)		Prélèvement isocinétique sur filtre (Ø25 mm) et analyse par fluorescence X	Prélèvement isocinétique avec impacteur à cascade et analyse par pesée et comptage par ou comptage par microscopie
STATION FIXE	Fluorescence UV (gamme de mesure 1 ppm V) (temps de réponse : 30 s)	Chimiluminescence (gamme de mesure 1 ppm V) (temps de réponse : 50 s)		Prélèvement sur filtre (Ø47 mm) (1,5 m3/h) et analyse par fluorescence X	Comptage par diffusion de lumière
STATION SEMI-MOBILE	Fluorescence UV (gamme de mesure 1 ppm V) (temps de réponse : 30 s)	Chimiluminescence (gamme de mesure 1 ppm V) (temps de réponse : 50 s)		Prélèvement sur filtre (Ø47 mm) (1,5 m3/h) et analyse par fluorescence X	Comptage par diffusion de lumière
LABORATOIRE MOBILE	Fluorescence UV (gamme de mesure 1 ppm V) (temps de réponse : 30 s)	Chimiluminescence (gamme de mesure 1 ppm V) (temps de réponse : 50 s)	Absorption UV (gamme de mesure 1 ppm V) (temps de réponse : 30 s)		
AVION	Fluorescence UV (gammes de mesure 5 et 50 ppm V) (temps de réponse : 4 s)	Chimiluminescence (gammes de mesure 5 et 50 ppm V) (temps de réponse : 4 s)	Absorption UV (gamme de mesure 1 ppm V) (temps de réponse : 4 s environ)	Prélèvement isocinétique sur filtre (Ø257 mm) (30 m3/heure) et analyse par fluorescence X	

Les nuits et fins de semaine, la station fonctionnait au poste central situé à 2 km des cheminées.

Pour les deux équipements décrits ci-dessus, la fréquence d'enregistrement des résultats sur cassette magnétique était de 10 minutes, la durée d'échantillonnage des particules de 3 heures.

Sa mobilité et son équipement, à la fois de localisation du panache et d'analyse au sol, font du laboratoire mobile l'outil central de l'expérience.

Sur la base des informations météorologiques fournies par les stations locales et celles recueillies par les radiosondages effectués depuis le poste central au moins deux fois par jour (en début et fin de journée), cet équipement sillonne le plus souvent les routes perpendiculaires à l'axe de panache. En s'éloignant progressivement de la source d'émission, il permet ainsi de connaître précisément la direction du panache et de localiser ses zones de retombées.

2 000 km ont ainsi été parcourus jusqu'à 30km du point d'émission.

Ces informations permettent de choisir un emplacement pour le poste semi-mobile et d'utiliser d'une manière optimale le Lidar et l'avion.

Au cours de la campagne, le Lidar a opéré à 3 emplacements différents choisi quotidiennement en fonction de la direction du panache de manière à pouvoir effectuer des coupes de panache dans l'espace depuis la cheminée jusqu'à 4 km environ.

Chronologiquement, l'avion est généralement le dernier équipement à intervenir au cours d'une journée de mesure ; en effet, les conditions météorologiques doivent être favorables et les données de localisation du panache disponibles.

Trois types de coupe ont été réalisés par l'avion (schéma II) :
- des coupes longitudinales effectuées à altitude constante depuis la cheminée jusqu'à dilution complète.
Ce type de coupe réalisé en début d'expérience permet de confirmer la direction du panache mais les données recueillies sur les gaz et surtout sur les particules ne sont pas toujours exploitables,
- des coupes en hippodrome dont les branches perpendiculaires à l'axe du panache se situent à la même altitude et à deux distances de la centrale. Ce type de coupe est intéressant à faible distance (moins de 5 km environ), mais requiert une durée de réalisation importante au-delà,
- des coupes en hélice. C'est le type qui a été le plus souvent réalisé. La coupe est effectuée en 50 minutes environ à une distance constante des cheminées et une altitude variable par pas de 300 à 500 pieds.

Les analyseurs de gaz fonctionnent en continu, depuis le décollage avant lequel ils ont été calibrés jusqu'à l'atterrissage.

En ce qui concerne l'échantillonnage des particules, quotidiennement, un filtre impacté en vol a permis la mesure du bruit de fond en sulfate hors panache.

Dans le panache, l'échantillonnage est cumulé sur un filtre par coupe, à distance constante et altitude variable, les temps de prélèvement hors panache et dans celui-ci étant comptabilisés.

Au cours de la campagne, l'avion a totalisé 35 heures de mesure en 11 vols (dont 1 de nuit). Il a évolué entre 500 m et 60 km de la centrale à une altitude comprise entre 300 et 4 000 pieds.

Il est à noter que la présence fréquente d'une inversion de température (comprise entre 500 et 800 mètres) a d'une part, facilité la localisation du panache, et d'autre part, permis son suivi sur plusieurs dizaines de kilomètres.

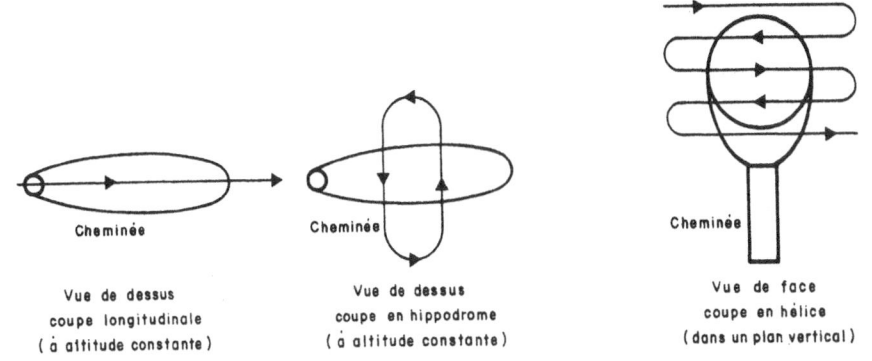

Cheminée   Cheminée   Cheminée

Vue de dessus   Vue de dessus   Vue de face
coupe longitudinale coupe en hippodrome coupe en hélice
(à altitude constante) (à altitude constante) (dans un plan vertical)

SCHÉMA II
EXEMPLES DE TYPES DE COUPE

   L'utilisation de l'ensemble des équipements nécessite une grande coordination basée sur la rapidité d'échanges d'informations.
   Des liaisons VHF et par radiotéléphones ont été utilisées à cet effet.
   En outre, les renseignements communiqués concernant les résultats relevés et leurs lieux d'obtention doivent être complets et précis.
   Les dispositifs de repérage automatique et de sortie en temps réel des résultats, équipant le laboratoire mobile, sont à ce titre, très utiles.
   L'avion possède également un système de sortie en temps réel. Sa localisation est basée en grande partie sur les qualités du navigateur ainsi que sur les indications fournies par le radar de poursuite, des photographies aériennes et une centrale à inertie.

4. PREMIERS RESULTATS ET REMARQUES SUR LES ESSAIS

   Les résultats actuellement disponibles sont ceux sortis en temps réel ; ils portent sur les analyses de gaz effectuées à partir de l'avion et du camion laboratoire pour lequel on dispose également des données enregistrées par le spectromètre de corrélation.
   Trois exemples de restitution sont donnés ; il s'agit tout d'abord de coupes de panache réalisées par le laboratoire mobile (schéma III).
Sur cet exemple, on notera plus particulièrement l'élargissement du panache au fur et à mesure que le laboratoire s'éloigne de la centrale, la rapidité de réalisation des coupes successives à partir du moment où l'infrastructure routière s'y prête et enfin la correspondance (en tenant compte du temps de réponse des appareils) entre les réponses du spectromètre de corrélation et le niveau en SO_2 au sol.
   En ce qui concerne l'avion, les teneurs représentées ci-après ont été obtenues au cours de coupes réalisées perpendiculairement à l'axe du panache respectivement à 8 et 28 km (schémas IV et V).
   Bien que ces deux résultats n'aient pas été obtenus le même jour, la comparaison de ces deux restitutions permet tout de même de noter qualitativement :
   - l'élargissement du panache et corrélativement la baisse des teneurs en SO_2 et NO_x entre 8 et 28 km,

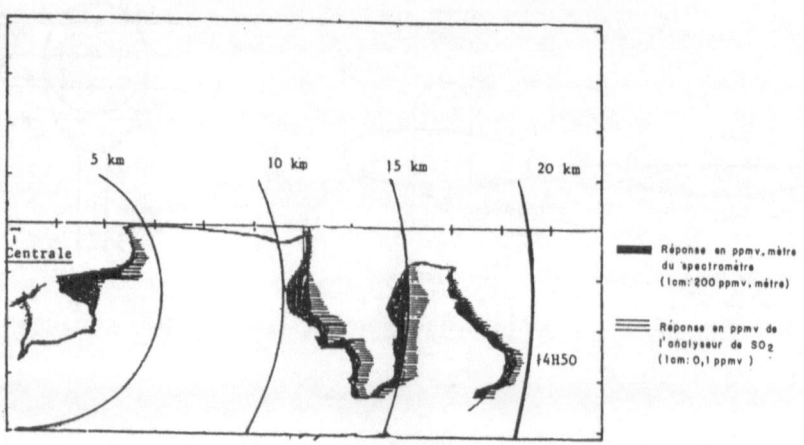

SCHEMA III
COUPES REALISEES PAR LE LABORATOIRE MOBILE

- la présence de bouffées de panache très marquées à 8 km,
- dans les deux cas, des relations très nettes entre concentrations en SO_2 et concentrations en NO et NO_x.
En ce qui concerne l'ozone, nous avons des doutes sur le bon fonctionnement de l'analyseur, du moins pendant une partie de la campagne, en l'absence actuelle de système de calibrage fiable. Des études sont d'ores et déjà prévues sur ce point en vue d'une prochaine campagne.
Les quelques résultats de teneurs en sulfate dépouillés ressortent à quelques microgrammes par m^3 pour les mesures réalisées hors panache et à 10-20 µg/m³ pour des prélèvements dans le panache.
Sur ce point également, des améliorations sont en cours ; elles portent principalement sur le choix d'un filtre mieux adapté, de plus petit diamètre et qui présente une teneur en sulfate initiale moindre.

5. CONCLUSION

Cette campagne s'inscrit dans un programme qui devrait s'étendre sur 3 ans et qui a pour but d'étudier les transformations physicochimiques des composés émis par une centrale thermique à combustible fossile, lors de la dispersion du panache dans l'atmosphère.
Les effets de certains produits issus de ces réactions d'oxydation (sulfates, nitrates) justifient une telle étude.
Deux approches complémentaires ont été retenues pour traiter ce problème : les mesures expérimentales et la modélisation. En ce qui concerne celle-ci, l'utilisation de modèle de panache réactif déjà existant semble préférable dans un premier temps.
Parmi les différents modèles réactifs de panache de type gaussien souvent utilisés, nous avons retenu celui développé par SAI (3) basé sur un simple équilibre de masse et utilisant un concept de panache gaussien "divisé". Pour faire intervenir les transformations chimiques, les coupes

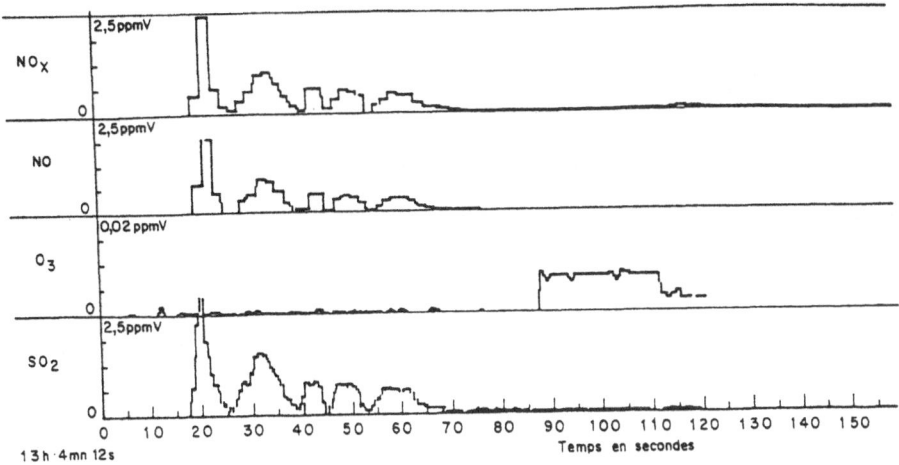

SCHEMA IV

RESTITUTION OBTENUE A 8 KM DE LA CENTRALE
ET A UNE ALTITUDE DE 2000 PIEDS

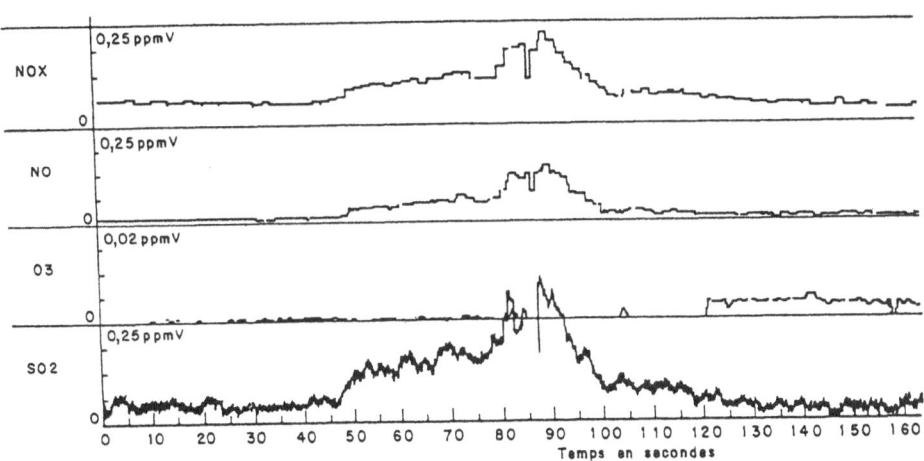

SCHEMA V

RESTITUTION OBTENUE A 28 KM DE LA CENTRALE
ET A UNE ALTITUDE DE 1200 PIEDS

des panaches sont divisées en un nombre déterminé de cellules. Celles-ci s'élargissent d'une manière prévue suivant une loi gaussienne lorsque le panache se déplace. Pour les espèces polluantes inertes du panache, il n'y a pas de flux massique à travers les parois des cellules au cours de leur dilatation.

En ce qui concerne les polluants réactifs du panache ou de l'atmosphère, un coefficient équivalent de diffusion est utilisé pour caractériser les flux massiques à travers les parois, dus à la turbulence.

Si ce modèle ne traite pas le problème du terme de corrélation, il tient en revanche compte du mélange inhomogène.

Il n'est cependant pas exclu qu'une prise en compte des fluctuations turbulentes au niveau de la cinétique de certaines réactions ne soit effectuée, par exemple au moyen d'une diminution de la constante de réaction.

Le développement d'un modèle dynamique de panache résolvant les équations de conservation des différentes espèces et incorporant un schéma réactionnel simplifié, pourrait également être envisagé dans les prochains mois.

Quant à la chimie, incorporée dans le modèle de panache sous forme d'un schéma réactionnel indépendant, la première étape indispensable consiste à traiter la phase gazeuse par l'intermédiaire d'un mécanisme cinétique le plus complet possible tout en lui conservant une taille raisonnable pour le calcul.

Les deux problèmes essentiels concernent la prise en compte des composés hydrocarbonés et des réactions photochimiques.

Le premier point a été résolu en utilisant le mécanisme carbon-bond développé par S.A.I.

Le second implique une réactualisation des valeurs des différents facteurs intervenant dans le calcul des constantes de réactions photolytiques : sections efficaces d'absorption et rendements quantiques mais aussi flux solaires.

Dans une deuxième étape, il paraît souhaitable d'inclure les réactions faisant intervenir la phase aqueuse, compte tenu de l'augmentation des taux de conversion SO_2 - sulfates et NO_x - nitrates, constatée pour des humidités relatives élevées.

Enfin, les transformations gaz - particules, principaux responsables de la formation des aérosols de sulfate et de nitrate pour des humidités relatives inférieures à 50 %, seront également incluses dans le modèle de panache réactif (MPR).

Le calibrage et la validation de MPR nécessitent de nombreuses mesures expérimentales que la campagne de 1983, ainsi que celles qui auront lieu en 1984-85, devraient fournir.

L'importance des transformations physicochimiques des composés d'un panache étant fonction de la qualité de l'air ambiant dans lequel celui-ci se disperse, différents cas de figure peuvent être examinés.

Ainsi, en ce qui concerne la campagne 03, la dilution du panache s'est effectuée dans un milieu de type rural à moyennement pollué. Les concentrations d'hydrocarbures non méthaniques n'ont malheureusement pas pu être mesurées et seules les mesures lidar et les analyses granulométriques au sol peuvent nous renseigner sur la pollution particulaire.

Les prochaines campagnes pourraient avoir lieu dans des environnements de type urbain où l'impact chimique des composés hydrocarbonés devrait être sensible.

De même, en phase gazeuse et surtout en phase aqueuse, l'importance de l'humidité relative dans les processus d'oxydation nous amèneront à étudier différentes situations : transport nocturne (un vol de nuit a déjà été effectué lors de cette campagne), passage dans les nuages, etc...

Compte tenu de cela, l'instrumentation aéroportée sera améliorée en ce qui concerne les particules. Un Knollenberg à faible diamètre minimum (< 0,1 µm) pour l'analyse en taille des aérosols devrait être installé.

Un système de prise d'échantillons d'air devrait également être aménagé afin d'effectuer, entre autres, des analyses de composés hydrocarbonés par chromatographie en phase gazeuse.

Enfin, des capteurs de rayonnement UV dans la gamme 295- 400 nm, particulièrement importante pour les réactions photochimiques, complèteront l'équipement du HD 34.

Le programme mentionné ci-dessus est bien entendu subordonné aux résultats du dépouillement qui, compte tenu du nombre de données collectées, requerra plusieurs mois.

REMERCIEMENTS : Les auteurs remercient les personnes de l'IGN, de l'INAG et d'EDF qui ont collaboré à cette étude.

REFERENCES BIBLIOGRAPHIQUES

[1] G. MAFFIOLO, Y. LERIQUIER, C. DUTRANNOY, J. DUBOIS : "Apport de la télédétection dans l'étude de l'influence d'une source ponctuelle sur les niveaux de pollution au sol". Congrès UIAPPA - BUENOS AIRES - 1980.

[2] G. MAFFIOLO, J. DUBOIS : "Stabilité des gaz comprimés en bouteille contenant de très faibles teneurs en polluants (SO_2, NO_x)". Analusis, 1978, vol 6, n° 4, p. 173.

[3] D.A. STEWART, M.K. LIU : "Development and Application of a Reactive Plume Model". Atmospheric Environment, vol 15, 2377, 1981.

THE ATMOSPHERIC SIGNIFICANCE OF LIQUID PHASE OXIDATION OF SO$_2$ TO SULPHATE BY O$_3$ AND H$_2$O$_2$

T. Bøhler* and I.S.A. Isaksen**
* Norwegian Institute for Air Research
**Institute of Geophysics, University of Oslo

Summary

A time dependent chemical model describing atmospheric oxidation of sulphur dioxide in the liquid phase, is presented. Oxidizing species considered are ozone (O$_3$) and hydrogen peroxide (H$_2$O$_2$). The model calculations include the effects of liquid water content, cloud water residence time and varying solubililty with acidity. While the SO$_2$ oxidation by H$_2$O$_2$ is nearly independent of the pH value, the oxidation by O$_3$ depends strongly on the pH value, which is affected by the presence of water soluble species like NH$_3$ and H$_2$SO$_4$. Conversion rates of SO$_2$ to H$_2$SO$_4$ from less than 1% h^{-1} in dry periods up to 30% in one hour in cases with high liquid water content, high pH value and short residence time of the droplets, are calculated. Comparison with measured sulphate concentrations indicate that sulphate formation through the liquid phase reactions with ozone and hydrogen peroxide can be an important mechanism in addition to the gas phase formation.

1. INTRODUCTION

The oxidation of sulphur dioxide in the atmosphere is important because of its contribution to the acidity of rain-, fog- and cloudwater. Several chemical mechanisms in the gas- and liquid phase are believed to be of significance for the sulphate formation.

In this paper is discussed the oxidation of SO$_2$ in atmospheric droplets by O$_3$ and H$_2$O$_2$. This mechanism involves multiphase (gas-liquid) processes, but the oxidation reaction takes place in a homogeneous aqueous medium. The model calculations were based on the assumption that equilibrium exists between the gas and the liquid phase in the atmosphere. This may be anticipated if the conversion rates are slow compared to the characteristic time for diffusion into the particle. Recent work has shown that this is true in most ambient situations since the characteristic time for mass transfer to and from an aerosol particle is of the order of a fraction of a second (Bassett and Seinfeld, 1983; Schwartz and Freiberg, 1981).

Liquid phase oxidation by O$_3$ and H$_2$O$_2$ seems to be the main mechanisms for sulphate formation in droplets on regional and global scales. Laboratory studies have shown that the oxidation by H$_2$O$_2$ is very fast, and the rate of this mechanism is limited by the formation of H$_2$O$_2$ in the gas- and the liquid phase (Isaksen, 1980; Chameides and Davis, 1982). Experiments with ozone as oxidizing specie have been performed by several authors (Penkett et al., 1979; Larsen et al., 1978; Ericksen et al., 1977). Ozone is important because of its rather uniform concentration distribution throughout the troposphere. In the unpolluted background troposphere the concentration of ozone varies between 50 and 100 µg/m$_3$. In polluted regions it often exceeds 100 µg/m during summer (OECD, 1976).

In the model calculations typical atmospheric situations were considered when polluted air from distant sources was brought to areas with highly concentrated local sources. The significance of atmospheric oxidation of SO_2 in droplets by O_3 and H_2O_2 in these situations, was estimated.

2. LIQUID PHASE SULPHUR CHEMISTRY

2.1 Equilibrium distribution in the atmosphere

In Table 1 is given the complete list of the equilibrium reactions and the equilibrium constants for the components used in the present model calculations of liquid phase sulphur oxidation.

Reaction	Equilibrium constant	Values (5^{0}C)
$H_2O \rightleftarrows H^+ + OH^-$	$K_w = [H^+]/[OH^-]$	$1.80 \times 10^{-15} M^2$
$SO_2(g) + H_2O(\iota) \rightleftarrows SO_2H_2O(\iota)$	$K_{HS} = [SO_2(\iota)]/P_{SO_2}$	2.64 M/atm
$SO_2H_2O(\iota) \rightleftarrows H^+ + HSO_3^-$	$K_{1S} = [H^+][HSO_3^-]/[SO_2(\iota)]$	$2.07 \times 10^{-2} M$
$HSO_3^- \rightleftarrows H^+ + SO_3^{2-}$	$K_{2S} = [H^+][SO_3^{2-}]/[HSO_3^-]$	$1.0 \times 10^{-7} M$
$HSO_4^- \rightleftarrows H^+ + SO_4^{2-}$	$K_{3S} = [H^+][SO_4^{2-}]/[HSO_4^-]$	$1.2 \times 10^{-2} M$
$CO_2(g) + H_2O(\iota) \rightleftarrows CO_2H_2O(\iota)$	$K_{HC} = [CO_2(\iota)]/P_{CO_2}$	$6.42 \times 10^{-2} M/atm$
$CO_2H_2O(\iota) \rightleftarrows H^+ + HCO_3^-$	$K_{1C} = [H^+][HCO_3^-]/[CO_2(\iota)]$	$3.03 \times 10^{-7} M$
$HCO_3^- \rightleftarrows H^+ + CO_3^{2-}$	$K_{2C} = [H^+][CO_3^{2-}]/[HCO_3^-]$	$2.78 \times 10^{-11} M$
$NH_3(g) + H_2O(\iota) \rightleftarrows NH_3H_2O(\iota)$	$K_{HA} = [NH_3(\iota)]/P_{NH_3}$	$1.41 \times 10^2 M/atm$
$NH_3H_2O(\iota) \rightleftarrows NH_4^+ + OH^-$	$K_{1A} = [NH_4^+][OH^-]/[NH_3(\iota)]$	$1.50 \times 10^{-5} M$
$O_3(g) + H_2O(\iota) \rightleftarrows O_3H_2O(\iota)$	$K_{HO_3} = [O_3(\iota)]/P_{O_3}$	$2.20 \times 10^{-2} M/atm$

Table 1: Equilibrium reactions with equilibrium constants and their values at 5^{0}C. (M = mole/l).

The droplet uptake of ammonium increases with increasing H^+ concentrations, otherwise the ion concentrations in the liquid solution are reduced with increasing H^+. We will assume that we have a neutral solution:

$$[H^+] + [NH_4^+] = [OH^-] + [HCO_3^-] + 2[CO_3^{2-}] + [HSO_3^-] + 2[SO_3^{2-}] + [HSO_4^-] + 2[SO_4^{2-}] \quad (1)$$

Since H_2SO_4 is assumed to be completely absorbed in the droplets, the equilibrium between the sulphate and bisulphate ions is given by the expression in column 2 in Table 1.

Equilibrium between the gas and the liquid phase can be expressed as:

$$C_T = C(g) + b\Sigma_i C_i(\iota) \qquad (2)$$

where C_T is the total concentration of a compound within a unit volume in the atmosphere; $C(g)$ and $\Sigma_i C_i(\iota)$ are the concentrations of the compounds in the gas and the liquid phases respectively, and b is the mixing ratio (by volume) of liquid water in the air mass considered. If w is the liquid water content in g/m^3, b is simply given by: $b = 10^{-6} w$.

From eqs. 1 and 2 it is clear that both the pH value and the amount of liquid water determine the partitioning of species between gas and liquid phase and between species in the droplets. These calculations mainly cover the pH range 3-6, where the bisulphite ion (HSO_3^-) is the dominant S(IV) compound in the droplets.

In Table 2 is given the amount of species in the gas phase relative to the liquid phase for different pH values, and different water vapour contents in the atmosphere. Ammonia is effectively absorbed in the droplets. The SO_2 absorption is strongly dependent on the pH value and liquid water content. For high pH and high liquid water content, the amount of sulphur in the droplets is comparable to the amount in the surrounding air. This clearly shows the importance of a proper representation of the acidity and the water vapour content for the oxidation of sulphur dioxide.

	w	pH	SO_2	CO_2	NH_3	O_3	H_2O_2	H_2SO_4
$c(g)/b\Sigma_i C_i(\iota)$	0.1	3.12	5.9×10^3	6.8×10^6	4.9×10^{-4}	2.0×10^7	$1/\infty$	$1/\infty$
		4.12	6.1×10^2	6.8×10^6	4.8×10^{-3}	"	"	"
		5.12	6.0×10^1	6.6×10^6	5.1×10^{-2}	"	"	"
	1.0	4.12	6.1×10^1	6.8×10^5	4.8×10^{-4}	2.0×10^6	"	"
		5.12	6.0	6.6×10^5	5.1×10^{-3}	"	"	"

Table 2: Variations in the ratio of gas- to liquid-phase concentrations for species in equilibrium. Liquid water content is given in g/m^3.

For ozone the gas phase concentrations dominate completely. H_2O_2 is effectively absorbed in droplets, and its concentration in the is droplets given by:

$$[H_2O_2(\iota)] = [H_2O_2(g)]/b \qquad (3)$$

2.2 Oxidation mechanisms for sulphur in the droplets

Away from the most polluted sources, oxidation proceeds predominantly through uncatalytical oxidation with O_3 and H_2O_2 as oxidating agents (Penkett et al., 1979). Although experimental results seem to differ substantially with respect to the actual mechanisms responsible for the oxidation by ozone, it is commonly believed that it is of first order with respect to O_3 in the pH range 3-6. The dependence on the dissolved sulphur species is more unclear. In this paper we will use the oxidation scheme suggested by Penkett et al. (1979). The contribution from O_2 oxidation is negligible, and is omitted in our model. A plausible oxidation scheme for the ozone oxidation is suggested, leading to a first order dependence on O_3 and HSO_3^-, and to a pH dependence which is proportional to $[H^+]^{-0.5}$. The oxidation reaction involving H_2O_2 is suggested to be of first order with respect to H_2O_2, HSO_3^-, and H^+.

The following equation describing sulphate formation in droplets is therefore obtained:

$$d/dt[S(VI)]=k_1[HSO_3^-][H^+]^{-0.5}[O_3(t)] + k_2[HSO_3^-][H^+][H_2O_2(t)] \quad (4)$$

The rate constants k_1 and k_2 are temperature dependent, and are adopted from Penkett et al. (1979). An atmospheric temperature of $5^{\circ}C$ is used.

Equation (4) also determines the chemical loss of S(IV) in the liquid phase as well. A similar expression is used to calculate O_3 and H_2O_2 changes due to droplet oxidation with S(IV) species, since they are removed in the S(IV) to S(VI) transition. The chemical loss rate is described through:

$$\frac{d}{dt}(C_T)_{ch}= -L_{ch}C_T \qquad (5)$$

where C_T is total concentration of a compound, and L_{ch} is the characteristic chemical loss rate. The total concentrations of ozone, sulphate, and hydrogen peroxide are either equal to their gas phase (O_3) or to their liquid phase concentrations (H_2O_2, H_2SO_4), while the partitioning of S(IV) and NH_3 species depends on the pH value and on liquid water content.

In Table 3 is shown the result of some calculations where the dependence of the ozone oxidation on the pH value was illustrated. The initial pH values were determined by the amounts of SO_2, NH_3, H_2SO_4, and liquid water present. Although NH_3 and H_2SO_4 are not involved in the oxidation process, they have a pronounced effect on the oxidation by O_3 through the pH value. Furthermore, other acids would also lower the pH value (HCl, HNO_3).

In mixtures with fast initial oxidation, the pH value was calculated to drop after a short time of oxidation. Further oxidation by ozone was then inhibited. At pH values below 4 the oxidation rate is less than 1% h^{-1}, even at large liquid water content (0.8 g/m^3). The marked variation with pH in sulphur oxidation by ozone is easily understood if we substitute HSO_3^- in eq. (4) by its equilibrium expression. This yields an oxidation rate of S(IV) with ozone which is proportional to $[H^+]^{-1.5}$ in the pH range 3-6.

H_2SO_4	NH_3	SO_2	w	pH	$S(IV) \to S(VI)$
$\mu g/m^3$	$\mu g/m^3$	$\mu g/m^3$	g/m^3	(t=0)	($\% \, h^{-1}$)
2	1.4	2.7	0.1	5.89	28.0
2	1.4	26.7	0.1	5.60	4.70
4	1.4	26.7	0.1	4.43	0.33
4	0.7	26.7	0.8	4.22	1.30
4	0.7	26.7	0.1	3.38	0.01
8	0.7	26.7	0.1	2.92	0.002

Table 3: $S(IV) \to S(VI)$ coversion due to the oxidation by dissolved O_3 in droplets in per cent during one hour of liquid phase oxidation, as a result of different initial distribution of H_2SO_4, NH_3, SO_2, and w.
Initial ozone mixing ratio: 60 $\mu g/m^3$. The calculated initial pH values are also given in the table.

Sulphate formation with H_2O_2 as oxidizing agent does not show the same pH dependence. Its rate is nearly independent of the H^+ ions. The droplet removal rate of H_2O_2 is:

$$L_{H_2O_2} = k_2 [H^+][HSO_3^-] \qquad (6)$$

As the time constant $\tau_{H_2O_2} = 1/L_{H_2O_2}$, the residence time of H_2O_2 when droplets are present is ~40 seconds and ~7 minutes for SO_2 concentrations of 26 and 2.6 $\mu g/m^3$ respectively.
    It is obvious that H_2O_2 is oxidized almost immediately to give sulphate, and any further oxidation depends on the rate at which H_2O_2 is formed. The formation in the gas phase proceeds via free hydrogen peroxy radicals:

$$HO_2 + HO_2 \xrightarrow{k_7} H_2O_2 + O_2 \qquad (7)$$

with the rate constant $k_7 = 3.4 \times 10^{-14} \exp(1245/T)$ $cm^3 molec^{-1}s^{-1}$ (Cox, and Tyndall, 1979). Recent work have shown that formation of H_2O_2 by free radial chemistry in the droplets can be of atmospheric significance (Chameides and Davis, 1982). In the model calculations presented in this paper we have adopted a H_2O_2 production rate in the gas phase from a photochemical gas phase model (Isaksen et al., 1978). The liquid phase formation is not included in this model.

3. SULPHUR OXIDATION DURING ATMOSPHERIC TRANSPORT

    In the atmosphere several processes determine the rate of sulphur oxidation. The mass continuity equation for the concentration C_T of a group of species is expressed as:

$$\frac{d}{dt} C_T = P_{ch} + P_s - L_{ch}C_T - L_d \, C_T - L_m(C-C_0) - L_t C_T \qquad (8)$$

P and LC are production and loss terms respectively. The subscript ch refers to chemical transformation, s to emissions, m to mixing with the surrounding air (concentration C_o), ι and d removal by wet and dry deposition respectively.

Note that C_T in eq. (8) refers to a sum of species which are in equililbrium, and they may exist both in the gas- and liquid-phase.

In this paper two atmospheric situations were considered:

a) A fog situation where droplets growth is very slow, and the precipitation is negligible ($\sim10^{-4}$ cm/h).
   Estimated values: w = 0.1 g/m^3, τ_w = 50 h.
b) A situation with moderate droplet growth and light rain (0.1 cm/h).
   Estimated values: w = 0.5 g/m^3, τ_w = 3 h.

Two very different removal rates for droplets were used to demonstrate the importance of acidity removal on the sulphate formation in solutions where ozone is the oxidizing agent.

3.1 Oxidation during long range transport

In the first model simulation a certain initial composition of gaseous species and particulates was assumed, with zero emissons during transport. The calculation should represent the long range transport from continental Europe to Scandinavia. The initial values used are given in Table 4, and are typical for those observed during the LRTAP project (OECD, 1977). Both cases have slow oxidation rates initially ($<1\%$ h^{-1}), due to low pH. The result of oxidation after 30 hours of long range transport is also shown in Table 4.

Cases	Oxidants	SO$_2$ μg/m^3	NH$_4^+$ μg/m^3	H$_2$SO$_4$ μg/m^3	O$_3$ μg/m^3	pH	Total oxid. % of S(IV)
Fog (case a)	Initial End distribution	13	0.5	5	60	3.12	
	O$_3$	9	0.27	2.8	60	3.82	0.31
	O$_3$ and H$_2$O$_2$	6.2	0.22	6.0	60	2.93	16.0
Moderate precipitation (case b)	Initial End distribution	13	0.5	5	60	3.38	
	O$_3$	3.9	2.7x10^{-5}	0.67	56	4.51	38.0
	O$_3$ and H$_2$O$_2$	2.7	3.1x10^{-5}	0.8	57	4.41	51.0

Table 4: Initial and end distribution of gases and particulates after 30 hours of oxidation in long range transported air. Liquid phase oxidation of S(IV) is given by O$_3$ alone and by O$_3$ and H$_2$O$_2$ together. The total loss of S(IV) during 30 hours of transport are also given.

The pH value remained low in the case with τ_w = 50 h, and hardly any sulphate was formed through ozone oxidation. During 30 h of transport only 0.3% of the initial SO_2 was converted to sulphate. In the case with τ_w = 3 h, the efficient liquid water removal caused a marked increase in the pH values during the transport, and 38% of SO_2 was oxidized in 30 h.

Liquid phase SO_2 oxidation by H_2O_2 is of potential importance in particular in strongly acid droplets, as the oxidation mechanism is nearly independent of pH values (Penkett et al., 1979). The H_2O_2 production rate adopted here gave rise to a SO_2 conversion of 16 and 13% during the transport period for cases a and b, respectively. H_2O_2 production rate in the gas phase of 3×10^5 molecules/$(cm^3 s)$ was used. To validate the importance of ozone oxidation the estimated pH values must be checked experimentally. Rain samples collected during the LRTAP Project (OECD, 1977) showed pH values typically in the range 4-4.5 for stations in Southern Scandinavia during air transport from continental Europe. These values agree rather well with the calculated pH values shown in Table 4. For comparison we give in Table 5 observed 3 years average concentrations from Birkenes, (see map in Figure 1), which is frequently affected by long range transport (SFT, 1977).

Species	SO_2	SO_4	NH_4^+	O_3
Concentrations $(\mu g/m^3)$	10.4	2.4	0.6	60.0

Table 5: Average atmospheric concentrations measured at Birkenes, Norway for the period 1973-1975. The corresponding pH value in precipitation was measured to 4.3 (SFT, 1977). The values are quite representative for long range transport situations.

3.2 Local oxidation in a polluted area

The effects of local industrial sources on long range transported air masses were considered. Initial values were taken from the calculations above. In Figure 1 is shown a map describing the location of the area and the observation sites referred to in Tables 5 and 6. The area covers approximately 2×5 km^2 and industrial releases of SO_2 and NH_3 were estimated to be 1.7×10^2 g(SO_2)/s and 2.3×10^1 g(NH_3)/s, respectivley.

Figures 2a and 2b show SO_2 loss rates in the two cases mentioned before with different w and τ_w. The pH values increased because of the high release rates of ammonia in this area. Both liquid phase S(IV) oxidation by ozone and washout through precipitation follow the change in pH values, as both processes become more efficient with increasing pH. The oxidation rate by H_2O_2 was low compared with the oxidation by ozone, and it was limited by the production rate of H_2O_2 in the gas phase. After leaving the source area, the pH values were reduced markedly and so did the conversion rates and washout through precipitation. The corresponding variations in gaseous and particulate distributions are given in Figures 3a and 3b. The fast conversion rates due to the reaction of S(IV) with ozone lead to a pronounced increase in the sulphate levels during transport over the industrial area. When the air mass left the polluted region, the sulphate concentrations reached 25 and 15 $\mu g/m^3$ in the cases with negligible and light precipitation, respectively. In the latter case sulphate was removed

efficiently through precipitation due to its complete adsorption in droplets. A similar fast removal will reduce NH_4^+ during transport outside the
source region, while SO_2, which is only partly absorbed in the droplets,
was removed much more slowly. In both situations nearly 20% of the released SO_2 was oxidized to sulphate inside the polluted region. This took
place during 40 min of transport time. It is interesting to notice that
ozone was markedly reduced as a result of the reaction with S(IV). During
the transport over the polluted area its concentration was estimated to be
reduced by approximately 20%. In Table 6 is given measured atmospheric
concentrations in a situation with negligible precipitation for one particular day in Southern Telemark (cfr. Figure 1). The observations agree
well with the model calculations.

Species	SO_4^{2-}	NH_4^+	O_3
Concentrations ($\mu g/m^3$)	31	20	58

Table 6: Measured atmospheric concentrations on 27 Oct. 1977 near
Porsgrunn, Norway (Bjørnstadjordet). The corresponding pH value
was in the range 5-5.5.

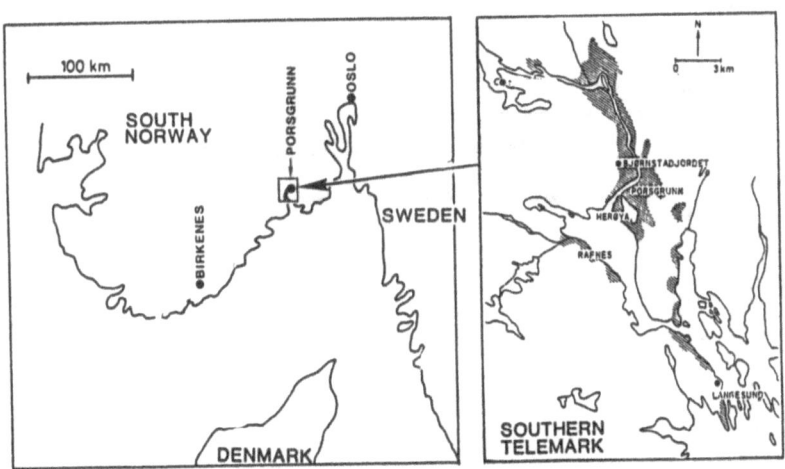

Figure 1: Map of South Norway and the highly industrialized coastal area
in Southern Telemark.

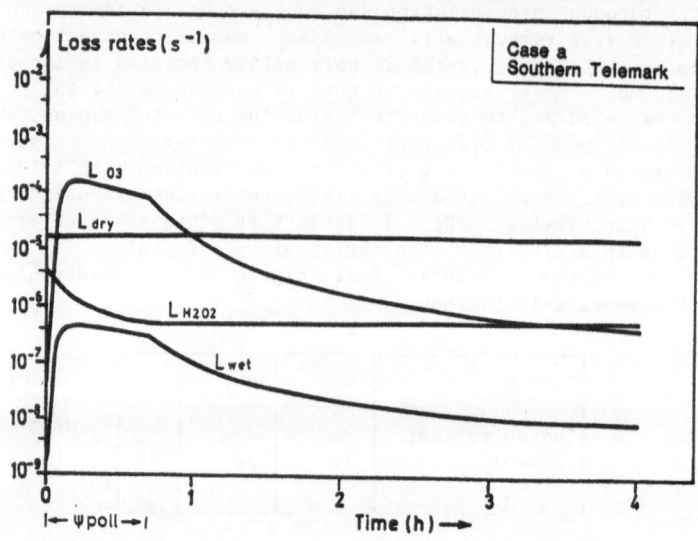

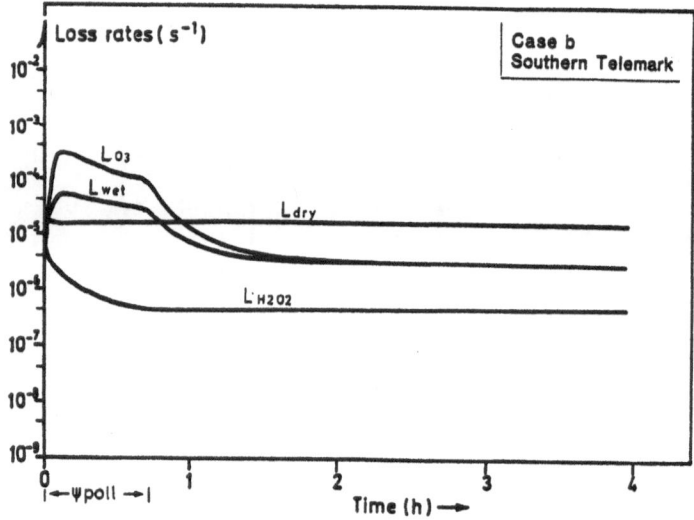

Figure 2: S(IV) loss rates (s^{-1}) versus time in an air mass which passes over an industrial area with emissions of sulphur dioxide and ammonia. L_{dry}, L_{wet}, L_{H2O2} and L_{O3} give the loss rates due to dry deposition, wet removal of SO_2, and removal due to liquid phase oxidations by H_2O_2 and O_3 respectively.
Case a) is the fog situation and case b) is the situation with light rain.

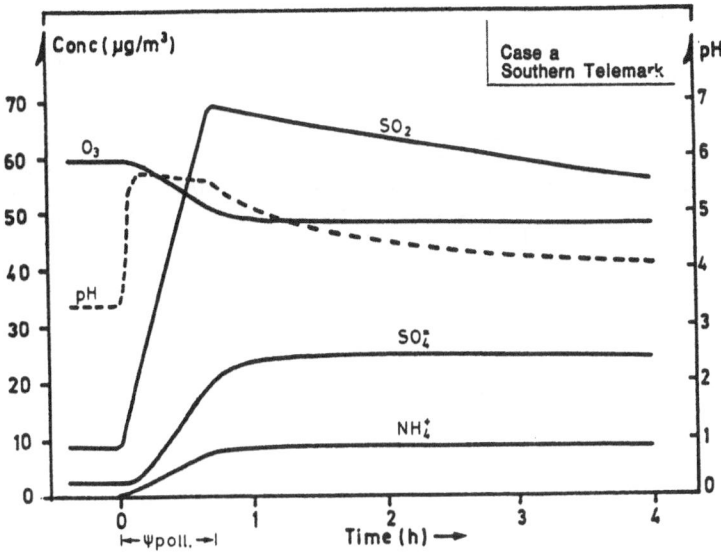

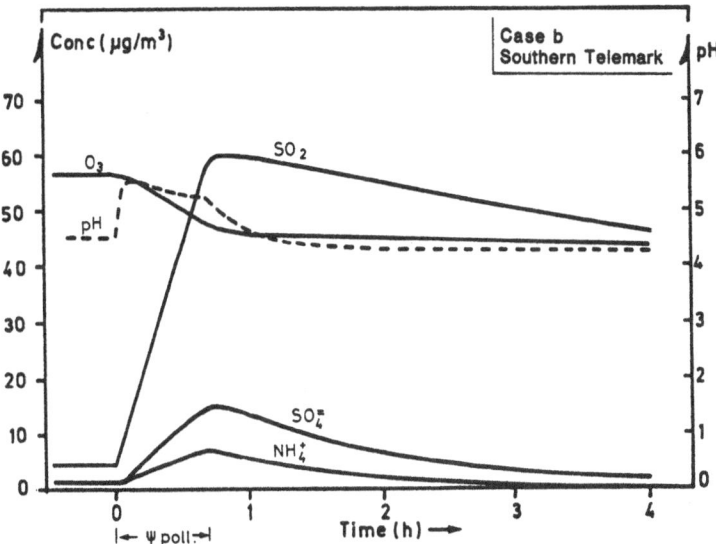

Figure 3: Variations of O_3, SO_2, SO_4^{2-}, NH_4^+ concentrations and pH values versus time for the same cases as shown in Figure 2.

4. DISCUSSION

The model calculations in this paper have shown that the oxidation of sulphur dioxide in the droplets can be of atmospheric significance. Estimates of the gas phase formation of sulphate (Rodhe and Isaksen, 1980) have indicated that the gas phase mechanisms are sufficiently fast during the summer months to explain the observed average conversion rates of approximately $1\% \; h^{-1}$ (OECD, 1977). During the fall and winter seasons, however, when the observed data still indicate $1\% \; h^{-1}$ as conversion rate, gas phase oxidation is estimated to be about a factor 10 slower. This points to oxidation processes in the liquid phase as major mechanism for sulphate formation. The oxidation of SO_2 by H_2O_2 in the droplets is limited by the production rate of the oxidant. In these calculations a simple gas phase production rate was adopted, because of the uncertainty in estimating the formation of H_2O_2 in the liquid phase. This can only be treated properly in models where the combined effects of gas and liquid phase chemistry are considered.

5. REFERENCES

Bassett, M. and Seinfeld, J.H. (1983) Atmospheric equilibrium models of sulfate and nitrate aerosols. Atmospheric Environment, 17, 2237-2252.

Beilke, S. and Gravenhorst, G. Heterogeneous SO_2-oxidation in the droplet phase, Atmospheric Environment, 13, 231-240, 1978.

Brosset, C. Water-soluble sulphur compounds in aerosols, Atmospheric Environment, 12, 25-38, 1978.

Calvert, J.G., Su, F., Bottenheim, J.W. and Strausz, O.P. Mechanism of the homogeneous oxidation of sulfur dioxide in the troposphere, Atmospheric Environment, 12, 197-226, 1978.

Chameides, W.L. and Davis, D.D. The free radical chemistry of cloud droplets and its impact upon the composition of rain. J. Geophys. Res. 87, 4863-4877, 1982.

Cox, R.A. and Tyndall, G. Rate constants for reactions of CH_3O_2 in the gas phase, Cem. Phys. Lett., 65, 357-360, 1979.

Erickson, R.E., Yates, L.M., Clark, R.L. and McEwen, D. The reaction of sulfur dioxide with ozone in water and its possible atmospheric significance, Atmospheric Environment, 11, 813-817, 1977.

Isaksen, I.S.A., Hov, Ø. and Hesstvedt, E. Ozone generation over rural areas, Environ Sci. Technol., 12, 1279-1284, 1978.

Larson, T.V., Horike, N.R. and Harrison, H. Oxidation of sulfur dioxide by oxygen and ozone in aqueous solution: A kinetic study with significance to atmospheric rate processes. Atmospheric Environment, 12, 1597-1611, 1978.

OECD, Program on long range transport of air pollutants. Measurements and findings, Paris, 1977.

Penkett,S.A., Jones,D.M.R., Brice,K.A. and Eggleton,A.E.J. The importance of atmospheric ozone and hydrogen peroxide in oxidizing sulphur dioxide in cloud and rainwater, Atmosperic Environment,13,123-137,1979.

Rodhe, H. and Isaksen, I.S.A. Global distribution of sulfur compounds in the troposphere estimated in a height latitude transport model, J. Geophys. Res., 85, 7401-7409, 1980.

Schwartz, S.E. and Freiberg, J.E. (1981) Mass-transport limitations to the rate of reaction of gases in liquid droplets; application to oxidation of SO_2 in aqueous solutions. Atmospheric Environment, 15, 1129-1144, 1981.

SFT, Annual report 1977 from Telemark district office of the Norwegian State Pollution Control Authority (in Norwegian), pp. 55-59, 1977.

Some possibilities of modelling data
from the "FOS" european field experiment (june 6/15 1983)

=-=-=-=-=-=-=

H. AUGUSTIN (*) P. BESSEMOULIN (**)

(*) Délégation aux Risques Majeurs (Paris)
(**) Etablissement d'Etudes et de Recherches Météorologiques (Toulouse).

Summary

Modelling of atmospheric pollution is often done with an insufficient set of field data.

On the other hand, the complexity of the atmospheric behaviour
is well described by a field experiment like the Fos-Etang de Berre european experiment performed from the 6th to the 15th of june 1983.

The whole set of corresponding data, may be "sliced" in space
and time, to perform various modelling tentatives either of the atmospheric flow, or of pollution phenomena, or else of both in a more
complete model.

The possibility to give to the interested modellers some more
elaborate data, for example wind fields or boundary layer thickness, is
being examinated.

The coordination of modelling research and a comparison of
results is proposed in the frame of COST 61 a bis action.

Perspectives de modélisation des données
de la campagne européenne du Fos (6 au 15 juin 1983)

Résumé

Pour modéliser la pollution atmosphérique, nombreux sont les
chercheurs qui ne disposent pas de mesures sur site suffisantes.

La complexité du comportement atmosphérique peut, par contre,
être décrite par une campagne de mesure telle que celle qui vient d'avoir
lieu sur le site de FOS - l'Etang de Berre du 6 au 15 juin 1983.

L'ensemble de ces mesures peut être découpé, dans l'espace et
dans le temps, au gré des chercheurs pour tenter diverses expériences de
modélisation, soit du comportement de l'atmosphère, soit de l'évolution
de la pollution, soit de la conjonction des deux.

Certaines données intermédiaires plus élaborées, par exemple,
champ de vent, épaisseur de la couche limite, peuvent être aussi mises
à la disposition des chercheurs intéressés.

Il est proposé d'examiner, dans le cadre de l'action COST 61
a bis, les modalités d'une coordination des modélisations et d'une con-
frontation des résultats.

I - Introduction

        Le rassemblement sur la zone industrielle de FOS-BERRE d'une
densité importante de mesures météorologiques et physico-chimiques, permet de proposer, dans le cadre de notre action COST 61 a bis, l'utilisation de ces mesures pour initialiser et vérifier divers types de modèles.
        L'objectif initial de cette campagne est l'intercomparaison
d'instruments de télédétection. Il s'agit d'une des campagnes effectuées
périodiquement, en des sites différents à l'initiative des Communautés
européennes.
        Ces campagnes comportent deux phases : la première est l'inter-
calibration des instruments au même point, et la seconde est l'estimation
des flux de polluants sur la zone étudiée.
        Effectuée en général à l'aide de moyens mobiles de télédétec-
tion, cette estimation ne peut être assurée que si l'on connaît le champ
de vent et le profil vertical de température (et, bien sûr, les caractéristiques des sources polluantes).
        L'Etablissement d'Etudes et de Recherches Météorologiques de la
Météorologie Nationale (EERM) et le Service Etudes et Recherches de l'E-
lectricité de France (EDF) sollicités pour fournir l'assistance météorologique correspondante, ont proposé, du fait de la présence de moyens importants sur le site, une campagne mésométéorologique. Ce deuxième objectif a permis d'obtenir des données très complètes, que les modélisateurs
européens devraient pouvoir utiliser.

II - Aspects météorologiques, reconstitution des champs

        Le site de FOS-BERRE se caractérise par sa géographie tourmen-
tée : présence de la mer, de l'Etang de Berre et de reliefs. Toutefois,
le mécanisme engendrant les fortes pollutions reste en gros le même que
sur site homogène.
        En cas de vent faible et de forte stabilité thermique des basses
couches, une accumulation de polluants se produit à différents niveaux. Un
mouvement de brassage ultérieur, dû par exemple à la convection thermique,
rabat l'air pollué jusqu'au sol.
        Deux effets principaux sont responsables de la stabilité de
l'air des très basses couches. L'un est le refroidissement nocturne radiatif de l'air au-dessus du continent, l'autre est l'advection d'air maritime plus frais au-dessus de la terre ; ce phénomène est souvent renforcé
par un effet de subsidence.
        Les couches stables observées sont très différentes d'un cas à
l'autre.
        L'inversion de rayonnement, plus basse en général, engendre des
épisodes de pollution plus intenses. C'est pour étudier ce dernier cas que
le plan d'expérience météorologique est conçu.
        Il s'agit d'une étude des circulations locales en situation sta-
ble, comprenant également la phase de disparition de l'inversion nocturne
(la transition matinale).
        Les écoulements d'air marin vers la terre sont également docu-
mentés mais le dispositif de mesure mis en place n'est pas optimum.
        Compte tenu de la finesse du relief, et de la stabilité de l'é-
coulement, il n'existe pas encore de modèle numérique adapté (non hydro-
statique).
        Pour un modèle de méso-échelle tel que CYBELE (BLONDIN, 1978)
(1), les critères de stabilités des schémas numériques et des impératifs

liés au temps de calcul ne permettent pas l'utilisation de mailles inférieures à quelques kilomètres (de l'ordre de cinq) en atmosphère à stra-
tification autre qu'adiabatique. On rappelle que ce modèle tridimensionnel développé à l'EERM, basé sur les équations dites primitives, travaille sur un domaine de taille limitée, typiquement 250 km x 250 km (maille
supérieure ou égale à 5 kilomètres).

L'étude reposera donc pour l'essentiel sur les résultats d'expériences et leur interprétation. Indiquons toutefois dès maintenant, que
l'on vise à obtenir des champs de vent au sol et en altitude sur une région de 50 km x 40 km centrée sur l'Etang de Berre. La structure verticale est déterminée grâce à deux systèmes de radio-sondages et un réseau
de cinq sondeurs acoustiques dont 3 sodars Doppler disposés autour de
l'Etang de Berre, et deux sodars simples.

Dans un certain nombre de situations jugées intéressantes, l'a-
vion du DFVLR, équipé d'un lidar visant vers le bas, effectue une exploration systématique de la topographie de l'inversion.

Ceci posé, l'analyse est conduite de la façon suivante :

a) phase nocturne

L'évolution du champ de vent est en général lente au cours de
la nuit. Ce champ est reconstitué par une méthode d'analyse optimale. Il
s'agit :

- soit de modèles diagnostiques permettant de calculer selon
une grille régulière un champ de vent tridimensionnel à partir des vents
de surface observés, et des profils verticaux obtenus à l'aide des sodars
Doppler et des radiosondages. Le relief h(x,y) est pris en compte par
l'utilisation d'une altitude réduite :

$$\sigma(Z) = \frac{Z - h(x,y)}{H(x,y) - h(x,y)}$$

où H(x,y) est l'altitude du sommet du domaine étudié. La méthode consiste
à utiliser un champ initial ("first guess"), qui est ensuite ajusté pour
satisfaire l'équation de continuité (P. RACHER, R. ROSSET (1980) ; R.M.
ENDLICH et Alii (1982)) (2), (3).

La structure thermique ne peut pas être connue d'une manière
aussi détaillée que la structure du champ de vent. Les radiosondages en
fixent les grandes lignes.

Pour le reste, l'inversion radiative n'est pas étudiée dans
tout le secteur, mais seulement au voisinage de l'Etang de Berre.

- soit des modèles de reconstitution de champ, avec un schéma
d'interprétation en $\dfrac{1}{a^2 + r^2}$ de type CRESSMAN (a : rayon d'influence,

r : distance à l'observation) dont un exemple est donné figure 1.

Divers modèles de ce type sont actuellement testés. En phase
diurne, sous réserve d'une variation suffisamment lente du champ de vent
ces deux schémas d'interpolation peuvent aussi être utilisés.

b) Transition matinale

L'écoulement prend un caractère instationnaire et donc l'analyse précédente se trouve en défaut.

Il apparaît nécessaire d'observer la transition en un nombre
limité de points (ceux où sont implantés les sondeurs acoustiques triples)
et de noter les heures de transition indiquées par le vent, la température et l'humidité au niveau du sol.

Par nuit claire, en l'absence de vent synoptique, la séquence
d'évènements est la suivante : écoulement d'air stable venant du Nord-Est

Fig. 1 - Vent au sol le 10 juin 83 à 03 TU.

(courant de drainage) nocturne, suivi d'une phase intermédiaire de durée une demi-heure environ, caractérisée par des circulations très localisées (par exemple une brise d'étang) ; ensuite c'est le contraste général terre-mer qui prédomine, donnant lieu dans ce cas à une brise de mer.

Afin de permettre le cas échéant une simulation numérique, quelques mesures supplémentaires sont effectuées sur le site, notamment la mesure de la température de la mer et de l'étang, ainsi que les termes du bilan d'énergie au sol, et en particulier le flux turbulent de chaleur sensible, Q_0, sur un site réputé représentatif (Berre la Fare-4M).

On sait, en effet, que la connaissance de ce flux turbulent est une donnée primordiale pour le calcul :

- de l'évolution de l'épaisseur de la couche stable nocturne (il faut connaître en outre la divergence du flux radiatif ; (NIEUWSTADT, 1980)) (4)

- de l'évolution de l'épaisseur de la couche limite convective, que l'on peut décrire de façon simple, en site homogène, à l'aide d'une équation du type

$$\frac{dh}{dt} = 1.4 \frac{Q_0}{h\gamma}$$

où :

- $\frac{dh}{dt}$ est la variation de hauteur de la couche, h, en fonction du temps, t,

- γ est la pente de la température potentielle en fonction de l'altitude dans la couche stable ; cette pente étant considérée, en première approximation, comme constante (ARTAZ et ANDRE (1980), (5)).

III - Dispositif météorologique expérimental et conditions rencontrées

Les emplacements retenus figurent sur la carte (figure 2). Sauf exceptions, le réseau fonctionne sur une base quart-horaire (ce qui est également l'échelle de temps du réseau local de surveillance de la pollution atmosphérique).

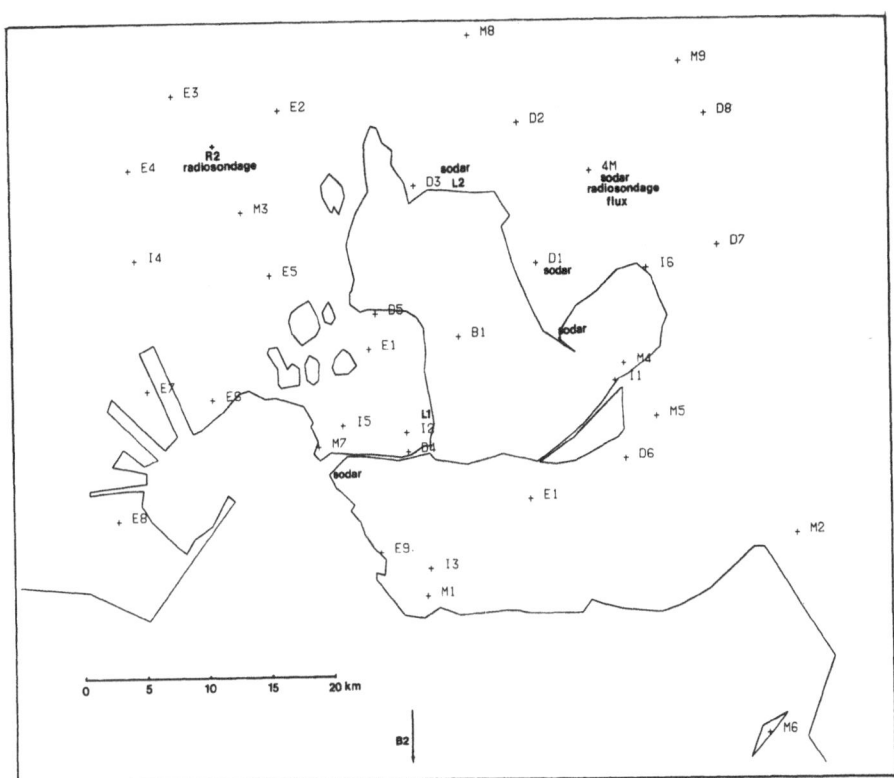

Fig. 2 - Réseau d'observations météorologiques au sol et en altitude.

Les moyens mis en oeuvre sont les suivants :
EERM : une station centrale (4M), où sont effectuées en particulier les mesures de flux, 8 stations automatiques périphériques (repérées Di), un ensemble de radiosondages, un sodar simple et deux bouées (repérées B1 et B2).
EDF : une station centrale, 10 stations automatiques (repérées Ei), un ensemble de radiosondages, deux sodars Doppler.
DII Marseille (*) : 6 stations automatiques (Ii), un sodar Doppler
JRC ISPRA (**) : une station de mesure des paramètres turbulents et un sodar simple (D1)
SMM (***) : neuf stations du réseau (Mi) figurent sur le site (Marignane Nord et Sud, Istres, Salon, Cap Couronne, Marseille Port, Pomègues, Port de Bouc, Saint-Cannat.
Les lidars du CEA (****) (L1) et d'EDF (L2) fournissent en certaines occasions des informations sur l'épaisseur de la couche de mélange, ainsi que l'avion FALCON 20 du DFVLR (*****).

(*) Direction Interdépartementale de l'Industrie
(**) Joint Research Center Ispra (Italie)
(***) Service Météorologique Métropolitain
(****) Deutschen Forschungs und Versuchsanstalt fur Luft und Raumfahrt.
Les situations météorologiques rencontrées sont résumées dans le tableau suivant :

DATE	SITUATION METEOROLOGIQUE	EVENEMENTS NOTABLES en heures TU	VENT MOYEN DD	m/s
L. 6	Marais barométrique	Brise de terre la nuit	NE÷E	2
		Brise de mer après 10 h	S	5/8
M. 7	Régime anticyclonique à faible gradient barométrique	Brise de terre le matin	E	3
		Brise de mer après 10 h	SSE	8/12
M. 8	Régime anticyclonique à faible gradient barométrique	Régime d'Est la nuit	E	2/5
		Brise de mer après 9 h	S-SW	5/7
J. 9	Régime anticyclonique à faible gradient barométrique	Brise de terre la nuit	NE-SE	1/3
		Brise d'étang pendant la transition		
		Brise de mer après 10 h	SW-N	5/8
V. 10	Marais barométrique Tendance orageuse	Brise de mer après 10 h	SW	4/6
		Passage d'un cumulonimbus au N de l'étang entre 12 et 13 h		
S. 11	Marais barométrique	Brise composée après 11 h	W	4
		Rotation au NW après 14 h	NW	5/8
D. 12	Marais barométrique	Faible courant de N en début de nuit	N-NE	1/3
		Brise de mer en milieu de journée	SSW	4/6
L. 13	Régime de NW	Brise de terre en début de nuit	E	2
		Passage d'un front froid vers 10 h		
		Vent se renforçant rapidement	NW	15/25
M. 14	Fort gradient de Nord au sol et en altitude	Vent fort	N-NNW	10/15
M. 15	Fort gradient de Nord au Sud et en altitude	Vent fort	N-NNW	10/15

Ce tableau montre qu'en première semaine certaines périodes nocturnes correspondent bien à l'étude météorologique souhaitée des circulations locales en phase stable.

Les différentes autres conditions rencontrées, brises de mer, mistral se révèlent par ailleurs valables pour les physico-chimistes.

IV - Aspects physico-chimiques

Les principales sources sont indiquées sur la figure 3 ainsi que leur débit horaire en tonnes de SO_2 par heure avec la fourchette des hauteurs de cheminées correspondante.

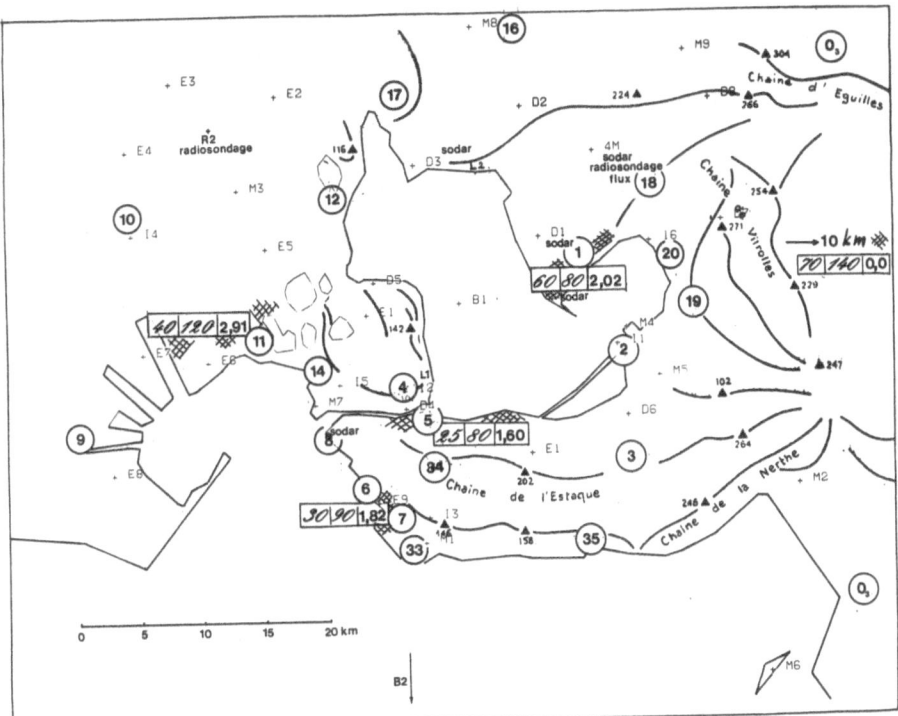

Fig. 3 :

◯ : Stations de mesure de pollution au sol,

(Hauteurs maxi. et mini. des cheminées (mètres)

| max / mini | E |

: E, émission de SO_2 en tonnes/heure le 6 juin 1983
(à 03 h TU.

Le réseau permanent de mesure au sol toutes les 15 minutes, porté figure 3, est complété par des mesures itinérantes par spectromètre à corrélation et par lidar dont les parcours et les mesures seront introduits dans la banque de données.

V - Possibilité de mise en oeuvre de modèles de diffusion

Les deux modèles de reconstitution de champ de vent décrits au paragraphe 2 peuvent être couplés à des modèles de diffusion :

- soit gaussiens : panaches ou bouffées auquel cas un simple calcul de trajectographie suffit,
- soit à coefficient d'échange, ce qui nécessite la conservation de la masse qui n'est assurée que par le modèle variationnel.

Pour le moment une exploitation est prévue par le CEA à l'aide d'un modèle à bouffées. Dans ce modèle les écarts types de la distribution des polluants sont paramétrés en fonction du temps de transfert (DOURY et Al.) (6).

VI - Disponibilité des données

De nombreuses autres expériences de trajectographie et de modélisation restent possibles. Les données sont actuellement centralisées et mises en forme de la façon suivante :

Le Centre National de Recherches Météorologiques de l'EERM assure la constitution de la banque de données météorologiques, qui se présente sous la forme de réseaux quart-horaires, classés par ordre chronologique.

Le Commissariat à l'Energie Atomique établit une banque regroupant l'ensemble des données physico-chimiques et météorologiques, ordonnées de manière chronologique quart-horaire pour obtenir un ensemble cohérent et facilement manipulable. Ce fichier doit être disponible à la fin du premier semestre 1984. Il sera alors utilisable par toute équipe désirant confronter des modèles de diffusion et/ou transformation de polluants avec des mesures.

Les modélisateurs de l'action COST intéressés peuvent demander dès maintenant le descriptif des données correspondantes mises sur support magnétique neuf pistes 1600 BPI. Celui-ci leur sera envoyé dès sa parution.

Une concertation ultérieure devrait permettre le choix des situations intéressantes, et ensuite aboutir soit à une répartition des tâches, soit surtout à une intercomparaison de modèles.

BIBLIOGRAPHIE

(5) M. A. ARTAZ, J.C. ANDRE (1980) : Similarity studies of entrainment in convective mixed layers. Boundary Layer Meteorology, Vol 19, n° 1, pp 51-66
(1) C. BLONDIN (1978) : Un modèle de méso-échelle : conception, utilisation, développement. Note EERM n° 416.
(6) A. DOURY, R. GERARD, M. PICOL (1980) : Abaques d'évaluation directe des transferts atmosphériques d'effluents gazeux. Rapport DSN n° 84 (Rev. 2)
(3) R.M. ENDLICH, F.L. LUDWIG, C.M. BHUMRALKAR (1982) : A diagnostic model for estimating winds at potential sites for wind turbines. Journal of Applied Meteorology, Vol 21, pp 1441-1454.
(4) F.T.M. NIEUWSTADT (1980) : A rate equation for the inversion height in a nocturnal boundary layer. Journal of Applied Meteorology, Vol 19 pp 1445-1447.
(2) P. RACHER, R. ROSSET (1980) : Un modèle variationnel d'ajustement du vent sur la Plaine d'Alsace : application au transport des polluants. Colloque "Météorologie et Environnement". EVRY, 6-7/10/1980.

ATMOSPHERIC TRACER DISPERSION EXPERIMENTS

(at the Mol Nuclear Energy Research Center)

B. Vanderborght, I. Mertens and J. Kretzschmar
Nuclear Energy Research Center, SCK/CEN
B-2400 Mol, Belgium

Summary

SF_6-tracer dispersion experiments have been executed for the validation of on-line and off-line mathematical models for the simulation of short-term emission-immission situations. Cross-wind measured concentration profiles are fitted with a Gaussian curve and the parameters of the fit (plume direction, horizontal and vertical dispersion parameters and maximum concentration) are compared with model values. The comparison of calculated and measured data is illustrated using one experiment as an example and the most important general validation results of all the experiments of the project are summarized.

1. INTRODUCTION

Mathematical models, describing the relation between the emission of a contaminant in the atmosphere, the meteorological conditions and the resulting concentration patterns, have become an indispensable instrument in the field of fundamental and applied air pollution and safety research. In fundamental research, many complex models of different kinds have been proposed, but when applications are involved the Gaussian distribution models have the highest score. The bi-Gaussian dispersion approach is basically a statistical model, relying on a statistical characterization of the atmosphere (the turbulence typing schemes) and using a statistical set of semi-empirical dispersion parameters for the simulation of the concentration patterns over an extended time period. While the bi-Gaussian dispersion principle remains unchanged, several classifications of the atmospheric turbulence and the corresponding dispersion parameters have been proposed. The influence of different schemes on the results of the model simulation of long-term concentration statistics has been the subject of task 1 of this project (KRETZSCHMAR e.a., 1983) and of several other sensitivity studies.

The objective of task 2 of the project is to evaluate the possibilities of a practical model - the bi-Gaussian dispersion model, a trajectory model and a tri-Gaussian diffusion bi-dimensional puff trajectory model - to simulate short-term emission and immission situations and to indicate the applicability of on-line models for assistance in emergency situations.

This study has three successive phases :
1. Execution of tracer dispersion experiments to construct a data
   base for the evaluation of model performances for individual
   emission-immission periods.
2. If the off-line versions of the models are capable to simulate
   the short-term immissions with reasonable accuracy, implement
   on-line versions on the computer system for assistance in emergency response.
3. Evaluate by means of tracer dispersion experiments the applicability and operationality of the on-line models to coordinate
   the field immission measurements in case of an accidental release.

2. EXPERIMENTAL

A computer code (the Computer Aided Emergency Response System :
CAERS) has been developed for the off-line and on-line simulation of the
short-term concentrations caused by an atmospheric emission. The concentration calculations are based on the bi-Gaussian theorem with the possibility of using twelve different sets of turbulence typing schemes and
dispersion parameters or the plume can be simulated with a bi-dimensional
puff trajectory model with tri-Gaussian diffusion of the puffs. With the
puff trajectory model the emission and the wind-conditions can be variable in time.

An experimental strategy has been optimized for the measurement of
the dispersion of pollutants in the atmosphere. SF_6-tracer gas is released
in the atmosphere under well controlled conditions, the plume is tracked
with mobile measurements of the tracer and the half hour average atmospheric SF_6-concentration is measured at 20 to 30 places under the plume.
SF_6-gas was released at 2 m above ground-level, at a rate of about
3 g.s^{-1} and at ambient temperature and a cap above the emission opening
to prevent plume rise. SF_6 was sampled at 20 to 30 locations on 1 to 3
arcs 2 to 7 km downwind the source during 1 to 3 consecutive 30 minutes
averaging sampling periods. Meanwhile the plume was tracked with instantaneous SF_6-concentration measurements using a gaschromatograph in a measuring van. Meteorological parameters were measured at 6 levels on the
120 m mast of the SCK.
Fifteen SF_6-tracer dispersion experiments were carried out in the period
September 1982 - September 1983 to validate the bi-Gaussian dispersion
model on short-time periods. The operationality of the on-line model
strategy was tested during a tracer dispersion experiment in November '83.

3. RESULTS AND DISCUSSION

It is obviously impossible to give a complete review of the results
of the experiments within the scope of this paper. The data reduction
principle will be presented using one experiment as an example and then
the general conclusions will be summarized.
The experiment of 17 January 1983 is a typical - not the best nor the
worst - example. $3.4 \text{ g } SF_6.s^{-1}$ was released at 2 m altitude, under WSW-
wind with 5.4 m.s^{-1} average speed and under neutral atmosphere, from
10.45 to 12.45. Half hour average air samples were taken at 25 places on
three arcs downwind the source from 11.30 to 12.00 and from 12.15 to 12.45.
A graphical representation of the emission and sampling places with the

measured concentrations of two consecutive half-hour sampling periods is
given in figure 1. The sampling places have been projected on a line
perpendicular to the plume axis, i.e. the line from the source to the
sampling place with the maximum concentration of the cross-wind profile.
The measured concentrations on every cross-wind arc have been fitted with
a Gaussian curve from which the parameters are : place of the profile
maximum, horizontal (σ_y) and vertical (σ_z) dispersion parameters. The
curve fitting on the 6 sampling lines of 17.01.1983 are given in figure 2.
From these curves the maximum concentration of the measured cross-wind
profiles can be deduced.
In most cases, the measured points can be fairly well fitted with a Gaussian curve. Aberrations of the measured concentrations from a Gaussian
curve are caused by sampling or analysis problems, important wind-direction shear during sampling or small erratic errors of minor importance.
        In table I, the direction of the plume axis, the horizontal and
vertical dispersion parameters and the maximum concentration of the different measured profiles are compared with wind-direction measurements and
model values. The model values were obtained from the bi-Gaussian model
with the SCK/CEN turbulence typing scheme.

Table I - Comparison of measured (M) and calculated (C) plume direction
          (dd), horizontal (σ_y) and vertical (σ_z) dispersion and maximum
          concentration (C_{max}) for the different cross-wind profiles of
          17.01.1983.

period	distance (m)	dd (°)		σ_y (m)		σ_z (m)		$C_{max}(\mu g.m^{-3})$	
		M	C	M	C	M	C	M	C
1	2260	251	254	138	195	135	126	11	8.2
2	2260	247	252	138	195	110	126	13	8.2
1	4380	252	254	210	330	250	200	3.8	3.0
2	4380	247	252	180	330	312	200	3.6	3.0
1	6240	251	254	340	440	187	260	2.2	1.8
2	6240	247	252	230	440	340	260	2.6	1.8

        The figures and table illustrate an operational problem for tracer
dispersion experiments : in order to obtain a measured profile with a
sufficient number of samplers under the plume, the samplers must be placed close to each other but a slight wind-direction change between the
start of the emission and the start of the sampling, or during the sampling can then easily move the plume out of the sampling network. Moreover
there is a small difference between measured wind-direction and plume axis
(on the average 5° difference, but in some cases up to 20° difference).
The dispersion parameters σ_y and σ_z of successive half hour sampling periods, under apparently equal meteo conditions, can be almost equal but
can also differ up to 50 % one from the next (see table I). The variability of the measured dispersion parameters are caused by real variations
of atmospheric dispersion and by measuring precision limitations. The
variability must be held in mind when measured values are compared with
model values. The same is true for the maximum concentration at a given
distance for successive sampling periods.

Averaged over all sampling arcs and periods, the average ratio of measured over model σ_y is 0.82 $\pm$ 0.22 for 41 half-hour periods. The small overestimation of the measured horizontal dispersion by the model can be due to the fact that the dispersion parameters in the SCK turbulence typing scheme have been deduced from wind-vector fluctuation measurements over wooded area while the tracer experiments have been done over open area with lower surface roughness.

The average ratio of measured over model σ_z is 1.25 $\pm$ 0.65. The scatter is much larger because σ_z is determined indirectly from mass conservation considerations and an assumption on the shape of the vertical concentration profile that is not measured directly. The apparent underestimation by the model is statistically not significant.

The measured σ values have also been compared with the σ values determined from the 12 "turbulence typing schemes – dispersion parameter sets" discussed in (KRETZSCHMAR and MERTENS, 1983). The comparison has been done through the relative difference between model σ and measured σ defined as

$$ r = \frac{(\sigma)_{model} - (\sigma)_{meas}}{1/2 \left[(\sigma)_{model} + (\sigma)_{meas} \right]} \times 100 $$

This parameter has the same absolute value but opposite sign when the model value is 1/n the measured value or vice versa. A positive value results from an overestimation of the measurement by the model. An average relative difference equal to zero corresponds with a perfect accuracy. The standard deviation is a measure for the precision of the turbulence typing scheme – dispersion parameter set.

The average of the relative differences of 41 sampling periods and arcs, independent of source receptor distance or atmospheric stability as well as the relative standard deviations are collected in table II for the horizontal dispersion and in table III for the vertical one.

This representation of the comparison between measured and model horizontal dispersion also reveals the overestimation by the SCK scheme with roughly 20 %. All the turbulence typing schemes using the Pasquill dispersion parameters have an average relative difference closer to 0 (Pas SYN, Pas NRC, Pas PRA, Briggs OPEN, Turner, Smith-Hosker and Klug). The other schemes (Briggs URBAN, Vogt SYN and GRAD, Doury) overestimate the measured horizontal dispersion very much.

Table II - Average relative difference between model σ_y and measured σ_y
and corresponding standard deviation for 41 measurements
and 12 turbulence typing schemes

scheme	% relative difference ($\bar{r}$)	standard deviation (s.d.)
SCK	22	25
Pas SYN	7.9	41
Pas NRC	2.2	34
Pas PRA	-5.3	30
Briggs OPEN	15	38
Briggs URBAN	59	37
Turner	-9.2	30
Smith-Hosker	20	37
Klug	-13	51
Vogt SYN	44	32
Vogt GRAD	61	40
Doury	54	49

For all schemes, the scatter of the relative difference between model and
measured horizontal dispersion parameter is rather high. The best precision is obtained by the SCK scheme (s.d. = 25). The precision is between
30 and 40 for the Pasquill based systems and around 50 for Klug, Vogt and
Doury.
The scatter of the fractional error of the vertical dispersion is for all
schemes very large, emphasizing the large uncertainty in the knowledge of
the vertical dispersion. The average relative difference is closest to
zero, for the SCK scheme. The average underestimation of the measured
vertical dispersion by the model is not significant for the SCK scheme,
but becomes more evident for the Pasquill based systems, with an average
relative difference between -23 and -76 %. The Doury system largely un-
derestimates while Vogt overestimates vertical dispersion.

Table III - Average relative difference between model σ_z and measured σ_z
and corresponding standard deviation for 41 measurements
and 12 turbulence typing schemes

scheme	% relative difference ($\bar{r}$)	standard deviation (s.d.)
SCK	-7.1	59
Pas SYN	-45	83
Pas NRC	-43	80
Pas PRA	-66	53
Briggs OPEN	-34	82
Briggs URBAN	69	62
Turner	-76	66
Smith-Hosker	-23	70
Klug	-66	96
Vogt SYN	61	57
Vogt GRAD	50	38
Doury	-84	40

The SCK bi-Gaussian model simulates the maximum concentration of the cross-wind profiles fairly well with an average ratio of calculated over measured maximum concentration equal to 1.02 ± 0.65.

The scatter diagram of calculated versus measured concentration on 17.01.1983 is given in figure 3. The parameters of the least square fit and of the emission and meteo conditions are reported in the figure caption. The average slope of all the scatter diagrams of the sampling days is very close to one : 1.03 and the average correlation coefficient is 0.92.

4. Conclusions

From this validation it is clear that the bi-Gaussian model can simulate individual immission periods with an acceptable accuracy and precision, when an appropriate set of dispersion parameters is used.
An on-line version of the model can be helpful in emergency situations : to calculate the area and the levels of contamination so that the necessary precautions can be taken and to coordinate the immission measurements that must be carried out in the contaminated area during the emergency situation. An off-line version of the validated model can help in the study of the measurement results in view of the evaluation of the global immission or to estimate the source-term.

Acknowledgments

The work is a part of a larger project and was mainly financed by the Commission of European Communities under contract No. 028 SRB.

References

KRETZSCHMAR J. and MERTENS I. (1983)
Influence of the turbulence typing scheme upon the cumulative frequency distributions of the calculated relative concentrations for different averaging times.
EUR 8478 EN

KRETZSCHMAR J., MERTENS I. and VANDERBORGHT B. (1984)
Development and evaluation of the possibilities of on-line dispersion models for use in an emergency plan of a nuclear installation.
Final report of contract No. SR-028-B (G)

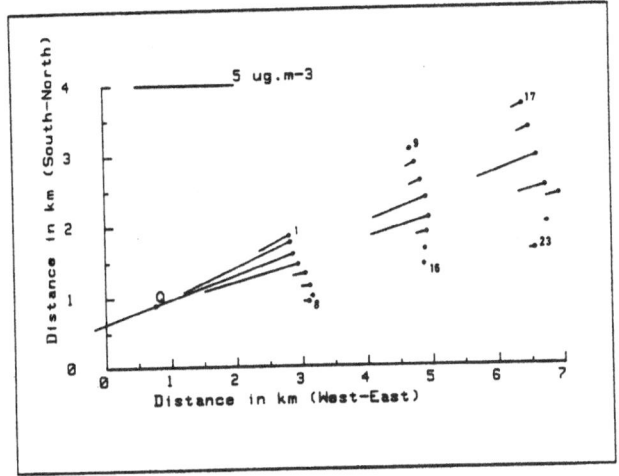

FIGURE   1a :SF6 CONCENTRATION PATTERN DAY 17/01/83
              PERIOD: 1a   SAMPLING POINTS 1 TO 23

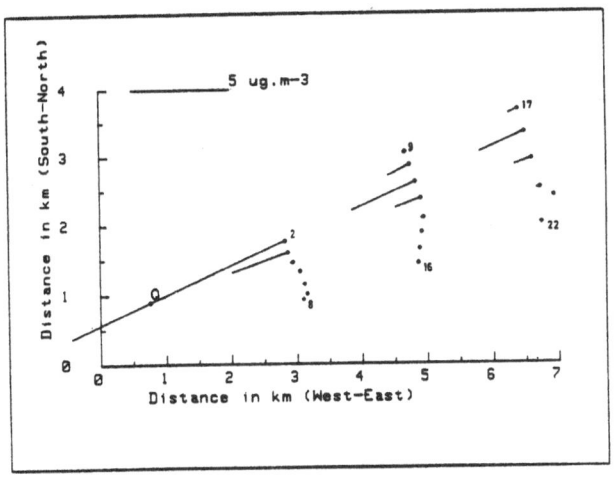

FIGURE   1b :SF6 CONCENTRATION PATTERN DAY 17/01/83
              PERIOD: 2   SAMPLING POINTS 2 TO 22

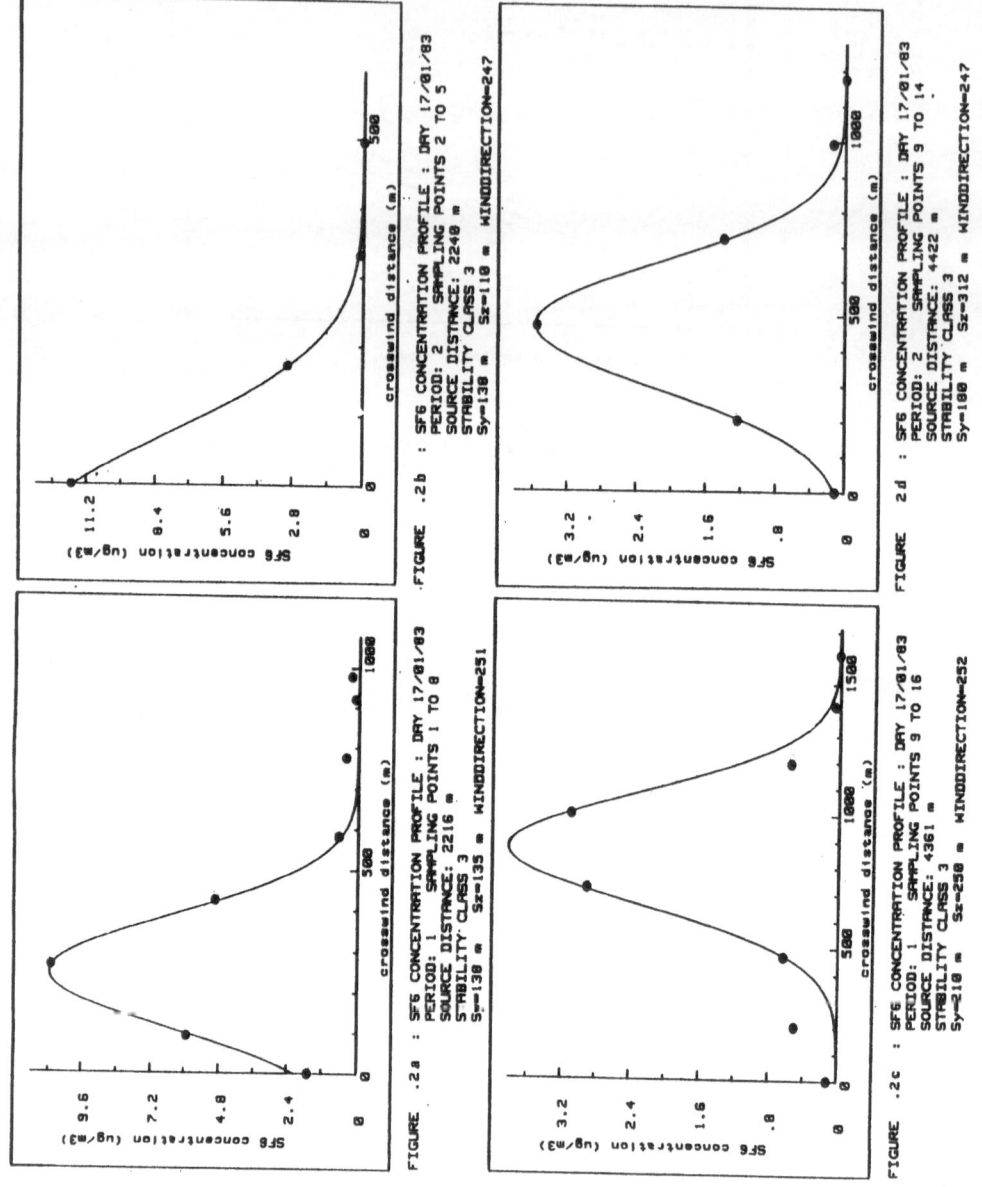

FIGURE .2a : SF6 CONCENTRATION PROFILE : DAY 17/01/83
PERIOD: 1   SAMPLING POINTS 1 TO 8
SOURCE DISTANCE: 2216 m
STABILITY CLASS 3
Sy=130 m   Sz=135 m   WINDDIRECTION=251

FIGURE .2b : SF6 CONCENTRATION PROFILE : DAY 17/01/83
PERIOD: 2   SAMPLING POINTS 2 TO 5
SOURCE DISTANCE: 2240 m
STABILITY CLASS 3
Sy=130 m   Sz=118 m   WINDDIRECTION=247

FIGURE .2c : SF6 CONCENTRATION PROFILE : DAY 17/01/83
PERIOD: 1   SAMPLING POINTS 9 TO 16
SOURCE DISTANCE: 4361 m
STABILITY CLASS 3
Sy=210 m   Sz=250 m   WINDDIRECTION=252

FIGURE 2d : SF6 CONCENTRATION PROFILE : DAY 17/01/83
PERIOD: 2   SAMPLING POINTS 9 TO 14
SOURCE DISTANCE: 4422 m
STABILITY CLASS 3
Sy=180 m   Sz=312 m   WINDDIRECTION=247

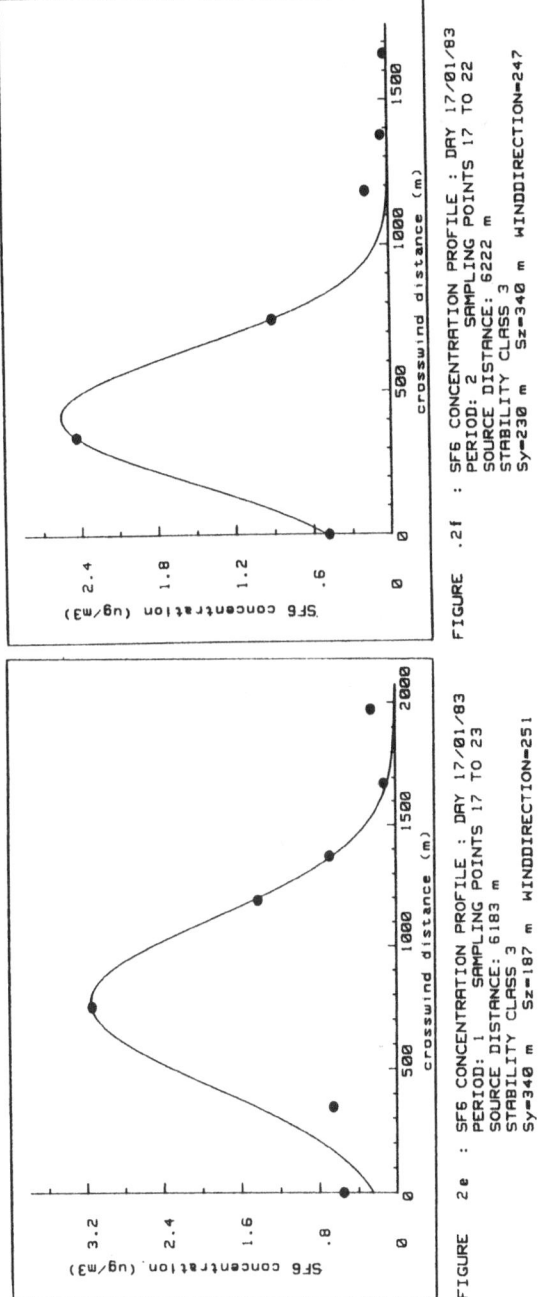

FIGURE 2 e : SF6 CONCENTRATION PROFILE : DAY 17/01/83
PERIOD: 1   SAMPLING POINTS 17 TO 23
SOURCE DISTANCE: 6183 m
STABILITY CLASS 3
Sy=340 m   Sz=187 m   WINDDIRECTION=251

FIGURE .2f : SF6 CONCENTRATION PROFILE : DAY 17/01/83
PERIOD: 2   SAMPLING POINTS 17 TO 22
SOURCE DISTANCE: 6222 m
STABILITY CLASS 3
Sy=230 m   Sz=340 m   WINDDIRECTION=247

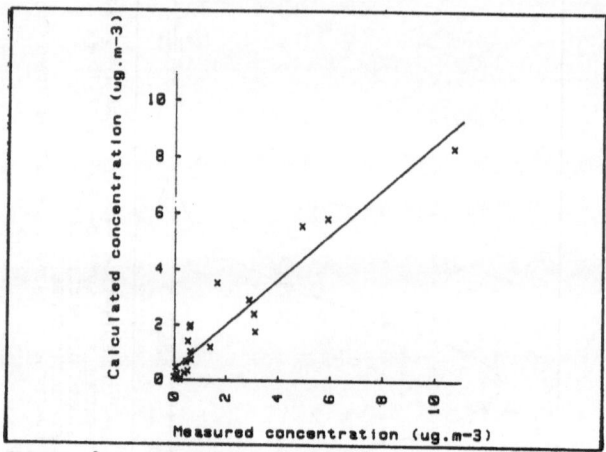

FIGURE 3a : CALCULATED vs. MEASURED SF6 CONCENTRATION
DAY: 17/01/83
PERIOD 1   POINTS 1 TO 24
n = 23   Q = 3.40 g/s
a = .814   WD = 251
b = .388   WS = 5.4 m/s
r = .944   Ei = 3

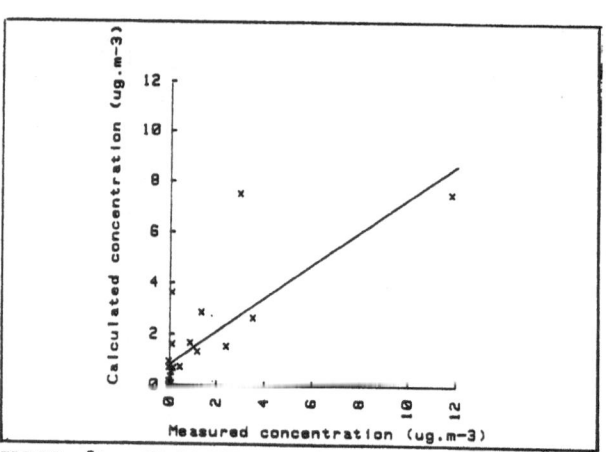

FIGURE 3b : CALCULATED vs. MEASURED SF6 CONCENTRATION
. DAY: 17/01/83
PERIOD 2   POINTS 1 TO 24
n = 22   Q = 3.40 g/s
a = .653   WD = 249
b = .812   WS = 5.4 m/s
r = .780   Ei = 3

A SAMPLING NETWORK FOR THE ASSESSMENT OF THE HEAVY METAL POLLUTION ORIGINATING FROM A MUNICIPAL INCINERATOR PLANT IN BELGIUM

F.CANDREVA and R.DAMS

Institute for Nuclear Sciences, University of Gent, Belgium

Summary

A two-weeks sampling campaign covering simultaneous measurements of particulate emission and ambient suspended particulates is briefly described. The heavy metals Pb, Cd, Cr, As and V in the different types of samples are determined by instrumental neutron activation or atomic absorption spectrophotometry. The composition of the particulate emission samples from stack off-gases is compared with that of the fly ash samples collected downstream the electrostatic precipitator. Results of deposition measurements performed at the different stations of the sampling network are also included. Measured ambient aerosol concentrations are compared with calculated values, using a bi-Gaussian dispersion model. Finally the validation of the mathematical model is briefly discussed.

1. INTRODUCTION

The growing concern with respect to solid waste disposal is receiving at present the deserved attention. Besides soil and groundwater pollution caused by land disposal of slags and fly ash, municipal incinerators may be generating a serious air pollution problem. The latter topic is investigated in the present paper.

Within the framework of a European Community research project, a network of 5 sampling stations is set up around a municipal incinerator plant, in order to study the impact of emissions of heavy metals on ambient air and fall-out. Total stack emissions - obtained after isokinetical sampling of particulates and stack flow measurements - ambient air concentrations and meteorological conditions are simultaneously measured during a two-weeks campaign. A simple mathematical model is applied and validated by comparison of measured and calculated ambient air concentrations. The results reported in this paper are limited to a one-day campaign, however, a global evaluation of the results of the entire campaign is briefly reported.

2. DESCRIPTION OF THE INCINERATOR PLANT

The incinerator is equipped with two furnaces each having a capacity of at least 5.5 metric ton per hour. The refuse burned in this plant consists of the daily collected waste from the residents of the city of Gent and vicinity, the daily or weekly collected waste from hospitals, restaurants, schools, stores and small industries, and finally the residual materials from the neighbouring municipal compostation plant. The waste material is deposited on a four-stage rocking-grate and burned at a furnace temperature of ca. 1000°C. Complete combustion is achieved by insufflation of primary and secondary air. Depending on the composition of the waste additional fuel is injected. After combustion, the hot residues are mixed with fly ash (removed by an electrostatic precipitator) and carried to a silo, awaiting further disposal. The combustion gases are first conducted into a cooling-tower, and cooled down to 400°C with water that is pumped out of a nearby river and is injected at a rate of 13 m^3 per hour. Further cooling down to

300°C is achieved with supplementary air. Subsequently, the larger fly ash particles are removed with an electrostatic precipitator in order to reduce the particulate load below 100 mg/Nm^3 at a CO_2 content of 7% (v/v). Finally the off-gases are ducted into a 65 m high exhaust stack.

3. EXPERIMENTAL

3.1. Sampling

It is common knowledge that it is waste of time to expend fine analytical skill using advanced analytical techniques when dealing with a doubtful sample. Therefore most of the attention was focused on obtaining valuable and representative samples, especially when dealing with isokinetical sampling of particulate emissions.

Emission measurements. The determination of heavy metals in off-gases originating from incineration of municipal waste gives rise to specific sampling problems. Corrosivity ($\pm$ 250 mg Cl$^-$/Nm^3,w) (1), high temperature ($\pm$ 280°C) and high water content ($\pm$ 15 v/v %) are three main characteristics of the off-gases to be sampled under isokinetical conditions. For these reasons special attention must be paid to obtain valuable samples free from contamination. In this context it must be emphasized that at present no commercially available apparatus meets the requirements for proper in-stack sampling and subsequent measuring the heavy metals in the particulate emission. Within the limits of this paper the applied procedure and arrangement is briefly described.

A BCURA-dust sampling equipment (Airflow Developments, High Wycombe, England) is thoroughly modified as pictured in Fig.I. The in-stack sampling head contains a cyclone (95% collection efficiency for coarse particles > 5 μm) and a microquartz filter housed in a stainless steel pot as pictured in the exploded view on Fig.I. Because of the high water content of the hot off-gases, adequate cooling and collection of the condensed water is necessary in order to on the one hand protect the suction unit and dry gasmeter, and on the other hand determine the water content of the off-gases. Some technical problems, however, may arise when dealing with tall stack dimensions (1.30 m stack wall and 2 m internal diameter) which necessitates the handling of probes and pitot tubes of almost 4 m. An attempt is also made to determine the heavy metals in the vapour phase. For this purpose a test arrangement with an Andersen in-stack cascade impactor was used as sampling head along with a heated probe.

Collection of electrostatic precipitator fly ash. Fly ash samples are taken downstream of a precipitator duct, from which a fraction of the ashes are tapped during the total sampling period at the stack site. In this way a comparison can be made between the precipitator fly ash (for land disposal) and the particulate emission (stack discharge). The one-day fly ash sample ($\pm$ 2 kg) is subsampled by means of a "chute splitter" until a representative subsample of $\pm$ 100 g is obtained.

Suspended particulate matter and deposition measurement. A sampling network with 5 stations is set up NE of the incinerator plant, 4 stations are in downwind and one in upwind direction with respect to the prevailing SW winds. Every station is equipped for measurement of suspended particulates and for deposition measurement. The suspended particulates are measured with LIB-type high volume samplers (2) with a sampling head mounted on a 7 m height stand. The suspended particulates are collected on a Whatman 41 cellulose filter. Samples are taken daily on a 21 h-basis ; the filters are changed between 14.00 h and 17.00 h.

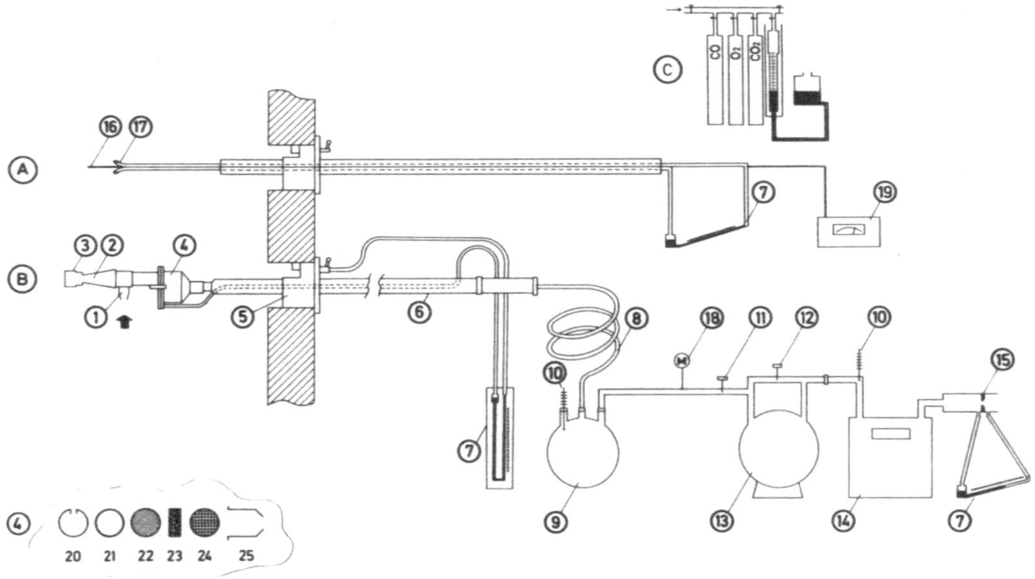

Fig. I : Arrangement for in-stack sampling of particulate emissions

A. Velocity measurement
B. Particulate sampling
C. Off-gas analysis (Orsat)

Legend

1. Sampling nozzle
2. Cyclone
3. Hopper
4. Filter housing
5. Probe holder
6. Probe with pressure tapping
7. Manometer
8. Copper spiral cooler
9. Condens catchpot
10. Thermometer
11. Coarse adjust valve
12. By-pass valve
13. Rotary vane pump
14. Dry gas meter
15. Orifice
16. Thermocouple
17. S-pitot tube
18. Vacuum gauge
19. Digital thermometer

Exploded view of filter housing

20. Spring clip (stainless steel)
21. Teflon ring
22. Microquartz filter
23. Glass wool fibre packing (1-4 μm)
24. Rooster (stainless steel)
25. Sampling pot (stainless steel)

Deposition measurements (total fall-out) are carried out with NILU-gauges (3) on a monthly basis. Duplicate samplings and analyses are executed in order to estimate the precision of the sampling method.

3.2. Meteorological observations – Mathematical model

Wind direction and speed are continuously measured in the immediate vicinity of the incinerator plant at 20 m height. These data are averaged over 1 h periods. The atmospheric stability class is estimated according to the original classification system of Pasquill (4), based on surface wind speed, insolation during daytime and cloud cover during nighttime.

Within the limits of the available meteorological data the mathematical model applied is based on the bi-Gaussian dispersion formula with reflection of the pollutant by the ground.

3.3. Analytical techniques

Instrumental neutron activation analysis (5) is used for the determination of Cr, As and V in the emission, fly ash and suspended particulate samples. Thermal atomic absorption spectrophotometry with Zeeman background correction is applied for the determination of Pb and Cd in the samples, after acid digestion.

4. RESULTS AND DISCUSSION

The results reported in the following are limited to the data obtained from a one-day campaign (22 March 1983) as previously stated.

Table I summarizes the emission data for 22 March 1983. Obviously the heavy metals Pb, Cd and As are strongly correlated with the fine particulate matter, whereas Cr and V are associated with the coarse particles in the off-gases. With respect to V it is remarkable but not surprising that the lithophilic fraction (associated with coarse particulates) largely exceeds the fine particulate fraction, commonly ascribed to combustion of fuel oil for space heating.

TABLE I : Emission data for 33 March 1983

STACK GAS FLOWRATE 74 000 Nm^3,w/h	TOTAL DUST mg/Nm^3,w	Pb	Cd	Cr	As	V
				μg/Nm^3,w		
Concentration	120	2600	106	107	8	14
Emission rate (g/h)	8900	190	8	8	0.6	1
Particulate ratio fine : coarse	0.1	8	8	0.3	3	0.1

From a comparison of fly ash and particulate emission data (Table II) the conclusions stated above are affirmed, since it is common knowledge that an electrostatic precipitator has a high collection efficiency to retain large particles.

TABLE II : Comparison between fly ash and particulate emission data

	Pollutant concentration (μg/g)				
	Pb	Cd	Cr	As	V
Fly ash	7800+700	250+15	700+20	57+2	124+5
Coarse particulate emission	12600+500	520+20	960+20	56+2	126+4
Fine particulate emission	101000+7000	4000+100	330+10	165+6	18+3

Table III compares the measured and calculated <u>ambient air concentra-</u><u>tions</u> for 22-23 March 1983. In the calculations Pasquill stability class D (neutral conditions) and an effective plume height of 85 m are applied. The plume height is based on the formula of Stümke (6). In general the aerosol concentrations appear to be at background level in this particular case. However, it is found that dealing with N to NE winds a highly increased ambient air concentration is measured. As a result the pollution roses as pictured on Figure II seem to assign the Gent city and the northern indus-tries as major and predominant pollution sources. This phenomenon was al-ready established by R.Heindryckx in 1974 (7).

TABLE III : Ambient air concentrations (ng/m^3) on 22-23 March 1983.
Measured and calculated values ( )

Element	Sampling station	I	II	III	IV	V upwind (background)
		downwind sampling stations				
Pb		28 (0)	36 (0)	38 (20)	5.2 (12)	82
Cd		0.2 (0)	2.2 (0)	4.0 (0.8)	1.3 (0.5)	0.6
Cr		2.74 (0)	0.44 (0)	0.60 (0.84)	0.74 (0.54)	1.86
As		0.17 (0)	0.26 (0)	0.03 (0.06)	0.25 (0.04)	1.04
V		8 (0)	44 (0)	29 (0)	30 (0)	32

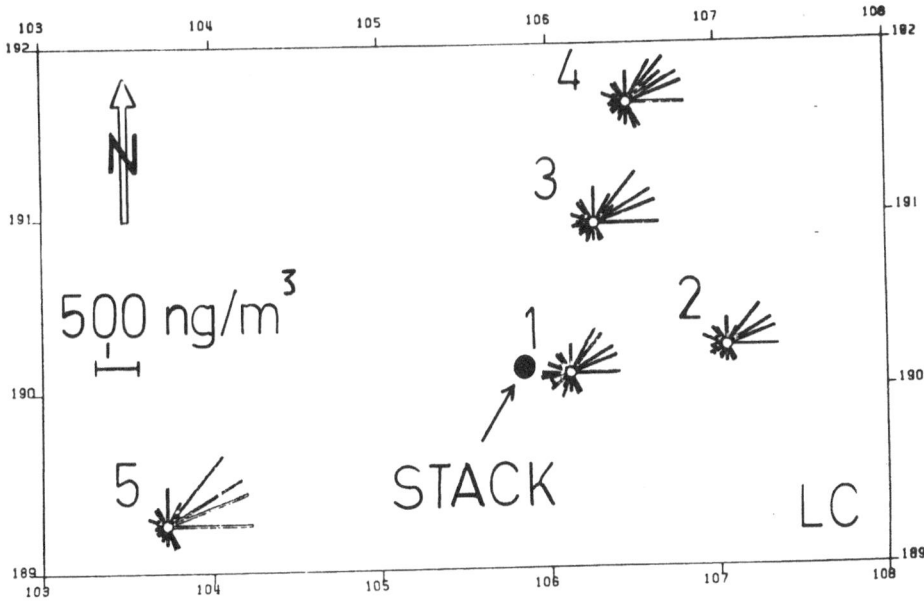

Fig. II : Configuration of the sampling network / Pb-pollution roses

At present <u>deposition measurements</u> are analysed for Pb and Cd as summarized in Table IV. The measurements carried out in duplo, refer to the sampling period of 15 March - 19 April 1983. Atmospheric deposition is obviously more important in the immediate vicinity of the incinerator plant. The duplicate samplings reveal an acceptable precision for the sampling technique and analytical method applied (generally better than 25%).

TABEL IV : Results of deposition measurements ($\mu g/m^2.d$)

Sampling station / Element	I	II	III	IV	V
Pb	88	44	35	37	36
	103	38	40	43	35
Cd	9.4	4.2	2.1	2.0	1.6
	11.6	2.8	1.8	1.5	1.6

5. GLOBAL EVALUATION AND VALIDATION OF THE MATHEMATICAL MODEL

As far as the present tentative results allow to be thoroughly evaluated, it must first be well-remarked that the overall measured ambient air concentrations are not strikingly high to directly pinpoint the municipal incinerator as a significant pollution source for the environment. The applied mathematical model does not include deposition velocities for the different elements, nor is the pollutant reflection at the mixing layer taken into account. Background concentrations are subtracted from the measured values ; to this purpose a uniform pollution background is assumed for all the sampling stations. Nevertheless these considerations some general conclusions can be drawn from the comparison of measured and calculated ambient air concentrations.

There is a reasonable agreement, except for V, between measured and calculated immmission data with respect to sampling stations III and IV. The example also demonstrates that the applied mathematical model (and probably most of the models) reveals a serious underestimation of the ambient air concentrations in the nearest vicinity of the point source.

With respect to V it is remarkable that the measured ambient air concentrations are far beyond the calculated values. This means on the one hand that no significant impact can be ascribed to the point source under investigation, and on the other hand it is obvious that the V concentration in ambient air can easily be ascribed to a more diffusive pollution source e.g. the combustion of fuel oil for space heating.

The highly variable pollutant emission rate is of utmost importance in the calculation of immission data; it is therefore required to control this input parameter as good as possible. Experience has clearly demonstrated the frequent occurrence of operational problems with the dust control system.

We may conclude from the present results that additional emission, immission and deposition data are required in order to fully evaluate the environmental impact of municipal incinerate emissions. The fate of land disposal of slags and toxic fly ash is beyond the interest of air pollution but may, however, not be separated from the global environmental impact of municipal incinerators.

ACKNOWLEDGMENTS

The authors are indebted to M.Nagels, G.Vermeir, J.Cluydts and G. Cosemans for technical assistance during sampling, analysis and data treatment. This work is fully supported by the Commission of the European Communities, under contract N° ENV.619-B(RS).

REFERENCES

(1) F.Candreva and R.Dams (1981). Determination of gaseous fluoride and chloride emissions in a municipal incinerator. Science of Total Env. 17, 155-163

(2) VDI Verein Deutscher Ingenieur (1976). Messen der Massenkonzentration von Partikelen in der Aussenluft/LIB verfahren. VDI 2463, Blatt 4, VDI-Verlag GmbH, Düsseldorf.

(3) International Standards Organisation (ISO) (1979). Air Quality - Measurement of atmospheric dustfall - Horizontal deposit gauge method. ISO/DIS 4222.2

(4) Pasquill F. (1961). The estimation of the dispersion of windborne material. Met.Magazine 90, 33-49

(5) R.Dams, J.A.Robbins, K.A.Rahn and J.W.Winchester (1970). Nondestructive neutron activation analysis of air pollutant particulates. Anal.Chem., 42, 861-869

(6) R.Heindryckx (1974). Anorganische verontreiniging door aerosolen in de Gentse industriezone. Ph.D. thesis, Rijksuniversiteit Gent (B).

(7) Stümke H. (1963). Vorschlag einer empirischen Formel für die Schornsteinüberhöhung in Auschuss und eine überprüfung bekannter Formeln mit zusätzlichem Beobachtungs-Material. Institut für Gasströmungen der Technischen Hochschule, Stuttgart (G).

EVALUATION OF THE INFORMATION FROM A CONTINUOUSLY WORKING

PRECIPITATION MONITOR

P. Winkler

Deutscher Wetterdienst, Meteorologisches Observatorium Hamburg
Frahmredder 95, D-2000 Hamburg 65

Summary

The discussion of the effects of acid deposition in the ecosystem has
shown that a statistical description of the frequency of critical values is needed rather than the knowledge of the average deposition. The
comparison of such frequency distributions of the pH-value based (1)
on daily averages and (2) on hourly values shows no difference in the
pH-values with the highest frequency but a higher standard deviation
for the hourly values by about 0.1 pH units. If the frequency distributions of the H^+-depositions of daily and hourly values are compared,
a shift of the pH-value at which H^+-deposition is highest, can be established.
In this context the relative contribution of acid to the total dissolved material in the course of single precipitation events is discussed. It can be shown that although the concentration of dissolved
material usually decreases during precipitation events of several
hours, the relative contribution of acid to the total dissolved material increases. This means that acid formation in clouds due to liquid
phase oxidation developes completely different as compared with the
removal of aerosol particles.

1. INTRODUCTION

Acid deposition is believed to cause most of the severe tree and water
damages observed in the environment. In order to discuss the water- or soil-
acidification it is sufficient to know the average values of the deposition
because the time constants of acidification are large. Plant damages at
their green parts can depend on critical values, however, and here the knowledge of average values is not sufficient. Even the derivation of a frequency
distribution of the pH-values of single precipitation avents brings not more
information, because the pH-changes during single events.
At the Meteorological Observatory of Hamburg an instrument has been developed (Winkler, 1977) continuously monitoring electrical conductivity,
pH-value, amount and intensity during precipitation events. The instrument
is a wet only collector free from perturbations due to dry deposition. The
records of this instrument are evaluated in order to derive frequency distributions. More-over the records can be evaluated with respect to the development of acidity during precipitation events. This in turn gives insight
into the scavenging process and some observational evidence can be presented
that the formation and scavenging of acid shows different behaviour than
that of other trace substances.

2. FREQUENCY DISTRIBUTIONS OF pH

The question: How often precipitation with a certain acidity is falling, can be answered by drawing up a frequency distribution of pH-values. In order to come to a conclusion with respect to the effects of acidity the question: At which pH-value do we find the maximum of H^+-deposition, becomes important. For this reason two types of frequency distributions are presented in fig. 1 and the comparison between a daily and hourly data basis is made.

The pH-frequency distributions for daily and hourly data basis do not differ very much. The maximal frequencies have been normalized to 100% for reasons of comparison. The most frequent pH-values for both distributions are nearly the same, the hourly pH-distribution is some what broader, however. Especially at the lower pH-values, the frequencies are up to 15% higher for the hourly pH-values than for the daily pH-values.

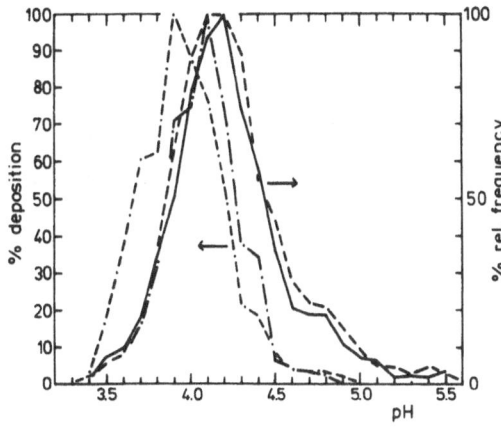

Fig. 1:
Frequency distribution of pH-values of precipitation of Hamburg.
——————— based on daily pH-average values (volume weighed),
------- based on hourly pH-average values,
—·—·— distribution of H^+-deposition of daily time basis,
—··—··— distribution of H^+-deposition of hourly time basis,
All distributions have been normalized to 100% at the maximum position.

On can conclude that the frequency distribution would become still broader, when a time basis of 1/4 hour for the pH would be taken. This is easily understood because in any case the frequency distribution of average values· is smaller than the distribution of the original values. The only slight differences between the daily and hourly distribution means that within a day no extreme pH-variation are to be expected. This is a consequence of the meteorological situation. The conditions leading to more or less acidification do not change dramatically within one day.

If we now ask for the deposition of H^+-ions as a function of the pH-value the sitiation changes because the distributions of precipitation amount and frequency of pH-values are not congruent. The deposition is defined as $[H^+]_i$. x RR_i, where RR is the precipitation amount and the subscript i means hourly resp. daily values. If we compare the two respective distributions in fig.1, the shift is more pronounced. Note that again the maximum deposition values have been normalized to 100%. We see that the maximum of the hourly distribution has shifted by 0.2 pH units and again the distribution is broader. Even the total distribution shifts to the acid side, when changing from a daily to an hourly time basis. If we compare absolute values of the deposition, the maximum of the daily deposition is higher by 20% than the maximum of the hourly deposition, while most of the other daily deposition values are lower than the corresponding hourly deposition values. Here we can say that the hourly deposition distribution of H^+-ions gives a more realistic picture than the daily deposition distribution.

3. ACID FRACTION OF DISSOLVED MATERIAL

The acid fraction is defined as the relative mass contribution of acid to the total dissolved matter. Its percentage is derived from the simultaneous measurement of the electrical conductivity and the pH-value of precipitation with the above mentioned precipitation monitor. The method has been described by Winkler (1980). In principle the amount of acid is estimated from the pH-measurement, assuming a composition of 70% H_2SO_4 and 30% HNO_3. The total dissolved material is derived from the electrical conductivity. Assuming an average equivalent weight of 35g and an average equivalent conductivity of 60 S/val, the amount of dissolved material is easily calculated. The influence of varying chemical composition is not very critical.

4. CASE STUDIES

With the continuously operating instrument the overall scavenging of H^+-ions, of the total amount of ions except H^+, the acid fraction and precipitation amount and intensity has been monitored. The instrument is free from influences due to dry deposition by opening its lid only during a precipitation event. The records have been evaluated every 0.05 mm of precipitation depth.

In the first case study the event was a frontal precipitation of several hours duration. Fig. 2 shows in the upper part the time variation of the concentration of H^+-ions and the non-H^+-ions. The concentration of the non-H^+-ions is high at the beginning and decreases rapidly at first and only slightly after the first hour. This is typical for aerosol scavenging as can be seen from the behaviour of heavy metals like Pb, Cd. Zn, Cu, Mn which have been analyzed by fractionated sampling (Kins, 1982; Nürnberg et al., 1982) and which are in the atmosphere bound to aerosol particles only. In contrast, the H^+-ions do not show such a pronounced decrease. Moreover, their

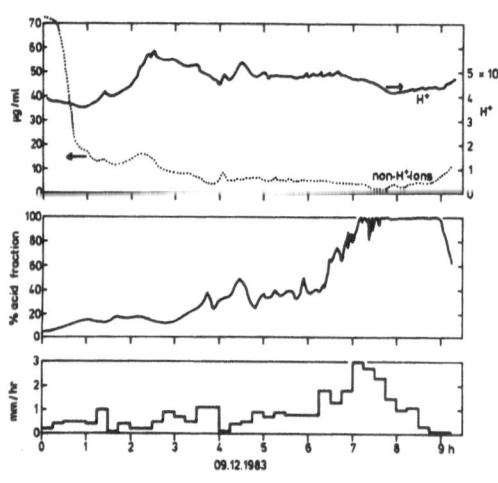

Fig. 2:
Frontal precipitation
Upper part: time variation of the H^+- and non-H^+-ions during a frontal precipitation event.
Middle: Time variation of the acid fraction of the dissolved material.
Lower part: Time variation of the precipitation intensity.

concentration increases during the second 1 1/2 hours and decreases only slightly for the rest of the time. Correspondingly the acid fraction shows low values at the beginning and increases towards the end of the event.

In fig. 3 another event but now with shower type precipitation is investigated.

During the two showers (1^{00} and 10^{00}), both, the H^+- and the non-H^+-ion concentrations decrease in the course of the event after a slight increase at the beginning. But again the non-H^+-ions decrease more quickly so that the acid fraction increases during the shower. The longer lasting precipitation between 4^{00} and 8^{00} is composed of several shower cells as can be seen at the marked intensity peaks. Correspondingly the concentration of H^+- and non-H^+-ions varies up and down more pronounced as in case 1. The acid fraction again shows more or less marked increases with each shower cell. After the cell has passed the acid fraction falls down to low values rather rapidly. So on a shorter time scale we see a similar development of the acid fraction as in case 1.

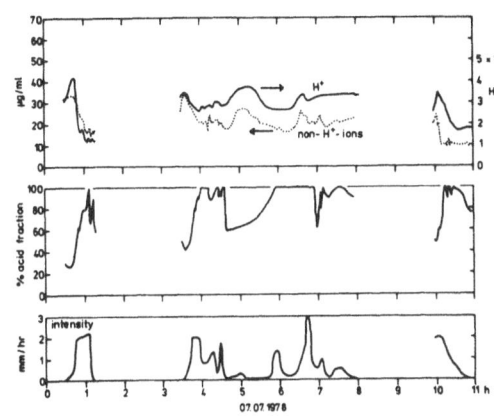

Fig. 3:
Showertype precipitation

(arrangement as in fig. 2)

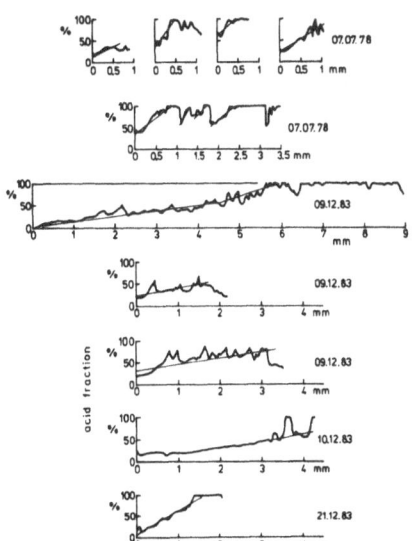

Fig. 4:
Development of the acid fraction as function of the precipitation depth for precipitation events from convective processes (7.7.78) and from upslide motion (9.-21.12.83).

In fig. 4 several cases are depicted with the development of the acid fraction as function of the precipitation depth during precipitation from convective processes or upslide motion..In almost all cases the acid fraction increases during the first time of the event. From the figs.2 and 3 it can be seen that in most cases the acid fraction becomes very high when the precipitation intensity reaches or exceeds values of about 2 mm/hr. Although cases with decreasing acid fraction have been observed, an increase with intensity seems to occur more frequently.

5. DISCUSSION

The observations of the behaviour of the acid fraction allows conclusions on the scavenging processes. From the figures 2 and 3 we have seen that the concentrations of H^+- and non-H^+-ions develope in different ways. If we make the rather likely assumption that SO_4 and NO_3 are the counterions of H^+ and if we assume the contribution of aerosol native H^+-ions to the measured H^+-ions to be small (Winkler, 1983) than the amount of non-H^+-ions corrected for the amount of the H^+-counterion is a good approach for the overall aerosol contribution to the dissolved material due to scavenging processes. This contribution generally is high at the beginning of a precipitation process and decreases rapidly as is confirmed by the heavy metal analysis during fractionated sampling (Kins, 1982; Nürnberg et al. 1982). This aerosol contribution can be approximated by the formula:

$$ac = \left[ec - (\alpha+\beta)10^{-pH} \right] \frac{ew}{\Lambda}$$

with ac = amount of aerosol native ions dissolved in precipitation; ec = electrical conductivity; ew = average equivalent weight of aerosol native ions (NH_4^+, CA^{++}, NA^+, Mg^{++}, K^+, NO_3^-, Cl^-, SO_4^{--}); Λ = average equivalent conductance of these ions; α = specific conductivity of H^+-ions; β = average specific conductivity of (60-70%) SO_4^- and (30-40%) NO_3^-.

It should be emphasized that this method does not hold for clean areas where the aerosol contribution predominantly determines the amount of the dissolved material in rain water.

The acid fraction most likely originates from scavenging of gas phase acids (HCl, HNO_3) and from liquid phase oxidation of SO_2 and NO_x. The increasing amount with time of the acid fraction can be interpreted in two ways: (a) The dissolved acid is partly neutralized due to uptake of NH_3 and the NH_3 amount of an air parcel is scavenged more rapidly than the acid forming gases. This may be due to the higher diffusion coefficient of NH_3 than of the other gases (Adewuyi and Carmichael 1982). (b) It is possible that liquid phase oxidation of gases and/or the gas phase oxidation of gases to acids with subsequent scavenging by cloud and rain drops has a longer time constant than the aerosol scavenging process. This means that aerosol particles are scavenged more rapidly than acid can be produced. Although the acidity forming gases decrease in their concentration during the precipitation event, this decrease is slower than aerosol scavenging. If this interpretation is right the influence of the intensity on the acid fraction can be understood. The precipitation intensity primarily is a function of the vertical velocity. Doubling the intensity means that an air parcel twice as large is lifted over a certain distance in the time unit. This means that the trace substances of a twice as large volume become available to scavenging. As was shown by Beilke (1970) the scavenging velocity for SO_2 and NO_2 is much higher than for aerosol particles, a result which can be transduced to the in-cloud scavenging. Assuming a constant oxidation rate for SO_2 the higher scavenging velocity for gases means an relative inscrease of the acid oxidation products over the removal of aerosol particles.

Literature

Adewuyi, Y.G., G.R. Carmichael: A theoretical investigation of gaseous absorption by water droplets from SO_2 - HNO_3 - NH_3 - HCl mixtures. Atm. Environment 16 (1982) 719-729.

Beilke, S.: Untersuchungen über das Auswaschen atmosphärischer Spurenstoffe durch Niederschläge. Ber. Inst. Met. Geoph. Univ. Frankfurt Nr. 19 (1970) 1-61.

Kins, L.: Temporal variation of chemical composition of rainwater during individual precipitation events. In: Deposition of Atmospheric Pollutants, eds. H.W. Georgii and J. Pankrath, Reidel Publishing Comp., Dordrecht 1983, pp. 87-96.

Nürnberg, H.W., P. Valenta, V.D. Nguyen: Wet deposition of toxic metals from the atmosphere in the Federal Rebublic of Germany. In: Deposition of Atmospheric Pollutants. eds. W. Georgii and G. Pankrath, Reidel Publishing Comp.,Dordrecht 1983,

Winkler, P.: Automatic analyser for pH and electrical conductivity of precipitation. In: Papers presented at the WMO Technical Conference on Instruments and Methods of Observation (TECIMO). Hamburg 1977, WMO-No. 480, Genf, pp. 191-196.

Winkler, P.: Observations on acidity in continental and in marine atmospheric aerosols and in precipitation. J. Geophys. Res. 85 (1980) 4481-4486.

Winkler, P.: Trend development of precipitation-pH in Central Europe. In: Acid Deposition 116-125; S. Beilke, A.J. Elshout (eds.), Reidel Publishing Company, Dordrecht, Holland 1983.

PRODUCTION ET TRANSFERT D'OZONE SUR LE BASSIN DE FOS-BERRE

P. PERROS et G. TOUPANCE
Laboratoire de Physico-Chimie de l'Environnement
Université Paris Val de Marne, Av. Gén. de Gaulle, F 94000 CRETEIL

Summary :

During the 1983 European International Campaign of Fos-Berre, the
local ozone pollution survey network (5 analysers) have been completed by measurements on 3 sites distant of about 30-50 km far from
the industrial area. A regional analysis of the set of data is performed with respect to meteorological parameters.
The Fos-Berre area appears to be one of the major source of photooxy-
dant pollution in the region ; however some other industrial or urban
sites are probably also involved. Medium range transport of O3 have
been evidenced and local peak concentrations have been tentatively
interpreted.

I. INTRODUCTION

La concentration industrialo-portuaire de FOS-BERRE (Sud de la
France) rassemble une forte capacité de raffinage, des pôles chimiques
et pétrochimiques, des aciéries, une cimenterie et une importante
centrale thermique. On trouvera par ailleurs une description du site
(1-6). La nature des effluents (SO2, NOx, hydrocarbures,...) et les
conditions climatiques sont très favorables à la formation d'oxydants
photochimiques: des concentrations atteignant 230 ppb de O3 ont ainsi pu
être observées dans l'est du bassin.
Lors de la VIème Campagne Européenne d'Etude de la Pollution
Atmosphérique qui s'est déroulée sur le bassin de FOS-BERRE du 6 au
14/06/83 nous nous sommes proposés d'évaluer le comportement de l'ozone à
l'échelle régionale. Nous voulions d'une part contribuer à la
compréhension des processus de formation de la pollution photo-oxydante
dans le bassin de FOS-BERRE, d'autre part vérifier dans quelle mesure les
zones industrielles du bassin agissent comme puits piégeant l'ozone
préexistant, ainsi que cela a été montré dans certains cas pour l'Europe
du Nord-Ouest (2 - 3), ou comme source d'ozone, formé au sein du panache
et entrainé sur les zones sous le vent, ainsi que cela a été bien étudié
dans le cas du corridor nord-est des USA (4 - 5) et enfin de mettre
en évidence d'éventuels phénomènes de transport.

II. CHOIX DES SITES D'ETUDE

Le programme de la campagne européenne était principalement axé sur
l'étude de l'étang de Berre et s'étendait sur une zone d'environ 30x30
km. Un important potentiel de moyens, tant météorologique que
physico-chimique, a été exploité sur ce secteur au cours de la campagne.
En outre le site comporte en permanence deux réseaux bien équipés de
surveillance de la pollution, AIRFOBEP sur le bassin Fos-Berre et
AIRMARAIX pour la région d'Aix-Marseille.

Nous avons choisi de compléter ce dispositif par trois sites de
mesure d'ozone de type rural, éloignés du centre du bassin, susceptibles
d'être influencés par les rejets de la zone industrielle et permettant
de suivre l'évolution de la composition en O3 du panache:
1. Baux de Provence: transferts par régime de S à SE. L'analyseur
était implanté dans la chapelle désaffectée du parc du château
historique.
2. Saint Cannat et 3. Sommet du Lubéron: transferts par régime de
SW. L'analyseur du Lubéron était implanté dans l'enceinte d'un relais de
télécommunications. L'analyseur de Saint Cannat était implanté à 20m par
rapport au sol dans le clocher de l'église. Les conséquences de ce choix
sont discutées plus loin (IV-2).
Le tableau suivant réunit les principales caractéristiques des 8
sites de mesure d'ozone exploités dans cette étude.

1	2	3	4	5	6
SAINTE BAUME	40	ESE	690	SBA	AIRMARAIX
LUBERON	40	NE	1125	LUB	UPVM
BAUX DE PROVENCE	40	NW	280	BAU	UPVM
ST CANNAT	15	NW	220	SCA	UPVM
VITROLLES	0	E	40	VIT	AIRFOBEP
ROGNAC	0	E	5	RGN	AIRFOBEP
PORT DE BOUC	0	SW	68	PDB	AIRFOBEP
MARSEILLE	15	SE	35	MAR	AIRMARAIX

Légende: 1: Nom du site - 2 et 3: distance (km) et direction par
rapport aux bords de l'étang de Berre - 4: altitude du point de mesure /
niveau de la mer - 5: abréviation utilisée sur les figures - 6: organisme
(AIRMARAIX et AIRFOBEP = réseaux de surveillance; UPVM = Université Paris
Val de Marne).
La localisation exacte des sites est représentée sur les figures 3
et 4 (par ex. fig 3, le 6 à 6h).

III. METROLOGIE ET METEOROLOGIE

Toutes les mesures d'ozone ont été réalisées par absorption UV à
l'aide d'appareils ENVIRONNEMENT SA, type 1003 AH. Les appareils UPVM et
AIRFOBEP ont été intercalibrés par titration en phase gazeuse. Nous
n'avons pas vérifié s'il en avait été de même pour ceux d'AIRMARAIX.
Nous avons disposé des données météorologiques et métrologiques
(SO2, NO, NO2, NMHC, CH4, particules) fournies par les réseaux de surveillance
permanents et de celles des autres équipes participant à la campagne et
qui ont pu nous être transmises. Nous avons complété le jeu des données
météorologiques rassemblées dans le cadre de la campagne, par les
vitesses et directions de vent acquises en dehors de la zone d'étude par
le réseau permanent de la météorologie nationale et par le réseau de
surveillance des feux de forêts.
La caractérisation météorologique du site et la description des
situations rencontrées au cours de la campagne sont données par AUGUSTIN
et BESSEMOULIN (6).

IV. RESULTATS ET INTERPRETATION

Nous présentons ici une première analyse des résultats obtenus.

Les profils diurnes de O3 relevés sur la période du 3 au 14/06/83 sont donnés figure 1 et 2. Les heures sont en temps universel et les concentrations en ppb (1ppb = 2 microgrammes par m3 pour O3). On note l'existence de pics brefs et intenses le matin entre 10 et 11h dans l'est du bassin (Vitrolles, Rognac, St Cannat), de même que des pointes brèves sur les sites naturels des Baux et du Lubéron. Les figures 3 et 4 représentent quelques-unes des situations typiques relevées sur le site.

Ces résultats peuvent être, dans leurs grandes lignes, interprétés à travers quatre paramètres principaux.

1- Niveau de fond: Les sites naturels (BAU, LUB, SBA) mettent nettement en évidence un niveau de fond, durant la campagne, de 50 ppb environ. Ce niveau est rencontré ici couramment entre 300 et 1200m quelles que soient les conditions météorologiques. La situation observée le 6/06 à 6h (fig 3) est significative d'une situation nocturne: valeurs de fond sur les trois sites naturels; valeurs très faibles sur l'ensemble du bassin industriel. Sur ces trois sites, on n' observe pas de cycle diurne notable à l'exception d'accidents spécifiques qui seront discutés plus loin (IV-4).

2- Sources locales de pièges pour O3 (réducteurs comme NO): L'analyseur de Marseille est situé au coeur de l'agglomération. Le débit de NO rejeté par le trafic apparait être en permanence supérieur à celui de production photochimique de O3: le niveau de O3 reste faible, ainsi que cela a déjà été noté pour le coeur d'autres agglomérations (3 -7 -8). Nous n'avons pu controler si la valeur nocturne de 20 ppb (reproductible et probablement trop forte selon nous) était significative ou non.

Rognac est situé au niveau de la mer, à proximité immédiate de la zone industrielle de Berre. La forte stabilité nocturne de l'atmosphère en ce point, confine les rejets industriels dans les couches inférieures et conduit durant la nuit à la consommation totale du stock d'ozone existant: les concentrations nocturnes de O3 à Rognac sont souvent de l'ordre de 5 ppb seulement.

Au contraire, l'implantation du capteur de Port de Bouc à une altitude de plus de 60m le place dans une tranche d'atmosphère moins approvisionnée durant la nuit en composés pièges de O3: parmi les sites industriels, c'est celui qui connait les concentrations nocturnes de O3 les plus élevées. Vitrolles situé à 40m dans un site résidentiel, bien qu'à proximité immédiate de Rognac, connait une situation nocturne intermédiaire.

Les valeurs nocturnes très faibles enregistrées à St Cannat confirment cette analyse, bien que, paradoxalement, il s'agisse d'un site rural. L'expérience a montré en effet que, malgré les précautions prises, le point de prélèvement était soumis aux émanations réductrices du guano accumulé par les oiseaux nichant dans le clocher. La nuit, du fait de la forte stabilité, tout l'ozone était consommé avec en conséquence l'enregistrement de niveaux pratiquement nuls (fig. 1 et 2). Des mesures de contrôle effectuées de jour comme de nuit les 14 et 15/06 ont montré que la teneur observée au point de prélèvement était d'environ 20 ppb inférieure à celle mesurée sur la place de l'église. Il est probable qu'au cours des journées sans mistral la différence était plus importante. Devant la difficulté d'apporter une correction pertinente, nous avons préféré donner les résultats bruts obtenus à St Cannat. Les valeurs réelles étaient vraisemblablement supérieures de 20 à 40 ppb.

L'ensemble de ces résultats montre que l'intensité des minimums observés sur les profils de O3 sont pour une large part attribuables à l'intensité des sources locales de composés réducteurs susceptibles de piéger O3.

3- Convection; hauteur de la couche de mélange: La teneur au sol en O3 est pour une large part gouvernée par la convection (3 -2 -8 -10). C'est à une convection nocturne négligeable, associée à des sources locales de réducteurs, que nous venons d'attribuer les faibles niveaux d'O3 observés la nuit sur plusieurs sites. En cours de journée, la convection a deux effets contraires: le premier est de ramener au sol des masses d'air qui peuvent être riches en ozone; c'est ce qui se produit le matin. Le second est de diluer les polluants primaires émis par les sources industrielles dans une couche de mélange de hauteur croissante au cours de la journée et d'influer ainsi directement sur les débits volumiques de production photochimique de O3. Nous n'analyserons pas ici en détail ces processus; toutefois, nous les évoquerons en IV-4 pour interpréter les pics observés à Vitrolles et à Rognac.

Bien que la pertinence de l'utilisation de la hauteur de la couche de mélange pour interpréter les niveaux de pollution observés ait été récemment discutée (11), nous avons cherché si une relation raisonnable pouvait être dégagée sur le site étudié durant la campagne.

Nous avons ainsi porté sur la figure 5 la moyenne des concentrations de O3 relevées de 6h à 19h sur l'ensemble des sites industriels de la région (PDB, VIT, et ROG), en fonction de la hauteur de la couche, de mélange. Celle-ci a été obtenue à partir des données acquises lors de la campagne, à La Suzanne par l'équipe HASENJSEGER d'ISPRA. Nous avons calculé la moyenne arithmétique de la hauteur de la couche de mélange durant la période de mesure, en éliminant deux valeurs apparemment aberrantes (une valeur nulle le 6 et un pic bref le 12).

On observe un alignement raisonnable des points qui indique que l'essentiel de l'ozone observé est formé au cours de la journée au sein de la couche de mélange. Deux points (les 8 et 11) s'écartent de la relation générale. La situation du 8 est caractérisée par une valeur relativement trop forte de O3. Elle fait suite à un courant d'est de 3 à 5 m/s pendant la nuit du 7 au 8 et qui a gêné la formation de l'inversion de rayonnement (12). Ce phénomène, unique pendant la campagne, a pour effet de favoriser la dispersion du monoxyde d'azote émis pendant la nuit. L'ozone aurait ainsi pu atteindre des valeurs plus élevées sur tous les sites (fig. 1 et 2). La production photochimique au cours de la journée du 8 se cumulerait alors aux valeurs préexistantes. Au contraire, la situation nocturne du 11 est classique (advection faible, forte stabilité) tandis que la journée est marquée par l'absence de renverse de brise. Dès le matin on relève un flux permanent de NW (5 à 6 m/s) qui renouvelle en permanence l'air sur le site permanent : la concentration d'ozone est voisine de la concentration de fond et sans rapport direct avec la hauteur de la couche de mélange.

4- Transport: Les observations menées sur les trois sites naturels mettent nettement en évidence des transports d'ozone.

La figure 3 représente l'évolution des concentrations sur le site au cours de la journée du 6. Le régime général est établi au SW dès 10h. Le panache chargé d'ozone qui est observé à Vitrolles à 10h est entraîné en cours de journée vers St Cannat où il est observé vers 11h-12h pour atteindre le Lubéron vers 15h. A ce moment c'est au sommet du Lubéron qu'est observée la plus forte concentration d'ozone sur l'ensemble du site. Cette situation durera jusque vers 18h. Il faut noter que l'observation d'un transport au sommet du Lubéron suppose un développement de la couche de mélange sur 1000m. Dans tous les cas ce phénomène doit retarder l'heure d'observation du maximum au sommet par rapport à la durée normale du transfert. Dans certains cas il est possible que le panache reste bloqué à plus faible altitude.

Simultanément on observe un pic brutal de O3 aux Baux à midi (fig. 1). La finesse du pic et le fait que les Baux ne sont pas sous le vent de l'étang de Berre depuis 9-10h nous a amené à reconstituer la trajectoire des masses d'air par l'exploitation de l'ensemble des données de vitesse et de direction de vent au sol dont nous avons pu disposer. La figure 6 représente les trajectoires des masses d'air qui atteignent les Baux respectivement à 10h, 12h, 14h. On constate que la masse d'air qui a transité lentement au cours de la nuit sur les sites industriels du bassin de Fos-Berre, atteint les Baux à midi. L'ozone observé provient pour une large part des processus photochimiques au sein du panache tout au cours de la matinée. Du fait de la rotation rapide des vents lors de l'établissement du régime de brise de mer, cette situation est très éphémère: plus tôt les Baux sont alimentés en air provenant de l'est; plus tard c'est de l'air marin non pollué qui atteint le site.

La situation aux Baux le 7 est très différente (fig. 1).La direction moyenne du vent (6) est fixée au S.E. Le site est sous le vent du bassin Fos-Berre et des concentrations nettement supérieures à la concentration de fond sont observées toute la journée: le profil est large et des concentrations de 100 ppb et plus sont observées pendant plusieurs heures. Simultanément aucun accident marqué n'est observé au Lubéron. St Cannat n'est pas sous le vent du bassin Fos-Berre; les concentrations observées sont nettement plus faibles que les autres jours.

La figure 4 représente la situation observée le 9/06. Celle-ci est caractérisée par un régime d'ouest, un épisode bref de brise d'étang à 9h et une très forte concentration d'ozone à Vitrolles et Rognac à 10h.

Au cours de la période étudiée, le 9 est le seul jour où un régime d'ouest est observé. C'est aussi le seul jour où une concentration élevée d'O3 est observée à Sainte Baume. La figure 4 met nettement en évidence le transit de la masse d'air chargée en O3 depuis Vitrolles jusqu'à Sainte Baume. De 14 à 16h c'est à Sainte Baume que sont observées les concentrations d'O3 les plus élevées sur l'ensemble du site.

De même que ci-dessus, la figure 7 retrace l'histoire de la masse d'air observée à Sainte Baume à 14h. Malgré les aléas d'une telle reconstitution (brises très instables la nuit, mauvaise couverture météorologique dans la zone de transit nocturne) il semble que la masse d'air a transité une première fois sur l'étang de Berre dans la journée du 8, puis au cours de la nuit (vers 0h) au-dessus du centre industriel de Gardanne (entre Aix en Provence et Marseille), pour repasser une 2è fois sur l'étang de Berre au début de la matinée du 9. La brise d'étang observée à 9h (12) est nette sur la figure 4. Elle concerne tout l'étang. Il lui est vraisemblablement associée une subsidence au centre de l'étang. On peut supposer que la masse d'air qui arrive vers 8h en altitude au-dessus de l'étang se trouve réintroduite au niveau du sol par cette subsidence. La très forte concentration d'ozone observée à Vitrolles-Rognac pourrait avoir été photoproduite en altitude dès les premières heures de la matinée au sein de la couche d'air chargée en précurseurs une première fois le 8 sur l'étang et une deuxième fois la nuit au-dessus de Gardanne. La brièveté de l'épisode serait liée d'une part à la brièveté du phénomène de brise d'étang et d'autre part à la dilution dans la couche de mélange du fait d'une convection croissante au cours de la matinée.

Il semble que des situations semblables puissent être invoquées pour interpréter de nombreux pics observés à Vitrolles, Rognac et peut-être à St Cannat. La situation comparée des 9 et 10 montre en effet que l'épisode aigu est très localisé et concerne soit le groupe Vitrolles-Rognac, soit St Cannat.

Il faut enfin mentionner que des teneurs notables de 03 sont observées aux Baux dans la journée du 9 et du 10. Il ne semble pas que des transferts depuis le bassin de Fos-Berre puissent rendre compte des observations. Il faudrait sans doute étudier en détail les sources situées plus à l'ouest.

V. CONCLUSION

Les profils d'ozone observés du 3 au 14/06/83 sur l'aire d'étude résultent de la compétition de plusieurs facteurs: niveau de fond, sources de pièges à ozone, sources de précurseurs, photochimie, convection, transport.

En début de matinée les couches situées en altitude (typiquement 100 à 300 m) et chargées en précurseurs sont le siège d'une photochimie intense avec forte production d'oxydants. Au niveau du sol l'accumulation durant la nuit d'un important stock de réducteurs ralentit la production d'ozone. Les pics d'ozone se produisent lorsque l'augmentation de l'épaisseur de la couche de mélange injecte dans celle-ci l'ozone produit en altitude. L'homogénéisation ultérieure par convection ainsi que le déplacement des masses d'air sont responsables de la disparition du pic. Ces effets peuvent être accentués en fonction de l'histoire des masses d'air et de phénomènes spécifiques comme la brise d'étang.

Des transports complexes à longue distance ont été observés; ils montrent que des niveaux très élevés d'ozone peuvent être observés en des localités éloignées de toute source de précurseurs de polluants photochimiques.

Nous nous proposons d'approfondir l'analyse des processus de transfert lorsque les champs de vent détaillés sur l'ensemble de la zone seront disponibles.

Remerciements: Cette étude a été soutenue par le Ministère de l'Environnement, contrat No 83 133.

VI. REFERENCES

(1)- PERRIN M.L., MADELAINE G., FRAMBOURT C., Third European Symposium on Physico-chemical Behaviour of Atmospheric Pollutants, Varese, Italy, 10-12/04/84, PREPRINTS Volume, p 1.

(2)- SCHERER B., STERN R., in proceedings of Second Symposium on Behaviour of Atmospheric Pollutants, Varese, Italy, 29/09 -1/10/81, D. REIDEL Pub. Co edit., 1982, p 561.

(3)- Van DUREN H., ROMER F.G., in proceedings of Second Symposium on Behaviour of Atmospheric Pollutants, Varese, Italy, 29/09 -1/10/81, D.REIDEL Pub. Co edit., 1982, p 460.

(4)- CLEVELAND W.S., KLEINER B., Mc RAE J.E., WARNER J.L., Science, 191, 179-181, 1976.

(5)- CLARK T.L., CLARKE J.F., Atmospheric Environment, 17, No 2, 287-297, 1984.

(6)- AUGUSTIN H. et BESSEMOULIN P., Third European Symposium on Physico-chemical Behaviour of Atmospheric Pollutants, Varese, Italy, 10-12/04/84, PREPRINTS Volume, p 541.

(7)- BENARIE M., BENECH A., CHUONG B.T., MENARD T., Pollution Atm., No 81, 44-52, 1979.

(8)- TOUPANCE G., Rapport de contrat No 78 142, Ministère de l'Environment, France, 1981.

(9)- MIZUNO T., YOSHIKADO H., Atmospheric Environment, 17, No 12, 2575-2582, 1983.

(10)- LOPEZ A., PRIEUR S., FONTAN J., KIM P.S., in proceedings of Second European Symposium on Behaviour of Atmospheric Pollutants, Varese, Italy, 29/09 -1/10/81, D. REIDEL Pub. Co edit., 1982, p 362.

(11)- ARUN R., Atmospheric Environment, 17, No 11, 2193-2197, 1983.

(12)- SOL B., Résumé des conditions météorologiques au cours de la campagne Fos-Berre, polygraphié, 3p, juin 1983.

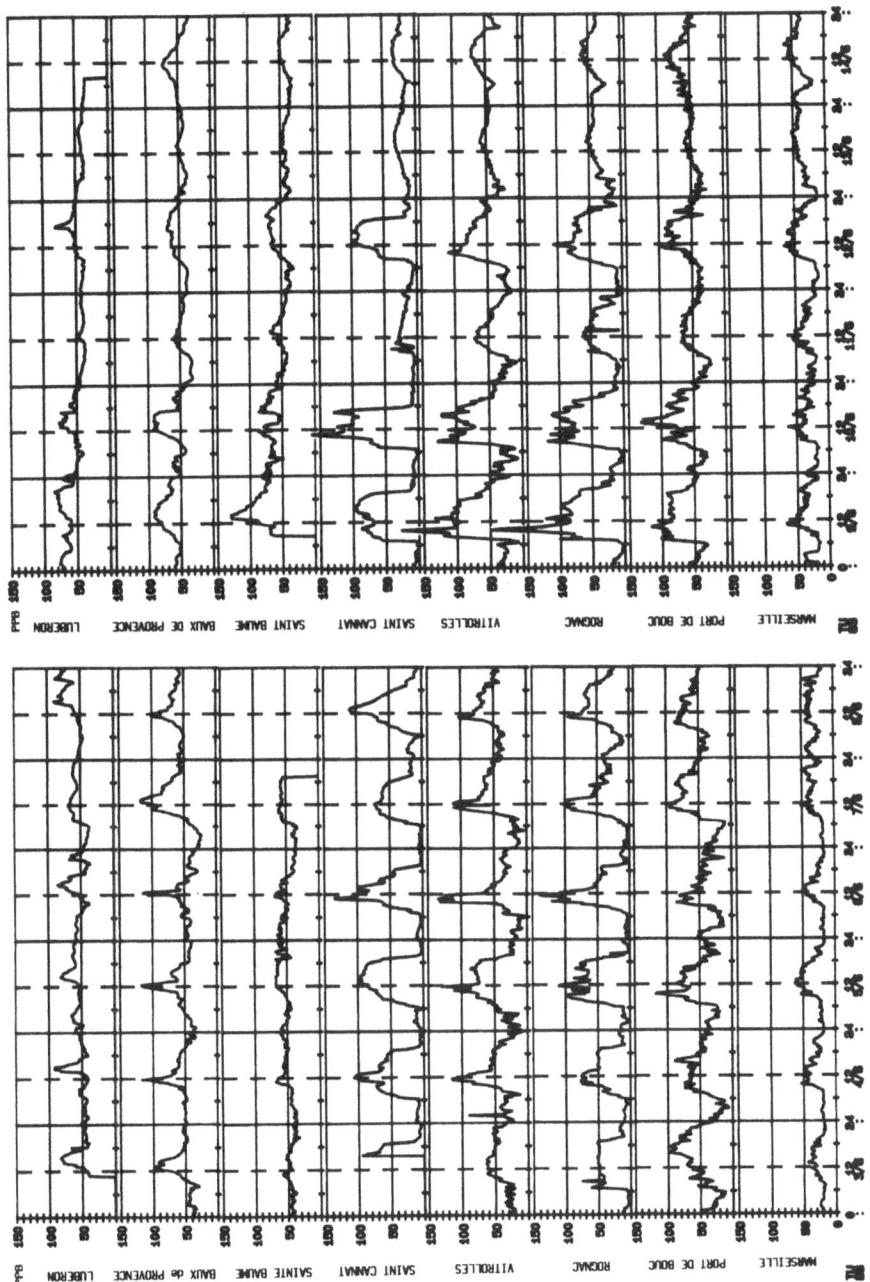

Fig.2

Fig.1

Figure 1 et figure 2 : Profils diurnes d'ozone du 3/6/83 au 14/6/83.

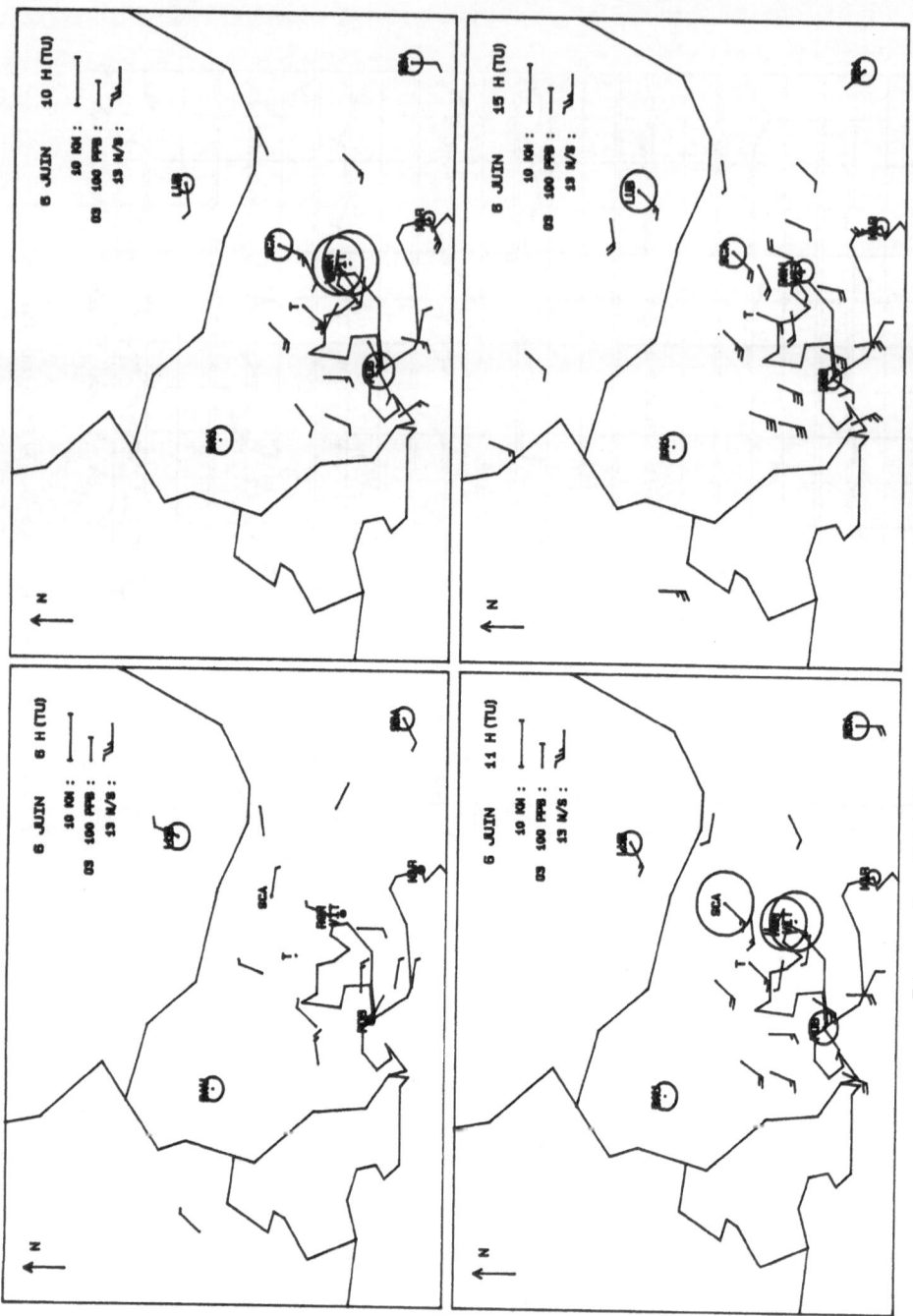

Figure 3 : Répartition spatiale de O3 le 6/06/83

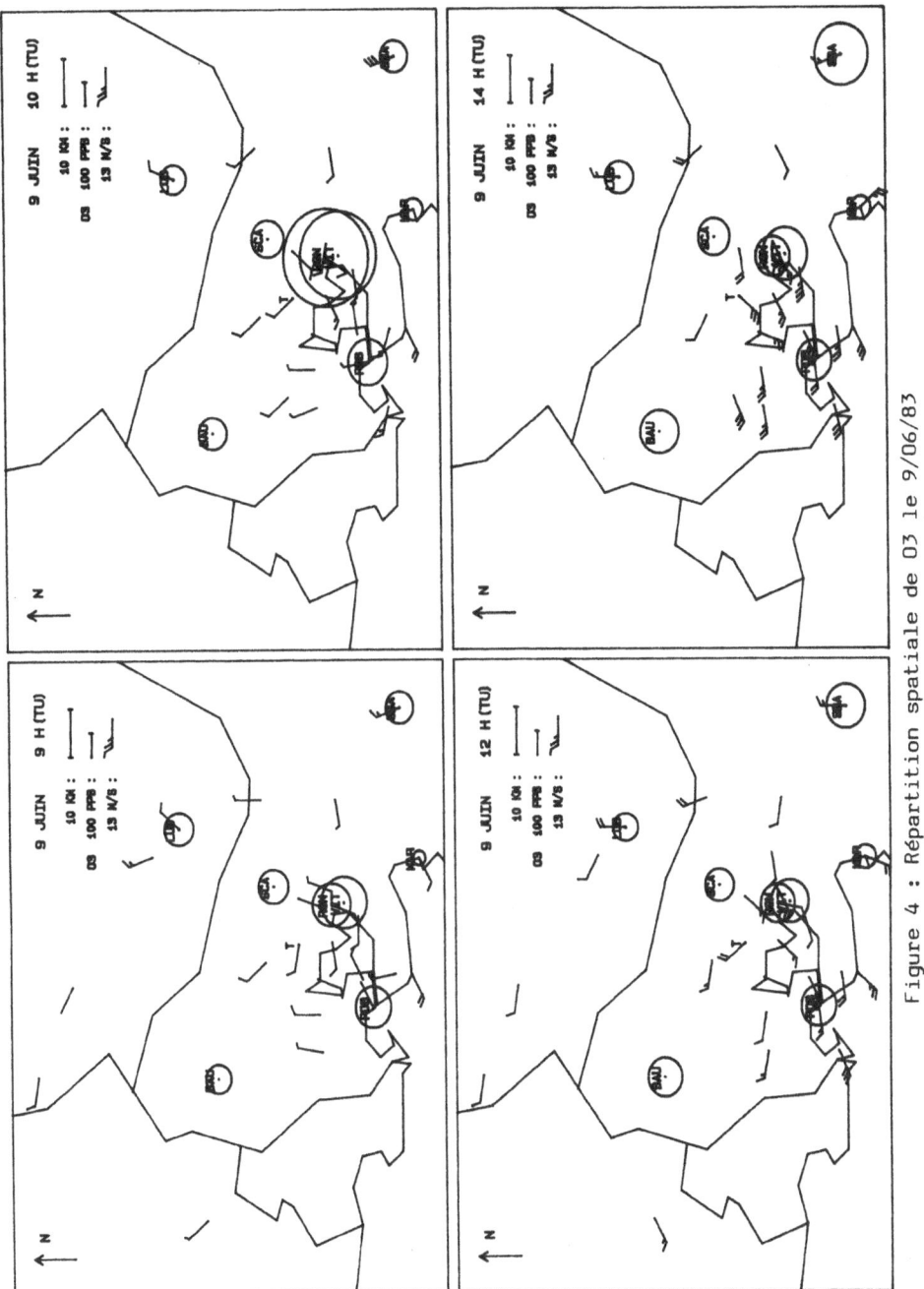

Figure 4 : Répartition spatiale de 03 le 9/06/83

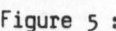

Figure 5 :

Relation entre la concen-
tration moyenne de O3 et
la hauteur moyenne de la
couche de mélange pour
les dates comprises
entre le 6 et le 12/6/83.

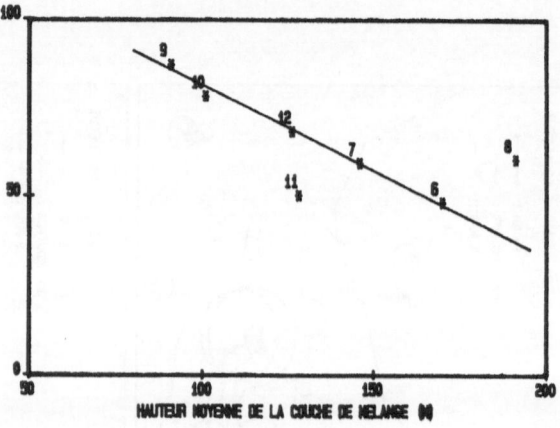

Figure 6 :

Trajectoires des masses
d'air qui parviennent aux
Baux-de-Provence à 10 h.
(← →, 12 h. (——),
14 h. (— - -) le 6/06/83.

**
* Zones industrielles

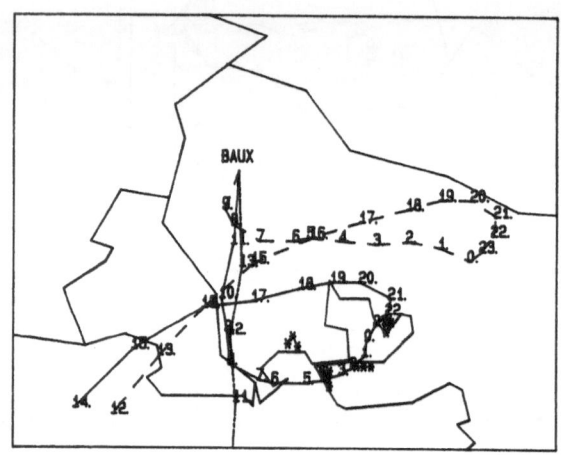

Figure 7 :

Trajectoire de la masse
d'air qui parvient le
9/06/83 à Vitrolles à
10 h. et à Sainte Baume
à 14 h.

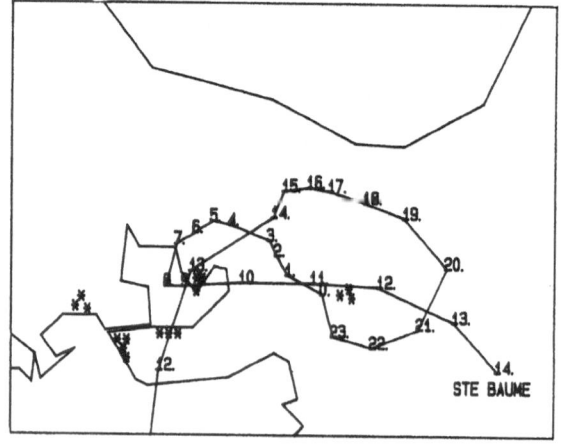

COMPARISON BETWEEN THE SCAVENGING RATIOS FOR NITRATE AND SULPHATE AT A RURAL SITE

M. Ferm
Swedish Environmental Research Institute
P.O. Box 5207, S-402 24 Gothenburg, Sweden

Summary

A study of the scavenging ratios during 35 days with precipitation was made at a remote station on the Swedish west coast. Ground level concentrations of HNO_3, SO_2 and particulate NO_3^- and SO_4^{2-} were measured on a diurnial basis. Precipitation was collected on the same time basis and analyzed for NO_3^-, SO_4^{2-} and Cl^- by ion chromatography. The concentration of H^+ was determined from coulometric titrations using Gran extrapolation and NH_4^+ was analyzed with an ion specific electrode. Different assumptions were used to calculate the scavenging ratios. The most consistent data were obtained when it was assumed that HNO_3 and particulate nitrate were scavenged with equal efficiency and the oxidation rate of SO_2 in rain was so low that its concentration in precipitation was mainly determined by its solubility. The average concentration of SO_2 in ground level air was higher on wet than on dry days. This supported the theory that the oxidation rate of SO_2 in aqeous phase is low. Simultaneously determined scavenging ratios for nitrate and particulate sulphate were for simultaneous data equal within a factor of two and varied with a factor of 30. During winter the precipitation amount was the most important factor determining the scavenging ratio. HNO_3 and SO_2 could not alone explain the acidity in rain. The missing part was well correlated with particulate sulphate.

1. INTRODUCTION

A one year monitoring programme of nitrogen and sulphur compounds has been performed on the Swedish west coast. The station is situated 40 km south of Gothenburg. Some results concerning the concentrations in air have previously been presented (1). HNO_3, NO_2, PAN, SO_2 and particulate NO_3^-, SO_4^{2-} and soot were measured on a diurnal basis (6.00-6.00 GMT). At this site precipitation measurements within the European Monitoring Network (EMEP) were performed on the same diurnal sampling time bases. The concentrations of H^+, NH_4^+, Cl^-, NO_3^- and SO_4^{2-} were determined togheter with measurements of pH and precipitation amount. The aim of this project was to see if there is any relationship between the ambient ground level concentrations and the concentrations in precipitation.

2. ANALYTICAL METHODS

Nitric acid and particulate nitrate were sampled with a combined denuder and impregnated filter technique (2). Na_2CO_3 was used as coating. After sampling the denuder and filter were leached in water and nitrate was analyzed using ion chromatography (IC). The method may give too low nitric acid values since the sampling flow used was adapted to the diffusion coefficient for HNO_3 in dry air. In humid air the coefficient may be lower due to agglomeration of water molecules on the HNO_3 molecule. The sum of the nitric acid and particualte nitrate concentrations will not be affected by this. NO_2 was analyzed with a modified Saltzman technique and PAN by gas chromato graphy. Sulphur dioxide was trapped in a bubbler containing acid hydrogen peroxide which was subsequently analyzed by IC. Particulate sulphate was collected on a paper filter mounted in front of the bubbler and was analyzed with X-ray fluorescense. Acidity in rain was analysed by using coulometric titration with Gran extrapolation (3). Ammonium was analyzed with a gas sensitive electrode (Orion 95-10). Chloride, nitrate and sulphate were analysed by IC.

3. SCAVENGING RATIOS

The scavenging ratio is defined as the concentration of a compound in precipitation divided by the concentration of its precursor in the air at ground level. Here the concentrations were measured on a diurnal basis. It implies that the concentrations in air and precipitation are not averaged over the same period of time. The air concentrations before, during and after a rainfall were thus frequently included in the mean value. This can be advantageous if the concentration in air drops drastically during precipitation. No information about such a concentration drop came out of the project, but a decrease on a 24 hour basis was observed and will be discussed later. The scavenging ratio is a dimension less feature but the value depends on the units chosen for the concentrations. In this context the "concentrations" are often given in ug/g. Here they are expressed on a volume basis (e.g. nmoles/m^3) which are more common in all other situations. To get the values here into the ug/g ratio they should be multiplied with the density of the air (1.2 kg/m^3) and divided by the density of the precipitation (1000 kg/m^3).

To define the scavenging ratio the precursors must be known. This is difficult because they often exist in different forms in precipitation and in air. There can also be more than one form in the air. Then their individual contribution to the concentrations in precipitation has to be estimated. In the hypothesis tested the following assumptions were made.

Nitrate in precipitation originates from both gaseous nitric acid and particulate nitrate. Sulphate comes from sulphur dioxide and particulate sulphate. Hydrogen ions come from nitric acid and sulphur dioxide and acid sulphate particles. This can be written

$$[NO_3^-]_p = R_{HNO_3} [HNO_3]_g + R_{NO_3^-} [NO_3^-]_a \tag{1}$$

$$[SO_4^{2-}]_{p,ex} = R_{SO_2} [SO_2]_g + R_{SO_4^{2-}} [SO_4^{2-}]_a \tag{2}$$

$$[H^+]_p = R_{HNO_3} [HNO_3]_g + 2R_{SO_2} [SO_2]_g \quad \delta R_{SO_4^{2-}} [SO_4^{2-}]_a \tag{3}$$

where index p stands for precipitation, ex for excess, g for gas and a for aerosol phase. δ denotes the molar fraction of hydrogen ions associated with particulate sulphate. There are three independent equations containing five unknown coefficients. Different assumptions have been made to solve them which will be discussed later.

The sulphate not originating from sea salt is usually called excess sulphate. It is difficult to estimate the concentration of sea sulphate by multiplying a measure of the salt concentration (e.g. Mg^{2+}) in precipitation by the relative abundance of sulphate in the sea, because sulphate to salt ratio in ambient aerosols may be different from that in the sea. The station is situated close to the sea and the salt concentration was often high. There was, however, no other alternative than to make correction from chloride which was the only sea salt component measured. Correction was made with the formula

$$[SO_4^{2-}]_{p,ex} = [SO_4^{2-}]_p - 0.0516 \times [Cl]_p \tag{4}$$

Other compounds than those discussed here may also appear in the air. There may for instance be hydrochloric acid, carbonate particles and sodium nitrate. Only data sets with a fairly good ion balance between H^+, NH_4^+, NO_3^- and excess SO_4^{2-} were therefore used to test the hypothesis. The condition can be described by eq. 5:

$$0.8 < ([H^+]_p + [NH_4^+]_p) / ([NO_3^-]_p + 2[SO_4^{2-}]_{p,ex}) < 1.2 \tag{5}$$

The acid in precipitation seemed to mainly come from strong acids because there was a very good relationship between the measured hydrogen ion concentration and 10^{-pH}.

Acid particles are very rare at the ground level. They have mainly been observed in connection with episodes of high particle concentration (5). Very little is know about their abundance at higher altitudes. In the first approximation it was assumed that δ in equation 3 was zero. If particulate nitrate mainly consists of NH_4NO_3 it can be scavenged out via the gas phase. When HNO_3 is dissolved in water droplets the gas phase concentration of HNO_3 in equilibrium with the droplet will be very samll (6). Particulate NH_4NO_3 will then dissociate to HNO_3 and NH_3. It was therefore assumed that R_{HNO_3} was equal to $R_{NO_3^-}$. Using these approximations there are now three coefficients to solve with three indeptendent equations. In many cases the HNO_3 sampling failed due to dissolution of the coating by small water droplets. This happened more frequently during days with rain. At 62 occasions the sampling and analysis of all parameters in equations 1-3 were successful and in 38 occations of those the condition in equation 5 was fulfilled. Using this model it was found that 64% of the excess sulphate was due to the dissolution and oxidation of sulphur dioxide. There was a very poor correlation between $R_{SO_4^{2-}}$ and $R_{NO_3^-}$ and the data seemed inconsistant in comparison with other parameters. A clooser examination of the data showed that there did not seem to be higher SO_4^{2-} concentrations in precipitation when there were high SO_2 concentrations when days with similar precipitation amounts and particulate sulphate concentrations were compared.

Table 1 gives the average concentrations of the different species in rain and ground level air. From the table it is obvious that SO_2 has a small influence on the excess sulphate. The next approach used was to assume that the oxidation of SO_2 in water is slow and that the SO_4^{2-} coming from SO_2 is equal to the solubility of SO_2 in rain water.

Pena et al. (7) measured the HSO_3^- concentration in rainwater using the "TCM" method and found that the observed concentration was close to the equilibrium concentration. They used the formula

$$[HSO_3^-] = \frac{K_H \times K_I \times P_{SO_2}}{[H^+]_p} \tag{6}$$

in which P_{SO_2} is the partial pressure of SO_2 in atm. K_H is the distribution constant calculated from

$$\log K_H = \frac{1373.9}{T} - 4.159 \tag{7}$$

where K_H is given in moles per kg H_2O and atm. and K_I is the thermodynamic dissociation constant for the first step.

$$\log K_I = \frac{868.3}{T} - 4.805 \tag{8}$$

where K_I is expressed in moles/liter.

The temperature at ground level during precipitation together with the bulk acidity for the whole rain were used. By replacing R_{SO_2} $[SO_2]$ in eq. 2 with $[HSO_3^-]$ calculated from eq. 6 and assuming that $R_{HNO_3} \approx R_{NO_3^-}$ the equation system 1 to 3 could be solved again. The mean HSO_3^- was 6.8 μM during winter and 0.7 μM during summer. At very low temperatures $R_{SO_4^{2-}}$ was negative. This only happened at three occasions. A plot of $R_{SO_4^{2-}}$--versus $R_{NO_3^-}$ is given in figure II. Assuming a log normal distribution a standard deviation of 58% was obtained if three outliers were excluded. This is shown as dashed lines in figure II. A standard deviation of 93% was obtained for the whole material.

It is a fairly good correlation considering that $R_{NO_3^-}$ varies with a factor 24 and $R_{SO_4^{2-}}$ with a factor 34. Both the concentrations in precipitation and ambient air could vary with more than one order of magnitude for the similar scavenging ratios because the ambient concentrations varied so much with wind trajectories.

During winter the precipitation amount was the most important factor which influenced the R values. Figure III shows the correlation between R and precipitation amount P. The empirical relations obtained from this plot were
$R_{NO_3^-} = 1.87 \times 10^6 \times P^{-0.87}$ and $R_{SO_4^{2-}} = 1.67 \times 10^6 \times P^{-0.82}$. During the summer this plot gave very scattered points indicating that other parameters had a big influence on R. High precipitation amounts could give high scavenging ratios and vice versa but $R_{SO_4^{2-}}$ and $R_{NO_3^-}$ followed each other quite well during the summer, see Figure II.

The arithmetic means of $R_{NO_3^-}$ and $R_{SO_4^{2-}}$ were during the winter 2.25×10^6 and 1.94×10^6, rescpectively and during the summer 2.24×10^6 and 2.39×10^6, respectively. δ was fairly constant and equal to 1.3 ± 0.6 during winter and 1.5 ± 0.6 during summer. It seems as if the acidity is to a great extent associated with the particulate sulphate. It may for instance be HCl coming from the same sources as sulphate.

However, there is not enough ammonium ions to cover two times the excess sulphate concentration and some NH_4^+ will probably come from gasous NH_3 near ground or from particulate NH_4NO_3. It therefore seems likely that the sulphate aerosol is acid at higher altitudes.

By applying multiple linear regression on the data obtained with the latter model the following equations were obtained

$$[NH_4^+]_p = 0.41 \times R_{NO_3^-} [NO_3^-]_a + 0.84 \times R_{SO_4^{2-}} [SO_4^{2-}]_a + 7.9 \ (r=0.83)$$

$$[H^+]_p = 2.02 \times R_{NO_3} [HNO_3]_g + 0.63 \times R_{SO_4^{2-}} [SO_4^{2-}]_a + 16.3 \ (r=0.90)$$

It seems like HNO_3 has a higher influence on the acidity in precipitation than assumed in the last model. As mentioned in the introduction the HNO_3 concnetrations may be a little too low due to its lower diffusion coefficient in ambient air. This was investigated in a field test and the apparent diffusion coefficient obtained was on the average $7 \cdot 10^{-6}$ m²/s. In laboratory experiments a value around $1.2 \cdot 10^{-5}$ m²/s was obtained for dry air. This can, however, not explain the high coefficient in front of the hydrogen ion contribution to the total acidity. A diffusion coefficient of $7 \cdot 10^{-6}$ m²/s corresponds to a sampling efficiency of about 85%. If the particulate nitrate at ground level mainly consists of NH_4NO_3 it may to a higher degree be dissociated to HNO_3 and NH_3 at higher altitudes because the ambient concentration of NH_3 is higher at ground level than higher up. In an earlier study it has been pointed out that the concentration of NH_3 at ground level usually decreased with height and depends on the NH_3 flux from the soil and the diffusitivity of the air (8). The NH_3 flux is a function of the equilibrium concentration between air and soil and the transfer coefficient for NH_3. This seems to be the case if there are no NH_3 sources nearby. No indication that gaseous NH_3 can be long range transported has been found. If equetion 3 is replaced by 9

$$[H^+]_p = 2R_{HNO_3} \times [HNO_3]_g + R_{SO_4^{2-}} \times [SO_4^{2-}]_a + 2[HSO_3^-] \qquad (9)$$

and equation 1,2 and 4 to 9 are applied to the whole data material, new values for δ will be obtained. The scavenging ratios will not be affected by this. During winter δ became 0.6 ± 0.5 and during summer 1.0 ± 0.6. Tang (9) presented a diagram over the equilibrium partial pressures of HNO_3 and NH_3 over acid ammonium sulphates at varying ammonium to sulphate ratios and nitrate to sulphate ratios at 25°C. If the sulphate aerosol consists mainly of NH_4HSO_4 and the partial pressure of HNO_3 is <40 nmole/m³ (1 ppb) the nitrate to sulphate ratio in such an aerosol (liquid) at 85% r.h. would be <0.001. It means that the main fraction of the nitrates will not be in the same system as the sulphates. The partial pressure of NH_3 over that droplet would be 0.8 nmole/m³ (0.02 ppb). Stelson and Seinfeld (10) have presented a graph of the partial pressure product of HNO_3 and NH_3 over NH_4NO_3 as a function of relative humidity. If the partial pressure of HNO_3 is less than 40 nmole/m³ the pressure of NH_3 above a solution of NH_4NO_3 at 85% relative humidity would be >360 nmole/m³ (>9 ppb). The NH_4^+ to NO_3^- ratio in an aerosol in equilibrium with a NH_4HSO_4 aerosol must therefore be much smaller than one. No diagram for the vapor pressures of HNO_3 and NH_3 over a solution with varying NH_4^+/NO_3^- ratio has, however, been found. From this discussion it seems likely that nitrate is to a great extent in the form of gaseous or concentrated nitric acid solution. The sum of particulate nitrate and nitric acid concentrations divided by the nitric acid concentration at ground level was on the average 2.9. If only a minor fraction of the nitrates is associated with other ions than NH_4^+ it seems possible that the acidity correlates with two times the ground level concentration of nitric acid. The figures used represent the equilibria at 25°C. At lower temperatures the partial pressure for the gas phase species will be lower in both the sulphate and nitrate system, but data for other temperatures are not yet available. If the nitrates are in HNO_3 form in the most part of the scavenged air mass it explains why more consistant data were obtained when R_{HNO_3} was assumed to be equal to $R_{NO_3^-}$.

If these assumptions are correct the average acid contribution from HNO_3 in comparison to the total acidity will be 54%.

All the events when the ion balance did not satisfy equation 5 were also investigated. The scavenging ratios for those fitted very well with the others when they were plotted in figure 2 and 3, but the calculated values were different and directly proportional to the ion balance. Therefore it seems as if other acid or basic components with other anions than NO_3^- or SO_4^{2-} were more abundant at these occasions.

5. THE PRECIPITATION EFFECT ON GROUND LEVEL CONCENTRATIONS

The ground level concentration is to a great extent dependent on the origin of the air mass. Back trajectories have been calculated by the Norwegian Meteorological Institute for the whole period of measurements. Wind rose diagrams of average concentrations have been evaluated both for dry days and for days with precipitation. By comparing the two wind roses a weighted mean value of the ratio between the concentration during wet and dry days (α) could be obtained. It was calculated from

$$\alpha = \sum_{j=1}^{j=8} \left(\frac{\bar{c}_{p,j}}{\bar{c}_{d,j}} \times N_j \right) / \sum_{j=1}^{j=8} N_j \qquad (10)$$

where j is a code for the trajectory sector. $\bar{c}_{p,j}$ is the mean concentration during days with precipitation and wind from a certain sector. $\bar{c}_{d,j}$ is the mean concentration for dry days with wind from the same sector. N_j is the smallest number of measurements with trajectory j of the dry and wet days. In all cases here there are smaller number of wet days for all trajectories. Table II shows $\bar{\alpha}$ together with the number of measurements for HNO_3, NO_3^-, NO_2, PAN, SO_2, SO_4^{2-} and soot. The average percent of dry time during days with precipitation events is also shown. Data for NO_2 and PAN were obtained from an eralier study (1).

The high α value for SO_2 supports the hypothesis that the oxidation rate in precipitation is low. When a droplet falls from high altitudes it will evaporate when it passes through unsaturated air masses near ground. The concentration of H^+ and HSO_3^- will then increase. The partial pressure of SO_2 in equilibrium with the droplet will then increase and cause a release of SO_2. The higher temperature in the lower troposphere will further increase the equilibrium concentration of SO_2.

If the particulate nitrate consists of NH_4NO_3 it will dissociate when the partial pressure of HNO_3 and NH_3 drops due to wash out. More HNO_3 and NH_3 will thus be produced while the concentration of particulate nitrate will drop. This can explain that the concentration of particulate nitrate drops more rapidly.

During wet days the air mass may have been scavenged by precipitation at other locations. The average concentration decrease during wet days may therefore be lower than the average ratio of dry to total time.

REFERENCES

Ferm. M., Samuelsson, U., Sjödin, Å. and Grennfelt, P. (1983)
Long range transport of gaseous and particulate oxidized nitrogen compounds.
Presented at the CACGP Symposium on trophospheric chemistry in Oxford, 28 Aug - 3 Sept 1983
The article is submitted to Atm. Environment

Ferm, M. (1982)
Method for determination of gaseous nitric acid and particulate nitrate in the atmosphere.
EMEP Expert meeting on chemical matters, Genéva, 10-12 March

Brosset, C. and Ferm, M. (1978)
Man-made airborne acidity and its determination
Atm. Environ. 12 909-916

Small, H., Stevens, T.S. and Bauman, W.C. (1975)
Novel ion exchange chromatographic method using conductimetric detection.
Anal. Chem. 47 1801-1809

Brosset, C., Andréasson, K. and Ferm, M. (1975)
The nature and possible origin of acid particles observed at the Swedish west coast.
Atm. Environ. 9 631-642

Levine, S.Z. and Schwartz, S.E. (1982)
In-cloud and below-cloud scavenging of nitric acid vapor.
Atm. Environ. 16 1725-1734

Pena, J.A., Pena, R.G., Bowersox, V.C. and Takacks, J.F. (1982)
SO_2 content in precipitation and its relationship with surface concentration of SO_2 in the air.
Atm. Environ. 16 1711-1715

Ferm. M. (1983)
Ammonia volatilization from arable land - an evaluation of the chamber technique.
Presented at the WMO technical conference on observation and measurement of atmospheric contaminants (Vienna 17-21 October 1983)

Tang, I.N. (1980)
On the equilibrium partial pressures of nitric acid and ammonia in the atmosphere.
Atm. Environ. 14 819-828

Stelson, A.W. and Seinfeld, J.H. (1982)
Relative humidity and pH dependence of the vapor pressure of ammonium nitrate - nitric acid solutions at 25°C
Atm. Enivron. 16 993-1000

Period	Concentrations in precipitation in µmol/litre							Concentrations in ground level air in nmol/m³			
	$[H^+]$	$[NH_4^+]$	$[NO_3^-]$	$[SO_4^{2-}]_{ex}$	$[Cl^-]$	P mm	Air temp. °C	$[HNO_3]$	$[NO_3^-]$	$[SO_2]$	$[SO_4^{2-}]$
November-April	108.3	70.7	87.8	56.2	192.8	2.9	1.1	26.9	54.8	148.6	44.7
April-October	104.8	70.4	69.4	53.2	182.6	4.2	13.1	19.4	35.6	33.1	45.2

<u>Table 1.</u> Average concentrations of different species in precipitation and air at ground level, when simultaneous data from all species were available and equation 5 was fulfilled.

Parameter	α	Number of wet days with defined trajectories	Number of dry days with defined trajectories
HNO_3	0.77	47	101
NO_3^-	0.56	47	101
NO_2	1.05	64	115
PAN	0.90	37	85
SO_2	1.29	66	114
SO_4^{2-}	0.73	66	117
soot	0.70	66	115
%dry time	0.72	62	117

<u>Table II.</u> Weighted mean ratios of average concentration during wet days to average concentration during dry days (α). The function is defined in eq 10.

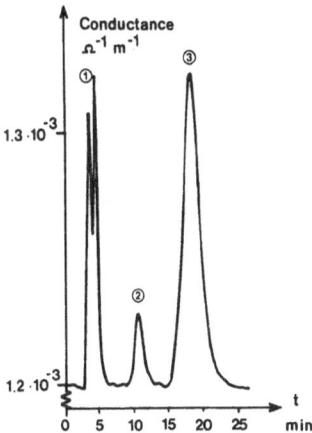

Figure 1. Chromatogram from analysis of the leaching solution from a denuder.
Peak 1 : $\sim 10^{-2}$ M CO_3^{2-}, 2 : $1.0 \cdot 10^{-5}$ M NO_3^{-}, 3 : $2.6 \cdot 10^{-5}$ M SO_4^{2-}.
The analysis represents a sample with a yearly mean HNO_3 and
SO_2 concentration (1982-10-28). $3 \cdot 10^{-3}$ M Na_2CO_3 at a flow of
1.7 ml/min and a 300 µl loop was used.

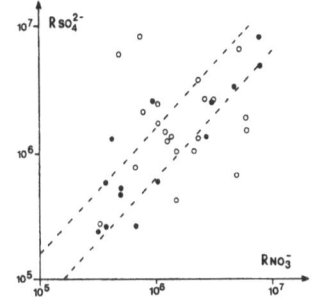

Figure 2. The scavenging ratio for SO_4^{2-} as a function of ratio for NO_3^{-} on
the assumption that $R_{HNO_3} = R_{NO_3^-}$ and $R_{SO_2} |SO_2| = [HSO_3^-]$ at
equilibrium.
● denotes values for the period November 1981 to April 1982
○ denotes values   the period May to October 1982

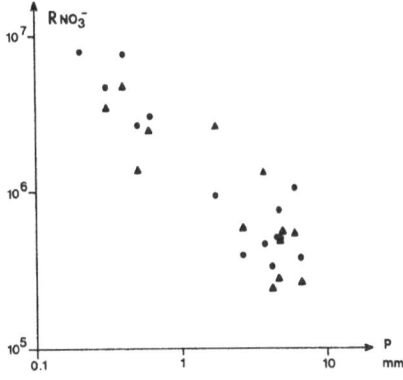

Figure 3. The scavenging ratios for nitrate (●) and sulphate (▲) as a
function of total precipitation amount during November 1981 to
April 1982. The assumptions are the same as in figure 2.

IDENTIFICATION DES SOURCES D'HYDROCARBURES AROMATIQUES

POLYCYCLIQUES PARTICULAIRES DANS L'ATMOSPHERE URBAINE.

P. MASCLET, K. NIKOLAOU et G. MOUVIER[*]
Laboratoire de Physico-Chimie Instrumentale – Université PARIS VII.
75251   PARIS CEDEX 05.

Summary

To analyze PAH, a method with short sampling (1 hour) on representative sites of various origins in urban areas was optimized : residential heating, gasoline and diesel-fuel powered vehicles, incinerators,
oil reffinery, casting work and coal-fired power plant. The current
methods for identifying : profiles or histograms and characteristic
ratio, were used and compared to the values of the scholarly literature. It is very important to count with PAH reactivity differences
to obtain significative ratio. Only data obtained with short times
and well determined samplings conditions must be considered.

INTRODUCTION

Plusieurs méthodes ont été utilisées pour identifier les sources anthropogéniques des hydrocarbures aromatiques polynucléaires (HAP) et leurs
rejets dans l'atmosphère ; les méthodes déjà employées utilisent les profils caractéristiques de concentrations de HAP (1) et les rapports caractéristiques de composés pris 2 à 2 (2).
Les mesures sont parfois anciennes et les évaluations ont été souvent
faites à partir de mesures de concentrations obtenues par des techniques
différentes, les résultats sont alors difficilement comparables et les
ordres de grandeur sont parfois surprenants (3). Une autre source d'erreur
provient de la non-prise en compte de la transformation physico-chimique
des HAP dans l'atmosphère et dans le système de prélèvement (4). Or, les
mesures peuvent être faites directement à l'émission, dans le panache des
émissions à une distance plus ou moins grande du point d'émission, ou même
dans l'atmosphère ambiante moyenne. Cette pluralité des sites de prélèvement
représentent des états physico-chimiques différents pour les HAP. A l'émission, la température est souvent élevée et le rapport HAP particulaire /
HAP gazeux est beaucoup plus faible que dans l'atmosphère. Dans l'atmosphère
l'influence des dégradations photochimiques par les agents oxydants peut
être fort importante. Il est donc tout-à-fait essentiel de bien définir les
conditions de prélèvement des HAP et de tenir compte de la stabilité relative de ces polluants dans l'atmosphère (4).
Pour notre part, nous n'avons pris en compte que la forme particulaire
des HAP, dans le panache des émissions (à moins de 1 ou 2 km de distance),
pour des températures toujours très proches de 10 °C (entre 8 et 11 °C),
pour des vents inférieurs à 3 m/sec et pour des luminosités et concentrations de NO_x et O_3 comparables. Les temps de prélèvement ont toujours été

courts (1 heure) ; nous pensons ainsi avoir minimisé les artefacts sur le filtre et même annuler cet effet dans l'évaluation des rapports de concentrations.

La méthode de préparation de l'échantillon et son analyse étant bien définies (5), nous pensons avoir minimisé les erreurs systématiques d'analyse, bien que le nombre de mesures, pour chaque source, soit relativement limité (entre 3 et 6 mesures).

Dans un premier temps, nous avons cherché à identifier les sources et à les caractériser par des HAP prépondérants ; dans un deuxième temps, il est envisagé d'établir un bilan global pour une région urbanisée des différentes contributions à partir d'une évaluation des flux émis.

I. MÉTHODE EXPÉRIMENTALE

Les particules ont été prélevées par un HIVOL sampler sur des filtres en fibre de verre, pendant 1 heure. Les échantillons ont été extraits au Soxhlet pendant 3 heures avec un mélange cyclohexane/dichlorométhane (2/3) à l'obscurité ; l'analyse a été réalisée par CLHP en phase inverse, par un gradient eau-méthanol (95 à 100 % à 0,25 % par mn). 16 HAP ont été dosés par fluorescence en utilisant 4 couples de longueur d'onde 297/340, 297/480, 313/390 et 306/390 nm. Les détails opératoires ont été publiés par ailleurs (5).

Lieux de prélèvement :

A : air ambiant Paris, 20 m hauteur - HIVER (- 3 °C)
B : air ambiant Paris, 20 m hauteur - ETE (28 °C)
C : autoroute, niveau sol   - HIVER (4 °C)
D : autoroute, niveau sol   - ETE (27 °C)
E : Parking souterrain de l'Université PARIS VII (11 °C)
F : garage de véhicules diesel   (8 °C)
G : usine d'incinération des déchets urbains(9 °C)
H : raffinerie de pétrole   (10 °C)
I : usine d'engrais   (10 °C)
J : usine fonderie automobile   (22 °C)
K : centrale électrique charbon   (9 °C)
L : zone rurale, à 50 km de Paris   (9 °C)

Les valeurs des concentrations moyennes sont données dans le tableau 1.

On remarque que les concentrations varient de plusieurs ordres de grandeur entre la zone rurale (L) et l'usine de fonderie automobile (J) ; ces valeurs absolues n'ont qu'une signification relative puisque nous ne tenons pas compte des flux émis pour les prélèvements effectués dans les panaches des émissions (G, H, I, K) et sur les lieux mêmes des émissions (C, D, E, F, J).

II. LES HAP EN ZONE URBAINE (Figure 1)

Les nombreuses mesures effectuées en zone urbaine (A et B), à 20 m de hauteur pour s'affranchir de la proximité des gaz d'échappement automobile, ont permis de mettre en évidence l'apport de la circulation automobile (véhicules à essence + diesel) et du chauffage domestique ; les deux séries de mesures ont été effectuées dans des conditions de température caractéristiques des mois les plus froids et les plus chauds. Pour les prélèvements A et B, seules les sources : véhicules et chauffage domestique, sont à prendre en considération, car les sources industrielles sont éloignées et constituent un "back ground" faible et constant.

Pour le prélèvement A (en hiver), la plupart des HAP sont abondants ; les HAP prédominants sont le fluoranthène, le pyrène, le BaAnthracène, le chrysène et le BaPyrène, c'est-à-dire essentiellement des composés tétra-cycliques que l'on peut considérer comme prépondérants dans les émissions dues au chauffage domestique.

Pour le prélèvement B (même site en été), l'absence totale de chauffa-ge domestique privilégie - en valeur relative - les composés lourds comme le coronène et l'indenopyrène ; on sait déjà que le coronène peut être considéré comme un traceur des véhicules à essence ; les concentrations en fluoranthène et pyrène ne sont pas non plus négligeables. En valeur absolue les concentrations sont toujours plus faibles en été qu'en hiver ; cette diminution n'est pas seulement due à des émissions plus faibles ; la trans-formation physico-chimique des HAP (volatilisation par élévation de tempé-rature et réactions photochimiques) est importante, comme nous l'avons montré par ailleurs. Ainsi, le coronène, d'origine purement automobile, a une concentration 3 fois plus faible en été qu'en hiver, pour une circula-tion quasiment identique. La cause en est exclusivement la transformation physico-chimique (réactivité importante).

Pour le fluoranthène (peu réactif), la baisse de concentration est due essentiellement à des émissions moindres. Pour le BaPyrène, les deux fac-teurs s'additionnent.

Nous avons tenté, à partir d'hypothèses simples, de connaitre la part du chauffage domestique et celle de la transformation physico-chimique dans la concentration mesurée de chaque HAP, en tenant compte de l'échelle de stabilité. Nous proposons une méthode simple de calcul de l'apport de cha-que source et de la transformation physico-chimique (6).

Les rapports caractéristiques de 2 HAP constituent une bonne représen-tation des diverses sources et permettent de les caractériser sans ambigui-té.

Sur la figure 2, nous avons représenté quelques rapports permettant de distinguer les émissions véhiculaires et les émissions du chauffage domes-tique. Les rapports les plus significatifs concernent le coronène toujours prépondérant dans les émissions des automobiles à essence. Les rapports BbF/BkF, PYR/CHR ou FLA/CHR sont moins différenciés, mais la bonne repro-ductibilité des mesures permet de les utiliser comme indicateurs de source; le pyrène et le fluoranthène sont présents après combustion des fuels domes-tiques comme dans les diesels, mais les émissions particulaires sont nette-ment plus abondantes dans les véhicules diesels. D'autres combinaisons de rapports peuvent être aussi utilisées.

Les prélèvements C et D (autoroute), pour lesquels les sources sont bien définies (véhicules à essence et véhicules diesels), mettent bien en évidence l'importance des transformations physico-chimiques.

Les prélèvements ont été effectués aux mêmes tranches horaires (moyen-ne de 5 mesures). Le trafic est tout-à-fait comparable en hiver et en été ; on remarque néanmoins que les concentrations sont en gros 10 fois plus for-tes en hiver qu'en été ; il est évident que cette diminution est due essen-tiellement à la dégradation photochimique et à la modification du rapport particules/gaz. Les mesures effectuées en phase gazeuse sont trop partielles pour chiffrer l'importance de ce phénomène ; néanmoins, il montre qualita-tivement l'absence quasi totale de HAP sous forme gazeuse en hiver.

Les HAP les plus abondants sont encore le fluoranthène (provenant des camions diesels) et l'indénopyrène et le coronène (essence). La valeur relativement basse observée pour le pyrène est assez surprenante, mais le pyrène est plus réactif que le fluoranthène.

Les prélèvements E et F, concernant respectivement un parking de véhi-cules automobiles et un parking d'autobus diesel (10), ont été effectués

dans des endroits clos à faible luminosité ; ils sont donc représentatifs
de sources pures pour lesquelles les transformations physico-chimiques doivent jouer un rôle mineur. Les valeurs absolues montrent l'abondance des
HAP particulaires pour les véhicules diesels ; ceux-ci sont bien caracté-
ristisés par les HAP légers : phénanthène, fluoranthène et pyrène, alors
que les véhicules à essence présentent de fortes concentrations en HAP
lourds : BaP, indenopyrène et coronène.

Il est évident que tous les rapports du type HAP lourds/HAP légers
donnent des valeurs supérieures à 1 pour les véhicules à essence et inférieures à 1 pour les véhicules diesels. Nous ne les avons pas mentionnés
dans la figure 2.

Les abondances relatives d'autres HAP sont peu mis en évidence dans
les profils ; les facteurs d'enrichissement de HAP pris 2 à 2 le montrent
beaucoup plus clairement : ainsi, le rapport FLA/PYR met en évidence que
les émissions de ces 2 composés caractéristiques des diesels et de l'essence ne se font pas dans les mêmes proportions relatives. Il en est de même
pour le BaA et le BghiPerylène, composés non prépondérants dans les émissions des véhicules. Les composés même mineurs peuvent donc aussi servir
d'indicateurs de sources de pollution.

Comparaison avec d'autres études

Le rapport BaP/BghiP est souvent utilisé afin d'identifier les sources
véhicules et chauffage domestique. A New York, durant l'été, on a évalué
un rapport de 1,3 caractéristique des véhicules (1). A Oslo, il est de 0,4
pour les véhicules et supérieur à 0,8 pour le chauffage (8). Une autre
étude (3) donne un rapport de 0,2 pour des émissions à l'échappement.

Pour les véhicules, nous observons un rapport de 1,25 en été, donc en
bon accord avec l'une des deux études. En revanche, nous observons un
rapport de 1,7 pour le chauffage domestique, supérieur à celui observé par
d'autres auteurs.

Les prélèvements effectués par ces auteurs ont toujours porté sur 24
heures ; sachant que la durée de vie du BaP est inférieure à celle du
BghiPérylène (1,9/4,2), il est évident que les concentrations de BaP sont
sous-estimées pour des prélèvements de 24 heures ; il n'en est pas de même
pour des prélèvements d'une heure, et nos rapports sont plus significatifs.

Quant aux mesures faites à l'émission directement, elles n'ont aucune
valeur, car seule la forme particulaire a été prise en compte ; à la température de l'échappement, une majorité de HAP (même lourds) se trouve sous
forme gazeuse.

Un autre rapport a été utilisé : pyrène/BaPyrène. Les valeurs sont
variables suivant les études : 0,3 à 0,8 pour le chauffage, 2 à 12 pour
les véhicules à essence et 50 à 100 pour les véhicules diesels (à l'échappement). Pour notre part, nous observons un rapport de 0,8 pour le chauffage et 1,5 pour les véhicules à essence et 11 pour les véhicules diesels.
La différence de réactivité du BaP et du pyrène explique encore les différences observées avec les autres auteurs.

Enfin, un dernier rapport est utilisé : COR/BaP = 1,7 (2) pour les
véhicules. Cette valeur est en accord avec la nôtre : 1,5 ; en effet, les
deux composés ont des réactivités comparables et l'erreur sur l'évaluation
des concentrations pour une mesure de 24 h. est sans influence sur le
rapport des concentrations.

III. EMISSIONS INDUSTRIELLES

5 émissions industrielles ont été caractérisées. Elles concernent les sources les plus fréquentes en zone urbaine. La figure 3 montre les concentrations relatives des divers HAP ; les concentrations observées en zone rurale y sont aussi présentées ; 3 gammes de concentrations sont représentées : 0 à 10 ng pour les prélèvements G, H, I et K, concernant des mesures dans le panache des émissions ; 0 à 1000ng pour un prélèvement à l'intérieur des locaux J (usine de fonderie) et 0 à 1 ng pour la zone rurale (L).

Comme nous l'avons fait remarquer précédemment, les mesures dans les panaches et en zone rurale ont été effectuées dans des conditions météorologiques et de température à peu près identiques. Néanmoins, seuls les ordres de grandeur sont significatifs. En revanche, les profils sont caractéristiques de chaque source.

Le prélèvement H, concernant les émissions d'une raffinerie de pétrole montre que les HAP légers (3 cycles) sont prépondérants ; les concentrations en pyrène et fluoranthène sont logiquement importantes, mais le HAP le plus caractéristique est le BeP.

Le prélèvement I (usine d'engrais) est caractérisé par d'importantes émissions de chrysène, pyrène, BeP et pérylène.

Les émissions concernant les centrales électriques à combustible fossile font l'objet d'une étude particulière, plus approfondie (7).

Le prélèvement K concerne une centrale électrique de moyenne puissance fonctionnant au charbon et située dans une zone éloignée des centres urbains. Les études ont montré que, suivant le type de chaudière, le type de combustible et même la provenance du combustible (pour le charbon par exemple), les émissions sont différenciées quantitativement et qualitativement ; néanmoins, dans tous les cas, la présence de HAP sous forme gazeuse est prépondérante. Sous forme particulaire, les HAP légers sont les plus abondants, et le profil diffère peu de celui observé pour la raffinerie de pétrole (12).

Le prélèvement J concerne une fonderie. Les concentrations à l'intérieur de l'usine sont en moyenne 100 fois plus fortes que dans les panaches et 1000 fois plus fortes que dans l'air ambiant des villes. Le profil est très caractéristique. Les composés tétra et pentacycliques y sont fortement prépondérants : fluoranthène, BaAnthracène, BkFluoranthène, BaP et indenopyrène, composés tous réputés comme cancérigènes.

Le prélèvement G (13,14) concerne une usine d'incinération des déchets urbains, située dans la zone centrale de l'agglomération parisienne ; le panache étudié ne constitue donc pas une source pure ; la proximité d'axes de circulation automobile est évidente : les fortes concentrations en indenopyrène et coronène ne sont pas significatives de l'usine d'incinération. Néanmoins, les concentrations élevées en chrysène, BeP et dibzahAnthracène (peu présents dans les émissions véhiculaires) doivent être représentatives de ce type d'émission.

Le prélèvement L, effectué en zone rurale loin de toutes usines et d'axes routiers importants, montre que si les concentrations sont faibles, elles ne sont pas négligeables : de 0,05 à 0,6 ng/m3 d'air. L'origine des HAP détectés est évidente : les composés les plus abondants sont le fluoranthène (présent dans toutes les émissions provenant d'une combustion des fuels), le BaP, l'indenopyrène et le coronène. La principale cause de la pollution en zone rurale est donc la circulation des véhicules.

La figure 2 représente quelques rapports caractéristiques de ces sources. Comme pour différencier les émissions du chauffage domestique et des véhicules à essence ou diesel, nous avons souvent utiliser des rapports avec le coronène, de façon à comparer l'apport des émissions industrielles des émissions par les véhicules.

Le rapport caractéristique permettant de différencier sans ambiguité toutes les industries étudiées est le rapport des concentrations fluoranthène/pyrène.

Il est de 0,5 dans les raffineries de pétrole, 1,7 dans les usines d'engrais, 65 dans les fonderies, 3 dans la centrale thermique au charbon, 0,15 dans l'usine d'incinération des déchets.

Remarquons que les centrales thermiques et les raffineries de pétrole peu différenciées par le profil, le sont sans ambiguité par le facteur d'enrichissement du fluoranthène par rapport au pyrène.

Comparaison avec d'autres études

Le rapport BaP/BghiP a été proposé pour l'identification des sources industrielles (8). Pour les incinérateurs, il varie de 0,14 à 0,6 ; pour les raffineries de 0,65 à 1,7 et pour les centrales au charbon de 0,9 à 7. Dans notre étude, ce rapport ne varie pas sensiblement pour les diverses industries mesurées : 0,8 à 1,1 ; il est donc en accord avec les autres études, mais peu différencié, donc sans intérêt quant à l'identification des sources.

CONCLUSION

L'analyse des profils des HAP pour différentes sources présentes en milieu urbain a montré que chacune d'elle pouvait être caractérisée par l'association d'un nombre restreint de HAP (Tableau 2). L'apport de la circulation automobile est facilement mis en évidence, car elle est seule responsable des hautes concentrations en HAP lourds (coronène et indéno-pyrène), donc que le fluoranthène et le pyrène caractérisent les fuels et les composés tricycliques (sauf anthracène) les émissions à haute température.

Le tableau 2 donne une synthèse des HAP caractéristiques.Mieux que les profils, les rapports caractéristiques ou facteurs d'enrichissement sont adéquates pour caractériser les sources. Nous avons sélectionné les plus significatifs, encore faut-il que l'évaluation de ces rapports résulte de mesures de concentration sur des temps courts (1 heure) afin de minimiser l'effet important de la transformation physico-chimique différenciée des divers HAP ; les rapports établis sur des mesures de 24 h. sont donc erronés, si on ne tient pas compte de la durée de vie des divers composés.

BIBLIOGRAPHIE

01 G. GRIMMER
   VDI Berichte 39, 1980, 358

02 R.P. HANGERBRAUCK, R.P. LAUCH, J.E. MEEKER
   Amer. Jud. Hyg. Assoc. J. 27, 1966, 47.

03 Particulate Polycyclic Organic Matter
   National Academy of Sciences, Washington D.C., 1972.

04 K. NIKOLAOU, P. MASCLET, G. MOUVIER
   *Proceedings of the 2nd. Int. Conf. on Carbonaceous Particles in the
   Atmosphere, Linz (Austria) - Sept. 1983, published in Science of the
   Total Environment 1984.*

05 P. MASCLET, K. NIKOLAOU, G. MOUVIER
   Poll. Atm. 95, 1982, 175.

06 A. KOUTSANDREAS, K. NIKOLAOU, P. MASCLET, G. MOUVIER
   Cahiers de l'Analyse des Données - Vol. 3 n° 3, 1983, 371.

07 M.A. BRESSON, P. MASCLET, G. MOUVIER
   à paraître.

08 M. MØLLER, I. ALFHEIM
   Atm. Env. 14(1), 1980, 83.

09 P. KOTTIN, H.L. FALK, M. THOMAS
   AMA Arch. Ind. Hyg. Occup. Med. 9, 1954, 164.

10 D.T. KASCHANI
   Erdoel. Koble, Erdgas, Petrochem. 32, 1979, 572.

11 R. BENNETT, K. KNAPP, P. JONES, J. WILKERSON, P. STRUP
   *3rd. Int. Symp on PAH - JONES P.W., LEBER P., Ed., Ann Arbor Science
   1979, p. 419.*

12 M.M. DUVAL, S.K. FRIEDLANDLER
   *"Source resolution of PAH in the Los Angeles Atmosphere" - US Environ-
   mental Protection Agency - Rapport EPA 600/2-81.161 - Sept. 1981 -
   NTIS PB 82 - 121 336.*

13 I.W. DAVIES, R.M. HARRISON, R. PERRY, P. RATNAYAKA, R.A. WELLINGS
   Env. Sci. Technol. 10(5), 1976, 451.

14 G.A. EICEMAN, R.E. CLEMENT, F.W. KARASEK
   Anal. Chem. 53(7), 1981, 955.

TABLEAU 1 : CONCENTRATIONS DE PAH EN NG/M3

(NM : non mesurable — ND : non détectable)

SITE DE PRELEVEMENT / HAP	A	B	C	D	E	F	G	H	I	J	K	L
Fluorène	6,2	NM	ND	1,5	35,8	36,9	2,5	2,4	0,8	NM	1,9	ND
Phénanthène	38,2	NM	ND	7,5	41,0	152,2	7,0	7,9	5,5	NM	5,2	ND
Anthracène	10,0	0,5	1,2	0,4	5,4	6,2	2,0	3,9	0,9	254	0,5	0,04
Fluoranthène	69,7	2,7	10,9	5,4	55,9	317	1,0	2,3	5,8	980	4,1	0,65
Pyrène	40,0	2,4	2,3	3,1	65,3	469	5,9	4,3	3,4	15	1,3	0,11
Bza Anthracène	53,7	0,9	2,5	1,7	25,2	50	3,3	1,6	1,5	832	0,6	0,02
Chrysène	40,8	1,3	1,8	2,9	40,0	25,8	10,0	1,7	5,0	113	1,8	0,06
BbFluoranthène	31,7	1,5	1,6	1,5	21,3	23,3	2,8	1,3	3,8	463	2,0	0,11
BePyrène	12,1	0,8	NM	1,1	13,1	33,9	5,8	2,8	2,9	NM	ND	0,04
BkFluoranthène	12,8	0,7	0,8	0,6	10,3	11,6	1,3	0,5	1,4	219	0,9	0,04
BaPyrène	46,2	1,8	1,9	2,5	44,4	43,5	2,8	0,4	1,6	567	1,3	0,27
DibzaAnthracène	13,5	1,1	0,7	0,6	15,5	12,8	4,1	1,6	1,3	NM	8,7	0,03
Perylène	11,0	ND	ND	2,4	11,5	57,6	ND	ND	ND	160	ND	ND
BzghiPerylène	27,4	1,4	1,0	1,3	21,9	19,0	3,7	0,7	2,7	133	1,2	0,10
Indenopyrène	30,1	4,0	4,2	4,7	55,8	46,7	24	1,2	2,3	370	1,7	0,21
Coronène	14,5	5,0	4,3	5,3	64,9	38,0	33,0	1,0	0,7	28	0,9	0,13

NM : non mesurable ; ND : non détectable

TABLEAU 2 : LES HAP PREDOMINANTS DANS LES EMISSIONS DE CHAQUE SOURCE.

	Chauffage Domestique	Véhicules à essence	Véhicules à moteur diesel	Usine d'engrais	Raffinerie de pétrole	Centrale électrique à charbon	Usine d'incinération des déchets	Fonderie
FLE					+	+		
PHE			+	+	+	+	+	
ANT					+			
FLA	+		+	+		+		+
PYR	+		+		+		+	
BaA	+							+
CHR	+			+			+	
BbF								+
BeP							+	
BkF								
BaP		+						+
DiBahA								
PER				+				
BghiP								
INPY		+						
COR		+						

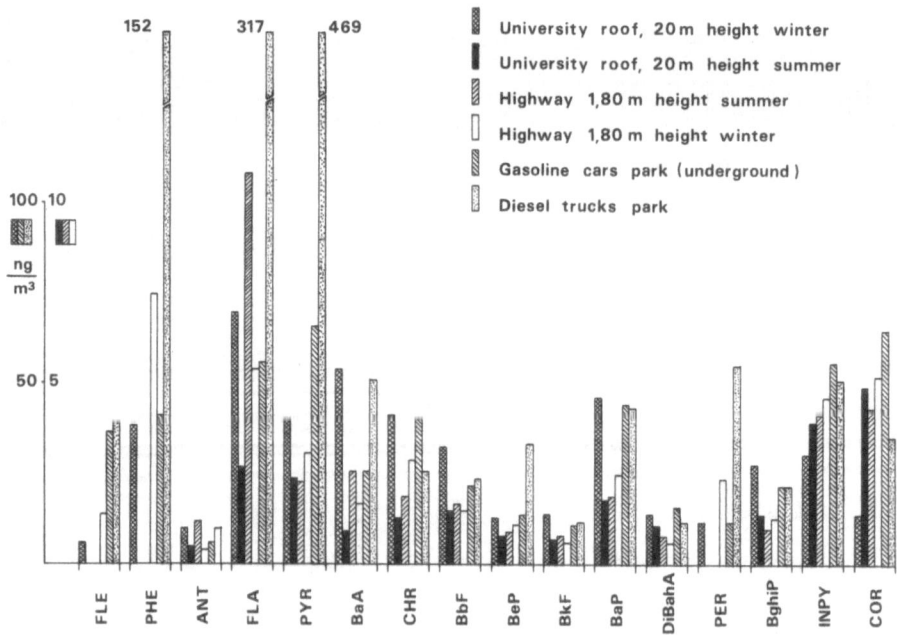

Fig. 1

University roof, 20 m height winter
University roof, 20 m height summer
Highway 1,80 m height summer
Highway 1,80 m height winter
Gasoline cars park (underground)
Diesel trucks park

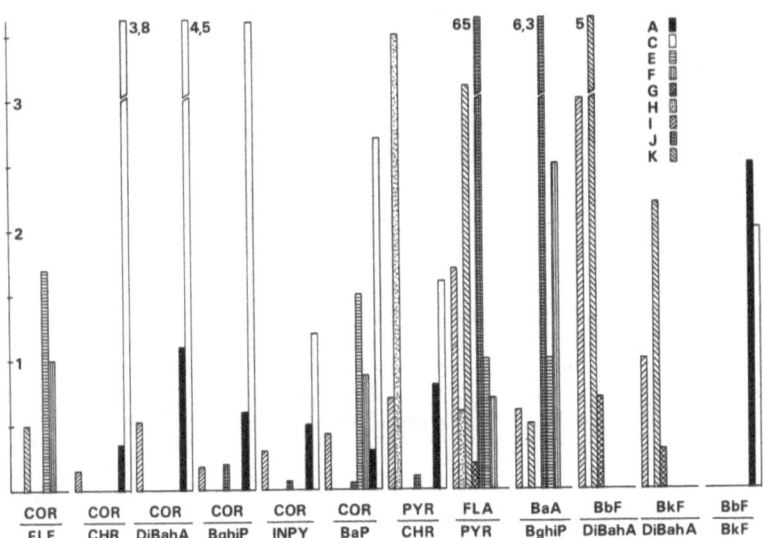

Fig. 2

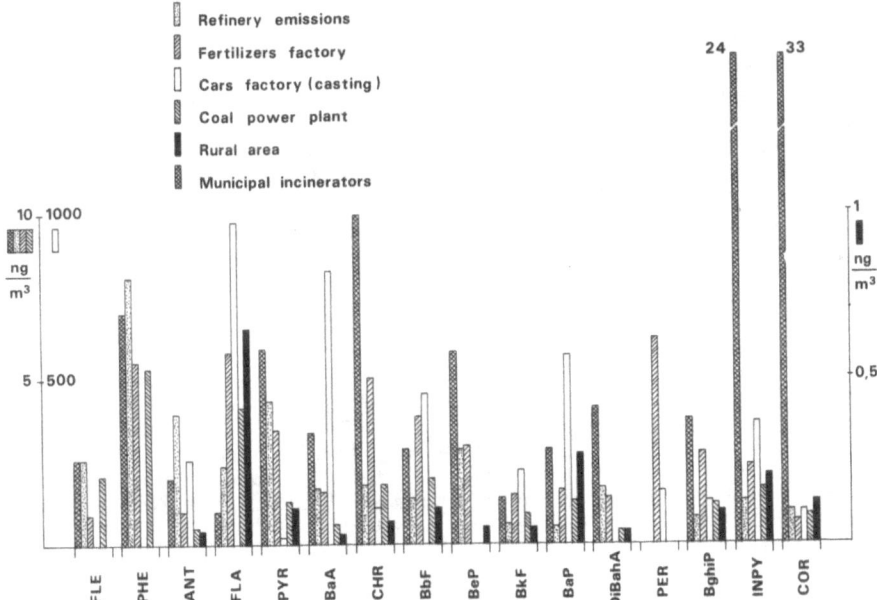

Fig. 3

PHOTOCHEMICAL AIR POLLUTION IN DENMARK.
WEEKDAY EFFECTS AND EVIDENCE OF LARGE-SCALE FORMATION

J. FENGER
Air Pollution Laboratory
National Agency of Environmental Protection
Risø National Laboratory
DK-4000 Roskilde, Denmark

SUMMARY

Ozone was monitored in Copenhagen and at Risø, about 30 km west of Copenhagen. The concentrations are independent of cloud-cover, and a dependence of temperature appears to be due to seasonal variation; therefore, local photochemical activity is assumed to be of minor importance. Although the ozone must thus be due to distant sources, no correlation with wind direction was found. The concentration, however, increases with wind-speed and is significantly lower in Copenhagen (urban) than at Risø (rural); this indicates that the net reaction is an ozone decomposition. Since concentrations in Copenhagen are higher during the weekend than on the weekdays the cause must be the emission of NO_x from urban traffic.

1.   INTRODUCTION

Elevated ozone concentrations at ground level are normally ascribed to photochemical reactions between nitrogenoxides and hydrocarbons. Consequently, one should expect the highest concentrations in urban areas during the day. This is also typical for so-called photochemical smog-episodes in highly polluted cities, situated not too far north.

Since, however, the reaction scheme is very complex, with a series of subsequent and competing reactions, and since both horizontal and vertical transport of ozone or primary pollutants are possible, the outcome may be completely different in less polluted areas. Elevated ozone concentrations simultaneously in larger rural and urban areas are often taken as indication of long-range transport (1), but it may be difficult to identify a specific source area (2).

Often an increase in ozone concentrations is observed during the night - typically with a maximum at about 3 a.m. For measurements in London (3) the phenomenon was ascribed to diffusion of stratospheric ozone. For Tel-Aviv it was assumed (4) that ozone formed during the day can be trapped under an inversion, whereas nightly maxima in Toronto and Montreal were explained (5) by a trapping of primary pollutants. German observations (6) were explained by assuming a brief increase in the vertical mixing.

Relatively high ozone concentrations at night are, in cities, often followed by a minimum in the morning; this is ascribed to reactions between ozone and nitrogenoxides emitted from the morning traffic. Consequently, the minimum may be less pronounced or absent in the weekend (7). This may also be the reason why in some cities the ozone concentration is lower than in the surrounding country side. The influence of urban traffic on ozone pollution in adjecent rural areas have been demonstrated i.a. for New York (8) where it is independent of the day in the week in the city, but in the weekend decreases downwind, due to a reduction in emssion of primary pollutants.

It is well established that elevated ozone concentrations can occur in the southern part of Scandinavia. In the southeastern Norway (59°-60° N) measurements have been carried out since 1975 (9) and concentrations up to about 200 ppb were observed in episodes. In Sweden ozone is measured at Rørvik, south of Gothenburg, where the earlier EPA standard of 80 ppb has been exceeded several times, with the largest value recorded being 200 ppb. In the central parts of Stockholm and Nyköping the concentrations are lower; the highest value being 110 ppb (10). Before the investigations described here, ozone concentrations had not been measured systematically in Denmark, although Ro-Poulsen et al. (11) using tobacco plants as monitors had demonstrated that elevated levels might occur during the summer. It is, however, not obvious where the ozone is formed.

Denmark is a flat country, partly composed of islands, with no points exceeding 200 m above sea level. The wind is mainly blowing from west or southwest giving mild winters and cool summers; typical monthly temperature averages are -1 °C in February and +17 °C in July. The sunlight intensity may during the summer be sufficient for local photochemical activity, but it may be counteracted by a rapid dispersion of the primary pollutants. The NO_x-level (as NO_2) is about 60 $\mu g/m^3$ (30 ppb) in Copenhagen and 30 $\mu g/m^3$ in provincial towns (12) and probably below 10 $\mu g/m^3$ in rural areas. The concentration of reactive hydrocarbons is not known better than that it must be below 1000 ppb. Assuming a ratio of 5-8 between HC and NO_x (10) yield an HCconcentration in Copenhagen of about 200 ppb.

2. DATA COLLECTION

The measurements described here were originally carried out in support of a study of plant damage by photochemical air pollution (13). Therefore, we do not have parallel measurements of the primary pollutants.

Ozone measurements were performed at the Royal Veterinary and Agricultural University, Copenhagen, at an open area approximately 200 m from several streets with normal city traffic. - And at the Air Pollution Laboratory, Risø, in a rural area approximately 1 km from a secondary highway with modest traffic. The positions are shown in figure 1, where also the Norwegian and Swedish monitoring stations are indicated.

The measurements were performed with Monitor Labs Inc., Ozone Analysers model 8410E. The instruments are based on detection of chemiluminescens from a reaction between ozone and ethylene. The ozone concentrations were registered as 1 hour averages; all times were indicated as normal meantime, although Denmark had summer time during the summer period.

Measurements in Copenhagen were carried out in the period June 1980–May 1981 with a coverage of about 80% ($\approx$ 7000 1 hour averages). At Risø measurements were made from June 1980 to September 1982, but only with a coverage of about 45% ($\approx$ 9000 1 hour averages) since the ozone analyzer was also used for other purposes. At both stations there was a reasonable coverage both in winter and summer.

Metorological data were received from air-base Værløse, situated between Copenhagen and Risø, and from Højbakkegård 15 km west of Copenhagen. They comprised i.a. wind speed and wind direction, temperature and cloud-cover.

All data were stored and processed in the Burroughs 6700 computer at Risø.

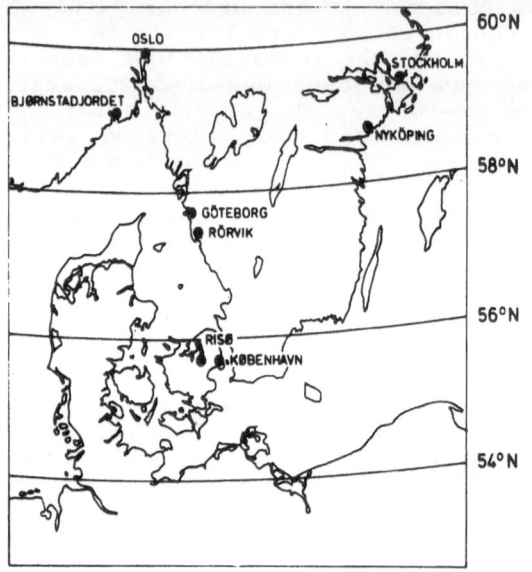

Figure 1. Denmark and the southern parts of Norway and Sweden with indication of monitoring stations.

3. OZONE CONCENTRATIONS AS A FUNCTION OF TIME

Raw plots of the ozone concentration showed that it was generally two times higher at Risø than in Copenhagen and two times higher in the summer than in the winter. Fractiles and maximum values are shown in table 1. the highest value (125 ppb) was the only one exceeding 120 ppb and was only recorded once.

As demonstrated with figure 2, a clearcut diurnal variation was not immediately apparent.

Table 1. Ozone concentrations in Copenhagen and at Risø indicated as 50% and 98% fractiles and the maximum value; 1 hour averages in ppb.

	50%	98%	max
Copenhagen (winter/summer)	8/23	26/52	40/72
Risø (winter/summer)	20/36	39/73	62/125

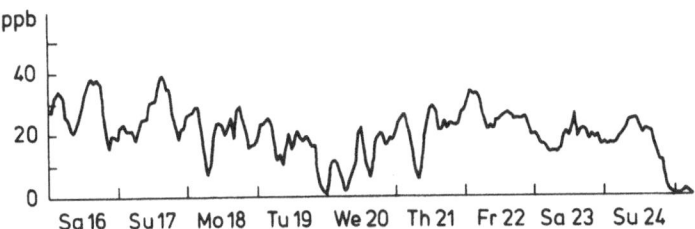

Figure 2. Typical time variation of the ozone concentration (Copenhagen Saturday 16th – Sunday 24th, August 1980).

Calculation of average diurnal ozone concentrations for each day in the week during the summer and winter period in Copenhagen and at Risø gave, however, the results presented in figure 3.
In the winter period the ozone concentration is low and nearly constant, but in the summer a pronounced maximum is seen in the afternoon. In Copenhagen there is apparently a nightly maximum of nearly equal magnitude during the weekdays; it is smaller on Saturday and nearly absent on Sunday. Closer inspection, however, reveal that this pattern is better described by assuming a mimimum (dip) in the morning hours in weekdays, since the ozone concentrations is generally higher during the weekend than on weekdays. The difference between Sunday and weekdays (figure 4) show a similar dip in the afternoon.

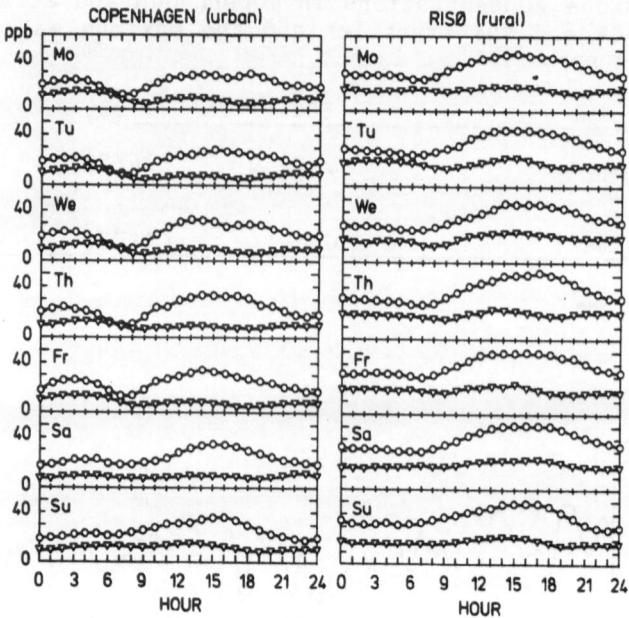

Figure 3. Average diurnal ozone concentration in Copenhagen and at Risø. Upper curves: summer (Apr.-Aug.). Lower curves: winter (Jan.-Mar. and Sep.-Dec.).

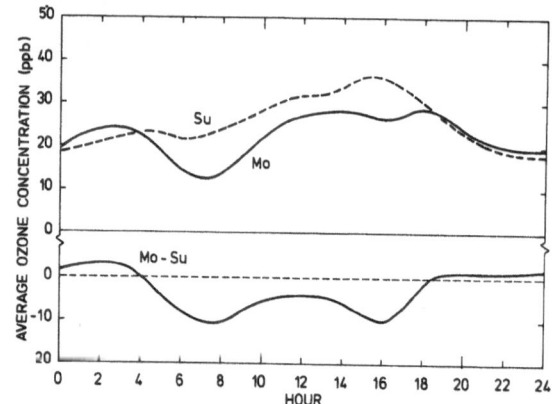

Figure 4. The upper plot compares the average ozone-concentration in Copenhagen on summer Sundays (full curve) and summer Sundays (dashed curve). The lower plot shows the difference between Monday and Sunday concentrations. (Data from figure 3).

4. DEPENDENCE OF METEOROLOGICAL PARAMETERS

On some occasions periods with elevated ozone concentrations occurred, although the EPA-standard (14) was never vio-
lated. One of these 'episodes' have been attributed to long-
range transport of photochemical pollutants (15), but generally no clearcut evidence of such phenomena was found. Neither
was it possible to demonstrate break-through of ozone from the
stratosphere (13). It was therefore attempted to correlate
the concentrations with various local meteorological parameters.

No correlation between cloud-cover and average ozone concentration was found (figure 5). There was a tendency to increase of ozone-concentrations with temperature, but not more
than could be attributed to the general seasonal variation.

At Risø the average concentration did not depend upon the
wind-direktion; in Copenhagen only slightly higher concentrations were observed in the summer when the wind was blowing
from the quadrant E-S, i.e. from the seaside (figure 6).

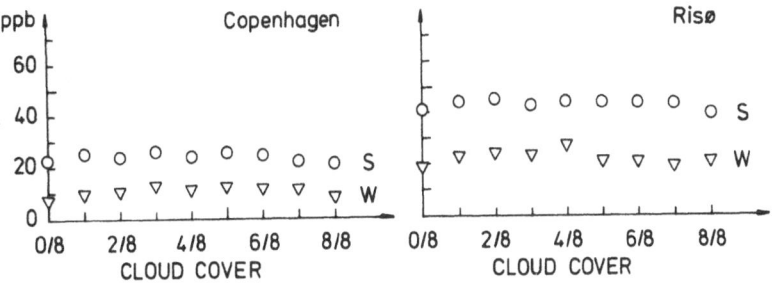

Figure 5. Average concentrations as a function of cloud cover
measured in '8ths'.

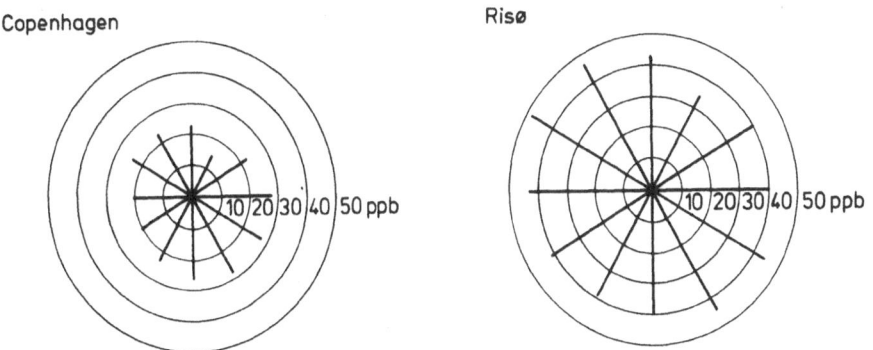

Figure 6. Average ozone-concentrations as a function of wind-
direction in the summer period (Apr.-Sep.).

Average concentrations, however, increased with wind-speed, most pronounced in Copenhagen, where the values were nearly zero for zero velocity (figure 7).

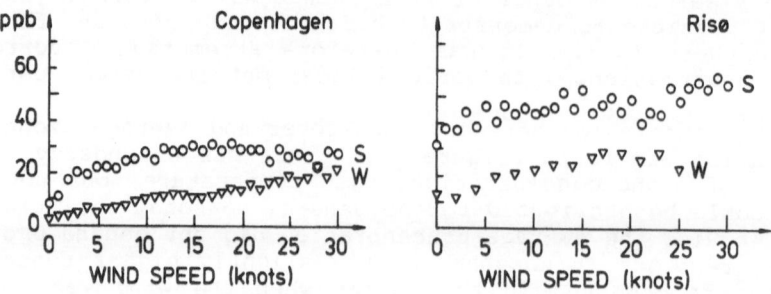

Figure 7. Average ozone-concentrations as a function of wind-speed. Upper points: summer; lower points: winter.

5. CONCLUSION

Ozone-concentrations in Denmark are generally low, and lower than found at the more northerly monitoring stations in Sweden and Norway.

Although, on the average, the highest concentrations are found during the day, there is no dependence of the cloud-cover. This indicates that local photochemical activity plays a minor role; which again is consistent with the low concentrations of the primarly pollutants, NO_x and HC.

The observed ozone must therefore arise from distant sources. However, practically no dependence on wind direction and no clearcut evidence of general long-range transport were observed. It must therefore be assumed that ozone is formed in large systems possibly covering major parts of northwest-Europe (2,10).

It has often been observed (e.g. 2) that low wind-speeds favour high ozone-concentrations, presumably because the primary pollutats (NO_x and CH) are kept together. Here, however, it was observed that the concentration - especially in Copenhagen - increases with wind-speed. This indicates that destruction of ozone is the dominant proces. Accordingly lower concentrations were found in Copenhagen (urban) that at Risø (rural).

In Copenhagen the overall ozone-concentration is higher in the weekend that on the weekdays. This is due to a reduction in concentration in the morning and in the afternoon - i. e. in the rush-hours with heavy traffic and increased emission of nitrogenoxides.

ACKNOWLEDGEMENT

Ozone measurements were carried out by Henrik Borgen and Jørgen Freundt, the Danish Boiler Association, and by Erling Lund Thomsen and Bjarne Westerberg, The Air Pollution Laboratory. Data processing was performed by Steen Rahbeck Petersen, Risø National Laboratory.
The work forms part of a larger project and was mainly financed by the National Agency of Environmental Protection and the Environmental Control, Copenhagen.

REFERENCES

(1) H. Nieboer and J. van Ham. - Peroxyacetyl nitrate (PAN) in relation to ozone and some meteorological parameters at Delft in the Netherlands. Atmos. Environ. 19, 115-120. (1976).

(2) R. Guicherit and H. van Dop. - Photochemical production of ozone in western Europe (1971-1975) and its relation to meteorology. Atmos. Environ. 11, 145-155. (1977).

(3) H.N.M. Stewart, E.J. Sullivan and M.L. Williams. - Ozone levels in central London. Nature 263, 582-584. (1976).

(4) E.H. Steinberger and E. Ganor. - High Ozone Concentrations at Night in Jerusalem and Tel-Aviv. Atmos. Environ. 14, 221-225. (1980).

(5) Y.S. Chung. - Ground-level ozone and regional transport of air pollutants. J. App. Meteorol. 16, 1127-1136. (1977).

(6) P. Winkler. - Störungen der nächtlichen Grenzschicht. - Meteorol. Rdsch. 33, 90-94. (1980).

(7) B. Elkus and K.R. Wilson. - Photochemical air pollution; Weekend-weekday differences. Atmos. Environ. 11, 509-515. (1977).

(8) W.S. Cleveland and J.E. McRae. - Weekday-weekend ozone concentrations in the Northeast United States. In: Air quality meteorology and atmospheric ozone. Proceedings of an syposium sponsored by ASTM, Boulder, Colorado, USA 31. July - 6. August 1977, 407-420 (1978).

(9) J. Schjoldager. - Ambient ozone measurements in Norway 1975-1979. J. Air Pollut. Contr. Ass. 31, 1187-1191. (1981).

(10) Naturvårdsverket. - Ozon, halt, effekt, åtgärd. SNV PM 1426, 27 pp. (1981).

(11) H. Ro-Poulsen, B. Andersen, L. Mortensen and L. Moseholm. - Elevated zone levels in ambient air in and around Copenhagen indicated by means of tobacco indicator plants. Oikos 36, 171-176 (1981).

(12) F. Palmgren Jensen. - The Danish Air Quality Monitoring Network, Air Quality Measurements, annual report 1982. MST LUFT-A78, 157 pp (1983) (in Danish).

(13) J. Fenger, I. Johnsen, L. Mortensen and H. Ro-Poulsen. - Photochemical air pollution: Formation, distribution, effects and control. Miljøprojekter 45 March 1983, National Agency of Environmental Protection, 152 pp. (In Danish).

(14) D.M. Costle. - Revisions to the national ambient air quality standards for photochemical oxidants. Fed. Reg. 44, 8202-8221. (1979).

(15) P. Grennfelt and U. Samuelsson and T. Nielsen. The presence of PAN in long-range transported polluted air masses. Proc. of the 2nd European Symposium, Varese, Italy 29. Sept. - 1. Oct. 1981, 619-624. (1982).

CLASSEMENT AUTOMATIQUE DES TRAJECTOIRES DU PANACHE DE L'ETNA

ETUDE CLIMATOLOGIQUE

D. MARTIN et D. CHEYMOL
Etablissement d'Etudes et Recherches Meteorologiques
78470 Magny les Hameaux   France
M. IMBARD et B. STRAUSS
Service Meteorologique Metropolitain
Avenue Rapp Paris

Summary

We establish a descriptive and statistical climatology of forward trajectories of Mount ETNA plume. A significant difference between the seasonal typical trajectories is pointed out. It is shown that the plume can reach a range of more than 2000 km, specially during summer. The paper presents a new methodology to assess the effect of the transport processes on the pollution, from a climatological point of view.

1. INTRODUCTION

La présente note a pour but l'établissement d'une climatologie descriptive et statistique des trajectoires du panache de l'ETNA, volcan sicilien culminant à 3300 m. Elle est essentiellement motivée par :

1°) Une action thématique programmée du CNRS (PIREN-PIRPSEV) dont l'objectif est de définir la contribution du volcanisme à la composition chimique de l'atmosphère.

2°) Le PIRPSEV, Programme Interdisciplinaire de Recherches sur la Prévision et la Surveillance des Eruptions Volcaniques (CNRS) a défini l'ETNA comme volcan laboratoire (schéma directeur du PIRPSEV 1983 - 1985 - Bulletin d'information n° 3). Rappelons que l'ETNA est une source importante de matière, (600.10^3 t/an de $SO2$, débit équivalent à la production de 10 centrales thermiques de 500 MW).[3],[4],[11].

3°) L'étude de la pollution de la Méditerranée par voie atmosphérique par les métaux lourds. Il s'agit là d'un programme prioritaire de WMO/PNUE. On a montré [2] que les flux de matières solides du volcan par rapport aux flux d'origine anthropogène à l'échelle du bassin méditerranéen étaient négligeables pour le plomb, comparables pour le cadmium, le mercure, le cuivre, le zinc et prédominant pour le selenium. L'ETNA peut cependant avoir un effet différent à l'échelle régionale [1]. La climatologie des trajectoires de masses d'air passant au-dessus de l'ETNA doit permettre de mettre en évidence une zone où cet effet, s'il existe, pourra être observé.

4°) L'étude de l'histoire des volcans fondée sur les analyses des sédiments prélevés par forages en mer peut également avoir besoin de cette climatologie afin de mieux orienter les investigations sur le terrain [5]. Une climatologie des trajectoires du panache doit donc permettre de mieux définir une campagne expérimentale d'étude du panache froid et devrait être une aide précieuse à l'interprétation de résultats de mesure.

5°) Enfin, cette étude permet le développement d'un outil d'investigation pour l'étude du transport longue distance, axe de recherche récemment défini par l'EERM et faisant partie du programme de recherche de l'équipe Physico Chimie du CRPA. L'établissement de cette climatologie a été possible grâce à l'existence d'un fichier d'archives des géopotentiels à 1000 mb, 850 mb et 700 mb, de 1963 à 1980 et d'un modèle de calcul des trajectoires de masses d'air.

2. METHODOLOGIE

Généralités :

Nous avons calculé les trajectoires des particules d'air, passant au-dessus de de l'ETNA à 00. TU et 12. TU à 700 mb. Nous avons constitué un fichier RESULTATS sur lequel figurent les coordonnées des points d'arrivée du panache correspondants à 1, 2 et 3 j d'échéance. L'utilisation de ces résultats doit tenir compte de la climatologie des précipitations sur le bassin méditerranéen et sur les pays riverains (fig. 1). Elle montre que les trajectoires à 3 j sont davantage significatives l'été que l'hiver, la fréquence d'occurrence d'une précipitation étant plus élevée en hiver. Pendant cette saison, les trajectoires de 1 jour restent cependant valables. Au-delà de 3 jours d'échéance, les incertitudes sur les trajectoires peuvent devenir importantes, et la distribution géographique des points d'arrivée trop dispersée. La climatologie au-delà de 3 j a été jugée trop peu significative et n'a donc pas été entreprise.

Le fichier HEMIS :

Le fichier HEMIS constitué par SMM/CLIM/DEV est un fichier chronologique, par niveaux isobars, de données de géopotentiel et température en points de grille sur domaine hémisphérique (grille 47*51). Cette zone correspond approximativement aux latitudes supérieures à 20 degrés nord. La période archivée débute le 1.1.63/00.TU à raison de deux réseaux par jour. Le fichier est organisé par niveau et par paramètres (géopotentiel et température). Les niveaux archivés sont : 1000 - 850 - 700 - 500 - 300 - 200 - 100 mb soit, au total 7 niveaux isobariques.

Le modèle de trajectoires :

Deux modèles sont actuellement opérationnels à la Météorologie pour calculer les trajectoires de particules d'air. Le premier a été créé en 1975, il utilise l'hypothèse géostrophique et les champs de géopotentiels archivés au SMM. Le deuxième, créé en 1981, utilise les champs de vents et de vitesses verticales analysés et archivés au CEPMMT (Centre Européen pour les Prévisions Météorologiques à Moyen Terme - READING - G.B.). Cette méthode est plus performante (4 échéances par jour au lieu de 2 et calcul des trajectoires à partir des vents analysés) mais n'est pas applicable sur le fichier HEMIS car celui-ci ne contient que les géopotentiels. On a donc utilisé le premier modèle. Il est bien évident que l'hypothèse géostrophique ne peut rendre compte des mouvements verticaux, toutefois, appliquée au niveau 700 mb où les forces de frottement sont négligeables elle donne des résultats satisfaisants surtout quand il s'agit d'une étude statistique des courants empruntés par le panache (9).

Résolution :

Le schéma numérique temporel consiste en des avances centrales effectuées toutes les 300 s. Pour ce faire, à partir des données de géopotentiel connues toutes les 12 heures on calcule le vent géostrophique à chaque pas dans le temps par une interpolation linéaire temporelle entre les deux réseaux d'analyse adjacents. Ce système permet d'éviter les discontinuités dues aux changements de réseau et l'incrémentation de 300 s assure une bonne finesse dans le calcul de la trajectoire. A chaque changement de réseau une avance latérale brise le rythme des avances centrales ; on stabilise ainsi le modèle, l'intégration ne s'étant pas révélée stable à longue échéance en l'absence de cette procédure surtout dans les zones où le champ d'isohypses présente de fortes courbures.

Calcul des vents :

Le vent utilisé est le vent géostrophique sur une surface isobare

$$u_g = -\frac{f}{g}\frac{\partial z}{\partial y} \qquad v_g = \frac{f}{g}\frac{\partial z}{\partial x}$$

avec z : géopotentiel de la surface isobare

f : paramètre de Coriolis : $2\omega \sin\varphi$

On calcule le vent géostrophique en connaissant les valeurs du géopotentiel uniquement sur les noeuds d'une maille carrée. Pour cela on utilise une méthode simple d'interpolation linéaire des géopotentiels dans une maille selon les directions O_x et O_y et on calcule le vent géostrophique par différences finies. Ce procédé très simple a le mérite d'exiger uniquement la connaissance de quatre valeurs de géopotentiel (sur les sommets du carré dans lequel on travaille) et donc de ne pas faire appel à des points trop éloignés spatialement comme l'exigent les méthodes consistant à calculer la structure polynomiale du champ de géopotentiel près du point considéré à partir des points de grille adjacents. Le désavantage de cette méthode est que l'on ne calcule pas exactement le vent au point de calcul.

3. RESULTATS

Nous avons ainsi pu calculer 10976 trajectoires à un jour d'échéance, 10327 trajectoires à deux jours d'échéance et 9707 trajectoires à trois jours d'échéance pendant une période allant du 1.1.63 au 31.10.80.

généralités:

L'ensemble des résultats peut être classiquement représenté par des histogrammes (figure 1). (7).

Les distributions de la direction du point d'arrivée à 1 j, 2 j, et 3 j d'échéance pour les mois de juillet et décembre sont données dans la figure 2. On peut ainsi comparer l'évolution de la distribution proprement dite. A un jour d'échéance en juillet elle a une forme gaussienne centrée en SE alors qu'à trois jours une bimodalité apparaît en ESE et SSW. On trouvera plus loin une trajectoire type caractérisant cette saison. Au contraire, pour le mois de décembre le maximum reste centré en E.

Par ailleurs, à partir de la distribution géographique annuelle et saisonnière, sur une maille de 2º par 2º, pondérée par la surface de la maille (celle-ci étant fonction de la latitude), nous avons tracé les lignes d'égale distribution (fig. 3).

Ces premiers résultats montrent que à un jour d'échéance, les particules d'air suivent les flux caractéristiques des vents zonaux, tandis qu'à trois jours, la composante NW du vent est prépondérante.

En faisant remarquer que les trajectoires à un jour peuvent caractériser l'effet "régional" et les trajectoires à trois jours l'effet "longue distance", nous allons analyser les résultats en suivant cette dichotomie.

Occurence régionale :

Nous rappelons que la fréquence des précipitations (1 j sur 2 à CATANIA en décembre) implique que l'effet régional de l'ETNA devrait être sensible en hiver.
L'analyse annuelle et mensuelle de la direction, de la distance du secteur, les plus probables est donnée dans les tableaux I et II. On retrouve les vents de secteurs WNW et forts en hiver, les distances parcourues sont comprises entre 1000 et 1500 km, alors que les mois de juillet et août sont caractérisés par des vents dominants NNW plus faibles (distances parcourues comprises entre 500 et 1000 km). Ces résultats sont bien concordants avec la climatologie des vents à 700 mb de Messine, Athènes et Malte . L'occurence régionale est caractérisée par un maximum net (19º E, 35º N) et une répartition circulaire autour de ce point (fig. 3).

Occurence longue distance :

Nous admettrons que les trajectoires ayant leur point d'arrivée à au moins 2000 km sont représentatives de l'effet éventuel "longue distance".(8).Les distances moyennes annuelles les plus probables, toutes directions confondues, sont comprises entre 1500 et 2000 km (21%) et entre 2000 et 2500 km et la fréquence d'occurence d'une distance supérieure à 2000 km est élevée (46%). La figure 4 représente une analyse mensuelle de cette fréquence. Les fréquences des vents d'hiver (DEC, JANV, FEV, MARS) sont les plus élevés(> 50%). Le minimum apparait en octobre (29.8% ce qui reste élevé).Ces remarques montrent que l'ETNA, en activité volcanique "normale", c'est-à-dire non éruptive, a un effet longue distance certain.Pendant les mois d'hiver (NOV, DEC, JANV, FEV) les maximum d'occurence apparaissent à des distances supérieures à 3000 km, avec une direction du vent W à WSW traduisant bien les flux zonaux sous ces latitudes.Les mois d'août est nettement caractérisé par un flux particulier de NNE. En fait, il faut tenir compte dans ce cas des points à 1 j et 2 j d'échéance qui indiquent l'esquisse d'une trajectoire type pour le mois d'août, présentant une courbure très nette. On verra au chapitre IV comment s'explique cette trajectoire type.Enfin, la figure 5 montre que les directions les plus probables pour les distances supérieures à 2000 km sont ENE (15,4%) et ESE (11,9%). L'évolution mensuelle de ces deux fréquences est représentée par la figure 6.

4. TYPOLOGIE DES TRAJECTOIRES

La typologie des trajectoires du panache de l'ETNA a été faite en adaptant un programme de classement automatique. Le principe s'inspire de la méthode des nuées dynamiques (6). Soient IFIN le nombre de données à classer et NET le nombre d'étalons. Pour classer nos IFIN données, on va tout d'abord, tirer aléatoirement NET nombres compris entre 1 et IFIN. On prend alors les données correspondantes et on regroupe les données du fichier autour de ces étalons. Pour cela, on calcule les distances curvilignes D_i entre le point d'arrivée au $i^{ième}$ jour de la donnée à classer et le point d'arrivée au $i^{ième}$ jour de l'étalon considéré. On définit alors la distance D :

$$D = \sum_{i=1}^{P} D_i$$

(p = 3 pour 3 jours)

On range l'élément dans la classe où D est minimum. On répète l'opération pour chaque donnée. On calcule alors le nouveau barycentre de chaque classe. Pour chaque barycentre aléatoire, on calcule les distances moyennes,DM, définies ci-dessus, entre l'étalon et les éléments de sa classe et, les distances, DE, entre les étalons entre eux.Si DE est plus grand et DM plus petit, on estime avoir fait un meilleur tirage. On a donc un tirage aléatoire optimisé. On calcule alors le nouveau barycentre, on regroupe et ainsi de suite jusqu'à obtenir une variation moyenne entre un regroupement et son suivant inférieur à un seuil défini par l'utilisateur. Le résultat de ce classement est reporté fig. 7, pour les saisons d'été et d'hiver. On peut comparer ces résultats à une classification des types de temps en EUROPE (10). On constate que les trajectoires types trouvées ont une situation météorologique correspondante. La correspondance est très bonne et prouve que les résultats obtenus ne sont pas des artifices numériques mais qu'ils représentent bien la réalité atmosphérique. Enfin, remarquons que si d'une manière générale les flux dominants sont bien d'W et de SW, les mois de juillet et août sont caractérisés par des flux de N liés à la remontée et au renforcement de l'anticyclone des ACORES.

5. CONCLUSIONS

Après adaptation d'outils existants (fichier Hémis, modèle de trajectoires, programme de classement automatique), nous avons pu établir une première climatologie descriptive et statistique des trajectoires du panache de l'ETNA. Des trajectoires types saisonnières et mensuelles ont été calculées et sont cohérentes avec une classification des types de temps.
Nous avons montré que l'ETNA pouvait avoir une influence à des distances supérieures à 2000 km surtout en été.
Enfin, nous avons ainsi défini une méthodologie pour l'approche de l'étude de l'influence d'une source. Ce travail devrait maintenant être complété par une étude

plus quantitative qui examinerait :
- l'homogénéïté des résultats
- les modes et les profils des distributions
- les corrélations entre les distances et les directions
- les tendances

6. BIBLIOGRAPHIE

1. M. ARNOLD et Alii, "Geochimie et transport des aérosols métalliques au-dessus de la Méditerranée occidentale" CIESM/PNUE Journées d'études sur les pollutions marines en Méditerranée - CANNES 2-4 DEC 1982.
2. P. BUAT MENARD et Alii, "The heavy metal chemistry of atmospheric particulate matter emitted by Mount Etna Volcano" Geophysical Res. letters 5 n° 4 p. 245-248 (1978).
3. R. FAIVRE PIERRET et Alii "Contribution des sondes aérologiques motorisées à l'étude de la physico chimie du panache volcanique. Bull. of volcanology 43-3 (1980).
4. W. JAESCHKE et Alii "Sulfur emissions from Mount ETNA" J. of Geophysical research vol. 87 n° C9 p. 7253-7261 august 20 (1982).
5. J. KELLER et Alii "Explosive volcanic activity in the Mediterraneen over the past 200.000 years as recorded in deep sea sediments" Geophysical society of American Bull. Vol. 89 p. 591-604 (1978).
6. L. LEBART et Alii "Traitement des données statistiques" DUNOD (1979).
7. D. MARTIN et Alii: "Climatologie des trajectoires du panache de l'Etna" NIT EERM Déc. 1983 à paraître.
8. J.M. MILLER "A five year climatology of five days back trajectories from BARROW, Alaska". Atmospheric Environment. Vol. 15 N° 8 pp 1401-1405 (1981).
9. R. SYKES "Computation of horizontal trajectories based on the surface geostrophic wind". Atmospheric Environment. Vol. 10 pp 925-934 (1976).
10. M. URBANI "Su una classificazione di typi di tempo in Europa" Rivista di Meteorologia Aeronautica. Anno 15 n. 3-4 p. 31 (1955).
11. P. ZETTWOGG et Alii "Experimental results on the SO2 transfer in the mediterranean obtained with remote sensing devices" Atmospheric Environment 12, 795-796 (1978).

1J/63.80	DIRECTION (%)	DISTANCE (%) X 100 KM		SECTEUR		
			DIR	DIR	DIS	%
annuelle	ESE 29.0	5-10 44.8		ESE	5-10	12.7
JANV	ESE 26.7	5-10 32.7		ESE	10-15	9.9
FEV	ESE 32.3	5-10 40.0		ESE	10-15	13.0
MARS	ESE 29.2	5-10 39.4		ESE	10-15	12.2
AVRIL	ESE 33.3	5-10 39.8		ESE	10-15	13.6
MAI	ESE 29.2	5-10 51.0		ESE	5-10	14.4
JUIN	ESE 36.6	5-10 47.7		ESE	5-10	15.1
JUILL	SSE 39.8	5-10 52.6		SSE	5-10	21.5
AOUT	SSE 33.3	5-10 58.1		SSE	5-10	22.1
SEPT	ESE 29.1	5-10 44.8		SSE	5-10	12.8
OCT	ESE 20.9	5-10 46.9		ESE	5-10	12.0
NOV	ENE 30.1	5-10 44.4		ENE	5-10	11.9
				ESE	5-10	11.9
DEC	ENE 26.6	5-10 38.2		ESE	10-15	10.1

Tableau I : Caractérisation de la direction, de la distance et du secteur les plus probables en fonction du mois à un jour d'échéance.

3J/63.80	DIRECTION (%)	DISTANCE (%) X 100 KM	DISTANCE >2000 KM %	SECTEUR DIR.	DIS.	%
annuelle	ENE 25.9	15-20 21.0	46	ESE	20-25	6.4
	ESE 24.7					
JANV.	ENE 28.4	20-25 20.8	57	ENE	>30	10.5
FEV.	ENE 34.0	20-25 24.8	64	ENE	>30	11.0
MARS	ENE 31.4	20-25 21.6	54.8	ENE	20-25	7.7
AVRIL	ENE 37.7	20-25 18.4	52	ENE	>30	8.3
MAI	ESE 27.1	10-15 23.1	39.6	ESE	20-25	7.5
				SSE	10-15	6.2
JUIN	ESE 30.2	15-20 24.7	45.6	ESE	20-25	9.7
JUILL	ESE 26.1	15-20 30.2	40.6	ESE	15-20	9.1
AOUT	SSW 26.9	15-20 29.2	34.2	SSW	15-20	8.9
SEPT	ESE 26.6	15-20 24.1	31.4	ESE	15-20	9.7
OCT	ENE 22.3	10-15 22.4	29.8	SSE	15-20	5.9
NOV	ENE 31.6	15-20 19.9	46.0	ENE	20-25	7.5
DEC	ENE 29.9	>30 23.9	59.8	ENE	>30	11.7

Tableau II : Caractérisation de la direction, de la distance et du secteur les plus probables en fonction du mois à trois jours d'échéance.

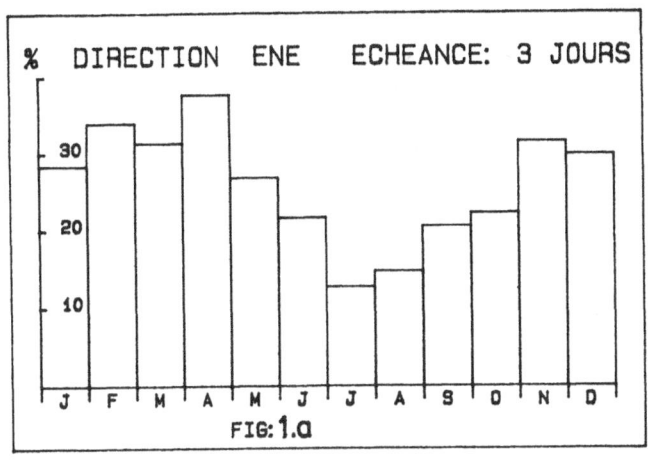

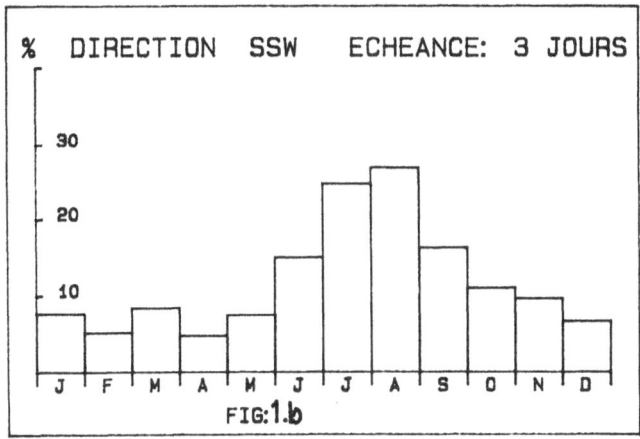

Fig.1:Histogrammes pour les directions ENE et SSW à 3 jours d'échéance.

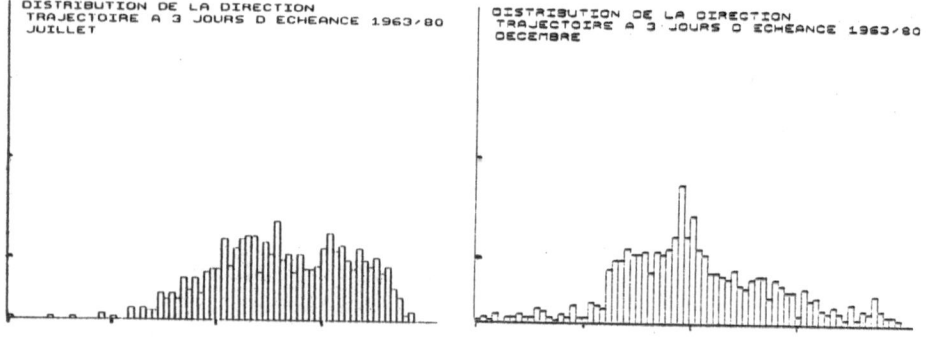

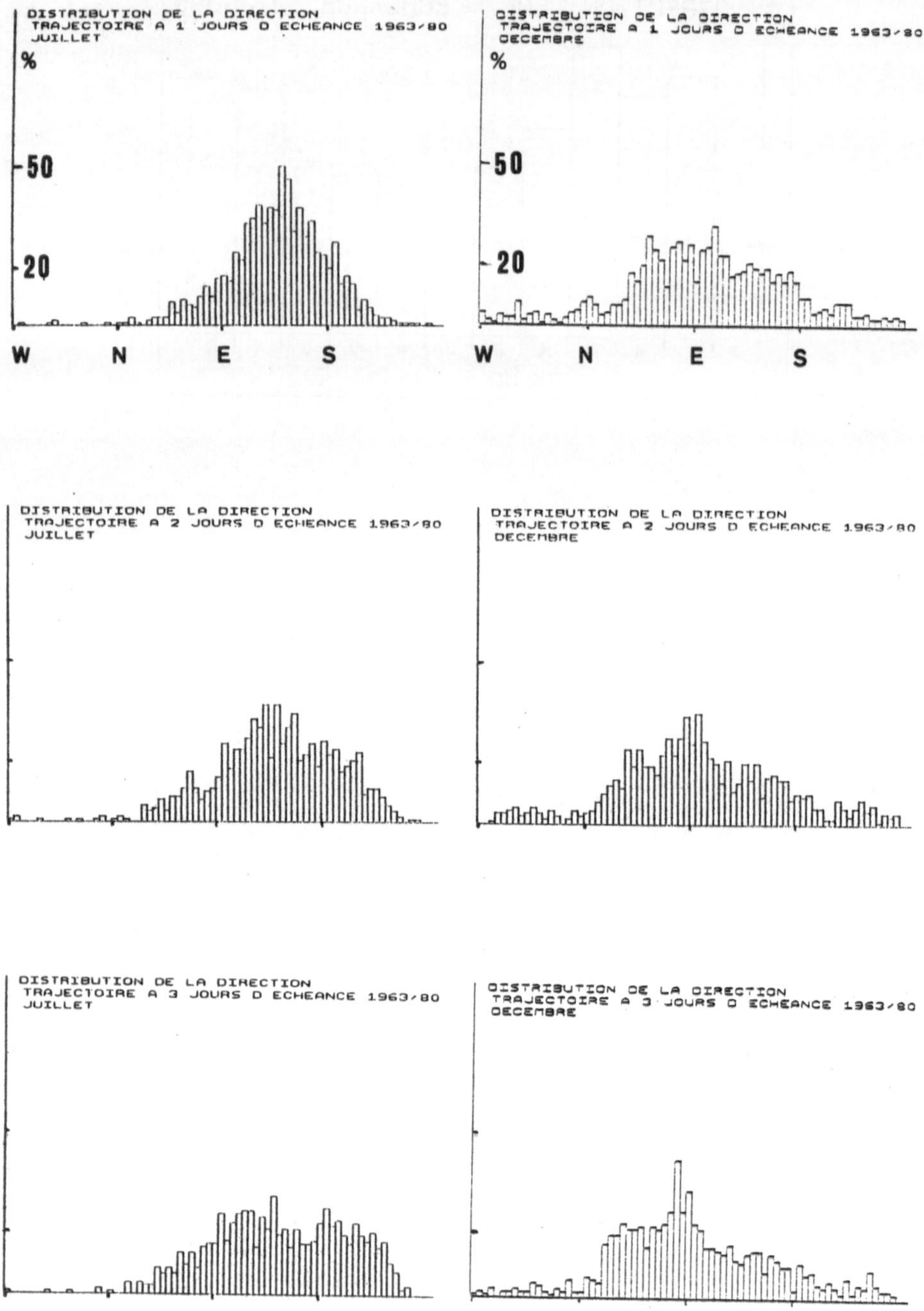

Fig.2:Distribution (5° X 5°) de la direction des points d'arrivée à 1, 2, 3 jours d'échéance pour les mois de juillet et de décembre.

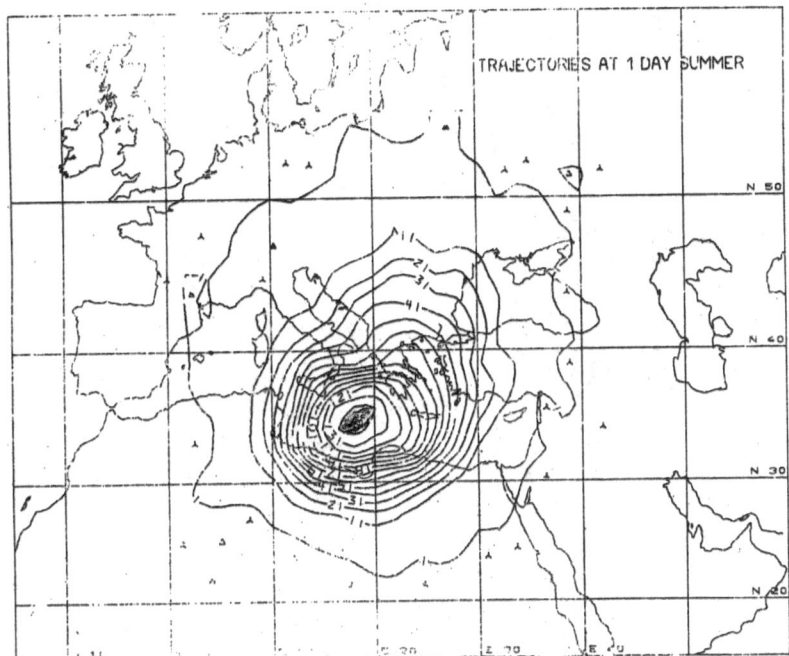

Fig.3A

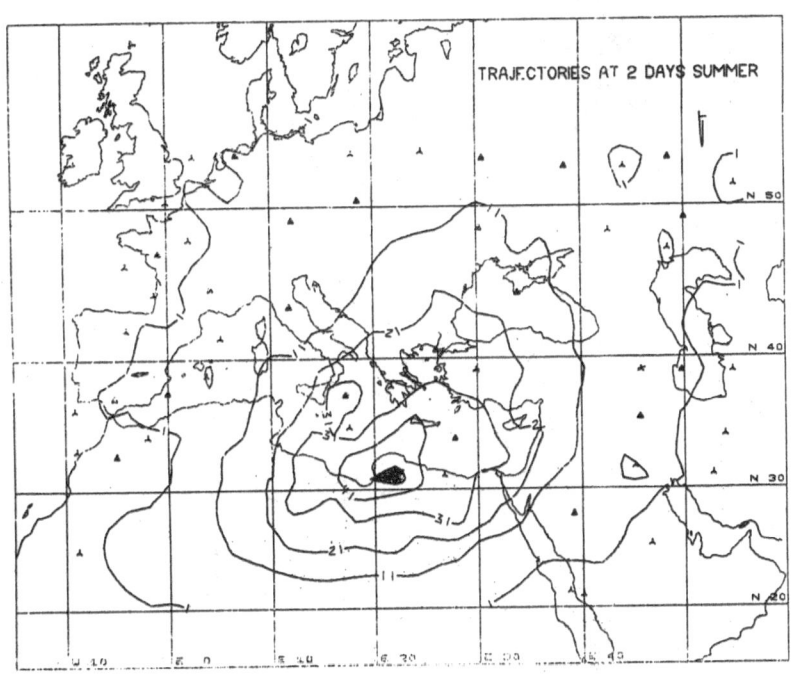

Fig.3B

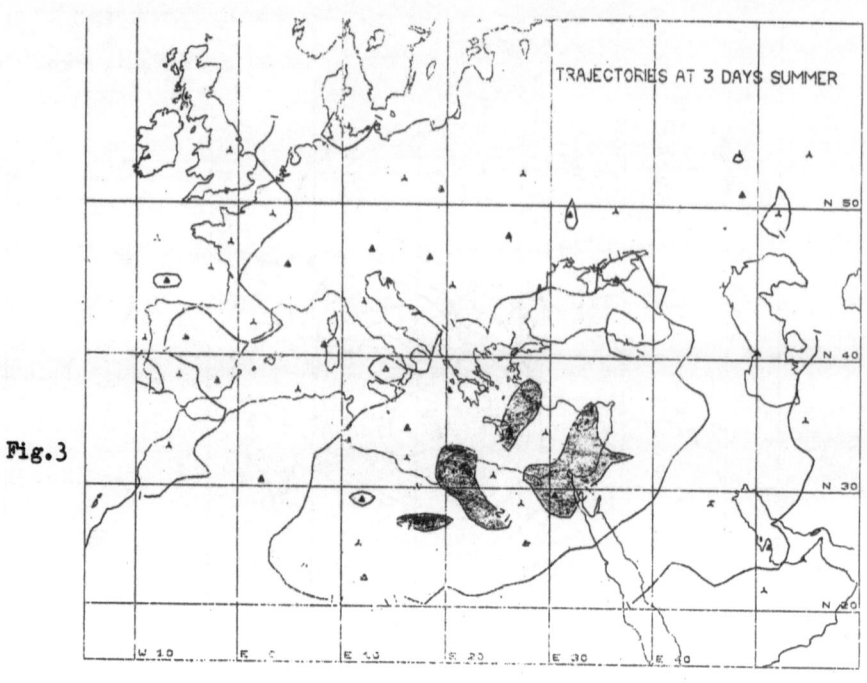

Fig.3

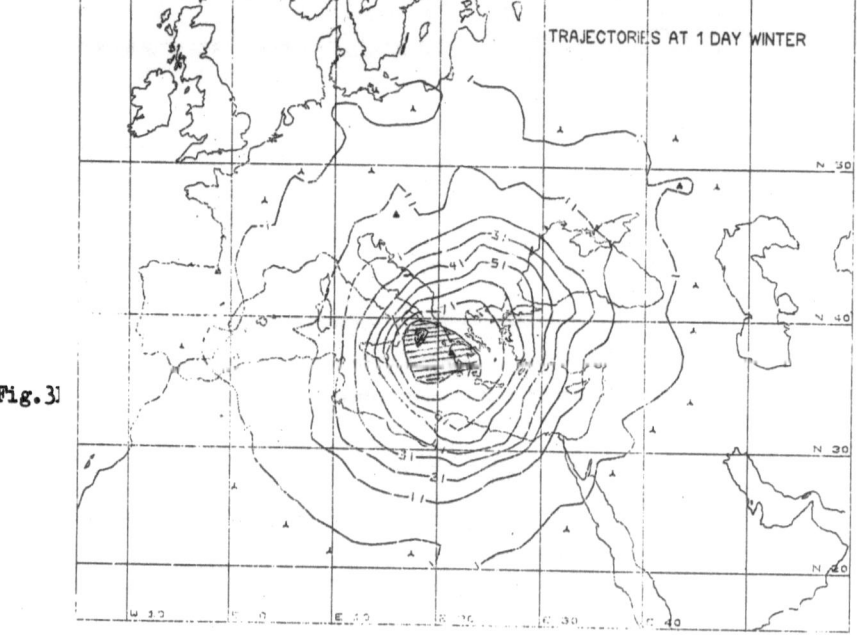

Fig.3]

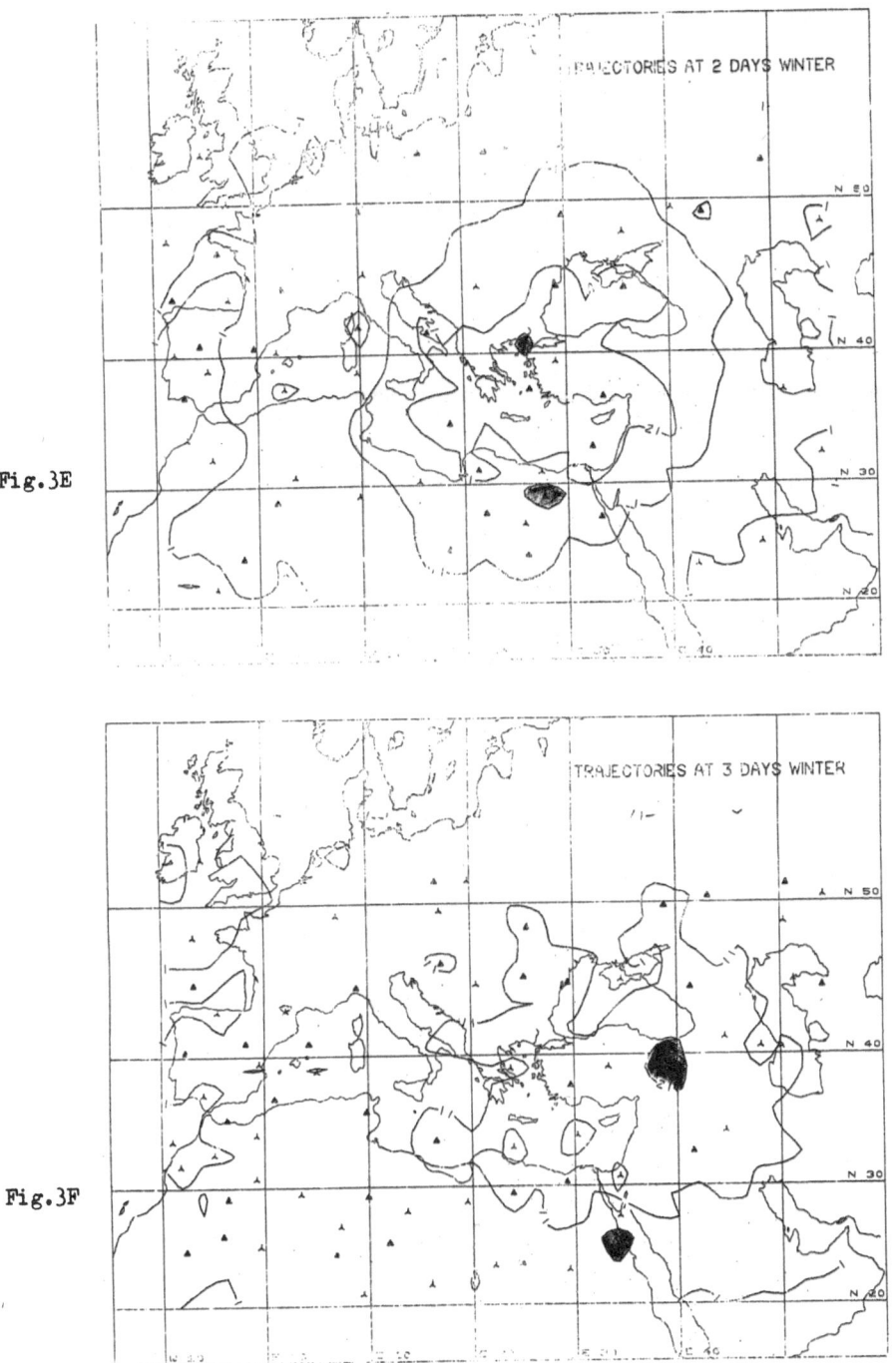

Fig.3E

Fig.3F

Fig.3:Distribution géographique des points d'arrivée à 1, 2, 3 j d'échéance en été et
hiver.

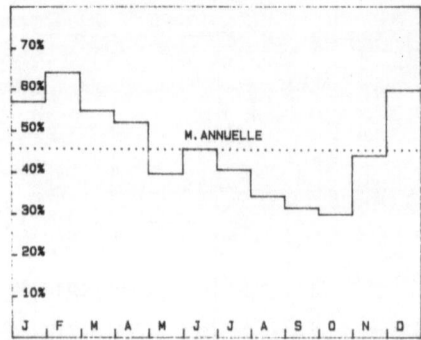

Fig.4:Fréquences d'occurence du point d'arrivée d'une trajectoire à 3 j d'échéance et à une distance supérieure à 2000 km, toutes distances confondues.

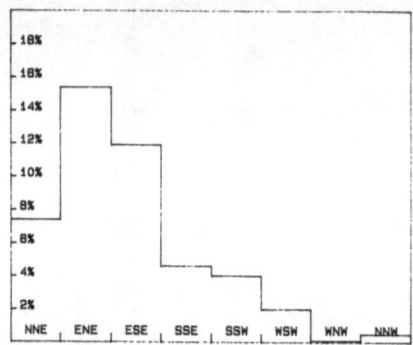

Fig.5:Fréquences d'occurence des points d'arrivée d'une trajectoire à 3 j d'échéance et à une distance supérieure à 2000 km en fonction de la DIRECTION.

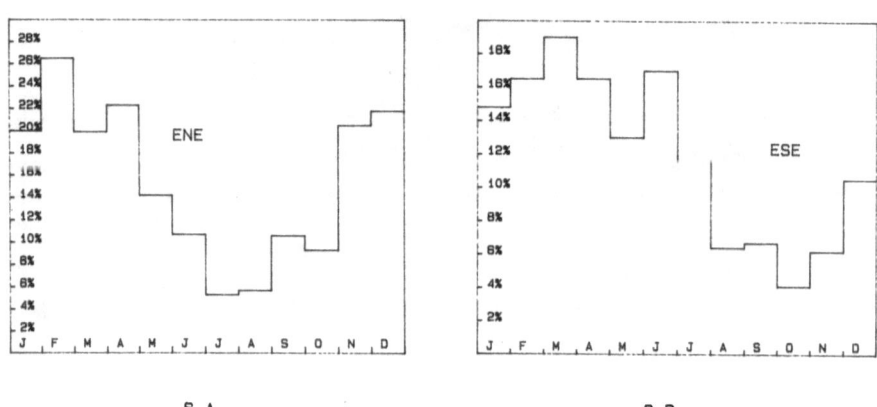

8.A

8.B

Fig.6: Analyse mensuelle des directions les plus probables (ENE et ESE) et dont les points d'arrivée à 3 j d'échéance sont à une distance supérieure à 2000 km.

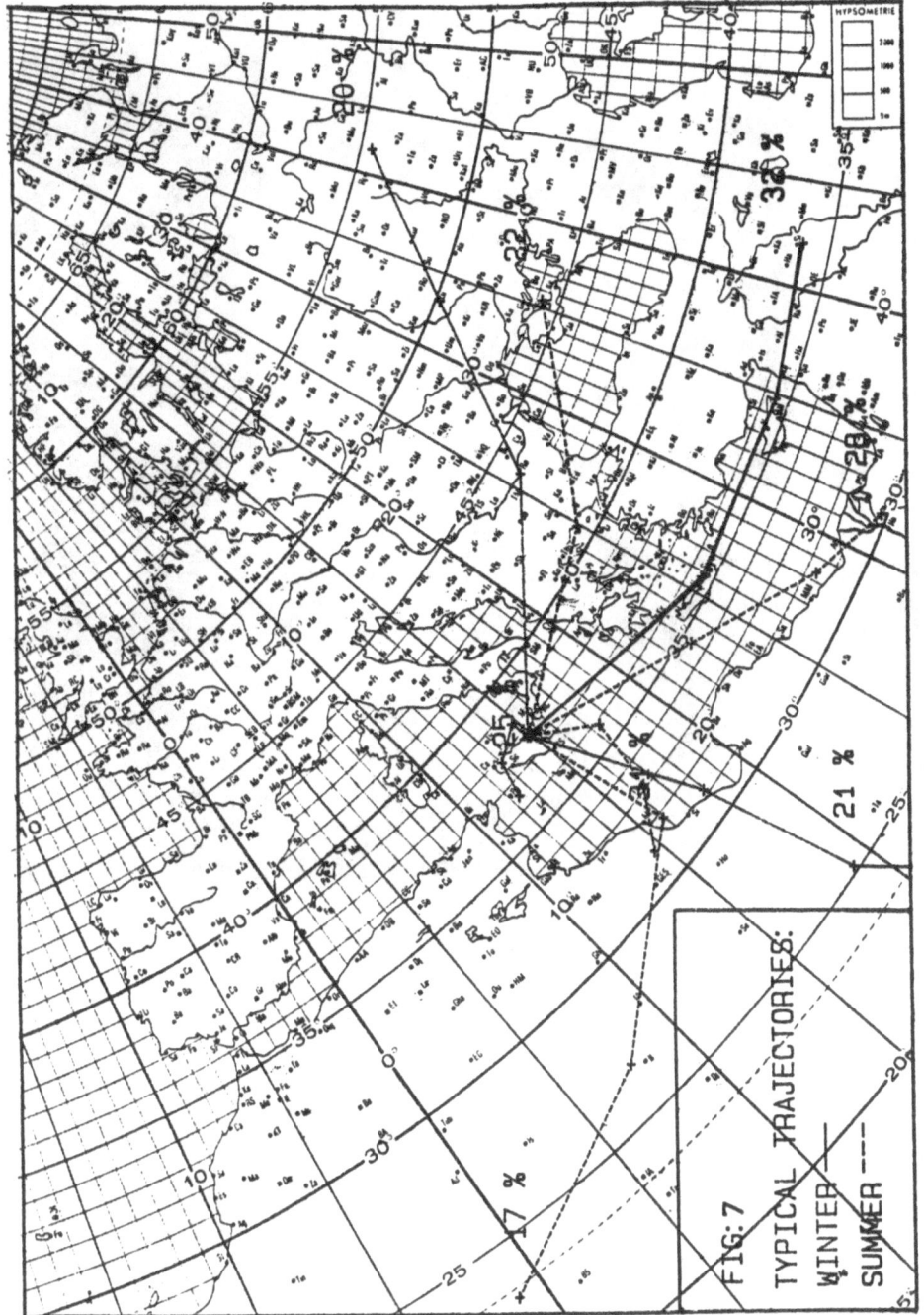

Fig.7:Trajectoires types en été et en hiver.

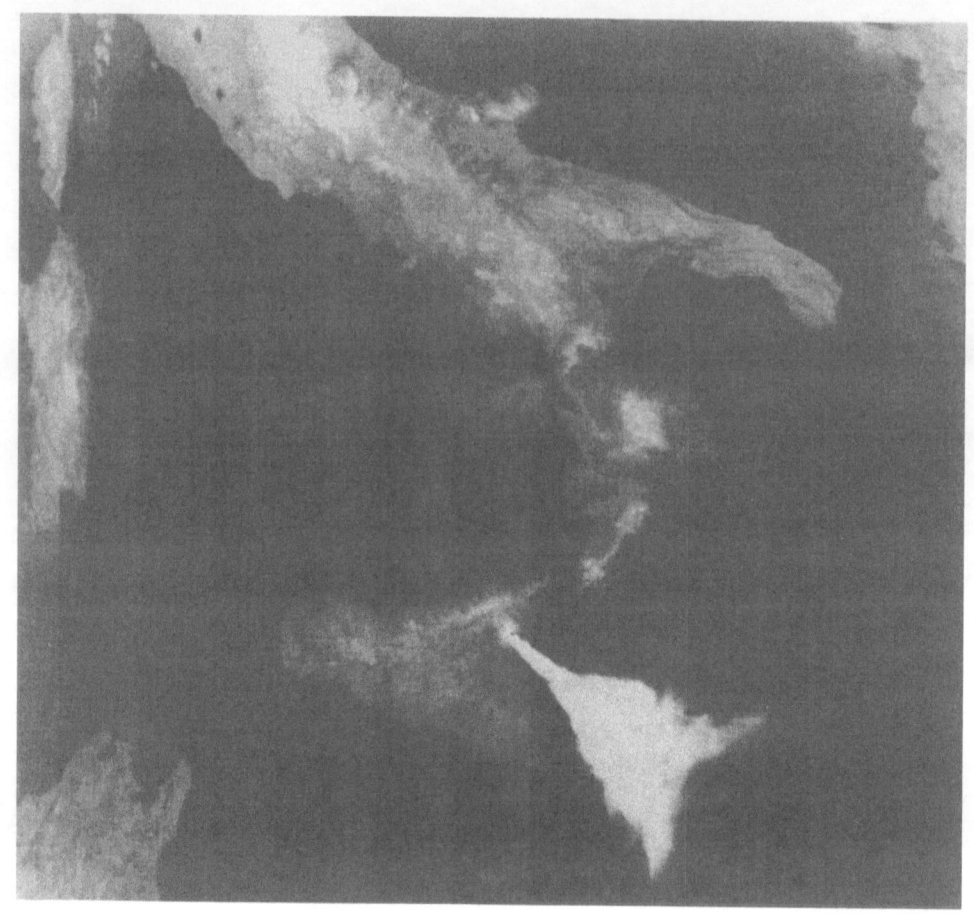

Satellite TIROS N

Orbite n° 4155 du 4 août 1979 à 02 heures TU

AVHRR Canal 4 (infra-rouge)

DIRECTION DE LA METEOROLOGIE
ETABLISSEMENT D'ETUDES
ET DE RECHERCHES METEOROLOGIQUES
CMS - LANNION

COMMENTAIRE DE LA PHOTOGRAPHIE SATELLITAIRE

Le panache de l'ETNA en activité normale n'est pas visible sur les photographies
satellitaires (visible ou I.R.). Cette photographie montre l'éruption phréatique de
l'ETNA du 4 août 1979, prise par le satellite TIROS N (Infra Rouge). Cette
spectaculaire émission s'accompagne encore d'une énigme : pourquoi le panache de
l'ETNA a-t-il la forme de la Sicile ?

WHAT IS THE SOURCE OF ACID IN CLOUDS?

F.G. Römer, Environmental Department, N.V. KEMA,
Arnhem, The Netherlands

H.F.R. Reijnders, Laboratory for Inorganic Chemistry,
National Institute of Public Health
and Environmental Hygiene,
Bilthoven, The Netherlands

Summary

Measurements of main components were performed in cloud water over
The Netherlands. It was found that besides nitrate and sulphate,
chloride may occur in considerable amounts. These three components
form the bulk of acid anions in clouds. It is shown that ammonia
emitted from ground level neutralizes the acids from nitrogen and
sulphur oxides, and hydrogen chloride even at high altitudes. The
free acid ratio decreases quite drastically above the Netherlands
when air masses are transported from the North Sea into the country.
This decrease is accompanied by an increase in ammonium content and
to a lesser extent in hydronium ion content.

1.   INTRODUCTION

Explorative flights are carried out regularly since the end of 1981.
The measurements were mainly intended to inventory the chemical composition of cloud water. A strong variability was found for the concentrations of main components e.g. nitrate and sulphate. This variability
turned out to be dependent on sample location in combination with the
origin of the air masses [1]. Differences in the molar nitrate-sulphate
ratio can partly be explained by differences in conversion rates for
sulphur dioxide and nitrogen oxides in the gas and/or liquid phase. With
respect to this the availability of oxidant is crucial. In water particularly the oxidation of sulphur dioxide may be important if hydrogen peroxide is present. For this substance the concentration varied with both
the sample location and the time of the year [1, 2].

It cannot be deduced solely from cloud water data to what extent the
acidity in cloud and rain water is determined by the absorption (rainout,
washout) of the acids (sulphuric and nitric acid) and its neutralized
(ammonium) salts or the in situ formation of acids by oxidation of dissolved oxides. The absorption and conversion rate of sulphur dioxide are
pH dependent [3, 4]. Therefore, substances like hydrogen chloride and
ammonia affect the oxidation processes indirectly. In this paper the
acidity of clouds is discussed on the basis of cloud water data only.

2. INSTRUMENTATION

The aircraft used were a Piper Navajo Chieftain and, on a few occasions, a Cessna 172 owned by Geosens B.V., Rotterdam. A description of the aircraft and instruments (monitors, cloud and rain samplers) is given elsewhere [1].

The water samples are stored at a temperature of 4°C directly after collection and analysed as soon as possible. The methods of analysis are described in references 1 and 2. The analysis include the determination of chloride, nitrate, sulphate, ammonium, hydronium ions and hydrogen peroxide. Absolute volumes needed for the separate analysis range from 50 tot 400 μl. To run the analysis automatically, a total volume of 5 ml is needed. To set up the ionic balance also the cations of sodium, potassium, calcium and magnesium were analysed. For the complete analysis of all anions and cations about 10 ml is needed. Using the ionic balance the accuracy of the analysis is checked.

3. THE ACIDITY OF CLOUD WATER

3.1. The contribution of NO_x and SO_2

A simplified scheme of the relative importance of NO_x and SO_2 conversion processes is given in table I.

Table I   Chemical conversion (%/h) of SO_2 and NO_x *

	gas phase	water phase
SO_2	+ (< 4)	+++ ($\leq$ 100)
NO_x	++ (< 25)	–

*   formation of H_2SO_4 aerosol (+), HNO_3 gas (++), H^+ and SO_4^{2-} (+++)

The conversion of SO_2 and NO_x in the different phases depend on several factors (e.g. concentration, oxidant concentration, light intensity, temperature). In general NO_x will be removed faster from the atmosphere than SO_2. Neglecting local differences in emission densities of SO_2 and NO_x, the value of the NO_3^-/SO_4^{2-} ratio in cloud water therefore indicates the origin of the samples (distance from source areas) and includes also information about atmospheric chemical behaviour of these gases.

Typical values of the molar NO_3^-/SO_4^{2-} ratio in industrial areas in The Netherlands range from 2 tot 1. In a relatively clean area (Southern Sweden) the effect of SO_2 is more pronounced resulting in values of 1 to 1/2 for NO_3^-/SO_4^{2-} [1].

3.2. The contribution of hydrogen chloride

It is found that in The Netherlands the contribution of hydrochloric acid to the acidity in rain collected at ground level is relatively unimportant [5]. However, from the data of four flights significant contributions of hydrochloric acid (excess chloride) could be calculated to the acidity in cloud water. The values were derived according to the next two procedures:

a   from seasalt (Na$^+$) corrected values and
b   from the difference $([H^+]+[NH_4^+])-([NO_3^-]+[SO_4^{2-}])$
     (concentration in equivalents per liter).

The method under a is probably not very reliable because considerable deviations from theoretical seasalt ratios were often observed. To achieve a high reliability, calculations were carried out on selected data sets. The limiting conditions are:

1   electroneutrality (e): $0,95 \leq e \leq 1,05$
2    $[Na^+] > 0,5$ µeq/l

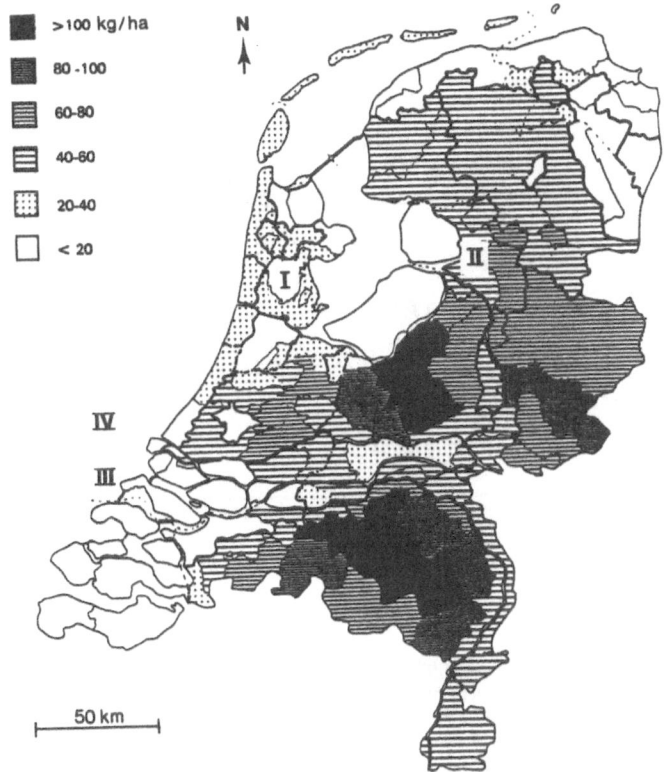

Figure 1   Map showing   a   locations (I, II, III, IV) where excess chloride concentrations were found[*]
                         b   emission density of ammonia (kg/ha) from agricultural activities in The Netherlands in 1980 according to Buijsman [6]

*   I   northern part of province of Noord-Holland, wind: East, altitude: 600-900 m
    II   northern part of province of Overijssel, wind: South South West, altitude: 1200 m
    III   North Sea at the coasts of provinces Zuid-Holland and Zeeland, wind: West, altitude: 800-900 m
    IV   background of North Sea, altitude: 2400 m

3   (|a-b|)/a ≤ 75%, with a(b) = percentage contribution following
       procedure a(b).

The results are summarized in table II. Excess chloride was found at
different locations I, II, III and IV (see figure 1) above the North Sea
and The Netherlands in cloud water. As the total quantities are not very
high, nevertheless higher concentrations may be expected near source
areas.

Tabel II The contribution of HCl to the acidification of cloud water
         using methods a and b (see text)

flight[**]	$(Cl^- \times 100)/(Cl^- + NO_3^- + SO_4^{2-})$ (%)				$(Cl^- + NO_3^- + SO_4^{2-})$ [*] (µeq/l)	
	cloud a	b	rain a	b	cloud	rain
83-05-16	11	7	21	6	261	163
	12	7	14	8	320	293
83-05-18			18	7		153
83-08-17	29	20	15	13	167	261
	28	16			188	
	35	28			157	
83-09-01			54	38		24

[*]   the sum of the anions equivalents contains the uncorrected
           chloride concentrations
[**]   for locations see figure 1: I, II, III, IV

3.3. The effect of ammonia on cloud acidity

     In The Netherlands high ammonia concentrations occur in the atmos-
phere. The areas with intensive ammonia emissions are indicated on the
map in figure 1. Ammonia is emitted mainly from agricultural sources [5,
6]. As a result relatively high ammonium ion concentrations are found
both in rain sampled at ground level and in cloud water collected at
different altitudes [1, 7].
     The free acid ratio defined as F = $[H^+]/([H^+]+[NH_4^+])$,
varies theoretically from 0 to 1 ($[H^+]$ concentration of free acid,
$[H^+]+[NH_4^+]$ potential acidity). In practice relatively high values
could be established in cloud water collected above sea and from westerly
origin: 0,70-0,80. In this situation pollutant concentration levels often
are rather low. Above the country pollutant concentrations increase and F
will depend on the amount of ammonia (or ammonium ions from aerosol) and
the amounts of nitrogen oxides, sulphur dioxide and acids (nitric acid,
sulphuric acid, hydrochloric acid) absorbed in the liquid phase. As far
as the present measurements concern F values in clouds from westerly
origin decrease in down wind direction to values as low as 0,2. At the
same time the pronounced effect of ammonia emissions is demonstrated.

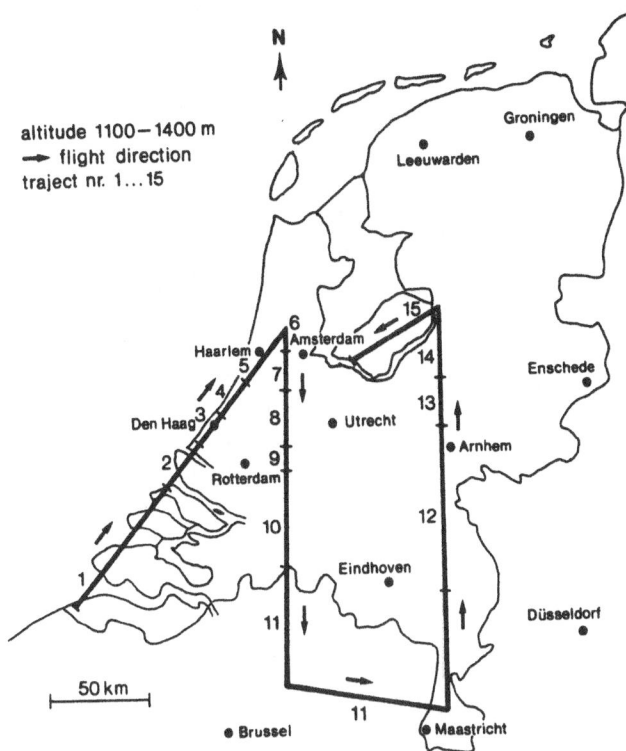

Figure 2   Flight pattern for a flight on 82-07-06

The effect of ammonia on the free acid ratio is illustrated in figure 3. During a flight on 82-07-06 (for flight scheme see figure 2) samples were taken in such a way that the increase in concentration could be traced in the air mass transport direction from west to east. From a meteorological point of view the samples were still taken in air parcels of similar origin. The total sampling time (for sample 1-15) amounted to 3,5h. Assuming that no significant differences in emission patterns during the flights occur, the data of figure 3 may be compared.

The same increasing concentrations and decreasing free acid ratios are found down wind from the Rijnmond industrial area. The flight pattern is given in figure 4. In this case the effect of ammonia emission on the free acid ratio is clearly more pronounced than the effects of acids from nitrogen and sulphur oxides.

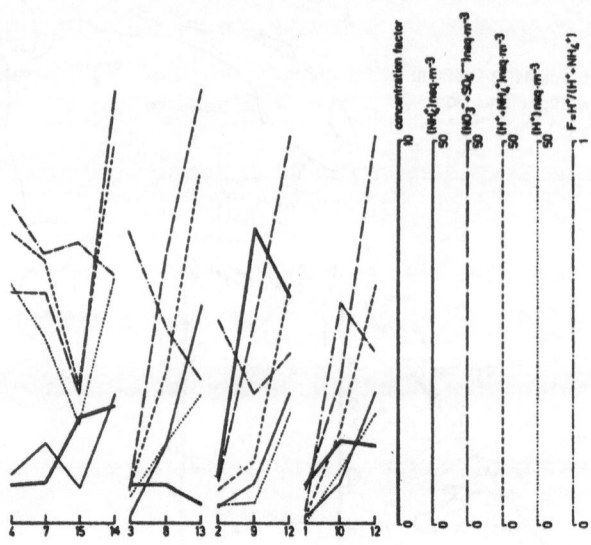

Figure 3 Change in free acid ratio (see text) and in concentrations of some main components in cloud water during transport (concentrations expressed in nano equivalents per m³ air; concentration factor i.e. cloud water density of a sample divided by the cloud water density of the upwind sample

4. CONCLUSIONS

Besides nitrate and sulphate also chloride is found in considerable concentrations in cloud water at higher altitudes. It is concluded that the major sources of acid in clouds are nitrogen and sulphur oxides; on the other hand hydrogen chloride contribution to the total amount of acid may not be neglected. Furthermore it is remarkable that ammonia, which is quite a reactive substance and is emitted mainly (more than 95%) by low sources, occurs in the same concentration range as nitrate, sulphate and chloride. Ammonia neutralizes the main part of hydronium ions. The neutralization process may proceed in the gas, liquid as well as in the solid phase. The increase in ammonium content in cloud water by emission of source areas in The Netherlands could be demonstrated quite convincingly. This increase in ammonium is greater than the increase in hydronium ion content. As a result the free acid ratio decreases in the transport direction while the potential quantity of acid increases.

ACKNOWLEDGEMENT

The authors thank mr. J.F. den Tonkelaar (KNMI, De Bilt) for delivering valuable meteorical information.

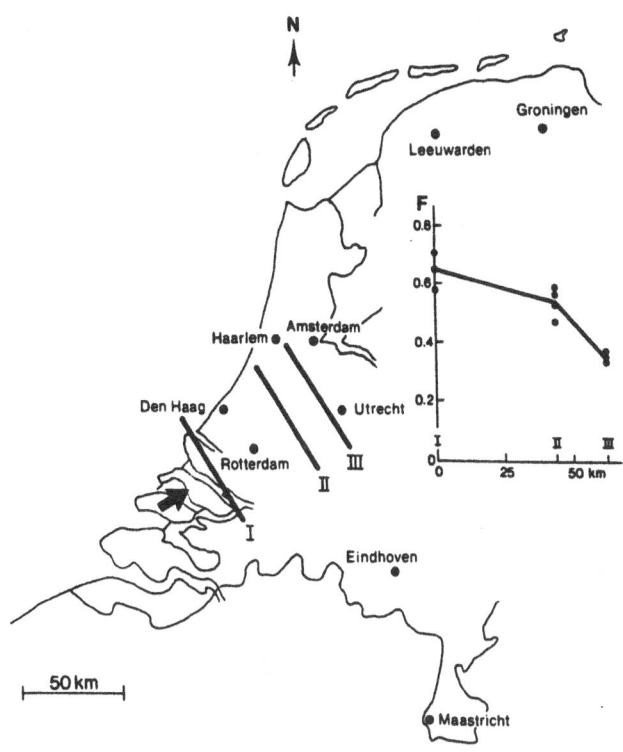

Figure 4   Flight pattern for a flight on 83-11-25 and change in free acid
ratio

REFERENCES

1   RÖMER, F.G., VILJEER, J.W., VAN DEN BELD, J., SLANGEWAL, H.J.,
    VELDKAMP, A.A. and REIJNDERS, H.F.R., (1984). The chemical composi-
    tion of cloud and rain water. Results of preliminary measurements
    from an aircraft. Atmos.Environ. (accepted for publication)

2   RÖMER, F.G., SLANGEWAL, H.J., VELDKAMP, A.A. and REIJNDERS, H.F.R.,
    (1983). Wolkenwateronderzoek in Nederland. In: Zure regen; oorzaken,
    effecten en beleid. Proc. Symp. november 1983, 's Hertogenbosch; ed.
    E.H. Adema and J. van Ham, Pudoc I II, Wageningen

3   PENKETT, S.A., Jones, B.M.R., BRICE, K.A. and EGGLETON, A.E.J.,
    (1979). The importance or atmospheric ozone and hydrogen peroxide in
    oxidizing sulphur dioxide at low pH. Atmos.Environ. 13, 123-137

4   MARTIN, L.R. and DAMSCHEN, D.E., (1981). Aqueous oxidation of sulphur
    dioxide by hydrogen peroxide at low pH. Atmos.Environ. 15, 1615-1621

5   VAN AALST, R.M., (1984). Verzuring door atmosferische depositie. Atmosferische processen en depositie. Publikatiereeks Milieubeheer. VROM 83656/1-84

6   BUIJSMAN, E., (1983). Ammoniakemissies in Nederland. IMOU-report V-83-3. Institute for Meteorology and Oceanography, Univ. Utrecht

7   RIDDER, T.B., REIJNDERS, H.F.R., VAN DEN ESHOF, A.J., WEGMAN, R.C.C. and MOOK, W.G., (1982). Chemical composition of precipitation over The Netherlands. Annual Report 1982. RIV-KNMI report, KNMI, De Bilt, The Netherlands

LIST OF PARTICIPANTS

ANGELETTI, G.
Commission of the European
Communities - D.G. Science,
Research and Development
200, rue de la Loi
B - 1049 BRUSSELS

AUGUSTIN, H.
Commissariat aux Risques Naturels
25, av. Charles Floquet
F - 75007 PARIS

BABELOT, J.F.
Commission of the European
Communities
Joint Research Centre
Karlsruhe Establishment
P.O. Box 2266
D - 7500 KARLSRUHE

BARNES, I.
Universität
D - 5600 WUPPERTAL 1

BATT, L.
Dept. of Chemistry, University
Meston Walk
GB - OLD ABERDEEN AB92UE

BAUER, H.
G.S.F. - U. Umweltforschung m.b.H.
Bereich Projektträgerschaften
Josephspitalstrasse, 15
D - 8000 MUENCHEN

BECKER, K.H.
Phys. Chem./FB9 - Universität
Gauss Str., 20
D - 5600 WUPPERTAL 1

BEHNKE, W.
Fraunhofer Institut ITA
Stadtfelddamm, 35
D - 3000 HANNOVER

BEILKE, S.
Umweltbundesamt
Feldbergstrasse, 45
D -6000 FRANKFURT

BESEMER, A.
Division of Technology
for Society
T.N.O.
P.O. Box 217
NL - 2600 AE DELFT

BESSEMOULIN, P.
Centre National de Recherches
Météorologiques
42, av. Coriolis
F - 31062 TOULOUSE

BOHLER, T.
N.I.L.U.
P.O. Box 130
N - 2001 LILLESTROM

BOURDEAU, P.
Commission of the European
Communities - D.G. Science,
Research and Development
200, rue de la Loi
B - 1049 BRUSSELS

BRESSON, M.A.
Université Paris VII
Lab. P.C.I.
2, place Jussien
F - 75251 PARIS CEDEX 05

BUCK, M.
Landesanstalt für Immissions-
schutz
D - 4300 ESSEN

BUIJS, K.
Commission of the European
Communities
Joint Research Centre
Karlsruhe Establishment
P.O. Box 2266
D - 7500 KARLSRUHE

BURROWS, J.P.
Max Planck Institut für Chemie
Saarstrasse, 23
D -6500 MAINZ

CANDREVA, F.
Ryksuniversiteit of Gent
86, Proeftuinstraat
B - 9000 GENT

CAPPELLANI, F.
Commission of the European
Communities
Joint Research Centre
Ispra Establishment
I - 21020 ISPRA - VARESE

CARBONNELLE, J.
C.E.A. DPT/SPIN
Bât. 93
F - 91191 GIF-SUR-YVETTE CEDEX

CARLIER, P.
Université Paris VII - Lab. P.C.I.
2, place Jussien
F - 75251 PARIS CEDEX 05

CICCIOLI, P.
Istituto Inquinamento Atmosferico
C.N.R.
Via Salaria, Km. 29.300
I - 00016 MONTEROTONDO

CIESLIK, St.
Commission of the European
Communities
Joint Research Centre
Ispra Establishment
I - 21020 ISPRA

CLARKE, A.
Dept. of Fuel and Energy - University
GB - L52 9JT LEEDS

COX, R.
UKAEA
B551 AERE HARWELL
GB - DIDCOT, OXON OX11 ORA

DAMS, R.
University
I.N.W.
86, Proeftuinstraat
B - 9000 GENT

DE LA SERNA, J.
Ministerio de Sanidad y Consumo
Escuola Nacional de Sanidad
Ciudad Universitaria
E - MADRID 3

DE LEEUW, F.
National Institute of Public
Health
P.O. Box 1
NL - 3270 BA BILTHOVEN

DELMAS, R.J.
Laboratoire de Glaciologie et
Géophysique de l'Environnement
B.P 68
F - 38402 ST. MARTIN D'HERES

DE SANTIS
Istituto Inquinamento Atmosferico
C.N.R.
Via Salria, Km. 29300
I - 00016 MONTEROTONDO

DESIATO, F
ENEA - DISP
Via Brancati, 48
I - 00148 ROMA

DEUMLING, D.
Hatzfelotsche Verwaltung
Schloss Schönstein
D - 5248 WISSEN

DE WISPELAERE, C.
Ministero Programmatie " Van HEP
Wetenschapsbeleid
8, Wetenschapstraat
B -1040 BRUSSELS

DUBOIS, J.
E.D.F.
6, quai Watier
F - 78401 CHATOU

ELSHOUT, A.
KEMA
Utrechtseweg, 310
NL - 6812 AR ARNHEM

FEDOROFF, M.
C.N.R.S. - C.E.C.M.
15, rue G. Urbain
F - 94400 VITRY

FENGER, J.
Air Pollutions lab. - Nat. Agency
of Envir. Protection-
Risø National Lab.
DK - 4000 ROSKILDE

FILBY, G.
Commission of the European
Communities
Joint Research Centre
Karlsruhe Establishment
P.O. Box 2266
D - 7500 KARLSRUHE

FINK, E.H.
University
D - 56 WUPPERTAL

FLYGER, H.
Air Pollutions Lab. - Nat. Agency
of Envir. Protection
Risø National Lab.
DK - 4000 ROSKILDE

FRAMBOURT, C.
L.P.M.A./SPIN CEN - FAR
B.P. 6
F - 92260 FONTENAY-AUX-ROSES

FRICKE, W.
Gesamtverband Steinkohle
Friedrichstrasse, 1
D - 4300 ESSEN 1

FUGAS, M.
Institute for Medical Research and
Occupational Health
P.O. Box 291 Mose Pijade, 158
YU - 41001 ZAGREB

FUGLSANG, K.
Air Pollution Lab. - Agency of
Envir. Protection
Risø National Lab.
DK - 4000 ROSKILDE

FUHRER,J.
University
Altenbergrain, 21
CH - 3013 BERN

FUNCKE, W.
Fraunhofer Inst. für Toxikologie
une Aerosolforschung
Nottulner Landweg 102
D - 4400 MUNSTER ROXEL

GEISS, F.
Commission of the European
Communities
Joint Research Centre
Ispra Establishment
I - 21020 ISPRA

GERVAT, G.P.
Meteorological Office
London Road
UK - BRACKWELL RG12 2SZ

GLAVAS, S.
Dept. of Chemistry - University
Alex. Ypsilantou St., 260
GR - 262 22 PATRAS

GUILLOT, P.
Commission of the European
Communities - D.G. Science,
Research and Development
200, rue de la Loi
B - 1049 BRUSSELS

HAGELE, J.
Institute of Physico Chemie
Tammanstrasse, 6
D - 03400 GOETTINGEN

HELAS, G.
Max Planck Institut für Chemie
Saarstrasse, 23
D - 6500 MAINZ

HOLLANDER, W.
Inst. für Toxicologie und
Aerosolforschung
Nottulner Landweg 104
D - 440 MUNSTER ROXEL

ISRAEL, G.
Technische Universitat
St. des 17 Juni, 135
D - 1000 BERLIN 12

JOOS, E.
E.D.F.
6, quai Watier
F - 78401 CHATOU

KALLEND, A.
Central Electr. Gen. Board (CERL)
Kelvin Av.
GB - LEATHERHEAD (SURREY)

KESSLER, C.
Institut für Chemie 3
Kernforschungsanlage
D - JUELICH

KLAIS, O.
Hoechst A.G. - Ang. Physik
Postbox
D - 6230 FRANKFURT 80

KLOEPFFER, W.
Battelle Institut E.V.
Am Römerhof, 35
D - 6000 FRANKFURT 90

KLOSE, A.
Commission of the European
Communities - D.G. Science,
Research and Development
200, rue de la Loi
B - 1049 BRUSSELS

KORNDOERFER, G.
Commission of the European
Communities
Joint Research Centre
Ispra Establishment I -
21020 ISPRA

LE BRAS, G.
C.N.R.S. - C.R.C.C.H.T.
F - 45045 ORLEANS CEDEX

LIBERTI, A.
Istituto Inquinamento Atmosferico
C.N.R.
Via Salaria, Km. 29.300
I - 00016 MONTEROTONDO

LINDNER, R.
Commission of the European
Communities
Joint Research Centre
Karlsruhe Establishment
P.O. Box 2266
D - 7500 KARLSRUHE

LINDSKOG, A.
Swedish Environmental Research Inst.
Box 5207
S - 40224 GOTHENBURG

LOEBEL, J.
Verein Deutscher Ingenieure
Schenkspfad, 9
D - 5000 KOELN 91

LORENZ, K.
Institut für Physik und
Chemie - Universität
Tammanstrasse, 6
D - 3400 GOETTINGEN

MADELAINE, G.
L.P.M.A./SPIN CEN-FAR
B.P. 6
F - 92260 FONTENAY AUX ROSES

MANNING, P.
Imperial College of
Science and Technology
Exhibition Road
GB - LONDON SW7 2BX

MARTIN, D.
Centre de Recherches en
Physique
de l'Atmosphère
F - 78470 MAGNY-LES-HAMEAUX

MASCLET, P.
Université Paris VII
Lab. P.C.I.
2, place Jussien
F - 75251 PARIS CEDEX 05

MASNIERE, P.
E.D.F.
6, quai Watier
F - 78400 CHATOU

METTERNICH, P.
Institut für Meteorologie
Goethe Universität
Feldbergstrasse, 47
D - 6000 FRANKFURT 1

MEYRAHN, H.
Max Planck Institut für Chemie
Saarstrasse, 23
D - 6500 MAINZ

MITCHELL, I.
Commission of the European
Communities - D.G. Science,
Research and Development
200, rue de la Loi
B - 1049 BRUSSELS

MOORTGAT, G.K.
Max Planck Institut für Chemie
Saarstrasse, 23
D - 6500 MAINZ

MORELLI, J.
Lab. Chemie Minérale des Milieux
Naturels, Université Paris VII
2, place Jussien
F - 75251 PARIX CEDEX 05

MOUVIER, G.
Université Paris VII - Lab. P.C.I.
2, place Jussien
F - 75251 PARIX CEDEX 05

MUELLER, J.
UBA - Pilotstation
Feldbergstr., 45
D - FRANKFURT

MUELLER, W.
Commission of the European
Communities
Joint Research Centre
Karlsruhe Establishment
P.O. Box 2266
D - 7500 KARLSRUHE

MULLER, M.
Ministère de l'Environnement
14, bvd du Général Leclerc
F - 92524 NEULLY sur SEINE CEDEX

NGUYEN, B.C.
Centre Faibles Radioactivités
C.N.R.S.
F - 91191 GIF SUR YVETTE

NIELSEN, O.
Risø National Lab.
DK - 4000 ROSKILDE

NIELSEN, T.
Chemistry Dept.
Risø National Lab.
DK - 4000 ROSKILDE

NOLTING, F.
Fraunhofer Institut ITA
Stadtfelddamm, 35
D - 3000 HANNOVER

OTT, H.
Commission of the European
Communities - D.G. Science,
Research and Development
200, rue de la Loi
B - 1049 BRUSSELS

PAFFRATH, D.
DFVLR NE-PA
D - 8031 OBERPFAFFENHAUSEN

PAYRISSAT, M.
Commission of the European
Communities
Joint Research Centre
Ispra Establishment
I - 21020 ISPRA

PEETERS, J.
Dept. of Chemistry - K.U. Leuven
Celestijnenlaan, 200 F
B - 3030 HEVERLEE

PENKETT, S.
UKAEA
AERE HARWELL
GB - DIDCOT, OXON OX11 ORA

PERRIN, M.L.
L.P.M.A./SPIN CEN FAR
B.P. 6
F - 92260 FONTENAY AUX ROSES

PERROS, P.
L.P.C.E. - Université
F - 94010 CRETEIL CEDEX

PERSSON, K.A.
Dept. of Analytical Chemistry
Arrhenius Lab. - University
S - 10691 STOCKHOLM

PICKERING, S.
Commission of the European
Communities
Joint Research Centre
Karlsruhe Establishment
P.O. Box 2266
D - 7500 KARLSRUHE

RAES, F.
Rijksuniversiteit
Lab. voor Kernfysica
Proeftuinstraat, 42
B - 9000 GENT

REITER, R.
Institute Atmospheric Env.
Research
Muellerstrasse, 54
D - 8100 GARMISCH

RESTELLI, G.
Commission of the European
Communities
Joint Research Centre
Ispra Establishment
I - 21020 ISPRA

ROEMER, F.
N.V. KEMA
Utrechtseweg, 310
NL - 6812 AR ARNHEM

RUDOLPH, J.
K.F.A. Jülich - ICH 3
P.O. Box 1913
D - 5170 JUELICH

SAAB, A.
E.D.F.
Direction des Etudes et Recherches
6, quai Watier
F - 78400 CHATOU

SANDRONI, S.
Commission of the European
Communities
Joint Research Centre
Ispra Establishment
I - 21020 ISPRA

SCHMIDT, H.E.
Commission of the European
Communities
Joint Research Centre
Karlsruhe Establishment
P.O. Box 2266
D - 7500 KARLSRUHE

Schmidt, V.
GHS Wuppertal
Gauss Strasse, 20
D - 5600 WUPPERTAL 1

Schubert, B.
K.F.A. - ICH 3
Postfach 1913
D - 5170 JUELICH

SCHURATH, U.
Inst. für Physikalische
Chemie der Universität
Wegelerstrasse, 12
D - 5300 BONN

SCOTT, J.A.
University College
Physics Dept.
Belfield
IRL - DUBLIN 4

SIDEBOTTOM, H.
Chemistry Dept.
univer. College
Belfield
IRL - DUBLIN 4

SJOEDIN, A.
Swedish Environmental Res. Inst.
P.O. Box 5207
S - 40224 GOTHENBURG

STANGL, H.
Commission of the European
Communities
Joint Research Centre
Ispra Establishment
I - 21020 ISPRA

STINGELE, A.
Commission of the European
Communities
Joint Research Centre
Ispra Establishment
I - 21020 ISPRA

TEN BRINK, H.
Netherlands Energy
Research Foundation
P.O. Box 1
NL - 1755 ZG PETTEN

TOUPANCE, G.
Université Paris VII
L.P.C.E.
F - 94010 CRETEIL CEDEX

TREACY, J.
Dept. of Chemistry
University College
Belfield
IRL - DUBLIN 4

TYNDALL, G.
Max-Planck-Institut
für Chemie
Saarstrasse, 23
D - 6500 MAINZ

VANDERBORGHT, B.
S.C.K./CEN
Boeretang, 200
B - 2400 MOL

VAN HAECKE
Ministero Programmatie van HEP
Wetenschapstraat, 8
B - 1040 BRUSSELS

VERSINO, B.
Commission of the European
Communities
Joint Research Centre
Ispra Establishment
I - 21020 ISPRA

VIERKORN-RUDOLPH, B.
K.F.A. Jülich
Institut für Chemie
Postfach 1913
D - 5170 JUELICH

VINCKIER, C.
K.U. Leuven
Celestijnenlaan, 200 F
B - 3030 HEVERLEE

WALSH, J.J.
Chemistry Engineering Dept.
University
IRL - DUBLIN 2

WAETJEN, U.
Commission of the European
Communities
Joint Research Centre
Geel Establishment
B - GEEL

WESTERHOLM, R.
Dept. of Analytical Chemistry
University
Arrhenius Lab.
S - 10691 STOCKHOLM

WILMES, R.
G.D.C.H. - Bayer AG
D - 5600 WUPPERTAL 1

WEBER, E.
Bundesministerium des Innern
D - 5300 BONN 1

WINKLER, P.
Meteorological Observatory
Frahmredder, 95
D - 2000 HAMBURG 65

WITTE, F.
Ruhr Universität
Phys. Chemie I
Postfach 102148
D - 4630 BOCHUM

ZELLNER, R.
Universität
Tammanstrasse, 6
D - 3400 GOETTINGEN

ZEPHORIS, M.
C.R.P.A. Ministère des Transports
F - 78470 MAGNY LES HAMEAUX

ZETZSCH, C.
Fraunhofer Institut ITA
Stadtfelddamm, 35
D - 3000 HANNOVER

INDEX OF AUTHORS

ALLEGRINI, I., 12
AUGUSTIN, H., 565

BÄCHMANN, K., 433
BAGNALL, G.N., 188
BAMBER, D.J., 517
BARNES, I., 149
BARTON, R.A., 205
BASTIAN, V., 149
BATIFOL, F.M., 471
BATT, L., 293
BAUER, H.W., 320
BECKER, K.H., 149,177
BEHNKE, W., 309
BEILKE, S., 380,535
BESSEMOULIN, P., 565
BEYNE, S., 53
BØHLER, T., 554
BOUTRON, C.F., 471
BRACHETTI, A., 62
BRANCALEONI, E., 62
BRESSON, M.A., 53
BROLL, A., 390
BROSTRÖM-LUNDEN, E., 101,264
BUCK, M., 83
BURROWS, J.P., 240,249

CADENE, S., 451
CANDREVA, F., 583
CAPPELLANI, F., 216
CARLIER, P., 451,461
CHEYMOL, D., 635
CICCIOLI, P., 62
CLARK, P.A., 517
CLARKE, A.G., 331
COX, R.A., 141,205

DAMS, R., 583
DAUM, P.H., 356
DE LA SERNA, J., 322
DE SANTIS, F., 12
DEBOVE, L., 461
DELMAS, R.J., 441
DI PALO, C., 62

EHHALT, D.H., 44
ELSHOUT, A.J., 504,535

FEBO, A., 12
FEDOROFF, M., 461
FENGER, J., 626
FERM, M., 607
FERNANDEZ PATIER, R., 322
FINK, E.H., 149,177
FONDERIE, V., 274
FRAMBOURT, C., 300
FUGAS, M., 348
FUGLSANG, K., 111
FUHRER, J., 423

GEORGII, H.W., 339
GERVAT, G.P., 506
GIRARD, R., 451
GIRARD-REYDET, L., 451,461
GLAVAS, S., 27
GLOVER, G.M., 517
GOEDE, H.J., 227
GRENNFELT, P., 401
GRIFFITH, D.W.T., 249
GROENEVELD, K.O., 339
GUI-YUN ZHU, 177

HÄGELE, J., 5
HAHN, J., 38
HANSEN, K.A., 111
HEALEY, P.G., 517
HELAS, G., 38,390
HJORTH, J., 216
HOLLÄNDER, W., 309
HRSAK, J., 348

IMBARD, M., 635
ISAKSEN, I.S.A., 554
ISRAËL, G.W., 320

JANSSENS, A., 364
JEBSEN, C., 492
JOHNEN, F.J., 492
JOOS, E., 544
JOURDAIN, J.L., 143

KALLEND, A.S., 506,517
KELLY, T.J., 20
KESSLER, C., 412
KHEDIM, A., 492
KOCH, W., 309
KORTMANN, U., 227
KRETZSCHMAR, J., 573

LE BRAS, G., 143
LEE, Y.N., 20
LEGRAND, M., 441
LEWIN, E.E., 111
LIBERTI, A., 2,12
LINDSKOG, A., 264
LJUNGSTRÖM, E., 205
LORENZ, K., 158
LUCE, C., 451

MAC LEOD, H., 143
MADELAINE, J.G., 298,300
MAES, D., 274
MAFFIOLO, G.., 544
MARCHAL, T., 451,461
MARSH, A.R.W., 506,517
MARTIN, D., 635
MASCLET, P., 53,616
MASNIERE, P., 461

MCQUIGG, R.D., 194
MEIXNER, F.X., 433
MERTENS, I., 573
METTERNICH, P., 339
MEYRAHN, H., 38
MITCHELL, I.V., 120
MOORTGAT, G.K., 194,240,249
MORELLI, J., 451,461
MOUVIER, G., 53,451,616
MÜLLER, J., 373
MUNZERT, K., 480

NELSON, L., 258
NICOLAOU, K., 616
NICOLLIN, B., 90
NIELSEN, O.J., 283

OTTOBRINI, G., 216

PAGSBERG, P., 283
PASCHKE, R., 5
PEETERS, J., 274
PENKETT, S.A., 38
PEREZ CARLES, F., 322
PERRIN, M.L., 300
PERROS, P., 596
PEYRISSAT, M., 90
PLATT, U., 412
POHLMANN, G., 309
POSSANZINI, M., 12,62
PÖTZL, K., 480
POULET, G., 143

RAES, F., 364
REITER, R., 480
RESTELLI, G., 216
RHÄSA, D., 158
RIEDEL, F., 373
RÖMER, F.G., 74
ROUCHAUD, J.C., 461
RUDOLPH, J., 433,492
RUMPEL, K.J., 390

SAAB, A., 544
SCHMIDT, U., 44
SCHMIDT, V., 177

SCHUBERT, B., 44
SCHURATH, U., 27,227
SCHWARTZ, S.E., 20,356
SCHWARZ, B., 433
SEGA, K., 348
SIDEBOTTOM, H.W., 188,258
SILLESEN, A., 283
SJÖDIN, A., 264,401
SKÄRBY, L., 101
SOUVENT, P., 348
STANGL, H., 90,216
STOCKER, D.W., 205
STRAUSS, B., 635

TEN BRINK, H.M., 20,356
TOUPANCE, G., 596
TREACY, J.J., 258
TUCK, A.F., 517
TYNDALL, G.S., 240,249

VAN GALEN, P., 74
VANDENBORGHT, B., 573
VAUGHAN, G., 517 -
VELDKAMP, A.A., 74
VIERKORN-RUDOLPH, B., 433

WARNECK, P., 38,390
WENGENROTH, K., 320
WILLISON, M.J., 331
WINKLER, P., 590
WITTE, F., 168

ZANOLINI, F., 441
ZEKI, E.M., 331
ZELLNER, R., 5,158
ZETZSCH, C., 168